PERGAMON INTERNATIONAL LIBRARY
of Science, Technology, Engineering and Social Studies

The 1000-volume original paperback library in aid of education, industrial training and the enjoyment of leisure

Publisher: Robert Maxwell, M.C.

INTRODUCTION TO HEALTH PHYSICS

Pergamon Titles of Related Interest

Bentel Treatment Planning and Dose Calculation in Radiation Oncology Third Edition
Gollnick Experimental Radiological Health Physics
ICRP Limits for Intakes of Radionuclides by Workers (#30)
Kase/Nelson Concepts of Radiation Dosimetry
Kathren Health Physics: A Backward Glance

Related Journals*

ANNALS OF THE ICR
HEALTH PHYSICS
INTERNATIONAL JOURNAL OF NUCLEAR MEDICINE AND BIOLOGY
INTERNATIONAL JOURNAL OF APPLIED RADIATION AND ISOTOPES
RADIATION PHYSICS AND CHEMISTRY

***Free specimen copies available upon request**

INTRODUCTION TO HEALTH PHYSICS

SECOND EDITION—REVISED AND ENLARGED

Herman Cember
Northwestern University

Pergamon Press
New York • Oxford • Toronto • Sydney • Paris • Frankfurt

Pergamon Press Offices:

U.S.A.	Pergamon Press Inc., Maxwell House, Fairview Park, Elmsford, New York 10523, U.S.A.
U.K.	Pergamon Press Ltd., Headington Hill Hall, Oxford OX3 0BW, England
CANADA	Pergamon Press Canada Ltd., Suite 104, 150 Consumers Road, Willowdale, Ontario M2J 1P9, Canada
AUSTRALIA	Pergamon Press (Aust.) Pty. Ltd., P.O. Box 544, Potts Point, NSW 2011, Australia
FRANCE	Pergamon Press SARL, 24 rue des Ecoles, 75240 Paris, Cedex 05, France
FEDERAL REPUBLIC OF GERMANY	Pergamon Press GmbH, Hammerweg 6, D-6242 Kronberg-Taunus, Federal Republic of Germany

Library of Congress Cataloging in Publication Data

Cember, Herman.

Introduction to health physics.

Includes bibliographies.
1. Radiation--Safety measures. 2. Radiation--
Toxicology. 3. Environmental health. I. Title.
RA569.C4 1983 616.07'57'0288 82-19051
ISBN 0-08-030129-0
ISBN 0-08-030936-4 (pbk.)

Second printing, 1985.

Printed in Great Britain by A. Wheaton & Co. Ltd., Exeter

This book is respectfully dedicated to the memory of

Dr. Elda E. Anderson

and

Dr. Thomas Parran

CONTENTS

PREFACE

A number of changes have occurred in the years since the first edition of *Introduction to Health Physics* was published. The rapid growth of the technology of lasers and microwaves and the widespread application of these two sources of radiation have focused the attention of the public on the health and safety of these technologies. Although laser and microwave radiation are both non-ionizing radiations, they nevertheless share many physical characteristics with X-rays. Furthermore, the general principles for the control of the hazards from laser and microwave sources are similar to those for the control of hazards from ionizing radiation. It is thus natural that the health physicist be concerned with non-ionizing as well as ionizing radiation. Accordingly, a chapter on lasers and microwaves has been added to introduce the student of health physics to the control of the potential hazards from these two sources of non-ionizing radiation.

Another change incorporated into the second edition is the introduction of the SI (Système International) units. Scientists and engineers have been using various systems, such as the English system and the metric system with its subsets, the cgs, the emu, and the MKS systems. Units have been confused as well as abused. Often the same name is given to fundamentally different physical concepts—such as kilogram (or pound) of force and kilogram (or pound) of mass. On the other hand, the same physical concept frequently is called by two or more different names. Thus, for example, an electric motor "converts" watts to horsepower, while an electrical cooling or heating system "converts" watts to Btu per hour! The SI system, which, although based on metric units is nevertheless a new system, is an attempt to reduce this plethora of measurement systems to a single system that will be understood by everyone. Because of the increasing use of SI units in science and technology and because the SI system of units was adopted by the International Commission on Radiological Protection, SI units are used in the second edition of *Introduction to Health Physics*. Since most of us "old-timers" are accustomed to thinking in terms of the old units, however, the equivalent old units for radiation and radioactivity quantities are given in parentheses after the new units. In several examples in the book, the numerical values are given in the old system of units, with the SI equivalents in parentheses.

The advent of the computer has made possible a more accurate method of dose calculation (based on Monte Carlo techniques) than the computational methods that had been available previously. This new computational technique and its application have been incorporated into the chapter on Radiation Dosimetry.

The major change in health physics that has occurred since the first edition of this book is the new basis on which the ICRP sets safety standards. Dose limits in the earlier ICRP recommendations were based mainly on consideration of possible genetic effects and on harm to a critical organ. The new ICRP recommendations are based on the total detriment from radiation exposure. This total detriment reflects the combined risk of fatal cancer and of genetic effects. The chapter on Radiation Protection Guides has thus been significantly changed to include the latest recommendations of the ICRP, with illustrative examples applied to the calculation of dose limits and to the calculation of allowable limits for the intake—by ingestion or by inhalation—of radionuclides. Since the regulations of many countries have not yet been adapted to the new ICRP recommendations, the chapter on

Radiation Protection Guides retains the methods for calculating maximum permissible environmental concentrations of radioisotopes based on the critical organ concept.

I wish to thank the many people, too numerous to mention by name, who made suggested changes and who pointed out errors in the first edition. I would also like to acknowledge the contribution of Dr. Allen Brodsky throughout the many "brainstorming" conversations that we had about the material in the book, and to thank Dr. Morris Brodwin for his critical review of the chapter on Non-ionizing Radiation. Finally, I owe a debt of gratitude to my wife Sylvia, for all her help in preparing the manuscript and for her continuous encouragement when my own efforts started to lag.

INTRODUCTION TO HEALTH PHYSICS

CHAPTER 1

INTRODUCTION

HEALTH physics, or radiological health, as it is frequently called, is that area of environmental health engineering that deals with the protection of the individual and population groups against the harmful effects of ionizing and non-ionizing radiation. The health physicist is responsible for the safety aspects in the design of processes, equipment, and facilities utilizing radiation sources, so that radiation exposure to personnel will be minimized, and will at all times be within acceptable limits; and he must keep personnel and the environment under constant surveillance in order to ascertain that his designs are indeed effective. If control measures are found to be ineffective, or if they breakdown, he must be able to evaluate the degree of hazard, and to make recommendations regarding remedial action.

The scientific and engineering aspects of health physics are concerned mainly with: (1) the physical measurements of different types of radiation and radioactive materials, (2) the establishment of quantitative relationships between radiation exposure and biological damage, (3) the movement of radioactivity through the environment, and (4) the design of radiologically safe equipment, processes, and environments. Clearly, health physics is a professional field that cuts across the basic physical, life, and earth sciences, as well as such applied areas as toxicology, industrial hygiene, medicine, public health, and engineering. The professional health physicist, therefore, in order to perform effectively, must be competent in the wide spectrum of disciplines that bridge the fields between industrial operations and technology on one hand, the modern health science on the other hand. He must have an appreciation of the complex interrelationships between man and the physical, chemical, biological, and even social components of the environment; as well as a quantitative understanding of group phenomena. In addition to these general prerequisites, he must be technically competent in the subject matter unique to his speciality.

The main purpose of this book is to lay the groundwork for attaining technical competency in health physics. Because of the nature of the subject matter and the topics covered, however, it is hoped that the book will be a useful source of information to workers in environmental health as well as to those who will use radiation as a tool. For the latter group, it is also hoped that this book will impart an appreciation for radiation safety as well as an understanding of the philosophy of environmental health.

CHAPTER 2

REVIEW OF PHYSICAL PRINCIPLES

Mechanics

Units and Dimensions

Health Physics is a science, and hence is a systematic organization of knowledge about the interaction between radiation and organic and inorganic matter. Quite clearly, the organization must be quantitative as well as qualitative, since control of radiation hazards implies a knowledge of the dose response relationship between radiation and the biological effects of radiation.

Quantitative relationships are based on measurements, which in reality are comparisons of the attribute under investigation to a standard. A measurement includes two components: a number and a unit. When measuring the height of a person, for example, the result is given as 70 inches if the British system of units is used, or as 177.8 centimeters if the metric system is used. The units "inches" in the first case and "centimeters" in the second tell us what the criterion for comparison is, and the number tells us how many of these units are included in the quantity being measured. Although 70 inches means exactly the same thing as 177.8 centimeters, it is clear that, without an understanding of the units, the information contained in the number above would be meaningless. The British system of units is used chiefly in engineering, while the metric system is widely used in science.

Three attributes are considered basic in the physical sciences: length, mass, and time. In the British system of units, these attributes are measured in feet, pounds, and seconds, respectively, while the metric system is divided into two subsystems—the mks, in which the three qualities are specified in meters, kilograms, and seconds—and the cgs, in which the centimeter, gram, and second are used to designate length, mass, and time. In health physics, the cgs system was used most frequently.

By international agreement, the metric system and the British system are being replaced by a third, and new, system called The International System or simply the SI System. Although many familiar metric units are employed in SI, it should be emphasized that SI is a new system and must not be thought of as a new form of the metric system. All the other units, such as force, energy, power, etc. are derived from the three basic units of mass in kilograms (kg), length in meters (m), and time in seconds (sec), plus the four additional basic units for electric current in amperes (A), temperature in degrees (C), amount of a substance in moles (mol) and, luminous intensity in candelas (cd). For example, the unit of force, the newton (N) is defined as follows:

> One newton is the unbalanced force that will accelerate a mass of 1 kilogram at a rate of 1 meter per second per second.

Expressed mathematically:

$$\text{force} = \text{mass} \times \text{acceleration},$$

$$f = ma, \tag{2.1}$$

and the units are

$$\text{newton} = \text{kg} \times \frac{\text{m}}{\text{sec}} \Big/ \text{sec}.$$

Since dimensions may be treated in exactly the same way as numbers, the dimension for acceleration is written as m/sec^2. The dimensions for force, therefore, are

$$\text{newton} = \frac{\text{kg-m}}{\text{sec}^2}.$$

Work and Energy

Energy is defined as the ability to do work. Since all work requires the expenditure of energy, the two terms are expressed in the same units, and consequently have the same dimensions. Work is done, or energy expended, when a force f is exerted through some distance r:

$$W = fr. \tag{2.2}$$

In the SI system, the joule is the unit of work and energy, and is defined as follows:

> One joule of work is done when a force of 1 newton is exerted through a distance of 1 meter.

Since work is defined as the product of a force and a distance, the dimensions for work and energy are:

$$\begin{aligned}\text{joules} &= \text{newtons} \times \text{meters}\\ &= \frac{\text{kg-m}}{\text{sec}^2} \times \text{m} = \frac{\text{kg-m}^2}{\text{sec}^2}.\end{aligned}$$

Although the joule is the basic unit of energy in the SI system, it is not too practical for many measurements in the field of health physics because it is an extremely large unit in terms of the energies encountered in the microscopic world of the atom. For many purposes, a more practical unit, called the electron volt (abbreviated eV), is used. The electron volt is a *unit of energy*, and is defined as follows:

$$1 \text{ electron volt} = 1.6 \times 10^{-19} \text{ joule}.$$

When work is done on a body, the energy expended in doing the work is added to the energy of the body. For example, if a mass is lifted from one elevation to another, the energy that was expended during the performance of the work is converted to potential energy. On the other hand, when work is done to accelerate a body, the energy that was expended appears as kinetic energy in the moving body. In the case where work was done in lifting a body, the mass possesses more potential energy at the higher elevation than it did before it was lifted. Work was done, in this case, against the force of gravity, and the total increase in potential energy of the mass is equal to its weight, which is merely the force with which the mass is attracted to the earth, multiplied by the height through which the mass was raised. *Potential energy* is defined as energy that a body possesses by virtue of its position in a force field. *Kinetic energy* is defined as energy possessed by a moving body as result of its motion. For bodies of mass m, moving with a velocity v (for the case where v is less than about 3×10^7 m/sec), the kinetic energy is given by

$$E_k = \tfrac{1}{2}mv^2. \tag{2.3}$$

Relativity Effects

According to the system of classical mechanics that was developed by Newton and the other great thinkers of the Renaissance period, mass is an immutable property of matter; it could be changed in size, shape, or state, but it could neither be created nor destroyed. Although this law of conservation of mass seems to be true for the world which we can perceive with our senses, it is in fact only a special case for conditions of large masses and slow speeds. In the sub-microscopic world of the atom, where masses are measured in units of 9×10^{-31} kg, where distances are measured in units of 10^{-10} m, and where velocities are measured in terms of the velocity of light, classical mechanics is not applicable. Einstein, in his special theory of relativity, postulated that the velocity of light in a vacuum, 3×10^8 m/sec, is an upper limit of speed that no material body can ever attain; the velocity of light may be asymptotically approached by any mass, but can never be reached. Furthermore, according to Einstein, the mass of a moving body is not constant, as was previously thought, but rather is a function of the velocity with which the body is moving. As the velocity increases, the mass increases, and when the velocity of the body approaches the velocity of light, the mass increases very rapidly. The mass, m, of a moving object whose velocity is v is related to its rest mass, m_0, by the equation

$$m = \frac{m_0}{\sqrt{\left(1 - \frac{v^2}{c^2}\right)}}, \tag{2.4}$$

where c is the velocity of light.

Example 2.1

Compute the mass of an electron moving at 10% of the speed of light. The rest mass of an electron is 9.11×10^{-31} kg. At $v = 0.1\,c$,

$$m = \frac{9.11 \times 10^{-31}\ \text{kg}}{\sqrt{\left(1 - \frac{(0.1\,c)^2}{c^2}\right)}} = 9.16 \times 10^{-31}\ \text{kg},$$

and at $v = 0.99\,c$

$$m = \frac{9.11 \times 10^{-31}\ \text{kg}}{\sqrt{\left(1 - \frac{(0.99\,c)^2}{c^2}\right)}} = 64.6 \times 10^{-31}\ \text{kg}.$$

This illustration shows that whereas an electron suffers a mass increase of only $\frac{1}{2}$% when it is moving at 10% of the speed of light, its mass increases about seven-fold when the velocity is increased to 99% of the velocity of light.

Kinetic energy of a moving body can be thought of as the income from work put into the body, or energy input, in order to bring the body up to its final velocity.

Expressed mathematically, we have

$$W = E_k = fr = \tfrac{1}{2}mv^2. \tag{2.5}$$

The expression for kinetic energy in equations (2.3) and (2.5) is a special case, however, since the mass is assumed to remain constant during the time that the body is undergoing acceleration from its initial to its final velocity. If the final velocity is sufficiently high to produce observable relativistic effects (this is usually taken as $v = 0.1\,c = 3 \times 10^7$ m/sec), then equations (2.3) and (2.5) are no longer valid.

As the body gains velocity under the influence of an unbalanced force, its mass continuously increases until it attains the value given by equation (2.4). This particular value for the mass is thus applicable only to one point during the time that the body was undergoing acceleration, that is, only after the body has completed its acceleration. The magnitude of the unbalanced force, therefore, must be continuously increased during the accelerating process to compensate for the increasing inertia of the body due to its continuously increasing mass. Equations (2.2) and (2.5) assume the force to be constant and therefore are not applicable to cases where relativistic effects must be considered. One way in which to overcome this difficulty is to divide the total distance r into many smaller distances, $\Delta r_1, \Delta r_2, \ldots, \Delta r_n$, as shown in Fig. 2.1, and then multiply each of these small distances by the average force exerted while traversing the small distance, and then summing the products. This process may be written as

$$W = f_1\Delta r_1 + f_2\Delta r_2 + \cdots + f_n\Delta r_n, \tag{2.6A}$$

and abbreviated as

$$W = \sum_{n=1}^{n} f_n\Delta r_n. \tag{2.6B}$$

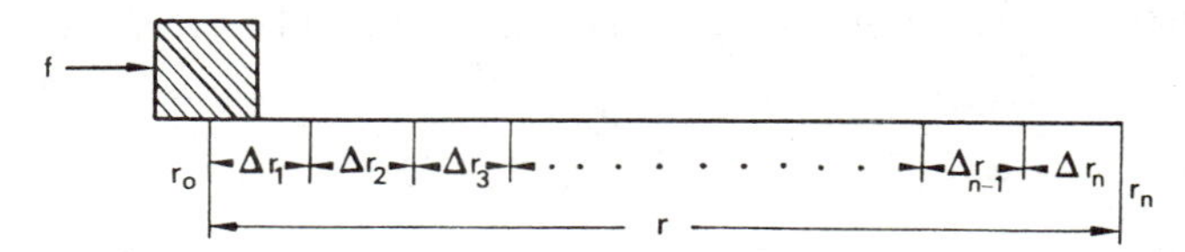

FIG. 2.1. Diagram illustrating that the total work done in accelerating a body is
$$W = \sum_{n=1}^{n} f_n\Delta r_n.$$

As r is successively divided into smaller and smaller lengths, the calculation, using equation (2.6), of the work done becomes more accurate. A limiting value for W may be obtained by letting each small distance, Δr_n, in equation (2.6) approach zero, that is, by considering such small increments of distance that the force remains approximately constant during the specified interval. In the notation of the calculus, such an infinitesimally small quantity is called a differential, and is specified by prefixing the symbol for the quantity with the letter "d". Thus, if r represents distance, dr represents an infinitesimally small distance, and the differential of work done, which is the product of the force and the infinitesimally small distance is

$$\mathrm{d}W = f\,\mathrm{d}r. \tag{2.7}$$

The total energy expended in going from the point r_0 to point r_n, then, is merely the sum of all the products of the force and the infinitesimally small distances through which it acted. This sum is indicated by the mathematical notation

$$W = \int_{r_0}^{r_n} f\,\mathrm{d}r. \tag{2.8}$$

The ratio of two differentials is called a derivative, and the process in which a derivative is obtained is called differentiating. Since acceleration is defined as the rate of change of velocity with respect to time,

$$a = \frac{v_2 - v_1}{t_2 - t_1} = \frac{\Delta v}{\Delta t}, \tag{2.9}$$

where v_1 and v_2 are the respective velocities at times t_1 and t_2 then equation (2.1) may be written as

$$f = m\frac{\Delta v}{\Delta t}, \tag{2.10}$$

and, by letting Δt approach zero, thereby obtaining the instantaneous rate of change of velocity, or the derivative of velocity with respect to time, we have, using the differential notation,

$$f = m\frac{\mathrm{d}v}{\mathrm{d}t}, \tag{2.11}$$

which is the expression of Newton's second law of motion for the case where the mass remains constant. Newton's second law states that the rate of change of momentum of an accelerating body is proportional to the unbalanced force acting on the body. For the general case, where mass is not constant, Newton's second law is therefore written as

$$f = \frac{\mathrm{d}(mv)}{\mathrm{d}t}. \tag{2.12}$$

Substitution of the value of f from equation (2.12) into equation (2.8) gives

$$W = \int_0^r \frac{\mathrm{d}(mv)}{\mathrm{d}t}\,\mathrm{d}r. \tag{2.13}$$

Since $v = \mathrm{d}r/\mathrm{d}t$, equation (2.13) can be written as

$$W = \int_0^r \frac{\mathrm{d}(mv)}{\mathrm{d}t}\cdot v\,\mathrm{d}t = \int_0^{mv} v\,\mathrm{d}(mv), \tag{2.14}$$

and substituting $m = m_0/\sqrt{(1 - v^2/c^2)}$, we have

$$W = \int_0^v v\,\mathrm{d}\left(\frac{m_0 v}{\sqrt{1 - v^2/c^2}}\right). \tag{2.15}$$

Differentiating the term in the parenthesis gives

$$W = m_0 \int_0^v \left[\frac{v}{(1 - v^2/c^2)^{1/2}} + \frac{v^3/c^2}{(1 - v^2/c^2)^{3/2}}\right]\mathrm{d}v. \tag{2.16}$$

Now, multiply the numerator and denominator of the first term in equation (2.16) by

$1 - v^2/c^2$ to give

$$W = m_0 \int_0^v \left[\frac{v - v^3/c^2}{(1 - v^2/c^2)^{3/2}} + \frac{v^3/c^2}{(1 - v^2/c^2)} \right] dv \tag{2.17}$$

$$= m_0 \int_0^v \frac{v}{(1 - v^2/c^2)^{3/2}} dv = m_0 \int v(1 - v^2/c^2)^{-3/2} dv. \tag{2.18}$$

The integrand in equation (2.18) is almost in the form

$$\int_a^b u^n du = \left. \frac{u^{n+1}}{n+1} \right|_a^b, \tag{2.19}$$

where

$$u^n = \left(1 - \frac{v^2}{c^2}\right)^{-3/2} \quad \text{and} \quad du = -\frac{2v}{c^2} dv.$$

To convert equation (2.18) into the form for integration given by equation (2.19), it is necessary only to complete du. This is done by multiplying the integrand by $-2/c^2$ and the entire expression by $-c^2/2$ in order to keep the total value of equation (2.18) unchanged. The solution of equation (2.18), which gives the kinetic energy of a body that was accelerated from zero velocity to a velocity v, is

$$E_k = W = m_0 c^2 \left(\frac{1}{\sqrt{(1 - v^2/c^2)}} - 1 \right) = m_0 c^2 \left(\frac{1}{\sqrt{(1 - \beta^2)}} - 1 \right), \tag{2.20}$$

where $\beta = v/c$.

Equation (2.20) is the exact expression for kinetic energy, and must be used whenever the moving body experiences observable relativistic effects.

Example 2.2

(a) What is the kinetic energy of the electron in Example 2.1 that travels at 99% of the velocity of light?

$$E_k = m_0 c^2 \left(\frac{1}{\sqrt{(1 - \beta^2)}} - 1 \right)$$

$$= 9.11 \times 10^{-31} \text{ kg} \left(3 \times 10^8 \frac{\text{m}}{\text{sec}} \right)^2 \left(\frac{1}{\sqrt{(1 - (0.99)^2)}} - 1 \right)$$

$$= 4.97 \times 10^{-13} \text{ joule.}$$

(b) How much additional energy is required to increase the velocity of this electron to 99.9% of the velocity of light, an increase in velocity of only 0.91%?

$$E_k = 9.11 \times 10^{-31} \text{ kg} \left(3 \times 10^8 \frac{\text{m}}{\text{sec}} \right)^2 \left(\frac{1}{\sqrt{(1 - (0.999)^2)}} - 1 \right)$$

$$= 17.55 \times 10^{-13} \text{ J.}$$

The additional work necessary to increase the kinetic energy of the electron from 99% to 99.9% of the velocity of light is

$$W = (17.55 - 4.97)\, 10^{-13} \text{ J}$$

$$= 12.58 \times 10^{-13} \text{ J.}$$

(c) What is the mass of the electron whose β is 0.999?

$$m = \frac{m_0}{\sqrt{(1-\beta^2)}}$$
$$= \frac{9.11 \times 10^{-31}\ \text{kg}}{\sqrt{(1-(0.999)^2)}} = 204 \times 10^{-31}\ \text{kg}.$$

The relativistic expression for kinetic energy given by equation (2.20) is rigorously true for particles moving at all velocities, while the non-relativistic expression for kinetic energy, equation (2.3), is applicable only to cases where the velocity of the moving particle is much less than the velocity of light. It can easily be shown that the relativistic expression reduces to the non-relativistic expression for low velocities by expanding the expression $1/\sqrt{(1-\beta^2)}$ in equation (2.20) according to the binomial theorem, and then dropping higher terms that become insignificant when $v \ll c$. According to the binomial theorem,

$$(a+b)^n = a^n + na^{n-1}b + \frac{n(n-1)a^{n-2}b^2}{2!} + \cdots. \tag{2.21}$$

The expansion of $1/\sqrt{(1-\beta^2)}$, or $(1-\beta^2)^{-1/2}$ according to equation (2.21) is accomplished by letting $a = 1$, $b = -\beta^2$, and $n = -\frac{1}{2}$.

$$(1-\beta^2)^{-1/2} = 1 + \tfrac{1}{2}\beta^2 + \tfrac{3}{8}\beta^4 + \cdots. \tag{2.22}$$

Since $\beta = v/v$, then, if $v \ll c$, terms from β^4 and higher will be insignificantly small, and may therefore be dropped. When this is done, and the first two terms from equation (2.22) are substituted into equation (2.20), we have

$$E_k = m_0c^2\left(1 + \frac{1}{2}\frac{v^2}{c^2} - 1\right)$$
$$= \tfrac{1}{2}m_0v^2, \tag{2.3}$$

which is the non-relativistic case. Equation (2.3) is applicable when $v \ll c$.

In Example 2.1 it was shown that, at a very high velocity ($\beta = 0.99$), a kinetic energy increase of 153% resulted in a velocity increase of the moving body by only 0.91%. In non-relativistic cases, the increase in velocity is directly proportional to the square root of the work done on the moving body or, in other words, to the kinetic energy of the body. In the relativistic case, the velocity increase due to additional energy is smaller than in the non-relativistic case because the additional energy serves to increase the mass of the moving body rather than its velocity. This equivalence of mass and energy is one of the most important consequences of Einstein's theory of relativity. According to Einstein, the relationship between mass and energy is

$$E = mc^2, \tag{2.23}$$

where E is the total energy of a piece of matter whose mass is m, and c is the velocity of light in a vacuum. The theory of relativity tells us that all matter contains potential energy by virtue of its mass. It is this energy source which is tapped to obtain nuclear energy. The main virtue of this energy source is the vast amount of energy that can be derived from conversion into its energy equivalent of small amounts of nuclear fuel.

Example 2.3

(a) How much energy can be obtained from one gram of nuclear fuel?

$$
\begin{aligned}
E &= mc^2 \\
&= 1 \times 10^{-3}\,\text{kg} \times \left(3 \times 10^8 \frac{\text{m}}{\text{sec}}\right)^2 \\
&= 9 \times 10^{13}\,\text{J}.
\end{aligned}
$$

Since there are 2.78×10^{-21} kilowatt hours per J, 1 g of nuclear fuel yeilds

$$
\begin{aligned}
E &= 9 \times 10^{13}\,\text{J} \times 2.78 \times 10^{-21} \frac{\text{kWh}}{\text{J}}, \\
&= 2.5 \times 10^7 \text{ kilowatt hours.}
\end{aligned}
$$

(b) How much coal, whose heat content is 13,000 Btu per pound, must be burned to liberate the same amount of energy as one gram of nuclear fuel?

$$1\,\text{Btu} = 2.93 \times 10^{-4}\,\text{kWh}.$$

∴ Amount of coal required is:

$$1.3 \times 10^4 \frac{\text{Btu}}{\text{lb}} \times 2.93 \times 10^{-4} \frac{\text{kWh}}{\text{Btu}} \times 2 \times 10^3 \frac{\text{lb}}{\text{ton}} \times C \text{ tons} = 2.5 \times 10^7\,\text{kWh}.$$

$$
\begin{aligned}
\therefore \quad C &= \frac{2.5 \times 10^7}{1.3 \times 10^4 \times 2.93 \times 10^{-4} \times 2 \times 10^3} \\
&= 3280 \text{ tons.}
\end{aligned}
$$

The loss in mass accompanying ordinary energy transformations is not detectable because of the very large amount of energy released per unit mass, and the consequent small change in mass for ordinary reactions. In the case of coal, for example, the above example shows a loss in mass of 1 g per 3280 tons. The fractional mass loss is

$$
\begin{aligned}
f &= \frac{\Delta m}{m} = \frac{1\,\text{g}}{3.28 \times 10^3\,\text{tons} \times 2 \times 10^3\,\text{lb/ton} \times 4.54 \times 10^2\,\text{g/lb}} \\
&= 3.3 \times 10^{-10}.
\end{aligned}
$$

Such a small fractional loss in mass is not detectable by any of our weighing techniques.

Electricity

Electrical Charge: The Coulomb

All matter is electrical in nature, and consists of extremely small charged particles called protons and electrons. The mass of the proton is 1.6723×10^{-24} g and the mass of the electron is 9.1085×10^{-28} g. These two particles have charges of exactly the same magnitude, but are qualitively different. A proton is said to have a positive charge and an electron a negative charge. Under normal conditions, matter is electrically neutral because the positive and negative charges are homogeneously (on a macroscopic scale) dispersed in equal numbers in a manner that results in no net charge. However, it is possible, by suitable treatment, to induce either net positive or negative charges on bodies. Combing the hair, for example, with a hard rubber comb transfers electrons to the comb from the hair, leaving a net negative charge on the comb.

Charged bodies exert forces on each other by virtue of their electrical fields. Bodies with like charges repel each other, while those with unlike charges attract each other. In the case of point charges, the magnitude of these electrical forces is proportional to the product of the charges and inversely proportional to the square of the distance between the charged bodies. This relationship was described by Coulomb, and is known as Coulomb's law. Expressed algebraically, it is

$$f = k\frac{q_1 q_2}{r^2}, \tag{2.24}$$

where k, the constant of proportionality, depends on the nature of the medium that separates the charges. In the SI system, the unit of electrical charge, called the *coulomb* (C) is defined in terms of electrical current rather than by Coulomb's law. For this reason, the constant of proportionality has a value not equal to one, but rather

$$k_0 = 9 \times 10^9 \frac{\text{N-m}^2}{\text{C}^2}, \tag{2.25}$$

when the two charges are in a vacuum or in air (air at atmospheric pressure exerts very little influence on the force developed between charges, and thus may be considered equivalent to a vacuum). The subscript 0 signifies the value of k in a vacuum. If the charges are separated by materials, other than air, that are poor conductors of electricity (such materials are called *dielectrics*), the value of k is different.

It is convenient to define k_0 in terms of another constant, ϵ_0, called the *permittivity*:

$$k_0 = \frac{1}{4\pi\epsilon_0} = 9 \times 10^9 \frac{\text{N-m}^2}{\text{C}^2} \tag{2.26}$$

$$\epsilon_0 = \frac{1}{4\pi k} = \frac{1}{4\pi \times 9 \times 10^9\ \text{N-m}^2/\text{C}^2}$$

$$\epsilon_0 = 8.85 \times 10^{-12}\ \text{C}^2/\text{N-m}^2.$$

ϵ_0 is the permittivity of a vacuum. The permittivity of any other medium is designated by ϵ. The relative permittivity of a substance is defined by

$$k_e = \frac{\epsilon}{\epsilon_0}, \tag{2.27}$$

and is called the *dielectric coefficient*.

For all dielectric materials, the dielectric coefficient has a value greater than 1. The permittivity, or the dielectric coefficient, is a measure of the amount of electrical energy that can be stored in a medium when the medium is placed into a given electric field. If everything else is held constant, a higher dielectric coefficient leads to a greater amount of stored electrical energy.

The smallest natural quantity of electrical charge is the charge on the electron or proton, $\pm 1.6 \times 10^{-19}$ C. The reciprocal of the electronic charge, 6.25×10^{18}, is the number of electrons whose aggregate charge is 1 C. In the cgs system, the unit of charge is the *statcoulomb* (sC), and the electronic charge is 4.8×10^{-10} sC. There are 3×10^9 sC in one C.

Example 2.4

Compare the electrical and gravitational forces of attraction between an electron and a proton separated by 5×10^{-11} m.

The electrical force is given by equation (2.24):

$$f = k_0 \frac{q_1 q_2}{r^2}$$

$$= 9 \times 10^9 \frac{\text{N-m}^2}{\text{C}^2} \times \frac{1.6 \times 10^{-19}\,\text{C} \times 1.6 \times 10^{-19}\,\text{C}}{(5 \times 10^{-11}\,\text{m})^2}$$

$$= 9.2 \times 10^{-8}\,\text{N}.$$

The gravitational force between two bodies follows the same mathematical formulation as Coulomb's law for electrical forces. In the case of gravitational forces, the force is always attractive. The gravitational force is given by

$$F = \frac{G m_1 m_2}{r^2}. \qquad (2.28)$$

G is a universal constant that is equal to 6.67×10^{-11} N-m²/kg², and must be used because the unit of force, the newton, was originally defined using "inertial" mass, according to Newton's second law of motion given by equation (2.1). The mass in equation (2.28), is commonly called "gravitational" mass. Despite the two different designations, however, it should be emphasized that inertial mass and gravitational mass are equivalent. It should also be pointed out that F in equation (2.28) gives the weight of an object of mass m_1 when m_2 represents the mass of earth and r is the distance from the object to the center of the earth. Weight is merely a measure of the gravitational attractive force between an object and the earth, and therefore varies from point to point on the surface of the earth, according to the distance of the point from the earth's center. On the surface of another planet, the weight of the same object would be different from that on earth because of the different size and mass of that planet and its consequent different attractive force. In outer space, if the object is not under the gravitational influence of any heavenly body, it must be weightless. Mass, on the other hand, is a measure of the amount of matter and its numerical value is therefore independent of the point in the universe where it is measured.

The gravitational force between the electron and the proton is

$$F = \frac{6.67 \times 10^{-11}\,\text{N-m}^2/\text{kg}^2 \times 9.11 \times 10^{-31}\,\text{kg} \times 1.67 \times 10^{-27}\,\text{kg}}{(5 \times 10^{-11}\,\text{m})^2}$$

$$= 4.05 \times 10^{-47}\,\text{N}.$$

It is immediately apparent that in the interaction between charged particles, gravitational forces are extremely small in comparison with the electrical forces acting between the particles, and may be completely neglected in most instances.

Electrical Potential: The Volt

If one charge is held rigidly, and another charge is placed in the electrical field of the first charge, it will have a certain amount of potential energy relative to any other point within the electric field. In the case of electrical potential energy, the reference point is taken at an infinite distance from the charge that sets up the electric field; that is, at a point far enough away from the charge so that its effect is negligible. As a consequence of the fact that these charges do not interact electrically, a value of zero is arbitrarily assigned to the potential energy in the system of charges; the charge at an infinite distance from the one that sets up the electric field has no electrical potential energy. If the two charges are of the same sign, bringing them closer together requires work, or the expenditure of energy, in order to overcome the repulsive force between the two charges. Since work was done in bringing the two charges together, the potential energy in the system of charges is now greater than it was initially. On the other hand, if the two charges are of opposite sign, then a decrease in distance between them occurs spontaneously because of the attractive forces, and work is done by the system. The potential energy of the system consequently decreases; that is, the potential energy of the freely moving charge, with respect to the rigidly held charge, decreases. This is exactly analogous to the case of a freely falling mass whose potential energy decreases as it approaches the surface of the earth. In the case of the mass in the earth's gravitational field, however, the reference point for potential energy of the mass is arbitrarily set on the surface of the earth. This means that the mass has no potential energy when it is lying right on the earth's surface. All numerical values for potential energy of the mass, therefore, are positive numbers. In the case of electrical potential energy, however, as a consequence of the arbitrary convention that the point of the zero numerical value is at an infinite distance from the charge that sets up the electric field, the numerical values for the potential energy of a charge, owing to attractive electrical forces, must be negative.

The quantitative aspects of electrical potential energy may be investigated with the aid of Fig. 2.2, which shows a charge $+Q$ that sets up an electric field which extends uniformly in all directions. Another charge, $+q$, is used to explore the electric field set up by Q. When

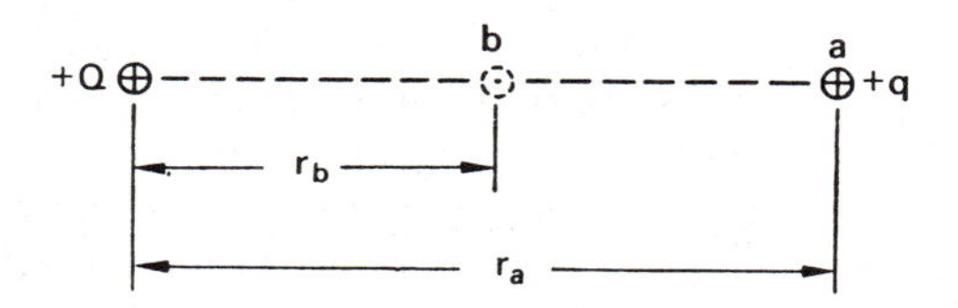

FIG. 2.2. Diagram illustrating work done in moving a charge between two points of different potential in an electric field.

the exploring charge is at point a, at a distance r_a cm from Q, it has an amount of potential energy that depends on the magnitudes of Q, q, and r_a. If the charge q is now to be moved to point b, which is closer to Q, then, because of the repulsive force between the two charges, work is done in moving the charge from point a to point b. The amount of work that is done in moving charge q from point a to point b may be calculated by multiplying the force exerted on the charge q by the distance through which it was moved, in accordance with equation (2.2). From equation (2.24), however, it is seen that the force is not constant, but varies inversely with the square of the distance between the charges. The magnitude of the force, therefore, increases rapidly as the charge q approaches Q, and increasingly greater amounts of work are done when the exploring charge q is moved a unit distance. The movement of the exploring charge may be accomplished by a series of infinitesimally small movements, during each of which an infinitesimally small amount of work is done. The total energy expenditure, or increase in potential energy of the exploring charge, is then merely equal to the sum of all the infinitesimal increments of work. This infinitesimal energy increment is given by

$$dW = -f\,dr \qquad (2.7)$$

(the minus sign is used here because an increase in potential energy results from a decrease in distance between the charges) and, if the value for f from equation (2.24) is substituted into equation (2.7), we have

$$dW = -k_0 \frac{Qq}{r^2} dr, \tag{2.29}$$

$$W = -k_0 Qq \int_{r_a}^{r_b} \frac{dr}{r^2}. \tag{2.30}$$

Integration of equation (2.30) gives

$$W = k_0 Qq \left(\frac{1}{r_b} - \frac{1}{r_a} \right). \tag{2.31}$$

If the distances a and b are measured in meters, and if the charges are given in coulombs, then the energy W is given in joules.

Example 2.5

If in Fig. 2.2, Q is $+44.4\ \mu C$, q is $5\ \mu C$, and r_a and r_b are 2 and 1 m, respectively, then the work done in moving the $5\ \mu C$ charge from point a to point b is, from equation (2.31),

$$W = 9 \times 10^9 \frac{\text{N-m}^2}{\text{C}^2} \times 44.4 \times 10^{-6}\,\text{C} \times 5 \times 10^{-6}\,\text{C} \left(\frac{1}{1\text{ m}} - \frac{1}{2\text{ m}} \right)$$

$$= 1\text{ N-m} = 1\text{ J}.$$

In this example, 1 joule of work was expended in moving $5\ \mu C$ of charge from a to b. The work per unit charge is

$$\frac{W}{C} = \frac{1\text{ J}}{5 \times 10^{-6}\,\text{C}} = 200{,}000 \frac{\text{J}}{\text{C}}.$$

We therefore say that the potential difference between points a and b is 200,000 volts, since

> One volt of potential difference exists between any two points in an electric field if one joule of energy is expended in moving a charge of one coulomb between the two points.

Expressed more concisely, the definition of a volt is

$$1\text{ V} = 1\text{ J/C}.$$

In example 2.5, point b is the point of higher potential with respect to point a, because work had to be done on the charge to move it to b from a.

The electrical potential at any point due to an electric field from a point charge Q is defined as the potential energy that a unit positive exploring charge, $+q$, would have if it were brought from a point at an infinite distance from Q to the point in question. The electrical potential at point b in Fig. 2.2 can be computed from equation (2.30) by setting distance r_a equal to infinity. The potential at point b, V_b, which is defined as the potential energy per unit positive charge at b, is, therefore,

$$V_b = \frac{W}{q} = k_0 \frac{Q}{r_b}. \tag{2.32}$$

Example 2.6

(a) What is the potential at a distance of 5×10^{-11} m from a proton?

$$V = k_0 \frac{Q}{r} = 9 \times 10^{-9} \frac{\text{N-m}^2}{\text{C}^2} \times \frac{1.6 \times 10^{-19}\,\text{C}}{5 \times 10^{-11}\,\text{m}}$$
$$= 28.8 \text{ volts.}$$

(b) What is the potential energy of another proton at this point? According to equation (2.32), the potential energy of the proton is equal to the product of its charge and the potential of its location.

$$\therefore \quad W = Vq = 28.8\ \text{V} \times 1.6 \times 10^{-19}\,\text{C}$$
$$= 4.6 \times 10^{-18}\,\text{J}.$$

The Electron Volt: A Unit of Energy

If two electrodes are connected to the terminals of a source of voltage, as shown in Fig. 2.3, then a charged particle anywhere in the electric field between the two plates will have an amount of potential energy given by equation (2.32).

$$W = Vq,$$

where V is the electrical potential at the point occupied by the charged particle. If, for example, the cathode in Fig. 2.3 is one volt negative with respect to the anode and the

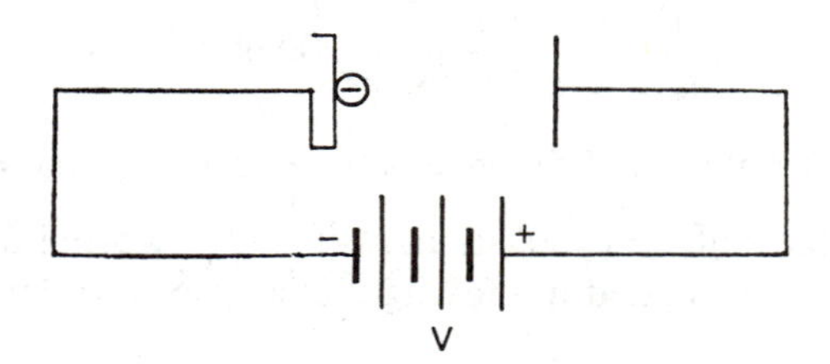

FIG. 2.3. Diagram showing the potential energy in an electric field.

charged particle is an electron on the surface of the cathode, then the potential energy of the electron with respect to the anode is

$$W = Vq$$
$$= -1\,\text{V}(-1.6 \times 10^{-19}\,\text{C})$$
$$= 1.6 \times 10^{-19}\,\text{J}.$$

This amount of energy, 1.6×10^{-19} joule, is called an *electron volt*, and is abbreviated eV. Since the magnitude of the electron volt is convenient when dealing with the energetics of atomic and nuclear mechanics, this quantity of energy is taken as a unit and, consequently, is frequently used in health physics. Multiples of the electron volt are the keV, 10^3 eV; the MeV, 10^6 eV; and the BeV, 10^9 eV.

Example 2.7

How many electron volts of energy correspond to the mass of a resting electron?

$$\begin{aligned} E &= mc^2 \\ &= 9.11 \times 10^{-31}\ \text{kg} \times (3 \times 10^8\ \text{m/sec})^2 \\ &= 81.99 \times 10^{-15}\ \text{joule.} \end{aligned}$$

Since there are 1.6×10^{-19} J/eV,

$$\begin{aligned} E &= \frac{81.99 \times 10^{-15}\ \text{J}}{1.6 \times 10^{-19}\ \text{J/eV}} \\ &= 0.51 \times 10^6\ \text{eV.} \end{aligned}$$

It should be emphasized that, although the numerical value for the electron volt was calculated by computing the potential energy of an electron at a potential of 1 volt, the electron volt is not a unit of electrons or volts; it is a unit of energy, and may be interchanged (after numerical correction) with any other unit of energy.

Example 2.8

How many electron volts (eV) of heat must be added to change 1 liter of water, whose temperature is 50°C, to completely dry steam?

The specific heat of water is 1 calorie per gram, and the heat of vaporization of water is 539 calories per gram.

$$\begin{aligned} \therefore \quad \text{Heat added} &= 1000\ \text{g}\left(1\,\frac{\text{cal}}{\text{g deg}}(100 - 50)\ \text{deg} + 539\,\frac{\text{cal}}{\text{g}}\right) \\ &= 589{,}000\ \text{calories.} \end{aligned}$$

Since there are $4.186\,\frac{\text{J}}{\text{cal}}$ and $1.6 \times 10^{-19}\,\frac{\text{J}}{\text{eV}}$, we have:

$$\begin{aligned} \text{heat added} &= \frac{5.89 \times 10^5\ \text{cal} \times 4.186\ \text{J/cal}}{1.6 \times 10^{-19}\ \text{J/eV}} \\ &= 1.54 \times 10^{25}\ \text{eV.} \end{aligned}$$

The answer to Example 2.8 is an astronomically large number (but not very much energy on the scale of ordinary physical and chemical reactions) and shows why the electron volt is a useful energy unit only for reactions in the atomic world.

Example 2.9

An alpha particle, whose charge is $+(2 \times 1.6 \times 10^{-19})$ C and whose mass is 6.601×10^{-27} kg, is accelerated across a potential difference of 100,000 V. What is its kinetic energy, in joules and in electron volts; and how fast is it moving?

The potential energy of the alpha particle at the moment it starts to accelerate is, from equation (2.32),

$$\begin{aligned} W &= Vq \\ &= 10^5\ \text{V} \times 2 \times 1.6 \times 10^{-19}\ \text{C} \\ &= 3.2 \times 10^{-14}\ \text{J.} \end{aligned}$$

In terms of electron volts

$$W = \frac{3.2 \times 10^{-14}\,\text{J}}{1.6 \times 10^{-19}\,\text{J/eV}}$$

$$= 200{,}000\ \text{eV}.$$

Since all the alpha's potential energy is converted into kinetic energy after the alpha particle falls through the 100,000 volt (100 kV) potential difference, the kinetic energy must then be 200,000 eV (200 keV).

The velocity of the alpha particle may be computed by equating its potential and kinetic energies,

$$Vq = \tfrac{1}{2}mv^2, \tag{2.33}$$

and solving for v:

$$v = \sqrt{\left(\frac{2\,Vq}{m}\right)}$$

$$= \sqrt{\left(\frac{2 \times 10^5\ \text{V} \times 3.2 \times 10^{-19}\ \text{C}}{6.601 \times 10^{-27}\ \text{kg}}\right)}$$

$$= 3.11 \times 10^6\,\frac{\text{m}}{\text{sec}}.$$

Electric Field

The term "electric field" was used in the preceding sections of the chapter without explicitly defining the term. Implicit in the use of the term, however, was the connotation by the context that an electric field is any region where electric forces act. An electric field is not merely a descriptive term; to define an electric field requires a number to specify the magnitude of the electric forces that act in the electric field and a direction in which these forces act. The strength of an electric field is called the electric field intensity, and may be

FIG. 2.4. The force on an exploring charge q in the electric field of charge Q.

defined in terms of the force (magnitude and direction) that acts on a unit-exploring charge which is placed into the electric field. Consider an isolated charge, $+Q$, that sets up an electric field, and an exploring charge, $+q$, that is used to investigate the electric field, as shown in Fig. 2.4. The exploring charge will experience a force in the direction shown and of a magnitude given by equation (2.24).

$$f = k_0\frac{Qq}{r^2}. \tag{2.24}$$

The force per unit charge at the point r meters from charge Q is the *electric field* intensity at that point, and is given by the equation

$$\mathscr{E} = \frac{f}{q}\ \text{N/C} = k_0\frac{\text{N-m}^2}{\text{C}^2} \times \frac{q}{r^2}\ \text{C/m}^2. \tag{2.34}$$

According to equation (2.34), electric field intensity is expressed in units of force per unit charge, i.e., in newtons per coulomb. It should be emphasized that $\mathscr{E}$ is a vector quantity, that is, that it has direction as well as magnitude.

Example 2.10

(a) What is the electric field intensity at point P due to the two charges $+6$ and $+3$ C, shown in Fig. 2.5a. The electric field intensity at point P due to the $+6$ C charge is

$$\mathscr{E}_1 = k_0 \frac{Q_1}{r_1^2} = 9.9 \times 10^9 \frac{\text{N-m}^2}{\text{C}^2} \times \frac{6\ \text{C}}{(2\ \text{m})^2} = 1.5 \times 10^{10}\ \text{N/C},$$

and acts in the direction shown in Fig. 2.5a. (The magnitude of the field intensity is shown graphically by a vector whose length is proportional to the field intensity. In Fig. 2.5a, the

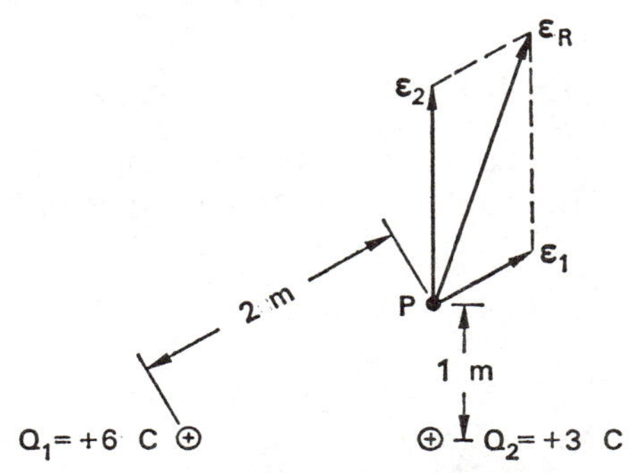

FIG. 2.5a. Resultant electric field from two positive charges.

scale is 1 cm $= 1 \times 10^{10}$ N/C. $\mathscr{E}_1$ therefore is drawn 1.5 cm long). $\mathscr{E}_2$, the electric field intensity at P due to the $+3$ C charge is

$$\mathscr{E}_2 = k_0 \frac{Q_2}{r_2^2} = 9.9 \times 10^9 \frac{\text{N-m}^2}{\text{C}^2} \times \frac{3\ \text{C}}{(1\ \text{m})^2} = 2.97 \times 10^{10}\ \text{N/C},$$

and acts along the line Q_2P_1, as shown in the illustration. The resultant electrical intensity at point P is the vector sum of $\mathscr{E}_1$ and $\mathscr{E}_2$. If these two vectors are accurately drawn in magnitude and in direction, the resultant may be obtained graphically by completing the parallelogram of forces and drawing the diagonal $\mathscr{E}_R$. The length of the diagonal is proportional to the magnitude of the resultant electric field intensity and its direction shows the direction of the electric field at point P. In this case since 1×10^{10} N/C is represented by 1 cm, the resultant electric field intensity is found to be about 4×10^{10} N/C, and it acts in a direction 30° clockwise from the vertical. The value of $\mathscr{E}_R$ may also be determined from the law of cosines

$$a^2 = b^2 + c^2 - 2bc \cos A, \tag{2.35}$$

where b and c are two adjacent sides of a triangle, A is the included angle, and a is the side opposite angle A. In this case, b is 2.97×10^{10}, c is 1.49×10^{10}, angle A is 120°, and a is the resultant, $\mathscr{E}_R$, the electric field intensity whose magnitude is to be found. From equation (2.35), we find

$$\mathscr{E}_R^2 = (2.97 \times 10^{10})^2 + (1.49 \times 10^{10})^2 - 2(2.97 \times 10^{10})(1.49 \times 10^{10}) \cos 120°,$$

$$\mathscr{E}_R = 3.97 \times 10^{10}\ \text{N/C}.$$

(b) What is the magnitude and direction of $\mathscr{E}$ if the 3C charge is negative and the 6C charge is positive?

In this case, the magnitudes of $\mathscr{E}_1$ and $\mathscr{E}_2$ would be exactly the same as in part (a) of this example; the direction of $\mathscr{E}_1$ would also remain unchanged, but the direction of $\mathscr{E}_2$

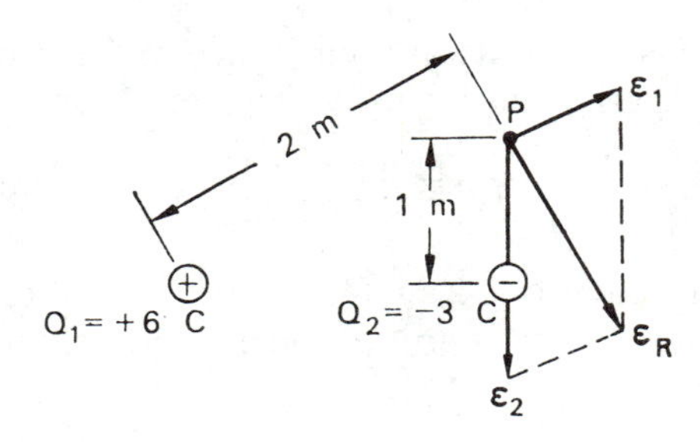

FIG. 2.5b. Resultant electric field from two opposite charges.

would be toward the -3C charge, as shown in Fig. 2.5b. From the geometrical arrangement, it is seen that the resultant intensity acts in a direction 120° clockwise from the vertical. The magnitude of $\mathscr{E}_R$, from equation (2.35), is

$$\mathscr{E}_R^2 = (2.97 \times 10^{10})^2 + (1.49 \times 10^{10})^2 - 2(2.97 \times 10^{10})(1.49 \times 10^{-10}) \cos 60°,$$

$$\mathscr{E}_R = 2.6 \times 10^{10} \text{ N/C}.$$

Point charges result in non-uniform electric fields. A uniform electric field may be produced by applying a potential difference across two large parallel plates made of electrical conductors separated by an insulator, as shown in Fig. 2.6. The electric intensity

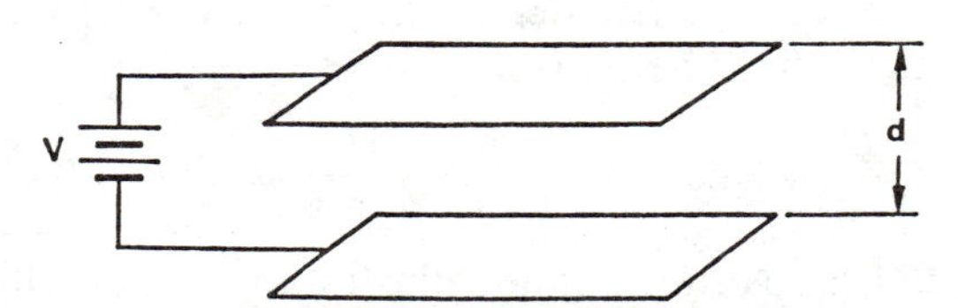

FIG. 2.6. Conditions for producing a relatively uniform electric field. The field will be quite uniform throughout the region between the two plates, but will be distorted at the edges of the plates.

throughout the region between the two plates is $\mathscr{E}$ newtons per coulomb. The force acting on any charge within this field, therefore, is

$$f = \mathscr{E}q \text{ newtons} \tag{2.36}$$

If the charge q happens to be positive, then, to move it across the distance d, from the negative to the positive plates, against the electrical force in the uniform field requires the expenditure of energy given by the equation

$$W = fd = \mathscr{E}qd. \tag{2.37}$$

However, since potential difference is defined as work per unit charge, equation (2.37) may be written as

$$V = \frac{W}{q} = \mathscr{E}d, \tag{2.38}$$

or

$$\mathscr{E} = \frac{V}{d} \frac{\text{V}}{\text{m}}. \tag{2.39}$$

Equation (2.39) expresses electric field intensity in the units most commonly used for this purpose: volts per meter.

A non-uniform electric field that is of interest to the health physicist (in instrument design) is that due to a potential difference applied across two coaxial conductors, as shown in Fig. 2.7. If the radius of the inner conductor is a meters, that of the outer conductor b meters, then the electric intensity at any point between the two conductors, r meters from the center, is given by

$$\mathscr{E} = \frac{1}{r} \frac{V}{\ln b/a} \frac{\text{V}}{\text{m}}, \tag{2.40}$$

where V is the potential difference between the two conductors.

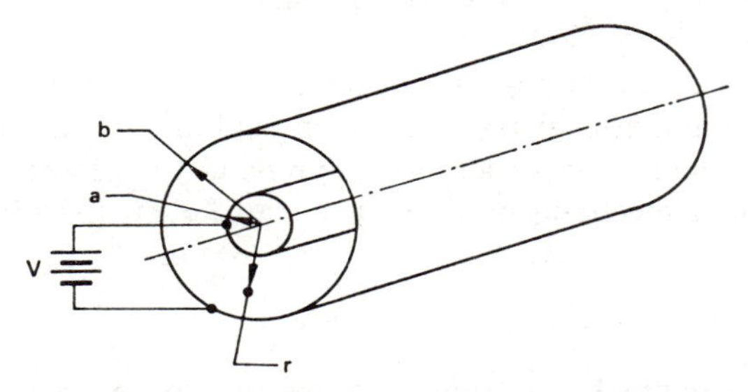

FIG. 2.7. Conditions for the non-uniform electric field between two coaxial conductors given by equation (2.40).

Example 2.11

A Geiger counter is constructed of a wire anode whose diameter is 0.1 mm and a cathode, coaxial with the anode, whose diameter is 2 cm. If the voltage across the tube is 1000 V, what is the electric field intensity (a) at a distance of 0.03 mm from the surface of the anode, and (b) at a point midway between the center of the tube and the cathode?

(a) $$\mathscr{E} = \frac{1}{r} \frac{V}{\ln b/a}$$

at $r = \frac{1}{2}(0.01) + 0.003 = 0.008$ cm

$= 8 \times 10^{-5}$ m, we have

$$\mathscr{E} = \frac{1}{8 \times 10^{-5}\ \text{m}} \times \frac{1000\ \text{V}}{\ln 1/0.005}$$

$$= 2.36 \times 10^{6} \frac{\text{V}}{\text{m}}$$

(b) At $r = 0.01$ m,

$$\mathscr{E} = \frac{1}{0.01\ \text{m}} \times \frac{1000\ \text{V}}{\ln 1/0.005}$$

$$= 1.89 \times 10^{4} \frac{\text{V}}{\text{m}}.$$

It should be noted that in the case of coaxial geometry, extremely intense electric fields may be obtained with relatively small potential differences. Such large fields require mainly a large ratio of outer to inner electrode radii.

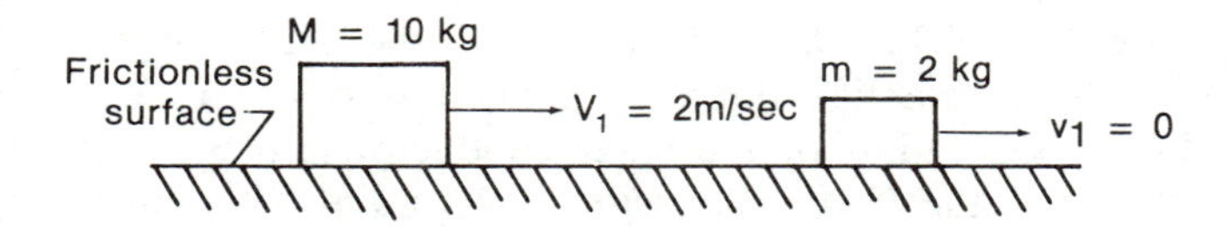

Fig. 2.8. Elastic collision between blocks *M* and *m*, in which the sum of both kinetic energy and momenta of the two blocks before and after the collision are the same.

Energy Transfer

In a quantitative sense, the biological effects of radiation depends on the amount of energy absorbed by living matter from a radiation field and by the spatial distribution, in tissue, of the absorbed energy. In order to comprehend the physics of tissue irradiation, therefore, some pertinent mechanisms of energy transfer must be understood.

Elastic Collision

An elastic collision is defined as a collision between two bodies in whcih kinetic energy and momentum are conserved; that is, the sum of the kinetic energy of the two bodies before the collision is equal to the sum after collision, and the sum of the momenta before and after the collision is the same. In an elastic collision, the total kinetic energy is redistributed between the colliding bodies; one body gains energy at the expense of the other. A simple case is illustrated in the example below.

Example 2.12

A block made of perfectly elastic material, whose mass is 10 kg, slides on a frictionless surface with a velocity of 2 m/sec, and strikes a stationary elastic block whose mass is 2 kg. How much energy was transferred from the large block to the small block during the collision?

If V_1, v_1, V_2, and v_2 are the respective velocities of the large and small blocks before and after the collision, then, according to the laws of conservation of energy and momentum, we have

$$\tfrac{1}{2}MV_1^2 + \tfrac{1}{2}mv_1^2 = \tfrac{1}{2}MV_2^2 + \tfrac{1}{2}mv_2^2, \tag{2.41}$$

$$MV_1 + mv_1 = MV_2 + mv_2. \tag{2.42}$$

Since $v_1 = 0$, equations (2.41) and (2.42) may be solved simultaneously to give $V_2 = 1\tfrac{1}{3}$ m/sec and $v_2 = 3\tfrac{1}{3}$ m/sec.

The kinetic energy transferred during the collision is

$$\tfrac{1}{2}MV_1^2 - \tfrac{1}{2}MV_2^2 = \tfrac{1}{2} \times 10\left(4 - \frac{16}{9}\right) = 11\tfrac{1}{9}\ \text{J},$$

and this, of course, is the energy gained by the smaller block:

$$\tfrac{1}{2}mv^2 = \tfrac{1}{2} \times 2 \times \frac{100}{9} = 11\tfrac{1}{9}\ \text{J}.$$

Note that the magnitude of the force exerted by the larger block on the smaller block during the collision was not considered in the solution of Example 2.12. The reason for not explicitly considering the force in the solution can be seen from equation (2.10), which may be written as

$$f\,\Delta t = m\,\Delta v. \tag{2.10A}$$

According to equation (2.10A), the force necessary to change the momentum of a block is dependent on the time during which it acts. The parameter of importance in this case is the product of the force and the time. This parameter is called the impulse; equation (2.10A) may be written in words as

$$\text{impulse} = \text{change of momentum}.$$

The length of time during which the force acts depends on the relative velocity of the system of moving masses and on the nature of the mass. Generally, the more "give" in the colliding blocks, the greater will be the time of application of the force and the smaller, consequently, will be the magnitude of the force. For this reason, for example, a baseball player who catches a ball moves his hand back at the moment of impact, thereby increasing the time during which the stopping force acts and decreasing the shock to his hand. For this same reason, a jumper flexes his knees as his feet strike the ground, thereby increasing the time that his body comes to rest and decreasing the force on his body. For example, a man who jumps down a distance of 1 meter is moving with a velocity of 4.43 m/sec at the instant that he strikes the floor. If he weighs 70 kg, and if he lands rigidly flat footed and is brought to a complete stop in 0.01 sec, then the stopping force, from equation (2.10), is 3.1×10^4 N, or 6980 lb. If, however, he lands on his toes, then lowers his heels, and flexes his knees as he strikes, thereby increasing his actual stopping time to 0.5 sec, the average stopping force is only about 140 lb.

In the case of the two blocks in Example 2.12, if the time of contact is 0.01 sec, then the average force of the collision during this time interval is

$$f = \frac{10\ \text{kg} \times 0.0067\ \text{m/sec}}{0.01\ \text{sec}} = 6.7\ \text{N}$$

The instantaneous forces acting on the two blocks vary from zero at the instant of impact to a maximum value at some time during the collision, then to zero again as the second block leaves the first one. This may be graphically shown in Fig. 2.9, a curve of force vs. time during the collision. The average force during the collision is the area under the curve divided by the time that the two blocks are in contact.

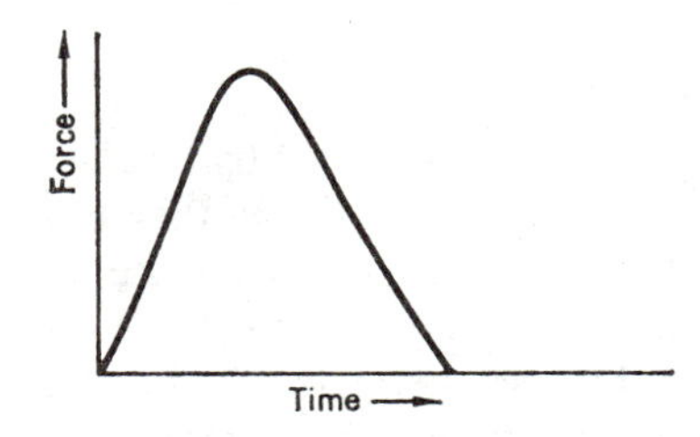

FIG. 2.9. The variation in time of the force between the colliding bodies.

In the case of a collision between two masses, such as that described above, one block exerts a force on the other only while the two blocks are in "contact." During "contact" the two blocks seem to be physically touching each other. Actually, however, the two blocks are merely very close together, too close in fact to perceive any space between them. Under this condition the two blocks repel each other by very short-range forces that are thought to be electrical in nature. (These forces will be discussed again in Chapter 3.) This concept of a "collision" without actual contact between the colliding masses may be easily demonstrated with the aid of magnets. If magnets are affixed to the two blocks in Example 2.12, as shown in Fig. 2.10, then the magnetic force, which acts over relatively long distances, will repel the two blocks, and block B will move. If the total mass of each block, including the magnet, remains the same as in example 2.12, then the calculations and results of Example 2.12 are applicable. The only difference between the physical "collision" and magnetic "collision" is that the magnitude of the force in the former case is greater than in the latter instance, but the time during which the forces are effective is greater in the case of the magnetic "collision." In both instances, the product of average force and time is exactly the same.

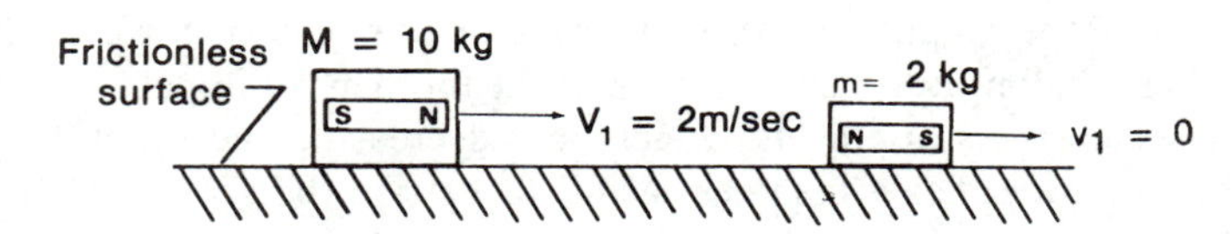

FIG. 2.10. "Collision" between two magnetic fields.

Inelastic Collision

If the conditions in Fig. 2.8 are modified by fastening block B to the floor with a rubber band, then, in order to break the rubber band and cause the block to slide freely, block A must transfer at least sufficient energy to break the rubber band. Any additional energy transferred would then appear as kinetic energy of Block B. If the energy necessary to break the rubber band is called the binding energy of block B, then the kinetic energy of block B after it is struck by block A is equal to the difference between the energy lost by A and the binding energy of B. Algebraically, this may be written as

$$E_b = E_a - \phi, \tag{2.43}$$

where E_a is the energy lost by block A and ϕ is the binding energy of block B. In a collision of this type, where energy is expended to free one of the colliding bodies, kinetic energy is not conserved, and the collision is therefore not elastic, i.e., it is inelastic.

Example 2.13

A stationary block B, whose mass is 2 kg is held by an elastic cord whose elastic constant is 10 N/m and whose ultimate strength is 5 N. Another block A, whose mass is 10 kg is moving with a velocity of 2 m/sec on a frictionless surface. If block A strikes block B, with what velocity will block B move after the collision?

From Example 2.12 it is seen that the energy lost by block A in this collision is $11\frac{1}{9}$ joules. The energy expended in breaking the rubber band may be calculated from the product of the force needed to break the elastic cord and the distance that the elastic cord stretches before breaking. In the case of a spring, rubber band, or any other substance that is elastically deformed, the deforming force is opposed by a restoring force whose magnitude is proportional to the deformation, that is,

$$f = kr, \tag{2.44}$$

where f is the force needed to deform the elastic body a distance r, and k is the "spring constant," or the force per unit deformation. Since equation (2.44) shows that the force is not constant, but rather is proportional to the deformation of the rubber band, the work done in stretching the rubber band must be computed by application of the calculus. The infinitesimal work done in stretching the rubber band through a distance dr is

$$\mathrm{d}W = f\,\mathrm{d}r, \tag{2.7}$$

and the total work done in stretching the rubber band from 0 to r is given by equation (2.8)

$$W = \int_0^r f\,\mathrm{d}r. \tag{2.8}$$

Substituting equation (2.44) for f, we have

$$W = \int_0^r kr\,\mathrm{d}r, \tag{2.45}$$

and solving equation (2.45) shows the work done in stretching the rubber band to be

$$W = \frac{kr^2}{2}. \tag{2.46}$$

Since in this example k is equal to 10 N/m, the ultimate strenght of the elastic cord, 5 N, is reached when the rubber is extended 0.5 m. With these numerical values, equation (2.46) may be solved:

$$\begin{aligned} W &= \frac{10\ \mathrm{N/m} \times (0.5\ \mathrm{m})^2}{2} \\ &= 1.25 \text{ newton meters,} \\ &= 1.25 \text{ joules.} \end{aligned}$$

Of the $11\frac{1}{9}$ J lost by block A in its collision with block B, 1.25 J are dissipated in breaking the elastic cord that holds block B (the binding energy). The kinetic energy of block B, therefore, is, from equation (2.43),

$$E_b = 11.11 - 1.25 = 2.89\ \mathrm{J}.$$

If block A would have had less than 1.25 J of kinetic energy, then the elastic cord would not have been broken; the restoring force in the elastic cord would have pulled block B back, and would have caused it to oscillate about its equilibrium position. (For this oscillation to actually occur, block A would have to be withdrawn immediately after the collision, otherwise block B, on its rebound, would transfer its energy back to block A, and send it back with the same velocity that it had before the first collision. The net effect of the two collisions, then, would have been only the reversal of the direction in which block A traveled.)

Waves

Energy may be transmitted by disturbing a "medium," permitting the disturbance to travel through the medium, and then collecting the energy with a suitable receiver. For example, if work is done in raising a stone, and the stone is dropped into water, the potential energy of the stone before being dropped is converted into kinetic energy, which is then transferred to the water when the stone strikes. The energy gained by the water disturbs the water and causes it to move up and down. This disturbance spreads out from the point of the initial disturbance at a velocity characteristic of the medium (in this case the water). The energy can be "received" at a remote distance from the point of the initial disturbance by a bob that floats on the water. The wave, in passing by the bob, will cause the bob to move up and down, thereby imparting energy to it. It should be noted here that the *water* moves only in a vertical direction; while the *disturbance* moves in the horizontal direction (Fig. 2.11).

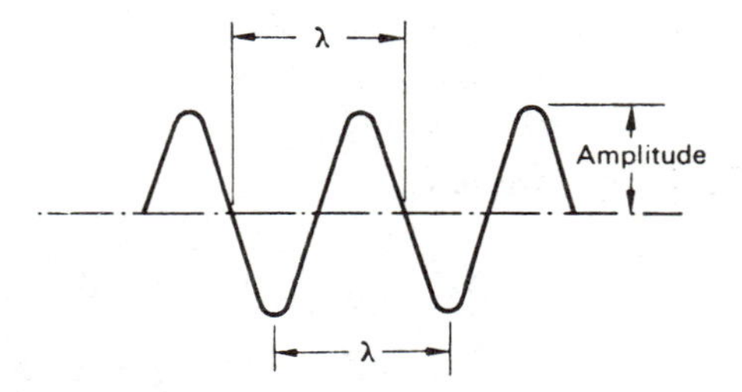

FIG. 2.11. Graphical representation of a wave.

Displacement of water upwards from the undisturbed surface produces a crest, while downward displacement results in a trough. The amplitude of a wave is a measure of the vertical displacement, and the distance between corresponding points on adjacent disturbances is called the wavelength. (The wavelength is usually represented by the Greek letter lambda, λ). The number of disturbances per second at any point in the medium is called the frequency. The velocity with which a wave (disturbance) travels is equal to the product of the wavelength and the frequency,

$$v = f\lambda. \tag{2.47}$$

Example 2.14

Sound waves, which are disturbances in the air, travel through air at a velocity of 344 meters per second. Middle C has a frequency of 264 Hz (cycles per second).

Calculate the wavelength of this note.

$$\lambda = \frac{v}{f}$$
$$= \frac{344 \text{ m/sec}}{264 \text{ sec}^{-1}}$$
$$= 1.3 \text{ m}.$$

If more than one disturbance passes through a medium at the same time, then, where the respective waves meet, the total displacement of the medium is equal to the algebraic sum of the two waves. For example, if two rocks are dropped into a pond, then, if the crests of the two waves should coincide as the waves pass each other, the resulting crest is equal to the height of the two separate crests, and the trough is as deep as the sum of the two individual troughs, as shown in Fig. 2.12. If, on the other hand, the two waves are exactly out of phase, that is, if a crest of one coincides with the trough of the other,

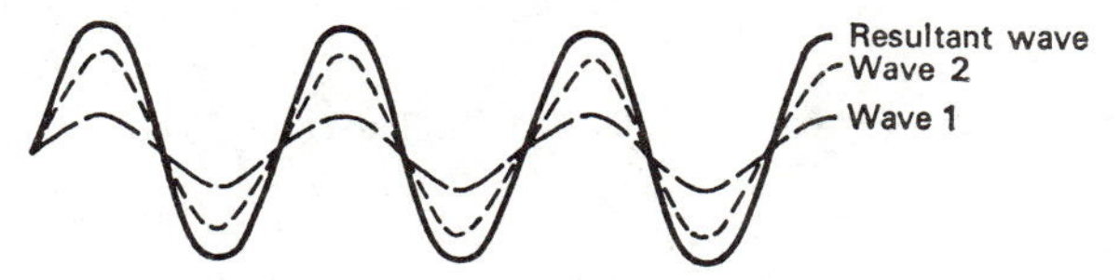

FIG. 2.12. The addition of two waves of equal frequency, and in phase.

then the positive and negative displacements cancel each other, as shown in Fig. 2.13. If, in Fig. 2.13, wave 1 and wave 2 are of exactly the same amplitude as well as the same frequency, there would be no net disturbance. For the more general case, in which the component waves are of different frequencies, different amplitudes, and only partly out of phase, complex wave forms may be formed, as seen below in Fig. 2.14.

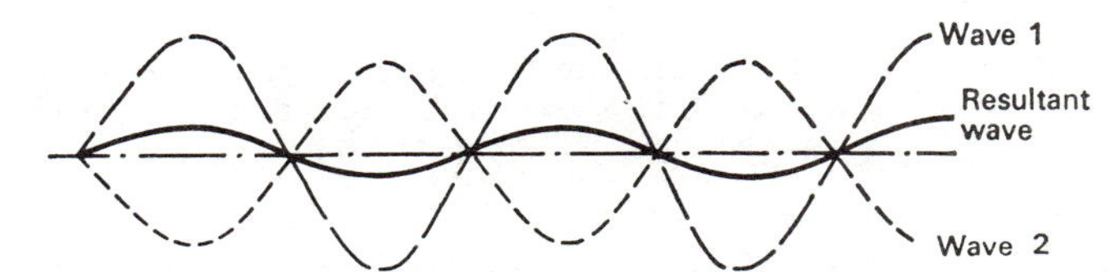

FIG. 2.13. The addition of two waves of equal frequency, but different amplitude, and 180° out of phase.

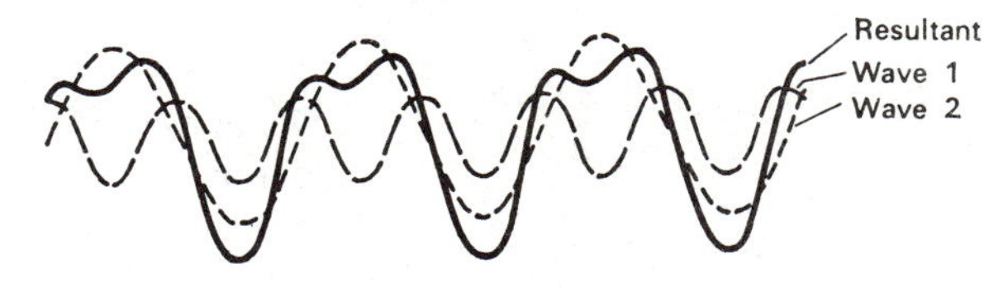

FIG. 2.14. Complex wave formed by the algebraic addition of two different pure waves.

Electromagnetic Waves

In 1820, Oersted, a Danish physicist, observed that a compass needle deflected whenever it was placed in the vicinity of a current-carrying wire. He thus discovered the intimate relationship between electricity and magnetism, and found that a magnetic flux coaxial with the wire is always induced in the space around a current carrying wire. Furthermore, he found that the direction of deflection of the compass needle depended on the direction of the electric current, thus showing that the induced magnetic flux has direction as well as magnitude. The direction of the induced magnetic flux can be determined by the "righthand rule": if the fingers of the right hand are curled around the wire, as though grasping the wire, with the thumb outstretched and pointing in the direction of conventional current flow, then the curled fingers point in the direction of the induced magnetic flux. If two parallel current carrying wires are near each other, they either attract or repel one another, depending on whether the currents flow in the same or in opposite directions. The attractive or repulsive force (F) per unit length (l) of wire, as shown in Fig. 2.15, is proportional to the product of the currents, and inversely proportional to the distance between the wires.

$$\frac{F}{l} \propto \frac{i_1 \times i_2}{r} = k_m \times \frac{i_1 \times i_2}{r}. \tag{2.48}$$

If the current-carrying wires are in free space (or in air), if i_1 and i_2 are 1 ampere each,

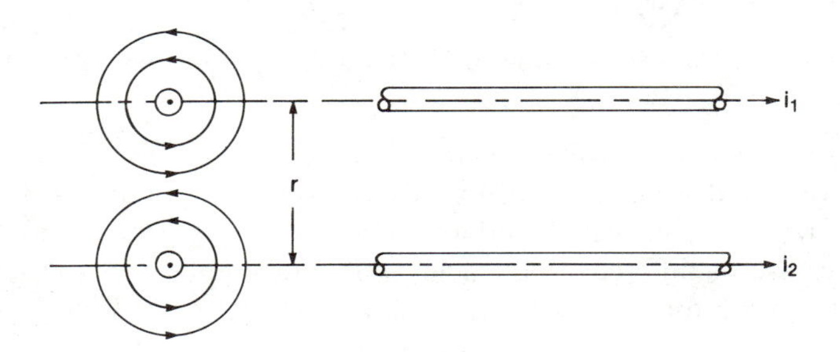

FIG. 2.15. Force between two parallel current-carrying wires. The force, in this case attractive, is due to the magnetic fields shown by the circular lines in the end view that are generated by the electric current.

and if the distance r between the wires is 1 meter, then the force per unit length of wire is found to be

$$\frac{F}{l} = 2 \times 10^{-7}\ \text{N/m}.$$

The constant of proportionality, k_m, therefore, is $k_m = 2 \times 10^{-7}\ \text{N/A}^2$. It is convenient to define k_m in terms of another constant, μ_0:

$$k_m = \frac{\mu_0}{2\pi} = 2 \times 10^{-7}\ \text{N/A}^2, \tag{2.49A}$$

$$\mu_0 = 4\pi \times 10^{-7}\ \text{N/A}^2. \tag{2.49B}$$

μ_0 is called the *permeability* of free space. Permeability is a property of the medium in which magnetic flux is established. The permeability of any medium other than free space is designated by μ. The *relative permeability* of any medium, K_m, is defined by

$$K_m = \frac{\mu}{\mu_0}. \tag{2.50}$$

Iron, cobalt, nickel, and gadolinium have high values of relative permeability, i.e., $K_m \gg 1$; these substances we called *ferromagnetic*. Those substances, such as silver, copper, and bismuth, whose relative permeability is less than one are said to be diamagnetic. Most substances including all biological materials, have relative permeabilities of one or slightly greater; these materials are called *paramagnetic*.

The unit of magnetic flux is called the *weber* (Wb):

$$1\ \text{weber} = 1\ \text{joule/ampere}, \tag{2.51}$$

and the unit of flux density, which is a measure of magnetic intensity, is called the tesla, T (after the Croatian born American electrical engineer Nikola Tesla):

$$1\ \text{tesla} = 1\ \text{weber/m}^2. \tag{2.52}$$

Since joule = newton × meter, the dimensions of μ_0 may be written as

$$\mu_0 = 4\pi \times 10^{-7}\ \mathrm{J/A^2\text{-}m}.$$

Furthermore, since weber × ampere = joule, μ_0 may also be expressed as

$$\mu_0 = 4\pi \times 10^{-7}\ \mathrm{Wb/A\text{-}m}.$$

The magnetic flux density at a distance r from a wire carrying a current i is given by

$$B = \frac{\mu_0}{2\pi} \times \frac{i}{r}\ \mathrm{Wb/m^2}. \tag{2.53}$$

Example 2.15

What is the magnetic flux density at a distance of 0.1 m from a wire that carries a current of 0.25 ampere? Substituting the numerical values into equation (2.53) yields

$$\begin{aligned} B &= \frac{4\pi \times 10^{-7}\ \mathrm{Wb/A\ m}}{2\pi} \times \frac{0.25\ \mathrm{A}}{0.1\ \mathrm{m}} \\ &= 5 \times 10^{-5}\ \mathrm{Wb/m^2} = 5 \times 10^{-5}\ \mathrm{T}. \end{aligned}$$

In comparison, the magnetic flux density at the equator is about 1.3×10^{-7} T.

Any region in which there is a magnetic flux is called a magnetic field, and the field intensity (or field strength) is directly proportional to the magnetic flux density. Since magnetic flux has direction as well as magnitude, magnetic field strength is a vector quantity. Faraday, a Scottish experimental physicist, found in 1831 that electricity could be generated from a magnetic field. In 1864, Maxwell, a Scottish theoretical physicist, published a general theory that related the experimental findings of Oersted and Faraday. His theory states that a changing electric field is always associated with a changing magnetic field, and that a changing magnetic field is always associated with a changing electric field. Together, these associated electric and magnetic fields constitute an electromagnetic wave.

In dealing with electromagnetic waves, it is more convenient to describe the magnetic component in terms of magnetic field strength H, rather than in terms of magnetic flux density, B. In free space, magnetic field strength is related to magnetic flux density by

$$H = B/\mu_0. \tag{2.54}$$

Since B has dimensions of Wb/m^2, and μ_0 has dimensions of Wb/A-m, the dimensions of magnetic field strength are A/m.

An electromagnetic wave may be initiated by accelerating a charged particle. When this happens, some of the energy of the charged particle is radiated as electromagnetic radiation. This phenomenon is the basis of radio transmission, in which electrons are accelerated up and down an antenna that is connected to an oscillator. The electromagnetic wave thus generated has a frequency equal to that of the oscillator, and a velocity of 3×10^8 m/sec in free space. The waves consist of oscillating electric and magnetic fields that are perpendicular to each other, and are mutually perpendicular to the direction of propagation of the wave, Fig. 2.16. The energy carried by the waves depends on the strenght of the associated electric and magnetic fields.

The relationship between the peak magnetic and electric field intensities, H_0 and E_0 depends on the magnetic permeability, μ, and the electrical permittivity, ϵ, of the medium

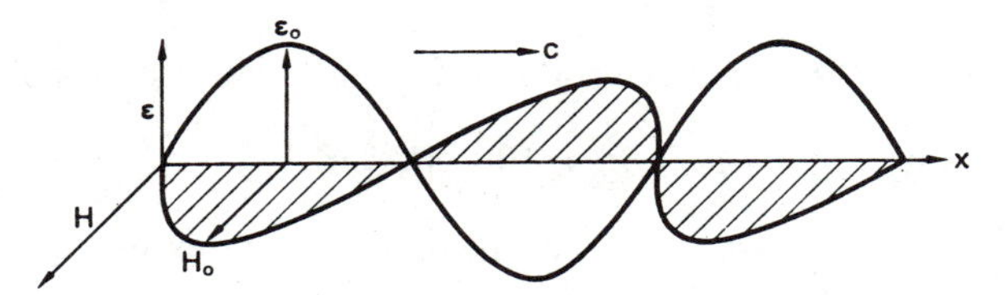

FIG. 2.16. Schematic representation of an electromagnetic wave. The electric intensity E and the magnetic intensity H are at right angles to each other, and the two are mutually perpendicular to the direction of propagation of the wave. The velocity of propagation is c, the electric intensity is $E = E_0 \sin 2\pi/\lambda(x - ct)$, and the magnetic intensity is $H = H_0 \sin 2\pi/\lambda(x - ct)$. The plane of polarization is the plane containing the electric field vector.

through which the electromagnetic wave is propagating. This relationship is given by

$$H_0\sqrt{\mu} = E_0\sqrt{\epsilon}. \tag{2.55}$$

If the wave is traveling through free space, then

$$H_0\sqrt{\mu_0} = E_0\sqrt{\epsilon_0}. \tag{2.56}$$

Radio waves, microwaves (radar), infrared radiation, visible light, ultraviolet light, and X-rays are all electromagnetic radiations. They are qualitatively alike, but differ in wavelength to form a continuous electromagnetic spectrum.

All these radiations are transmitted through the atmosphere (which may be considered, for this purpose, free space) at a speed very close to 3×10^8 meters per second.

Since the speed of all electromagnetic waves in free space is a constant, equation (2.45), when applied to electromagnetic waves in free space, becomes

$$c = 3 \times 10^8 \frac{\text{m}}{\text{sec}} = f\lambda. \tag{2.57}$$

Specifying either the frequency or wavelength of an electromagnetic wave in free space is equivalent to specifying both. Free space wavelengths may range from 5×10^6 m for 60-Hz electric waves through visible light (green light has a wavelength about 500 nanometers and a frequency of 6×10^{14} Hz to short wavelength X- and gamma radiation (whose wavelengths are of the order of 10 nm or less). There is no sharp cut-off in wavelength at either end of the spectrum, nor is there a sharp dividing line between the various portions of the electromagnetic spectrum. Each portion blends into the next, and the lines of demarcation shown in Fig. 2.17 are arbitrarily placed to show the approximate wavelength span of the regions of the electromagnetic spectrum.

Generally, the speed of an electromagnetic wave in any medium depends on the electrical and magnetic properties of that medium: on the permittivity and the permeability. The speed is given by

$$v = \sqrt{\frac{1}{\epsilon\mu}}. \tag{2.58}$$

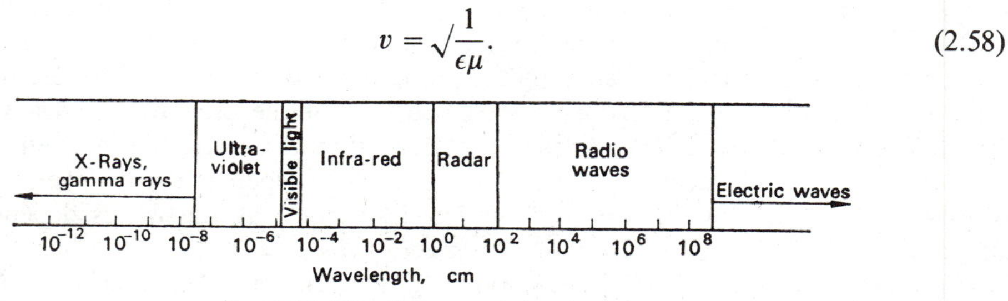

FIG. 2.17. The electromagnetic spectrum.

In free space,

$$\mu = \mu_0 = 4\pi \times 10^{-7}\,\text{N/A}^2$$

$$\epsilon = \epsilon_0 = \frac{1}{4\pi \times 9 \times 10^9\,\text{N-m}^2/\text{C}^2}.$$

Substituting the values above into equation (2.58) gives the speed, c, of an electromagnetic wave in free space.

$$v = c = \sqrt{\frac{4\pi \times 9 \times 10^9\,\text{N-m}^2/\text{C}^2}{4\pi \times 10^{-7}\,\text{N/A}^2}} = 3 \times 10^8\,\text{m/sec}.$$

An electromagnetic wave travels slower through a medium than through free space. The frequency of the electromagnetic wave is independent of the medium through which it travels. The wavelength is decreased, however, so that the relationship

$$v = f\lambda \tag{2.47}$$

is maintained. The wavelength in a medium is given by

$$\lambda = \lambda_0 \sqrt{\frac{1}{K_e K_m}}, \tag{2.59}$$

where λ_0 is the free space wavelength and K_e and K_m are the relative permittivity (dielectric coefficient) and relative permeability of the medium. Since the relative permeability is ≈ 1 for most biological materials and dielectrics, we can approximate equation (2.59) by

$$\lambda = \lambda_0 \sqrt{\frac{1}{K_e}}. \tag{2.60}$$

If the medium is a *lossy dielectric* (a lossy dielectric is one that absorbs energy from an electromagnetic field; all biological media are lossy), the wave length of our electromagnetic wave within the medium is given by

$$\lambda = \frac{\lambda_0}{\sqrt{K_e}}\left[\frac{1}{2} + \frac{1}{2}\sqrt{1 + \left(\frac{\sigma}{\omega\epsilon}\right)^2}\right]^{-1/2}, \tag{2.61}$$

or, in terms of the *loss tangent*,

$$\lambda = \frac{\lambda_0}{\sqrt{K_e}}\left(\frac{1}{2} + \frac{1}{2}\sqrt{1 + \tan^2\delta}\right)^{-1/2}. \tag{2.62}$$

where λ_0 = free space wavelength, K_e = dielectric coefficient, σ = resistivity, in ohm meters, $\omega = 2\pi \times$ frequency, $\epsilon = K_e\epsilon_0$, and

$$\tan^2\delta = \left(\frac{\sigma}{\omega\epsilon}\right)^2 = \text{loss tangent}. \tag{2.63}$$

Example 2.16

The dielectric coefficient for brain tissue is $K_e = 82$, and the resistivity, $\rho = 1.88$ ohm meters at 100 MHz. What is the wavelength of this radiation in the brain?

$$\lambda_0 = \frac{c}{f} = \frac{3 \times 10^8\,\text{m/sec}}{100 \times 10^6\,\text{sec}^{-1}} = 3\,\text{m}$$

$$\tan^2\delta = \left(\frac{\sigma}{\omega\epsilon}\right)^2 = \left(\frac{\sigma}{\omega k_e \epsilon_0}\right)^2$$

$$= \left(\frac{1.88}{2\pi \times 100 \times 10^6 \times 82 \times \dfrac{1}{4\pi \times 9 \times 10^9}}\right)^2 = 17.03$$

$$\lambda = \frac{\lambda_0}{\sqrt{K_e}}\left[\frac{1}{2} + \frac{1}{2}\sqrt{(1 + \tan^2\delta)}\right]^{-1/2}$$

$$\lambda = \frac{3}{\sqrt{82}}\left[\frac{1}{2} + \frac{1}{2}\sqrt{(1 + 17.03)}\right]^{-1/2} = 0.20 \text{ m}.$$

The loss tangent is a measure of energy absorption by a medium through which an electromagnetic wave passes; energy absorption by the medium is directly proportional to the loss tangent. The loss tangent is defined as

$$\text{loss tangent} = \frac{\text{conduction current}}{\text{displacement current}}. \tag{2.64}$$

Conduction current is the ordinary current that consists of a flow of electrons across a potential difference in a circuit. Since no dielectric material is a perfect insulator, some small conduction current will flow through any insulating material under the influence of a potential difference. *Displacement* is a concept proposed by Maxwell to account for the apparent flow of current through an insulator—even a perfect insulator—under the action of a changing voltage. Consider the circuit shown in Fig. 2.18, a capacitor whose dielectric

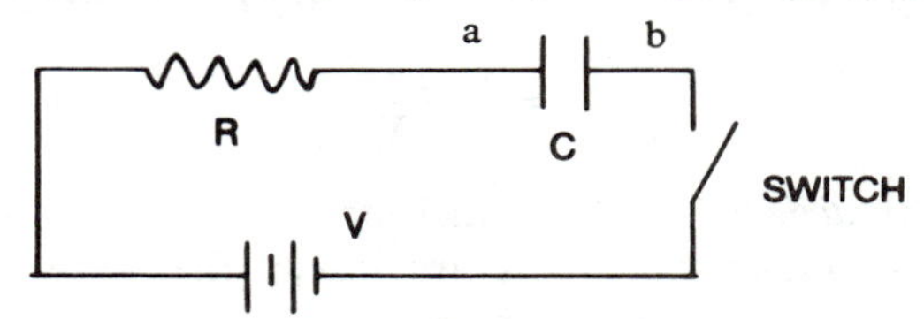

FIG. 2.18. A circuit having capacitance, C, and resistance, R, in series with a voltage source, V.

is a perfect insulator, a resistor, a switch, and a battery connected in series. While the switch is open, plate *a* of the capacitor is at the same potential as the positive terminal of the battery to which it is connected through the resistor. When the switch is closed, plate *b* of the capacitor is connected to the negative pole of the battery, and the capacitor begins to charge under the influence of the potential across the plates. While the capacitor is charging, a current given by

$$i = \frac{V}{R} e^{-t/RC}, \tag{2.65}$$

where t = time after closing the switch, in seconds; R = resistance, in ohms; C = capacitance, in farads; and V = voltage across capacitor; flows through the curcuit. The product RC is called the *time constant* of the circuit, and represents the time, in seconds, after closing the switch, until the current decreases to $1/e$, or 37% of the initial current. Since the dielectric in this circuit is a perfect insulator, then clearly no current can flow through the dielectric. However, connecting the plates of the capacitor to the battery terminals causes electrons to pile up on the negative plate, and to be drained off from the positive plate. That is, electrons flow onto the plate connected to the negative terminal, and flow from the other plate to the positive terminal until an equilibrium is reached when no more charge flows. At that time, the charge on each plate of the capacitor is given by

$$Q = CV. \tag{2.66}$$

Thus, during the time that the capacitor is charging, the circuit behaves *as if* current were flowing through every portion of the circuit, including the dielectric. This apparent current flowing through the dielectric is called the *displacement* current. Since no real dielectric is a perfect insulator, there always is a conduction current through the dielectric in addition to the displacement current. Furthermore, since a conduction current is always accompanied by energy loss through Joule heating, there always is some loss of energy when a dielectric is placed in an electric field.

Impedance

Whereas electric current in a circuit is transmitted as electron flow through conducting wires, electromagnetic waves are transmitted as disturbances in the electromagnetic fields established in various media, including a vacuum (free space). The electric and magnetic components of the field may be considered the analogs of voltage and current in an electrical circuit. The impedance of an alternating current, Z, is given by Ohm's Law:

$$Z = \frac{V}{i}, \tag{2.67}$$

where i is the current flow due to the voltage V. By analogy, the impedance of a medium through which an electromagnetic wave is propagated is given by

$$Z = \frac{E_0}{H_0}, \tag{2.68}$$

where E_0 and H_0 are the electric and magnetic field strengths respectively. Using the relationship given by equation (2.55),

$$H_0\sqrt{\mu} = E_0\sqrt{\epsilon},$$

we can rewrite equation (2.68) as

$$Z = \sqrt{\frac{\mu}{\epsilon}}. \tag{2.69}$$

In free space,

$$\mu = \mu_0 = 4\pi \times 10^{-7}\ \text{N/A}^2$$

$$\epsilon = \epsilon_0 = \frac{1}{4\pi \times 9 \times 10^9\ \text{N-m}^2/\text{C}^2}.$$

Using these values in equation (2.69), one obtains

$$Z = \left(\frac{4\pi \times 10^{-7}\ \text{N/A}^2}{\dfrac{1}{4\pi \times 9 \times 10^9\ \text{N-m}^2/\text{C}^2}} \right)^{1/2} = 377\ \text{ohms},$$

the impedance of free space.

In an electric circuit the power dissipated in a load whose impedance is Z ohms, across which the voltage drop is V volts, is given by

$$P = \frac{V^2}{Z}. \tag{2.70}$$

The analogous quantities in an electromagnetic field of mean power density $\bar{W}$ watts/m^2 and effective electric field strength, $\bar{E}$ volts/m are related by

$$\bar{W} = \frac{\bar{E}^2}{Z} = \frac{(E_0/\sqrt{2})^2}{Z}, \tag{2.71}$$

where E_0 is the maximum value of the sinusoidally varying electric field strength. $E_0\sqrt{2}$ is the effective, or root-mean-square (rms) value of the electric field strength. In free space, where $Z = 377$ ohms, the power density corresponding to $E_0 = 1$ V/m is

$$\bar{W} = \frac{(1\sqrt{2})^2}{377} = 1.33 = 10^{-3} \text{ watt/m}^2 = 1.33 \text{ mW/cm}^2.$$

Energy in an electromagnetic wave is transported in a direction mutually perpendicular to the electric field vector and the magnetic field vector, as shown in Fig. 2.16. The instantaneous rate of energy flow in an electromagnetic wave, through a unit area perpendicular to the direction of propagation, is given by the vector product

$$\mathbf{W} = \mathbf{E} \times \mathbf{H}. \tag{2.72}$$

The vector $\vec{\mathbf{W}}$, which represents the flow of energy, is called the Poynting vector. Since the instantaneous values of E and H vary sinusoidally from 0 to their maxima, their effective values are $E_0/\sqrt{2}$ and $H_0/\sqrt{2}$. Since the magnetic and electric field vectors are always at 90° to each other, equation (2.72) may be rewritten as

$$\bar{\mathbf{W}} = \frac{E_0}{\sqrt{2}} \times \frac{H_0}{\sqrt{2}} = \frac{1}{2} E_0 H_0. \tag{2.73}$$

Example 2.17

The electric field intensity E_0, of a plane electromagnetic wave is 1 V/m. What is the average power density perpendicular to the direction of propagation of the wave?

From equation (2.73), the average power density is

$$\bar{W} = \tfrac{1}{2} E_0 H_0, \tag{2.74}$$

and from equation (2.56), H_0 is found to be

$$H_0 = E_0 \sqrt{\frac{\epsilon_0}{\mu_0}}. \tag{2.56A}$$

If we substitute the expression for H_0 into equation (2.67) and insert the numerical values for ϵ_0 and μ_0, we have

$$\begin{aligned}\bar{W} &= \tfrac{1}{2}(1 \text{ V/m})^2(4\pi \times 9 \times 10^9 \text{ N-m}^2/\text{C}^2 \times 4\pi \times 10^{-7} \text{ N/A}^2)^{-1/2} \\ &= 1.33 \times 10^{-3} \text{ watt/m}^2 \\ &= 1.33 \times 10^{-4} \text{ mW/cm}^2.\end{aligned}$$

For health physics purposes, the electromagnetic spectrum may be divided into two portions: one portion, called *ionizing radiation*, extends from the very shortest wavelengths until about several nanometers. Electromagnetic radiation of these wave-lengths include X-rays and gamma rays. (X-rays and gamma rays are exactly the same radiations; they differ only in their manner of origin. Once created, it is impossible to distinguish between the two types.) Electromagnetic radiation whose wavelength exceeds that of ionizing radiation is called collectively *non-ionizing radiation*. In this region, the health physicist is interested mainly in laser radiation, which includes ultraviolet, visible, and infrared light, and in the high frequency bands from about 3 MHz to 300 GHz. Although ultraviolet light is capable of producing ions, it nevertheless is considered as a non-ionizing radiation in the context of health physics.

Quantum Theory

The representation of light and other electromagnetic radiations as a continuous succession of periodic disturbances or as a "wave train" in an electromagnetic field is a satisfactory model which can be used to explain many physical phenomena, and which can serve as a basis for the design of appparatus for transmitting and receiving electromagnetic energy. According to this wave theory, or the classical theory as it is often called, the energy transmitted by a wave is proportional to the square of the amplitude of the wave. This model, despite its wide usefulness, nevertheless fails to predict certain phenomena in the field of modern physics. Accordingly, therefore, a new model for electromagnetic radiation was postulated, one which could be used to "explain" certain phenomena which are not amenable to explanation by the wave theory. It should be emphasized that to hypothesize a new theory is not synonymous with abandonment of the old theory. Models or theories are useful in so far as they describe observed phenomena and permit prediction of consequences of certain acts. Philosophically, most scientists subscribe to the school of thought known as "logical positivism." According to this philosophy, there is no way to discover or verify an absolute truth. Science is not concerned with absolute truth or with reality—it is concerned with giving the simplest possible unified description of as many experimental findings as possible. According to this philosophy of science, the acceptance of several different theories on the nature of electromagnetic radiations (or for the nature of matter, electricity, etc.) is perfectly acceptable, provided each theory is capable of describing experimental facts that the others cannot describe.

The more recent theory of the nature of electromagnetic radiation is called the quantum theory. According to the quantum theory, electromagnetic radiation consists of "corpuscles" or "particles" of energy which travel at a velocity of 3×10^8 m/sec. Each particle, or "quantum" as it is called, contains a discrete quantity of electromagnetic energy. The energy contained in a quantum is proportional to the "color" of the radiation, or to its frequency when it is considered as a wave, and is given by the relationship

$$E = hf. \tag{2.75}$$

The symbol h is called Planck's constant, and is a fundamental constant of nature whose magnitude in the S.I system is $h = 6.614 \times 10^{-34}$ J-sec. E, in equation (2.75), is given in joules and f, the frequency, is in hertz.

By substituting the value of f from equation (2.57) into equation (2.75), we have

$$E = h\frac{c}{\lambda}. \tag{2.76}$$

A quantum of electromagnetic energy is also called a photon. Equations (2.75) and (2.76) show that a photon is completely described when either its energy, or its frequency, or its wavelength is given.

Example 2.18

(a) Radio station KDKA in Pittsburgh, Pennsylvania, broadcasts on a carrier frequency of 1020 kilocycles per second.

(1) What is the wavelength of the carrier frequency?

$$\lambda f = c,$$

$$\lambda = \frac{c}{f} = \frac{3 \times 10^8 \text{ m/sec}}{1.02 \times 10^3 \text{ kc/sec} \times 10^3 \text{ cycle/kc}}$$

$$= 294 \text{ m/cycle}.$$

(2) What is the energy of a KDKA photon, in joules and in electron volts?

$$E = hf$$

$$= 6.614 \times 10^{-34} \text{ J sec} \times 1.02 \times 10^6 \text{ sec}^{-1}$$

$$= 6.75 \times 10^{-28} \text{ J}$$

or

$$E = \frac{6.75 \times 10^{-28} \text{ J}}{1.6 \times 10^{-19} \text{ J/eV}}$$

$$= 4.14 \times 10^{-9} \text{ eV}.$$

(b) What is the energy, in electron volts, of an X-ray photon whose wavelength is 1×10^{-10} m?

$$E = \frac{hc}{\lambda}$$

$$= \frac{6.614 \times 10^{-34} \text{ J-sec} \times 3 \times 10^8 \text{ m/sec}}{10^{-10} \text{ m} \times 1.6 \times 10^{-19} \text{ J/eV}}$$

$$= 1.24 \times 10^4 \text{ eV}.$$

The wavelengths of X-rays and gamma-rays are very short; of the order of 10^{-10} m or less. Because of this, and in order to avoid writing the factor 10^{-8} repeatedly, another unit, called the angstrom unit, is used in X-ray work and in health physics. The angstrom unit, which is symbolized by Å, is a unit of length which is equal to 1×10^{-10} m.

It may seem strange that, having found the wave theory of electromagnetic radiation inadequate to explain certain physical phenomena, we should incorporate part of the wave model into the quantum model of electromagnetic radiation. This dualism, however, seems to be inherent in the "explanations" of atomic and nuclear physics. Mass and energy, particle and wave in the case of electromagnetic energy, and, as will be shown below in subsequent paragraphs, wave and particle in the case of sub-atomic particulates, all seem to be part of a dualism in nature; either aspect of this dualism can be demonstrated in the laboratory by appropriate experiments.

In the case of the photon, some degree of correspondence with the classical picture of electromagnetic radiation can be demonstrated by a simple thought experiment. It is conceivable that, given a very large number of waves that differ from each other in frequency and amplitude, a wave packet, or quantum, could result from reinforcement of the waves over a very limited region, and complete interference ahead and behind the region of reinforcement. Fig. 2.19 is an attempt to portray graphically such a phenomenon.

The model of a photon shown in Fig. 2.19 combines wave properties and particle properties. Furthermore, this model suggests that a photon may be considered a moving

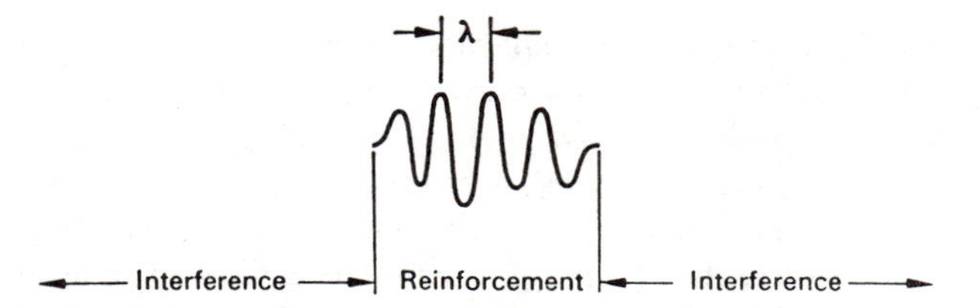

FIG. 2.19. Possible combination of electromagnetic waves to produce a wave packet, a quantum of electromagnetic energy called a *photon*. The energy content of the photon is $E = hc/\lambda$.

particle that is guided in its path by the waves that combine to produce the particle. The "mass" of a photon may be found by equating its energy with the relativistic energy of a moving particle:

$$E = hf = mc^2, \tag{2.76A}$$

$$\therefore \quad m = \frac{hf}{c^2}. \tag{2.76B}$$

The "momentum" p, of the photon, therefore, is

$$p = mc = \frac{hf}{c}, \tag{2.77}$$

and, if the value of f from equation (2.57) is substituted into equation (2.77), we have

$$p = \frac{h}{\lambda}. \tag{2.78}$$

The duality of nature was further emphasized when two experimenters in the Bell Telephone Laboratories, Davisson and Germer, found a beam of electrons to behave like a wave. When they bombarded a nickel crystal with a fine beam of electrons whose kinetic energy was 54 eV, they found the electrons to be reflected only in certain directions, rather than isotropically as expected. Only at angles of 50 and 0° (backscatter) were scattered electrons detected. Such a behavior was unexplainable if the bombarding electrons were considered to be particulate in nature. By assuming them to be waves, however, the observed distribution of scattered electrons could easily be explained. The answer simply was that the electron "waves" underwent destructive interference at all angles except those at which they were observed; there the electron waves reinforced each other.

Matter Waves

In 1924 the French physicist, Louis de Broglie, suggested that not only electrons, but all moving particles were associated with wave properties. The length of these waves, according to de Broglie, was inversely proportional to the momentum of the moving particle and, furthermore, that the constant of proportionality was Planck's constant, h. The length of these matter waves is given by equation (2.78). Since momentum p is equal to mv, equation (2.78) may be rewritten as

$$\lambda = \frac{h}{mv}. \tag{2.79}$$

Here, too, as in the case of the photon, we have in the same equation properties characteristic of particles and properties characteristic of waves. The mass of the moving particle, m in equation (2.79) represents a particle concept, while λ, the wavelength of the "matter" wave associated with the moving particle, is, quite clearly, a wave concept.

The fact that moving particles possess wave properties is the basis of the electron microscope. In any kind of a microscope, whether used with beams of light waves or beams of de Broglie "matter" waves, the resolving power (the ability to separate two points that are close together, or the ability to see the edges of a very small object sharply and distinctly) is an inverse function of the wavelength of the probing beam; a shorter wavelength permits better resolution than a longer wavelength. For this reason, optical microscopes are usually illuminated with blue light, since blue is near the short wavelength end of the visible spectrum. Under optimum conditions, the limit of resolution of an optical microscope, using blue light whose wavelength is 4000 Å, is of the order of 100 nm. Since high velocity electrons are associated with very much shorter wavelengths than blue light, an electron microscope, which uses a beam of electrons instead of a beam of light, has a much greater resolving power than even the best optical microscope. Since useful magnification is limited by resolving power, the increased resolution possible with an electron microscope permits much greater useful magnification than could be obtained with the optical microscope.

Example 2.19

What is the de Broglie wavelength of an electron that is accelerated across a potential difference of 100,000 V?

According to equation (2.20), the kinetic energy of a particle moving with a velocity $v = \beta c$ is

$$E_k = m_0c^2\left(\frac{1}{\sqrt{(1-\beta^2)}} - 1\right),$$

from which it follows that

$$\frac{m_0c^2}{E_k + m_0c^2} = \sqrt{(1-\beta^2)}. \tag{2.80}$$

In Example 2.8 it was shown that m_0c^2 for an electron is 5.1×10^5 eV. We therefore have, after substituting the appropriate numerical values into equation (2.80)

$$\frac{5.1 \times 10^5 \text{ eV}}{1 \times 10^5 \text{ eV} + 5.1 \times 10^5 \text{ eV}} = \sqrt{(1-\beta^2)},$$

$$0.837 = \sqrt{(1-\beta^2)},$$

$$(0.837)^2 = 1 - \beta^2,$$

$$\therefore \quad \beta = \sqrt{[1-(0.837)^2]}$$

$$= 0.55.$$

The momentum of the electron is

$$p = mv = \frac{m_0}{\sqrt{(1-\beta^2)}}\beta c$$

$$= \frac{9.11 \times 10^{-31} \text{ g} \times 0.55 \times 3 \times 10^8 \text{ m/sec}}{0.837}$$

$$= 1.80 \times 10^{-22} \text{ kg-m/sec}.$$

and the de Broglie wavelength, consequently, is

$$\begin{aligned}\lambda &= \frac{h}{p} \\ &= \frac{6.614 \times 10^{-34}\ \text{J sec}}{1.80 \times 10^{-22}\ \text{kg-m/sec}} \\ &= 3.69 \times 10^{-12}\ \text{m} \\ &= 0.0369\ \text{Å}.\end{aligned}$$

The wave-particle dualism may seem especially abstract when it is extended to include particles of matter whose existence may be confirmed by our experience, by our senses, and by our intuition. At first it may seem that the wave properties of particles are purely mathematical figments of the imagination whose only purpose is to quantitatively describe experimental phenomena that are not otherwise amenable to theoretical analysis. In this regard, of course, the wave properties of matter serve a useful purpose. However, it is possible to give a physical interpretation of matter waves. According to the physical theory of waves, the intensity of a wave is proportional to the square of the amplitude of the wave. In 1926, Max Born applied this concept to the wave properties of matter. In the case of a beam of electrons, the square of the amplitude of the electron waves was postulated to be proportional to the intensity of the beam, or to the number of electrons per square centimeter per second incident on a plane perpendicular to the direction of the beam. If this beam strikes a crystal, as in the case shown in Fig. 2.20, then the reflected electron

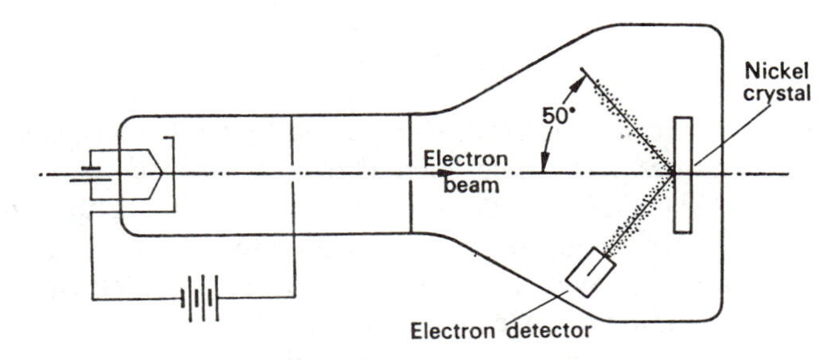

FIG. 2.20. Experiment of Davisson and Germer suggesting the wave nature of electrons.

waves either reinforce or interfere with each other, and cause the observed interference pattern. The waves are reinforced most strongly at certain points, and at other points are exactly out of phase, thereby cancelling each other out. Where reinforcement occurs, a maximum electron density is observed, whereas interference results in a decrease in electron density. The exact distribution of the interference pattern is determined by the crystalline structure of the scatterer, and therefore is uniquely representative of the scatterer. This method of "finger-printing" substances is the basis of electron diffraction methods used by physical chemists to identify unknowns.

For a beam of electrons, this relationship between wave and particle properties seems reasonable. In the case of a single electron, however, the electron-wave must be interpreted differently. If instead of bombarding the crystal in Fig. 2.20 with a beam of electrons, the electrons were fired at the crystal one at a time and each separate scattered electron was detected, it would be found that the single electrons would be scattered through the same angles as the beam of electrons. The exact coordinates of any particular scattered electron, however, would not be known until it is "seen" by the electron detector. After firing the same total number of single electrons as those in a beam, and plotting the positions of each scattered electron, exactly the same "interference" pattern would be observed as in the case of a beam of electron waves. This experiment shows that, although the behavior of a single electron cannot be precisely predicted, the behavior of a group of electrons can be predicted. From this it may be inferred that the square of the amplitude at any point of the curve of position versus electron intensity gives the probability of any single electron being scattered through that point.

Uncertainty Principle

It should be noted that the implications of Born's probability interpretation of the wave properties of matter were truly revolutionary. According to classical physics, if the mass and velocity of a particle, as well as the external forces acting on the particle, are known, then, in principle at least, all of its future actions could be precisely predicted. According to the wave-mechanical model, however, precise predictions are not possible—probability replaces certainty. Heisenberg, in 1927, developed these ideas still further, and postulated his uncertainty theory, in which he said that it is impossible, in principle, to know both the exact location and momentum of a moving particle at any point in time. Either one of these two quantities could be determined to any desired degree of accuracy. The accuracy of the other quantity, however, decreases as precision of the first quantity increases. The porduct of the two uncertainties Heinsenberg showed to be proportional to Planck's constant h, and to be given by the relationship

$$\Delta x \times \Delta p \geq \frac{h}{2\pi}. \tag{2.81}$$

It should be emphasized that the uncertainty expressed by Heisenberg is not due to faulty measuring tools or techniques, or to experimental errors; it is a fundamental limitation of nature, which is due to the fact that any measurement must disturb the object being measured. Precise knowledge about anything can therefore never be attained. In many instances, this inherent uncertainty can be understood intuitively. When a student is being tested, for example, it is common knowledge that he may become tense or suffer some other physio-psychological stress which may cast some doubt on the accuracy of the test results. In any case, the tester cannot be absolutely certain that the psychological strain of the examination did not influence the actions of the student and thereby influence the results of the examination. The uncertainty expressed by equation (2.81) can be illustrated by an example in which we try to locate a particle by looking at it. To "see" a particle means that light is reflected from the particle into the eye. When the quantum of light strikes the particle and is reflected toward the eye, some energy is transferred from the quantum to the particle, thereby changing both the position and momentum of the particle. The reflected photon tells the observer where the particle *was*, not where it *is*.

Problems

1. Two blocks, of mass 0.1 kg and 0.2 kg, approach each other along a frictionless surface, at velocities 0.4 and 1 m/sec, respectively. If the blocks collide, and remain together, calculate their joint velocity after the collision.

2. A bullet whose mass is 50 g travels at a velocity of 500 m/sec. It strikes a rigidly fixed wooden block, and penetrates a distance of 20 cm before coming to a stop.

(a) What is the deceleration of the bullet?

(b) What was the decelerating force?
(c) What was the initial momentum of the bullet?
(d) What was the impulse of the collision?

3. Compute the mass of the earth, assuming it to be a sphere of 25,000 miles circumference, if at its surface it attracts a mass of 1 g with a force of 980 dynes.

4. An automobile weighing 2000 kg, and going at a speed of 60 km/hr, collides with a truck weighing 5 metric tons that was moving at right angles to the direction of the auto at a speed of 4 km/hr. If the two vehicles become joined in the collision, what is the magnitude and direction of their velocity after the collision?

5. A small electrically charged sphere of mass 0.1 g hangs by a thread 100 cm long between two parallel vertical plates spaced 6 cm apart. If 100 volts are across the plates, and if the charge on the sphere is 10^{-9} coulombs, what angle does the thread make with the vertical direction?

6. A capacitor has a capacitance of $10\,\mu$F. How much charge must be removed to cause a decrease of 20 volts across the capacitor?

7. A small charged particle whose mass is 0.01 g remains stationary in space when it is placed in an upward directed electric field of 10 V/cm. What is the charge on the particle?

8. A 1-micron diameter droplet of oil, whose specific gravity is 0.9, is introduced into an electric field between two large parallel plates, separated by 5 mm, across which is placed a potential difference V volts. If the oil droplet carries a net charge of 100 electrons, how many volts must be across the plates if the droplet is to remain suspended in the space between the plates?

9. A diode vacuum tube consists of a cathode and an anode spaced 5 mm apart. If 300 volts are placed across the electrodes,
(a) What is the velocity of an electron midway between the electrodes, and at the instant of striking the plate, if the electrons are emitted by the cathode with zero velocity?
(b) If the plate current is 20 mA, what is the average force exerted on the anode?

10. Calculate the ratios v/c and m/m_0 for a 1-MeV electron and for a 1-MeV proton.

11. Assuming an uncertainty in the momentum of an electron equal to one half its momentum, calculate the uncertainty in position of a 1-MeV electron.

12. If light quanta have mass, they should be attracted by the earth's gravity. To test this hypothesis a parallel beam of light is directed horizontally at a receiver 10 miles away. How far would the photons have fallen during their flight, to the receiver, if quanta have mass?

13. The maximum wavelength of U.V. light observing the photoelectric effect in tungsten is 2730 Å. What will be the kinetic energy of photoelectrons produced by U.V. radiation of 1500 Å?

14. Calculate the uncertainty in position of an electron that was accelerated across a potential difference of $100{,}000 \pm 100$ volts.

15. (a) What voltage is required to accelerate a proton from zero velocity to a velocity corresponding to a de Broglie wavelength of 0.01 Å?
(b) What would be the kinetic energy of an electron with this wavelength?
(c) What is the energy of an X-ray photon whose wavelength is 0.01 Å?

16. A current of 25 mA flows through 25-gauge wire, 0.0179 in. (17.9 mils) in diameter. If there are 5×10^{22} free electrons per cm^3 in copper, calculate the average speed with which electrons flow in the wire.

17. An electron starts at rest on the negative plate of a parallel capacitor, and is accelerated by a potential of 1000 volts across a gap of 1 cm.
(a) With what velocity does the electron strike the positive plate?
(b) How long does it take the electron to travel the 1 cm distance?

18. A cylindrical capacitor is made of two coaxial conductors—the outer one has a diameter of 20.2 mm and the diameter of the inner one is 0.2 mm. The inner conductor is 1000 volts positive with respect to the outer conductor. Repeat parts (a) and (b) of problem 17, and compare the results to those of problem 17.

19. Two electrons are initially at rest, separated by 0.1 nm. After both electrons are released, they repel each other. What is the kinetic energy of each electron when they are 1.0 nm apart?

20. A cyclotron produces a 100-microamp beam of 15 MeV deuterons. If the cyclotron were 100% efficient in converting electrical energy into kinetic energy of the deuterons, what is the minimum required power input, in kilowatts?

21. A 1-μF capacitor is fully charged to 100 V by connecting it across the terminals of a 100-volt battery. Then, a 2-μF capacitor is charged to 100 V in the same manner. The two charged capacitors are then removed from the batteries, and connected as shown below. What is the charge on each capacitor?

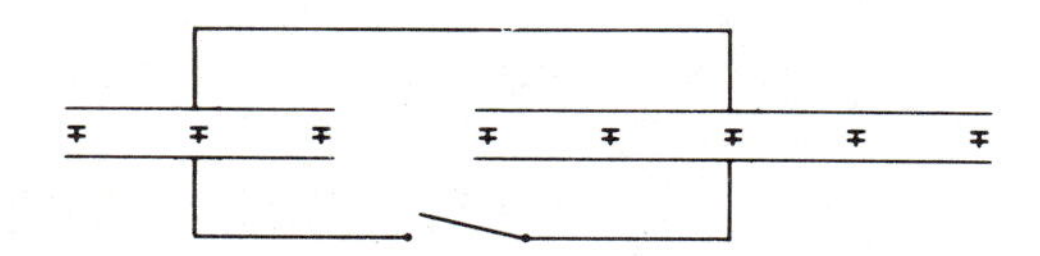

22. (a) What voltage must be applied across two oppositely charged parallel plates, 2 cm apart, in order to have an electron, starting from rest on the negative plate, strike the opposite plate in 10^{-8} second?
(b) With what speed will the electron strike the plate?

23. When hydrogen "burns," it combines with oxygen according to

$$2H_2 + O_2 \rightarrow 2H_2O,$$

and emits about 2.3×10^5 J of heat energy in the production of 1 mole water. By what fraction is the mass of the reactants decreased in this reaction?

24. (a) A 1000-MW(e) nuclear power plant operates at a thermal efficiency of 33% and at 75% capacity for 1 year. How many kilograms of nuclear fuel are consumed?

(b) If a coal fired plant operates at the same efficiency and capacity factor, how many kilograms of coal must be burned during the year if the heat content of the coal is 27 J/kg (11,700 Btu/pound)?

25. The solar constant is defined as the rate at which solar radiant energy falls on the earth's atmosphere on a surface normal to the incident radiation. The mean value for the solar constant is 1353 W/m^2, and the mean distance of the earth from the sun is 1.5×10^8 km.

(a) At what rate is energy being emitted from the sun?

(b) At what rate in tonnes per second, is the sun's mass being converted into energy?

26. What is the energy of a photon whose momentum is equal to that of a 10-MeV electron?

27. What is the wavelength of

(a) an electron whose kinetic energy is 1000 eV?

(b) a 10^{-8} kg oil droplet falling at a rate of 0.01 m/sec?

(c) a 1-MeV neutron?

28. (a) The specific heat of water in the English system of units is 1 Btu per pound per degree F; in the cgs. system, it is 1 calorie per gram per degree C. Calculate the number of joules per Btu, if there are 4.186 joules per calorie.

(b) What is the specific heat of water, in J/kg-°C?

29. The maximum amplitude of the electric vector in a plane wave in free space is 275 V/m.

(a) What is the amplitude of the magnetic field vector?

(b) What is rms value of the electric vector?

(c) What is the power density, in mW/cm^2, in this electromagnetic field?

30. (a) What is the free space power density, in milliwatts per cm^2, of a 2450-MHz electromagnetic wave whose maximum electric intensity is 100 millivolts/m?

(b) What is the maximum magnetic field intensity in this wave?

31. A radiostation transmits at a power of 50,000 W. Assuming the electromagnetic energy to be isotropically radiated (in the case of a real radio transmitter, emission is not isotropic),

(a) What is the mean power density at a distance of 50 km?

(b) What is the maximum electric field strength at that distance?

(c) What is the maximum magnetic field strength at that distance?

32. The mean value for the solar constant is 1.94 calories per minute per cm^2. Calculate the electric and magnetic field strengths corresponding to the solar constant.

Suggested References

ALONSO, M., and FINN, E. J. *Fundamental University Physics.* Addison-Wesley, Reading, Mass., 1967.

BORN, M. *Atomic Physics*, 8th ed. Hafner, Darien, Conn., 1970.

FEYNMAM, R. P., LEIGHTON, R. B., and SANDS, M. *The Feynman Lectures on Physics.* Addison-Wesley, Reading, Mass., 1965.

HALLIDAY, D., and RESNICK, R. *Physics*, 3rd ed. John Wiley & Sons, New York, 1977.

HOLTON, G. J. *Thematic Origins of Scientific Thought, Kepler to Einstein.* Harvard U. Press, Cambridge, 1973.

LAPPS, R. E., and ANDREWS, H. L. *Nuclear Radiation Physics*, 4th ed. Prentice Hall, Englewood Cliffs, N.J., 1972.

LINDSAY, R. B. *Basic Concepts of Physics.* Van Nostrand Reinhold, New York, 1971.

MAGID, L. M. *Electromagnetic Fields, Energy, and Waves.* John Wiley & Sons, New York, 1972.

MCGERVEY, J. D. *Introduction to Modern Physics.* Academic Press, New York, 1971.

RIPLEY, J. A., JR. *The Elements and Structure of the Physical Sciences.* John Wiley & Sons, New York, 1964.

ROGERS, E. M. *Physics for the Inquiring Mind.* D. van Nostrand, Princeton, 1960.

SEARS, F. W., and ZEMANSKY, M. W. *University Physics*, 4th ed. Addison-Wesley, Reading, Mass., 1970.

SEMAT, H., and ALBRIGHT, J. R. *Introduction to Atomic and Nuclear Physics*, 5th ed. Holt, Rinehart, and Winston, New York, 1972.

CHAPTER 3

ATOMIC AND NUCLEAR STRUCTURE

Atomic Structure

Matter, as we ordinarily know it, is electrically neutral. Yet the fact that matter can be easily electrified—by walking with rubber-soled shoes on a carpet, by sliding across a plastic auto-seat cover when the atmospheric humidity is low, and by numerous other commonplace means—testifies to the fact that matter is electrical in nature. The manner in which the positive and negative electrical charges were held together was a matter of concern to the physicists of the early twentieth century.

Rutherford's Nuclear Atom

The British physicist Rutherford had postulated, in 1911, that the positive charge in an atom was concentrated in a central massive point called the nucleus, and that the negative electrons were situated at some remote points, about one angstrom unit distant from the nucleus. In one of the all-time classical experiments of physics, two of Rutherford's students, Geiger and Marsden, in 1913 tested the validity of this hypothesis by bombarding an extremely thin (6×10^{-5} cm) gold foil with highly energetic, massive, positively charged projectiles called alpha particles. These projectiles, whose kinetic energy was 7.68 MeV, were emitted from the radioactive substance polonium. If Rutherford's idea had merit, it was expected that most of the alpha particles would pass straight through the thin gold foil. Some of the alpha particles, however, those that would pass by a gold nucleus sufficiently close to permit a strong interaction between the electric field of the alpha particle and of the positive point charge in the gold nucleus, would be deflected as a result of the repulsive force between the alpha particle and the gold nucleus. An angular scan with an alpha particle detector about the point where the beam of alpha particles traversed the gold foil, as shown in Fig. 3.1, permitted Geiger and Marsden to measure the alpha particle intensity at various scattering angles. The experimental results verified Rutherford's hypothesis. Although most of the alpha particles passed undeflected through the gold foil, a continuous distribution of scattered alpha particles was observed as the alpha particle detector traversed a scattering angle from 0° to 150°. Similar results were obtained with other scatterers. The observed angular distributions of the scattered alpha particles agreed with those predicted by Rutherford's theory, thereby providing experimental evidence for the nuclear atom. Matter was found to consist mainly of open space. A lattice of atoms, consisting of positively charged nuclei about 5×10^{-15} m in diameter, and separated by distances on the order of 10^{-10} m, was inferred from the scattering data. Detailed analyses of many experimental data later showed the nucleus to have radius of

$$r = 1.2 \times 10^{-15} A^{1/3} \text{ m}, \tag{3.1}$$

where A is the atomic mass number. The number of unit charges in the nucleus (1 unit charge is 1.6×10^{-19} coulombs) was found to be approximately equal to the atomic

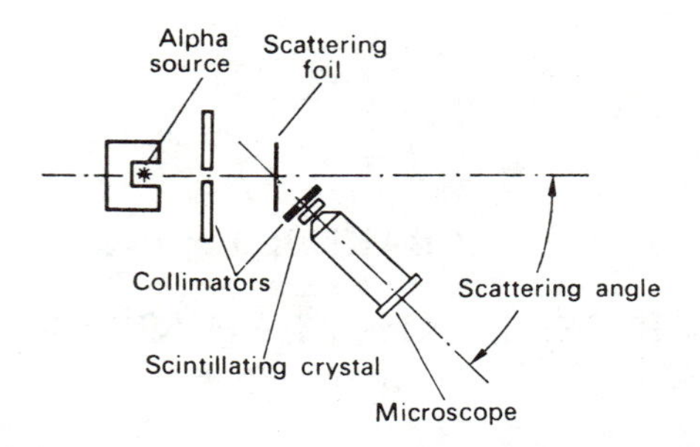

FIG. 3.1. Diagram showing principle of Rutherford's alpha-particle-scattering experiment. The alpha source, its collimator, and the scattering foil are fixed; the alpha particle detector, consisting of a collimator, a ZnS scintillating crystal, and a microscope rotates around the point where the alpha beam strikes the scattering foil.

number of the atom, and to about one-half the atomic weight. Later work in Rutherford's laboratory by Moseley and by Chadwick in 1920 showed the number of positive charges in the nucleus to be exactly equal to the atomic number. These data implied that the proton, which carries one charge unit, is a fundamental building block of nature. The outer periphery of the atom, at a distance of about 5×10^{-11} m from the nucleus, was thought to be formed by electrons, equal in number to the proton within the nucleus, distributed around the nucleus. However, no satisfactory theory to explain this structure of the atom was postulated by Rutherford. Any acceptable theory must answer two questions: first, how are the electrons held in place outside the nucleus despite the attractive electrostatic forces, and, second, what holds the positive charges in the nucleus together in the face of the repulsive electrostatic forces?

Bohr's Atomic Model

A simple solar system-like model, with the negative electrons revolving about the positively charged nucleus, seemed inviting. According to such a model, the attractive force between the electrons and the nucleus could be balanced by the centrifugal force due to the circular motion of the electrons. Classical electromagnetic theory, however, predicted that such an atom is unstable. The electrons revolving in their orbits undergo continuous radial acceleration. Since classical theory predicts that charged particles radiate electromagnetic energy whenever they experience a change in velocity (either in speed or in direction), it follows that the orbital electrons should eventually spiral into the nucleus as they lose their kinetic energy by radiation. (The loss of kinetic energy by this mechanism is called *Bremsstrahlung*, and is very important in health physics. This point will be taken up in more detail in later chapters.) The objection to the solar system-like atomic model, based on the argument of energy loss due to radial acceleration, was overcome in 1913 by the Danish physicist Niels Bohr simply by denying the validity of classical electromagnetic theory in the case of motion of orbital electrons.

Although this was a radical step, it was by no means without precedent. The German physicist Max Planck had already shown that a complete description of black body radiation could not be given with classical theory. To do this, he postulated a quantum theory of radiation, in which electromagnetic radiations are assumed to be particles whose energy depends only on the frequency of the radiation. Bohr adopted Planck's quantum theory, and used it to develop an atomic model that was consistent with the known atomic phenomena. The main source of experimental data from which Bohr inferred his model was atomic spectra. Each element, when excited by the addition of energy, radiates only certain colors that are unique to it. (This is the basis of "neon" signs. Neon, sealed in a glass tube, emits red light as a consequence of electrical excitation of the gas. Mercury vapor is used in the same way to produce blue light.) Because of the discrete nature of these colors, atomic spectra are called "sharp-line" spectra to distinguish them from white light or black body radiation, which has a continuous spectrum. Hydrogen, for example, emits electromagnetic radiation of several distinct frequencies when it is excited, as shown in Fig. 3.2. Some of these radiations are in the ultraviolet region, some are in the visible light region, and some are in the infrared region. The spectrum of hydrogen consists of

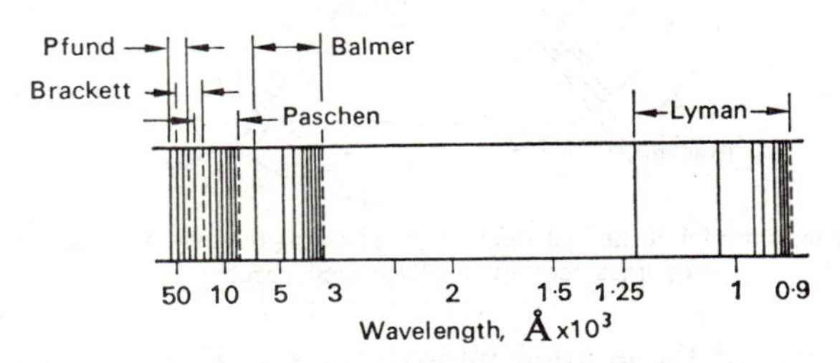

FIG. 3.2. Hydrogen spectrum.

several well defined series of lines whose wavelengths were described empirically by physicists of the late nineteenth century by the equation

$$\frac{1}{\lambda} = R\left(\frac{1}{n_1^2} - \frac{1}{n_2^2}\right), \tag{3.2}$$

where R is a constant, named after Rydberg, whose numerical value is $1.097 \times 10^{-2}\ \text{nm}^{-1}$, n_1 is any whole number equal to or greater than 1, and n_2 is a whole number equal to or greater than $n_1 + 1$. The Lyman series, which lies in the ultraviolet region, is the series in which $n_1 = 1$, and $n_2 = 2, 3, 4, \ldots$. The longest wavelength in this series, obtained by setting n_2 equal to 2 in equation (3.2), is 121.5 nm. Succeeding lines, when n_2 is 3 and 4, are 102.6 nm and 97.2 nm, respectively. The shortest line, called the series limit, is obtained by solving equation (3.2) with n_2 equal to infinity; in this case, the wavelength of the most energetic photon is 91.1 nm.

Bohr's atomic model, Fig. 3.3, is based on two fundamental postulates:

1. The orbital electrons can revolve around the nucleus only in certain fixed radii, called stationary states. These radii are such that the angular momentum of the revolving electrons must be integral multiples of $h/2\pi$,

$$mvr = \frac{nh}{2\pi}, \tag{3.3}$$

where m is the mass of the electron, v is its linear velocity, r is the radius of revolution, h is Planck's constant, and n is any positive integer.

2. A photon is emitted only when an electron falls from one orbit to another orbit of lower energy. The energy of the photon is equal to the difference between the energy levels of the electrons in the two orbits.

$$hf = E_2 - E_1, \tag{3.4A}$$

$$f = \frac{E_2}{h} - \frac{E_1}{h}, \tag{3.4B}$$

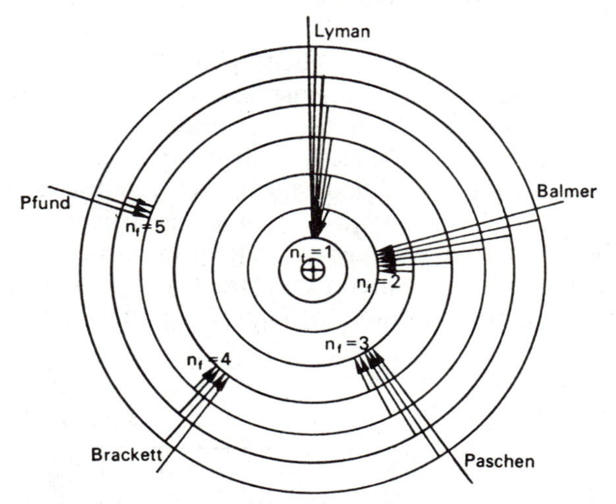

FIG. 3.3. Bohr's model of the hydrogen atom, showing the origin of the various series of lines seen in the hydrogen spectrum.

where f is the frequency of the emitted photon, E_2 and E_1 are the high- and low-energy orbits, respectively.

When the electron revolves around the nucleus, the electrostatic force of attraction between the electron and the nucleus is balanced by the centrifugal force due to the revolution of the electron:

$$k_0 \frac{Ze \cdot e}{r^2} = \frac{mv^2}{r}, \tag{3.5}$$

where k_0 is Coulomb's law constant, 9×10^9 N-m^2/C^2, Z is the atomic number of the atom, and e is the electronic and protonic charge; Ze, therefore, is the charge on the nucleus. Substituting the value for v from equation (3.3) into equation (3.5) and solving for r, we have

$$r = \frac{n^2h^2}{4\pi^2me^2Zk_0}. \tag{3.6}$$

Equation (3.6) gives the radii of the electronic orbits that will satisfy the condition for the stationary states when whole numbers are substituted for n. The normal condition of the atom, or the ground state, is that state for which n is 1. In the ground state the atom is in its lowest possible energy state and therefore in its most stable condition. Transitions from the ground state to higher energy orbits are possible through the absorption of sufficient energy to raise the electron to a larger radius.

The energy in any orbit may be calculated by considering the kinetic energy of the electron due to its motion around the nucleus and the potential energy due to its position in the electric field of the nucleus. Since the kinetic energy of the electron is equal to $\frac{1}{2}mv^2$ (the electron does not revolve rapidly enough to consider relativistic effects), then, from equation (3.5), we have

$$E_k = \tfrac{1}{2}mv^2 = k_0 \frac{Ze^2}{2r}. \tag{3.7}$$

The potential energy is, from equation (2.32),

$$E_p = k_0 \frac{Ze}{r}(-e) = -k_0 \frac{Ze^2}{r}. \tag{3.8}$$

$$E = E_k + E_p$$
$$= \frac{Ze^2}{2r} - \frac{k_0 Ze^2}{r} = -\frac{k_0 Ze^2}{2r}. \qquad (3.9)$$

The total energy given by equation (3.9) is negative simply as a result of the convention discussed in Chapter 2. By definition, the point of zero potential energy was set at an infinite distance from the nucleus. Since the force between the nucleus and the electron is attractive, it follows that at any point closer than infinity, the potential energy must be less than that at infinity, and therefore must have a negative numerical value.

The total energy in any permissible orbit is found by substituting the value for the radius, equation (3.6), into equation (3.9):

$$E = \frac{-2\pi^2 k_0^2 m Z^2 e^4}{h^2} \times \frac{1}{n^2}. \qquad (3.10)$$

Equation (3.10) may now be substituted into equation (3.4B) to get the frequency of the "light" that is radiated from an atom when an electron falls from an excited state into one of lower energy. Letting n_f and n_i represent respectively the orbit numbers for the lower and higher levels, we have for the frequency of the emitted radiation:

$$f = \frac{2\pi k_0^2 m Z^2 e^4}{h^3}\left(\frac{1}{n_f^2} - \frac{1}{n_i^2}\right), \qquad (3.11)$$

and the reciprocal of the wavelength of this radiation is

$$\frac{1}{\lambda} = \frac{2\pi k_0^2 m Z^2 e^4}{ch^3}\left(\frac{1}{n_f^2} - \frac{1}{n_i^2}\right). \qquad (3.12)$$

When numerical values are substituted into the first term of equation (3.12), the numerical value for Rydberg's constant is obtained.

Excitation and Ionization

The Bohr equation may be illustrated for the case of hydrogen by substituting $Z = 1$ into the equations. The radius of the ground state is found to be, from equation (3.6), 0.526×10^{-8} cm. The wavelength of the light emitted when the electron falls from the first excited state, $n_i = 2$, to the ground state, $n_f = 1$, may be calculated by substituting these values into equation (3.12), and is found to be 121.5 nm. The energy of this photon is

$$E = \frac{hc}{\lambda} = \frac{6.6 \times 10^{-34}\ \text{J-sec} \times 3 \times 10^8\ \text{m/sec}}{1.215 \times 10^{-7}\ \text{m} \times 1.6 \times 10^{-19}\ \text{J/eV}} = 10.2\ \text{eV}.$$

This same amount of energy, 10.2 eV, is necessary to excite hydrogen to the first excited state. Precisely this amount of energy, no more and no less, may be used.

When a sufficient amount of energy is imparted to raise it to an infinitely great orbit, that is, to remove the electron from the electrical field of the nucleus, the atom is said to be *ionized*, and the negative electron together with the remaining positively charged atom, is called an *ion pair*. This process is called *ionization*. Ionization or excitation may occur when either a photon or a charged particle, such as an electron, a proton, or an alpha particle, collides with an orbital electron. This mechanism is of great importance in health physics because it is the avenue through which energy is transferred from radiation to matter. When living matter is irradiated, the primary event in the sequence of events leading to biological damage is either excitation or ionization. The *ionization potential* of an element is the amount of energy necessary to ionize the least tightly bound electron in an atom of that element. To remove a second electron requires considerably more energy than removal of the first electron. For most elements, the first ionization potential is on the order of several electron volts. In the case of hydrogen, the ionization potential may be calculated from equation (3.11) by setting n_i equal to infinity:

$$I = hf = \frac{2\pi^2 m Z^2 e^4}{1.6 \times 10^{-19}\ \text{J/eV}\, h^2}\left(\frac{1}{1} - \frac{1}{\infty}\right)$$
$$= \frac{2\pi^2 \times 9.11 \times 10^{-31}\ \text{kg} \times 1 \times (1.6 \times 10^{-19}\ \text{C})^4}{1.6 \times 10^{-19}\ \text{J/eV}\,(6.625 \times 10^{-34}\ \text{J-sec})^2}$$
$$= 13.6\ \text{eV}.$$

A collision in which a rapidly moving particle transfers much more than 13.6 electron volts to the orbital electron of hydrogen results in the ionization of the hydrogen. The excess energy above 13.6 eV that is transferred in this collision appears as kinetic energy of the electron and of the resulting positive ion which recoils under the impact of the collision, in accordance with the requirements of the conservation of momentum. Such inelastic collisions occur only if the incident particle is sufficiently energetic, about 100 eV or greater, to meet this requirement. In those instances where the energy of the particle is insufficient to meet this requirement, an elastic collision with the atom as a whole occurs.

When a photon whose energy is great enough to ionize an atom collides with a tightly bound orbital electron, the photon disappears and the electron is ejected from the atom with a kinetic energy equal to the difference between the energy of the photon and the ionization potential. This mechanism is called the *photoelectric effect*, and is described by the equation

$$E_{pe} = hf - \phi, \tag{3.13}$$

where E_{pe} is the kinetic energy of the photoelectron (the ejected electron), hf is the photon energy, and ϕ is the ionization potential (commonly called the work function). Einstein won the Nobel prize in 1921 for his work on the theoretical aspects of the photoelectric effect.

Example 3.1

An ultraviolet photon whose wavelength is 2000 Å strikes the outer orbital electron of sodium; the ionization potential of the atom is 5.41 eV. What is the kinetic energy of the photoelectron?

The energy of the photon is

$$\frac{hf\ \text{J}}{1.6 \times 10^{-19}\ \text{J/eV}} = \frac{hc}{1.6 \times 10^{-19}\,\lambda}$$
$$= \frac{6.625 \times 10^{-34}\ \text{J-sec} \times 3 \times 10^{8}\ \text{m/sec}}{1.6 \times 10^{-19}\ \text{J/eV} \times 2 \times 10^{-7}\ \text{m}}$$
$$= 6.2\ \text{eV}.$$

From equation (3.13), the kinetic energy of the photoelectron is found to be

$$E_{pe} = 6.20 - 5.41 = 0.79\ \text{eV}.$$

Modifications of the Bohr Atom

The atomic model proposed by Bohr "explains" certain atomic phenomena for hydrogen and for hydrogen-like atoms, such as singly ionized helium, He^+ and doubly ionized lithium, Li^{++}. Calculation of spectra for other atoms according to the Bohr model is complicated by the screening effect of the other electrons, which effectively reduces the electrical field of the nucleus, and by electrical interactions among the electrons. The simple Bohr theory described above is inadequate even for the hydrogen atom. Examination of the spectral lines of hydrogen with spectroscopes of very high resolving power shows the lines to have a fine structure. The spectral lines are in reality each made of several lines very close together. This observation implies the existence of sub-levels of energy within the principal energy levels, and that these sub-levels were very close together. These sub-levels can be explained by assuming the orbits to be elliptical instead of circular, with the nucleus at one of the foci, and that ellipses of different eccentricities have slightly different energy level. For any given principal energy level, the major axes of these ellipses are the same; eccentricity varies only by changes in the length of the minor axes. The eccentricity of these ellipses is restricted by quantum conditions. The angular momentum of an electron rotating in an elliptical orbit is an integral multiple of $h/2\pi$, as in the case of the circular Bohr orbit. However, the numerical value for this multiple is not the same as that for the circular orbit. In the case of the circular orbit, this multiple, or quantum number, is called the principal quantum number and is given the symbol n. For the elliptical orbit, the multiple is called the azimuthal quantum number, usually symbolized by the letter l, and may be any integral number between 0 and $n - 1$, inclusive.

Elliptical orbits alone are insufficient to account for observed spectral lines. To explain the very fine structure, it was necessary to postulate that each orbital electron spins about its own axis in the same manner as the earth spins about its axis as it revolves around the sun. The angular momentum due to this spin also is quantized, and can only have a value equal to one-half a unit of angular momentum;

$$s = \pm\frac{1}{2}\frac{h}{2\pi}. \tag{3.14}$$

The orbital electron can spin in only one of two directions with respect to the direction of its revolution about the nucleus; either in the same direction or in the opposite direction. This accounts for the plus and minus signs in equation (3.14). The total angular momentum of the electron is therefore equal to the vector sum of the orbital and spin angular momenta.

One more fact must be considered before the description of the atom is complete. A closed loop through which an electric current flows has a magnetic moment that is proportional to the product of the current and the area of the loop. The electron revolving in its orbit is equivalent to a closed current-carrying loop, and therefore has a magnetic moment. The fact that the electron spins results in an additional magnetic moment, which may be either positive or negative, depending on the direction of spin relative to the direction of the orbital motion. The total magnetic moment is therefore equal to the sum of the orbital and spin magnetic moments. If the atoms of any substance are placed in a strong magnetic field, the orbital electrons, because of their magnetic moments, will orient themselves in definite directions relative to the applied magnetic field. These directions are such that the component of the vector representing the orbital angular momentum, l, that is parallel to the magnetic field must have an integral value of angular momentum. This integral number, which is given the symbol m, is called the magnetic quantum number; it can have numerical values ranging from l, $l-1$, $l-2, \ldots, 0 \ldots, -(l-2)$, $-(l-1)$, $-l$.

To describe an atom completely, it is necessary to specify four quantum numbers, which have the values given below, for each of the orbital electrons.

Symbol	Name	Value
n	principal quantum number	$1, 2, \ldots$
l	azimuthal quantum number	0 to $n-1$
m	magnetic quantum number	$-l$ to 0 to $+l$
s	spin quantum number	$-\frac{1}{2}, +\frac{1}{2}$

By using Bohr's atomic model, and by assigning all possible numerical values to these four quantum numbers according to certain rules, it is possible to construct the periodic table of the elements.

Periodic Table of the Elements

The period table of the elements may be constructed with Bohr's atomic model by application of the *Pauli Exclusion Principle*. This principle states that no two electrons in any atom may have the same set of four quantum numbers. Hydrogen, the first element, has a nuclear charge of $+1$, and therefore has only one electron. Since the principal quantum number of this electron must be 1, l and m must be 0, and the spin quantum number s may be either plus or minus $\frac{1}{2}$. If now we go to the second element, helium, we must have two orbital electrons, since helium has a nuclear charge of $+2$. The first electron in the helium atom may have the same set of quantum numbers as the electron in the hydrogen atom. The second electron, however, must differ. This difference can be only in the spin, since we may have two different spins for the set of quantum numbers $n = 1$, $l = 0$, and $m = 0$. This second electron exhausts all the possibilities for $n = 1$. If now a third electron is added when we go to atomic number 3, lithium, it must have the principal quantum number 2. In this principal energy level, the orbit may be either circular or elliptical, that is, the azimuthal quantum number l may be either 0 or 1. In the case of $l = 0$, the magnetic quantum number m can only be equal to 0; when $l = 1$, m may be either -1, 0, or $+1$. Each of these quantum states may contain two electrons, one each with spins of plus and minus $\frac{1}{2}$. Eight different electrons, each with its own unique set of quantum numbers, are therefore possible in the second principal energy level. These eight different possibilities are utilized by the elements Li, Be, B, C, N, O, F, and Ne, atomic-numbered elements 3–10 inclusive. The additional electron for sodium, atomic number 11, must have the principal quantum number $n = 3$. By assigning all the possible combinations of the four quantum numbers to the electrons in the third principal energy level, it is found that eighteen electrons are possible. These energy levels are not filled successively as were those in the K and L shells. (The principal quantum levels corresponding to $n = 1$, 2, 3, 4, 5, 6, and 7 are called the K, L, M, N, O, P, and Q shells, respectively.) No outermost electron shell contains more than eight electrons. After the M shell contains eight electrons, as in the case of argon, the next element in the periodic table, potassium, atomic number 20, starts another principal energy level with one electron in the N shell. Subsequent elements then may add electrons either in the M or in the N shells, until the M shell contains its full complement of eighteen electrons. No electrons appear in the O shell until the N shell has eight electrons. The maximum number of electrons that may exist in any principal energy level is given by the product $2n^2$, where n is the principal quantum number. Thus, the O shell may have a maximum of $2 \times 5^2 = 50$ electrons.

TABLE 3.1. ELECTRONIC STRUCTURE OF THE ELEMENTS

Period	Element	Atomic no.	K Shell 1	L Shell 2	M Shell 3	N Shell 4	O Shell 5	P Shell 6	Q Shell 7
1	H	1	1						
	He	2	2						
2	Li	3	2	1					
	Be	4	2	2					
	B	5	2	3					
	C	6	2	4					
	N	7	2	5					
	O	8	2	6					
	F	9	2	7					
3	Ne	10	2	8					
	Na	11	2	8	1				
	Mg	12	2	8	2				
	Al	13	2	8	3				
	Si	14	2	8	4				
	P	15	2	8	5				
	S	16	2	8	6				
	Cl	17	2	8	7				
4	A	18	2	8	8				
	K	19	2	8	8	1			
	Ca	20	2	8	8	2			
	Sc	21	2	8	9	2			
	Ti	22	2	8	10	2			
	V	23	2	8	11	2			
	Cr	24	2	8	13	1			
	Mn	25	2	8	13	2			
	Fe	26	2	8	14	2			
	Co	27	2	8	15	2			
	Ni	28	2	8	16	2			
	Cu	29	2	8	18	1			
	Zn	30	2	8	18	2			
	Ga	31	2	8	18	3			
	Ge	32	2	8	18	4			
	As	33	2	8	18	5			
	Se	34	2	8	18	6			
	Br	35	2	8	18	7			
5	Kr	36	2	8	18	8			
	Rb	37	2	8	18	8	1		
	Sr	38	2	8	18	8	2		
	Y	39	2	8	18	9	2		
	Zr	40	2	8	18	10	2		
	Nb	41	2	8	18	12	1		
	Mo	42	2	8	18	13	1		
	Tc	43	2	8	18	14	1		
	Ru	44	2	8	18	15	1		
	Rh	45	2	8	18	16	1		
	Pd	46	2	8	18	18	0		
	Ag	47	2	8	18	18	1		
	Cd	48	2	8	18	18	2		
	In	49	2	8	18	18	3		
	Sn	50	2	8	18	18	4		
	Sb	51	2	8	18	18	5		

TABLE 3.1 (*cont.*)

Period	Element	Atomic no.	K Shell 1	L Shell 2	M Shell 3	N Shell 4	O Shell 5	P Shell 6	Q Shell 7
	Te	52	2	8	18	18	6		
	I	53	2	8	18	18	7		
6	Xe	54	2	8	18	18	8		
	Cs	55	2	8	18	18	8	1	
	Ba	56	2	8	18	18	8	2	
	La	57	2	8	18	18	9	2	
	Ce	58	2	8	18	19	9	2	
	Pr	59	2	8	18	20	9	2	
	Nd	60	2	8	18	22	8	2	
	Pm	61	2	8	18	23	8	2	
	Sm	62	2	8	18	24	8	2	
	Eu	63	2	8	18	25	8	2	
	Gd	64	2	8	18	25	9	2	
	Tb	65	2	8	18	26	9	2	
	Dy	66	2	8	18	28	8	2	
	Ho	67	2	8	18	29	8	2	
	Er	68	2	8	18	30	8	2	
	Tm	69	2	8	18	31	8	2	
	Yb	70	2	8	18	32	8	2	
	Lu	71	2	8	18	32	9	2	
	Hf	72	2	8	18	32	10	2	
	Ta	73	2	8	18	32	11	2	
	W	74	2	8	18	32	12	2	
	Re	75	2	8	18	32	13	2	
	Os	76	2	8	18	32	14	2	
	Ir	77	2	8	18	32	15	2	
	Pt	78	2	8	18	32	17	1	
	Au	79	2	8	18	32	18	1	
	Hg	80	2	8	18	32	18	2	
	Tl	81	2	8	18	32	18	3	
	Pb	82	2	8	18	32	18	4	
	Bi	83	2	8	18	32	18	5	
	Po	84	2	8	18	32	18	6	
	At	85	2	8	18	32	18	7	
7	Rn	86	2	8	18	32	18	8	
	Fr	87	2	8	18	32	18	8	1
	Ra	88	2	8	18	32	18	8	2
	Ac	89	2	8	18	32	18	9	2
	Th	90	2	8	18	32	19	9	2
	Pa	91	2	8	18	32	20	9	2
	U	92	2	8	18	32	21	9	2
	Np	93	2	8	18	32	22	9	2
	Pu	94	2	8	18	32	23	9	2
	Am	95	2	8	18	32	25	8	2
	Cm	96	2	8	18	32	25	9	2
	Bk	97	2	8	18	32	26	9	2
	Cf	98	2	8	18	32	27	9	2
	Es	99	2	8	18	32	28	9	2
	Fm	100	2	8	18	32	29	9	2
	Md	101	2	8	18	32	30	9	2
	No	102	2	8	18	32	31	9	2

The fact that no outermost electron shell contains more than eight electrons is responsible for the periodicity of the chemical properties of many elements, and is the physical basis for the periodic table. Since chemical reactions involve the outer electrons, it is not surprising that atoms with similar outer electronic structures should have similar chemical properties. For example, Li, Na, K, Rb, and Cs behave chemically similarly because each of these elements has only one electron in its outermost orbit. The inert gases He, Ne, A, Kr, Xe, and Rn have similar electronic structures too; they all have eight electrons in their outermost shells. Because their outermost electron shells are filled, these elements do not undergo chemical reactions. The elements are thought to have the electronic configurations given in table 3.1.

Examination of table 3.1 reveals certain interesting points. The first twenty elements successively add electrons to their outermost shells. The next eight elements, scandium to nickel, have four shells, but add successive electrons to the third shell unit it is filled with the maximum number of 18. These elements are called *transition elements*. The same thing happens with elements 39 to 46 inclusive. Electrons are added to the fourth shell until they number 18, then the fifth shell increases until it contains eight electrons. In element number 55, cesium, the sixth principal electron orbit, the *P* shell, starts to fill. Instead of continuing, however, the *N* level starts to fill. Beginning with cerium and continuing through lutetium, electrons are successively added to the fourth electron shell, while the two outermost shells remain about the same. This group of elements is usually called the *rare earths*, and sometimes called the *lanthanides* because they begin immediately after lanthanum. The rare earths differ from the transition elements in the depth of the electronic orbit which is filling. While the transition elements fill the second outer orbit, the rare earths fill the third electron shell, which is deeper in the atom. Since, in the case of the rare earths, the two outermost electron shells are alike, it is extremely difficult to separate them by chemical means. They ere of importance to the health physicist because they include a great number of the fission products. The concern of the health physicist with the rare earths is aggravated by the fact that the analytical chemistry of the rare earths is very difficult, and also by the relative dearth of knowledge regarding their metabolic pathways and toxicological properties. Despite their name, the rare earths are not rare; they are found to be widely distributed in nature, albeit in small concentrations. Another group of rare earths is found. In the elements starting with thorium, and continuing to lawrencium, the *O* shell fills while the *P* and *Q* shells remain about the same. These rare earths are frequently called the *actinide* elements. They are of importance to the health physicist because they are all naturally radioactive and because they include the fuel used in nuclear reactors.

Characteristic X-rays

Some virtues of the solar system type of atomic model , in which electrons rotate about the nucleus in certain radii corresponding to unique energy levels, are the simple explanations that it allows for transfer of energy to matter by excitation and ionization, for the photoelectric effect, and for the origin of certain X-rays called characteristic X-rays. It was pointed out that optical and ultraviolet spectra of elements are due to excitation of outer electrons to levels up to several electron volts, and that spectral lines represent energy differences between excited states. As more and more electron shells are added the energy differences between the principal levels increases greatly. For the high atomic numbered elements, they reach tens of thousands of electron volts. In th case of lead, for example, the energy difference between the *K* and *L* shells is 72,000 eV. If this *K* electron

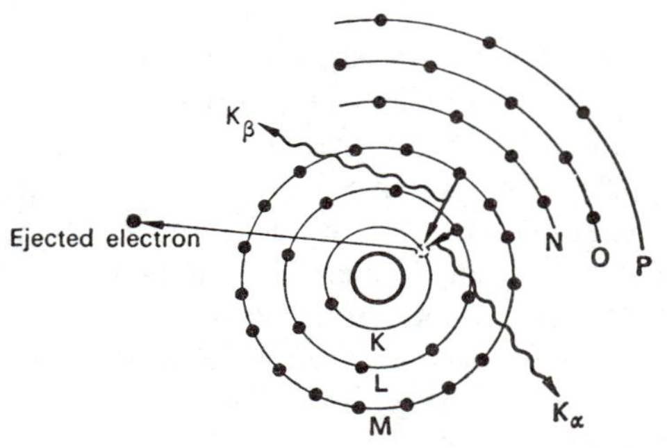

FIG. 3.4. Schematic representation of the origin of characteristic X-rays.

is struck by a photon whose energy exceeds 87.95 keV, the binding energy of the K electron, the electron is ejected from the atom, and leaves an empty slot in the K shell, as shown schematically in Fig. 3.4. Instantaneously, one of the outer electrons falls down into the vacant slot left by the photoelectron. When this happens, a photon is emitted whose energy is equal to the difference between the initial and final energy levels, in accordance with equation (3.4A). For the lead atom, when an electron falls from the L to the K levels, the emitted photon has a quantum energy of 72,000 eV. A photon of such high energy is an X-ray. When produced in this manner, the photon is called a *characteristic X-ray*, because the energy differences between electron orbits are unique for the different atoms, and the X-rays that represent these differences are "characteristic" of the elements in which they originate. This process is repeated until all the inner electron orbits are re-filled. It is possible, of course, that the first transition is from the M level, or even from the outermost electronic orbit. The most likely origin of the first electronic transition, however, is the L shell. When this happens, the resulting X-ray is called a K_α photon; if an electron falls from the M level to the K level, we have a K_β photon. When the vacancy in the L orbit is filled by an electron that falls from the M level, we have an L_α X-ray; if the L vacancy is filled by an electron originally in the N level, then an L_β X-ray results, and so on. These characteristic X-rays are sometimes called *fluorescent radiation*, since they are emitted when matter is irradiated with X-rays. Characteristic radiation is useful as a tool to the analytical chemist for identifying unknown elements. Obviously, characteristic radiation is of importance to the health physicist who must consider the fluorescent radiation that may be produced in radiation absorbers and in certain other cases where inner electrons are ejected from high-atomic-numbered elements.

The Wave Mechanics Atomic Model

The atomic model described above is sufficiently useful to explain most phenomena encountered in health physics. For the study of atomic physics, however, a more abstract concept of the atom was proposed by the Austrian physicist Schrödinger (for which he and the British physicist Dirac shared the Nobel prize in 1933). Instead of working with particulate electrons as Bohr had done, he treated them as de Broglie waves, and developed the branch of physics known as "wave mechanics." Starting with the de Broglie equation for the associated electron wave, Schrödinger derived a general differential equation that must be satisfied by an electron within an atom. The present-day atomic theory consists of solutions of this equation subject to certain conditions. A number of different solutions, corresponding to different energy levels, is possible. However, whereas Bohr pictured an atom with electrons at precisely determined distances from the nucleus, the Schrödinger wave equation gives the probability of finding an electron at any given distance from the nucleus. The two atomic pictures coincide to the extent that the most probable radius for the hydrogen electron is exactly the same as the first Bohr radius. Similarly, the second Bohr radius corresponds to the most probable distance from the nucleus of the electron in the first excited state. Furthermore, the four quantum numbers arbitrarily introduced into the Bohr atom falls naturally out of the solutions of the Schrödinger wave equation. Although the wave model has replaced the Bohr system of atomic mechanics for highly theoretical considerations, the older atomic model is still considered a very useful tool in helping to interpret atomic phenomena.

The Nucleus

The Neutron and Nuclear Force

It has already been pointed out that the positive charges in the atomic nucleus are due to protons, and that hydrogen is the simplest nucleus—it consists of only a single proton. If succeeding nuclei merely were multiples of the proton, then the mass numbers of the nuclei, if a mass number of 1 is assigned to the proton, should be equal to the atomic numbers of the nuclei. This was not found to be the case. Except for hydrogen, the nuclear mass numbers, were found to be about twice as great as the corresponding atomic numbers, and to become relatively greater as the atomic numbers increased. Furthermore, it was necessary to account for the stability of the nucleus in the face of the repulsive coulombic forces among the nuclear protons. A simple calculation shows that the gravitational force of attraction is insufficient to overcome the repulsive electrical forces. Both these problems were solved by the discovery, in 1932, by the Brithish physicist Chadwick, of the third basic building block in nature: the neutron. (Chadwick won the Nobel prize in 1935 for this discovery.) This particle, whose mass is about the same as that of a proton, 1.67474×10^{-27} kg, is electrically neutral. Its presence in the nucleus accounts for the differnce between the atomic number and the atomic mass number; it also supplies the cohesive force that holds the nucleus together. This force is called the nuclear force. It is thought to act over an extremely short range—about 2 to 3×10^{-15} m. By analogy to the ordinary case of charged particles, it may be assumed that the neutron and the proton carry certain nuclear charges and that force fields due to these nuclear charges are established around the nucleons (particles within the nucleus are called nucleons). Nuclear forces are all attractive, and the interaction between the nuclear force fields supplies the cohesive forces which overcome the repulsive electrical forces. However, since the range of the nuclear force is much shorter than the range of the electrical force, neutrons can interact only with those nucleons to which they are immediately adjacent, whereas protons interact with each other even though remotely located within the nucleus. For this reason, the number of neutrons must increase more rapidly than the number of protons.

Isotopes

It has been found that, for any particular element, the number of neutrons within the nucleus is not constant. Oxygen, for example, consists of three nuclear species; one whose nucleus has eight neutrons, one of nine neutrons, and one of ten neutrons. In these three cases, of course, the nucleus contains eight protons. The atomic mass numbers of these three species are 16, 17, and 18, respectively. These three nuclear species of the same element are called *isotopes* of oxygen. Isotopes of an element are atoms that contain the same number of positive nuclear charges and have the same extra-nuclear electronic structure, but differ in the number of neutrons. Most elements contain several isotopes. The atomic weight of an element is the weighted average of the weights of the different isotopes of which the element is composed. Isotopes cannot be distinguished chemically, since they have the same electronic structure and therefore undergo the same chemical reactions. An isotope is identified by writing the chemical symbol with a subscript to the left giving the atomic number and a superscript giving the atomic mass number, or the total number of nucleons. Thus the three isotopes of oxygen may be written as ${}^{16}_{8}O$, ${}^{17}_{8}O$, and ${}^{18}_{8}O$. Since the atomic number is synonomous with the chemical symbol, however, the subscript is usually omitted and the isotope is written as ${}^{16}O$. It should be pointed out that not all isotopes are equally abundant. In the case of oxygen 99.975% of the naturally occurring atoms are ${}^{16}O$, whereas ${}^{17}O$ and ${}^{18}O$ include 0.037% and 0.204%, respectively. In other elements the distribution of isotopes may be quite different. Chlorine, for example, consists of two naturally occurring isotopes, ${}^{35}Cl$ and ${}^{37}Cl$, whose respective abundances are 75.4% and 24.6%.

The Atomic Mass Unit

Atomic masses may be given either in grams or in relative numbers called atomic mass units. Since one mole of any substance contains 6.03×10^{23} molecules (Avogadro's number), and since the weight in grams of one mole is equal numerically to its molecular weight, the weight of a single atom can easily be computed. In the case of ${}^{12}C$, for example, 1 mole weighs 12.0000 g. One atom, therefore, weighs

$$\frac{12.0000 \text{ g/mole}}{6.027 \times 10^{23} \text{ molecules/mole}} = 1.9910 \times 10^{-23} \text{ g}.$$

Because the weight of one mole of carbon 12 was found to be a whole number, carbon was chosen as the reference standard for the system of relative weights known as atomic weights. (Actually the atomic weight of an element is the weighted average of all the isotopic weights. The physical scale is based only on the weight of ${}^{12}C$.) Since carbon 12 was assigned an atomic weight of 12.0000, one atomic weight unit is

$$1 \text{ amu} = \frac{1.9910 \times 10^{-23} \text{ g}}{12} = 1.6592 \times 10^{-24} \text{ g}.$$

On this basis, the weight of a neutron, m_n, is 1.008987 amu, that of a proton, m_p, is 1.007593 amu, and the atomic weight of an electron is 0.000552 amu. The energy equivalent of one atomic mass unit is

$$E = \frac{mc^2 \text{ joules}}{1.6 \times 10^{-13} \text{ J/MeV}}$$

$$= 931 \text{ MeV}.$$

Binding Energy

At this point, it is interesting to compare the sum of the weights of the constituent parts of an isotope, W, with the measured isotopic weight, M. For the case of ${}^{17}O$, whose atomic mass is 17.004533 amu, we have

$$W = Zm_p + (A - Z)m_n + Zm_e, \tag{3.15}$$

where Z is the atomic number and A is the atomic mass number, which is equal to the number of nucleons within the nucleus.

$$W = 8(1.007593) + (17 - 8)(1.008987) + 8(0.000552)$$

$$= 17.146043 \text{ amu}.$$

The weight of the sum of the parts is seen to be much greater than the actual weight of the entire atom. This is true not only for ^{17}O, but for all nuclei. The difference between the atomic weight and the sum of the weights of the parts is called the mass defect, and is defined by δ as follows:

$$\delta = W - M. \tag{3.16}$$

The mass decrement represents the mass equivalent of the work that must be done in order to separate the nucleus into its individual component nucleons, and is therefore called the binding energy. In energy units, the binding energy, *BE*, is

$$BE = (W - M)\,\text{amu} \times 931\,\frac{\text{MeV}}{\text{amu}}. \tag{3.17}$$

Quite clearly, the binding energy is a measure of the cohesiveness of a nucleus. Since the total binding energy of a nucleus depends on the number of nucleons within the nucleus, a more useful measure of the cohesiveness is the binding energy per nucleon, E_b, as given below:

$$E_b = \frac{931(W - M)}{A}\,\frac{\text{MeV}}{\text{nucleon}}, \tag{3.18}$$

where A, the atomic mass number, represents the number of nucleons within the nucleus. For the case of ^{17}O, the binding energy is 131.7 MeV, and the binding energy per nucleon is 7.74 MeV.

The binding energy per nucleon is very low for the low atomic numbered elements, but rises rapidly to a very broad peak at binding energies in excess of 8 MeV per nucleon, and then decreases very slowly until a value of 7.58 MeV per nucleon is reached for ^{238}U. Figure 3.5, in which the binding energy per nucleon is plotted against the number of nucleons in the various isotopes, shows that, with a very few exceptions, there is a systematic variation of binding energy per nucleon with the number of nucleons within the nucleus.

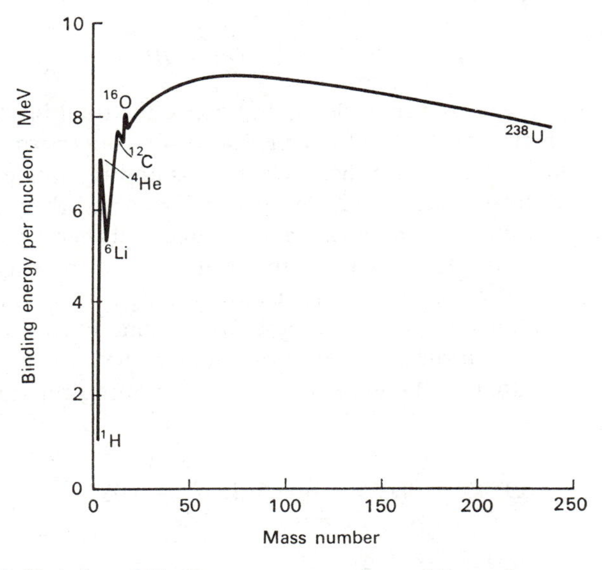

FIG. 3.5. Variation of binding energy per nucleon with atomic mass number.

The most notable departures from the smooth curve are the isotopes ^{4}He, ^{12}C, and ^{16}O. Each of these isotopes lies above the curve, indicating that they are very strongly bound. The ^{12}C and ^{16}O isotopes, as well as ^{20}Ne, which has more binding energy per nucleon than either of the isotopes that flank it, may be thought of as containing three, four, and five sub-units of ^{4}He respectively. The exceptional binding energies in these nuclei, together with the fact that ^{4}He nuclei, as alpha particles, are emitted in certain modes of radioactive transformation, suggest that nucleons tend to form stable subgroups of two protons and two neutrons within the nucleus.

The fact that the binding energy curve, Fig. 3.5, has the shape that it does, explains why it is possible to release energy by splitting the very heavy elements and by fusing two very light elements. Since the binding energy per nucleon is greater for nuclei in the center of the curve than for nuclei at both extremes, any change in nuclear structure that drives the nucleons towards the center of the curve must release the energy difference between the final and the initial states.

Nuclear Models

(a) *Liquid drop*

Although the nuclear building blocks seem to be well known, no definite structure for the nucleus has yet been established. However, two different nuclear models have been postulated. According to one of these, the nucleus is thought to be a homogenous mixture of nucleons in which all the nucleons interact strongly with each other. As a result, the internal energy of the nucleus is about equally distributed among the constituent nucleons, while surface tension forces tend to keep the nucleus spherical. This is analogous to a drop of liquid, and hence is called the *liquid drop model* of the nucleus. This model, which was proposed by Bohr and Wheeler, is particularly successful in explaining nuclear fission and in permitting the calculation of the atomic masses of isotopes whose atomic mass is very difficult to measure.

From the preceding discussion, we see that the mass of a nucleus of atomic number A and atomic mass number Z is

$$M = (A - Z)m_n + Zm_p - BE. \tag{3.19}$$

Furthermore, the binding energy per nucleon, and hence the total binding energy, is seen from Fig. 3.5 to be a function of Z and A. The nuclear drop model permits a semi-empirical equation to be formulated that relates the nuclear mass and binding energy to A and Z. According to the liquid drop model, the intra-nuclear forces and the potential energy due to these forces are due to the short-range attractive forces between adjacent nucleons, the long-range repulsive Coulomb forces among the protons, and the surface tension effect, in which nucleons on the surface of the nucleus are less tightly bound than those in the nuclear interior. The binding energy due to these forces is modified according to whether the numbers of neutrons and protons are even or odd. On the basis of this reasoning, the following equation was fitted to the experimental data relating nuclear mass, in atomic mass units, with A and Z:

$$M = 0.99389\ A - 0.00081\ Z + 0.014\ A^{2/3} + 0.083\frac{(A/2 - Z)^2}{A} + 0.000627\frac{Z^2}{A^{1/3}} + \Delta, \tag{3.20}$$

where: $\Delta = 0$ for odd A,

$\Delta = -0.036/A^{3/4}$ for even A, even Z,

$\Delta = +0.036/A^{3/4}$ for even A, odd Z.

(b) *Shell model*

The alternate nuclear model is called the *shell* model. According to this model of the nucleus, the various nucleons exist in certain energy levels within the nucleus, and interact weakly among themselves. Many observations and experimental data lend support to such a nuclear structure. Among the stable isotopes, the "even–even" nuclei that is, nuclei with even numbers of protons and neutrons, are most numerous, with a total of 162 isotopes. Even–odd nuclei, in which either the protons or neutrons are even in number, and the other one is odd, are second in abundance with a total of 108 isotopes. Odd–odd nuclei are the fewest in number; only four such stable isotopes are found in nature. Furthermore, so-called "magic numbers" have been found to recur among the stable isotopes. These magic numbers include 2, 8, 20, 50, 82, and 126. Isotopes containing these numbers of protons or neutrons or both are most abundant in nature, suggesting unusual stability in their structures. Nuclei containing these magic numbers are relatively inert in a nuclear sense, that is, they do not react easily when bombarded with neutrons. This is analogous to the case of chemically inert elements that have filled electron energy levels. All these observed facts are compatible with an energy level model of the nucleus similar to the electronic energy level model of the atom. Each nucleon in a nucleus is identified by its own set of four quantum numbers, as in the case of the extra-nuclear electrons. By application of the Pauli exclusion principle to nucleons, it is possible to construct energy levels which contain successively 2, 8, 20, 50, 82, and 126 nucleons.

As in the case of the extra-nuclear electrons, nucleons too may be excited by raising them to higher energy levels. When this occurs, the nucleon falls back into its ground state, and emits a photon whose energy is equal to the energy difference between the excited and ground states. This is the same type phenomenon as in the case of optical and characteristic X-ray spectra. The photon in this case is called a *gamma ray*. Because nuclear energy levels are usually much further apart than electronic energy levels, gamma rays are usually (though not necessarily) more energetic than X-ray photons. It should be emphasized that from the practical health physics point of view X-rays and gamma rays are identical. They differ only in their place of origin—X-rays in the extra-nuclear structure and gamma rays within the nucleus. Once produced, it is impossible to distinguish between X-rays and gamma rays.

Nuclear Stability

If a plot is made of the number of protons versus the number of neutrons for the stable isotopes, the curve shown in Fig. 3.6 is obtained. The stable isotopes lie within a relatively narrow range, indicating that the neutron to proton ratio must lie within certain limits if a nucleus is to be stable. Most radioactive nuclei lie outside this range of stability. The plot also shows that the slope of the curve, which initially has a value of unity, gradually increases as the atomic number increases, thereby showing the continuously increasing ratio of neutrons to protons.

Since all nuclear forces are attractive, it may appear surprising to find unstable nuclei with an excessive number of neutrons. This apparent anomaly may be explained simply in terms of the shell model of the nucleus. Acording to the Pauli exclusion principle, like

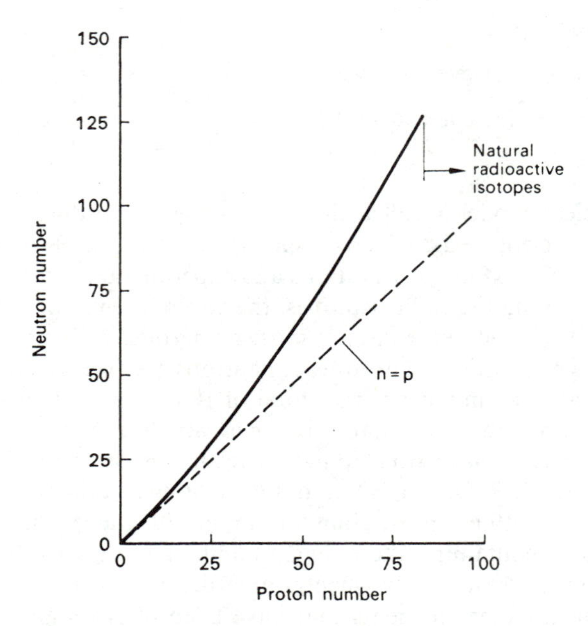

FIG. 3.6. Nuclear stability curve. The line represents the best fit to the neutron-proton coordinates of stable isotopes.

nucleons may be grouped in pairs, with each pair having all quantum numbers the same, except the spin quantum number. Since nuclei with completely filled energy levels are more stable than those with unfilled inner levels, additional neutrons in the case of nuclei with unfilled proton levels but filled neutron levels results in unstable nuclei. To achieve stability, the nucleus may undergo an internal rearrangement in which the additional neutron transforms itself into a proton by emitting an electron. The new proton then pairs off with a proton in one of the unfilled proton levels. As an example of this possible mechanism, consider the consequences of the addition of a neutron to $^{31}_{15}P$. This is the stable isotope of phosphorous that occurs naturally. According to the shell model, the fifteen protons inside the nucleus may be distributed among several pairs plus one, while the neutrons may be paired off into eight groups. If now an additional neutron is added to the nucleus to make $^{32}_{15}P$, the additional neutron may go into another energy level. This condition, however, is unstable. The additional neutron may therefore become a proton and an electron, with the electron being ejected from the nucleus, and the proton pairing off with the single proton, thereby forming stable $^{32}_{16}S$. This internal nuclear transformation is called a *radioactive transformation*.

Problems

1. What is the closest approach that a 5.3-MeV alpha particle can make to a gold nucleus?

2. Calculate the number of atoms per cm^3 of lead, given that the density of lead is 11.3 g/cm^3 and its atomic weight is 207.21.

3. A μ^- meson has a charge of -4.8×10^{-10} sC and a mass 207 times that of a resting electron. If a proton should capture a μ^- to form a "mesic" atom, calculate

(a) the radius of the first Bohr orbit, and

(b) the ionization potential.

4. Calculate the ionization potential of a singly ionized ^4He atom.

5. Calculate the current due to the hydrogen electron in the ground state of hydrogen.

6. Calculate the ratio of the velocity of a hydrogen electron in the ground state to the velocity of light.

7. Calculate the Rydberg constant for deuterium.

8. What is the uncertainty in the momentum of a proton inside a nucleus of ^{27}Al? What is the kinetic energy of this proton?

9. A sodium ion is neutralized by capturing a 1-eV electron. What is the wavelength of the emitted radiation if the ionization potential of Na is 5.41 volts?

10. (a) How much energy would be released if one gram deuterium were fused to form helium according to the equation $^{2}H + {}^{2}H {}^{4}He + Q$?

(b) How much energy is necessary to drive the two deuterium nuclei together?

11. The density of beryllium, atomic number 4, is 1.84 g/cm^3, and the density of lead, atomic number 82, is 11.3 g/cm^3. Calculate the density of a ^{9}Be and a ^{208}Pb nucleus.

12. Determine the electronic shell configuration for aluminum, atomic number 13.

13. What is the difference in mass between the hydrogen atom and the sum of the masses of a proton and an electron? Express the answer in energy equivalent (eV) of the mass difference.

14. If the heat of vaporization of water is 540 calories per gram at atmospheric pressure, what is the binding energy of a water molecule?

15. The ionization potential of He is 24.5 eV.

(a) What is the maximum velocity with which an electron is moving before it can ionize an unexcited He atom?

(b) What is the minimum wavelength of a photon in order that it ionize the He atom?

16. In a certain 25-watt mercury-vapor ultraviolet lamp, 0.1% of the electrical energy input appears as U.V. radiation of wavelength 2537 Å. What is the photon emission rate, per second, from this lamp?

17. The atomic mass of tritium is 3.017005 amu. How much energy in MeV is required to dissociate the tritium into its component parts?

18. Compute the frequency, wave length, and energy (electron volts) for the second and third lines in the Balmer series.

19. Using the Bohr atomic model, calculate the velocity of the ground state electrons in hydrogen and in helium.

20. The heat of combustion when H_2 combines with O_2 to form water is 60 kcal/mole water. How much energy (electron volts) is liberated per molecule of water produced?

Suggested References

BORN, M. *Atomic Physics*, 8th ed. Hafner, Darien, Conn., 1970.

COHEN, B. L. *Concepts of Nuclear Physics*. McGraw-Hill, New York, 1971.

EVANS, R. D. *The Atomic Nucleus*. McGraw-Hill, New York, 1955.

FRIEDLANDER, G., KENNEDY, J. W., MACIAS, E. S. and MILLER, J. M. *Nuclear and Radiochemistry*, 2nd ed. John Wiley & Sons, New York, 1981.

GLASSTONE, S. *Sourcebook on Atomic Energy*, 3rd ed. D. Van Norstrand, Princeton, 1967.

HOLLIDAY, D., and RESNICK, R. *Physics*, 3rd ed. John Wiley & Sons, New York, 1977.

KAPLAN, I. *Nuclear Physics*, 2nd ed. Addison-Wesley Publ. Co., Reading, Mass., 1963.

LAPP, R. E., and ANDREWS, H. L. *Nuclear Radiation Physics*, 4th ed. Prentice-Hall, Englewood Cliffs, N.J., 1972.

ROGERS, E. M. *Physics for the Inquiring Mind*. D. Van Nostrand, Princeton, 1960.

SEARS, F. W., and ZEMANSKY, M. W. *University Physics*, 4th ed. Addison-Wesley, Reading, Mass., 1970.

SEMAT, H., and ALBRIGHT, J. R. *Introduction to Atomic and Nuclear Physics*, 5th ed. Holt, Rinehart, and Winston, New York, 1972.

CHAPTER 4

RADIOACTIVITY

Radioactivity and Transformation Mechanisms

Radioactivity may be defined as spontaneous nuclear transformations that result in the formation of new elements. These transformations are accomplished by one of several different mechanisms, including alpha particle emission, beta particle and positron emission, and orbital electron capture. Each of these reactions may or may not be accompanied by gamma radiation. Radioactivity and radioactive properties of nuclides are determined by nuclear considerations only, and are independent of the chemical and physical states of the radioisotope. Radioactive properties of radioisotopes therefore cannot be changed by any means, and are unique to the respective radionuclides. The exact mode of radioactive transformation depends on two factors: the particular type of nuclear instability—that is, whether the neutron to proton ratio is either too high or too low for the particular nuclide under consideration, and on the mass–energy relationship among the parent nucleus, daughter nucleus, and the emitted particle.

Alpha Emission

An alpha particle is a highly energetic helium nucleus that is emitted from the nucleus of the radioactive isotope when the neutron to proton ratio is too low. It is a positively charged, massive particle, consisting of an assembly of two protons and two neutrons. Since atomic numbers and mass numbers are conserved in alpha transitions, it follows that the result of alpha emission is a daughter whose atomic number is two less than that of the parent, and whose atomic mass number is four less than that of the parent. In the case of ^{210}Po, for example, the reaction is

$$^{210}_{84}Po \longrightarrow {}^{4}_{2}He + {}^{206}_{82}Pb.$$

In this example, ^{210}Po has a neutron to proton ratio of 126 to 84, or 1.5 to 1. After decaying by alpha particle emission, a stable daughter nucleus, $^{206}_{82}Pb$, is formed whose neutron to proton ratio is 1.51 to 1. With one exception, $^{147}_{62}Sm$, naturally occurring alpha emitters are found only among elements of atomic number greater than 82. The explanation for this is two-fold; first is the fact that the electrostatic repulsive forces in the heavy nuclei increase much more rapidly than the cohesive nuclear forces, and the magnitude of the electrostatic forces, consequently, may closely approach or even exceed that of the nuclear force; the second part of the explanation is concerned with the fact that the emitted particle must have sufficient energy to overcome the high potential barrier at the surface of the nucleus resulting from the presence of the positively charged nucleons. This potential barrier may be graphically represented by the curve in Fig. 4.1. The inside of the nucleus, because of the negative potential there, may be thought of as a potential well that is surrounded by a wall whose height is about 25 MeV for an alpha particle inside a high atomic numbered nucleus. According to quantum mechanical theory, an alpha

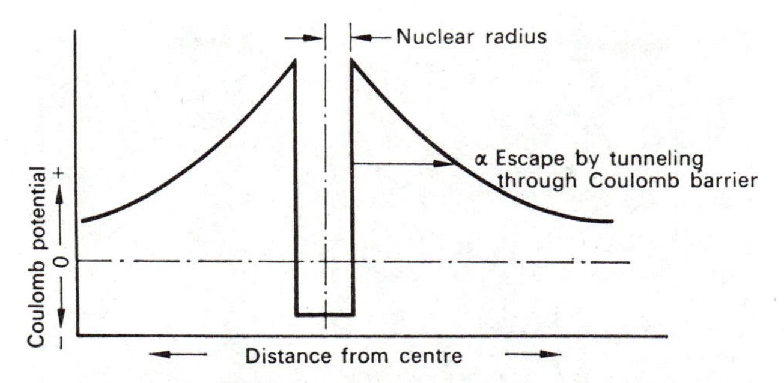

FIG. 4.1. Potential inside and in the vicinity of a nucleus.

particle may escape from the potential well by tunneling through the potential barrier. For alpha emission to be observed from the high atomic numbered naturally occurring elements, theoretical considerations demand that an alpha particle have a kinetic energy greater than 3.8 MeV. This condition is verified by the experimental finding that the lowest energy alpha particle from the high atomic numbered elements is 3.93 MeV. This alpha particle originates in ^{232}Th. (Samarium-147 emits an alpha particle whose energy is only 2.18 MeV. This low energy is consistent, however, with the theoretical calculations mentioned above if the low atomic number, 62, of samarium is considered.) The question regarding the source of this kinetic energy naturally arises. This energy results from the net decrease in mass following the formation of the alpha particle. Generally, for alpha emission to occur, the following conservation equation must be satisfied:

$$M_p = M_d + M_\alpha + 2M_e + Q, \tag{4.1}$$

where M_p, M_d, M_α, and M_e are respectively equal to the masses of the parent, the daughter, the emitted alpha particle, and the two orbital electrons that are lost during the transition to the lower atomic numbered daughter, while Q is the total energy release associated with the radioactive transformation. In the case of the decay of ^{210}Po, for example, we have, from equation (4.1),

$$\begin{aligned} Q &= M_{\text{Po}} - M_{\text{Pb}} - M_\alpha - 2M_e \\ &= 210.04850 - 206.03883 - 4.00277 - 2 \times 0.00055 \\ &= 0.0058 \text{ atomic mass units.} \end{aligned}$$

In energy units,

$$\begin{aligned} Q &= 0.0058 \text{ amu} \times 931 \frac{\text{MeV}}{\text{amu}} \\ &= 5.4 \text{ MeV.} \end{aligned}$$

This Q value represents the total energy associated with the transformation of ^{210}Po. Since no gamma ray is emitted in this transition, the total released energy appears as kinetic energy, and is divided between the alpha particle and the daughter, which recoils after the alpha particle is emitted. The exact energy division between the alpha and recoil nucleus depends on the mass of the daughter, and may be calculated by application of the laws of conservation of energy and momentum. If M and m are the masses, respectively, of the recoil nucleus and the alpha particle, and if V and v are their respective velocities, then

$$Q = \tfrac{1}{2}MV^2 + \tfrac{1}{2}mv^2. \tag{4.2}$$

FIG. 4.2. Tracks in a Wilson cloud chamber of alpha particles from thorium C (^{212}Bi), energy = 6.05 MeV, and thorium C (^{212}Po), energy = 8.78 MeV. (E. Rutherford, J. Chadwick, and C. D. Ellis, *Radiation from Radioactive Substances*, Macmillan, New York, 1930.)

We have, according to the law of conservation of momentum,

$$MV = mv, \tag{4.3}$$

or,

$$V = \frac{mv}{M}.$$

When the value for V from equation (4.3) is substituted into equation (4.2), we have

$$Q = \tfrac{1}{2}M\frac{m^2v^2}{M^2} + \tfrac{1}{2}mv^2. \tag{4.4}$$

If we let E represent the kinetic energy of the alpha particle, $\frac{1}{2}mv^2$, then equation (4.4) may be rewritten as

$$Q = E\left(\frac{m}{M} + 1\right),$$

or

$$E = \frac{Q}{1 + m/M}. \tag{4.5}$$

According to equation (4.5), the kinetic energy of the alpha particle emitted in the decay of ^{210}Po is:

$$E = \frac{5.4}{(1 + 4/206)}$$
$$= 5.3 \text{ MeV}.$$

The kinetic energy of the recoil nucleus, therefore, is 0.1 MeV.

Alpha particles are essentially monoenergetic. However, alpha-particle spectrograms do show discrete energy groupings, with small energy differences among the different groups. These small differences are attributed to differences in the energy level of the daughter nucleus. That is, a nucleus that emits one of the lower energy alpha particles is left in an excited state, while the nucleus that emits the highest energy alpha particle for any particular isotope is usually left in the "ground" state. A nucleus left in an excited state emist its energy of excitation in the form of a gamma ray. It should be pointed out that most of the alpha particles are usually emitted with the maximum energy. Very few nuclei, consequently, are left in excited states, and gamma radiation, therefore, accompanies only a small fraction of the alpha rays. Radium may be cited as an example of an alpha emitter with a complex spectrum. In the overwhelming majority of transformations of ^{226}Ra, 94.3%, alphas are emitted with a kinetic energy of 4.777 MeV. The balance of the alpha particles, 5.7%, have kinetic energies of only 4.591 MeV. In that instance where a lower energy alpha is emitted, the daughter nucleus is left in an excited state, and rids itself of its energy of excitation by emitting a gamma-ray photon whose energy is equal to the difference between the energies of the two alpha particles: $4.777 - 4.591 = 0.186$ MeV. (About 35% of these gamma-ray photons are internally converted: see section below on internal conversion.) The ^{226}Ra spectrum is the least intricate of all the complex alpha spectra. Most alpha emitters have several groups of alphas, and therefore more gammas. All alpha spectra, however, show the same consistent relationship among the various nuclear energy levels.

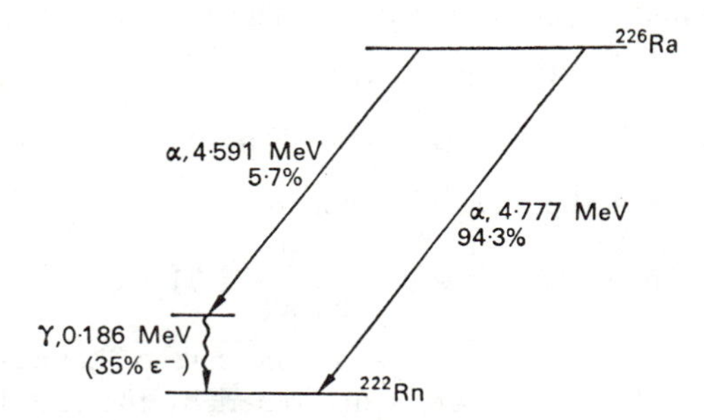

FIG. 4.3. Radium-226 transformation (decay) scheme.

Alpha particles are extremely limited in their ability to penetrate matter. The dead outer layer of skin is sufficiently thick to absorb all alpha radiations from radioactive materials. As a consequence, alpha radiation from sources outside the body do not constitute a radiation hazard. In the case of internally deposited alpha-emitting isotopes, however, the shielding effect of the dead outer layer of skin is absent, and the energy of the alpha radiation is dissipated in living tissue. For this reason, and others to be discussed in Chapter 7, alpha radiation is highly toxic when it irradiates the inside of the body from internally deposited radioisotopes.

Beta Emission

A beta particle is an ordinary electron that is ejected from the nucleus of a beta-unstable radioactive atom. The particle has a single negative electrical charge (1.6×10^{-19} C) and a very small mass (0.00055 atomic mass units). Since theoretical considerations preclude the independent existence of an intra-nuclear electron, it is postulated that the beta particle is formed at the instant of emission by the transformation of a neutron into a proton and an electron according to the equation

$$^{1}_{0}n \rightarrow {}^{1}_{1}\mathrm{H} + {}_{-1}^{0}e. \tag{4.6}$$

This transformation shows that beta decay occurs among those isotopes that have a surplus of neutrons. For beta emission to be energetically possible, the exact nuclear mass of the parent must be greater than the sum of the exact masses of the daughter nucleus plus the beta particle.

$$M_p = M_d + M_e + Q. \tag{4.7}$$

This restriction, of course, is analogous to the corresponding restriction on alpha emitters. Because a unit negative charge is lost during beta decay, and because the mass of the beta particle is very much less than 1 atomic mass unit, the daughter nucleus is one atomic number higher than its parent but retains the same atomic mass number as the parent. For example, radioactive phosphorous decays to stable sulfur according to the equation

$$^{32}_{15}\mathrm{P} \rightarrow {}^{32}_{16}\mathrm{S} + {}_{-1}^{0}e + 1.71\ \mathrm{MeV}.$$

The transformation energy, in this instance 1.71 MeV, is the energy equivalent of the difference in mass between the ^{32}P nucleus and the sum of the ^{32}S nucleus plus the beta particle, and appears as kinetic energy of the beta particle. If neutral atomic masses are used to complete the mass-energy equation, then, of course, the mass of the electron shown in the right hand side of equation (4.7) is not considered, since it is implicitly included in the extra-nuclear electronic structure of the ^{32}S. The mass difference is

$$31.98403 = 31.98224 + Q,$$

$$Q = 0.00179\ \mathrm{amu},$$

and the energy equivalent of the mass difference is

$$0.00179\ \mathrm{amu} \times 931\,\frac{\mathrm{MeV}}{\mathrm{amu}} = 1.71\ \mathrm{MeV}.$$

Examination of equation (4.5) shows that in the case of beta emission, an extremely small part of the energy of the reaction is dissipated by the recoil nucleus, since m/M, where m is now the mass of the beta particle, and M is the mass of the daughter nucleus, is very small. In the example given above,

$$\frac{m}{M} = \frac{0.00055}{32} = 0.000017,$$

and Q is only 1.000017 times greater than the kinetic energy of the beta particle.

On the basis of the above analysis, one might expect beta particles to be monoenergetic, as in the case of alpha radiation. This expectation is not confirmed by experiment. Instead, beta particles are found to be emitted with a continuous energy distribution ranging from zero to the theoretically expected value based on mass-energy considerations for the particular beta transition. In the case of ^{32}P, for example, although the maximum energy of the beta particle may be 1.71 MeV, most of the beta-rays have considerably smaller kinetic energies, as shown in Fig. 4.4. The average energy of a ^{32}P beta particle is 0.7 MeV, or about 41% of the maximum energy. Generally, the average energy of the beta radiation from most beta-active radioisotopes is about 30%–40% of the maximum energy. Unless otherwise specified, when the energy of a beta emitter is given, it is the maximum energy.

The fact that beta radiation is emitted with a continuous energy distribution up to a definite maximum seems to violate the established energy-mass conservation laws. To

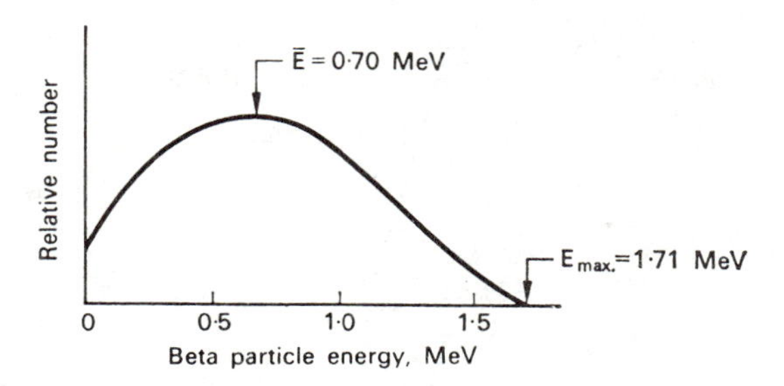

FIG. 4.4. Phosphorous-32 beta ray spectrum.

prevent violation of the conservation laws, it was postulated that the beta particle is accompanied by another particle, called a neutrino, whose energy is equal to the difference between the kinetic energy of the accompanying beta particle and the maximum energy of the spectral distribution. The neutrino, as postulated, has no electrical charge and a vanishingly small mass. Because of these two characteristics, detection of the neutrino is extremely difficult. Recent experimental work, however, has confirmed the validity of the neutrino hypothesis. Equation (4.6) should therefore be modified to

$$ {}^{1}_{0}n \rightarrow {}^{1}_{1}\mathrm{H} + {}^{0}_{-1}e + \nu, \tag{4.8}$$

where ν represents the neutrino.

Phosphorous-32, like several other beta emitters including ^{3}H, ^{14}C, ^{90}Sr, and ^{90}Y, emits no gamma rays. These isotopes are known as pure beta emitters. The opposite of a pure beta emitter is a beta–gamma emitter. In this case, the beta particle is followed (instantaneously) by a gamma ray. The explanation for the gamma ray here is the same as that in the case of the alpha ray. The daughter nucleus, after the emission of a beta ray, is left in an excited condition, and rids itself of the energy of excitation by the emission of a gamma ray. Mercury-203 may be given as an example. It emits a 0.21-MeV beta ray and a 0.279-MeV gamma photon, as seen in the transformation scheme shown in Fig. 4.5.

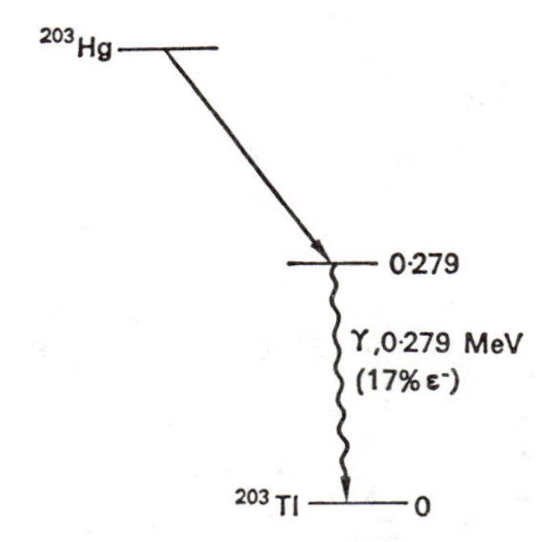

FIG. 4.5. Transformation (decay) scheme of ^{203}Hg.

Both illustrations given above (^{32}P and ^{203}Hg) are for beta emitters with simple spectra, that is, for emitters with only one group of beta rays. Complex beta emitters are those isotopes whose beta ray spectra contains more than one distinct group of beta rays. Potassium-42, for example, in about 82% of its transformations, decays to stable ^{42}Ca by emission of a beta particle from a group whose maximum energy is 3.55 MeV and in 18%

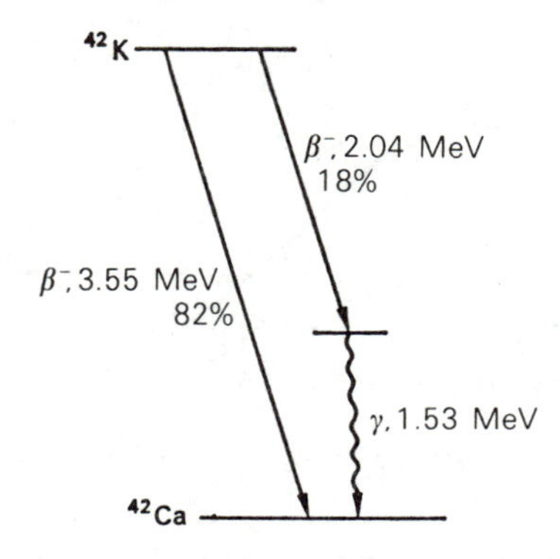

FIG. 4.6. Potassium-42 transformation (decay) scheme.

of its transformations by emitting a 2.04-MeV beta (Fig. 4.6). In this case, however, the excited ^{42}Ca immediately emits a gamma-ray photon whose energy is 1.53 MeV. A commonly used isotope which has an even more complex beta–gamma spectrum is ^{131}I. This isotope decays to stable ^{131}Xe by emission of a beta particle. In 85% of the transformations, however, the beta particle is a member of a group whose maximum energy is 0.6 MeV, while the remainder of beta particles belong to a group whose maximum energy is 0.315 MeV. In both instances, the xenon daughter nucleus is left in an excited state, and rids itself of its energy of excitation by the emission of gamma radiation. In the case of the nucleus resulting from the emission of the lower energy beta particle, a gamma photon whose energy is 0.638 MeV is emitted. The nucleus in the lower energy level, the one resulting from the emission of the 0.6-MeV beta particle, rids itself of its excitation energy by two competing gamma ray transitions. About 93% of these nuclei (corresponding to 79% of the ^{131}I transformations) emit 0.364-MeV gamma rays and the balance of the excited nuclei emit two gamma photons in cascade, one of 0.284 MeV and one of 0.080 MeV. The decay scheme for ^{131}I is shown in Fig. 4.7.

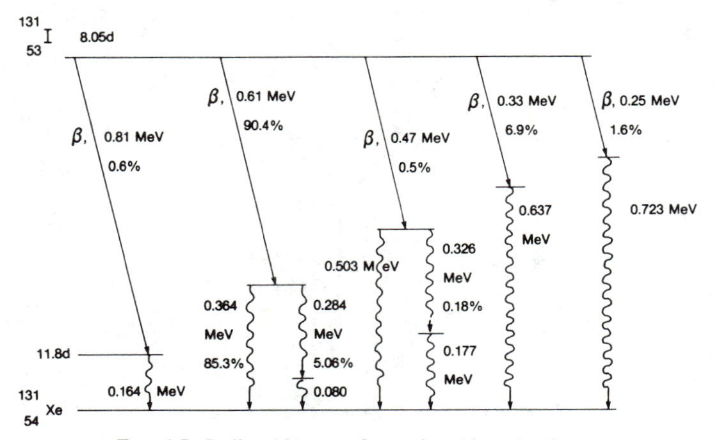

FIG. 4.7. Iodine-131 transformation (decay) scheme.

Beta radiation, because of its ability to penetrate tissue to varying depths, depending on the energy of the beta particle, may be an external radiation hazard. The exact degree of hazard, of course, depends on the beta-emitting isotope, and must be evaluated in every case. Generally, however, beta rays whose energies are less than 200 keV and therefore have very limited penetrability, such as those from tritium, ^{35}S, and ^{14}C, are not considered as external radiation hazards. It should be noted, however, that beta rays give rise to highly penetrating X-rays called *Bremsstrahlung* when they are stopped by shielding. (This interaction will be more fully discussed later, in Chapter 5.) Unless shielding is properly designed, and proper precautionary measures adopted, beta radiation may indirectly result in an external radiation hazard through the production of *Bremsstrahlung*. Any beta-emitting isotope, of course, is potentially hazardous when it is deposited in the body in amounts exceeding those thought to be safe.

Positron Emission

In those instances where the neutron to proton ratio is too low and alpha emission is not energetically possible, the nucleus may, under certain conditions, attain stability by emitting a positron. A positron is a beta particle whose charge is positive. In all other respects it is the same as the negative beta particle, or an ordinary electron. Its mass is 0.000548 atomic mass units and its charge is $+1.6 \times 10^{-19}$ C. Because of the fact that the nucleus loses a positive charge when a positron is emitted, the daughter product is one atomic number less than the parent. The mass number of the daughter remains unchanged, as in all nuclear transitions involving electrons. In the case of ^{22}Na, for example, we have

$$^{22}_{11}\text{Na} \rightarrow {}^{22}_{10}\text{Ne} + {}^{0}_{1}e + \nu. \tag{4.9}$$

Whereas negative electrons occur freely in nature, positrons have only a transitory existence. They occur in nature only as the result of the interaction between cosmic rays and the atmosphere, and disappear in a matter of microseconds after formation. The manner of disappearance is of interest and great importance to the health physicist. The positron combines with an electron, and the two particles are annihilated, giving rise to two gamma-ray photons whose energies are equal to the mass equivalent to the positron and electron. This interaction will be discussed more fully in Chapter 5. The positron is not thought to exist independently within the nucleus. Rather, it is believed that the positron results from a transformation, within the nucleus, of a proton into a neutron according to the reaction

$$^{1}_{1}\text{H} \rightarrow {}^{1}_{0}n + {}^{0}_{1}e + \nu. \tag{4.10}$$

For positron emission, the following conservation equation must be satisfied:

$$M_p = M_d + M_e + Q, \tag{4.11}$$

where M_p, M_d, and M_e are the masses of the parent nucleus, daughter nucleus, and positron, respectively, and Q is the mass equivalent of the energy of the reaction. Since the daughter is one atomic number less than the parent, it must also lose an orbital electron immediately after the nuclear transition. In terms of atomic masses, therefore, the conservation equation is

$$M'_p = M'_d + 2M_e + Q. \tag{4.12}$$

Sodium-22, a useful isotope for biomedical research, is transformed into ^{22}Ne by two competing mechanisms, positron emission and K capture (which is discussed in the following section), according to the decay scheme shown below in Fig. 4.8. Positrons are emitted in 89.8% of the transformations, while the competing decay mode, K capture, occurs in 10.2% of the nuclear transformations. Both modes of decay result in ^{22}Ne, which is in an excited state; the excitation energy instantly appears as a 1.277 MeV gamma ray.

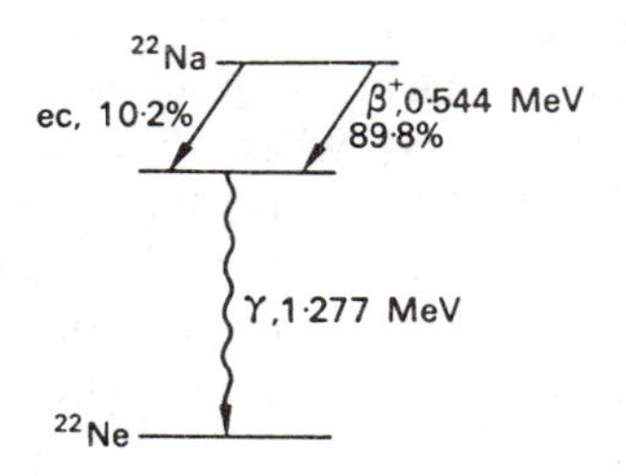

FIG. 4.8. Sodium-22 transformation (decay) scheme.

The exact atomic mass of the neon may be calculated from the positron transformation data with the aid of equation (4.12):

$$M(^{22}\text{Ne}) = M(^{22}\text{Na}) - 2M_e - Q_m$$

$$= 22.001404 - 2 \times 0.000548 - \frac{(0.544 + 1.277)\ \text{MeV}}{931\ \text{MeV/amu}}$$

$$= 21.998352\ \text{amu}.$$

Since positrons are electrons, the radiation hazard from the positrons themselves is very similar to the hazard from beta particles. The gamma radiation resulting from the annihilation of the positron, however, makes all positron emitting isotopes potential external radiation hazards.

Orbital Electron Capture

Equation (4.12) shows that, if a neutron-deficient atom is to attain stability by positron emission, it must exceed the weight of its daughter by at least two electron masses. If this requirement cannot be met, then the neutron deficiency is overcome by the process known as orbital electron capture or, alternatively, as *K* capture. In this radioactive transformation, one of the extra-nuclear electrons is captured by the nucleus, and unites with an intra-nuclear proton to form a neutron according to the equation

$$_{-1}^{0}e + _{1}^{1}\text{H} \rightarrow _{0}^{1}n + \nu. \tag{4.13}$$

Since the electrons in the *K* shell are much closer to the nucleus than those in any other shell, the probability that the captured orbital electron will be from the *K* shell is much greater than that for any other shell, hence the alternate (and most common name) for this mechanism. In the case of *K* capture, as in positron emission, the atomic number of the daughter is one less than that of the parent, while the atomic mass number remains unchanged. The energy conservation requirements for *K* capture are much less rigorous than for positron emission. It is merely required that the following conservation equation be satisfied:

$$M_p + M_e = M_d + \phi + Q, \tag{4.14}$$

where M_p and M_d are the exact atomic masses of the parent and daughter, M_e is the mass of the captured electron, ϕ is the binding energy of the captured electron, and Q is the energy of the reaction.

Equation (4.14) may be illustrated by the *K* capture mode of transformation of ^{22}Na. The binding energy, ϕ, of the sodium *K* electron is 1.08 keV. The energy of decay, Q, may therefore be calculated as follows:

$$Q = M(^{22}\text{Na}) + M_e - M(^{22}\text{Ne}) - \phi$$
$$= 22.001404 + 0.000548 - 21.998352 - \frac{0.00108 \text{ MeV}}{931 \text{ MeV/amu}}$$
$$= 0.003600 \text{ amu}.$$

In terms of MeV, we have

$$Q = 0.0036 \text{ amu} \times 931 \frac{\text{MeV}}{\text{amu}}$$
$$= 3.352 \text{ MeV}.$$

Since a 1.277-MeV gamma ray is emitted, we are left with an excess of $3.352 - 1.277 = 2.075$ MeV. The recoil energy associated with the emission of the gamma-ray photon is insignificantly small. The excess energy, therefore, must be carried away by a neutrino. Although the example given above is for a specific reaction, it is nevertheless typical of all reactions involving *K* capture; a neutrino is always emitted when an orbital electron is captured. It is thus seen that in all types of radioactive decay involving either the capture or emission of an electron, a neutrino must be emitted in order to conserve energy. In contrast to positron and negatron (ordinary beta) decay, however, in which the neutrino carries off the difference between the actual kinetic energy of the particle and the maximum observed kinetic energy, and therefore has a continuous energy distribution, the neutrino in orbital electron capture is necessarily monoenergetic.

Whenever an atom is transformed by orbital electron capture, an X-ray characteristic of the daughter element is emitted as an electron from an outer orbit falls into the energy level occupied by the electron which had been captured. That characteristic X-rays of the daughter should be observed follows from the fact that the X-ray photon is emitted after the nucleus captures the orbital electron and is thereby transformed into the daughter. These low energy characteristic X-rays must be considered by the health physicist when he computes absorbed radiation doses from internally deposited isotopes which decay by orbital electron capture.

Gamma rays

Gamma rays are monochromatic electromagnetic radiations that are emitted from nuclei of excited atoms following radioactive transformation; they provide a mechanism for ridding excited nuclei of their excitation energy. Since the health physicist is concerned with all radiations which come from radioactive substances, and since X-rays are indistinguishable from gamma rays, characteristic X-rays that arise in the extra-nuclear structure of many isotopes must be considered by the health physicist when he evaluates radiation hazards. However, because of the low energy of characteristic X-rays, they are of importance mainly in the case of internally deposited radioisotopes. Annihilation radiation, the gamma rays resulting from the mutual annihilation of positrons and negatrons, are usually associated, for health physics purposes, with the isotope that emits the positrons. When considering the radiation hazard from ^{22}Na, for example, two photons from the annihilation process, which are not shown on the decay scheme, must be considered together with the 1.277-MeV gamma photon that is shown on the decay scheme. The general rule in health physics, therefore, is automatically to associate positron emission with gamma radiation in all problems involving shielding, dosimetry, and radiation hazard evaluation.

Internal Conversion

Internal conversion is an alternative mechanism by means of which an excited nucleus of a gamma-emitting isotope may rid itself of the excitation energy. It is an interaction in which a tightly bound electron interacts with its nucleus, absorbs the excitation energy from the nucleus, and is ejected from the atom. Internally converted electrons appear in monoenergetic groups. The kinetic energy of the converted electron is always found to be equal to the difference between the energy of the gamma-ray photon emitted by the radioisotope and the binding energy of the converted electron of the daughter element. Since electrons in the *L* energy level of high atomic-numbered elements are also tightly bound, internal conversion in those elements results in two groups of electrons which differ in energies by the difference between the binding energies of the *K* and *L* levels. Because of these experimental findings, internal conversion may be thought of as an internal photoelectric effect, that is, an interaction in which the gamma photon collides with the tightly bound electron and transfers all of its energy to the electron. The energy of the photon is divided between work done to overcome the binding energy of the electron and kinetic energy imparted to the electron. Mathematically, this may be expressed by the equation

$$E_\gamma = E_e + \phi, \tag{4.15}$$

where E_γ is the energy of the gamma ray, E_e is the kinetic energy of the conversion electron, and ϕ is the binding energy of the electron. Since conversion electrons are monoenergetic, they appear as line spectra superimposed on the continuous beta-ray spectra of the isotope. An interesting example of internal conversion is given by ^{137}Cs. This isotope is transformed by beta emission to an excited state of ^{137}Ba. The ^{137}Ba then emits a 0.661-MeV photon which undergoes internal conversion in 11% of the transitions. The internal conversion coefficient α, which is defined as the ratio of the number of conversion electrons per gamma-ray photon,

$$\alpha = \frac{N_e}{N_\gamma} \tag{4.16}$$

is equal to 0.11 in this instance. The conversion electrons, which seem to come from the ^{137}Cs, are found to be superimposed on the beta spectrum of the cesium, as shown below

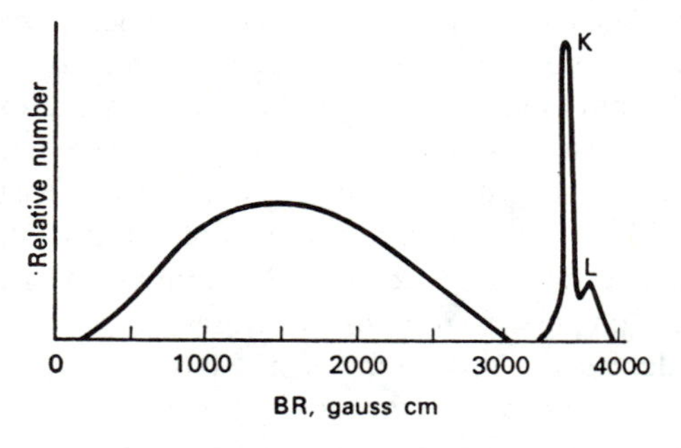

FIG. 4.9. Cesium-137 beta spectrum, showing conversion electrons from the *K* and *L* energy levels.

in Fig. 4.9. After internal conversion characteristic X-rays are emitted as outer orbital electrons fill the vacancies left in the deeper energy levels by the conversion electrons. These characteristic X-rays may themselves be absorbed by an internal photoelectric effect on the atom from which they were emitted, a process that is of the same nature as internal conversion. The ejected electrons from this process are called Auger electrons, and they possess very little kinetic energy.

Transformation Kinetics

Half-life

Different isotopes are transformed at different rates, and each isotope has its own characteristic transformation rate. For example, when the activity of ^{32}P is measured daily over a period of about 3 months, and the percentage of the initial activity is plotted as a function of time, the curve shown in Fig. 4.10 is obtained. The data show that one-half of the ^{32}P is gone in 14.3 days, half of the remainder in another 14.3 days, half of what is left during the following 14.3 days, and so on.

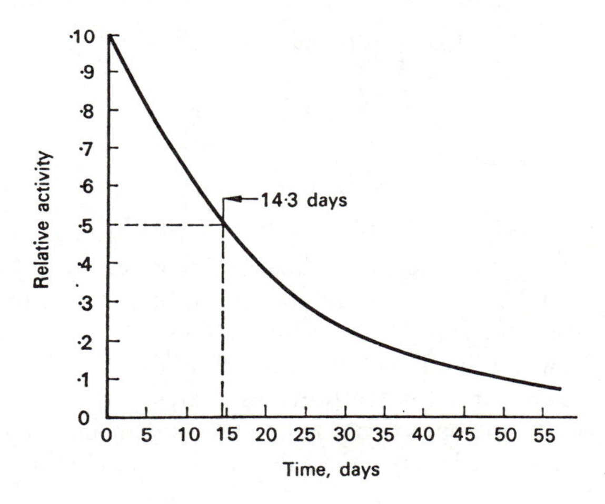

FIG. 4.10. Decrease of ^{32}P activity.

If a similar series of measurements were made on ^{131}I, it would be observed that the iodine would disappear at a faster rate. One-half would be gone after 8 days, and three-fourths of the initial activity would be gone after only 16 days, while seven-eighths of the iodine would be gone after 24 days.

The time required for any given radioisotope to decrease to one-half of its original quantity is a measure of the speed with which the isotope undergoes radioactive transformation. This period of time is called the half-life, and is characteristic of the particular radioisotope. Each radioisotope has its own unique rate of transformation, and no operation, either chemical or physical, is known that will change the transformation rate; the half-life of a radioisotope is an unalterable property of the isotope. Half-lives of radioisotopes range from microseconds to billions of years.

From the definition of the half-life, it follows that the fraction of a radioisotope remaining after n half-lives is given by the relationship

$$\frac{A}{A_0} = \frac{1}{2^n}, \tag{4.17}$$

where A_0 is the original quantity of activity, and A is the activity left after n half-lives.

Example 4.1

Cobalt-60, a gamma-emitting radioisotope whose half-life is 5.3 years, is used as a radiation source for radiographing pipe welds. Because of the decrease in radioactivity with increasing time, the exposure time for a radiograph will be increased annually. Calculate the correction factor to be applied to the exposure time in order to account for the decrease in the strength of the source. Equation (4.17) may be rewritten as

$$\frac{A_0}{A} = 2^n.$$

By taking the logarithm of each side of the equation, we have

$$\log \frac{A_0}{A} = n \log 2,$$

where n, the number of ^{60}Co half-lives in one year, is $1/5.3 = 0.189$.

$$\log \frac{A_0}{A} = 0.189 \times 0.301,$$

$$\frac{A_0}{A} = \text{antilog } 0.0569$$

$$= 1.14.$$

The ratio of the initial quantity of cobalt to the quantity remaining after 1 year is 1.14. The exposure time after 1 year, therefore, must be increased by 14%. It should be noted that this ratio is independent of the actual amount of activity at the beginning and end of the year. After the second year, the ratio of the cobalt at the beginning of the second year to that at the end will be 1.14. The same correction factor, 1.14, therefore, is applied every year to the exposure time for the previous year.

If the decay data for any isotope are plotted on semi-logarithmic paper, with the activity measurements recorded on the logarithmic axis and time on the linear axis, a straight line

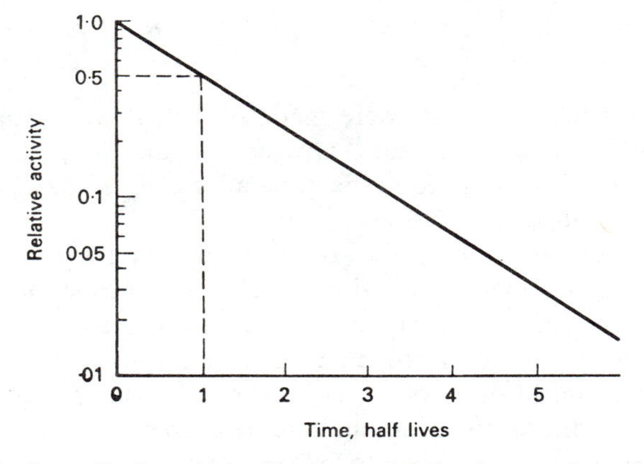

FIG. 4.11. Generalized semi-logarithmic plot of the decrease in activity due to radioactive transformation.

results. If time is measured in units of half-lives, the generally useful curve shown in Fig. 4.11 results.

The illustrative example given above could have been solved graphically with the aid of this curve. The ordinate at which the time in units of half-life, 0.189, intersects the curve shows that 87.7% of the original activity is left. The correction factor, therefore, is the reciprocal of 0.877:

$$\text{Correction factor} = \frac{1}{0.877} = 1.14.$$

The fact that the graph of activity vs. time, when drawn on semi-logarithmic paper, is a straight line tells us that the quantity of activity left after any time interval is given by the following equation:

$$A = A_0 e^{-\lambda t}, \tag{4.18}$$

where A_0 is the initial quantity of activity, A is the amount left after time t, λ is the transformation constant, and e is the base of the system of natural logarithms. The transformation constant is the fractional decrease in activity per unit time, and is defined as

$$\lim_{\Delta t \to 0} \frac{\Delta N / N}{\Delta t} = -\lambda, \tag{4.19}$$

where N is a number of radioactive atoms, and ΔN is the number of these atoms that are transformed during a time interval Δt. The fraction $\Delta N / N$ is the fractional decrease in the number of radioactive atoms during the time interval Δt. A negative sign is given to λ to indicate that the quantity N is decreasing. For a short-lived radioisotope, λ may be determined from the slope of an experimentally determined transformation curve. For long-lived isotopes, the transformation constant may be determined by measuring N, then by counting the number of transformations per second, and then calculating the numerical value of λ from equation (4.19).

Example 4.2

One microgram radium is found to emit 3.7×10^4 alpha particles per second. If each of these alphas represents a radioactive transformation of radium, what is the transformation constant for radium?

In this case, ΔN is 3.7×10^4, Δt is one second, and N, the number of radium atoms per microgram, may be calculated as follows:

$$N = \frac{6.03 \times 10^{23} \text{ atoms/mole}}{A \text{ g/mole}} \times W \text{ g}, \tag{4.20}$$

where A is the atomic weight and W the weight of the radium sample.

$$N = \frac{6.03 \times 10^{23} \times 10^{-6}}{2.26 \times 10^2} = 2.66 \times 10^{15} \text{ atoms}.$$

The transformation constant, therefore, is

$$\lambda = \frac{\Delta N / N}{\Delta t} = \frac{3.7 \times 10^4 \text{ atoms}/2.66 \times 10^{15} \text{ atoms}}{1 \text{ sec}} = 1.39 \times 10^{-11} \text{ sec}^{-1}.$$

On a per year basis, the transformation rate is

$$\lambda = 1.39 \times 10^{-11} \text{ sec}^{-1} \times 8.64 \times 10^4 \frac{\text{sec}}{\text{day}} \times 3.65 \times 10^2 \frac{\text{day}}{\text{yr}}$$

$$= 4.38 \times 10^{-4} \text{ yr}^{-1}.$$

This value of λ may be used in equation (4.18) to compute the amount of radium left after any given time period.

Example 4.3

What percentage of a given amount of radium will decay during a period of 1000 years? The fraction remaining after 1000 years is given by

$$\frac{A}{A_0} = e^{-\lambda t} = e^{-4.38 \times 10^{-4} \text{yr}^{-1} \times 10^3 \text{yr}} = e^{-0.438} = 0.645 = 64.5\%.$$

The percentage that was transformed during the 1000-year period, therefore, is

$$100\% - 64.5\% = 35.5\%.$$

The quantitative relationship between half-life, T, and decay constant, λ, may be found by setting A/A_0 in equation (4.18) equal to $\frac{1}{2}$, and solving the equation for t. In this case, of course, the time is the half-life:

$$\frac{A}{A_0} = \tfrac{1}{2} = e^{-\lambda T}$$
$$T = \frac{0.693}{\lambda}. \tag{4.21}$$

Example 4.4

Given that the transformation constant for ^{226}Ra is 4.38×10^{-4} per year, calculate the half-life for radium.

$$T = \frac{0.693}{\lambda}$$
$$= \frac{0.693}{4.38 \times 10^{-4}\ \text{yr}^{-1}}$$
$$= 1.6 \times 10^3 \text{ years.}$$

Average Life

Although the half-life of an isotope is a unique, reproducible characteristic of that isotope, it is nevertheless a statistical property, and is valid only because of the very large number of atoms involved. (One microgram of radium contains 2.79×10^{15} atoms.) Any particular atom of a radioisotope may be transformed at any time, from zero to infinity, after it is observed. For some applications, such as in the case of dosimetry of internally deposited radioisotopes (to be discussed in Chapter 6), it is convenient to use the average life of the radioisotope. The average life is defined simply as the sum of the lifetimes of the individual atoms divided by the total number of atoms originally present.

The instantaneous transformation rate of a quantity of radioisotope containing N atoms is λN. During the time interval between t and $t + dt$, the total number of transformations is $\lambda N\,dt$. Each of the atoms that decayed during this interval, however, had existed for a total lifetime t since the beginning of observation on them. The sum of the lifetimes, therefore, of all the atoms that were transformed during the time interval between t and $t + dt$, after having survived since time $t = 0$, is $t\lambda N\,dt$. The average life of the radioactive species, τ, is

$$\tau = \frac{1}{N_0} \int_0^\infty t\lambda N \, dt, \tag{4.22}$$

where N_0 is the number of radioactive atoms in existence at time $t = 0$. Since

$$N = N_0 e^{-\lambda t},$$

we have

$$\tau = \frac{1}{N_0} \int_0^\infty t\lambda N_0 e^{-\lambda t} \, dt. \tag{4.23}$$

This expression, when integrated by parts, shows the value for the mean life of a radioisotope to be

$$\tau = \frac{1}{\lambda}. \tag{4.24}$$

If the expression for the transformation constant in terms of the half-life of the radioisotope,

$$\lambda = \frac{0.693}{T},$$

is substituted into equation (4.22), the relationship between the half-life and the mean life is found to be

$$\tau = \frac{T}{0.693} = 1.45T. \tag{4.25}$$

Activity

The Becquerel

Uranium-238 and its daughter ^{234}Th each contain about the same number of atoms per gram; approximately 2.5×10^{21}. Their half-lives, however, are greatly different; ^{238}U has a half-life of 4.5×10^9 years, while ^{234}Th has a half-life of 24.1 days (or 6.63×10^{-2} years). Thorium-234, therefore, is transforming 6.8×10^{10} times faster than ^{238}U. Another example of greatly different rates of transformation that may be cited is ^{35}S and ^{32}P. These two radioisotopes, which have about the same number of atoms per gram, have half-lives of 87 and 14.3 days respectively. The radiophosphorous, therefore, is decaying about 6 times faster than the ^{35}S. When radioisotopes are used, the radiations are the center of interest. In this context, therefore, $\frac{1}{6}$ of a gram of ^{32}P is about equivalent to 1 g of ^{35}S in radioactivity, while 15 micromicrograms of ^{234}Th is about equivalent in activity to 1 g of ^{238}U. Obviously, therefore, when interest is centered on radioactivity, the gram is not a very useful unit of quantity. To be meaningful, the unit for quantity of radioactivity must be based on activity. Such a unit is called the *becquerel* (symbolized by Bq) and is defined as follows:

> The becquerel is that quantity of radioactive material in which one atom is transformed per second.

$$1 \text{ Bq} = 1 \text{ tps}. \tag{4.26}$$

It must be emphasized that, although the becquerel is defined in terms of a number of atoms transformed per second, it is not a measure of rate of transformation. *The becquerel is a measure only of quantity of radioactive material.* The phrase "one atom transformed per second" as used in the definition of the bequerel is not synonymous with the number of particles emitted by the radioactive isotope in one second. In the case of a pure beta emitter, for example, 1 Bq, or one transformation per second, does in fact result in one beta particle per second. In the case of a more complex radioactive isotope, however, such as ^{60}Co (Fig. 4.12), each transformation releases one beta particle and two gamma photons; the total number of radiations, therefore is 3 per second per Bq. In the case of ^{42}K (Fig. 4.6), on the other hand, 20% of the beta transformations are accompanied by a single

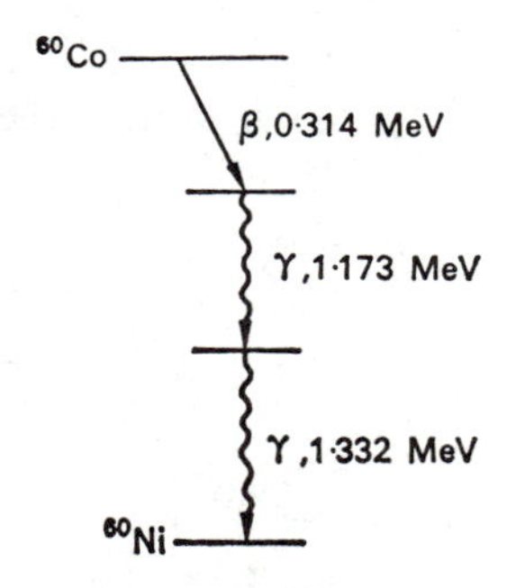

FIG. 4.12. Cobalt-60 decay scheme.

quantum of gamma radiation. The total number of emissions from 1 Bq ^{42}K, therefore is 1.2 per second. For many purposes, the becquerel is a very small quantity of activity, and multiples of the becquerel are commonly used:

1 kilobecquerel (kBq) $= 10^3$ Bq,

1 megabecquerel (MBq) $= 10^6$ Bq,

1 gigabecquerel (GBq) $= 10^9$ Bq,

1 terabecquerel (TBq) $= 10^{12}$ Bq.

The Curie

The *curie*, symbolized by Ci, is the unit for quantity of radioactivity that was used before the adoption of the SI units and the becquerel. The curie, which originally was defined as the activity of 1 gram ^{226}Ra, is more explicitly defined as:

> The curie is the activity of that quantity of radioactive material in which 3.7×10^{10} atoms are transformed per second.

The curie is related to the becquerel by

$$1 \text{ Ci} = 3.7 \times 10^{10} \text{ Bq}. \tag{4.27}$$

For health physics, as well as for many other purposes, the curie is a very large amount of activity. Sub-multiples of the curie, as listed below, are therefore used:

1 millicurie (mCi) $= 10^{-3}$ Ci,

1 microcurie (μCi) $= 10^{-6}$ Ci,

1 nanocurie (nCi) $= 10^{-9}$ Ci,

1 picocurie (pCi) $= 10^{12}$ Ci,

1 femtocurie (fCi) $= 10^{15}$ Ci.

Specific Activity

Note that the becquerel (or curie), although used as a unit of quantity, does not mention anything about the mass or volume of the radioactive material in which the specified number of transformations occur. The concentration of radioactivity, or the relationship between the mass of radioactive material and the activity is called the specific activity. Specific activity is the number of becquerels (or curies) per unit mass or volume. The specific activity of a carrier-free radioisotope, that is a radioisotope that is not mixed with any other isotope of the same element, may be calculated as follows:

If λ is the transformation constant in units of reciprocal seconds, then the number of transformations per second (tps), and hence the number of becquerels in an aggregation of N atoms, is simply given by λN.

If the radioisotope under consideration weighs 1 gram, then, according to equation (4.20), the number of atoms is given by

$$N = \frac{6.03 \times 10^{23} \text{ atoms/mole}}{A \text{ g/mole}},$$

where A is the atomic weight of the isotope. The activity per unit weight, or the specific activity, therefore is:

$$SA = \lambda N = \frac{\lambda \times 6.03 \times 10^{23}}{A} \text{ Bq/g}. \quad (4.28)$$

Equation (4.28) gives the desired relationship between the specific activity and weight of an isotope, and can be calculated in terms of the isotope's half life by substituting equation (4.21) for λ in equation (4.28):

$$\begin{aligned} \lambda N &= \frac{0.693}{T} \times \frac{6.03 \times 10^{23}}{A}, \\ SA &= \frac{4.18 \times 10^{23}}{A \times T} \text{ Bq/g}. \end{aligned} \quad (4.29)$$

Note that equations (4.28) and (4.29) are valid only if λ and T are given in time units of seconds. A more convenient form for calculating specific activity may be derived by making use of the fact that there are 3.7×10^{10} transformations per second in 1 gram ^{226}Ra. The specific activity, therefore, of ^{226}Ra is 3.7×10^{10} Bq per gram. The ratio of the specific activity of any isotope, SA_i, to that of ^{226}Ra is

$$\begin{aligned} \frac{SA_i, \text{ Bq/g}}{3.7 \times 10^{10} \text{ Bq/g}} &= \frac{4.18 \times 10^{23}/A_i \times T_i}{4.18 \times 10^{23}/A_{Ra} \times T_{Ra}}, \\ SA_i &= 3.7 \times 10^{10} \frac{A_{Ra} \times T_{Ra}}{A_i \times T_i} \text{ Bq/g}, \end{aligned} \quad (4.30)$$

where A_{Ra}, the atomic weight of ^{226}Ra is 226, A_i is the atomic weight of the radioisotope whose specific activity is being calculated, and T_{Ra} and T_i are the half-lives of the radium and the isotope *i*. The only restriction on equation (4.30) is that both half-lives must be in the same units of time. Analogously, the specific activity in units of Ci/g is given by

$$SA_i = \frac{A_{Ra} \times T_{Ra}}{A_i \times T_i}. \tag{4.31}$$

Example 4.5

Calculate the specific activities of ^{14}C and ^{35}S, given that their half-lives are 5,730 years and 87 days, respectively.
For ^{14}C

$$SA = 3.7 \times 10^{10} \times \frac{226 \times 1600 \text{ yr}}{14 \times 5730 \text{ yr}}$$
$$= 1.7 \times 10^{11} \text{ Bq/g} \quad (4.6 \text{ Ci/g}),$$

and, for ^{35}S,

$$SA = 3.7 \times 10^{10} \times \frac{226 \times 1600 \text{ yr} \times 365 \text{ days/yr}}{35 \times 87 \text{ days}}$$
$$= 1.6 \times 10^{15} \text{ Bq/g} \quad (4.3 \times 10^4 \text{ Ci/g}).$$

The specific activities calculated above are for the carrier-free isotopes of ^{14}C and ^{35}S. Very frequently, especially when radioisotopes are used to label compounds the radioisotope is not carrier free, but rather it constitutes an extremely small fraction, either by weight or number of atoms, of the element that is labeled. In such cases, it is customary to refer to the specific activity either of the element or the compound that is labeled. Generally, the exact meaning of the specific activity is clear from the context.

Example 4.6

A solution of $Hg(NO_3)_2$ tagged with ^{203}Hg has a specific activity of 1.5×10^5 Bq/ml (4μCi/ml). If the concentration of Hg in the solution is 5 mg/ml,

(a) What is the specific activity of the Hg?
(b) What fraction of the Hg in the $Hg(NO_3)_2$ is ^{203}Hg?
(c) What is the specific activity of the $Hg(NO_3)_2$?

(a) The specific activity of the Hg is

$$SA(\text{Hg}) = \frac{1.5 \times 10^5 \text{ Bq/ml}}{5 \text{ mg Hg/ml}} = 0.3 \times 10^5 \text{ Bq/mg Hg}.$$

(b) The weight-fraction of mercury that is tagged is given by

$$\frac{SA(\text{Hg})}{SA(^{203}\text{Hg})},$$

and the specific activity of ^{203}Hg is calculated from equation (4.30):

$$SA(^{203}\text{Hg}) = 3.7 \times 10^{10} \times \frac{226 \times 1.6 \times 10^3 \text{ yr} \times 3.65 \times 10^2 \text{ days/yr}}{203 \times 46.5 \text{ days}}$$

$$= 5.2 \times 10^{14} \text{ Bq/gm} \quad (1.4 \text{ Ci/g}).$$

The weight fraction of ^{203}Hg, therefore, is

$$\frac{0.3 \times 10^8 \text{ Bq/g Hg}}{5.2 \times 10^{14} \text{ Bq/g } ^{203}\text{Hg}} = 5.8 \times 10^{-8} \frac{\text{g } ^{203}\text{Hg}}{\text{g Hg}}.$$

(c) Since an infinitesimally small fraction of the mercury is tagged with ^{203}Hg, it may be assumed that the formula weight of the tagged $Hg(NO_3)_2$ is 324.63, and that the concentration of $Hg(NO_3)_2$ is

$$\frac{324.63 \text{ mg Hg(NO}_3)_2}{200.61 \text{ mg Hg}} \times \frac{5 \text{ mg Hg}}{\text{ml}} = 8.1 \frac{\text{mg Hg(NO}_3)_2}{\text{ml}}.$$

The specific activity, therefore, of the $Hg(NO_3)_2$ is

$$\frac{1.5 \times 10^5 \text{ Bq/ml}}{8.1 \text{ mg Hg(NO}_3)_2\text{/ml}} = 1.9 \times 10^4 \text{ Bq/mg Hg(NO}_3)_2 \quad (0.5\,\mu\text{Ci/mg Hg(NO}_3)_2).$$

Example 4.7

Can commercially available ^{14}C-tagged absolute ethanol, $CH_3—C^*H_2—OH$, whose specific activity is 1 mCi/mole, be used in an experiment that requires a minimum specific activity of 10^7 transformations per minute per milliliter? The density of the alcohol is 0.789 g/cm^3.

The specific activity of ^{14}C, as calculated from equation (4.31), is 4.61 Ci/g. One millicurie ^{14}C, therefore, weighs

$$\frac{10^{-3} \text{ Ci}}{4.61 \text{ Ci/g}} = 2.2 \times 10^{-4} \text{ g},$$

and the number of radioactive atoms represented by 0.22 mg ^{14}C is

$$\frac{6.03 \times 10^{23} \text{ atoms/mole}}{14 \text{ g/mole}} \times 2.2 \times 10^{-4} \text{ g} = 9.3 \times 10^{18} \text{ atoms}.$$

Since one mole contains Avogadro's number of molecules, and each tagged molecule contains only one carbon atom, there are

$$\frac{9.3 \times 10^{18}}{6.03 \times 10^{23}} = 15 \text{ per million}$$

ethanol molecules that are tagged. For all practical purposes, therefore, the additional mass due to the isotopic carbon may be neglected when calculating the molecular weight of the labeled ethanol, and the accepted molecular weight of ethanol, 46.078, may be used to compute the activity per cubic centimeter of the alcohol:

$$1 \frac{\text{mCi}}{\text{mole}} \times \frac{1 \text{ mole}}{46.078 \text{ g}} \times 2.22 \times 10^9 \frac{\text{dis/min}}{\text{mCi}} \times 0.789 \frac{\text{g}}{\text{cm}^3}$$

$$= 3.8 \times 10^7 \frac{\text{t/min}}{\text{cm}^3}.$$

The commercially available ethanol may be used.

Naturally Occurring Radioactivity

The naturally occurring radioactive substance which Becquerel discovered in 1896 was a mixture of several isotopes which were later found to be related to each other. They were members of long series of isotopes of various elements, all of which were radioactive but the last. Uranium, the most abundant of the radioactive elements in this mixture, consists of three different isotopes: about 99.3% of naturally occurring uranium is ^{238}U, about 0.7% is ^{235}U, and a trace quantity (about 5×10^{-3}%) is ^{234}U. The ^{238}U and ^{234}U belong to one family, the uranium series, while the ^{235}U isotope of uranium is the first member of another series called the actinium series. The most abundant of all naturally occurring radioisotopes, ^{232}Th, is the first member of still another long chain of successive radioisotopes. All of the isotopes that are members of radioactive series are found in the upper portion of the periodic table; the lowest atomic number in these groups is 81, while the lowest mass number is 207. All the radioactive series have several common characteristics. First is the fact that the first member of each series is very long lived, with half-lives that may be measured in geological time units. That the first member of each must be very long lived is obvious, since, if the time since the creation of the world is considered, relatively short-lived isotopes would have decayed away during the several billions of years that the earth is believed to be in existence. This point is well illustrated by the artificially produced series of isotopes called the neptunium series. In this case, the first member is the transuranium element ^{241}Pu which is produced in the laboratory by neutron irradiation of reactor produced ^{239}Pu. The half-life of ^{241}Pu, however, is only 13 years. Because of the short half-life, even a period of a century is long enough to permit most of the ^{241}Pu to decay away. Even the longest-lived member of this series, ^{237}Np, whose half-life is 2.2×10^6 years, is sufficiently short to have completely disappeared if it had been created at the same time as all the other elements of the earth.

A second characteristic common to all three naturally occurring series is that each has a gaseous member, and furthermore that the radioactive gas in each case is a different isotope of the element radon. In the case of the uranium series, the gas, $^{222}_{83}Rn$, is called radon; in the thorium series the gas, $^{220}_{83}Rn$, is called thoron, while in the actinium series it is called actinon, $^{219}_{83}Rn$. It should be noted that the artificial neptunium series has no gaseous member. The existence of the radioactive gasses in the three chains is one of the chief reasons for the presence of naturally occurring environmental radioactivity. The radon gas diffuses out of the earth into the air, and the radioactive radon daughters, which are solids under ordinary conditions, attach themselves to atmospheric dust. Atmospheric concentrations of radioactivity from this source vary widely around the earth, and are dependent on the local concentrations of uranium and thorium in the earth. Although the probable atmospheric radon concentration is in the order of 2×10^{-6} Bq/ml (5×10^{-11} μCi/ml), concentrations 10 times greater are not uncommon. Since the radioactive radon daughters are found on the surface of atmospheric particulates, and since air-borne particulates are washed out of the atmosphere by rain, it is reasonable to expect increased background radiation during periods of rain. This phenomenon is in fact observed, and must be considered by health physicists and others when interpreting routine monitoring data. Fallout caused by rain in Upton, N.Y., is shown in Fig. 4.13, a set of curves giving the beta and gamma activity during rainy and dry periods. When the ground is covered with snow, however, a decrease in airborne radioactivity occurs because of the filtering action of the snow blanket on the effusing radon and its daughters. Increased environmental radioactivity from this source also occurs during temperature inversions, when vertical mixing of the air and consequent dilution of radon and its daughters temporarily ceases. Because of this naturally occurring airborne activity, certain correction factors must be applied when computing the atmospheric concentration, from dust samples, of air-borne radioactive contaminants. These corrections will be discussed in more detail in Chapter 13.

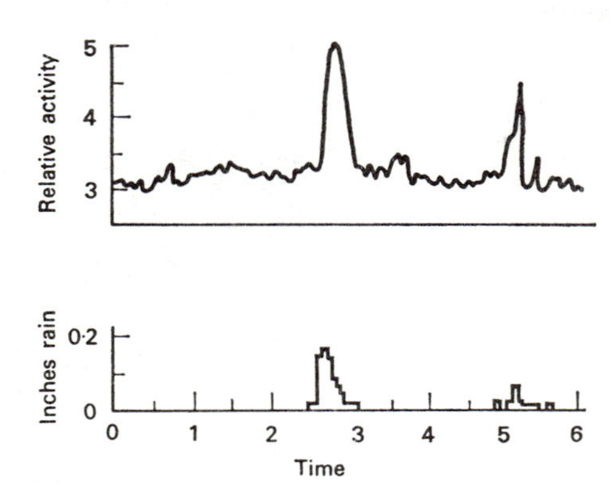

FIG. 4.13. Washout of atmospheric radioactivity by rain (M. M. Weiss, *Area Survey Manual*, BNL 344 (T-61), 15 June, 1955).

A third common characteristic among the three natural radioactive series is that the end product in each case is lead. In the case of the uranium series, the final member is stable ^{206}Pb, in the actinium series it is ^{207}Pb, and in the thorium series it is ^{208}Pb. The artificial neptunium series differs in this characteristic too from the natural series; the terminal member is stable bismuth, ^{209}Bi.

These four radioactive chains, the three naturally occurring ones and the artificially produced neptunium series, are often designated as the $4n$, $4n + 1$, $4n + 2$, and $4n + 3$ series. These identification numbers refer to the divisibility of the mass numbers of each of the series by 4. The atomic mass number of ^{232}Th, the first member of the thorium series, is exactly divisible by 4. Since all disintegrations in the series are accomplished by the emission of either an alpha particle of 4 atomic mass units, or a beta particle of 0 atomic mass units, it follows that the mass numbers of all members of the thorium series are exactly divisible by 4. The uranium series, whose first member is ^{238}U, consists of isotopes whose mass numbers are divisible by 4, and leave a remainder of 2 ($238 \div 4 = 59 + 2/4$). This series, therefore, is called the $4n + 2$ series. The actinium series, whose first member is ^{235}U (actinouranium), is the $4n + 3$ series. The "missing" series, $4n + 1$, is the artificially produced neptunium series, which begins with ^{241}Pu.

Radioactive isotopes that are found in nature are not restricted to the thorium, uranium, and actinium series. Several of the elements among the lower atomic numbered members of the periodic table also have radioactive isotopes. The most important of these low-atomic-numbered natural emitters are listed below in Table 4.5.

Of these naturally radioactive isotopes, ^{40}K, by virtue of the widespread distribution of potassium in the environment (the average concentration of potassium in crustal rocks is about 27 g/kg and in the ocean is about 380 mg/l, and in plants and animals, including man (the average concentration of potassium in man is about 1.7 g/kg), is the most important from the health physics point of view. Estimates of body burden of many radioisotopes, from which the degree of exposure to environmental contaminants may be inferred, is made from radiochemical analysis of urine from persons suspected of overexposure. Potassium, whose concentration in urine is about 1.5 g/l may interfere with the determination of the suspected contaminant unless special care is taken to remove the potassium from the urine or unless allowance is made for the ^{40}K activity. That this interfering activity must be considered is clearly shown by a comparison of the ^{40}K activity in urine with that of certain isotopic concentrations which are thought to be indicative of a significant body burden.

TABLE 4.1. THORIUM SERIES (4*n*)

Nuclide	Half-life	Energy, MeV		
		Alpha[a]	Beta	Gamma (photons/ trans.)[b]
$^{232}_{90}Th$	1.39 10^{10} y	3.98		
$^{228}_{88}Ra$(MsThl)	6.7 y		0.01	
$^{228}_{89}Ac$(MsTh2)	6.13 h		Complex decay scheme. Most intense beta group is 1.11 MeV	1.59(n.v.) 0.966(0.2) 0.908(0.25)
$^{228}_{90}Th$(RdTh)	1.91 y	5.421		0.084(0.016)
$^{224}_{88}Ra$(ThX)	3.64 d	5.681		0.241(0.038)
$^{220}_{86}Em$(Tn)	52 s	6.278		0.542(0.0002)
$^{216}_{84}Po$(ThA)	0.158 s	6.774		
$^{212}_{82}Pb$(ThB)	10.64 h		0.35, 0.59	0.239(0.40)
$^{212}_{83}Bi$(ThC)	60.5 m	6.086 (33.7%)[c]	2.25(66.3%)[c]	0.04 (0.034 branch)
$^{212}_{84}Po$(ThC′)	3.04×10^{-7} s	8.776		
$^{208}_{81}Tl$(ThC″)	3.1 m		1.80, 1.29, 1.52	2.615(0.997)
$^{208}_{82}Pb$(ThD)	Stable			

[a]Only the highest energy alpha is given. Complete information on alpha energies may be obtained from Sullivan's *Trilinear Chart of Nuclides*, Government Printing Office, Washington, D.C., 1957.

[b]Only the most prominent gamma photons are listed. For the complete gamma ray information, consult T. P. KOHMAN, "Natural radioactivity," in *Radiation Hygiene Handbook*, pp. 6–6 to 6–13, H. BLATZ, editor, McGraw-Hill, 1959.

[c]Indicates branching. The percentage enclosed in the parentheses gives the proportional decay by the indicated mode.

TABLE 4.2. NEPTUNIUM SERIES (4n + 1)

Nuclide	Half-life	Energy, MeV		
		Alpha[a]	Beta	Gamma (photons/trans.)[b]
$^{241}_{94}Pu$	13.2 y		0.02	
$^{241}_{95}Am$	462 y	5.496		0.060(0.4)
$^{237}_{93}Np$	2.2×10^6 y	4.77		
$^{233}_{91}Pa$	27.4 d		0.26, 0.15, 0.57	0.31(very strong)[d]
$^{233}_{92}U$	1.62×10^5 y	4.823		0.09(0.02) 0.056(0.02) 0.042(0.15)
$^{229}_{90}Th$	7.34×10^3 y	5.02		
$^{225}_{88}Ra$	14.8 d		0.32	
$^{225}_{89}Ac$	10.0 d	5.80		
$^{221}_{87}Fr$	4.8 m	6.30		0.216(1)
$^{217}_{85}At$	0.018 s	7.02		
$^{213}_{83}Bi$	47 m	5.86(2%)[c]	1.39(98%)[c]	
$^{213}_{84}Po$	4.2×10^{-6} s	8.336		
$^{209}_{81}Tl$	2.2 m		2.3	0.12 (weak)[d]
$^{209}_{82}Pb$	3.32 h		0.635	
$^{209}_{83}Bi$	Stable			

[a],[b],[c]See footnotes under Table 4.1.
[d]Exact intensity not known.

TABLE 4.3. URANIUM SERIES ($4n + 2$)

Nuclide	Half-life	Energy, MeV		
		Alpha[a]	Beta	Gamma (photons/trans.)[b]
$^{238}_{92}U$	4.51×10^{9} y	4.18		
$^{234}_{90}Th(UX_1)$	24.10 d		0.193, 0.103	0.092(0.04) 0.063(0.03)
$^{234m}_{91}Pa(UX_2)$	1.175 m		2.31	1.0(0.015) 0.76(0.0063), I.T.
$^{234g}_{91}Pa(UZ)$	6.66 h		0.5	Many weak
$^{234}_{92}U(UII)$	2.48×10^{5} y	4.763		
$^{230}_{90}Th(I_0)$	8.0×10^{4} y	4.685		0.068(0.0059)
$^{226}_{88}Ra$	1,622 y	4.777		
$^{222}_{86}Em(Rn)$	3.825 d	5.486		0.51 (very weak)
$^{218}_{84}Po(RaA)$	3.05 m	5.998 (99.978%)[c]	Energy not known (0.022%)[c]	0.186(0.030)
$^{218}_{85}At(RaA')$	2 s	6.63 (99.9%)[c]	Energy not known (0.1%)[c]	
$^{218}_{86}Em(RaA'')$	0.019 s	7.127		
$^{214}_{82}Pb(RaB)$	26.8 m		0.65	0.352(0.036) 0.295(0.020) 0.242(0.07)
$^{214}_{83}Bi(RaC)$	19.7 m	5.505 (0.04%)[c]	1.65, 3.7 (99.96%)[c]	0.609(0.295) 1.12(0.131)
$^{214}_{84}Po(RaC')$	1.64×10^{-4} s	7.680		
$^{210}_{81}Tl(RaC'')$	1.32 m		1.96	2.36(1) 0.783(1) 0.297(1)
$^{210}_{82}Pb(RaD)$	19.4 y		0.017	0.0467(0.045)
$^{210}_{83}Bi(RaE)$	5.00 d		1.17	
$^{210}_{84}Po(RaF)$	138.40 d	5.298		0.802(0.000012)
$^{206}_{82}Pb(RaG)$	Stable			

[a],[b],[c] See footnotes under Table 4.1.

TABLE 4.4. ACTINIUM SERIES $(4n + 3)$

Nuclide	Half-life	Energy, MeV		
		Alpha[(a)]	Beta	Gamma (photons/trans.)[(b)]
$^{235}_{92}U$	7.13×10^8 y	4.39		0.18(0.7)
$^{231}_{90}Th$(UY)	25.64 h		0.094, 0.302, 0.216	0.022(0.7) 0.0085(0.4) 0.061(0.16)
$^{231}_{91}Pa$	3.43×10^4 y	5.049		0.33(0.05) 0.027(0.05) 0.012(0.01)
$^{227}_{89}Ac$	21.8 y	4.94(1.2%)[(c)]	0.0455(98.8%)[(c)]	
$^{227}_{90}Th$(RdAc)	18.4 d	6.03		0.24(0.2) 0.05(0.15)
$^{223}_{87}Fr$(AcK)	21 m		1.15	0.05(0.40) 0.08(0.24)
$^{223}_{88}Ra$(AcX)	11.68 d	5.750		0.270(0.10) 0.155(0.055)
$^{219}_{86}Em$(An)	3.92 s	6.824		0.267(0.086) 0.392(0.048)
$^{215}_{84}Po$(AcA)	1.83×10^{-3} s	7.635		
$^{211}_{82}Pb$(AcB)	36.1 m		1.14, 0.5	Complex spectrum, 0.065 to 0.829 MeV
$^{211}_{83}Bi$(AcC)	2.16 m	6.619 (99.68%)[(c)]	Energy not known (0.32)%[(c)]	0.35(0.14)
$^{211}_{84}Po$(AcC′)	0.52 s	7.434		0.88(0.005) 0.56(0.005)
$^{207}_{81}Tl$(AcC″)	4.78 m		1.47	0.87(0.005)
$^{207}_{82}Pb$	Stable			

[(a),(b),(c)]See footnotes under Table 4.1.

TABLE 4.5. SOME LOW-ATOMIC-NUMBERED NATURALLY OCCURRING RADIOISOTOPES

Nuclide	Isotopic abundance (%)	Half-life (years)	Principal radiations	
			Particles	Gamma
^{40}K	0.0119	1.3×10^{9}	1.35 MeV	1.46 MeV
^{87}Rb	27.85	5×10^{10}	0.275 MeV	None
^{138}La	0.089	1.1×10^{11}	1.0 MeV	0.80, 1.43 MeV
^{147}Sm	15.07	1.3×10^{11}	2.18 MeV	None
^{176}Lu	2.6	3×10^{10}	0.43 MeV	0.20, 0.31 MeV
^{187}Re	62.93	5×10^{10}	0.043 MeV	None

Example 4.8

The specific activity of urine, with respect to ^{40}K, is

$$\frac{\text{t/min}}{\text{liter}} = \frac{3.7 \times 10^{10} \times 1600 \times 226}{1.3 \times 10^{9} \times 40} \frac{\text{Bq}}{\text{g}\ ^{40}\text{K}} \times 1.19 \times 10^{-4} \frac{\text{g}\ ^{40}\text{K}}{\text{g K}} \times \frac{1.5\ \text{g K}}{\text{liter}} \times 60 \frac{\text{t/min}}{\text{Bq}} = 2.8 \times 10^{3} \frac{\text{dis/min}}{\text{liter}}.$$

A gross beta activity in excess of 200 transformations per minute per liter of urine following possible exposure to mixed fission products is considered indicative of internal deposition of the fission products. This activity is less than 10% of that due to the naturally occurring potassium.

Another naturally occurring radioisotope of importance is ^{14}C. It may appear surprising that this radioisotope is found in our environment, since its half-life is only 5600 years. It is not a "natural" isotope in the same sense as the very long-lived isotopes listed in Table 4.5; ^{14}C was not created at the beginning of time, as those are thought to have been. The production of ^{14}C is still going on. This isotope of carbon is the result of a nuclear transformation induced by the cosmic-ray bombardment of ^{14}N. The environmental burden of ^{14}C before the advent of nuclear bombs was about 1.5×10^{11} MBq (4 MCi) in the atmosphere, 4.8×10^{11} MBq (13 MCi) in plants, and 9×10^{12} MBq (240 MCi) in the oceans. It should be pointed out that testing of nuclear weapons has resulted in a significant increase in the atmospheric level of radiocarbon. It is estimated that about 1.1×10^{11} MBq (3 MCi) ^{14}C have been injected into the air by all weapon tests which have been conducted until 1960. The atmospheric radiocarbon exists as $^{14}CO_2$. It is therefore inhaled by all animals, and utilized by plants in the process of photosynthesis. Because only living plants continue to incorporate ^{14}C along with non-radioactive carbon, it is possible to determine the age of organic matter by measuring the specific activity of the carbon present. If it is assumed that the rate of production of ^{14}C, as well as its concentration in the air, has remained constant during the past several tens of thousands of years, then a simple correction of specific activity data for half-life permits the estimation of the age of ancient samples of organic matter.

Example 4.9

If 2 g of carbon from a piece of wood found in an ancient temple is analyzed and found to have an activity of 10 transformations per minute per gram, what is the age of the wood if the current specific activity of ^{14}C in carbon is assumed to have been constant at 15 trans/min per gram:

Fraction of the original ^{14}C that remains today is, according to equation (4.18),

$$\frac{A}{A_0} = \frac{10}{15} = e^{-\lambda t}.$$

Since the half-life for ^{14}C is 5730 years,

$$\lambda = \frac{0.693}{5730} \text{ yr} = 1.21 \times 10^{-4} \text{ yr}^{-1},$$

$$\frac{10}{15} = e^{-1.21 \times 10^{-4} t},$$

$$t = 3.35 \times 10^3 \text{ years}.$$

Serial Transformation

In addition to the four chains of radioactive isotopes described above, there are a number of other groups of sequentially transforming isotopes which are important to the health physicist and the radiobiologist. Most of these series are associated with nuclear fission, and the first member of each series is a fission fragment. One of the most widely known fission products, for example, ^{90}Sr, is the middle member of a five-member series that starts with ^{90}Kr, and finally terminates with stable ^{90}Zr according to the following sequence:

$$^{90}_{36}\text{Kr} \xrightarrow[33\text{ s}]{\beta} {}^{90}_{37}\text{Rb} \xrightarrow[2.74\text{m}]{\beta} {}^{90}_{38}\text{Sr} \xrightarrow[19.9\text{y}]{\beta} {}^{90}_{39}\text{Y} \xrightarrow[64.2\text{h}]{\beta} {}^{90}_{40}\text{Zr}$$

The quantitative relationship among the various members of the series is of great significance, and must be considered when dealing with any of the group's members. Intuitively, it can be seen that any amount of ^{90}Kr will, in a time period of 10–15 min, have been transformed to such a degree that for practical purposes, the ^{90}Kr may be assumed to have been completely transformed. Rubidium-90, the ^{90}Kr daughter, because of its 2.74-min half-life, will suffer the same fate after about an hour. Essentially, all the ^{90}Kr is, as a result, converted into ^{90}Sr within about an hour after its formation. The buildup of ^{90}Sr is therefore very rapid. The half-life of ^{90}Sr is 20 years, and its transformation, therefore, is very slow. The ^{90}Y daughter of ^{90}Sr, with a half-life of 64.2 hours, transforms rapidly to stable ^{90}Zr. If initially pure ^{90}Sr is prepared, its radioactive transformation will result in an accumulation of ^{90}Y. Because the ^{90}Y transforms very much faster than ^{90}Sr, however, a point is soon reached at which the instantaneous amount of ^{90}Sr that transforms is equal to that of ^{90}Y. Under these conditions, the ^{90}Y is said to be in secular equilibrium. The quantitative relationship between isotopes in secular equilibrium may be derived in the following manner for the general case

$$A \xrightarrow{\lambda_A} B \xrightarrow{\lambda_B} C,$$

where the half-life of isotope A is very much greater than that of isotope B. The decay constant of A, λ_A, is therefore much smaller than λ_B, the decay constant for isotope B. Isotope C is stable and is not transformed. Because of the very long half-life of A relative to B, the rate of formation of B may be considered to be constant, and equal to K. Under these conditions, the net rate of change of isotope B with respect to time, if N_B is the number of atoms of isotope B in existence at any time t after an initial number N_{B_0}, is given by:

rate of change = rate of formation − rate of transformation

$$\frac{dN_B}{dt} = K - \lambda_B N_B, \tag{4.32}$$

or,

$$\int_{N_{B_0}}^{N_B} \frac{dN_B}{K - \lambda_B N_B} = \int_0^t dt. \tag{4.33}$$

The integrand can be changed to the form

$$\int_a^b \frac{dv}{v} = \ln v \Big|_a^b \tag{4.34}$$

if it is multiplied by $-\lambda_B$, and if the entire integral is multiplied by $-1/\lambda_B$ in order to keep the value of the integral unchanged. Equation (4.33) therefore may be solved to yield

$$\ln\left(\frac{K - \lambda_B N_B}{K - \lambda_B N_{B_0}}\right) = -\lambda_B t. \tag{4.35}$$

Equation (4.35) may be written in the exponential form

$$\frac{K - \lambda_B N_B}{K - \lambda_B N_{B_0}} = e^{-\lambda_B t}, \tag{4.36}$$

and then solved for N_B:

$$N_B = \frac{K}{\lambda_B}(1 - e^{-\lambda_B t}) + N_{B_0} e^{-\lambda_B t}. \tag{4.37}$$

If we start with pure A, that is, if $N_{B_0} = 0$, then equation (4.37) reduces to

$$N_B = \frac{K}{\lambda_B}(1 - e^{-\lambda_B t}). \tag{4.38}$$

The rate of formation of B from A is equal to the rate of transformation of A. K, therefore, is simply equal to $\lambda_A N_A$. An alternative way of expressing equation (4.38), therefore, is:

$$N_B = \frac{\lambda_A N_A}{\lambda_B}(1 - e^{-\lambda_B t}). \tag{4.39}$$

Note that the quantity of both the parent, isotope A, and the daughter, isotope B, is given in the same units, namely, λN, or transforming atoms per unit time. This is a reasonable unit, since each parent atom that is transformed changes into a daughter. Any other unit that implicitly states this fact is equally usable in equation (4.39). If λN represents transforming atoms per second, then division of both sides of the equation by the proper factor to convert the activity to curies, or multiples thereof is permissible, and converts equation (4.39) into a slightly more usable form. For example, if the activity of the parent is given in becquerels or millicuries, then the activity of the daughter must also be in units of becquerels or millicuries, and equation (4.39) may be written as

$$Q_B = Q_A(1 - e^{-\lambda_B t}), \tag{4.40}$$

where Q_A and Q_B are the respective activities in becquerels or millicuries of the parent and daughter.

Example 4.10

If we have 500 mg radium, how much ^{222}Rn will be collected after 1 day, after 3.8 days, after 10 days, and after 100 days?

Since the specific activity of radium is 3.7×10^{10} Bq/g, (1 Ci/g), 500 mg = 1.85×10^{10} Bq (500 mCi). The half-life of radon is 3.8 days; its decay constant therefore is:

$$\lambda_{\text{Rn}} = \frac{0.693}{(T_{1/2})_{\text{Rn}}} = \frac{0.693}{3.8 \text{ days}},$$

$$\lambda_{\text{Rn}} = 0.1825 \text{ day}^{-1}.$$

From equation (4.40), we have:

$$Q_{\text{Rn}} = Q_{\text{Ra}}(1 - e^{-\lambda_{\text{Rn}} t}),$$

$$Q_{\text{Rn}} = 1.85 \times 10^{10} \text{ Bq}(1 - e^{-0.1825 \text{ day}^{-1} \times t \text{ days}})$$

Substituting the respective time in days for t in the equation above gives 3.1×10^9 Bq (83.5 mCi) Rn after 1 day, 9.25×10^9 Bq (250 mCi) after 3.8 days, 1.55×10^{10} Bq (419.5 mCi) after 10 days, and 1.85×10^{10} Bq (500 mCi) after 100 days. Equation (4.40), as well as the illustrative example given above, shows a buildup of radon from 0 to a maximum activity that is equal to that of the parent from which it was derived. This buildup of daughter activity may be shown graphically by plotting equation (4.40). A generally useful curve showing the buildup of daughter activity under conditions of secular equilibrium may be obtained if t is plotted in units of daughter half-life, as shown in Fig. 4.14. As time increases, $e^{-\lambda t}$ decreases, and Q_B approaches Q_A. For practical purposes,

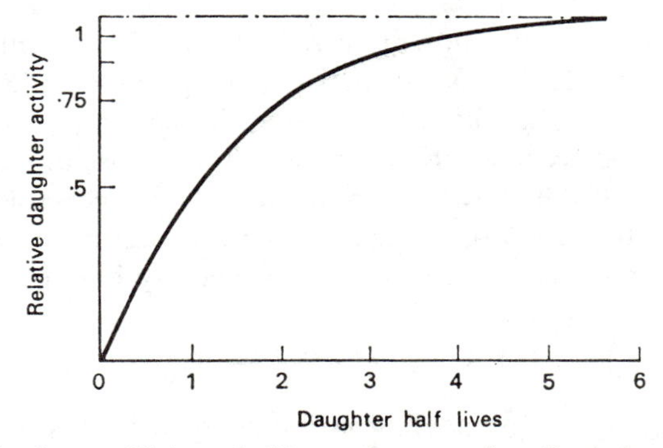

FIG. 4.14. Secular equilibrium: build-up of a very short-lived daughter from a long-lived parent. The activity of the parent remains constant.

equilibrium may be considered established after 7 daughter half-lives. At equilibrium, it should be noted that

$$\lambda_A N_A = \lambda_B N_B. \tag{4.41}$$

Equation (4.41) tells us that, at equilibrium, the activity of the parent is equal to that of the daughter, and that the ratio of the decay constants of the parent and daughter are in the inverse ratio of the equilibrium concentrations of the parent and daughter.

Example 4.11

Deduce the transformation constant and half-life of ^{226}Ra if the radon gas in secular equilibrium with 1 g Ra exerts a partial pressure of 4.8×10^{-4} mm Hg in a one-liter flask, and if the half-life of radon is 3.8 days.

The amount of radon in equilibrium with the radium is computed by:

$$\frac{\text{moles/liter}}{4.8 \times 10^{-4}\ \text{mm}} = \frac{1\ \text{mole}/22\ \text{liters}}{760\ \text{mm}}$$
$$= 2.88 \times 10^{-8}\ \text{moles},$$

and the number of atoms of radon is

$$6.03 \times 10^{23}\ \frac{\text{atoms}}{\text{mole}} \times 2.88 \times 10^{-8}\ \text{moles} = 1.736 \times 10^{16}\ \text{atoms}.$$

Since the radon is in equilibrium with 1 g radium, and since there are

$$\frac{1\ \text{g}}{226\ \text{g/mole}} = 4.42 \times 10^{-3}\ \text{moles radium}$$

$$\lambda_{\text{Ra}} \times N_{\text{Ra}} = \lambda_{\text{Rn}} \times N_{\text{Rn}},$$

$$\lambda_{\text{Ra}} \times 4.42 \times 10^{-3}\ \text{moles} = \frac{0.693}{3.8\ \text{days}} \times 2.88 \times 10^{-8}\ \text{moles}$$

$$\lambda_{\text{Ra}} = 1.19 \times 10^{-6}\ \text{day}^{-1}$$
$$= 4.35 \times 10^{-4}\ \text{year}^{-1},$$

$$T_{1/2} = \frac{0.693}{4.35 \times 10^{-4}} = 1600\ \text{years}.$$

In the case of secular equilibrium discussed above, the quantity of parent remains substantially constant during the period that it is being observed. Since it is required, for secular equilibrium, that the half-life of the parent be very much longer than that of the daughter, it is evident that secular equilibrium is a special case of a more general situation in which the half-life of the parent may be of any conceivable magnitude, and no restrictions are applied to the relative magnitudes of the decay constants of the parent and daughter. For the general case, where the parent activity is not relatively constant,

$$A \xrightarrow{\lambda_A} B \xrightarrow{\lambda_B} C,$$

the time rate of change of the number of atoms of species B is given by the differential equation

$$\frac{dN_B}{dt} = \lambda_A N_A - \lambda_B N_B. \tag{4.42}$$

In this equation, $\lambda_A N_A$ is the rate of transformation of species A, and is exactly equal to the rate of formation of species B. The rate of transformation of isotope B is $\lambda_B N_B$, and the difference between these two rates at any time is the instantaneous rate of growth of species B at that time.

Since we have, from equation (4.18),

$$N_A = N_{A_0} e^{-\lambda_A t}, \tag{4.43}$$

equation (4.42) may be rewritten, after substituting the expression above for N_A and transposing $\lambda_B N_B$, as

$$\frac{dN_B}{dt} + \lambda_B N_B = \lambda_A N_{A_0} e^{-\lambda_A t}. \tag{4.44}$$

Equation (4.44) is a first-order linear differential equation of the form

$$\frac{dy}{dx} + P(x)y = Q(x), \tag{4.45}$$

and may be integrable by multiplying both sides of the equation by

$$e^{\int P dx} = e^{\int \lambda_B dt} = e^{\lambda_B t},$$

and the solution to equation (4.45) is

$$y e^{\int P dx} = \int e^{\int P dx} \cdot Q \, dx. \tag{4.46}$$

Since N_B, λ_B, and $\lambda_A N_{A_0} e^{-\lambda_A t}$ from equation (4.44) are represented in equation (4.46) by y, P, and Q, respectively, the solution of equation (4.44) is

$$N_B e^{\lambda_B t} = \int e^{\lambda_B t} \lambda_A N_{A_0} e^{-\lambda_A t} \, dt + C, \tag{4.47}$$

or, if the two exponentials are combined, we have

$$N_B e^{\lambda_B t} = \int \lambda_A N_{A_0} e^{(\lambda_B - \lambda_A)t} \, dt + C. \tag{4.48}$$

If the integrand in equation (4.48) is multiplied by the integrating factor $\lambda_B - \lambda_A$, then equation (4.48) is in the form

$$\int e^v \, dv = e^v + C, \tag{4.49}$$

and the solution is

$$N_B e^{\lambda_B t} = \frac{1}{\lambda_B - \lambda_A} \lambda_A N_{A_0} e^{(\lambda_B - \lambda_A)t} + C. \tag{4.50}$$

The constant C may be evaluated by applying the boundary conditions

$$N_B = 0 \text{ when } t = 0.$$

$$0 = \frac{1}{\lambda_B - \lambda_A} \cdot \lambda_A N_{A_0} + C,$$

$$C = -\frac{\lambda_A N_{A_0}}{\lambda_B - \lambda_A}. \tag{4.51}$$

If the value for C, from equation (4.51), is substituted into equation (4.50), the solution for N_B is found to be

$$N_B = \frac{\lambda_A N_{A_0}}{\lambda_B - \lambda_A} (e^{-\lambda_A t} - e^{-\lambda_B t}). \tag{4.52}$$

For the case in which the half-life of the parent is very much greater than that of the daughter, that is, when $\lambda_A \ll \lambda_B$, equation (4.52) approaches the condition of secular equilibrium, the limiting case described by equation (4.41). Two other general cases should be considered. The case where the parent half-life is slightly greater than that of the daughter, $\lambda_A < \lambda_B$, and the case in which the parent half-life is less than that of the daughter, $\lambda_B < \lambda_A$. In the former case, where the half-life of the duaghter is slightly smaller than that of the parent, the daughter activity, if the parent is initially pure and free of any daughter activity, starts from zero, rises to a maximum and then seems to decay with the same half-life as the parent. When this occurs, the daughter is undergoing transformation at the same rate as it is being produced, and the two isotopes are said to be in a state of transient equilibrium. The quantitative relationships prevailing during transient equilibrium may be inferred from equation (4.52). If both sides of that equation are multiplied by λ_B, we have an explicit expression for the activity of the daughter:

$$\lambda_B N_B = \frac{\lambda_B \lambda_A N_{A0}}{\lambda_B - \lambda_A}(e^{-\lambda_A t} - e^{-\lambda_B t}). \tag{4.53}$$

Since λ_B is greater than λ_A, then, after a sufficiently long period of time, $e^{-\lambda_B t}$ will become much smaller than $e^{-\lambda_A t}$. Under this condition, equation (4.53) may be rewritten as

$$\lambda_B N_B = \frac{\lambda_B \lambda_A N_{A0}}{\lambda_B - \lambda_A} e^{-\lambda_A t}. \tag{4.54}$$

By application of equation (4.43) the mathematical expression for transient equilibrium, equation (4.54), may be rewritten as

$$\lambda_B N_B = \frac{\lambda_B \lambda_A N_A}{\lambda_B - \lambda_A}, \tag{4.55}$$

or in terms of activity units (becquerels, curies, etc.)

$$Q_B = \frac{\lambda_B}{\lambda_B - \lambda_A} Q_A. \tag{4.56}$$

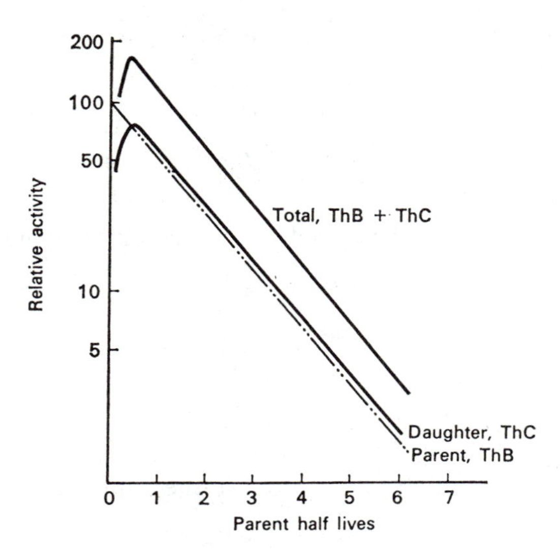

FIG. 4.15. Transient equilibrium: growth and decay of 60.5-min ThC from 10.6-hr ThB.

An example of this equilibrium that is of importance to the health physicist is the ThB to ThC to ThC′ and ThC″ chain which occurs near the end of the thorium series. In this sequence of transformations, ThB (^{212}Pb), whose half-life is 10.6 h, decays by beta emission to 60.5 min ThC (^{212}Bi), which then branches, 35.4% of the transformations going by alpha emission to ThC″ (^{208}Te) and 64.6% of the transitions being accomplished by beta emission to form ThC′ (^{212}Po). The ThC′ and ThC″ half-lives are very short, 3×10^{-7} sec and 3.1 min, respectively, and both decay to stable ^{208}Pb. Since ThB is a naturally occurring atmospheric isotope, correction for its activity and its daughter activity must be made before data on air samples can be accurately interpreted. The growth and decay of ThC is shown in Fig. 4.15. This curve graphically emphasizes the fact that, at transient equilibrium, the daughter activity seems to decrease at the same rate as the parent activity.

Figure 4.15 also shows that the daughter activity, which starts from zero, rises to a maximum value and then decreases. It is also seen that the total activity, daughter plus parent, reaches a maximum value that does not coincide in time with that of the daughter. The time after isolation of the parent that the daughter reaches its maximum activity may be computed by differentiating equation (4.53), the daughter activity, setting the derivative equal to zero, and then solving for t_{md}.

$$\lambda_B N_B = \frac{\lambda_B \lambda_A N_{A0}}{\lambda_B - \lambda_A}(e^{-\lambda_A t} - e^{-\lambda_B t}), \tag{4.53}$$

$$\frac{d(\lambda_B N_B)}{dt} = \frac{\lambda_B \lambda_A N_{A0}}{\lambda_B - \lambda_A}(-\lambda_A e^{-\lambda_A t} + \lambda_B e^{-\lambda_B t}) = 0,$$

$$\lambda_A e^{-\lambda_A t} = \lambda_B e^{-\lambda_B t},$$

$$\frac{\lambda_B}{\lambda_A} = \frac{e^{-\lambda_A t}}{e^{-\lambda_B t}} = e^{(\lambda_B - \lambda_A)t},$$

$$\ln \frac{\lambda_B}{\lambda_A} = (\lambda_B - \lambda_A)t,$$

$$t = t_{md} = \frac{\ln(\lambda_B/\lambda_A)}{\lambda_B - \lambda_A} = \frac{2.3 \log(\lambda_B/\lambda_A)}{\lambda_B - \lambda_A}. \tag{4.57}$$

For the case illustrated in Fig. 4.15,

$$\lambda_A = \frac{0.693}{10.6} = 0.065 \text{ h}^{-1},$$

and

$$\lambda_B = \frac{0.693}{1.01} = 0.686 \text{ h}^{-1}.$$

The time at which the daughter, ThC, will reach its maximum activity is:

$$t_{md} = \frac{2.3 \log(0.686/0.065)}{0.686 - 0.065} = 3.78 \text{ h}.$$

The time when the total activity is at its peak may be determined in a similar manner. In this case, the total activity, $A(t)$, must be maximized. The total activity at any time is given by

$$A(t) = \lambda_A N_A + \lambda_B N_B. \tag{4.58}$$

Substituting equations (4.43) and (4.52) for N_A and N_B, respectively, in equation (4.58), we have

$$A(t) = \lambda_A N_{A_0} e^{-\lambda_A t} + \frac{\lambda_B \lambda_A N_{A_0}}{\lambda_B - \lambda_A}(e^{-\lambda_A t} - e^{-\lambda_B t}), \tag{4.59}$$

and, differentiating, yields

$$\frac{dA(t)}{dt} = -\lambda_A^2 N_{A_0} e^{-\lambda_A t} + \frac{\lambda_B \lambda_A N_{A_0}}{\lambda_B - \lambda_A}(-\lambda_A e^{-\lambda_A t} + \lambda_B e^{\lambda_B t}) = 0. \tag{4.60}$$

Expanding and collecting terms, we have

$$-\lambda_A^2 N_{A_0} e^{-\lambda_A t}\left(1 + \frac{\lambda_B}{\lambda_B - \lambda_A}\right) + \frac{\lambda_A \lambda_{B^2}}{\lambda_B - \lambda_A} N_{A_0} e^{-\lambda_B t} = 0,$$

and solving for t_{mt}, the time when the total activity is a maximum, we get

$$t = t_{mt} = \frac{1}{\lambda_B - \lambda_A} \ln\left(\frac{\lambda_{B^2}}{2\lambda_A \lambda_B - \lambda_{A^2}}\right) = \frac{2.3}{\lambda_B - \lambda_A} \log\left(\frac{\lambda_{B^2}}{2\lambda_A \lambda_B - \lambda_{A^2}}\right). \tag{4.61}$$

Referring once more to the example in Fig. 4.15, one finds the maximum total activity to occur at

$$t_{mt} = \frac{2.3}{0.686 - 0.065} \log\left(\frac{(0.686)^2}{2 \times 0.065 \times 0.686 - 0.065^2}\right)$$
$$= 2.78 \text{ h}$$

after the initial purification of the parent. It should be noted that the maximum total activity occurs earlier than that of the daughter alone.

The time when the parent and daughter isotopes may be considered to be equilibrated depends on their respective half-lives. The shorter the half-life of the daughter, relative to the parent, the more rapidly will equilibrium be attained.

In the case where the half-life of the daughter exceeds that of the parent, no equilibrium is possible. The daughter activity reaches a maximum, at a time which can be calculated from equation (4.57), and then reaches a point where it decays at its own characteristic

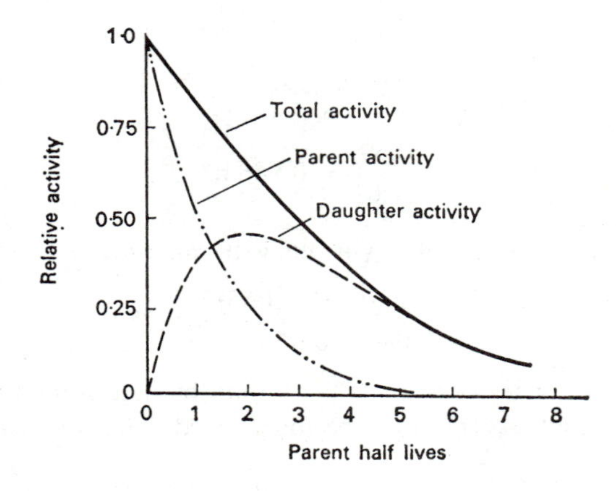

FIG. 4.16. No equilibrium: growth and decay of 24 min ^{140}Pr from 14 min ^{146}Ce.

rate. The parent, in the meantime, because of its shorter half-life, decays away. The total activity in this case does not increase to a maximum; it decreases continuously. Equation (4.58), which gives the time after isolation of a parent at which the total activity is maximum, cannot be solved for the case in which the half-life of the daughter exceeds that of the parent, or if $\lambda_B < \lambda_A$. These points are all illustrated in Fig. 4.16, which shows the course in time of the growth and decay of ^{146}Pr, whose half-life is 24.4 min, from the fission product ^{146}Ce, whose half-life is only 13.9 min. Solution of equation (4.57) shows the ^{146}Pr activity to reach its peak 26.2 min after isolation of ^{146}Ce.

Problems

1. Carbon-14 is a pure beta emitter that decays to ^{14}N. If the exact atomic masses of the parent and daughter are 14.007687 and 14.007520 atomic mass units, respectively, calculate the kinetic energy of the most energetic beta particle.

2. If 1.0 MBq (27 μCi)^{131}I is needed for a diagnostic test, and if 3 days elapse between shipment of the radioiodine and its use in the test, how many Bq must be shipped? To how many μCi does this correspond?

3. The gamma radiation from 1 ml of a solution containing 370 Bq (0.01 μCi) ^{198}Au and 185 Bq (0.005 μCi) ^{131}I is counted daily over a 16-day period. Assume an equal detection efficiency of the scintillation counter of 10% for all the quantum energies involved. What will be the relative counting rates of the ^{131}I and ^{198}Au at time $t = 0$, $t = 3$ days, $t = 8$ days, and $t = 16$ days. Plot the daily total counting rates on semi-log paper, and write the equation of the curve of total count rate vs. time.

4. The decay constant for ^{235}U is 9.72×10^{-10} per year. Compute the number of transformations per second in a 500-mg sample of ^{235}U.

5. Two hundred MBq (5.4 mCi) ^{210}Po are necessary for a certain ionization source. How many grams ^{210}Po does this represent?

6. How long would it take for 99.9% of ^{137}Cs to decay, if its half-life is 30 years?

7. For use in carcinogenesis studies, benzo(a)pyrene is tagged with ^{3}H to a specific activity of 4×10^{11} Bq/millimole. If there is only 1 tritium atom on a tagged molecule, what percentage of the benzo(*a*)pyrene molecules is tagged with ^{3}H?

8. How many alpha particles are emitted per minute by 1 cm^3 ^{222}Rn at a temperature of 27°C and a pressure of 100,000 Pa?

9. Calculate the number of beta particles emitted per minute by 1 kg KCl, if ^{40}K emits 1 beta particle per transformation.

10. Iodine 125, a widely used isotope in the practice of nuclear medicine, has a half life of 60 days.
(a) How long will it take for 4 MBq ($\sim$ 1 μCi) to decrease to 0.1% of its initial activity?
(b) What is the mean life of ^{125}I?

11. If uranium ore contains 10% U_3O_8, how many metric tons are necessary to produce 1 g radium if the extraction process is 90% efficient?

12. How much ^{234}U is there in 1 ton of the uranium ore containing 10% U_3O_8?

13. Compare the activity of the ^{234}U to that of the ^{235}U and the ^{238}U in the ore of problems 11 and 12.

14. What will be the temperature rise after 24 hours in a well insulated 100-ml aqueous solution containing 1 gram Na_2 $^{35}SO_4$, if the specific activity of the sulfur is 3.7×10^{12} Bq/gram (100 Ci/gram)?

15. The mean concentration of potassium in crustal rocks is 27 g/kg. If ^{40}K constitutes 0.012% of potassium, what is the ^{40}K activity in 1 ton of rock?

16. A solution of ^{203}Hg is received with the following assay: 1 MBq/ml on 1 March, 1981 at 8:00 a.m. It is desired to make a solution whose activity will be 0.1 MBq/ml on 1 April, 1981. Calculate the dilution factor to give the desired activity. $T_{\frac{1}{2}}$ ^{203}Hg = 46 days.

17. In a mixture of two radioisotopes, 99% of the activity is due to ^{24}Na and 1% is due to ^{32}P. At what subsequent time will the two activities be equal?

18. ThB is transformed to ThC at a rate of 6.54% per hour, and ThC is transformed at a rate of 1.15% per min. How long will it take for the two isotopes to reach their equilibrium state?

19. How many grams of ^{90}Y are there when ^{90}Y is equilibrated with 10 mg ^{90}Sr?

20. Radiogenic lead constitutes 98.5% of the element as found in lead ore. The isotopic constitution of lead in nature is: ^{204}Pb, 1.5%; ^{206}Pb, 23.6%; ^{207}Pb, 22.6%; ^{208}Pb, 52.3%. How much uranium and thorium decayed completely to produce 985 mg radiogenic lead?

21. How long after 1 kg of ^{241}Pu, is isolated will the ^{241}Am activity be at its maximum? What will the activity be at that time?

22. How long after 14-minute ^{146}Ce is isolated will the activity of the 24-minute ^{146}Pr daughter be equal to that of the parent?

23. Calculate the specific activity of ^{85}Kr ($T_{\frac{1}{2}}$ = 10.7 years) in Bq/m^3 and μCi/cm^3 at 25°C and 760 mm Hg.

24. Calculate the specific power of ^{35}S and of ^{14}C in

(a) watts per MBq,
(b) watts per kg.
25. Calculate the specific power of ^{90}Sr in
(a) watts per MBq,
(b) watts per kg.
27. How many joules of energy are released in 3 hours by an initial volume of 1 liter ^{41}A at 0°C and 760 mm Hg?
26. (a) Calculate the specific power, in watts/kg, of ^{41}A.
(b) What is the specific power of ^{41}A 4 hours after the ^{41}A is isolated in a bottle?
28. What volume of radon 222 (at 0°C and 760 torr) is in equilibrium with 0.1 gram radium 226?
29. One hundred milligrams radium as $RaBr_2$ (specific gravity = 5.79) is in a platinum capsule whose inside dimensions are 2 mm diameter × 4 cm long. What will be the gas pressure, at body temperature, inside the capsule 100 years after manufacture if it originally contained air at atmospheric pressure at room temperature (25°C)?
30. A volume of 10 cm^3 tritium gas $^3H_2$ at NTP dissipates 3.11 joules per hour.
(a) What is the activity of the tritium?
(b) What is the mean beta ray energy, if one beta particle is emitted per transition?
31. Barium 140 decays to ^{140}La with a half-life of 12.8 days, and the ^{140}La decays to stable ^{140}Ce with a half-life of 40.5 hours. A radiochemist, after precipitating ^{140}Ba, wishes to wait until he has a maximum amount of ^{140}La before separating the ^{140}La from the ^{140}Ba. (a) How long must he wait? (b) If he started with 1000 MBq (27 mCi) ^{140}Ba, how many micrograms ^{140}La will he collect?
32. Strontium 90 is to be used as a heat source for generating electrical energy in a satellite, (a) How much ^{90}Sr activity is required to generate 50 watts of electrical power, if the conversion efficiency from heat to electricity is 30%? (b) Weight of the isotopic heat source is an important factor in design of the power source. If weight is to be kept at a minimum, and if the source is to generate 50 watts after 1 year of operation, would there be an advantage to using ^{210}Po?
33. Carbon 14 is produced naturally by the ^{14}N(n, p) ^{14}C interaction of cosmic radiation with the nitrogen in the atmosphere at a rate of about 1.4×10^{15} Bq/year. If the half-life of ^{14}C is 5700 years, what is the steady state global inventory of ^{14}C?
34. The global steady state inventory of naturally produced tritium from the interaction of cosmic rays with the atmosphere is estimated by the United Nations Scientific Committee on the Effects of Atomic Radiation to be 1.26×10^{18} Bq (34×10^6 Ci). If the half-life of tritium is 12.3 years, what is the annual production of natural tritium?

Suggested References, Chapter 4

BORN, M. *Atomic Physics*, 8th ed. Hafner, Darien, Conn., 1970.
Radiological Health: *Radiological Health Handbook*, Revised ed. U.S. Dept. of Health, Education and Welfare, Public Health Service, Rockville, Md., 1970.
COHEN, B. L. *Concepts of Nuclear Physics*. McGraw-Hill, New York, 1971.
ETHERINGTON, H., ed. *Nuclear Engineering Handbook*. McGraw-Hill, New York, 1958.
EVANS, R. D. *The Atomic Nucleus*. McGraw-Hill, New York, 1955.
FRIEDLANDER, G., KENNEDY, J. W. MACIAS, E. S., and MILLER, J. M. *Nuclear and Radiochemistry*, 2nd ed. John Wiley & Sons, New York, 1981.
GLASSTONE, S. *Sourcebook on Atomic Energy*, 3rd ed. D. Van Nostrand, Princeton, 1967.
HAISSINSKY, M., and ADLOFF, J. P. *Radiochemical Survey of the Elements*. Elsevier, Amsterdam, 1965.
KAPLAN, I. *Nuclear Physics*, 2nd ed. Addison-Wesley Publ. Co., Reading, Mass., 1963.
LAPP, R. E., and ANDREWS, H. L. *Nuclear Radiation Physics*, 4th ed. Prentice-Hall, Englewook Cliffs, N.J., 1972.
National Council on Radiation Protection and Measurements. *Natural Background Radiation in the United States*. NCRP Report No. 45, 1975.
RUTHERFORD, E., CHADWICK, J., and ELLIS, C. D. *Radiations from Radioactive Substances*. Cambridge University Press, Cambridge, 1930.
SEMAT, H., and ALBRIGHT, J. R. *Introduction to Atomic and Nuclear Physics*, 5th ed. Holt, Rinehart, and Winston, New York, 1972.

CHAPTER 5

INTERACTION OF RADIATION WITH MATTER

Introduction

In order for the health physicist to understand the physical basis for radiation dosimetry and the theory of radiation shielding, he must understand the mechanisms by which the various radiations interact with matter. In most instances, these interactions involve a transfer of energy from the radiation to the matter with which it interacts. Matter consists of atomic nuclei and extra-nuclear electrons. Radiation may interact with either or both of these constituents of matter. The probability of occurrence of any particular category of interaction, and hence the penetrating power of the several radiations, depends on the type and energy of the radiation as well as on the nature of the absorbing medium. In all instances, excitation and ionization of the absorber atoms results from their interaction with the radiation. Ultimately, the energy transferred either to tissue or to a radiation shield is dissipated as heat.

Beta Rays

Range–Energy Relationship

The attenuation of beta rays by any given absorber may be measured by interposing successively thicker absorbers between a beta-ray source and a suitable beta-ray detector, such as a Geiger counter, as shown in Fig. 5.1, and counting the beta particles that penetrate the absorbers. When this is done with a pure beta emitter, it is found that the beta-particle counting rate decreases rapidly at first, and then slowly as the absorber thickness increases. Eventually, a thickness of absorber is reached that stops all the beta particles; the Geiger counter then registers only background counts due to environmental radiation. If semi-log paper is used to plot the data, and if the counting rate is plotted on the logarithmic axis while absorber thickness is plotted on the linear axis, the data approximate a straight line, as shown in Fig. 5.2. The end point in the absorption curve, where no further decrease in the counting rate is observed, is called the *range* of the beta rays in the material of which the absorbers are made. As a rough rule of thumb, a useful relationship is that the absorber half-thickness (that thickness of absorber that stops one-half of the beta particles) is about one-eighth the range of the beta rays. Since the maximum beta-ray energies for the various isotopes are known, then by measuring the beta-ray ranges in different absorbers, the systematic relationship between range and energy shown in Fig. 5.3. is established. Inspection of Fig. 5.3 shows that the required thickness of absorber for any given beta-ray energy decreases as the density of the absorber increases. Detailed analyses of experimental data show that the ability to absorb energy from beta rays depends mainly on the number of absorbing electrons in the path of the

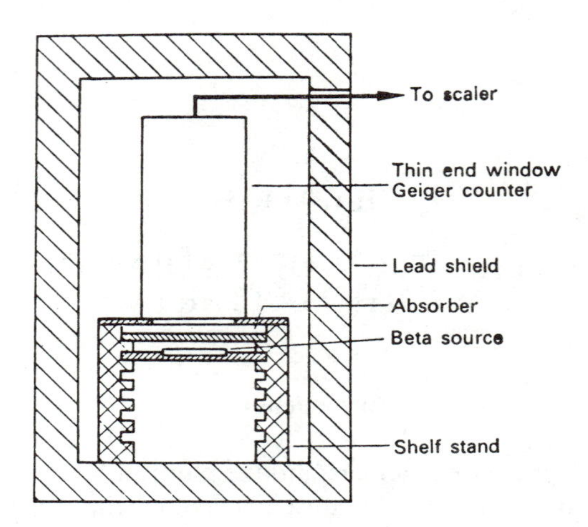

FIG. 5.1. Experimental arrangement for absorption measurements on beta particles.

beta ray, that is, on the areal density (electrons per cm^2) of electrons in the absorber; and, to a very much lesser degree, on the atomic number of the absorber. For practical purposes, therefore, in the calculation of shielding thickness against beta-rays, the effect of atomic number is neglected. Areal density of electrons is approximately proportional to the product of the density of the absorber material and the linear thickness of the absorber, thus giving rise to the unit of thickness called the *density thickness*. Mathematically, density thickness, t_d, is defined as

$$t_d \text{ g/cm}^2 = \rho \text{ g/cm}^3 \times t_l \text{ cm}. \tag{5.1}$$

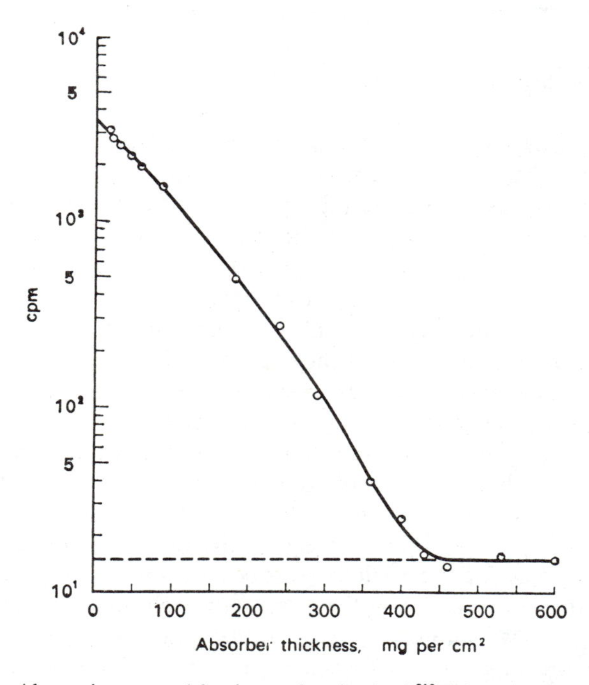

FIG. 5.2. Absorption curve (aluminum absorbers) of ^{210}Bi beta particles, 1.17 MeV. The broken line represents the mean background counting rate.

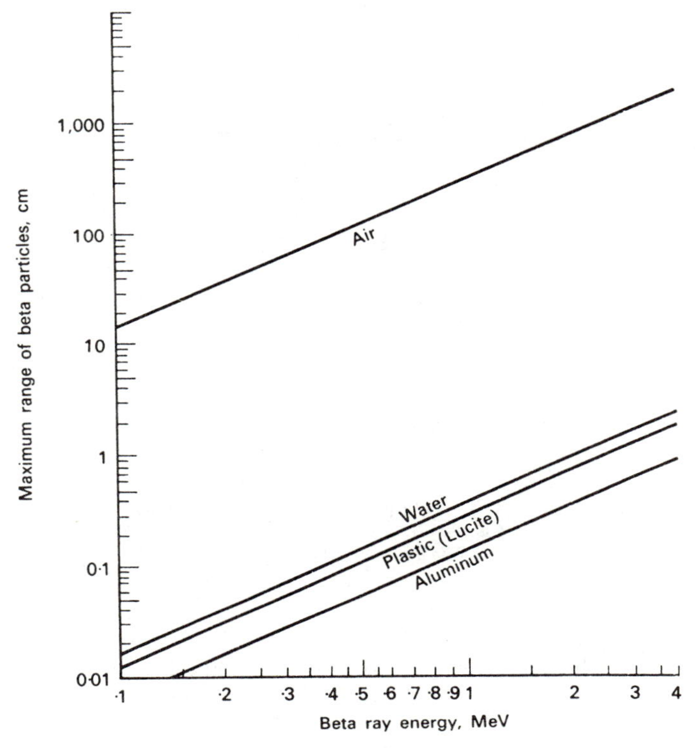

FIG. 5.3. Range-energy curves for beta rays in various substances. (Adapted from *Radiological Health Handbook*, Office of Technical Services, Washington, 1960.)

The units of density and thickness in equation (5.1), of course, need not be grams and centimeters, they may be any consistent set of units. Use of the density thickness unit, such as g/cm^2 or mg/cm^2 for absorber materials makes it possible to specify such absorbers independently of the absorber material. (It should be pointed out, that, for reasons to be given later, beta-ray shields are almost always made from low atomic-numbered materials.) For example, the density of aluminum is 2.7 g/cm^3. From equation (5.1), a sheet of aluminum 1-cm thick, therefore, has a density thickness of

$$t_d = 2.7\ \text{gm/cm}^3 \times 1\ \text{cm} = 2.7\ \text{gm/cm}^2.$$

If a sheet of Plexiglass, whose density is 1.18 g/cm^3 is to have a beta-ray absorbing quality very nearly equal to that of the 1-cm thick sheet of aluminum, i.e. 2.7 g/cm^2, its linear thickness is found, from equation (5.1), to be

$$t_l = \frac{t_d}{\rho} = \frac{2.7\ \text{g/cm}^2}{1.18\ \text{g/cm}^3} = 2.39\ \text{cm}.$$

Another practical advantage of using this system of thickness measurement is that it allows the addition of thicknesses of different materials in a radiologically meaningful way. A universal curve of beta-ray range (in units of density thickness) versus energy is given in Fig. 5.4. This curve, which is based on experimental range–energy measurements, is fitted by the following equations:

$$R = 412\,E^{1.265-0.0954 \ln E} \tag{5.2}$$

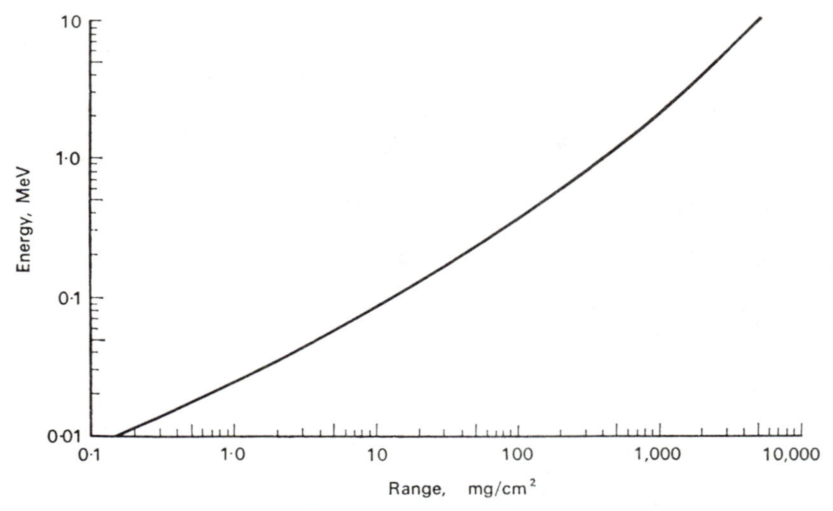

FIG. 5.4. Range-energy curve for beta particles. The range is expressed in units of density thickness (From *Radiation Health Handbook*, Office of Technical Services, Washington, 1960).

for $0.01 \leq E \leq 2.5$ MeV,

$$\ln E = 6.63 - 3.2376(10.2146 - \ln R)^{\frac{1}{2}} \tag{5.3}$$

for $R \leq 1200$,

$$R = 530E - 106 \tag{5.4}$$

for $E > 2.5$ MeV, $R > 1200$,

where R = range, mg/cm²

E = maximum beta-ray energy, MeV.

Example 5.1

What must be the minimum thickness of a shield made of (a) Plexiglass, and (b) aluminum in order that no beta rays from a ^{90}Sr source pass through?

Strontium 90 emits a 0.54-MeV beta particle. However, its daughter, ^{90}Y, emits a beta particle whose maximum energy is 2.27 MeV. Since ^{90}Y beta particles always accompany ^{90}Sr beta rays, the shield must be thick enough to stop these more energetic betas. From Fig. 5.4, the range of a 2.27-MeV beta particle is found to be 1.1 g/cm². The density of Plexiglass is 1.18 g/cm³. From equation (5.1), the required thickness is found to be

$$t_l = \frac{t_d}{\rho} = \frac{1.1 \text{ g/cm}^2}{1.18 \text{ g/cm}^3} = 0.932 \text{ cm}.$$

Plexiglass may suffer radiation damage and crack if exposed to very intense radiation for a long period of time. Under these conditions, aluminum is a better choice for a shield. Since the density of aluminum is 2.7 g/cm³, the required thickness is found to be 0.41 cm.

The range–energy relationship is often used by the health physicist as an aid in identifying an unknown beta-emitting containment. This is done by measuring the range of the beta-radiation, then finding energy of the beta ray, then looking up in a table of isotopes the isotope that emits a beta particle of that energy.

Example 5.2

Using the counting set-up shown in Fig. 5.5, the range of an unknown beta particle was found to be 0.111 mm aluminum. No measurable decay of the isotope was observed during a period of 1 month, and no other radiation was emitted from the isotope.

(a) What was the energy of the beta particle?
(b) What is the isotope?

The total range of the beta ray is

$$\text{Range} = 1.7\frac{\text{mg}}{\text{cm}^2}\text{ mica} + 1\text{ cm air} + 0.111\text{ mm aluminum.}$$

These different absorbing media may be added together if their thickness are expressed in density thickness. The density of air is 1.293 mg/cm^3 at STP. With equation (5.1) the density thicknesses of the air and aluminum are computed, and the range of the unknown beta particle is found to be

$$\text{Range} = 1.7\frac{\text{mg}}{\text{cm}^2} + 1.29\frac{\text{mg}}{\text{cm}^2} + 30\frac{\text{mg}}{\text{cm}^2} = 32.99\frac{\text{mg}}{\text{cm}^2}.$$

In Fig. 5.4, the energy corresponding to this range is seen to be about 0.17 MeV. The unknown isotope is therefore likely to be ^{14}C, a pure beta emitter whose maximum beta-ray energy is 0.155 MeV and whose half-life is 5700 years.

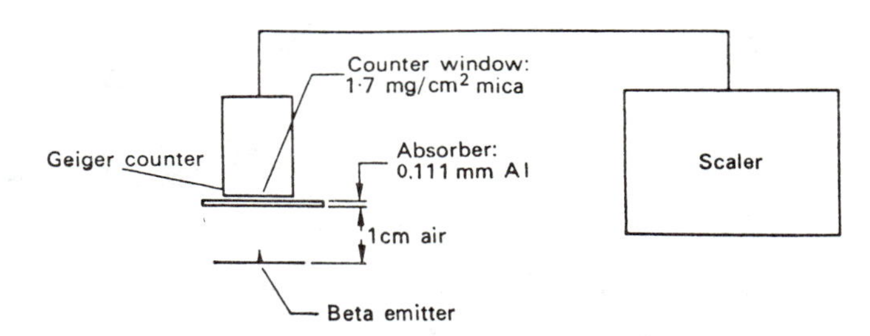

FIG. 5.5. Measuring the range of an unknown beta particle to identify the isotope.

Mechanisms of Energy Loss

Ionization and excitation

Interaction between the electric fields of a beta particle and the orbital electrons of the absorbing medium leads to electronic excitation and ionization. Such interactions are inelastic collisions, analogous to that described in Example 2.13. The electron is held in the atom by electrical forces, and energy is lost by the beta particle in overcoming these forces. Since electrical forces act over long distances, the "collision" between a beta particle and an electron occurs without the two particles coming into actual contact—as in the case of the collision between like poles of two magnets. the amount of energy lost by the beta particle depends on its distance of approach to the electron and on its kinetic energy. If ϕ is the ionization potential of the absorbing medium and E_t is the energy lost by the beta particle during the collision, the kinetic energy of the ejected electron, E_k, is

$$E_k = E_t - \phi. \tag{5.5}$$

In many ionizing collisions, only one ion pair is produced. In other cases, the ejected electron may have sufficient kinetic energy to produce a small cluster of several ionizations; and in a small proportion of the collisions the ejected electron may receive a considerable amount of energy, enough to cause it to travel a long distance and to leave a trail of ionizations. Such an electron, whose kinetic energy may be on the order of 1000 eV, is called a *delta ray*.

Beta particles have the same mass as orbital electrons, and hence are easily deflected during collisions. For this reason, beta particles follow tortuous paths as they pass through absorbing media. Figure 5.6 shows the path of a beta particle through a photographic emulsion. The ionizing events expose the film at the points of ionization, thereby making them visible after development of the film.

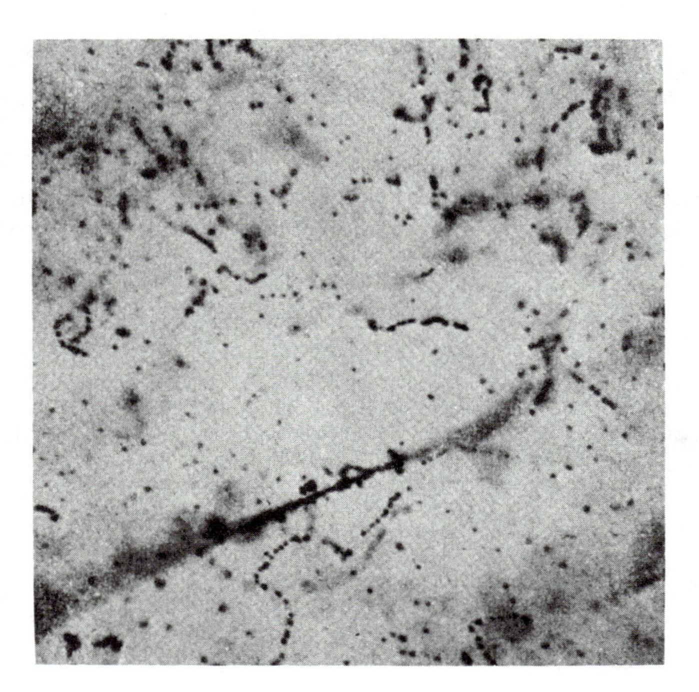

FIG. 5.6. Electron tracks in photographic emulsion. The tortuous lines are the electron tracks; the heavy line near the bottom was made by an oxygen nucleus in primary cosmic radiation. (From H. Yagoda, The tracks of nuclear particles, *Scientific American*, May 1956.)

By using a cloud chamber or a film to visualize the ionizing events, and by counting the actual number of ionizations due to a single primary ionizing particle of known energy, it was learned that the average energy expended in the production of an ion pair is about two to three times greater than the ionization potential. The difference between the energy expended in ionizing collisions and the total energy lost by the ionizing particle is attributed to electronic excitation. For oxygen and nitrogen, for example, the ionization potentials are 13.6 and 14.5 eV, respectively, while the average energy expenditure per ion pair in air is 34 eV. Table 5.1 shows the ionization potential and mean energy expenditure, w, for several gases of practical importance.

TABLE 5.1. AVERAGE ENERGY LOST BY A BETA PARTICLE IN THE PRODUCTION OF AN ION PAIR

Gas	Ionization potential	Mean energy expenditure per ion pair
H_2	13.6 eV	36.6 eV
He	24.5	41.5
N_2	14.5	34.6
O_2	13.6	30.8
Ne	21.5	36.2
A	15.7	26.2
Kr	14.0	24.3
Xe	12.1	21.9
Air		33.7
CO_2	14.4	32.9
CH_4	14.5	27.3
C_2H_2	11.6	25.7
C_2H_4	12.2	26.3
C_2H_6	12.8	24.6

Specific ionization. The linear rate of energy loss of a beta particle due to ionization and excitation, which is an important parameter in health physics instrument design and in the biological effects of radiation, is usually expressed by the *specific ionization*. Specific ionization is the number of ion pairs formed per unit distance travelled by the beta particle. Generally, the specific ionization is relatively high for low energy betas; it decreases rapidly as the beta particle energy increases, until a broad minimum is reached around 1 MeV. Further increase in beta energy results in slowly increasing specification, as shown in Fig. 5.7.

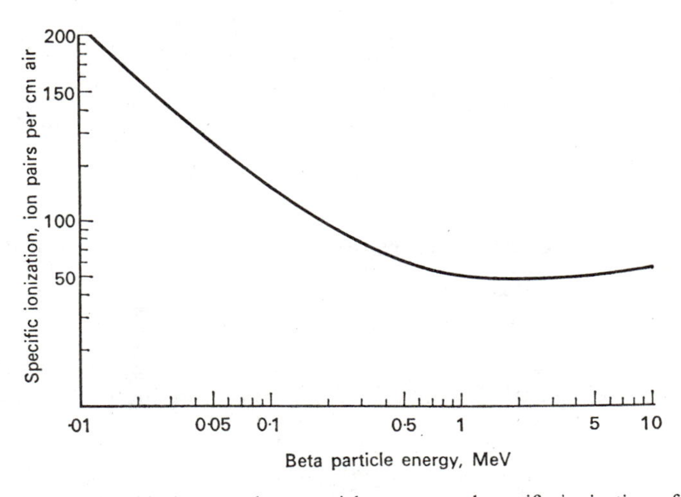

FIG 5.7. Relationship between beta particle energy and specific ionization of air.

The linear rate of energy loss due to excitation and ionization may be calculated from the equation,

$$\frac{dE}{dx} = \frac{2\pi q^4 \text{NZ} \times (3 \times 10^9)^4}{E_m \beta^2 (1.6 \times 10^{-6})^2} \left\{ \ln \left[\frac{E_m E_k \beta^2}{I^2(1-\beta^2)} \right] - \beta^2 \right\} \frac{\text{McV}}{\text{cm}} \tag{5.6}$$

where q = charge on the electron, 1.6×10^{-19} C,
N = number of absorber atoms per cm^3,
Z = atomic number of the absorber,
NZ = number of absorber electrons per cm^3 = 3.88×10^{20} for air at 0° and 76 cm Hg,
E_m = energy equivalent of electron mass, 0.51 MeV,
E_k = kinetic energy of the beta particle, MeV,
β = v/c,
I = mean ionization and excitation potential of absorbing atoms, MeV,
I = 8.6×10^{-5} for air; for other substances, $I = 1.35 \times 10^{-5} Z$.

If the mean energy expended in the creation of an ion pair, w, is known, then the specific ionization may be calculated from the equation below:

$$\text{S.I.} = \frac{dE/dx \text{ eV/cm}}{w \text{ eV/ip}}. \tag{5.7}$$

Example 5.3

What is the specific ionization resulting from the passage of a 0.1-MeV beta particle through standard air?

β^2 is found from equation (2.20):

$$E_k = m_0 c^2 \left(\frac{1}{\sqrt{(1-\beta^2)}} - 1 \right),$$

$$0.1 = 0.51 \left(\frac{1}{\sqrt{(1-\beta^2)}} - 1 \right),$$

$$\beta^2 = 0.3010.$$

Substituting the respective values into equation (5.6), we have

$$\frac{dE}{dx} = \frac{2\pi (1.6 \times 10^{-19})^4 \times 3.88 \times 10^{20} \times (3 \times 10^9)^4}{0.51 \times 0.3010 \times (1.6 \times 10^{-6})^2} \times \left\{ \ln \left[\frac{0.51 \times 0.1 \times 0.3025}{(8.6 \times 10^{-5})^2 (1 - 0.3010)} \right] - 0.3010 \right\} \frac{\text{MeV}}{\text{cm}},$$

$$\frac{dE}{dx} = 4.75 \times 10^{-3} \frac{\text{MeV}}{\text{cm}}.$$

For air, $w = 34$ eV/ip. The specific ionization, therefore, from equation (5.7), is

$$\text{S.I} = \frac{4750 \text{ eV/cm}}{34 \text{ eV/ip}} = 140 \text{ ip/cm}.$$

Very often, the unit of length used in expressing rate of energy loss is density thickness, that is, in units of MeV/g cm^2. This is called the mass stopping power, and is defined by the equation

$$S = \frac{dE/dx}{\rho}. \tag{5.8}$$

Since the density of standard air is 1.293×10^{-3} g/cm^3, the mass rate of energy loss, or the mass stopping power, in Example 5.3 is

$$S = \frac{4.75 \times 10^{-3}\ \text{MeV/cm}}{1.293 \times 10^{-3}\ \text{g/cm}^3} = 3.67\ \frac{\text{MeV}}{\text{g/cm}^2}.$$

Linear energy transfer. The term specific ionization is used when attention is focused on the energy lost by the radiation. When attention is focused on the absorbing medium, as is the case in radiobiology and radiation effects, we are interested in the linear rate of energy absorption by the absorbing medium as the ionizing particle traverses the medium. As a measure of the rate of energy absorption, we use the *linear energy transfer*, abbreviated LET, which is defined by the equation

$$\text{LET} = \frac{dE_L}{dl}, \tag{5.9}$$

where dE_L is the average energy locally imparted to the absorbing medium by a charged particle of specified energy in traversing a distance of dl. In health physics and radiobiology, LET is usually expressed in units of keV per micron. As used in the definition above, the term "locally imparted" may refer either to a maximum distance from the track of the ionizing particle or to a maximum value of discrete energy loss by the particle beyond which losses are no longer considered local. In either case, LET refers to energy imparted within a limited volume of absorber.

Relative mass stopping power. The relative mass stopping power is used to compare quantitatively the energy absorptive power of different media. It will be shown later that the mass stopping power of different absorbers relative to that of air is important in the practice of health physics. Relative mass stopping power is defined by

$$\rho_m = \frac{S_{\text{medium}}}{S_{\text{air}}}. \tag{5.10}$$

Example 5.4

What is the relative (to air) mass stopping power of graphite, density = 2.25 g/cm^3, for a 0.1-MeV beta particle?

The mass rate of energy loss in graphite is found by substituting the appropriate values into equations (5.6) and (5.8):

$$NZ = \frac{6.03 \times 10^{23}\ \text{atoms/mole} \times 2.25\ \text{g/cm}^3 \times 6\ \text{electrons/atom}}{12\ \text{g/mole}}$$

$$= 6.77 \times 10^{23}\ \text{electrons/cm}^3$$

$$I = 1.35 \times 10^{-5} \times 6 = 8.1 \times 10^{-5},$$

therefore

$$S_{\text{graphite}} = 3.85\ \frac{\text{MeV}}{\text{g/cm}^2}.$$

For air, the mass stopping power is 3.67 MeV/g/cm^2. From equation (5.10), the relative mass stopping power of graphite for a 0.1-MeV electron is

$$\rho_m = \frac{3.85}{3.67} = 1.05.$$

Bremsstrahlung

Bremsstrahlung are X-rays that are emitted when high-speed charged particles suffer rapid acceleration. When a beta particle passes close to a nucleus, the strong attractive coulomb force causes the beta particle to deviate sharply from its original path. The change in direction is due to radial acceleration, and the beta particle, in accordance with classical theory, loses energy by electromagnetic radiation at a rate proportional to the square of the acceleration. This means that the bremsstrahlung photons have a continuous energy distribution that ranges downward from a theoretical maximum equal to the kinetic energy of the beta particle. The exact shape of the bremsstrahlung spectrum, as well as the intensity of bremsstrahlung resulting from beta-ray absorption in any given configuration of beta source and absorber, is difficult to calculate and relatively simple to measure. For example, the bremsstrahlung dose rate at a distance of 10 cm from an aqueous solution of 4×10^9 Bq (~ 100 mCi) ^{32}P in a 25-ml volumetric flask is about 0.03 mGy/h (3 mrad/h) for 4×10^9 Bq (~ 100 mCi) ^{90}Sr in a small brass container, the bremsstrahlung dose rate at a distance of 10 cm is about 1 mGy/h (100 mrad/h). (The mGy and mrad will be formally introduced in the next chapter. At this point, it is sufficient to know that they are units for measuring radiation dose.)

For purposes of *estimating* the bremsstrahlung hazard, the following approximate relationship may be used:

$$f = 3.5 \times 10^{-4} ZE, \tag{5.11}$$

where f = the fraction of the incident beta energy converted into photons,
Z = atomic number of the absorber,
E = maximum energy of the beta particle, MeV.

Because the likelihood of bremsstrahlung production increases with atomic number of the absorber, beta-ray shields are made with material of the minimum practicable atomic number. In practice, beta-ray shields of higher atomic number than 13, aluminum, are seldom if ever used.

Example 5.5

A very small source (physically) of 3.7×10^{10}Bq (1 Ci) of ^{32}P is inside a lead shield just thick enough to prevent any beta particles from emerging. What is the bremsstrahlung flux at a distance of 10 cm from the source?

Since Z for lead is 82, and the maximum energy of the ^{32}P beta particle is 1.71 MeV, we have, from equation (5.11), the fraction of the beta energy converted into photons

$$f = 3.5 \times 10^{-4} \times 82 \times 1.71 = 0.049.$$

Since the average beta-ray energy is about one-third of the maximum energy, the energy in the beta particles incident on the shield, E_β, is

$$E_\beta = \tfrac{1}{3}E \times 3.7 \times 10^{10} \text{ MeV/sec}.$$

For health physics purposes, it is assumed that all the bremsstrahlung photons are of the maximum energy. The flux, ϕ, of bremsstrahlung photons at a distance r cm from a beta source of 3.7×10^{10} Bq (1 Ci) therefore is

$$\begin{aligned}\phi &= \frac{fE_\beta}{4\pi r^2 E} \qquad (5.12)\\ &= \frac{0.049 \times \frac{1}{3} \times 1.71 \text{ MeV}/\beta \times 3.7 \times 10^{10}\ \beta/\text{sec}}{4\pi \times 10^2 \text{ cm}^2 \times 1.71 \text{ MeV/photon}}\\ &= 4.8 \times 10^5 \text{ photons/cm}^2\text{/sec.}\end{aligned}$$

Alpha Rays

Range–Energy Relationship

Alpha rays are the least penetrating of the radiations. In air, even the most energetic alphas from radioactive substances travel only several centimeters, while in tissue, the range of alpha radiation is measured in mircons ($1\ \mu = 10^{-4}$ cm). The term range , in the case of alpha particles, may have two different definitions: mean range and extrapolated range. The difference between these two ranges can be seen in the alpha-particle absorption curve, Fig. 5.8.

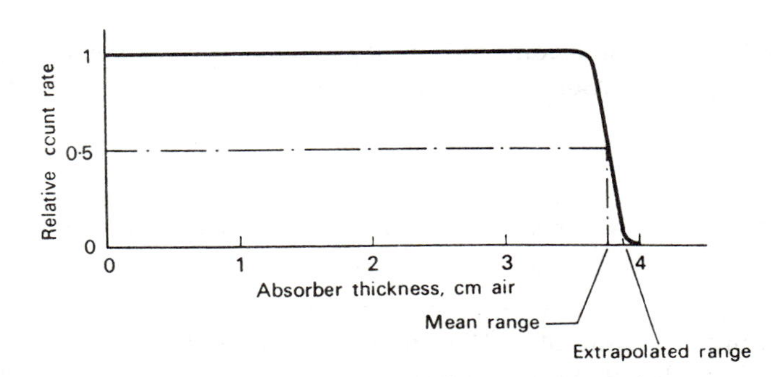

FIG. 5.8. Alpha-particle absorption curve.

An alpha-particle absorption curve is flat because alpha radiation is essentially monoenergetic. Increasing thickness of absorbers serves merely to reduce the energy of the alphas that pass through the absorbers; the number of alphas is not reduced until the approximate range is reached. At this point, there is a sharp decrease in the number of alphas that pass through the absorber. Near the very end of the curve, absorption rate decreases due to straggling, or the combined effects of the statistical distribution of the "'"average" energy loss per ion and the scattering by the absorber nuclei. The mean range is the range most accurately determined, and corresponds to the range of the "average" alpha particle. The extrapolated range is obtained by extrapolating the absorption curve to zero alpha particles transmitted.

Air is the most commonly used absorbing medium for specifying range–energy relationships of alpha particles. For energies less than 4 MeV, and for 4–8 MeV, the range in air at 0°C and 760 mm pressure is closely approximated (within 10%) by the equations

$$R \text{ cm} = 0.56E \text{ MeV} \qquad \text{for } E < 4 \text{ MeV}, \qquad (5.13)$$

$$R \text{ cm} = 1.24E \text{ MeV} - 2.62 \quad \text{for } 4 < E < 8 \text{ MeV}. \qquad (5.14)$$

The range of alpha particles in any other medium may be computed from the following relationship

$$R_m, \text{mg/cm}^2 = 0.56A^{1/3} R, \tag{5.15}$$

where A = atomic number of the medium, and R = range of the alpha particle in air, cm.

Example 5.6

What thickness of aluminum foil, density 2.7 g/cm³, is required to stop the alpha particles from ^{210}Po?

The energy of the ^{210}Po alpha particle is 5.3 MeV. From equation (5.14), the range of the alpha particle in air is

$$R = 1.24 \times 5.3 - 2.62 = 3.95 \text{ cm}.$$

Substituting this value for R into equaltion (5.15), and 27 for the atomic weight, A, we have

$$R_m = 0.56 \times 27^{1/3} \times 3.95 = 6.64 \text{ mg/cm}^2.$$

The thickness of this foil, in centimeters, is calculated from equation (5.1), and is found to be 0.00246 cm.

Because the effective atomic composition of tissue is not very much different from that of air, the following relationship may be used to calculate the range of alpha particles in tissue:

$$R_a \times \rho_a = R_T \times \rho_t, \tag{5.16}$$

where R_a and R_t = range in air and tissue, ρ_a and ρ_t = density of air and tissue.

Example 5.7

What is the range in tissue of a ^{210}Po alpha particle?

The range in air of this alpha was found in Example 5.6 to be 3.95 cm. Assuming tissue to have unit density, the range in tissue is, from equation (5.16),

$$R_t = \frac{3.95 \text{ cm} \times 1.293 \times 10^{-3} \text{ g/cm}^3}{1 \text{ g/cm}^3} = 5.1 \times 10^{-3} \text{ cm}.$$

Energy Transfer

The major energy loss mechanism for alpha particles, and the only one considered significant in health physics is electronic excitation and ionization. In passing through air or soft tissue, an alpha particle loses, on the average, 35 eV per ion pair that it creates. Because of its high electrical charge and relatively low velocity due to its great mass, the specific ionization of an alpha particle is very high, on the order of tens of thousands of ion pairs per centimeter in air, Fig. 5.9.

The linear rate of energy loss for all charged particles more massive than an electron is

$$\frac{\mathrm{d}E}{\mathrm{d}x} = \frac{4\pi z^2 q^4 NZ \times (3 \times 10^9)^4}{Mv^2 \times 1.6 \times 10^{-6}}\left[\ln\frac{2Mv^2}{I} - \ln\left(1 - \frac{v^2}{c^2}\right) - \frac{v^2}{c^2}\right]\frac{\text{MeV}}{\text{cm}}, \tag{5.17}$$

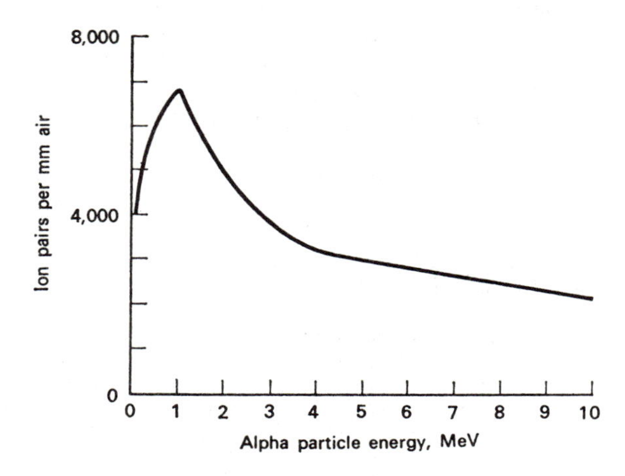

FIG. 5.9. Bragg curve of specific ionization by alpha particles in air at standard temperature and pressure.

where z = atomic number of the ionizing particle,
q = unit electrical charge, 1.6×10^{-19} C,
zq = electrical charge on the ionizing particle,
M = rest mass of the ionizing particle, grams,
v = velocity of the ionizing particle, cm/sec,
N = number of absorber atoms per cm^3,
Z = atomic number of absorber,
NZ = number of absorber electrons per cm^3,
c = velocity of light, 3×10^{10} cm/sec,
I = mean excitation and ionization potential of absorber atoms; for air, $I = 1.38 \times 10^{-10}$ ergs, for other substances, $I = 2.16 \times 10^{-11}Z$.

For the case where the ionizing particle is an alpha particle, $z = 2$ and $M = 6.60 \times 10^{-24}$ g. Equation (5.17) is valid for ionizing particle velocities that are greater than the velocity of the orbital electrons of the absorber.

The mass stopping power of any substance for an alpha particle is defined in the same way as for a beta particle; the same thing is true for the relative mass stopping power of any absorber. Equations (5.8) and (5.10) are therefore used to define these two properties.

Gamma Rays

Exponential Absorption

The attenuation of gamma radiation is qualitatively different from that of either alpha or beta radiation. Whereas both these corpuscular radiations have definite ranges in matter, and therefore can be completely absorbed, gamma radiation can only be reduced in intensity by increasingly thicker absorbers; it can not be completely absorbed. If gamma-ray attenuation measurements are made under conditions of *good geometry*, that is, with a well collimated, narrow beam of radiation, as shown in Fig. 5.10, and if the data are plotted on semi-log paper, a straight line results, as shown in Fig. 5.11, if the gamma

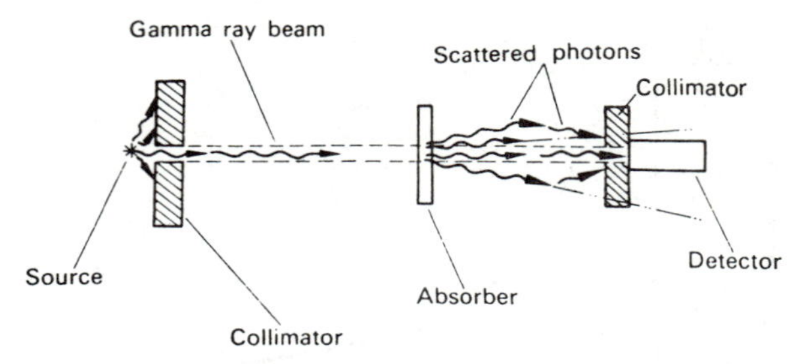

FIG. 5.10. Measuring attenuation of gamma rays under conditions of good geometry. Ideally, the beam should be well collimated, and the source should be as far away as possible from the detector; the absorber should be midway between the source and the detector, and it should be thin enough so that the likelihood of a second interaction between a photon already scattered by the absorber and the absorber is negligible; and there should be no scattering material in the vicinity of the detector.

rays are monoenergetic. If the gamma-ray beam is heterochromatic, a curve results, as shown by the dotted line in Fig. 5.11.

The equation of the straight line in Fig. 5.11 is

$$\ln I = -\mu t + \ln I_0 \tag{5.18A}$$

or

$$\ln I/I_0 = -\mu t. \tag{5.18B}$$

Taking the anti-logs of both sides of equation (5.18), we have

$$I/I_0 = e^{-\mu t}, \tag{5.19}$$

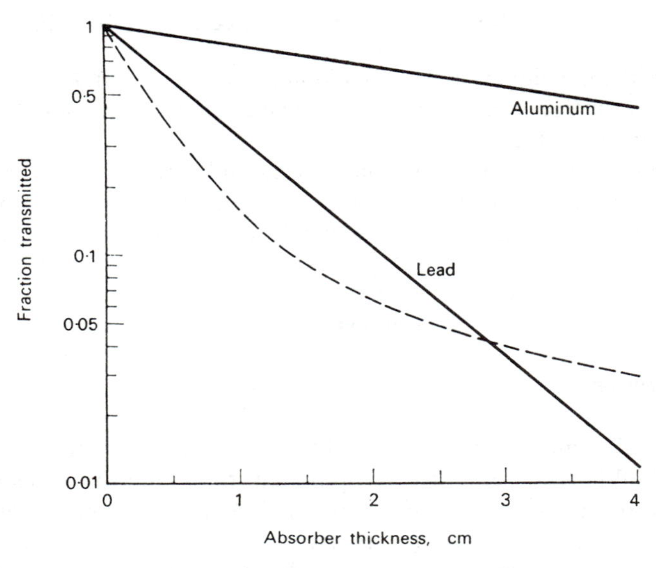

FIG. 5.11. Attenuation of gamma rays under conditions of good geometry. The solid lines are the attenuation curves for 0.662 MeV (monoenergetic) gamma rays. The dotted line is the attenuation curve for a heterochromatic beam.

where I_0 = gamma-ray intensity at zero absorber thickness,
t = absorber thickness,
I = gamma-ray intensity transmitted through an absorber of thickness t,
e = base of the natural logarithm system,
μ = slope of the absorption curve = the attenuation coefficient.

Since the exponent in an exponential equation must be dimensionless, μ and t must be in reciprocal dimensions; that is, if the absorber thickness is measured in centimeters, then the attenuation coefficient is called the *linear attenuation coefficient*, μ_l, and must have dimensions of "per cm". If t is in g/cm^2, then the absorption coefficient is called the *mass attenuation coefficient*, μ_m, and must have dimensions of $(g/cm^2)^{-1}$, or cm^2/g. The numerical relationship between μ_l and μ_m is given by the equation

$$\mu_l \text{ cm}^{-1} = \mu_m \text{ cm}^2/\text{g} \times \rho \text{ g/cm}^3, \tag{5.20}$$

where ρ is the density of the absorber.

The attenuation coefficient is the fraction of the gamma-ray beam attenuated per unit thickness of absorber, as defined by the equation below:

$$\lim_{\Delta t \to 0} \frac{\Delta I/I}{\Delta t} = -\mu, \tag{5.21}$$

where $\Delta I/I$ is the fraction of the gamma-ray beam attenuated by an absorber of thickness Δt. The attenuation coefficient thus defined is sometimes called the *total attenuation coefficient*. Values of the attenuation coefficients for several materials are given in Table 5.2.

For some purposes, it is useful to use the *atomic attenuation coefficient*, μ_a. The atomic attenuation coefficient is the fraction of an incident gamma-ray beam that is attenuated by a single atom. Another way of saying the same thing is that the atomic attenuation coefficient is the probability that an absorber atom will interact with one of the photons in the beam. The atomic attenuation coefficient may be defined by the equation

$$\mu_a \text{ cm}^2 = \frac{\mu_l \text{ cm}^{-1}}{N \text{ atoms/cm}^3}, \tag{5.22}$$

where N = the number of absorber atoms per cm^3. Note that the dimensions of μ_a are cm^2, the units of area. For this reason, the atomic attenuation coefficient is almost always referred to as the "cross section" of the absorber. The unit in which the cross section is specified is the *barn*.

$$1 \text{ barn} = 10^{-24} \text{ cm}^2.$$

The atomic attenuation coefficient is also called the microscopic cross section, and is symbolized by σ, while the linear absorption coefficient is often called the *macroscopic cross section*, and is given the symbol Σ. This nomenclature is almost always used when dealing with neutrons. Equation (5.22) can thus be written as

$$\Sigma \text{ cm}^{-1} = \sigma \frac{\text{cm}^2}{\text{atom}} \times N \frac{\text{atoms}}{\text{cm}^3}. \tag{5.23}$$

Using the relationship given in equation (5.22), equation (5.19) may be rewritten as

$$I/I_0 = e^{-\mu_a N t}, \tag{5.24A}$$

or

$$I/I_0 = e^{-\sigma N t}. \tag{5.24B}$$

TABLE 5.2. LINEAR ATTENUATION COEFFICIENTS, CM^{-1}

	ρ, g/cm^3	Quantum energy, MeV												
		0.1	0.15	0.2	0.3	0.5	0.8	1.0	1.5	2	3	5	8	10
C	2.25	0.335	0.301	0.274	0.238	0.196	0.159	0.143	0.117	0.100	0.080	0.061	0.048	0.044
Al	2.7	0.435	0.362	0.324	0.278	0.227	0.185	0.166	0.135	0.117	0.096	0.076	0.065	0.062
Fe	7.9	2.72	1.445	1.090	0.838	0.655	0.525	0.470	0.383	0.335	0.285	0.247	0.233	0.232
Cu	8.9	3.80	1.830	1.309	0.960	0.730	0.581	0.520	0.424	0.372	0.318	0.281	0.270	0.271
Pb	11.3	59.7	20.8	10.15	4.02	1.64	0.945	0.771	0.579	0.516	0.476	0.482	0.518	0.552
Air	1.29×10^{-3}	1.95×10^{-4}	1.73×10^{-4}	1.59×10^{-4}	1.37×10^{-4}	1.12×10^{-4}	9.12×10^{-5}	8.45×10^{-5}	6.67×10^{-5}	5.75×10^{-5}	4.6×10^{-5}	3.54×10^{-5}	2.84×10^{-5}	2.61×10^{-5}
H_2O	1	0.167	0.149	0.136	0.118	0.097	0.079	0.071	0.056	0.049	0.040	0.030	0.024	0.022
Concrete	2.35	0.397	0.326	0.291	0.251	0.204	0.166	0.149	0.122	0.105	0.085	0.067	0.057	0.054

Ordinary concrete of the following composition: 0.56% H, 49.56%O, 31.35% Si, 4.56% Al, 8.26% Ca, 1.22% Fe, 0.24% Mg, 1.71% Na, 1.92% K, 0.12% S.
From National Bureau of Standards Report No, 1003 (1952).

The numerical values for μ_a have been published for many elements and for a wide range of quantum energies.*

With the aid of atomic cross sections, it is possible to compute the attenuation coefficient of an alloy or a compound containing several different elements.

Example 5.8

Aluminum bronze, an alloy containing 90% Cu (atomic weight = 63.57) and 10% Al (atomic weight = 27) by weight, has a density of 7.6 g/cm³. What are the linear and mass attenuation coefficients for 0.4-MeV gamma rays, if the cross sections for Cu and Al for this quantum energy are 9.91 and 4.45 barns?

From equation (5.22) the linear attenuation coefficient of aluminum bronze is

$$\mu_l = (\mu_a)_{Cu} \times N_{Cu} + (\mu_a)_{Al} \times N_{Al}.$$

The number of Cu atoms per cm³ in the alloy is

$$N_{Cu} = \frac{6.03 \times 10^{23} \text{ atoms/mole}}{63.57 \text{ g/mole}} (7.6 \times 0.9) \text{ g/cm}^3 = 6.49 \times 10^{22} \text{ atoms/cm}^3,$$

and for aluminum, N_{Al} is 1.7×10^{22} atoms/cm³. The linear attenuation coefficient therefore is

$$\begin{aligned}\mu_l &= 9.9 \times 10^{-24} \text{ cm}^2 \times 6.49 \times 10^{22} \text{ atoms/cm}^3 + 4.45 \times 10^{-24} \text{ cm}^2 \\ &\quad \times 1.7 \times 10^{22} \text{ atoms/cm}^3, \\ &= 0.705 \text{ cm}^{-1}.\end{aligned}$$

The mass attenuation coefficient is, from equation (5.20)

$$\mu_m = \frac{0.705 \text{ cm}^{-1}}{7.6 \text{ g/cm}^3} = 0.0927 \text{ cm}^2/\text{g}.$$

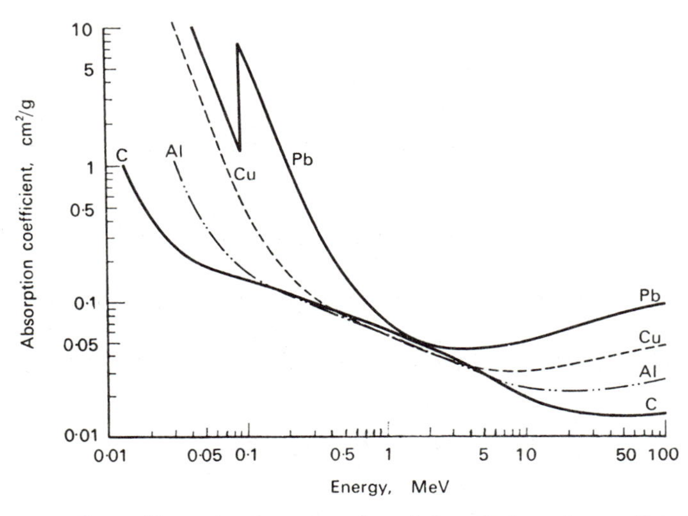

FIG. 5.12. Curves illustrating the systematic variation of attenuation coefficient with atomic number of absorber and with quantum energy.

* Gladys White Groodstein, *X-ray Attenuation Coefficients from 10 KeV to 100 MeV*. NBS Circular 583. U.S. Government Printing Office, Washington, 1957.

The attenuating properties of matter vary systematically with the atomic number of the absorber and with the energy of the gamma radiation, as shown in Fig. 5.12. It should be noted, however, that in the region where the Compton effect (this effect is more fully discussed below) predominates, the mass attenuation coefficient is almost independent of the atomic number of the absorber.

Example 5.9

(a) Compute the thickness of aluminum and lead to transmit 10% of a narrow beam of 0.1-MeV gamma radiation.

From Table 5.2, μ_l for Al is 0.435 cm^{-1} and for lead it is 59.7. From equation (5.19) we have for aluminum

$$\frac{1}{10} = e^{-(0.435\ \text{cm}^{-1})(t\ \text{cm})},$$

$$\ln 10 = 0.435t,$$

$$\frac{2.3}{0.435} = t = 5.3\ \text{cm}.$$

In a similar manner, we have for lead

$$\frac{1}{10} = e^{-(59.7\ \text{cm}^{-1})(t\ \text{cm})},$$

$$t = 0.0385\ \text{cm}.$$

(b) Repeat part (a) for a 1.0 MeV-gamma ray. μ_l for Al = 0.166, μ_l for Pb = 0.771. For Al we have

$$\frac{1}{10} = e^{-(0.166\ \text{cm}^{-1})(t\ \text{cm})},$$

$$t = 13.86\ \text{cm},$$

and for lead,

$$\frac{1}{10} = e^{-(0.771\ \text{cm}^{-1})(t\ \text{cm})},$$

$$t = 2.97\ \text{cm}.$$

(c) Compare the density thicknesses of the Al and Pb in each part of the illustrative example above.

The density thickness for the Al in the case of the 0.1 MeV photon is, from equation (5.1)

$$t_{d\text{Al}} = 2.7\ \text{g/cm}^3 \times 5.3\ \text{cm} = 14.3\ \text{g/cm}^2,$$

and for lead,

$$t_{d\text{Pb}} = 11.34\ \text{g/cm}^3 \times 0.0385\ \text{cm} = 0.435\ \text{g/cm}^2.$$

In the case of the 1.0-MeV quantum energy, the density thickness for the Al is computed as 37.4 g/cm^2 and for lead it is 33.6 g/cm^2.

Example 5.9 shows that for high-energy gamma ray, lead is only a slightly better absorber, on a mass basis, than aluminum. For the low-energy photon, on the other hand, lead is a very much better absorber than aluminum. Generally, for energies between about 0.75 and 5 MeV almost all materials have, on a mass basis, about the same gamma-ray attenuating properties. To a first approximation in this energy range, therefore, shielding properties are approximately proportional to the density of the shielding material. For lower and higher quantum energies, absorbers of high atomic number are more effective than those of low atomic number. To understand this behavior, we must examine the microscopic mechanisms of the interaction between gamma rays and matter.

Interaction Mechanisms

For radiation-protection purposes, four major mechanisms for the interaction of gamma-ray energy are considered significant. Two of these mechanisms, photoelectric absorption and Compton scattering, which involve interactions only with the orbital electrons of the absorber, predominate in the case where the quantum energy of the photons does not greatly exceed 1.02 MeV, the energy equivalent of the rest mass of two electrons. In the case of higher energy photons, pair production, which is a direct conversion of electromagnetic energy into mass, occurs. These three gamma-ray interaction mechanisms result in the emission of electrons from the absorber. Very high energy photons, $E \gg 2m_0c^2$, may also be absorbed into the nuclei of the absorber atoms, and then initiate nuclear reactions which result in the emission, from the excited nuclei, of other radiations.

Pair production

A photon whose energy exceeds 1.02 MeV may, as it passes near a nucleus, spontaneously disappear, and its energy may reappear as a positron and an electron, as pictured in Fig. 5.13. Each of these two particles has a mass of m_0c^2, or 0.51 MeV, and the total kinetic energy is nearly equal to $hf - 2m_0c^2$. This transformation of energy into mass must take place near a particle, such as a nucleus, in order that momentum be conserved. The

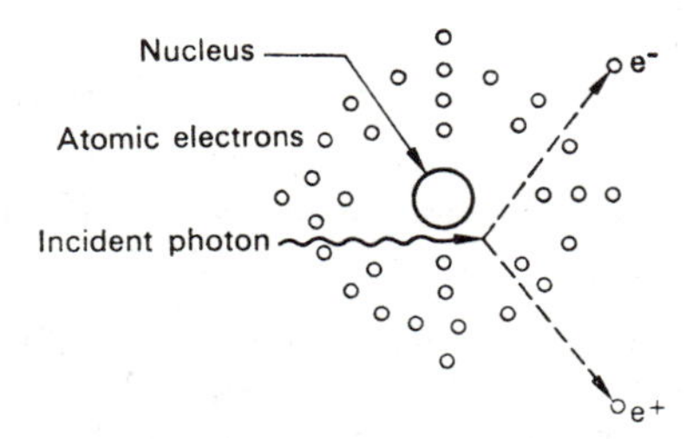

FIG. 5.13. Schematic representation of pair production. The positron–electron pair is generally projected in the forward direction (relative to the direction of the photon). The degree of forward projection increases with increasing photon energy.

kinetic energy of the recoiling nucleus is very small. For practical purposes, therefore, all the photon energy in excess of that needed to supply the mass of the pair appears as kinetic energy of the pair. This same phenomenon may also occur in the vicinity of an electron, but the probability of occurrence near a nucleus is very much greater. Furthermore, the threshold energy for pair production near an electron is $4m_0c^2$. This higher threshold energy is necessary because the recoil electron, which conserves momentum, must be projected back with a very high velocity, since its mass is the same as that of each of the newly created particles. The cross section, or probability of the production of a positron–electron pair, is approximately proportional to $Z^2 + Z$, and is therefore important for high-atomic-numbered absorbers. The cross section increases slowly with increasing energy between the threshold of 1.02 MeV and about 5 MeV. For higher energies, the cross section is proportional to the logarithm of the quantum energy. It is this increasing cross section that accounts for the increasing attenuation coefficient for high energy photons.

After production of a pair, the positron and electron are projected in a forward direction (relative to the direction of the photon) and lose their kinetic energy by excitation, ionization, and bremsstrahlung, as any other high-energy electron. When the positron has expended all of its kinetic energy, it combines with an electron to produce two quanta of 0.51 MeV each of annihilation radiations. Thus, a 10-MeV photon may, in passing through a lead absorber, be converted into a positron–electron pair in which each particle has about $4\frac{1}{2}$ MeV of kinetic energy. This kinetic energy is then dissipated in the same manner as beta particles. The positron then is annihilated by combining with an electron in the absorber, and two photons of 0.51 MeV each may emerge from the absorber (or they may undergo Compton scattering or photoelectric absorption). The net result of the pair production interaction in this case was the conversion of a single 10 MeV photon into two photons of 0.51 MeV each, and the dissipation of 8.98 MeV of energy.

Compton scattering

Compton scattering is an elastic collision between a photon and a "free" electron (a "free" electron is one whose binding energy to an atom is very much less than the energy of the photon), as shown diagramatically in Fig. 5.14.

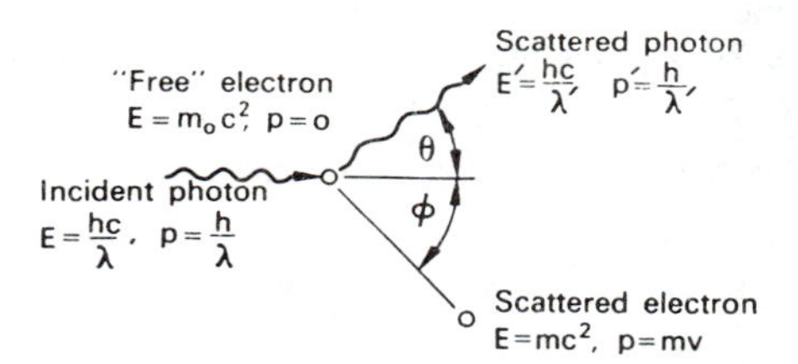

FIG. 5.14. Compton scattering: an elastic collision between a photon and an electron.

In a collision between a photon and a free electron, it is impossible for all the photon's energy to be transferred to the electron if momentum and energy are to be conserved. This can be shown by assuming that such a reaction is possible. If this were true, then, according to the conservation of energy, all the enrgy of the photon is imparted to the electron, and we have

$$E = mc^2.$$

According to the law of conservation of momentum, all the momentum of the photon, p, must be transferred to the electron if the photon is to disappear:

$$p = \frac{E}{c} = mv.$$

Eliminating m from these two equations and solving for v, we find $v = c$, an impossible solution. The original assumption, that the photon transferred all of its energy to the electron, must therefore be false.

Since all the photon's energy cannot be transferred, the photon must be scattered, and the scattered photon must have less energy—or a longer wavelength—than the incident photon. Only the energy difference between the incident and scattered photons is transferred to the free electron. The amount of energy transferred in any collision can be calculated by applying the laws of conservation of energy and momentum to the situation pictured in Fig. 5.14. To conserve energy, we must have

$$\frac{hc}{\lambda} + m_0c^2 = \frac{hc}{\lambda'} + mc^2, \tag{5.25}$$

and to conserve momentum in the horizontal and vertical directions respectively, we have

$$\frac{h}{\lambda} = \frac{h}{\lambda'} \cos\theta + mv \cos\phi, \tag{5.26}$$

$$0 = \frac{h}{\lambda'} \sin\theta - mv \sin\phi. \tag{5.27}$$

The solution of these equations shows the change in wavelength of the photon to be

$$\Delta\lambda = \lambda' - \lambda = \frac{h}{m_0c}(1 - \cos\theta) \text{ cm}, \tag{5.28}$$

and the relation between the scattering angles of the photon and the electron to be

$$\cot\frac{\theta}{2} = \left(1 + \frac{h}{\lambda m_0 c}\right)\tan\phi. \tag{5.29}$$

When the numerical values are substituted for the constants, and centimeters are converted into angstrom units, equation (5.28) reduces to

$$\Delta\lambda = 0.0242(1 - \cos\theta) \text{ Å}. \tag{5.30}$$

Equation (5.29) shows that the electron cannot be scattered through an angle greater than 90°. This scattered electron is of great importance in radiation dosimetry, because it is the vehicle by means of which energy from the scattered photon is transferred to an absorbing medium. The Compton electron dissipates its kinetic energy in the same manner as a beta particle, and is one of the primary ionizing particles produced by gamma radiation. Compton scattering is also important in health physics engineering because of the fact that a high-energy photon loses a greater fraction of its energy when it is scattered than does a low-energy photon. By taking advantage of this fact, required shielding thickness can be reduced and economic savings thereby effected.

Example 5.10

What fractions of their energies do 1 MeV and 0.1 MeV photons lose if they are scattered through an angle of 90°?

Substituting $E = hc/\lambda$ into equation (5.28), and solving for the energy of the scattered photon, we have

$$E' = \frac{E}{1 + (E/m_0c^2)(1 - \cos\theta)}, \tag{5.31A}$$

and the fraction of the incident energy carried by the scattered photon is

$$\frac{E'}{E} = \frac{1}{1 + (E/m_0c^2)(1 - \cos\theta)}. \tag{5.31B}$$

Substituting the values for the incident photon and the scattering angle into equation (5.31A), we have

$$E' = \frac{1 \text{ MeV}}{1 + (1/0.51)(1)} = 0.338 \text{ MeV},$$

and the fractional energy loss is

$$1 - \frac{E'}{E} = \frac{1 - 0.338}{1} = 0.662 = 66.2\%.$$

In the case of 0.1-MeV gamma ray, the energy of the scattered photon is, from equation (5.30), 0.0835 MeV, and the fractional energy loss is only 0.165, or 16.5%.

The probability of a Compton interaction decreases with increasing quantum energy and with increasing atomic number of the absorber. In light atomic numbered elements, Compton scattering is the main mechanism of interaction. In Compton scattering every electron acts as a scattering center, and the bulk scattering properties of matter depends mainly on the electronic density per unit mass. Probabilities for Compton scattering are

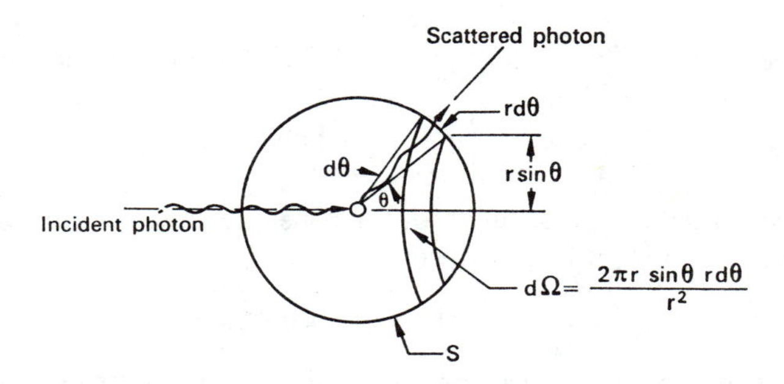

FIG. 5.15. Compton scattering diagram to illustrate differential scattering cross section. S is a sphere of unit radius whose center is the scattering electron.

therefore given on a per-electron basis. The theoretical cross sections for Compton scattering were derived by Klein and Nishina. For scattering into a differential solid angle $d\Omega$ at an angle θ to the direction of the incident photon, Fig. 5.15, they give the differential total scattering coefficient as

$$\frac{d\sigma_t}{d\Omega} = \frac{e^4}{2m_0^2c^4}\left[\frac{1}{1 + a(1 - \cos\theta)}\right]^2\left[\frac{1 + \cos^2\theta + a^2(1 - \cos\theta)^2}{1 + a(1 - \cos\theta)}\right], \tag{5.32}$$

where e, m_0, and c have the usual meaning, and $a = hf/m_0c^2$. Equation (5.32) and Fig. 5.16 give the probability of scattering a photon into a solid angle $d\Omega$ through an angle θ. The total probability of scattering, Ω_t, can be obtained by substituting $d\Omega = 2\pi \sin\theta \, d\theta$ and integrating the differential scattering coefficient over the entire sphere. The result of this calculation, for quantum energies up to 10 MeV, is presented graphically in Fig. 5.17.

The cross section for any reaction, wehn viewed from the point of view of an individual particle or a single photon, is the probability that the photon will undergo that reaction.

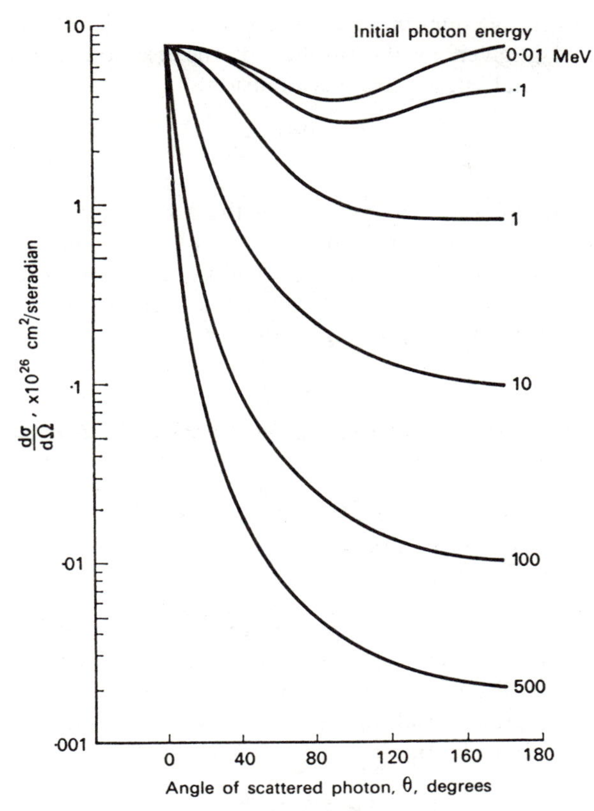

FIG. 5.16. Differential scattering coefficient showing the probable angular distribution of Compton scattered photons.

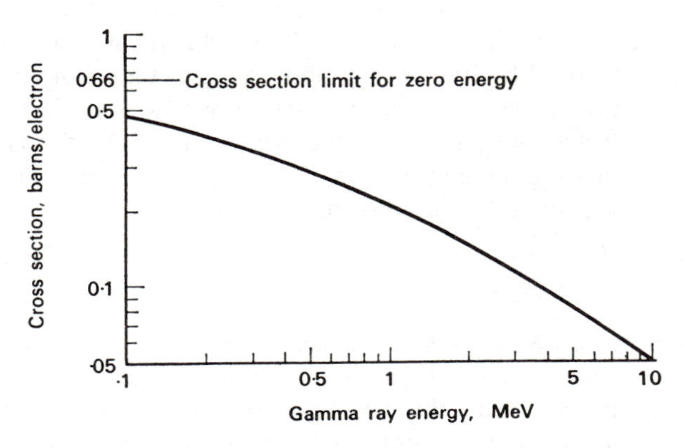

FIG. 5.17. Total Compton cross sections for a free electron.

From the point of view of a beam of radiation, the cross section gives the fraction of the particles in the beam that reacts in a given manner. In Compton scattering, energy is transferred from the photon to the scattered electron, and the scattered photon has less energy than the incident photon. The fraction of the incident energy that is carried by the scattered photon is given by equation (5.13B). The fraction of the energy in a gamma-ray beam that is scattered is given by the product of the fraction of the photons that are scattered and the fractional energy loss per collision:

$$\frac{d\sigma_s}{d\Omega} = \frac{d\sigma_t}{d\Omega} \cdot \frac{E'}{E}. \tag{5.33}$$

Of interest to the health physicist in calculating the absorbed dose from X- or gamma radiation is the fraction of the energy in the incident beam that is transferred to the Compton electron. This energy represents the energy absorbed from the beam due to Compton scattering. The Compton energy absorption cross section, σ_e, is merely the difference between the total and scattering cross sections

$$\sigma_e = \sigma_t - \sigma_s, \tag{5.34A}$$

and the Compton energy absorption coefficient, μ_{ce}, is the difference between the Compton total and scattering coefficients

$$\mu_{ce} = \mu_{ct} - \mu_{cs}. \tag{5.34B}$$

Photoelectric absorption

The photoelectric effect, in which the photon disappears, is an interaction between a photon and a tightly bound electron whose binding energy is equal to or less than the energy of the photon, as discussed in Chapter 3.

The primary ionizing particle resulting from this interaction is the photoelectron, whose energy is given by equation (3.13),

$$E_{pe} = hf - \phi.$$

The photoelectron dissipates its energy in the absorbing medium mainly by excitation and ionization. The binding energy ϕ is transferred to the absorber by means of the fluorescent radiation that follows the initial interaction. These low-energy photons are absorbed by outer electrons, in other photoelectric interactions, not far from their points of origin. The photoelectric effect is favored by low-energy photons and high-atomic-numbered absorbers. The cross section for this reaction varies approximately as $Z^4\lambda^3$. It is this very strong dependence of photoelectric absorption on the atomic number, Z, that makes lead such a good material for shielding against X-rays. For very low atomic numbered absorbers, the photoelectric effect is relatively unimportant.

Photodisintegration

In photodisintegration, the absorber nucleus captures a gamma ray and, in most instances, emits a neutron. This is a threshold reaction in which the quantum energy must exceed a certain minimum value that depends on the absorbing nucleus. This is a high-energy reaction, and with few exceptions is not an absorption mechanism for gamma rays from radiosotopes. An important exception is the case of ^{9}Be, in which the threshold energy is only 1.666 MeV. The reaction $^9\text{Be}(\gamma, n)^8\text{Be}$ is useful as a laboratory source of monoenergetic neutrons. Photodisintegration is an important reaction in the case of very high energy photons from electron accelerators such as betatrons and synchrotrons. Here, too, interest is centered on the fact that photodisintegration results in neutron production. Generally, the cross sections for photodisintegration are very much smaller than the total cross section given in equation (5.35). In shielding calculations, therefore, the photodisintegration cross sections are usually considered insignificant, and are neglected.

Photodisintegration is a threshold reaction because the energy added to the absorber nucleus must be at least equal to the binding energy of a nucleon. Furthermore, a neutron is preferentially emitted rather than a proton because it has no coulombic potential barrier to overcome in order to escape from the nucleus, and hence has a lower threshold. The range of energy thresholds for photodisintegration by neutron emission varies from 1.66 MeV for beryllium to about $8\frac{1}{2}$ MeV. For light nuclei, the thresholds fluctuate unsystematically; in the range of atomic mass numbers 20–130, the thresholds increase slowly to about $8\frac{1}{2}$ MeV, and then decreases slowly to about 6 MeV as the atomic mass numbers increase. Quantum energies greater than the threshold appear as kinetic energy of the emitted neutrons or, if great enough, may cause the emission of charged particles from the absorber nucleus.

Combined effects

The attenuation coefficients or cross sections give the probabilities of removal of a photon from a beam under conditions of good geometry, where it is assumed that any of the possible interactions removes the photon from the beam. The total attenuation coefficient, therefore, is the sum of the coefficients for each of the three reactions discussed above:

$$\mu_t = \mu_{pe} + \mu_{cs} + \mu_{pp}, \tag{5.35}$$

where the three right-hand terms are the attenuation coefficients, respectively, for the photoelectric effect, for Compton scattering, and for pair production. In computing attenuation of radiation for purposes of shielding design, the total attenuation coefficient as defined in equation (5.35) is used.

Equation (5.35) gives the fraction of the energy in a beam that is removed, per unit distance in an absorber, by the absorber. The fraction of the beam's energy that is deposited in the absorber considers only the energy transferred to the absorber by the photoelectron, by the Compton electron, and by the electron pair. Energy carried away by the scattered photon in a Compton interaction and the energy carried off by the annihilation radiation after pair production is not included. The *energy absorption coefficient*, which is also called the *true absorption coefficient*, is given by

$$\mu_e = \mu_{pe} + \mu_{ce} + \mu_{pp}\left(\frac{hf - 1.02}{hf}\right), \tag{5.36}$$

and is used in calculation of radiation dose. The total attenuation and true absorption coefficients for air are shown in Fig. 5.18, and the energy absorption coefficients for water, air, compact bone, and muscle are listed in Table 5.3.

Neutrons

Production

The most prolific source of neutrons is the nuclear reactor. The splitting of a uranium or a plutonium nucleus in a nuclear reactor is accompanied by the emission of several

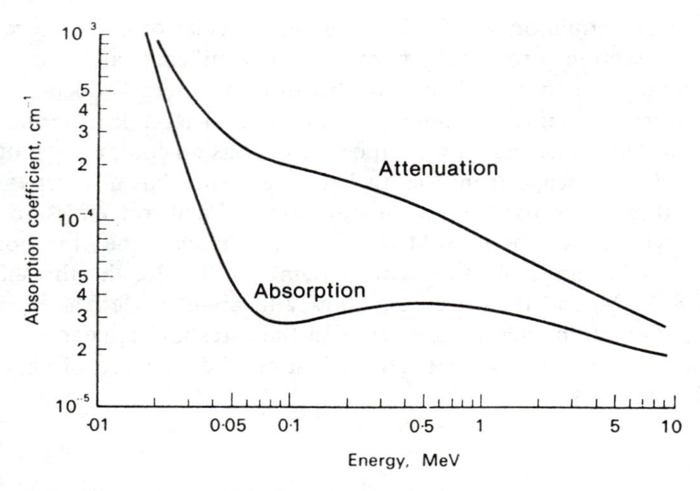

FIG. 5.18. Attenuation coefficients and absorption coefficients for gamma rays as a function of energy.

TABLE 5.3. VALUES OF THE MASS ENERGY-ABSORPTION COEFFICIENTS

Photon energy (MeV)	Mass Energy-absorption Coefficient, (μ_{en}/ρ), cm^2/g			
	Water	Air	Compact bone	Muscle
0.010	4.89	4.66	19.0	4.96
0.015	1.32	1.29	5.89	1.36
0.020	0.523	0.516	2.51	0.544
0.030	0.147	0.147	0.743	0.154
0.040	0.0647	0.0640	0.0305	0.0677
0.050	0.0394	0.0384	0.158	0.0409
0.060	0.0304	0.0292	0.0979	0.0312
0.080	0.0253	0.0236	0.0520	0.0255
0.10	0.0252	0.0231	0.0386	0.0252
0.15	0.0278	0.0251	0.0304	0.0276
0.20	0.0300	0.0268	0.0302	0.0297
0.30	0.0320	0.0288	0.0311	0.0317
0.40	0.0329	0.0296	0.0316	0.0325
0.50	0.0330	0.0297	0.0316	0.0327
0.60	0.0329	0.0296	0.0315	0.0326
0.80	0.0321	0.0289	0.0306	0.0318
1.0	0.0311	0.0280	0.0297	0.0308
1.5	0.0283	0.0255	0.0270	0.0281
2.0	0.0260	0.0234	0.0248	0.0257
3.0	0.0227	0.0205	0.0219	0.0225
4.0	0.0205	0.0186	0.0199	0.0203
5.0	0.0190	0.0173	0.0186	0.0188
6.0	0.0180	0.0163	0.0178	0.0178
8.0	0.0165	0.0150	0.0165	0.0163
10.0	0.0155	0.0144	0.0159	0.0154

From: *Physical Aspects of Irradiation.* NBS Handbook No. 85. Washington, D.C. Supt. of Docs., U.S. Government Printing Office, March 1964.

neutrons. These fission neutrons have a wide range of energies, as shown in Fig. 5.19. The distribution peaks at 0.7 MeV, and has a mean value of 2 MeV.

Except for several fission fragments of very short half-life, there are no radioisotopes that emit neutrons. Californium 252, however, an alpha emitter, undergoes spontaneous nuclear fission at an average rate of 10 fissions for every 313 alpha transformations. Since the half life of ^{252}Cf due to alpha emission is 2.73 years, its effective half-life, including spontaneous nuclear fission, is 2.65 years. Californium 252 thus simulates a neutron-emitting radioisotope. The neutron emission rate has been found to be 2.31×10^6 neutrons per second per microgram ^{252}Cf. The emitted neutrons span a wide range of energies. The most probable energy is about 1 MeV, while the average value of the energy distribution is about 2.3 MeV.

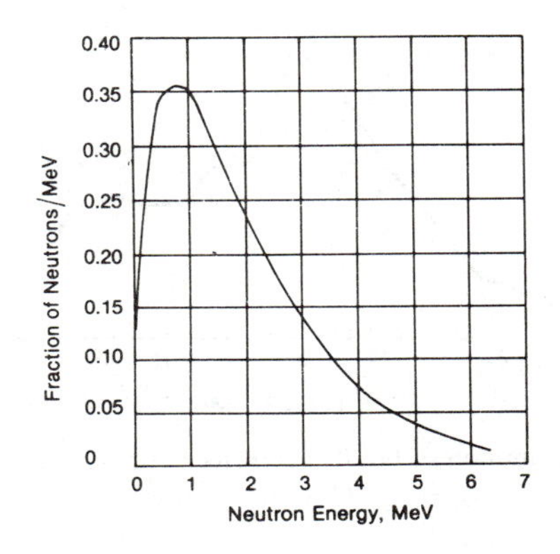

FIG. 5.19. Energy distribution of fission neutrons. The most probable energy is 0.7 MeV and the average energy is 2 MeV.

All other neutron sources depend on nuclear reactions for the emission of neutrons. Copious neutron beams may be produced in accelerators by many different reactions. For example, bombardment of beryllium by high-energy deuterons in a cyclotron produces neutrons according to the reaction

$$ {}^{9}_{4}\mathrm{Be} + {}^{2}_{1}\mathrm{D} \rightarrow ({}^{11}_{5}\mathrm{B})^* \rightarrow {}^{10}_{5}\mathrm{B} + {}^{1}_{0}n. \tag{5.37} $$

The term in the parenthesis is called a *compound nucleus*, and the asterisk shows that it is in an excited state. The compound nucleus rids itself of its excitation energy instantaneously ($<10^{-8}$ sec) by proceeding to the next step in the reaction. For small laboratory sources of neutrons, the photodisintegration of beryllium may be used. Another commonly used neutron source depends on the bombardment of beryllium with alpha particles. The reaction, in this case, is

$$ {}^{9}_{4}\mathrm{Be} + {}^{4}_{2}\mathrm{He} \rightarrow ({}^{13}_{6}\mathrm{C})^* \rightarrow {}^{12}_{6}\mathrm{C} + {}^{1}_{0}n. \tag{5.38} $$

For the source of the alpha particles, radium, polonium, and plutonium are used. The alpha emitter, as a powder, is thoroughly mixed with finely powdered beryllium, and the mixture is sealed in a capsule, as shown in Fig. 5.20. The neutrons that are produced are all high energy. In all cases of neutrons based on this reaction, the neutron energy is spread over a broad spectrum, as shown in Fig. 5.21. This spread of energies from a ^{9}Be $(\alpha, n)^{12}$C source is in sharp contrast to the monoenergetic neutrons from a photodisintegration source using monoenergetic photons. In the α, n reaction, the energy equivalent of the

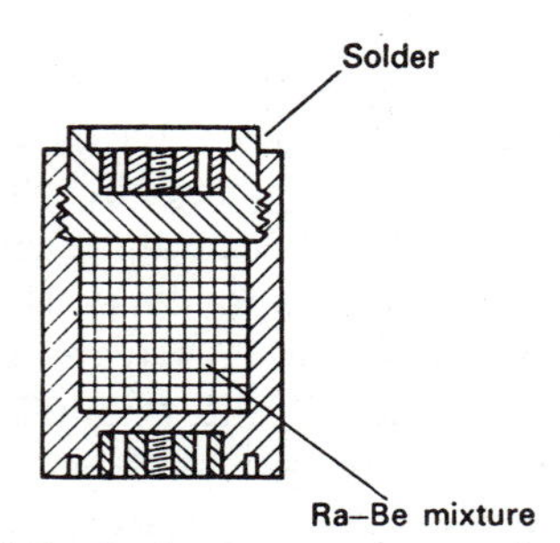

FIG. 5.20. Typical Ra–Be (α, n) neutron source in a sealed container.

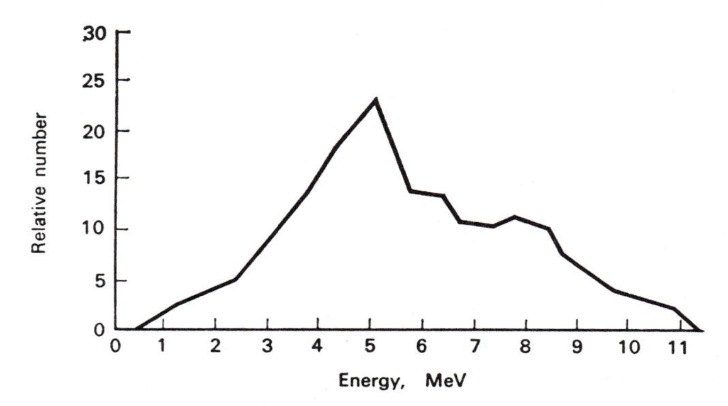

FIG. 5.21. Energy distribution of Po–Be neutrons. (Technical Bulletin NS-2, AECL, Ottawa.)

difference in mass between the reactants and the products plus the kinetic energy of the bombarding particle is divided between the neutron and the recoil nucleus. In practical α, n sources, some of the alpha particle energy is dissipated by self-absorption within the source. As a consequence, the alphas that initiate the reaction have a wide range of energy, thereby contributing to the spectral spread of the neutrons. The neutron yield for an a, n source increases with increasing alpha energy because of the greater ease with which higher-energy alphas penetrate the coulomb barrier at the nucleus. Tables 5.4 and 5.5 list some γ, n and α, n neutron sources, respectively.

Classification

Neutrons are classified according to their energy because the type of reaction that a neutron undergoes depends very strongly on its energy. High-energy neutrons, those whose energies exceed about 0.1 MeV, are called *fast neutrons. Thermal neutrons*, on the other hand, have the same average kinetic energy as gas molecules in their environment. In this respect, thermal neutrons are indistinguishable from gas molecules at the same temperature. The kinetic energies of gas molecules are related to temperature by the Maxwell–Boltzman distribution:

$$f(E) = \frac{2\pi}{(\pi kT)^{3/2}} e^{-E/kT} E^{1/2}, \tag{5.39}$$

where $f(E)$ is the fraction of the gas molecules (or neutrons) of energy E per unit energy interval; k is the Boltzmann constant, 1.38×10^{-23} J/°K or 8.6×10^{-5} eV/°K; and T is the absolute temperature of the gas, °K.

The most probable energy, represented by the peak of the curve in Fig. 5.22., is given by:

$$E_{mp} = kT, \tag{5.40}$$

while the average energy of gas molecules at any given temperature is:

$$\bar{E} = \tfrac{3}{2}kT. \tag{5.41}$$

For neutrons at a temperature of 293°K, the most probable energy is 0.025 eV. This is the energy often implied in the term "thermal" neutrons. The velocity corresponding to this energy, which is given by

$$\tfrac{1}{2}mv^2 = kT, \tag{5.42}$$

is 2200 m/sec.

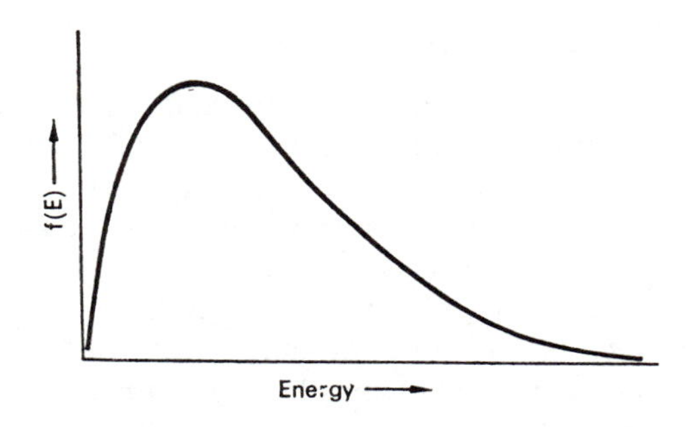

Fig. 5.22. Maxwell–Boltzmann distribution of energy among gas molecules.

In the region of energy between thermal and fast, neutrons are called by various names, including *intermediate neutrons*, *resonance neutrons*, and *slow neutrons*. All these descriptive adjectives are used loosely, and their exact meaning must be inferred from the context in which they are used.

Table 5.4. γ, n Photoneutron Sources

Source	Half-life	Average neutron energy MeV	Yield $\frac{n}{sec}$/Ci	Yield $\frac{n}{sec}$/MBq
$^{24}Na + Be$	15 hr	0.83	1.3×10^5	3.5
$^{24}Na + D_2O$	15 hr	0.22	2.7×10^5	7.3
$^{56}Mn + Be$	2.58 hr	0.1(90%), 0.3(10%)	2.9×10^4	0.8
$^{56}Mn + D_2O$	2.58 hr	0.22	3.1×10^3	0.08
$^{72}Ga + Be$	14.2 hr	0.78	5×10^4	1.4
$^{72}Ga + D_2O$	14.2 hr	0.13	6×10^4	1.6
$^{88}Y + Be$	88 d	0.16	1×10^5	2.7
$^{88}Y + D$	88 d	0.31	3×10^3	0.08
$^{116}In + Be$	54 min	0.30	8.2×10^3	0.2
$^{124}Sb + Be$	60 d	0.024	1.9×10^5	5.1
$^{140}La + Be$	40 hr	0.62	3×10^3	0.08
$^{140}La + D_2O$	40 hr	0.15	8×10^3	0.2
$Ra + D_2O$	1600 yr	0.12	1×10^3	0.03

TABLE 5.5. α, n NEUTRON SOURCES

Source	Half-life	Average neutron energy MeV	Yield	
			$\frac{n}{\text{sec}}/\text{Ci}$	$\frac{n}{\text{sec}}/\text{MBq}$
Ra + Be	1600 yr	5	1.7×10^7	459
Ra + B	3.8 d	3	6.8×10^6	184
^{222}Em + Be	3.8 d	5	1.5×10^7	405
^{210}Po + Be	138 d	4	3×10^6	81.1
^{210}Po + B	138 d	2.5	9×10^5	24.3
^{210}Po + F	138 d	1.4	4×10^5	10.8
^{210}Po + Li	138 d	0.42	9×10^4	2.4
^{239}Pu + Be	24,000 yr	4	10^6	27

Interaction

All neutrons, at the time of their birth, are fast. Generally, fast neutrons lose energy by colliding elastically with atoms in their environment, and then, after being slowed down to thermal or near thermal energies, they are captured by nuclei of the absorbing material. Although a number of possible neutron reaction types exist, for the health physicist the chief reactions are elastic scattering and capture followed by the emission of a photon or another particle from the absorber nucleus.

When absorbers are placed in a collinated beam of neutrons, and the transmitted neutron intensity is measured, as was done for gamma rays in Fig. 5.10, it is found that neutrons, too, are removed exponentially from the beam. Instead of using linear or mass absorption coefficients to describe the ability of a given absorber material to remove neutrons from the beam, it is customary to designate only the microscopic cross section, σ, for the absorbing material. The product σN, where N is the number of absorber atoms per cm^3, is the macroscopic cross section Σ. The removal of neutrons from the beam is thus given by

$$I = I_0 e^{-\sigma N t}. \tag{5.43}$$

Neutron cross sections are strongly energy dependent. If removal of a neutron from the beam may be effected by more than one mechanism, the total cross section is the sum of the cross sections for the various possible reactions.

Example 5.11

In an experiment designed to measure the total cross section of lead for 10 MeV neutrons, it was found that a 1-cm-thick lead absorber attenuated the neutron flux to 84.5% of its initial value. The atomic weight of lead is 207.21, and its specific gravity is 11.3. Calculate the total cross section from these data.

The atomic density of lead is

$$\frac{6.03 \times 10^{23}\ \text{atoms/mole}}{207.21\ \text{g/mole}} \times 11.3\ \text{g/cm}^3 = 3.29 \times 10^{22}\ \text{atoms/cm}^3,$$

$$\frac{I}{I_0} = e^{-\sigma N t},$$

$$0.845 = e^{-\sigma \times 3.29 \times 10^{22} \times 1},$$

$$\ln \frac{1}{0.845} = 3.29 \times 10^{22}\sigma,$$

$$\sigma = \frac{0.168}{3.29 \times 10^{22}} = 5.1 \times 10^{-24}\ \text{cm}^2,$$

$\sigma = 5.1$ barns, and the macroscopic cross section is

$$\Sigma = \sigma N = 5.1 \times 10^{-24}\ \text{cm}^2 \times 3.29 \times 10^{22}\ \text{cm}^{-3} = 0.168\ \text{cm}^{-1}.$$

Scattering

Neutrons may collide with nuclei, and undergo either inelastic or elastic scattering. In the former case, some of the kinetic energy that is transferred to the target nucleus excites the nucleus, and the excitation energy is emitted as a gamma-ray photon. This interaction is best described by the compound nucleus model, in which the neutron is captured, then re-emitted by that target nucleus together with the gamma photon. This is a threshold phenomena; the neutron energy threshold varies from infinity for hydrogen (inelastic scattering cannot occur) to about 6 MeV for oxygen to less than 1 MeV for uranium. Generally, the cross section for inelastic scattering is small, on the order of 1 barn or less, for low energy fast neutrons, but increases with increasing energy, and approaches a value corresponding to the geometrical cross section of the target nucleus.

Elastic scattering is the most likely interaction between fast neutrons and low-atomic-numbered absorbers. This interaction is a "billard ball" type collision, in which kinetic energy and momentum are conserved. By applying these conservation laws, it can be shown that the energy, E, of the scattered neutron after a head-on collision is

$$E = E_0\left(\frac{M - m}{M + m}\right)^2, \tag{5.44}$$

where E_0 = energy of the incident neutron,
m = mass of the incident neutron,
M = mass of the scattering nucleus.

The energy transferred to the target nucleus is $E_0 - E$. From equation (5.44), we have

$$E_0 - E = E_0\left[1 - \left(\frac{M - m}{M + m}\right)^2\right]. \tag{5.45}$$

According to equations (5.44) and (5.45), it is possible, in a head-on collision with a hydrogen nucleus, for a neutron to transfer all its energy to the hydrogen nucleus. With heavier nuclei, all the kinetic energy of the neutron cannot be transferred in a single collision. In the case of oxygen, for example, equation (5.45) shows that the maximum fraction, $(E_0 - E)/E_0$, of the neutron's kinetic energy that can be transferred during a single collision is only 22.2%. This shows that nuclei with small mass numbers are more effective, on a "per collision" basis, than nuclei with high mass numbers for slowing down neutrons.

Equations (5.44) and (5.45) are valid only for head-on collisions. Most collisions are not head-on, and the energy transferred to the target nuclei are consequently less than the maxima given by the two equations above.

In the course of the successive collisions suffered by a fast neutron as it passes through a slowing down medium, the average decrease, per collision, in the logarithm of the neutron energy (which is called the average logarithmic energy decrement) remains constant. It is independent of the neutron energy, and is a function only of the mass of scattering nuclei. The average logarithmic energy decrement is defined as

$$\xi = \overline{\Delta \ln E} = \overline{\ln E_0 - \ln E} = \overline{\ln \frac{E_0}{E}} = -\overline{\ln \frac{E}{E_0}}, \tag{5.46}$$

and can be shown to be given by

$$\xi = 1 + \frac{\alpha \ln \alpha}{1 - \alpha}, \tag{5.47}$$

where $\alpha = [(M - m)/(M + m)]^2$, as used in equation (5.44). If the slowing-down medium contains n kinds of nuclides, each of microscopic scattering cross section σ_s and average logarithmic energy decrement ξ, then the mean value of ξ for the n species is

$$\xi = \frac{\sum_{i=1}^{n} \sigma_{si} N_i \xi_i}{\sum_{i=1}^{n} \sigma_{si} N_i}. \tag{5.48}$$

Since

$$\overline{\ln \frac{E}{E_0}} = -\xi,$$

$$\frac{E}{E_0} = e^{-\xi},$$

and the median fraction of the incident neutron's energy tht is transferred to the target nucleus during a collision is

$$f = 1 - \frac{\bar{E}}{E_0} = 1 - e^{-\xi}. \tag{5.49}$$

Thus for hydrogen, $\xi = 1$, the median energy transfer during a collision with a fast neutron is 63% of the kinetic energy of the neutron. In the case of carbon, $\xi = 0.159$, an average of only 14.7% of the neutron's kinetic energy is absorbed by the struck nucleus during an elastic collision. The struck nucleus, as a result of the kinetic energy imparted to it by the neutron, becomes an ionizing particle, and dissipates its kinetic energy in the absorbing medium by excitation and ionization.

The distance travelled by a fast neutron between its introduction into a slowing down medium and its thermalization depends on the number of collisions made by the neutron and the distance between collisions. Although the actual path of the neutron is tortuous because of deflections due to collisions, the average straight line distance covered by the neutron can be determined; it is called the *fast-diffusion length*, or the *slowing-down length*. (The square of the fast-diffusion length is called the Fermi age of the neutron.) The distance traveled by the thermalized neutron until it is absorbed is measured by the *thermal diffusion length*. The thermal diffusion length is defined as the thickness of a slowing down medium that attenuates a beam of thermal neutrons by a factor of e. Thus, attenuation of a beam of thermal neutrons by a substance of thickness t cm whose thermal diffusion length is L cm is given by

$$n = n_0 e^{-t/L}. \tag{5.50}$$

(The terms fast diffusion length and thermal diffusion length are applicable only to materials in which the absorption cross section is very small. When this condition is not met, as in the case of boron or cadmium, the attenuation of a beam of thermal neutrons is given by equation (5.43).) Although fast and thermal diffusion lengths may be calculated, the assumptions inherent in the calculations make it preferable to use measured values for these parameters. Values for fast and thermal diffusion lengths for fission neutrons in certain slowing down media are given in Table 5.6.

TABLE 5.6. FAST AND THERMAL DIFFUSION LENGTHS OF SELECTED MATERIALS

Substance	Fast diffusion length	Thermal diffusion length	Diffusion coefficient
H_2O	5.75 cm	2.88 cm	0.16 cm
D_2O	11	171	0.87
Be	9.9	24	0.50
C(graphite)	17.3	50	0.84

For the case of a point source of n_0 thermal neutrons per second in a spherically shaped non-multiplying medium (a medium which contains no fissile material) of radius R, thermal diffusion length L, and diffusion coefficient D, the flux of neutrons escaping from the surface is

$$\phi = \frac{n_0 e^{-R/L}}{4\pi RD}. \tag{5.51}$$

Example 5.12

A Pu–Be neutron source that emits 10^6 neutrons per second is in the center of a spherical water shield whose diameter is 50 cm. How many thermal neutrons are escaping per cm^2/sec from the surface of the shield?

Since the radius of the water shield is much greater than the fast diffusion length given in Table 5.6, we may assume (for the purpose of this calculation) that essentially all the fast neutrons are thermalized, and that the thermal neutrons are diffusing outward from the centre. Substituting the appropriate numbers into equation (5.51), we have

$$\phi = \frac{10^6}{4\pi \times 25 \times 0.16} e^{-25/2.88} = 3.4 \frac{\text{neutrons}}{\text{cm}^2/\text{sec}}.$$

Absorption

From the discussion above, it is seen that fast neutrons are rapidly degraded in energy by elastic collisions if they interact with low-atomic-numbered substances. As neutrons reach thermal or near thermal energies, their likelihood of capture by an absorber nucleus increases. The absorption cross section of many nuclei, as the neutron energy becomes very small, has been found to be inversely proportional to the square root of its kinetic energy, and thus to vary inversely with its velocity:

$$\sigma \propto \frac{1}{\sqrt{E}} \propto \frac{1}{v}. \tag{5.52}$$

Equation (5.52) is called the "one-over-v law" for slow neutron absorption. For ^{10}B, this relationship is valid for the span of energies from 0.001 to 1000 eV, as shown in Fig. 5.23. Thermal neutron cross sections are usually given for neutrons whose most probable energy is 0.025 eV. If the cross section at energy E_0 is σ_0, then the cross section for any other energy within the range of validity of the $1/v$ law is given by

$$\frac{\sigma}{\sigma_0} = \sqrt{\frac{E_0}{E}}. \tag{5.53}$$

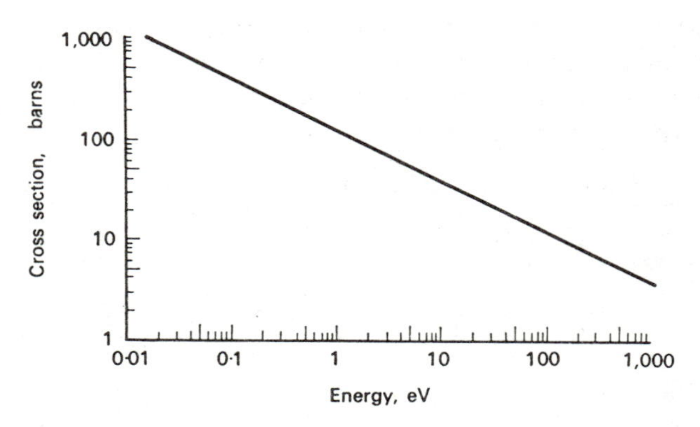

FIG. 5.23. Neutron absorption cross section for boron, showing the validity of the $1/v$ law for neutrons from 0.02 to 1000 eV in energy. The equation of the curve is $\sigma = 116\sqrt{\text{eV}}$ barns.

Example 5.13

The cross section of boron for the ^{10}B (n, α) ^{7}Li reaction is 753 barns for 0.025 eV neutrons. What is the boron cross section for 50 eV neutrons? Substituting into equation (5.54) gives

$$\sigma = 735\sqrt{\left(\frac{0.025}{50}\right)} = 16.8 \text{ barns.}$$

Some capture reactions of practical importance in health physics include the following:

$$^{1}\text{H}\,(n, \gamma)\,^{2}\text{H} \qquad \sigma = 0.33 \text{ barns,} \tag{5.54}$$

$$^{14}\text{N}\,(n, p)\,^{14}\text{C} \qquad \sigma = 1.70 \text{ barns,} \tag{5.55}$$

$$^{10}\text{B}\,(n, \alpha)\,^{7}\text{Li} \qquad \sigma = 4.01 \times 10^{3} \text{ barns,} \tag{5.56}$$

$$^{113}\text{Cd}\,(n, \gamma)\,^{114}\text{Cd} \qquad \sigma = 2.1 \times 10^{4} \text{ barns.} \tag{5.57}$$

Equations (5.54) and (5.55) are important in neutron dosimetry, since H and N are major constituents of tissue. Equation (5.56) is important in the design of instruments for measuring neutrons as well as neutron shielding, while the last equation is important mainly in shielding. It should be noted that the neutron reactions with hydrogen and with cadmium results in the emission of high-energy gamma rays, while the capture of a thermal neutron by ^{10}B releases a low-energy (0.48 MeV) gamma ray in 93% of the reactions. When a thermal neutron is captured by ^{14}N, a 0.6 MeV proton is emitted.

Neutron activation

Neutron activation is the production of a radioactive isotope by absorption of a neutron, such as the n, p reaction of equation (5.55). In that instance, ^{14}C is produced. Activation by neutrons is important to the health physicist for several reasons. First, it means that any substance that was irradiated by neutrons may be radioactive; a radiation hazard may therefore persist after the irradiation by neutrons is terminated. Secondly, it provides a convenient tool for measuring neutron flux. This is done simply by irradiating a known amount of the material to be activated, measuring the induced activity, and then, with a knowledge of the activation cross section, computing the neutron flux. In case of a criticality accident (an accidental attainment of an uncontrolled chain reaction), the measurement of induced radioactivity due to neutron irradiation permits calculation of the neutron dose. This same principle is applied by the chemist in neutron activation analysis. This method, which for many elements is more sensitive than other physical or chemical procedures, involves irradiation of that unknown sample in a neutron field of known intensity, measurement of the induced activity, and then calculation of the amount of the unknown in the sample. Furthermore, by spectroscopic examination of the induced radiation, qualitative analysis of the unknown is also possible.

If a radionuclide is being made by neutron irradiation, and is decaying at the same time, the net number of radioactive atoms present in the sample at any time is the difference between the rate of production and the rate of decay. This may be expressed mathematically by the equation: net rate of increase of radioactive atoms = rate of production–rate of decay.

$$\frac{dN}{dt} = \phi\sigma n - \lambda N, \tag{5.58}$$

where ϕ = flux, neutrons per cm² per sec,
σ = activation cross section, cm²,
λ = transformation constant of the induced activity,
N = number of radioactive atoms,
n = number of target atoms.

Equation (5.58) is a linear differential equation which may be integrated to yield

$$\lambda N = \phi\sigma n(1 - e^{-\lambda t}). \tag{5.59}$$

In equation (5.59), $\phi\sigma n$ is sometimes called the saturation activity; for an infinitely long irradiation time, it represents the maximum obtainable activity with any given neutron flux.

Example 5.14

A sample containing an unknown quantity of chromium is irradiated for 1 week in a thermal neutron flux of 10^{11} n/cm²/sec. The resulting ^{51}Cr gamma rays give a counting rate of 600 counts per minute in a scintillation counter whose overall efficiency is 10%. How many grams of chromium were there in the original sample? The reaction in this case is

$$^{50}\text{Cr} + {}^1_0n \rightarrow {}^{51}\text{Cr} + \gamma.$$

The thermal neutron activation cross section for ^{50}Cr is 13.5 barns, and ^{50}Cr forms 4.31% by number of the naturally occurring chromium atoms. Chromium 51 decays by orbital electron capture with a half-life of 27.8 days, and emits a 0.323-MeV gamma-ray in 9.8% of the decays. The atomic weight of Cr is 52.01.

The activity is given by λN in equation (5.59). This equation may therefore be solved for n, the number of target atoms. substituting the numerical values into equation (5.59), we have

$$10\,\frac{\text{counts}}{\text{sec}} \times 10\,\frac{\text{trnsf.}}{\text{count}} = 10^{11}\,\frac{1}{\text{cm}^2\text{sec}} \times 1.35 \times 10^{-23}\,\frac{\text{cm}^2}{\text{atom}}$$
$$\times\, 0.098 \times 0.0431\, n \text{ atoms } (1 - e^{-0.693/27.8 \times 7}),$$

$$n = \frac{10^2 \times 10^{23}}{10^{11} \times 1.35 \times 1.6 \times 10^{-1} \times 9.8 \times 10^{-2} \times 4.31 \times 10^{-2}}$$
$$= 1.095 \times 10^{18} \text{ atoms Cr.}$$

Since there are 52.01 g Cr/mole, the weight of chromium in the unknown is

$$\frac{1.095 \times 10^{18} \text{ atoms}}{6.027 \times 10^{23} \text{ atoms/mole}} \times 52.01 \frac{\text{g}}{\text{mole}} = 9.46 \times 10^{-5} \text{ g}.$$

Problems

1. The density of Hg is 13.6 grams/cm^3 and its atomic weight is 200.6. Calculate the number of Hg atoms/cm^3.

2. The density of quartz (SiO_2) crystals is 2.65 gm/cm^3. What is the atomic density (atoms/cm^3) of silicon and oxygen in quartz.

3. Compare the electronic densities of a piece of aluminum 5 mm thick and a piece of iron of the same density thickness.

4. In surveying a laboratory, a health physicist wipes a contaminated surface, and runs an absorption curve using a thin end window counter and aluminum absorbers. The range of the beta rays (no gammas were found) was found to be 800 mg/cm^2 aluminum. What could the contaminant be? What further studies could be done in the smear sample to help verify the identification of the comtaminant?

5. A health physicist finds an unknown contaminant that proves to be a pure beta emitter. To help identify the contaminant, he runs an absorption curve to determine the maximum energy of the beta rays. He uses an end window G.M. counter whose mica window (density = 2.7 gm/cm^3) is 0.1 mm thick, and he finds that 1.74 mm Al stops all the beta particles. The distance between the sample and the G.M. counter was 2 cm. What was the energy of the beta particle? What is the contaminant?

6. A Compton electron that was scattered straight forward ($\phi = 0°$) was completely stopped by an aluminum absorber 460 mg/cm^2 thick.

(a) What was the kinetic energy of the Compton electron?

(b) What was the energy of the incident photon?

7. Monochromatic 0.1-MeV gamma rays are scattered through an angle of 120° by a carbon block.

(a) What is the energy of the scattered photon?

(b) What is the kinetic energy of the Compton electron?

8. A 1.46-MeV gamma from naturally occurring ^{40}K is scattered two times: first through an angle of 30° and then through an angle of 150°.

(a) What is the energy of the photon after the second scattering?

(b) What is the energy of the scattered photon if the angular sequence is reversed?

9. What is the energy of the Compton edge for the 0.661-MeV gamma from ^{137}Cs?

10. The energy of a scattered photon is 0.2 MeV after it was scattered through an angle of 135°. What was the photon's energy before the scattering collision?

11. What is the energy of the Compton edge for the following gammas?

(a) 0.136 MeV from ^{57}Co,

(b) 0.811 MeV from ^{58}Co,

(c) 1.33 MeV from ^{60}Co.

12. The following gamma-ray absorption data were taken with lead absorbers:

Absorber thickness, mm	0	2	4	6	8	10	15	20	25
Counts per minute	1000	880	770	680	600	530	390	285	210

(a) Determine the linear and mass, and atomic absorption coefficients.

(b) What was the energy of the gamma ray?

13. The folowing absorption data were taken with aluminum absorbers:

Absorber thickness, mm	0	0.02	0.04	0.06	0.08	0.1	0.12	0.14	0.16	0.2	0.4	0.8	1.5	2	2.8
Counts per minute	1000	576	348	230	168	134	120	107	96	95	90	82	68	60	50

(a) Plot the data. What types of radiation does the curve suggest?

(b) If a beta particle is present, what is its energy?

(c) If a gamma ray is present, what is its energy?

(d) What isotope is compatible with the absorption data?

(e) Write the equation that fits the absorption data.

14. A collimated gamma ray beam consists of equal numbers of 0.1-MeV and 1.0-MeV photons. If the beam enters a 15-cm-thick concrete shield, what is the relative portion of 1-MeV photons to 0.1-MeV photons in the emergent beam?

15. Three collimated gamma-ray beams of equal flux, whose quantum energies are 2, 5, and 10 MeV, respectively, pass through a 5-cm thickness of lead. What is the ratio of the emergent fluxes?

16. Calculate the thickness of Al and Cu required to attenuate narrow, collimated, monochromatic beams of 0.1-MeV and 0.8-MeV gamma rays to

(a) one-half the incident intensity (HVL),

(b) one-tenth the incident intensity (TVL). Express the answers in cm and in grams/cm^2.

(c) What is the relationship between a half-value layer and a tenth-value layer?

17. A laminated shield consists of two layers each of alternating thickness of aluminum and lead, each layer having a density thickness of 1.35 g/cm^2. The shield is irradiated with a narrow collimated beam of 0.2-MeV photons.

(a) What is the overall thickness of the laminated shield, in cm?

(b) Calculate the shield attenuation factor when the (1) aluminum layer is first, (2) lead layer is first.

18. Calculate the gamma-ray threshold energy for the reaction $^{11}C(n, \gamma)^{12}C$.

19. X-rays are generated as bremsstrahlung by causing high-speed electrons to be stopped by a high atomic numbered target, as shown in the figure below. If the electrons are accelerated by a constant high voltage of 250 kV, and if the electron beam current is 10 mA, calculate the X-ray energy flux at a distance of 1 m from the tungsten target. Neglect absorption by the glass tube, and assume that the bremsstrahlung are emitted isotropically.

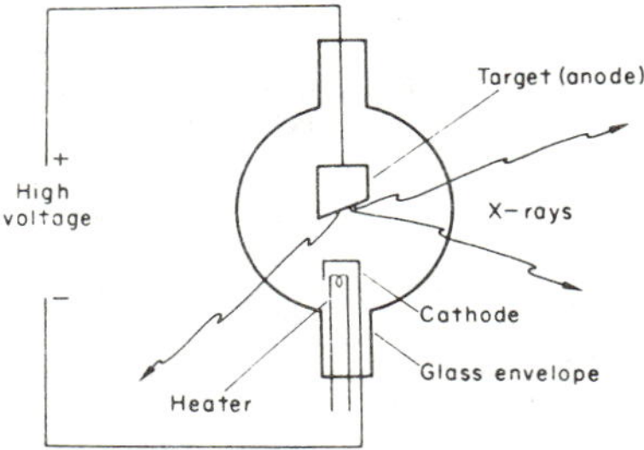

20. If the most energetic photon results from the instantaneous stopping of an electron in a single collision, what voltage must be applied across an X-ray tube in order to generate X-rays whose shortest wavelength approaches 0.124 Å?

21. A beta particle whose kinetic enegy is 0.159 MeV passes through a 4 mg/cm^2 window into a helium-filled Geiger tube. How many ion pairs will the beta particle produce inside the tube?

22. If the neutron emission rate from ^{252}Cf is 2.31×10^6 neutrons per second per μg, and the transformation rate constant for alpha emission is 0.25 per year, what is the neutron emission rate per MBq and per μCi?

23. Calculate the speed of a "slow" neutron whose kinetic energy is 0.1 eV. To what temperature does this energy correspond?

24. When ^{9}Be is irradiated with deuterons, neutrons are produced according to the reaction $^9_4$Be and (d, n) $^{10}_5$B. The cross section for this reaction for 15-MeV deuterons is 0.12 barns. What is the neutron flux at a distance of 25 cm from a 1 g beryllium target that is irradiated with a 100 μA beam of deuterons, 1.13 cm diameter, assuming an isotropic distribution of neutrons?

25. What is the thickness of Cd that will absorb 50% of an incident beam of thermal neutrons? The capture cross section for the element Cd is 2550 barns for thermal neutrons; the specific gravity of Cd is 8.65 and its atomic weight is 112.4.

26. A small ^{124}Sb gamma-ray source, whose activity is 3.7×10^{10} Bq (1 Ci), is completely surrounded by a 25 g sphere of beryllium. Calculate the number of neutrons per second from the $^9Be(\gamma, n)\ ^8Be$ reaction if the cross section is 1 millibarn, and if the diameter of the spherical cavity enclosing the gamma ray source is 1 cm.

27. Cadmium is used as a thermal neutron shield in an average flux of 10^{12} neutrons per cm^2/sec. How long will it take to use up 10% of the ^{113}Cd atoms?

28. The cross section for the $^{32}S(n, P)\ ^{32}P$ reaction is 300 millibarns for neutron energies greater than 2.5 MeV. How many microcuries of ^{32}P activity can we expect if 100 mg ^{32}S is irradiated in a fast flux of 10^2 neutrons/cm^2 sec for 1 week?

29. If the absorption coefficient of the high-energy component of cosmic radiation is 2.5×10^{-3} per meter water, calculate the reduction in intensity of these cosmic rays at the bottom of the ocean, at the depth of 10,000 m.

30. If deuterium is irradiated with 2.62-MeV gamma rays from ^{208}Tl(ThC″), the nucleus disintegrates into its component parts of 1 proton and 1 neutron. If the neutron and proton each has 0.225 MeV of kinetic energy, and if the proton has a mass of 1.007593 atomic mass units, calculate the mass of the neutron.

31. A beam of fast neutrons includes two energy groups. One group, of 1-MeV neutrons, includes 99% of the total neuton flux. The remaining 1% of the neutrons have an energy of 10 MeV.

(a) What will be the relative proportions of the two groups after passing through 25 cm of water?

(b) What would be the relative proportion of the two groups after passing through a slab of lead of the same density thickness?

The removal cross sections are:

	1 MeV	10 MeV
H	4.2 barns	0.95
O	8	1.5
Pb	5.5	5.1

32. Boral is an aluminum boron carbide alloy used as a shield against thermal neutrons. If the boron content is 35% by weight, and if the density of boral is 2.7 gm/cm^3, calculate the half-thickness of boral for thermal neutrons at room temperature. The capture cross sections are: boron = 755 barns, aluminum = 230 millibarns, carbon = 3.2 millibarns.

33. The scattering cross sections for N and O for thermal neutrons are 10 and 4.2 barns, respectively.

(a) Calculate the macroscopic cross section for air at STP. Air consists of 79 volume percent nitrogen and 21 volume percent oxygen.

(b) What is the scattering mean free path of thermal neutrons in air?

34. How many scattering collisions in graphite are required to reduce the energy of 2.5 MeV neutrons to

(a) 0.1% of the initial energy?

(b) 0.25 eV?

35. A cobalt foil, 1 cm diameter × 0.1 mm thick, is irradiated in a mean thermal flux of 1×10^{11} neutrons/cm^2 per second for a period of 7 days. If the activation cross section is 36 barns, and if the density of cobalt is 8.9 grams/cm^3, what is the activity, in Bq and in microcuries, at the end of the irradiation period? Note that natural cobalt is 100% ^{59}Co.

36. Type 304 stainless steel consists of 71 weight percent Fe, 19% Cr, and 10% Ni. The isotopic abundance, percentage, and the respective 2200 m/s capture cross sections (barns) are given below:

Fe			Cr			Ni		
A	Abund	σ_c	A	Abund	σ_c	A	Abund	σ_c
54	5.84	2.9	50	4.31	17	58	67.76	4.4
56	91.68	2.7	52	83.76	0.8	60	26.16	2.6
57	2.17	2.5	53	9.55	18	61	1.25	2
58	0.31	1.1	54	2.38	0.38	62	3.66	15
						64	1.16	1.5

(a) Calculate the macroscopic capture cross section.

(b) If a 1-cm-diameter collimated beam of 2200 m/s neutrons is incident on a 2-mm-thick sheet of type 304 stainless steel, how many neutrons per sec will be captured if the flux is 5×10^{11} n/cm^2 per second?

Suggested References

Attix, F. H., and Roesch, W. C., eds. *Radiation Dosimetry*, 2nd ed. Vol. 1, *Fundamentals*. Academic Press, New York, 1968.

Blatz, H., ed. *Radiation Hygiene Handbook*. McGraw-Hill, New York, 1959.

Born, M. *Atomic Physics*, 8th ed. Hafner, Darien, Conn., 1970.

Brodsky, A., ed. *Handbook of Radiation Measurement and Protection*, Vol. 1, *Physical Science and Engineering Data*. CRC Press, West Palm Beach, Fla., 1978.

Bureau of Radiological Health. *Radiological Health Handbook*, Revised ed. U.S. Dept. of Health, Education, and Welfare, Public Health Services, Rockville, Md., 1970.

Cohen, B. L. *Concepts of Nuclear Physics*. McGraw-Hill, New York, 1971.

Compton, A. H., and Allison, S. K. *X-rays in Theory and Experiment*. D. Van Nostrand, Princeton, 1935.

Etherington, H., ed. *Nuclear Engineering Handbook*. McGraw-Hill, New York, 1958.

Evans, R. D. *The Atomic Nucleus*. McGraw-Hill, New York, 1955.

Glasstone, S. *Sourcebook on Atomic Energy*, 3rd ed. D. Van Nostrand, Princeton, 1967.

Hurst, G. S., and Turner, J. E. *Elementary Radiation Physics*, Wiley, New York, 1969.

Johns, H. E., and Cummingham, J. R. *The Physics of Radiology*, 3rd ed. Charles Thomas, Springfield, Ill., 1973.

Kaplan, I. *Nuclear Physics*, 2nd ed. Addison-Wesley, Reading, Mass., 1963.

Lapp, R. E., and Andrews, H. L. *Nuclear Radiation Physics*, 4th ed. Prentice-Hall, Englewood Cliffs, N.J., 1972.

National Nuclear Data Center. *Neutron Cross Sections, Report BNL*-325 *and Supplements*. Brookhaven National Laboratory, Upton, New York.

Semat, H., and Albright, J. R. *Introduction to Atomic and Nuclear Physics*, 5th ed. Holt, Rinehart, and Winston, New York, 1972.

The Following reports of the International Commission on Radiation Units and Measurements, Washington, D.C.:

10b. *Physical Aspects of Irradiation* (1964).

16. *Linear Energy Transfer* (1970).

28. *Basic Aspects of High Energy Particle Interactions and Radiation Dosimetry* (1978).

CHAPTER 6

RADIATION DOSIMETRY

Units

During the early days of radiological experience, there was no precise unit of radiation dose that was suitable either for radiation protection or for radiation therapy. For purposes of radiation protection, a common "dosimeter" was a piece of dental film with a paper clip attached. A daily exposure great enough to just produce a detectable shadow was considered a maximum permissible dose. For greater doses and for therapy purposes, the dose unit was frequently the "skin erythema unit." Because of the great energy dependence of these dose units, as well as other inherent defects, neither of these two units could be biologically meaningful or useful either in the quantitative study of the biological effects of radiation or for radiation protection purposes. Furthermore, since the fraction of the energy in a radiation field that is absorbed by the body is energy dependent, it is necessary to distinguish between radiation *exposure* and radiation *absorbed dose*.

Absorbed Dose

Gray

Radiation damage depends on the absorption of energy from the radiation, and is approximately proportional to the concentration of absorbed energy in tissue. For this reason, the basic unit of radiation dose is expressed in terms of absorbed energy per unit mass of tissue. This unit is called the *gray* (Gy) and is defined as:

One gray is an absorbed radiation dose of one joule per kilogram.

$$1 \text{ Gy} = 1 \text{ J/kg}. \tag{6.1}$$

The gray is universally applicable to all types of ionizing radiation dosimetry—irradiation due to external fields of gamma rays, neutrons, or changed particles, as well as that due to internally deposited radioisotopes.

Rad

Before the universal adoption of the SI units, radiation dose was measured by a unit called the *rad* (Radiation Absorbed Dose).

One rad is an absorbed radiation dose of 100 ergs per gram.

$$1 \text{ rad} = 100 \text{ ergs/g}. \tag{6.2}$$

Since $1 \text{ J} = 10^7$ ergs, and since $1 \text{ kg} = 1000$ g,

$$1 \text{ Gy} = 100 \text{ rads}. \tag{6.3}$$

Although the gray is the newer unit, and will eventually replace the rad, the rad nevertheless continues to be widely used.

Exposure

Exposure unit

For external radiation of any given energy flux, the absorbed dose to any point within an organism depends on the type and energy of the radiation, the depth within the organism of the point at which the absorbed dose is required, and elementary constitution of the absorbing medium at this point. For example bone, consisting of higher atomic numbered elements (Ca and P) than soft tissue (C, O, H, and N) absorbs more energy from an X-ray beam, per unit mass of absorber, than soft tissue. For this reason, the X-ray fields to which an organism may be exposed are frequently specified in *exposure units*. The exposure unit is a measure of the photon flux, and is related to the amount of energy transferred from the X-ray field to a unit mass of *air*. One exposure unit is defined as that quantity of X- or gamma radiation that produces in air, ions carrying 1 coulomb of charge (of either sign) per kg air.

$$1 \text{ X unit} = 1 \text{ C/kg air}. \tag{6.4}$$

The exposure unit is based on ionization of air because of the relative ease with which radiation induced ionization can be measured. At quantum energies less than several keV and more than several MeV, however, it becomes difficult to fulfill the requirements for measuring the exposure unit. Accordingly the use of the exposure unit is limited to X- or gamma rays whose quantum energies do not exceed 3 MeV. For higher energy photons, exposure is expressed in units of watt-seconds per m^2 and exposure rate is expressed in units of watts per m^2. The operational definition of the exposure unit may be converted into the more fundamental units of energy absorption per unit mass of air by using the fact that the charge on a single ion is 1.6×10^{-19} coulombs and that the average energy dissipated in the production of a single ion pair in air is 34 eV. Therefore:

$$\begin{aligned} 1 \text{ X unit} &= 1 \frac{\text{C}}{\text{kg air}} \times \frac{1 \text{ ion}}{1.6 \times 10^{-19} \text{ C}} \times 34 \frac{\text{eV}}{\text{ion}} \times 1.6 \times 10^{-19} \text{ J/eV} \times 1 \frac{\text{Gy}}{\text{J/kg}} \\ &= 34 \text{ Gy (in air)}. \end{aligned} \tag{6.5}$$

It should be noted that the exposure unit is an integrated measure of exposure, and is independent of the time over which the exposure occurs. The strength of an X-ray or gamma-ray field is usually expressed as an exposure rate, such as coulombs per kg per hour. The total exposure, of course, is the product of exposure rate and time.

Roentgen

Formerly, before the SI system was adopted, the unit of X-ray exposure was called the roentgen, and was symbolized by R. The roentgen was defined as that quantity of X- or gamma radiation that produces ions carrying one statcoulomb of charge of either sign per cubic centimeter of air at 0°C and 760 mm Hg.

$$1 \text{ R} = 1 \text{ SC/cm}^3. \tag{6.6}$$

Since 1 ion carries a charge of 4.8×10^{-10} SC, and the mass of 1 cm^3 of standard air is 0.001293 g, an exposure of 1 R corresponds to an absorption of 87.7 ergs per gram of air, or to a dose to the air of 0.877 rad. When exposure is measured in roentgens, X-ray or gamma-ray field strength is measured in units such as roentgens per minute or milliroentgens per hour. (A milliroentgen, which is symbolized by "mR" is equal to 0.001 roentgen.)

The relationship between the exposure unit and the roentgen may be calculated as follows:

$$\frac{34\,\frac{\text{J}}{\text{kg}}\Big/\frac{\text{C}}{\text{kg}} \times 10^{7}\,\frac{\text{ergs}}{\text{J}} \times \frac{1\text{ kg}}{1000\text{ g}}}{87.7\,\frac{\text{ergs}}{\text{g}}\Big/\text{R}} = 3881\text{ R}\Big/\frac{\text{C}}{\text{kg}}.$$

or

$$1\text{ X unit} = 3881\text{ R}. \tag{6.7}$$

Example 6.1

Medical X-ray shielding design is based on maximum weekly exposures of 100 mR for controlled areas and 10 mR for uncontrolled areas. What are the corresponding exposures expressed in SI units?

From equation (6.7)

$$\frac{1\text{ X unit}}{3881\text{ R}} = 2.58 \times 10^{-4}\text{ X units/R} \times 1\,\frac{\text{C/kg}}{\text{X unit}} = 2.58 \times 10^{-4}\,\frac{\text{C/kg}}{\text{R}}.$$

For controlled areas

$$2.58 \times 10^{-4}\,\frac{\text{C/kg}}{\text{R}} \times 0.1\text{ R} = 25.8 \times 10^{-6}\text{ C/kg} = 25.8\ \mu\text{C/kg},$$

and for uncontrolled areas

$$2.58 \times 10^{-4}\,\frac{\text{C/kg}}{\text{R}} \times 0.01\text{ R} = 2.58 \times 10^{-6}\text{ C/kg} = 2.58\ \mu\text{C/kg}.$$

Exposure Measurement: The Free Air Chamber

The operational definition of the exposure unit can be satisfied by the instrument shown in Fig. 6.1. The X-ray beam enters through the portal and interacts with the cylindrical column of air defined by the entry port diaphragm. The ions resulting from interactions between the X-rays and the volume of air A B C D, which is determined by the intersection of the X-ray beam with the electric lines of force from the edges of the collector plate C, is collected by the plates, causing current to flow in the external circuit. The guard ring, G, and the guard wires, W, help to keep these electric field lines straight and perpendicular to the plates. The electric field intensity between the plates is on the order of 100 V/cm—high enough to collect the ions before they recombine, but not great enough to cause secondary ionization by the electrons released by the primary ionizing particles. The guard wires are connected to a voltage-dividing network to ensure a uniform potential drop across the plates. The number of ions collected because of X-ray interactions in the collecting volume is calculated from the current flow, and the dose rate, in roentgens per unit time, is then computed. For the exposure unit to be measured in this way, all the

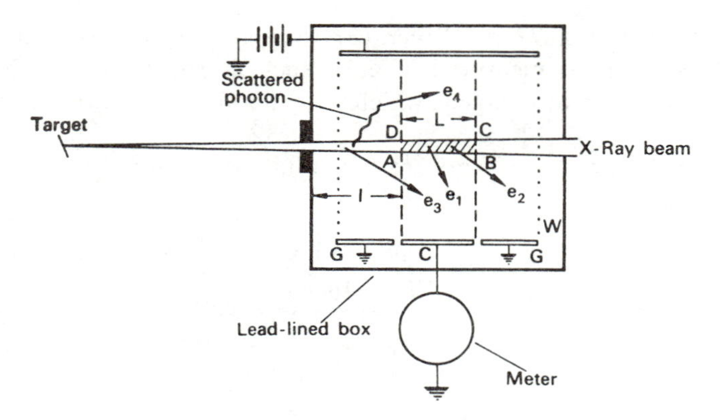

FIG. 6.1. Schematic diagram of a parallel plate free air ionization chamber. (From N.B.S. Handbook 64, *Design of Free Air Ionization Chamber*, 1957.)

energy of the primary electrons must be dissipated in the air within the meter. This condition can be satisfied by making the air chamber larger than the maximum range of the primary electrons. (For 300-keV X-rays, the spacing between the collector plates is about 30 cm, and the overall box is a cube about 50 cm on edge.) The fact that most of the ions produced as a consequence of X-ray interactions within the sensitive volume are not collected is of no significance if as many electrons from interactions elsewhere in the X-ray beam enter the sensitive volume as leave it. This condition is known as *electronic equilibrium*. When electronic equilibrium is attained, an electron of equal energy enters into the sensitive volume for every electron that leaves. A sufficient thickness of air, dimension 1 in Fig. 6.1, must be allowed between the beam entrance port and the sensitive volume in order to attain electronic equilibrium. For highly filtered 250-kV X-rays, 9 cm air is required, for 500-kV X-rays, the air thickness required for electronic equilibrium in the sensitive volume increases to 40 cm. Under conditions of electronic equilibrium, and assuming negligible attenuation of the X-ray beam by the air in length 1, the ions collected from the senstiive volume result from primary photon interactions at the beam entrance port; and the measured exposure, consequently, is at that point, and not in the sensitive volume. Free air chambers are in use that measure quantity of X-rays whose quantum energies reach as high as 500 keV. Higher-energy radiation necessitates much greater size free air chambers. The technical problems arising from the use of such large chambers makes it impractical to use the free air ionization chamber as a primary measuring device for quantum energies in excess of 500 keV.

The use of the free air ionization chamber to measure X-ray exposure rate in coulombs per kg per second may be illustrated by the following example:

Example 6.2

The opening of the diaphragm in the entrance port of a free air ionization chamber is 1 cm in diameter, and the length AB of the sensitive volume is 5 cm. A 200-kV X-ray beam projected into the chamber produces a steady current in the external circuit of 0.01 μA. The temperature at the time of the measurement was 20°C and the pressure was 750 mm Hg. What is the exposure rate from this beam of X-rays?

A current of 0.01 μA corresponds to a flow of electrical charge of 10^{-8} C/sec. The sensitive volume in this case is 3.927 cm^3. When the pressure and temperature are corrected to standard conditions, we have

$$\dot{X} = \frac{10^{-8}\ \text{C/sec}}{3.927\ \text{cm}^3 \times 1.293 \times 10^{-6}\ \text{kg/cm}^3} \times \frac{293}{273} \times \frac{760}{750}$$

$$= 2.14 \times 10^{-3}\ \text{X units/sec (8.31 R/sec)}.$$

Exposure Measurement: The Air Wall Chamber

The free air ionization chamber described above is practical only as a primary laboratory standard. For field use, a more portable instrument is required. Such an instrument could be made by compressing the air around the measuring cavity. If this were done, then the conditions for defining the exposure unit would continue to be met. In practice, of course, it would be quite difficult to construct an instrument whose walls are made of compressed air. However, it is possible to make an instrument with walls of "air equivalent" material, that is, a wall material whose X-ray absorption properties are very similar to those of air. Such a chamber can be built in the form of an electrical capacitor; its principle of operation can be explained with the aid of the following diagram.

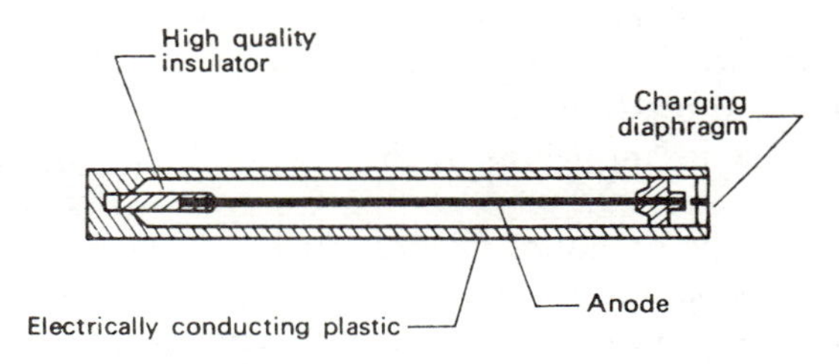

FIG. 6.2. Non self-reading condenser-type pocket ionization chamber.

The instrument consists of an outer cylindrical wall, about 4.75 mm thick, made of electrically conducting plastic. Coaxial with the outer wall, but separated from it by a very high quality insulator, is a center wire. This center wire, or central anode, is positively charged with respect to the wall. When the chamber is exposed to X- or to gamma radiation, the ionization produced in the measuring cavity as a result of interactions between photons and the wall, discharges the condenser, thereby decreasing the potential of the anode. This decrease in the anode voltage is directly proportional to the ionization produced in the cavity, which in turn is directly proportional to the radiation exposure. For example, consider the following instance:

Example 6.3

Chamber volume = 2 cm^3.
Chamber filled with air at S.T.P.
Electrical capacity = 5 $\mu\mu$F.
Voltage across chamber before exposure to radiation = 180 V.
Voltage across chamber after exposure to radiation = 160 V.
Exposure time = $\frac{1}{2}$ hr.
Calculate the radiation exposure and the exposure rate.

The exposure is calculated as follows:

$$C \times \Delta V = \Delta Q \tag{6.8}$$

$$5 \times 10^{-12} \text{ farads} \times (180 - 160) \text{ volts} = 1 \times 10^{-10} \text{ coulombs.}$$

Since one exposure unit is equal to 1 C/kg, the exposure measured by this chamber is

$$\frac{1 \times 10^{-10}\text{ C}}{2\text{ cm}^3 \times 1.293 \times 10^{-6}\text{ kg/cm}^3} = 3.867 \times 10^{-5}\text{ C/kg},$$

which corresponds to

$$3.867 \times 10^{-5}\text{ C/kg} \times 3881 \frac{\text{R}}{\text{C/kg}} = 0.150\text{ R},$$

or 150 mR. The exposure rate was

$$\frac{150\text{ mR}}{0.5\text{ hr}} = 300\text{ mR/hr},$$

or 77.3 μC/kg per hour.

A chamber built according to this principle is called an "air wall" chamber. When such a chamber is used, care must be taken that the walls are of the proper thickness for the energy of the radiation being measured. If the walls are too thin, an insufficient number of photons will interact to produce primary electrons; if too thick, the primary radiation will be absorbed to a significant degree by the wall, and an attenuated primary electron flux will result.

The determination of the optimum thickness may be illustrated by an experiment in which the ionization produced in the cavity of an ionization chamber is measured as the wall thickness is increased from a very thin wall until it becomes relatively thick. In performing this experiment, care must be taken to prevent secondary electrons that are formed outside the chamber walls and beta rays from the gamma ray source from reaching the sensitive volume of the chamber. When this is done, and the cavity ionization is plotted against the wall thickness, the curve shown in Fig. 6.3, results.

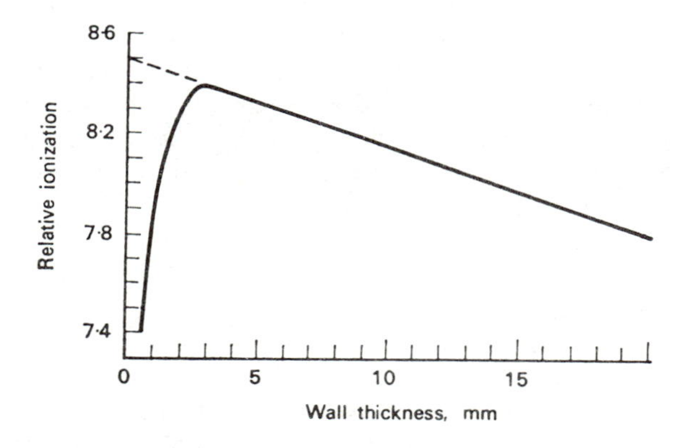

FIG. 6.3. Ion pairs per unit volume as a function of wall thickness. The ionization chamber in this case was made of pure carbon, and was a cylinder 20 mm inside diameter and 20 mm long. (W. V. Mayneord and J. E. Roberts, *British Journal of Radiology* **10**, 365, 1937.)

Since the cavity ionization is caused mainly by primary electrons resulting from gamma-ray interactions with the wall, increasing the wall thickness allows more photons to interact, thereby producing more primary electrons which ionize the gas in the chamber as they traverse the cavity. However, when the wall thickness reaches a point where a primary electron produced at the outer surface of the wall is not sufficiently energetic to pass through the wall into the cavity, the ionization in the cavity begins to decrease. The wall thickness at which this just begins is the *equilibrium wall thickness.*

As the wall material departs from air equivalence, the response of the ionization chamber becomes energy dependent. By proper choice of wall material and thickness, the maximum in the curve of Fig. 6.3 can be made quite broad, and the ionization chamber, as a consequence, made relatively energy independent over a wide range of quantum energies. In practice, this approximately flat response spans the energy range from about 200 keV to about 2 MeV. In this range of energies, the Compton effect is the predominant mechanism of energy transfer. For lower energies, the probability of a Compton interaction increases approximately as the wavelength, while the probability of a photoelectric interaction is approximately proportional to the cube of the quantum wavelength.

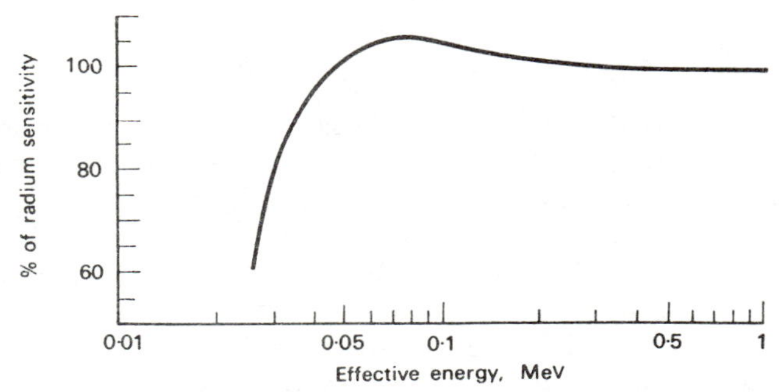

FIG. 6.4. Energy-dependence characteristics of the pocket dosimeter shown in Fig. 9.18.

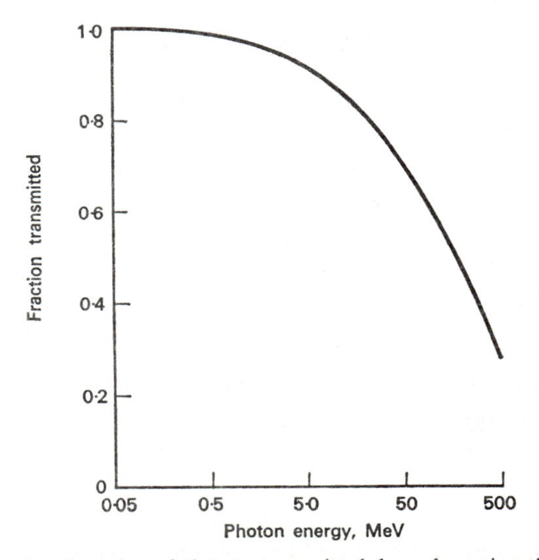

FIG. 6.5. Fractional number of photons transmitted through an air wall of thickness equal to the maximum range of the secondary electrons. (From N.B.S. Handbook 55, 1954.)

The total number of primary electrons therefore increases, and the sensitivity of the chamber consequently increases; the increased sensitivity, however, reached a peak as the quantum energy decreases and then, because of the severe attenuation of the incident radiation by the chamber wall, the sensitivity rapidly decreases. These effects are shown in Fig. 6.4, a curve showing the energy correction factor for a pocket dosimeter.

For quantum energies greater than 3 MeV, neither the C/kg nor the roentgen is used as the unit of measurement of exposure. This is due to the fact that the high energy, and consequently the long range of the primary electrons produced in the wall, makes it impossible to build an instrument that meets the criteria for measuring the roentgen. Because of the long range of the primary electrons, very thick walls are necessary. However, when the walls are sufficiently thick, on the basis of the range of the primary electrons, they attenuate the gamma radiation to a significant degree, as shown in Fig. 6.5. Under these conditions, it is not possible to attain electronic equilibrium, since the radiation intensity within the wall is not constant, and the primary electrons, consequently, are not produced uniformly throughout the entire volume of wall from which they may reach the cavity.

Exposure-Dose Relationship

The air wall chamber, as the name implies, measures the energy absorption in air. In most instances we are interested in the energy absorbed in tissue. Since energy absorption is approximately proportional to the electronic density of the absorber in the energy region where exposure units are valid, it can be shown that the tissue dose is not necessarily equal to the air dose for any given radiation field. For example, if we consider muscle tissue to have a specific gravity of 1, and to have an elementary composition of 5.98×10^{22} hydrogen atoms per gram, 2.75×10^{22} oxygen atoms per gram, 0.172×10^{22} nitrogen atoms per gram, and 6.02×10^{21} carbon atoms per gram, then the electronic density is 3.28×10^{23} electrons per gram. For air, whose density is 1.293×10^{-3} g/cm^3, the electronic density is 3.01×10^{23} electrons per gram. The energy absorption, in joules per kilogram of tissue, corresponding to an exposure of 1 coulomb per kilogram air is, therefore,

$$\frac{3.28}{3.01} \times 34 = 37 \text{ J/kg tissue.}$$

This value agrees very well with calorimetric measurements of energy absorption by soft tissue from an exposure of 1 C/kg air. By analogy, an exposure of 1 R, which corresponds to 87.8 ergs per gram air, leads to an absorption of 95 ergs per gram muscle tissue. This tissue dose from a 1-R exposure is very close to the tissue dose of 100 ergs per gram, which corresponds to 1 rad. For this reason, an exposure of 1 roentgen is frequently considered approximately equivalent to an absorbed dose of 1 rad, and the unit "roentgen" is loosely (but incorrectly) used to mean "rad".

The exposure unit bears a simple quantitative relationship to the gray that permits the calculation of absorbed dose in any medium whose exposure, in C/kg, is known. This relationship may be illustrated by the following example.

Example 6.4

Consider a gamma-ray beam of quantum energy 0.3 MeV. If the photon flux is 1000 quanta per cm^2/sec, what is the exposure rate at a point in this beam and what is the absorbed dose rate for soft tissue at this point?

From Fig. 5.18, the energy absorption coefficient for air, μ_a, at 20°C, for 300 keV photons is found to be $3.46 \times 10^{-5}\,\text{cm}^{-1}$. The exposure rate in C/kg per second is given by

$$\dot{X} = \frac{\phi \text{ photons/cm}^2\text{-sec} \times E \text{ MeV/photon} \times 1.6 \times 10^{-13}\text{ J/MeV} \times \mu_a\,\text{cm}^{-1}}{\rho_a\,\text{kg/cm}^3 \times 34\,\dfrac{\text{J/kg}}{\text{C/kg}}} \tag{6.9}$$

Substituting the appropriate numerical values into equation (6.9), we have

$$\dot{X} = \frac{10^3 \times 0.3 \times 1.6 \times 10^{-13} \times 3.46 \times 10^{-5}}{(1.293 \times 10^{-6} \times 273/293) \times 34}$$

$$= 4 \times 10^{-11}\,\frac{\text{C}}{\text{kg}}\Big/\text{sec.}$$

The absorbed dose rate, in grays per second, is given by

$$\dot{D} = \frac{\dfrac{\phi \text{ photons/cm}^2}{\text{sec}} \times E \text{ MeV/photon} \times 1.6 \times 10^{-13}\text{ J/MeV} \times \mu_m\,\text{cm}^{-1}}{\rho_m\,\text{kg/cm}^3 \times 1\,\dfrac{\text{J/kg}}{\text{Gy}}}. \tag{6.10}$$

When the value for the energy absorption coefficient for tissue for 0.3 MeV photons, $\mu_m = 0.0312\,\text{cm}^{-1}$, and a tissue density $\rho_m = 0.001\,\text{kg/cm}^3$ are substituted into equation (6.10), the absorbed dose rate is found to be 1.5×10^{-9} Gy/sec.

The relationship between exposure and dose is obtained from the ratio of equation (6.10) to (6.9):

$$\frac{\dot{D}}{\dot{X}} = \frac{(\phi \times E \times 1.6 \times 10^{-13} \times \mu_m)/\rho_m}{(\phi \times E \times 1.6 \times 10^{-13} \times \mu_a)/(\rho_a \times 34)},$$

$$\dot{D} = 34 \times \frac{\mu_m/\rho_m}{\mu_a/\rho_a} \times \dot{X}\text{ Gy/s.} \tag{6.11}$$

To obtain a tissue dose in rads when the exposure is given in roentgens, we use the analogous expression

$$\text{rads} = \frac{87.7}{100} \times \frac{\mu_m/\rho_m}{\mu_a/\rho_a} \times \text{roentgens.} \tag{6.12}$$

Equations (6.11) and (6.12) show that the radiation dose absorbed from any given exposure is determined by the ratio of the mass absorption coefficient of the medium to that of air. In the case of tissue, the ratio of dose to exposure remains approximately constant over the quantum energy range of about 0.1–10 MeV because the chief means of interaction between the tissue and the radiation is Compton scattering, and the cross section for Compton scattering depends mainly on electronic density of the absorbing medium. In the case of lower energies, photoelectric absorption becomes important, and the cross section for this mode of interaction increases with atomic number of the absorber. As a consequence of this dependence on atomic number, bone, which contains approximately 10% by weight of calcium absorbs much more energy than does soft tissue from a given air dose of low energy X-rays. This point is illustrated in Fig. 6.6, which shows the number of joules per kg absorbed per coulomb per kg of exposure for fat, muscle, and bone as a function of quantum energy.

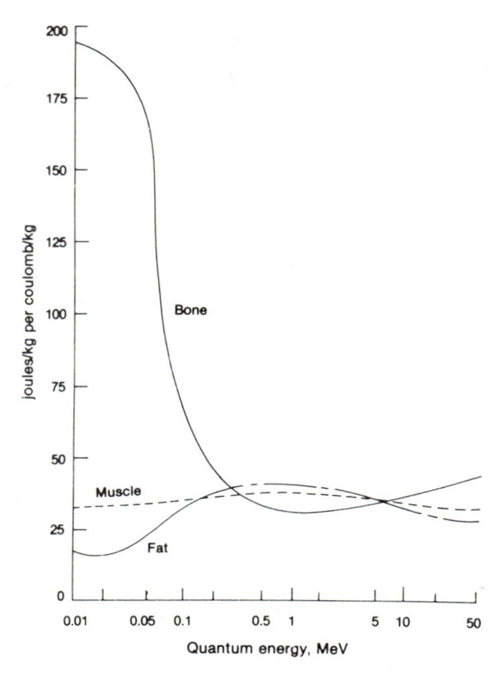

FIG. 6.6. Energy absorption per X unit (coulomb per kilogram) exposure for several tissues. (Adapted from O. Glasser, *Medical Physics*, Vol. II.)

Absorbed Dose Measurement: Bragg–Gray Principle

If a cavity ionization chamber is built with wall material whose radiation absorption property is similar to that of tissue, then, by taking advantage of the *Bragg–Gray principle*, an instrument can be built to measure tissue dose directly. According to the Bragg–Gray principle, the amount of ionization produced in a small gas-filled cavity surrounded by a solid absorbing medium is proportional to the energy absorbed by the solid. Implicit in the practical application of this principle is that the gas cavity be small enough relative to the mass of the solid absorber to leave unchanged the angular and velocity distributions of the primary electrons. This requirement is fulfilled if the primary electrons lose only a very small fraction of their energy in traversing the gas-filled cavity. If the cavity is surrounded by a solid medium of proper thickness to establish electronic equilibrium, then the energy absorbed per unit mass of wall, dE_m/dM_m, is related to the energy absorbed per unit mass of gas in the cavity, dE_g/dM_g, by

$$\frac{dE_m}{dM_m} = \frac{S_m}{S_g} \times \frac{dE_g}{dM_g}, \tag{6.13}$$

where S_m is the mass stopping power of the wall material and S_g is the mass stopping power of the gas. Since the ionization per unit mass of gas is a direct measure of dE_g/dM_g, equation (6.13) can be rewritten as

$$\frac{dE_m}{dM_m} = \rho_m \times w \times J, \tag{6.14}$$

where ρ_m is the ratio of the mass stopping powers of the solid relative to that of the gas,

S_m/S_g, w is the mean energy dissipated in the production of an ion pair in the gas, and J is the number of ion pairs per unit mass of gas. Using the appropriate equations for stopping power given in Chapter 5, it becomes a relatively simple matter to compute ρ_m for electrons of any given energy. For those cases where the gas in the cavity is the same substance as the chamber wall, such as methane and paraffin, ρ_m is equal to unity. Table 6.1 shows stopping power ratios, relative to air, of several substances for monoenergetic

TABLE 6.1. MEAN MASS STOPPING POWER RATIOS, RELATIVE TO AIR, FOR ELECTRONIC EQUILIBRIUM SPECTRA GENERATED BY INITIALLY MONOENERGETIC ELECTRONS

Initial energy, MeV	Element and state of molecular binding								
	Hydrogen saturated	Hydrogen unsaturated	Carbon saturated	Carbon unsaturated	Carbon highly chlorinated	Nitrogen, amines, nitrates	Nitrogen ring	Oxygen, —O—	Oxygen, O=
0.1	2.52	2.59	1.016	1.021	1.047	0.976	1.018	0.978	0.994
0.2	2.52	2.59	1.015	1.019	1.043	0.978	1.016	0.979	0.995
0.3	2.48	2.55	1.014	1.018	1.040	0.979	1.016	0.981	0.995
0.327	2.48	2.54	1.014	1.018	1.040	0.979	1.015	0.981	0.995
0.4	2.46	2.53	1.014	1.018	1.038	0.980	1.015	0.981	0.996
0.5	2.44	2.51	1.013	1.017	1.037	0.980	1.015	0.982	0.996
0.6	2.44	2.50	1.012	1.016	1.035	0.980	1.013	0.981	0.995
0.654	2.43	2.49	1.011	1.014	1.034	0.979	1.012	0.981	0.994
0.7	2.42	2.48	1.010	1.013	1.033	0.978	1.011	0.980	0.993
0.8	2.40	2.46	1.009	1.012	1.031	0.978	1.010	0.979	0.992
1.0	2.39	2.44	1.004	1.008	1.026	0.975	1.005	0.977	0.988
1.2	2.37	2.42	1.001	1.004	1.022	0.973	1.002	0.974	0.985
1.308	2.36	2.42	0.999	1.002	1.019	0.971	1.000	0.972	0.983
1.5	2.35	2.39	0.995	0.998	1.015	0.967	0.996	0.969	0.980

From N.B.S. Handbook 85, *Physical Aspects of Irradiation*, 1964.

electrons. For gamma radiation, however, the problem of evaluating ρ_m is more difficult; the relative fraction of the gamma rays that will interact by each of the competing mechanisms, as well as the spectral distribution of the primary electrons (Compton, photoelectric, and pair produced electrons) must be considered, and a mean value for relative stopping power must be determined. For the equilibrium electron spectra generated by gamma rays from ^{198}Au, ^{137}Cs, and ^{60}Co, the values for the mean mass relative stopping powers are given in Table 6.2. For air, w, the mean energy loss for the production

TABLE 6.2. MEAN MASS STOPPING POWER RATIOS, S_m/S_{air} FOR EQUILIBRIUM ELECTRON SPECTRA GENERATED BY ^{198}Au, ^{137}Cs, AND ^{60}Co, ON THE ASSUMPTION THAT THE ELECTRONS SLOW DOWN IN A CONTINUOUS MANNER

Energy, MeV	Medium		
	Graphite	Water	Tissue
0.411 (^{198}Au)	1.032		
0.670 (^{137}Cs)	1.027	1.162	1.145
1.25 (^{60}Co)	1.017	1.155	1.137

From N.B.S. Handbook 78, *Report of the International Commission on Radiological Units and Measurements*, 1959.

of an ion pair in air, has a value of 34 eV. To determine the radiation absorbed dose, it is necessary only to measure the ionization per unit mass of gas, J.

Example 6.5

Calculate the absorbed dose rate from the following data on a tissue equivalent chamber with walls of equilibrium thickness embedded within a phantom and exposed to ^{60}Co gamma-rays for 10 min. The volume of the air cavity in the chamber is 1 cm^3, the capacitance is 5 $\mu\mu$F, and the gamma-ray exposure results in a decrease of 72 V across the chamber. The charge collected by the chamber is

$$\begin{aligned} Q &= C\Delta V \\ &= 5 \times 10^{-12}\ \mathrm{F} \times 72\ \mathrm{V} \\ &= 3.6 \times 10^{-10}\ \mathrm{C}. \end{aligned}$$

The number of electrons collected, which corresponds to the number of ion pairs formed in the air cavity is

$$\frac{3.6 \times 10^{-10}\ \mathrm{C}}{1.6 \times 10^{-19}\ \mathrm{C/electron}} = 2.25 \times 10^{9}\ \text{electrons}.$$

Since 34 electron volts are expended per ion pair formed in air, and since the stopping power of tissue relative to air is 1.137, we have from the Bragg–Gray relationship of equation (6.14)

$$\begin{aligned} \frac{\mathrm{d}E_m}{\mathrm{d}M_m} &= \rho_m \times w \times J \\ &= \frac{1.137 \times 34\ \mathrm{eV/ip} \times 2.25 \times 10^{9}\ \mathrm{ip/cm^3} \times 1.6 \times 10^{-19}\ \mathrm{J/eV}}{1.293 \times 10^{-6}\ \mathrm{kg/cm^3} \times 1\ \mathrm{J/kg\ Gy^{-1}}} \\ &= 0.0108\ \mathrm{Gy}. \end{aligned}$$

The exposure time was 10 min, and the dose rate therefore is 1.08 mGy/min.

Kerma

In the case of indirectly ionizing radiation, such as X-rays and fast neutrons, we sometimes are interested in the initial kinetic energy of the primary ionizing particles (the photoelectrons, Compton electrons, or positron–negatron pairs in the case of photon radiation and the scattered nuclei in the case of fast neutrons) produced by the interaction of the incident radiation per unit mass of interacting medium. This quantity is called the kerma, and in the SI system is measured in units of joules per kilogram or grays (rads in the former system of units). The kerma decreases continuously with increasing depth in an absorbing medium because of the continuous decrease in the flux of the indirectly ionizing radiation. The absorbed dose, however, increases with increasing depth as the density of the primary ionizing particles and the secondary particles that they produce increases, until a maximum is reached, after which the absorbed dose decreases with continuing increase in depth. The maximum dose occurs at a depth approximately equal to the maximum range of the primary ionizing particles. The relationship between kerma and dose for photon radiation or for fast neutrons is shown in Fig. 6.7.

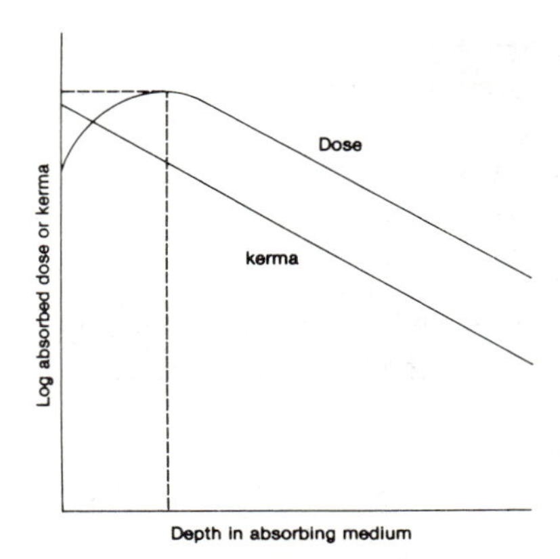

FIG. 6.7. Relationship between kerma and dose for photon radiation and for fast neutrons.

Source Strength: Specific Gamma-Ray Emission

The radiation intensity from any given gamma-ray source is used as a measure of the strength of the source. The gamma radiation exposure rate from a point source of unit activity at unit distance is called the specific gamma-ray emission, and is given in units of coulombs per kilogram per hour at 1 meter from a 1-MBq point source (or, in the older system, R per hour at 1 meter from a 1 Ci point source). The source strength may be calculated if the decay scheme of the isotope is known. In the case of ^{131}I, for example, whose gamma rays are shown in Fig. 4.7, and whose corresponding true absorption coefficients are found in Fig. 5.18, we have the following:

Quantum energy, MeV	Photons per disintegration	Energy absorption coefficient for air
0.723	0.016	$3.8 \times 10^{-3}\ m^{-1}$
0.637	0.069	3.9×10^{-3}
0.503	0.003	3.8×10^{-3}
0.326	0.002	3.8×10^{-3}
0.177	0.002	3.4×10^{-3}
0.365	0.853	3.8×10^{-3}
0.284	0.051	3.7×10^{-3}
0.080	0.051	3.2×10^{-3}
0.164	0.006	3.3×10^{-3}

The gamma-radiation exposure level is calculated by considering the energy absorbed per unit mass of air at the specified distance from the 1-MBq point source due to the photon flux at that distance, as shown in equation (6.15).

$$\dot{X} = \frac{f\ \text{phot/t} \times E\ \text{MeV/phot} \times 1.6 \times 10^{-13}\ \text{J/MeV} \times 1 \times 10^{6}\ \text{tps/MBq} \times 3.6 \times 10^{3}\ \text{s/hr} \times \mu\text{m}^{-1}}{4\pi (1\ \text{m})^2 \times \rho\ \text{kg/m}^3 \times 34 \dfrac{\text{J}}{\text{kg}} \Big/ \dfrac{\text{C}}{\text{kg}}}, \tag{6.15}$$

where $\dot{X}$ = exposure rate, C per kg per hour,
f = fraction of transformations that result in a photon of the quantum energy under consideration,
E = quantum energy, MeV,
μ = linear energy absorption coefficient, m^{-1},
ρ = density of air, kg/m^3.

This calculation is made for each different quantum energy, and the results are summed to obtain the source strength. For the 0.080-MeV gamma ray, we have

$$\dot{X} = \frac{5.1 \times 10^{-2} \times 8 \times 10^{-2} \times 1.6 \times 10^{-13} \times 1 \times 10^{6} \times 3.6 \times 10^{3} \times 3.2 \times 10^{-3}}{4\pi(1)^2 \times 1.293 \times 34}$$

$$= 1.36 \times 10^{-11} \frac{\text{C/(kg-hr)}}{\text{MBq}}.$$

The exposure rate for each of the other quanta emitted by ^{131}I is calculated in a similar manner, except that the corresponding frequency and absorption coefficient is used for each of the quanta of different energy. The results of this calculation are tabulated below:

Quantun energy, MeV	C/(kg hr) at 1 meter
0.723	4.583×10^{-11}
0.637	17.787×10^{-11}
0.503	0.598×10^{-11}
0.326	0.258×10^{-11}
0.177	0.126×10^{-11}
0.365	123.400×10^{-11}
0.284	5.569×10^{-11}
0.080	1.361×10^{-11}
0.164	0.339×10^{-11}
	Total = $1.540 \times 10^{-9} \frac{\text{C/(kg hr)}}{\text{MBq}}$ at 1 meter

Equation (6.15) contains several constants: 1×10^6 transformations per second per MBq, 3.6×10^3 seconds/hr, 1.6×10^{-13} J/MeV, $4\pi(1)^2$, 1.293 kg/m³, and 34 J/kg per C/kg. If all these constants are combined, the source strength, Γ, in C/kg per MBq at 1 meter is given by

$$\Gamma = 1.043 \times 10^{-6} \sum_i f_i \times E_i \times \mu_i \frac{\text{(C/kg) m}^2}{\text{MBq hr}}, \tag{6.16}$$

where f_i is the photons per transformation of the ith photon, E_i is the energy of the ith photon, and μ_i is the linear energy absorption coefficient in air of the ith photon. For many practical purposes, equation (6.16) may be simplified. For quantum energies from 60 keV to 2 MeV, the linear absorption coefficient varies little with energy; over this range μ is about

3.5×10^{-3} per meter. With this value, equation (6.16) may be approximated as

$$\Gamma = 3.65 \times 10^{-9} \sum_i f_i \times E_i \frac{(\mathrm{C/kg})\,\mathrm{m}^2}{\mathrm{MBq\text{-}hr}}. \quad (6.17)$$

When exposure is measured in roentgens and activity in curies, the specific gamma ray emission is approximated by

$$\Gamma = 0.5 \sum_i f_i \times E_i \frac{\mathrm{R\text{-}m}^2}{\mathrm{Ci\text{-}hr}}. \quad (6.18)$$

Table 6.3 lists the specific gamma ray emission of some isotopes that are frequently encountered by health physicists.

TABLE 6.3. SPECIFIC GAMMA-RAY EMISSION OF SOME RADIOISOTOPES

Isotope	Γ	
	$\frac{\mathrm{R\text{-}m}^2}{\mathrm{Ci\text{-}hr}}$ *	$\frac{\mathrm{X\text{-}m}^2}{\mathrm{MBq\text{-}hr}}$ **
Antimony-122	0.24	1.67E—09
Cesium-137	0.33	2.30E—09
Chromium-51	0.016	1.11E—10
Cobalt-60	1.32	9.19E—09
Gold-198	0.23	1.60E—09
Iodine-125	0.07	4.87E—10
Iodine-131	0.22	1.53E—09
Iridium-192	0.48	3.34E—09
Mercury-203	0.13	9.05E—10
Potassium-42	0.14	1.39E—09
Radium-226	0.825	5.75E—09
Sodium-22	1.20	8.36E—09
Sodium-24	1.84	12.80E—09
Zinc-65	0.27	1.88E—09

*From *Radiological Health Handbook*, revised edition, 1970.
**1 X-unit = 1 C/kg.

Internally Deposited Radioisotopes

Corpuscular Radiation

The calculation of the absorbed dose from internally deposited radioisotopes follows directly from the definition of the gray. For an infinitely large medium containing a uniformly distributed radioisotope, the concentration of absorbed energy must be equal to the concentration of energy emitted by the isotope. The energy absorbed per unit mass per transformation is called the *specific effective energy* (SEE). For practical health physics purposes, "infinitely large" may be approximated by a tissue mass whose dimensions exceed the range of the radiation from the distributed isotope. For the case of alpha and most beta radiation, this condition is easily met in practice, and the SEE is simply the average energy of the radiation divided by the mass of the tissue in which it is distributed:

$$\text{SEE } (\alpha \text{ or } \beta) = \frac{\bar{E}(\alpha \text{ or } \beta)}{m} \frac{\text{MeV}}{t} \Big/ \text{kg}. \tag{6.19}$$

The computation of the radiation absorbed dose from a uniformly distributed beta emitter within a tissue may be illustrated with the following example:

Example 6.6

Calculate the daily dose rate to a testis that weighs 18 grams and has 6660 Bq of ^{35}S uniformly distributed throughout the organ.

Sulfur is a pure beta emitter, whose maximum energy beta particle is 0.1674 MeV, and whose average energy is 0.0488 MeV. The beta-ray dose rate from q Bq uniformly dispersed in m kg of tissue, if the specific effective energy is SEE MeV per transformation per kg, is:

$$\dot{D}(\beta) = \frac{q \text{ Bq} \times 1 \text{ tps/Bq} \times \text{SEE} \frac{\text{MeV}}{t} \Big/ \text{kg} \times 1.6 \times 10^{-13} \text{ J/MeV} \times 8.64 \times 10^{4} \text{ sec/day}}{1 \frac{\text{J}}{\text{kg}} \Big/ \text{Gy}}. \tag{6.20}$$

Substituting equation (6.19) into equation (6.20) yields

$$\dot{D}(\beta) = \frac{q \text{ Bq} \times 1 \text{ tps/Bq} \times \bar{E} \text{ MeV}/t \times 1.6 \times 10^{-13} \text{ J/MeV} \times 8.64 \times 10^{4} \text{ sec/day}}{m \text{ kg} \times 1 \frac{\text{J}}{\text{kg}} \Big/ \text{Gy}} \tag{6.21}$$

$$= \frac{6.66 \times 10^{3} \times 1 \times 4.88 \times 10^{-2} \times 1.6 \times 10^{-13} \times 8.64 \times 10^{4}}{0.018 \times 1}$$

$$= 2.5 \times 10^{-4} \text{ Gy/day } (0.025 \text{ rad/day}).$$

Effective Half-life

The total dose absorbed during any given time interval after the deposition of the sulfur in the testis may be calculated by integrating the dose rate over the required time interval. In making this calculation, two factors must be considered, viz.

1. *In situ* radioactive decay of the isotope.
2. Biological elimination of the isotope.

In most instances, biological elimination follows first-order kinetics. In this case, the equation for the quantity of radioisotope within an organ at any time t after deposition of a quantity Q_0 is given by

$$Q = (Q_0 e^{-\lambda_R t})(e^{-\lambda_B t}), \tag{6.22}$$

where λ_R is the radioactive decay constant, and λ_B is the biological elimination constant. The two exponentials in equation (6.22) may be combined

$$Q = Q_0 e^{-(\lambda_R + \lambda_B)t}, \tag{6.23}$$

and, if $\lambda_E = \lambda_R + \lambda_B$, we have (6.24)

$$Q = Q_0 e^{-\lambda_E t}, \tag{6.25}$$

where λ_E is called the *effective* elimination constant. The effective half-life is then

$$T_E = \frac{0.693}{\lambda_E}. \tag{6.26}$$

From the relationship among λ_E, λ_R, and λ_B, we have

$$\frac{1}{T_E} = \frac{1}{T_R} + \frac{1}{T_B}, \tag{6.27}$$

or

$$T_E = \frac{T_R \times T_B}{T_R + T_B}. \tag{6.28}$$

In Example 6.6, for ^{35}S, $T_R = 87.1$ days and T_B, the biological half-life in the testis, is 623 days. The effective half-life, therefore, is

$$T_E = \frac{87.1 \times 623}{87.1 + 623} = 76.4 \text{ days},$$

and the effective elimination constant is

$$\lambda_E = \frac{0.693}{76.4} = 0.009 \text{ day}^{-1}.$$

Total Dose: Dose Commitment

The dose, dD, during an infinitesimally small period of time, dt, at a time interval t after an initial dose rate $\dot{D}_0$ is

$$\begin{aligned} dD &= \text{instantaneous dose rate} \times dt \\ &= \dot{D}_0 e^{-\lambda_E t}\, dt. \end{aligned} \tag{6.29}$$

The total dose during a time interval τ after deposition of the isotope is

$$D = \dot{D}_0 \int_0^{\tau} e^{-\lambda_E t}\, dt, \tag{6.30}$$

which, when integrated, yields

$$D = \frac{\dot{D}_0}{\lambda_E}(1 - e^{-\lambda_E \tau}). \tag{6.31}$$

For an infinitely long time—that is, when the isotope is completely gone—equation (6.31) reduces to

$$D = \frac{\dot{D}_0}{\lambda_E}. \tag{6.32}$$

For practical purposes, an "infinitely long time" corresponds to about six effective half-lives. It should be noted that the dose due to total decay is merely equal to the product of the initial dose rate, $\dot{D}_0$, and the average life of the radioisotope within the organ, $1/\lambda_E$. For the case in Example 6.6, the total absorbed dose during the first 5 days after deposition of the radiosulfur in the testis is, according to equation (6.31),

$$\begin{aligned} D &= \frac{2.5 \times 10^{-4}\,\text{Gy/day}}{0.009\,\text{day}^{-1}}(1 - e^{-0.009\,\text{d}^{-1} \times 5\text{d}}) \\ &= 1.2 \times 10^{-3}\,\text{Gy (0.12 rad)}, \end{aligned}$$

and the dose from complete decay, is, from equation (6.32)

$$D = \frac{2.5 \times 10^{-4}\,\text{Gy/day}}{0.009\,\text{day}^{-1}} = 0.028\,\text{Gy (2.8 rads)}.$$

The 0.028-Gy total dose absorbed from the deposition of the radiosulfur is called the dose commitment to the testes from this incident. The *dose commitment* is defined as the absorbed dose from a given practice or from a given exposure. Although the dose commitment in this example was due to an internally deposited radioisotope, the dose commitment concept is applicable to external radiation as well as to radiation from internally deposited radioisotopes.

In the example cited above, the testis behaved as if the radioisotope were stored in a single compartment. In many cases, an organ or tissue behaves as if the radioisotope is stored in more than one compartment. Each compartment follows first order kinetics, and is emptied at its own clearance rate. Thus, for example, cesium is found to be uniformly distributed throughout the body, although the body behaves as if the cesium is stored in two compartments. One compartment contains 10% of the total body burden and has a retention half-time of 2 days, while the second compartment contains the other 90% of the body's cesium content, and has a clearance half-time of 110 days. The retention curve for cesium, therefore, is given by the equation

$$q(\tau) = 0.1\, q_0\, e^{-(0.693\tau/2\,\text{days})} + 0.9\, q_0\, e^{-(0.693\tau/110\,\text{days})}, \tag{6.33}$$

where $q(\tau)$ is the body burden at time τ after deposition of q_0 amount of cesium in the body. Ten per cent of the total is deposited in compartment 1 and 90% is deposited in compartment 2. Generally, if there is more than one compartment, the body burden at any time τ after deposition of q amount of the isotope is given by

$$q(\tau) = q_{10}\, e^{-\lambda_1 \tau} + q_{20}\, e^{-\lambda_2 \tau} + \cdots + q_{n0}\, e^{-\lambda_n \tau} \tag{6.34}$$

where q_{10}, $q_{20} \ldots q_{n0}$ = amount deposited in compartments 1, 2, ... n and λ_1, $\lambda_2, \ldots \lambda_n$ = effective clearance rates for compartments 1, 2, ... n.

Since the activity in each compartment contributes to the dose to that organ or tissue, equation (6.31) becomes, for the multi-compartment case,

$$D = \frac{\dot{D}_{10}}{\lambda_{1E}}(1 - e^{-\lambda_{1E}\tau}) + \frac{\dot{D}_{20}}{\lambda_{2E}}(1 - e^{-\lambda_{2E}\tau}) + \cdots \frac{\dot{D}_{n0}}{\lambda_{nE}}(1 - e^{-\lambda_{nE}\tau}), \tag{6.35}$$

and when the isotope has completely been eliminated, equation (6.35) reduces to:

$$D = \frac{\dot{D}_{10}}{\lambda_{1E}} + \frac{\dot{D}_{20}}{\lambda_{2E}} + \cdots \frac{\dot{D}_{n0}}{\lambda_{nE}}. \tag{6.36}$$

Gamma Emitters

For gamma-emitting isotopes, we cannot simply calculate the absorbed dose by assuming the organ to be infinitely large because gammas, being penetrating radiations, may travel great distances within tissue, and leave the tissue without interacting. Thus, only a fraction of the energy carried by photons originating in the radioisotope containing tissue is absorbed within that tissue.

For a uniformly distributed gamma emitting isotope, the dose rate at any point p due to the isotope in the infinitesimal volume dV at any other point at a distance r from point p, as shown in Fig. 6.8, is

$$\mathrm{d}D = C\Gamma \frac{e^{-\mu r}}{r^2} \mathrm{d}V, \tag{6.37}$$

where C is the concentration of the isotope, Γ is the specific gamma-ray emission, and μ is the linear energy absorption coefficient. The dose rate at point p due to all the isotope in the tissue is computed by the contributions from all the infinitesimal volume elements

$$D = C\Gamma \int_0^V \frac{e^{-\mu r}}{r^2} \mathrm{d}V. \tag{6.38}$$

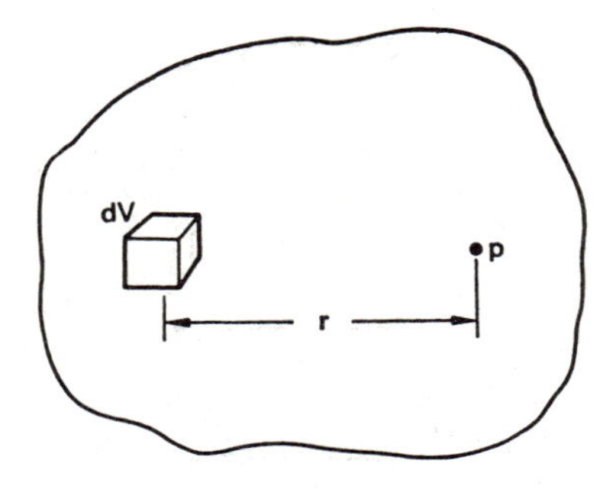

FIG. 6.8. Diagram for calculating dose at point p from the gamma rays emitted from the volume element dV in a tissue mass containing a uniformly distributed isotope.

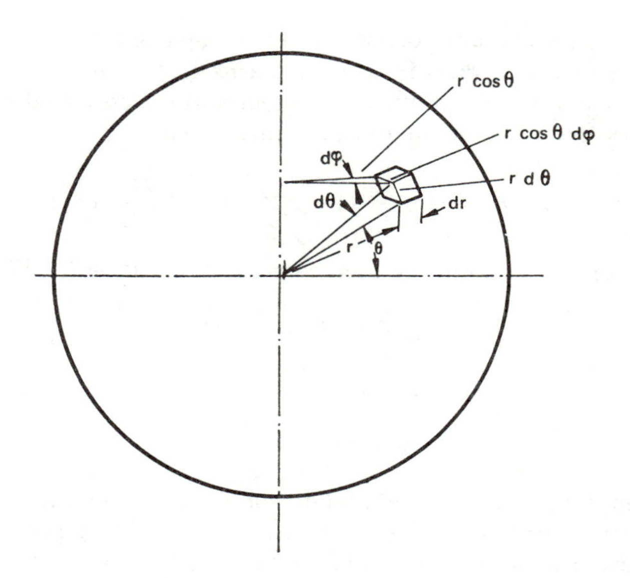

FIG. 6.9. Geometry for evaluating equation (6.38) for the center of a sphere.

For the case of a sphere, the dose rate at the center is

$$D = 4C\Gamma \int_{r=0}^{r=R} \int_{\theta=0}^{\theta=\pi/2} \int_{\varphi=0}^{\varphi=\pi} \frac{e^{-\mu r}}{r^2} \cdot r\,\mathrm{d}\theta \cdot r \cos\theta\,\mathrm{d}\varphi \cdot \mathrm{d}r. \tag{6.39}$$

Integrating with respect to each of the variables, we have, for the dose rate at the center of the sphere,

$$D \doteq C\Gamma \cdot \frac{4\pi}{\mu}(1 - e^{-\mu R}). \tag{6.40}$$

From an examination of equations (6.37), (6.38), and (6.39), it is seen that the factor that multiplies $C\Gamma$ depends only on the geometry of the tissue mass, and hence is called the geometry factor. The geometry factor, g, is defined by

$$g = \int_0^V \frac{e^{-\mu r}}{r^2}\,\mathrm{d}V. \tag{6.41}$$

Equation (6.38) may therefore be rewritten as

$$D = C\Gamma g. \tag{6.42}$$

The definition of g in equation (6.41) applies to a given point within a volume of tissue. In many health physics instances, we are interested in the average dose rate rather than the dose rate at a specific point. For this purpose, we may define an average geometry factor

$$\bar{g} = \frac{1}{V}\int g\,\mathrm{d}V. \tag{6.43}$$

For a sphere,

$$\bar{g} = \tfrac{3}{4}(g)_{\text{center}}. \tag{6.44}$$

At any other point in the sphere at a distance d from the center, the geometrical factor is given by

$$g_p = (g)_{\text{center}}\left[0.5 + \frac{1-(d^2/R)}{4(d/R)}\ln\frac{1+d/R}{|1-d/R|}\right]. \tag{6.45}$$

For a cylinder, the average geometry factor depends on the radius and height. Table 6.4 gives the numerical values of $\bar{g}$ for cylinders of various heights and radii.

Example 6.7

A spherical tank, capacity 1 cubic meter and radius 0.62 m is filled with aqueous ^{137}Cs waste containing a total activity of 37,000 MBq (1 Ci). What is the dose rate at the tank surface, if we neglect absorption by the tank wall?

TABLE 6.4. AVERAGE GEOMETRY FACTORS FOR CYLINDERS CONTAINING A UNIFORMLY DISTRIBUTED GAMMA EMITTER

Cylinder height, cm	Radius of cylinder, cm							
	3	5	10	15	20	25	30	35
2	17.5	22.1	30.3	34.0	36.2	37.5	38.6	39.3
5	22.3	31.8	47.7	56.4	61.6	65.2	67.9	70.5
10	25.1	38.1	61.3	76.1	86.5	93.4	98.4	103
20	25.7	40.5	68.9	89.8	105	117	126	133
30	25.9	41.0	71.3	94.6	112	126	137	146
40	25.9	41.3	72.4	96.5	116	131	143	153
60	26.0	41.6	73.0	97.8	118	134	148	159
80	26.0	41.6	73.3	98.4	119	135	150	161
100	26.0	41.6	73.3	98.5	119	136	150	162

From HINE and BROWNELL, *Radiation Dosimetry*, Academic Press.

From Table 6.3 we find $\Gamma = 2.3 \times 10^{-9}$ X units per hour per MBq at 1 meter. Since water absorbs 38 Gy per X unit, the dose rate is 8.74×10^{-8} Gy per hour per MBq at 1 meter. The absorption coefficient of water for the 0.661 MeV gammas from ^{137}Cs is listed in Table 5.3 as 0.0327 cm²/g. Since the density of water is 1 g/cm³, the linear absorption coefficient is 0.0327 per cm, or 3.27 per meter. The dose rate at the center of the sphere is found by substituting the respective values into equation (6.40):

$$\begin{aligned}\dot{D}_0 &= 3.7 \times 10^4\ \text{MBq/m}^3 \times 8.74 \times 10^{-8}\ \frac{\text{Gy-m}^2}{\text{MBq-hr}} \times \frac{4\pi}{3.27\ \text{m}^{-1}}(1 - e^{-3.27 \times 0.62})\\ &= 1.08 \times 10^{-2}\ \text{Gy/hr} \quad (1.08\ \text{rad/hr}).\end{aligned}$$

From equation (6.45), we see that the surface dose rate is $0.5 \times D_0$.

$$\therefore \quad \dot{D}_{\text{surface}} = 0.5 \times 1.08 \times 10^{-2} = 0.54 \times 10^{-2}\ \text{Gy/hr} \quad (0.54\ \text{rad/hr}).$$

MIRD Method

To account for the partial absorption of gamma ray energy in organs and tissues, the Medical Internal Radiation Dose Committee of the Society of Nuclear Medicine (MIRD) developed a formalized system for calculating the dose to a "target" organ or tissue (T) from a "source" organ (S) containing a uniformly distributed radioisotope. This system is based on the concept of *absorbed fraction*, that is, the fraction of the energy radiated by the source organ which is absorbed by the target organ. S and T may be either the same organ or two different organs bearing any of the possible relationships to each other shown in Fig. 6.10. These absorbed fractions are calculated by the application of Monte Carlo methods to the interactions and fate of photons or electrons following their emission from the deposited radionuclide.

Monte Carlo methods are useful in the solution of problems where events such as the interaction of photons with matter are governed by probabilistic rather than deterministic laws. In Monte Carlo solutions, individual simulated photons (or other corpuscular radiation) are "followed" in a computer from one interaction to the next. Since the radioisotope is assumed to be uniformly distributed throughout a given volume, and since radioactive transformation is a random process occurring at a mean rate that is an inherent characteristic of given isotope, we can start the process by randomly (in both space and time within the constraints of the boundaries of the known volume and the transformation rate constant of the radionuclide) initiating a radioactive transformation. For any of these transformations, we know the energy of the emitted radiation, its starting point, and its initial direction. The probability of each possible type of interaction within the organ, and the energy transferred during each interaction are also known. A situation is simulated by starting a very large number of such nuclear transformations, following the history of each particle as it traverses the target tissue and summing the total amount of energy that the particles dissipate within the target tissue. For a concentration of 1 Bq per cm^3 of tissue, for example, there would be 1 such start per cm^3 per second. Since the sum of the initial energies of these particles is known, the fraction of the emitted energy absorbed by the target tissue can be calculated:

$$\text{absorbed fraction} = \varphi = \frac{\text{energy absorbed by target}}{\text{energy emitted by source}}. \tag{6.46}$$

Since the mean free paths of photons usually are large relative to the dimensions of the organ in which the photon emitting isotope is distributed, the absorbed fraction for photons is less than 1. For nonpenetrating radiation, the absorbed fraction usually is either 1 or 0, depending on whether the source and target organs are the same or different.

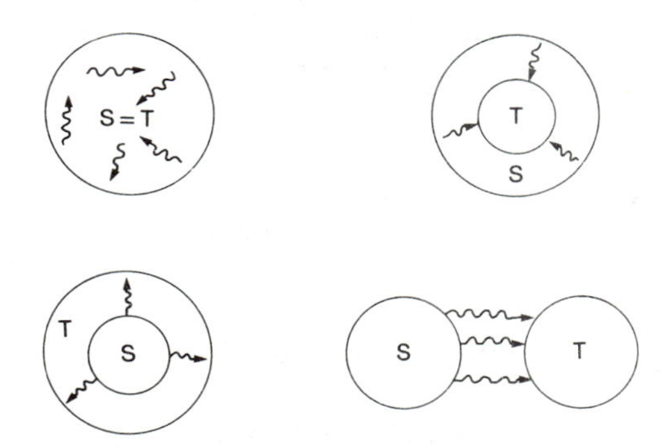

FIG. 6.10. Possible physical relationships beteen source organ and target organ.

Absorbed fractions for photons of various energies for point isotropic sources and for uniformly distributed sources in tissue and in water for spheres, cylinders, and ellipsoids have been calculated and published by MIRD in several Supplements to the *Journal of Nuclear Medicine*. Tables 6.5, 6.6, and 6.7 show some of these absorbed fractions.

TABLE 6.5. ABSORBED FRACTIONS FOR UNIFORM DISTRIBUTION OF ACTIVITY IN SMALL SPHERES* AND THICK ELLIPSOIDS*

	ϕ, MeV										
Mass	0.020	0.030	0.040	0.060	0.080	0.100	0.160	0.364	0.662	1.460	2.750
0.3	0.684	0.357	0.191	0.109	0.086	0.085	0.087	0.099	0.096	0.092	0.077
0.4	0.712	0.388	0.212	0.121	0.096	0.093	0.097	0.108	0.108	0.099	0.083
0.5	0.731	0.412	0.229	0.131	0.104	0.099	0.104	0.116	0.117	0.104	0.089
0.6	0.745	0.431	0.244	0.140	0.111	0.105	0.111	0.122	0.124	0.109	0.093
1.0	0.780	0.486	0.289	0.167	0.135	0.125	0.130	0.142	0.144	0.125	0.106
2.0	0.818	0.559	0.360	0.212	0.173	0.160	0.162	0.174	0.173	0.153	0.127
3.0	0.840	0.600	0.405	0.245	0.201	0.188	0.186	0.197	0.195	0.174	0.143
4.0	0.856	0.629	0.438	0.271	0.222	0.209	0.205	0.216	0.213	0.190	0.156
5.0	0.868	0.652	0.464	0.294	0.241	0.227	0.222	0.231	0.228	0.204	0.167
6.0	0.876	0.671	0.485	0.312	0.258	0.241	0.236	0.245	0.240	0.216	0.177

*The principal axes of the small spheres and thick ellipsoids are in the ratios of 1/1/1 and 1/0.667/1.333.
From G. L. Brownell, W. H. Ellett, and A. R. Reddy: Absorbed Fractions for Photon Dosimetry, *J. Nuclear Medicine*, Supplement No. 1, MIRD Pamphlet No. 3, 1968.

TABLE 6.6. ABSORBED FRACTIONS FOR CENTRAL POINT SOURCES IN RIGHT CIRCULAR CYLINDERS*

	ϕ, MeV					
Mass, kg	0.040	0.080	0.160	0.364	0.662	1.460
2	0.528	0.258	0.224	0.240	0.229	0.200
4	0.645	0.336	0.290	0.295	0.288	0.253
6	0.712	0.391	0.335	0.333	0.326	0.286
8	0.757	0.435	0.370	0.363	0.354	0.311
10	0.789	0.471	0.399	0.387	0.376	0.332
20	0.878	0.593	0.501	0.472	0.453	0.401
30	0.917	0.668	0.568	0.528	0.504	0.446
40	0.940	0.721	0.618	0.571	0.543	0.480
50	0.954	0.761	0.658	0.605	0.575	0.509
60	0.964	0.792	0.691	0.633	0.602	0.533
70	0.971	0.818	0.719	0.658	0.625	0.553
80	0.977	0.838	0.743	0.679	0.646	0.572
90	0.981	0.856	0.763	0.698	0.664	0.588
100	0.984	0.871	0.781	0.714	0.680	0.603
120	0.989	0.894	0.811	0.742	0.708	0.629
140	0.992	0.911	0.834	0.765	0.730	0.651
160	0.994	0.924	0.852	0.784	0.749	0.669
180	0.994	0.933	0.866	0.800	0.765	0.685
200	0.994	0.939	0.877	0.813	0.777	0.698

*The principal axes of the right circular cylinders are in the ratio of 1/1/0.75.
From G. L. Brownell, W. H. Ellet, and A. R. Reddy: Absorbed Fractions for Photon Dosimetry, *J. Nuclear Medicine*, Supplement No. 1, MIRD Pamphlet No. 3, 1968.

TABLE 6.7. ABSORBED FRACTIONS FOR CENTRAL POINT SOURCES IN SPHERES

	ϕ, MeV									
Mass, kg	0.020	0.030	0.040	0.060	0.100	0.140	0.160	0.279	0.662	2.750
2	0.989	0.794	0.537	0.322	0.243	0.233	0.234	0.241	0.235	0.168
4	0.996	0.878	0.658	0.421	0.317	0.301	0.297	0.302	0.293	0.209
6	0.999	0.916	0.727	0.488	0.370	0.348	0.342	0.344	0.330	0.238
8	0.999	0.938	0.772	0.540	0.413	0.386	0.379	0.377	0.359	0.259
10	0.999	0.952	0.806	0.581	0.448	0.418	0.409	0.405	0.382	0.277
20	0.999	0.982	0.894	0.709	0.569	0.529	0.517	0.500	0.461	0.339
30	0.999	0.991	0.932	0.780	0.644	0.600	0.587	0.562	0.514	0.380
40	0.999	0.995	0.954	0.826	0.698	0.652	0.639	0.608	0.554	0.411
50	0.999	0.996	0.966	0.859	0.738	0.692	0.679	0.644	0.586	0.436
60	0.999	0.997	0.974	0.882	0.770	0.725	0.712	0.675	0.613	0.457
70	0.999	0.998	0.980	0.900	0.796	0.752	0.739	0.700	0.637	0.476
80	0.999	0.998	0.983	0.915	0.818	0.775	0.762	0.722	0.657	0.492
90	0.999	0.999	0.986	0.926	0.836	0.794	0.781	0.741	0.675	0.507
100	0.999	0.999	0.988	0.935	0.851	0.811	0.799	0.758	0.691	0.520
120	0.999	0.999	0.991	0.948	0.876	0.839	0.827	0.786	0.719	0.544
140	0.999	0.999	0.993	0.958	0.895	0.860	0.849	0.809	0.742	0.564
160	0.999	0.999	0.995	0.965	0.910	0.878	0.867	0.829	0.761	0.582
180	0.999	0.999	0.996	0.971	0.923	0.892	0.882	0.845	0.778	0.598
200	0.999	0.999	0.998	0.976	0.933	0.904	0.894	0.858	0.792	0.612

From G. L. Brownell, W. H. Ellett, and A. R. Reddy: Absorbed Fractions for Photon Dosimetry, *J. Nuclear Medicine*, Supplement No. 1, MIRD Pamphoet No. 3, 1968.

Example 6.8

The use of these absorbed dose fractions may be illustrated by their application to calculations of the dose rate to a 0.6-kg sphere made of tissue equivalent material in which 1 MBq of ^{131}I is uniformly distributed.

In this case, energy will be absorbed from the beta particles and from the gamma rays. Since the range in tissue of the betas is very small, we can assume that all the beta ray energy is absorbed. For the gammas however, only a fraction of the energy will be absorbed. The total energy absorbed from the ^{131}I is simply the sum of the emitted beta ray energy plus the fraction of the emitted gamma ray energy that is absorbed by the sphere. This sum is called the effective energy per transformation. The absorbed fraction of the gamma rays depends on the size of the absorbing medium and on the photon energy.

Using the absorbed dose fractions given in Table 6.5 and interpolating for gamma-ray energies lying between those listed in the table, we calculate the absorbed gamma ray energy per ^{131}I transformation as follows:

$$E_e(\gamma) = \sum_i E_{\gamma i} \times n_i \times \varphi_i, \tag{6.47}$$

where $E_e(\gamma)$ = absorbed gamma ray energy, MeV/transformation,
$E_{\gamma i}$ = energy of the ith gamma photon, MeV,
n_i = number of photons of ith energy per transformation,
φ_i = absorbed fraction of the ith's photon energy.

Photon energy, $E_{\gamma i}$, MeV	×	Photons per transformation, n_i	Absorbed fraction, φ	=	Absorbed energy, MeV/t
0.723		0.016	0.123		0.0014
0.637		0.069	0.124		0.0055
0.503		0.003	0.123		0.0002
0.326		0.002	0.120		0.0001
0.177		0.002	0.112		0.0000
0.365		0.853	0.122		0.0380
0.284		0.051	0.118		0.0017
0.080		0.051	0.111		0.0005
0.164		0.006	0.111		0.0001
					$E_e(\gamma) = 0.0474$ MeV/t

The mean beta ray energy per ^{131}I transformation, which corresponds to the effective energy for the betas, may be calculated by substituting the mean beta-ray energies for ^{131}I listed in the Output Data in Fig. 6.11 into equation (6.48).

$$E_e(\beta) = \sum \bar{E}_{\beta i} \times n_{\beta i} \tag{6.48}$$

$$= (0.0701 \times 0.016) + (0.0955 \times 0.069) + (0.1428 \times 0.005) + (0.1917 \times 0.904) + (0.285 \times 0.006) = 0.183 \text{ MeV/t}.$$

The effective energy per transformation, E_e, that is, the amount of energy absorbed by the 0.6-kg tissue equivalent sphere per ^{131}I transformation, is

$$E_e = E_e(\gamma) + E_e(\beta) \tag{6.49}$$

$$= 0.047 + 0.183 = 0.230 \text{ MeV/t}.$$

The dose rate to a mass of m kg that absorbs E_e MeV per transformation from q Bq of activity within the mass is given by

$$\dot{D} = \frac{q \text{ Bq} \times 1 \text{ tps/Bq} \times E_e \text{ MeV/t} \times 1.6 \times 10^{-13} \text{ J/MeV} \times 8.64 \times 10^4 \text{ sec/day}}{m \text{ kg} \times 1 \frac{\text{J}}{\text{kg}} \Big/ \text{Gy}}. \tag{6.50}$$

If we substitute $q = 1 \times 10^6$ Bq,
$E_e = 0.230$ MeV/t,
$m = 0.6$ kg,

into equation (6.50), we find the dose rate to be

$$\dot{D} = 5.30 \times 10^{-3} \text{ Gy/day} \quad (0.530 \text{ rad/day}).$$

Let us now return to the formalism of the MIRD method for internal dose calculation. Let us consider two organs in the body, one of which contains the distributed radioisotope and is called the source, S; and the organ of interest T, the target which is being irradiated by S. S and T may be either the same organ or two different organs bearing any of the possible geometric relationships shown in Fig. 6.10.

The rate of energy emission by the radionuclide in the source at any time that is carried by the ith particle is given by

$$\chi_{ei} = A_s \text{ Bq} \times 1 \text{ tps/Bq} \times \bar{E}_i \text{ MeV/part} \times n_i \text{ part/t} \times 1.6 \times 10^{-13} \text{ J/MeV} \tag{6.51}$$

$$= 1.6 \times 10^{-13} A_s \times \bar{E}_i \times n_i \text{ J/sec}, \tag{6.52}$$

INPUT DATA

Radiation	%/disintegration	Transition energy (MeV)	Other nuclear parameters
Beta-1	1.6	0.25 *	Allowed
Beta-2	6.9	0.33 *	Allowed
Beta-3	0.5	0.47 *	Allowed
Beta-4	90.4	0.606 *	Allowed
Beta-5	0.6	0.81 *	First forbidden unique
Gamma-1	5.06	0.0802	M1, $\alpha_K = 1.7$, $\alpha_L = 0.17$
Gamma-2	0.6	0.1640	M4, $\alpha_K = 29$, K/L = 2.3
Gamma-3	0.18	0.1772	E2, $\alpha_K = 0.189$ (T), K/L = 4.0
Gamma-4	5.06	0.2843	E2, $\alpha_K = 0.052$, K/(L + M) = 4.0
Gamma-5	0.18	0.3258	M1, $\alpha_K = 0.0285$ (T), K/L = 6.0
Gamma-6	85.3	0.3645	E2 + 2% M1, $\alpha_K = 0.02$, K/L = 6.0
Gamma-7	0.32	0.5030	E2, $\alpha_K = 0.00749$ (T), $\alpha_L = 0.0011$ (T)
Gamma-8	6.9	0.6370	E2, $\alpha_K = 0.0039$, $\alpha_L = 0.000563$ (T)
Gamma-9	1.6	0.7229	M1, $\alpha_K = 0.004$, $\alpha_L = 0.000515$ (T)

Ref.: Lederer, C. M. et al, *Table of Isotopes*, 6th ed.
* Endpoint energy (MeV). (T) = Theoretical value.

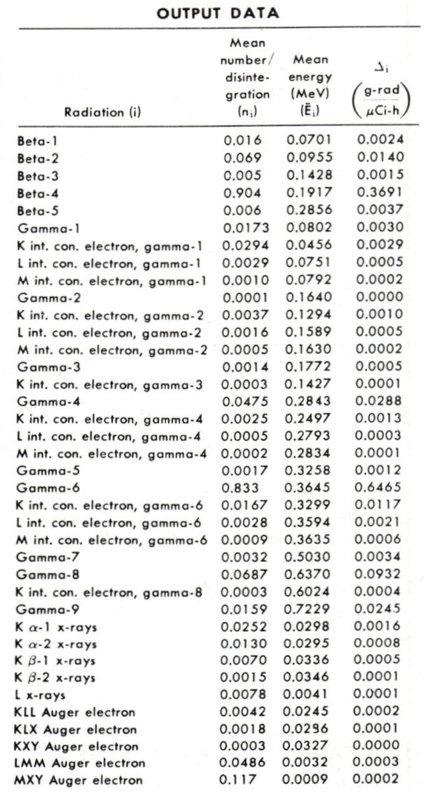

OUTPUT DATA

Radiation (i)	Mean number/disintegration (n_i)	Mean energy (MeV) ($\bar{E}_i$)	Δ_i $\left(\frac{\text{g-rad}}{\mu\text{Ci-h}}\right)$
Beta-1	0.016	0.0701	0.0024
Beta-2	0.069	0.0955	0.0140
Beta-3	0.005	0.1428	0.0015
Beta-4	0.904	0.1917	0.3691
Beta-5	0.006	0.2856	0.0037
Gamma-1	0.0173	0.0802	0.0030
K int. con. electron, gamma-1	0.0294	0.0456	0.0029
L int. con. electron, gamma-1	0.0029	0.0751	0.0005
M int. con. electron, gamma-1	0.0010	0.0792	0.0002
Gamma-2	0.0001	0.1640	0.0000
K int. con. electron, gamma-2	0.0037	0.1294	0.0010
L int. con. electron, gamma-2	0.0016	0.1589	0.0005
M int. con. electron, gamma-2	0.0005	0.1630	0.0002
Gamma-3	0.0014	0.1772	0.0005
K int. con. electron, gamma-3	0.0003	0.1427	0.0001
Gamma-4	0.0475	0.2843	0.0288
K int. con. electron, gamma-4	0.0025	0.2497	0.0013
L int. con. electron, gamma-4	0.0005	0.2793	0.0003
M int. con. electron, gamma-4	0.0002	0.2834	0.0001
Gamma-5	0.0017	0.3258	0.0012
Gamma-6	0.833	0.3645	0.6465
K int. con. electron, gamma-6	0.0167	0.3299	0.0117
L int. con. electron, gamma-6	0.0028	0.3594	0.0021
M int. con. electron, gamma-6	0.0009	0.3635	0.0006
Gamma-7	0.0032	0.5030	0.0034
Gamma-8	0.0687	0.6370	0.0932
K int. con. electron, gamma-8	0.0003	0.6024	0.0004
Gamma-9	0.0159	0.7229	0.0245
K α-1 x-rays	0.0252	0.0298	0.0016
K α-2 x-rays	0.0130	0.0295	0.0008
K β-1 x-rays	0.0070	0.0336	0.0005
K β-2 x-rays	0.0015	0.0346	0.0001
L x-rays	0.0078	0.0041	0.0001
KLL Auger electron	0.0042	0.0245	0.0002
KLX Auger electron	0.0018	0.0296	0.0001
KXY Auger electron	0.0003	0.0327	0.0000
LMM Auger electron	0.0486	0.0032	0.0003
MXY Auger electron	0.117	0.0009	0.0002

BETA-MINUS DECAY

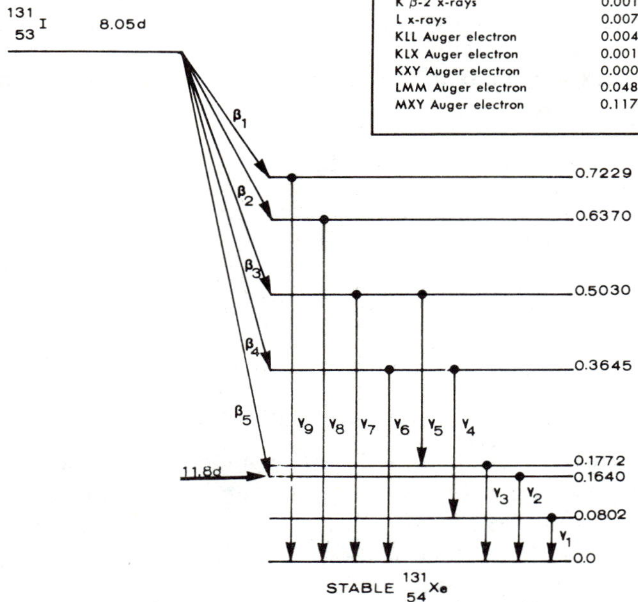

FIG. 6.11. Transformation scheme and input and output data for ^{131}I dosimetry. From Dillman, L. T.: Radionuclide Decay Schemes and Nuclear Parameters for Use in Radiation Dose Estimation, *J. Nuclear Medicine*, Vol. 10, Suppl. No. 2, MIRD Pamphlet No. 4, 1969.

where χ_{ei} = energy emission rate, J/sec,

A_s = activity in source, Bq,

$\bar{E}_i$ = energy of the ith particle, MeV,

n_i = number of particles of the ith kind per decay.

If the fraction of this emitted energy that is absorbed by the target is called φ_i, then the amount of energy absorbed by the target due to emission from the source is given by

$$\chi_{ai} = \chi_{ei} \times \varphi_i = 1.6 \times 10^{-13} \times A_s \times \bar{E}_i \times n_i \times \varphi_i \text{ J/sec.} \tag{6.53}$$

Since 1 gray corresponds to the absorption of 1 joule per kilogram, the dose rate from the ith particle to a target that weighs m kilograms is given by

$$\dot{D}_i = \frac{1.6 \times 10^{-13} \times A_s \times \bar{E}_i \times n_i \times \varphi_i \text{ J/sec}}{1 \frac{\text{J}}{\text{kg}} \Big/ \text{Gy} \times m \text{ kg}}. \tag{6.54}$$

If we let

$$\Delta_i = 1.6 \times 10^{-13} \times n_i \times \bar{E}_i \frac{\text{kg Gy}}{\text{Bq sec}}, \tag{6.55}$$

then equation (6.54) can be written as

$$\dot{D}_i = \frac{A_s}{m} \times \varphi_i \times \Delta_i \text{ Gy/sec.} \tag{6.56}$$

Δ_i is the dose rate in an infinitely large homogeneous mass of tissue containing a uniformly distributed radioisotope at a concentration of 1 Bq/kg. Numerical values for Δ_i for each of the radiations generated by radioisotopes in infinitely large masses of tissue are included in

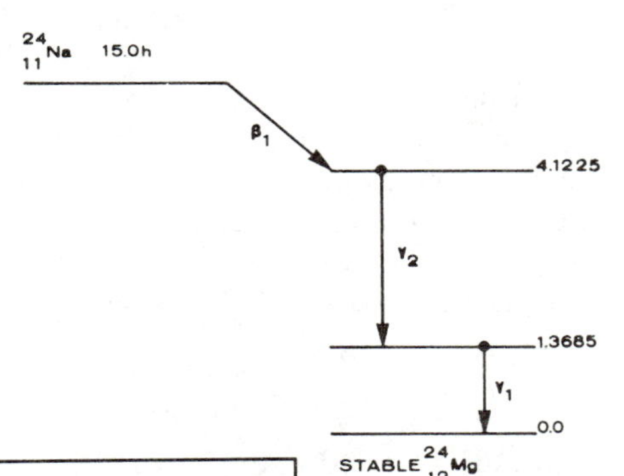

INPUT DATA

Radiation	%/disin-tegration	Transition energy (MeV)	Other nuclear parameters
Beta-1	99.9	1.392*	Allowed
All other betas	<0.1	—	—
Gamma-1	100.	1.3685	E2, α_K < 0.00001 (T)
Gamma-2	99.9	2.7539	E2, α_K < 0.00001 (T)
All other gammas	<0.1	—	—

Ref. Lederer, C. M. et al, *Table of Isotopes*, 6th ed.
* Endpoint energy (MeV). (T) = Theoretical value.

OUTPUT DATA

Radiation (i)	Mean number/disinte-gration (n_i)	Mean energy (MeV) ($\bar{E}_i$)	Δ_i $\left(\frac{\text{g-rad}}{\mu\text{Ci-h}}\right)$
Beta-1	0.999	0.5547	1.1803
Gamma-1	0.999	1.3685	2.9149
Gamma-2	0.999	2.7539	5.8599

FIG. 6.12. Transformation scheme and input and output data for ^{24}Na dosimetry.

From L. T. Dillman, Radionuclide Decay Schemes and Nuclear Parameters for Use in Radiation-Dose Estimation, *J. Nuclear Medicine*, Vol. 10, Supplement No. 2, MIRD Pamphlet No. 4, 1969.

the Output Data section of the decay schemes and nuclear parameters for use in radiation dose estimation that have been published by the Medical Internal Radiation Dose (MIRD) Committee of the Society of Nuclear Medicine. Considering all types of the particles emitted from the source, the dose rate to the target organ is:

$$\dot{D} = \frac{A_s}{m} \sum \varphi_i \Delta_i. \tag{6.57}$$

Since $\dot{D}$ is a function of A_s, which is a function of time, $\dot{D}$ too is a function of time. The dose commitment, that is, the total dose due to the complete decay of the deposited radioisotope, is given by integrating the dose rate with respect to time:

$$D = \int_0^\infty \dot{D}(t)\,\mathrm{d}t = \frac{\sum \varphi_i \Delta_i}{m} \int_0^\infty A_s(t)\,\mathrm{d}t. \tag{6.58}$$

If we call the time integral of the deposited radioactivity the cumulated activity $\tilde{A}$,

$$\tilde{A} = \int_0^\infty A_s(t)\,\mathrm{d}t, \tag{6.59}$$

then the total dose to the target organ is given by

$$D = \frac{\tilde{A}}{m} \sum \varphi_i \Delta_i. \tag{6.60}$$

Example 6.9

Calculate the total dose and initial dose rate to a 70-kg, 160-cm-tall reference man who is intravaneously injected with 1 MBq ^{24}NaCl. Assume the ^{24}NaCl to become uniformly distributed within a very short time, and to have a biological half-life of 11 days (264 hours). The decay scheme, and tables of input and output data are shown on page 161. The decay scheme and the accompanying table of input data show one beta (actually, >0.999) particle whose maximum energy is 1.392 MeV, and one 1.3685 MeV gamma per decay. The output data list the integral dose in an infinite medium, per unit of cumulated activity, in units of gm–rads/μCi–hr for each radiation. To convert from the old system of units found in the MIRD publications to the SI system, that is, to go

$$\text{from } \frac{\text{gm-rad}}{\mu\text{Ci-hr}} \text{ to } \frac{\text{kg-gray}}{\text{Bq-sec}}:$$

$$\frac{\text{kg-Gy}}{\text{Bq-sec}} = \frac{\text{gm-rad}}{\mu\text{Ci-hr}} \times \frac{10^{-3}\,\text{kg/g} \times 10^{-2}\,\text{Gy/rad}}{3.7 \times 10^4\,\text{Bq}/\mu\text{Ci} \times 3.6 \times 10^3\,\text{sec/hr}}$$

$$\frac{\text{kg-Gy}}{\text{Bq-sec}} = \frac{\text{gm-rad}}{\mu\text{Ci-hr}} \times 7.51 \times 10^{-14}. \tag{6.61}$$

Since there are 10^{15} femtograys (fGy) per gray, the conversion factor in equation (6.61) may be written as

$$\frac{\text{kg-fGy}}{\text{Bq-sec}} = \frac{\text{gm-rad}}{\mu\text{Ci-hr}} \times 75.1. \tag{6.62}$$

Furthermore, since

$$1\,\text{Gy} = 1\,\text{J/kg}$$

and since 1 Bq = 1 transformation/sec, the dimensions of equation (6.62) may also be given as

$$\frac{\text{gm rad}}{\mu\text{Ci hr}} \times 75.1 = \text{fJ/transformation}, \tag{6.63}$$

which is simply the amount of energy absorbed per nuclear transformation.

Now let us return to the problem. Since the ^{24}Na is cleared exponentially at an effective rate λ_E, the amount of activity in the source organ is given by

$$A_s(t) = A_s(0)\,e^{-\lambda_E t}, \tag{6.64}$$

where $A_s(0)$ is the initial activity in the source.

$$\tilde{A} = \int^{\infty} A_s(t)\,dt = A_s(0)\int_0^{\infty} e^{-\lambda_E t}\,dt = \frac{A_s(0)}{\lambda_E}. \tag{6.65}$$

Since

$$\lambda_E = \frac{0.693}{T_E} = \frac{0.693}{(T_R \times T_B)/(T_R + T_B)},$$

and the biological half-life, T_B, is found in ICRP Publication 2 to be 264 hours, and the radioactive half-life, T_R, is 15 hours,

$$\therefore \quad \tilde{A} = \frac{10^6\ \text{Bq}}{1.36 \times 10^{-5}\ \text{sec}^{-1}} = 7.35 \times 10^{10}\ \text{Bq-sec}.$$

Now we must calculate $\Sigma\, \varphi_i \Delta_i$.

The absorbed fractions, φ_i, in a number of target organs and tissues, for photons ranging in energy from 0.01 to 4 MeV that originate in a number of different source organs and tissues, are tabulated in Appendix A of the *Journal of Nuclear Medicine* Supplement No. 3, August, 1969. Table 6.8 shows the absorbed fractions from a photon emitter that is uniformly distributed throughout the body, as in the case of ^{24}Na. The values of φ_i for the 1.369 MeV and 2.754 MeV gammas were found by interpolation between values in Table 6.8, and are listed below, together with Δ_i, which was found in the Output Data listing in Fig. 6.12 and was converted to kg fGy/Bq sec through equation (6.62).

Radiation	E_i	φ_i	$\Delta_i\ \frac{\text{kg-fGy}}{\text{Bq-sec}}$	$\varphi_i \Delta_i$	
Beta 1	0.555	1.000	88.64	88.64	
Gamma 1	1.369	0.31	218.91	67.86	
Gamma 2	2.754	0.265	440.08	116.62	
				273.12	$\frac{\text{kg-fGy}}{\text{Bq-sec}}$

Substituting the value for $\tilde{A} = 7.35 \times 10^{10}$ Bq sec, $\Sigma\, \varphi_i \Delta_i = 273.12$ kg fGy/Bq sec, and $m = 70$ kg into equation (6.60) yields

$$D = \frac{7.35 \times 10^{10}\ \text{Bq-sec}}{70\ \text{kg}} \times 273.12\,\frac{\text{kg-fGy}}{\text{Bq-sec}}$$

$$= 2.868 \times 10^{11}\ \text{fGy (29 mrads)}.$$

TABLE 6.8. ABSORBED FRACTIONS (AND COEFFICIENTS OF VARIATION), GAMMA EMITTER UNIFORMLY DISTRIBUTED THROUGHOUT THE BODY

	Photon energy, MeV												
	0.010		0.015		0.020		0.030		0.050		0.100		
Target organ	ϕ	$\frac{100\sigma_\phi}{\phi}$	ϕ	$\frac{100\sigma_\phi}{\phi}$	ϕ	$\frac{100\sigma_\phi}{\phi}$	ϕ	$\frac{100\sigma_\phi}{\phi}$	ϕ	$\frac{100\sigma_\phi}{\phi}$	ϕ	$\frac{100\sigma_\phi}{\phi}$	Target organ
Adrenals	0.270E—03	35.	0.228E—03	34.	0.175E—03	37.	0.209E—03	28.	0.131E—03	23.	0.101E—03	26.	Adrenals
Bladder	0.757E—02	6.6	0.762E—02	6.5	0.683E—02	6.6	0.625E—02	6.1	0.445E—02	5.6	0.352E—02	5.2	Bladder
Gi (stom)	0.570E—02	7.6	0.507E—02	8.0	0.573E—02	7.1	0.560E—02	6.4	0.391E—02	5.8	0.273E—02	5.9	GI (stom)
GI (SI)	0.254E—01	3.6	0.236E—01	3.7	0.234E—01	3.6	0.209E—01	3.4	0.163E—01	3.1	0.120E—01	3.2	GI (SI)
GI (ULI)	0.541E—02	7.8	0.561E—02	7.5	0.647E—02	6.6	0.533E—02	5.9	0.374E—02	5.4	0.262E—02	5.7	GI (ULI)
GI (LLI)	0.350E—02	9.7	0.441E—02	8.5	0.457E—02	7.7	0.285E—02	7.9	0.256E—02	6.2	0.187E—02	6.3	GI (LLI)
Heart	0.756E—02	6.6	0.804E—02	6.3	0.769E—02	6.2	0.635E—02	6.0	0.469E—02	5.4	0.420E—02	5.0	Heart
Kidneys	0.410E—02	9.0	0.446E—02	8.5	0.412E—02	8.3	0.338E—02	7.4	0.233E—02	6.4	0.183E—02	6.6	Kidneys
Liver	0.260E—01	3.5	0.244E—01	3.6	0.249E—01	3.5	0.221E—01	3.3	0.154E—01	3.2	0.120E—01	3.2	Liver
Lungs	0.127E—01	5.1	0.142E—01	4.7	0.138E—01	4.4	0.122E—01	3.8	0.808E—02	3.4	0.551E—02	3.6	Lungs
Marrow	0.560E—01	1.4	0.594E—01	1.4	0.655E—01	1.3	0.740E—01	1.1	0.613E—01	1.1	0.329E—01	1.3	Marrow
Pancreas	0.134E—02	16.	0.103E—02	18.	0.828E—03	17.	0.780E—03	14.	0.567E—03	12.	0.449E—03	12.	Pancreas
Sk. (rib)	0.168E—01	4.4	0.206E—01	3.9	0.247E—01	3.4	0.263E—01	2.9	0.176E—01	2.9	0.764E—02	3.3	Sk. (rib)
Sk. (pelvis)	0.147E—01	4.7	0.160E—01	4.5	0.163E—01	4.3	0.224E—01	3.4	0.199E—01	3.0	0.103E—01	3.3	Sk. (pelvis)
Sk. (spine)	0.186E—01	4.2	0.190E—01	4.1	0.234E—01	3.7	0.253E—01	3.3	0.229E—01	3.0	0.144E—01	3.2	Sk. (spine)
Sk. (skull)	0.103E—01	5.6	0.115E—01	5.3	0.123E—01	5.	0.128E—01	4.6	0.722E—02	5.1	0.313E—02	6.0	Sk. (skull)
Skeleton (total)	0.144	1.4	0.153	1.3	0.167	1.2	0.188	1.1	0.153	1.1	0.810E—01	1.3	Skeleton (total)
Skin	0.258E—01	3.5	0.227E—01	3.5	0.169E—01	3.7	0.116E—01	3.3	0.758E—02	2.9	0.585E—02	3.1	Skin
Spleen	0.260E—02	11.	0.237E—02	12.	0.242E—02	11.	0.223E—02	9.1	0.149E—02	8.5	0.111E—02	8.7	Spleen
Thyroid	0.265E—03	35.	0.263E—03	34.	0.602E—04	48.	0.111E—03	36.	0.114E—03	27.	0.873E—04	29.	Thyroid
Uterus	0.999E—03	18.	0.109E—02	17.	0.122E—02	15.	0.924E—03	13.	0.712E—03	12.	0.611E—03	11.	Uterus
Trunk	0.604	0.47	0.589	0.48	0.566	0.50	0.500	0.55	0.358	0.67	0.245	0.79	Trunk
Legs	0.309	0.86	0.299	0.88	0.285	0.90	0.242	0.97	0.171	1.1	0.113	1.3	Legs
Head	0.488E—01	2.5	0.474E—01	2.5	0.440E—01	2.6	0.342E—01	2.7	0.200E—01	3.1	0.127E—01	3.1	Head
Total body	0.959	0.11	0.933	0.15	0.892	0.19	0.774	0.27	0.548	0.43	0.370	0.56	Total body

TABLE 6.8. continued

Target organ	0.200 ϕ	0.200 $\frac{100\sigma_\phi}{\phi}$	0.500 ϕ	0.500 $\frac{100\sigma_\phi}{\phi}$	1.000 ϕ	1.000 $\frac{100\sigma_\phi}{\phi}$	1.500 ϕ	1.500 $\frac{100\sigma_\phi}{\phi}$	2.000 ϕ	2.000 $\frac{100\sigma_\phi}{\phi}$	4.000 ϕ	4.000 $\frac{100\sigma_\phi}{\phi}$	Target organ
Adrenals	0.352E—04	36.	0.138E—03	35.	0.100E—03	42.	0.107E—03	43.	0.114E—03	43.			Adrenals
Bladder	0.327E—02	5.0	0.341E—02	6.6	0.274E—02	8.3	0.291E—02	8.4	0.231E—02	9.6	0.147E—02	12.	Bladder
GI (stom)	0.218E—02	7.0	0.258E—02	7.7	0.181E—02	9.8	0.199E—02	10.	0.212E—02	10.	0.119E—02	14.	GI (stom)
GI (SI)	0.106E—01	3.4	0.114E—01	3.8	0.109E—01	4.2	0.915E—02	4.8	0.820E—02	5.2	0.409E—02	7.3	GI (SI)
GI (ULI)	0.256E—02	6.3	0.306E—02	7.0	0.228E—02	8.9	0.209E—02	9.4	0.197E—02	10.	0.160E—02	12.	GI (ULI)
GI (LLI)	0.151E—02	7.6	0.184E—02	8.8	0.178E—02	9.7	0.181E—02	11.	0.157E—02	12.	0.673E—03	18.	GI (LLI)
Heart	0.337E—02	5.8	0.372E—02	6.6	0.301E—02	8.1	0.345E—02	7.8	0.312E—02	8.3	0.145E—02	13.	Heart
Kidneys	0.171E—02	7.4	0.142E—02	9.7	0.161E—02	10.	0.152E—02	11.	0.154E—02	12.	0.904E—03	16.	Kidneys
Liver	0.111E—01	3.4	0.101E—01	4.1	0.896E—02	4.7	0.912E—02	4.9	0.847E—02	5.1	0.560E—02	6.4	Liver
Lungs	0.507E—02	4.3	0.496E—02	5.2	0.466E—02	6.1	0.466E—02	6.5	0.427E—02	6.9	0.568E—02	6.4	Lungs
Marrow	0.221E—01	1.5	0.194E—01	1.0	0.182E—01	2.0	0.164E—01	2.2	0.156E—01	2.3	0.969E—02	3.0	Marrow
Pancreas	0.444E—03	14.	0.382E—03	17.	0.534E—03	19.	0.348E—03	22.	0.358E—03	24.	0.142E—03	39.	Pancreas
Sk. (rib)	0.505E—02	4.1	0.435E—02	5.6	0.421E—02	6.3	0.405E—02	7.0	0.350E—02	7.7	0.338E—02	8.0	Sk. (rib)
Sk. (pelvis)	0.668E—02	3.9	0.569E—02	5.0	0.562E—02	5.7	0.511E—02	6.3	0.422E—02	7.0	0.256E—02	9.3	Sk. (pelvis)
Sk. (spine)	0.910E—02	3.6	0.763E—02	4.5	0.751E—02	5.1	0.610E—02	5.7	0.606E—02	5.9	0.341E—02	8.1	Sk. (spine)
Sk. (skull)	0.277E—02	6.3	0.304E—02	7.2	0.280E—02	8.0	0.254E—02	9.0	0.292E—02	8.8	0.224E—02	10.	Sk. (skull)
Skeleton (total)	0.550E—01	1.4	0.488E—01	1.7	0.456E—01	2.0	0.413E—01	2.2	0.396E—01	2.3	0.252E—01	3.0	Skeleton (total)
Skin	0.677E—02	3.5	0.757E—02	4.2	0.745E—02	4.8	0.759E—02	5.0	0.664E—02	5.5	0.123E—01	4.3	Skin
Spleen	0.798E—03	11.	0.116E—02	11.	0.914E—03	14.	0.903E—03	16.	0.740E—03	17.	0.368E—03	24.	Spleen
Thyroid	0.418E—04	42.							0.810E—04	46.			Thyroid
Uterus	0.408E—03	15.	0.473E—03	16.	0.517E—03	18.	0.323E—03	23.	0.364E—03	25.	0.238E—03	33.	Uterus
Trunk	0.223	0.81	0.225	0.84	0.210	0.92	0.198	0.99	0.186	1.0	9.156	1.2	Trunk
Legs	0.102	1.3	0.101	1.4	0.965E—01	1.5	0.917E—01	1.6	0.846E—01	1.6	0.710E—01	1.8	Legs
Head	0.134E—01	3.2	0.147E—01	3.5	0.145E—01	3.8	0.130E—01	4.1	0.139E—01	4.1	0.127E—01	4.4	Head
Total body	0.338	0.57	0.340	0.60	0.321	0.67	0.302	0.73	0.284	0.77	0.240	0.90	Total body

The digits following the symbol E indicate the powers of ten by which each number is to be multiplied.

A blank in the table indicates that the coefficient of variation was greater than 50%.

Total body = head + trunk + legs.

From W. S. Snyder, M. R. Ford, G. G. Warner, and H. L. Fisher, Jr.: Estimates of Absorbed Fractions for Monoenergetic Photon Sources uniformly Distributed in Various Organs of a Heterogeneous Phantom, *J. Nuclear Medicine*, Vol. 10, Supplement No. 3, MIRD Pamphlet No. 5, 1969.

The initial dose rate may be found by substituting 10^6 Bq for A_s in equation (6.57):

$$\dot{D} = \frac{10^6\ \text{Bq}}{70\ \text{kg}} \times 273.12 \frac{\text{kg-fGy}}{\text{Bq-sec}}$$

$$= 3.9 \times 10^6\ \text{fGy/sec (1.4 mrad/hr)}.$$

The physiological kinetics on which the calculated dose from internally deposited radioisotopes is based are contained in the term for the cumulated activity, $\tilde{A}$, while the balance of the right hand side of equation (6.60) deals with physical data and measurements. The absorbed fraction φ_i represents the fraction of the energy that is absorbed by the total organ or tissue. According to equation (6.60), we must divide the total absorbed energy, $\Sigma\ \varphi_i\Delta_i$ by the mass of the target organ, m. Rather than consider the fraction of energy absorbed by the target organ and then divide by the organ weight, it may be more convenient to use the *specific absorbed fraction*, Φ_i, which is the fraction of the absorbed energy per unit mass of target tissue from the ith particle emitted in the source organ. Specific absorbed fractions of photons of several energies for reference man, which were calculated by Monte Carlo methods, are tabulated in the appendix. Specific absorbed fractions for beta or alpha radiation are easily calculated. In a large medium containing a uniformly distributed beta or alpha emitter, essentially all of the emitted energy is absorbed. For the case where the target and source are the same organ, and where the range of the radiation from the deposited radioisotope is less than the smallest dimension of the organ in which it is deposited, the specific absorbed fraction in an organ of mass m may be closely approximated by

$$\Phi = \frac{\varphi}{m} = \frac{1}{m}. \tag{6.66}$$

When the target organ is widely separated from the source organ, that is, when the distance between them is greater than the range of beta or alpha particles, then the target absorbs no energy from the source, and $\Phi = 0$. For the case where the target tissue is a region surrounded by the source, the specific absorbed fraction in the target is

$$\Phi = \frac{1}{m\ \text{(source)}}. \tag{6.67}$$

For example, if a beta emitter is uniformly distributed throughout the body, then the specific absorbed fraction to the liver from the radioactivity outside the liver, if the liver weighs 1.8 kg and the person weighs 70 kg, is

$$\Phi = \frac{1}{70 - 1.8} = 1.47 \times 10^{-2}\ \text{per kg}.$$

When the specific absorbed fraction is used, the absorbed dose is given by

$$D = \tilde{A} \sum \Delta_i \Phi_i. \tag{6.68}$$

Since every organ in the body is a target for radiation from the source organ, the exact target-source relationship is identified explicitly by the symbol

$$(r_k \leftarrow r_h),$$

where r_k represents the target organ and r_h represents the source organ. Thus, the dose to target organ r_k from activity $\tilde{A}_h$ in source organ r_h is written as

$$D(r_k \leftarrow r_h) = \tilde{A}_h \sum \Delta_i \Phi_i(r_k \leftarrow r_h). \tag{6.69}$$

Using the specific absorbed fraction, we can define the quantity

$$S(r_k \leftarrow r_h) = \sum \Delta_i \Phi_i (r_k \leftarrow r_h). \tag{6.70}$$

Since S depends only on physical factors, such as the geometrical relationship between the source, we can calculate the value of S for all of the target-source relationships of interest and for any radioisotope in the source organ. The dose to any target organ, r_k, from a source organ r_h, is then

$$D(r_k \leftarrow r_h) = \tilde{A}_h \times S(r_k \leftarrow r_h). \tag{6.71}$$

Furthermore, since the radioactivity is usually widespread within the body, a target organ may be irradiated by several different source organs. The dose to the target, therefore is

$$D(r_k) = \sum_h D(r_k \leftarrow r_h). \tag{6.72}$$

Tables of $S(r_k \leftarrow r_h)$ per unit cumulated activity for numerous target and source organs and for numerous radioisotopes of interest, are published in MIRD Pamphlet No. 11. Table 6.9 which is excerpted from Pamphlet No. 11, gives the values of S for ^{203}Hg. The use of the "S" tables in calculating internal dose is illustrated by the following example:

Example 6.10

The retention of mercury in the kidney was found, after an accidental inhalation (K. W. Brown, J. C. McFarlane, and D. E. Bernhardt, *Health Physics*, **28**, 1–4, 1975), to be given by single exponential function with an effective turnover rate of 2.6% per day. If 0.5 MBq of ^{203}Hg is deposited in the kidneys as a result of an acute accidental inhalation of tagged mercury vapor, what dose commitment to the kidneys resulted from this accidental exposure? Assume, for the purpose of this calculation, that the mercury is uniformly distributed throughout the kidneys, and that 50% of the body's Hg burden is in the kidneys, 20% is in the liver, and the balance is uniformly distributed throughout the rest of the body.

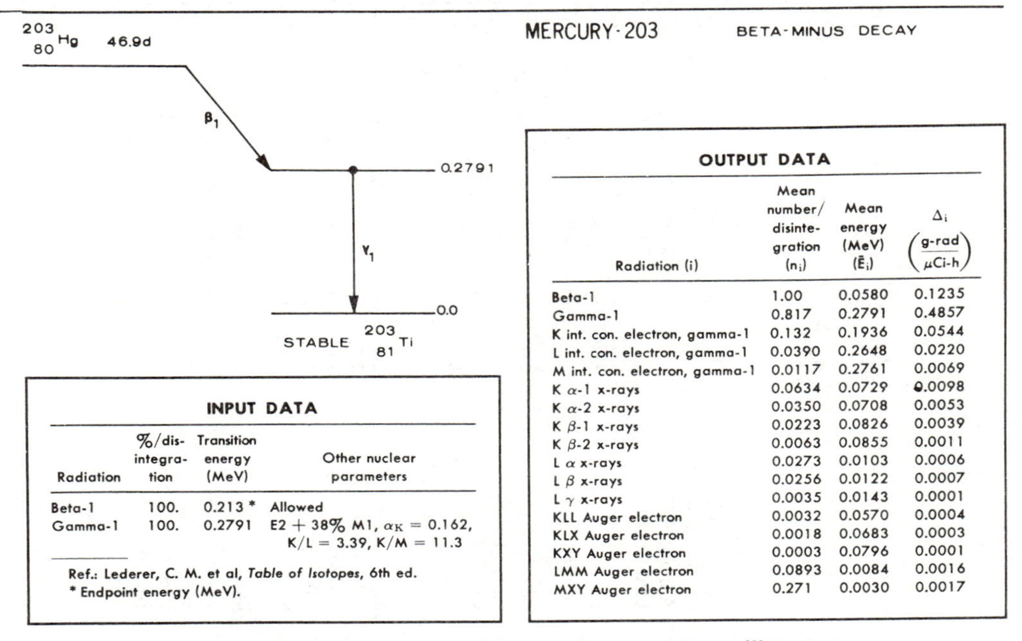

INPUT DATA

Radiation	%/dis-integra-tion	Transition energy (MeV)	Other nuclear parameters
Beta-1	100.	0.213 *	Allowed
Gamma-1	100.	0.2791	E2 + 38% M1, α_K = 0.162, K/L = 3.39, K/M = 11.3

Ref.: Lederer, C. M. et al, *Table of Isotopes*, 6th ed.
* Endpoint energy (MeV).

OUTPUT DATA

Radiation (i)	Mean number/disintegration (n_i)	Mean energy (MeV) ($\bar{E}_i$)	Δ_i (g-rad/μCi-h)
Beta-1	1.00	0.0580	0.1235
Gamma-1	0.817	0.2791	0.4857
K int. con. electron, gamma-1	0.132	0.1936	0.0544
L int. con. electron, gamma-1	0.0390	0.2648	0.0220
M int. con. electron, gamma-1	0.0117	0.2761	0.0069
K α-1 x-rays	0.0634	0.0729	0.0098
K α-2 x-rays	0.0350	0.0708	0.0053
K β-1 x-rays	0.0223	0.0826	0.0039
K β-2 x-rays	0.0063	0.0855	0.0011
L α x-rays	0.0273	0.0103	0.0006
L β x-rays	0.0256	0.0122	0.0007
L γ x-rays	0.0035	0.0143	0.0001
KLL Auger electron	0.0032	0.0570	0.0004
KLX Auger electron	0.0018	0.0683	0.0003
KXY Auger electron	0.0003	0.0796	0.0001
LMM Auger electron	0.0893	0.0084	0.0016
MXY Auger electron	0.271	0.0030	0.0017

FIG. 6.13. Transformation scheme and input and output data for ^{203}Hg dosimetry.

From L. T. Dillman, Radionuclide Decay Schemes and Nuclear Parameters for Use in Radiation-Dose Estimation, *J. Nuclear Medicine*, Vol. 10, Supplement No. 2, MIRD Pamphlet No. 4, 1969.

TABLE 6.9. *S*, ABSORBED DOSE PER UNIT CUMULATED ACTIVITY (RAD/UCI-H) MERCURY-203 (HALF-LIFE 1.12E 03 HOURS)

	Source organs									
			Intestinal tract							
Target organs	Adrenals	Bladder contents	Stomach contents	SI contents	ULI contents	LLI contents	Kidneys	Liver	Lungs	Other tissue (muscle)
Adrenals	1.6E—02	3.6E—07	4.2E—06	2.7E—06	1.7E—06	8.4E—07	1.9E—05	9.0E—06	4.4E—06	2.7E—06
Bladder wall	2.1E—07	6.6E—04	5.0E—07	5.0E—06	3.8E—06	1.1E—05	5.9E—07	3.6E—07	1.0E—07	3.2E—06
Bone (total)	2.9E—06	1.3E—06	1.3E—06	1.8E—06	1.6E—06	2.3E—06	2.1E—06	1.6E—06	2.1E—06	1.9E—06
GI (stom. wall)	5.3E—06	5.2E—07	5.1E—04	6.5E—06	6.7E—06	3.2E—06	6.1E—06	3.5E—06	3.3E—06	2.5E—06
GI (SI)	1.6E—06	5.1E—06	4.7E—06	3.2E—04	3.0E—05	1.7E—05	5.1E—06	3.0E—06	3.9E—07	2.8E—06
GI (ULI wall)	1.7E—06	4.2E—06	6.3E—06	4.2E—05	5.5E—04	7.7E—06	5.2E—06	4.5E—06	4.9E—07	2.9E—06
GI (LLI wall)	4.6E—07	1.3E—05	2.3E—06	1.3E—05	5.4E—06	8.7E—04	1.5E—06	4.6E—07	1.8E—07	3.1E—06
Kidneys	2.1E—05	5.5E—07	6.3E—06	5.5E—06	5.0E—06	1.7E—06	8.1E—04	6.9E—06	1.7E—06	2.5E—06
Liver	8.9E—06	4.1E—07	3.6E—06	3.3E—06	4.6E—06	5.3E—07	7.0E—06	1.6E—04	4.4E—06	2.0E—06
Lungs	4.4E—06	5.6E—08	3.1E—06	4.8E—07	5.2E—07	1.6E—07	1.6E—06	4.5E—06	2.4E—04	2.4E—06
Marrow (red)	5.2E—06	3.0E—06	2.3E—06	5.8E—06	5.0E—06	6.9E—06	5.3E—06	2.3E—06	2.7E—06	2.9E—06
Other tissues (musc.)	2.7E—06	3.2E—06	2.5E—06	2.8E—06	2.7E—06	3.1E—06	2.5E—06	2.0E—06	2.4E—06	1.0E—05
Ovaries	1.0E—06	1.3E—05	8.1E—07	1.8E—05	2.1E—05	3.4E—05	2.2E—06	7.6E—07	2.3E—07	3.6E—06
Pancreas	1.5E—05	5.2E—07	3.2E—05	3.7E—06	4.1E—06	1.3E—06	1.2E—05	7.4E—06	4.7E—06	3.2E—06
Skin	1.1E—06	1.1E—06	9.0E—07	8.5E—07	8.6E—07	9.8E—07	1.1E—06	9.8E—07	1.1E—06	1.5E—06
Spleen	1.2E—05	3.3E—07	1.8E—05	2.8E—06	2.5E—06	1.5E—06	1.6E—05	1.7E—06	4.1E—06	2.6E—06
Testes	9.2E—08	8.4E—06	9.9E—08	6.3E—07	6.5E—07	3.7E—06	2.2E—07	1.6E—07	2.8E—08	2.1E—06
Thyroid	3.0E—07	9.0E—09	2.3E—07	4.8E—08	5.2E—08	2.0E—08	1.3E—07	3.8E—07	1.7E—06	2.4E—06
Uterus (nongrvd)	3.3E—06	2.8E—05	1.5E—06	1.7E—05	8.5E—06	1.2E—05	1.8E—06	7.2E—07	1.7E—07	4.0E—06
Total body	6.0E—06	3.7E—06	4.1E—06	6.0E—06	4.9E—06	5.2E—06	6.0E—06	6.1E—06	5.6E—06	5.5E—06

TABLE 6.9. continued

Target organs	Source organs									
	Ovaries	Pancreas	Skeleton			Skin	Spleen	Testes	Thyroid	Total body
			Red marrow	Cort. bone	Tra. bone					
Adrenals	7.7E—07	1.5E—05	4.4E—06	2.5E—06	2.5E—06	1.4E—06	1.2E—05	9.2E—08	3.0E—07	6.5E—06
Bladder wall	1.2E—05	3.0E—07	1.4E—06	9.3E—07	9.3E—07	1.0E—06	2.9E—07	8.8E—06	9.1E—09	6.2E—06
Bone (total)	2.1E—06	2.0E—06	1.0E—05	4.9E—05	4.0E—05	1.6E—06	1.6E—06	1.4E—06	1.5E—06	6.0E—06
GI (stom. wall)	1.4E—06	3.3E—05	1.9E—06	1.0E—06	1.0E—06	1.0E—06	1.7E—05	1.2E—07	1.1E—07	6.4E—06
GI (SI)	2.1E—05	3.3E—06	4.7E—06	1.4E—06	1.4E—06	9.0E—07	2.5E—06	8.0E—07	2.4E—08	6.6E—06
GI (ULI wall)	2.0E—05	4.0E—06	3.8E—06	1.3E—06	1.3E—06	8.9E—07	2.3E—06	6.4E—07	2.0E—08	6.3E—06
GI (LLI wall)	2.6E—05	1.0E—06	5.3E—06	1.9E—06	1.9E—06	9.4E—07	1.2E—06	5.1E—06	1.6E—08	6.2E—06
Kidneys	1.8E—06	1.1E—05	4.1E—06	1.6E—06	1.6E—06	1.2E—06	1.6E—05	1.2E—07	7.3E—08	6.1E—06
Liver	1.1E—06	7.8E—06	1.8E—06	1.2E—06	1.2E—06	1.1E—06	1.9E—06	7.7E—08	2.3E—07	6.0E—06
Lungs	1.4E—07	4.5E—06	2.2E—06	1.8E—06	1.8E—06	1.2E—06	4.0E—06	2.2E—08	1.8E—06	5.6E—06
Marrow (red)	7.2E—06	3.8E—06	1.3E—04	6.5E—06	3.5E—05	1.5E—06	2.5E—06	1.1E—06	1.7E—06	6.4E—06
Other tissues (musc.)	3.6E—06	3.2E—06	2.3E—06	1.9E—06	1.9E—06	1.5E—06	2.6E—06	2.1E—06	2.4E—06	5.5E—06
Ovaries	2.1E—02	6.0E—07	4.7E—06	1.4E—06	1.4E—06	8.0E—07	1.2E—06	0.0	1.9E—08	6.3E—06
Pancreas	9.1E—07	2.5E—03	3.0E—06	1.9E—06	1.9E—06	1.0E—06	3.4E—05	1.2E—07	1.8E—07	6.7E—06
Skin	1.5E—07	8.1E—07	1.2E—06	1.4E—06	1.4E—06	8.4E—05	9.6E—07	2.7E—06	1.5E—06	4.5E—06
Spleen	1.0E—06	3.5E—05	1.6E—06	1.3E—06	1.3E—06	1.1E—06	1.4E—03	8.6E—08	2.2E—07	6.2E—06
Testes	0.0	1.3E—07	6.1E—07	1.1E—06	1.1E—06	1.8E—06	1.3E—07	6.6E—03	2.8E—09	5.3E—06
Thyroid	1.9E—08	2.9E—07	1.4E—06	1.7E—06	1.7E—06	1.4E—06	2.2E—07	2.8E—09	1.1E—02	5.3E—06
Uterus (nongrvd)	3.7E—05	1.1E—06	4.0E—06	1.1E—06	1.1E—06	7.8E—07	7.4E—07	0.0	1.8E—08	6.7E—06
Total body	6.7E—06	6.6E—06	5.9E—06	5.6E—06	5.6E—06	4.5E—06	6.1E—06	5.5E—06	5.3E—06	5.6E—06

From W. S. Snyder, M. R. Ford, G. G. Warner, and S. B. Watson: "S," Absorbed Dose per Unit Cumulated Activity for selected Radionuclides and Organs, MIRD Pamphlet No. 11, 1975.

The decay scheme for ^{203}Hg, Fig. 6.13 shows that the mercury emits a single group of beta particles whose maximum energy is 0.212 MeV and whose mean energy is listed in the output data as 0.0577 MeV. A 0.279-MeV gamma ray is emitted after each beta transformation. The gamma ray, however, is internally converted in 18.3% of the transformations, thus leading to conversion electrons from the K, L, or M energy levels and, effectively, gamma ray emission in only 81.7% of the transformations.

Table 6.9 lists the absorbed dose per unit cumulated ^{203}Hg activity. For the kidneys as the source, S (kidneys←kidneys) = 8.1×10^{-4} rad per μCi-hr, and for the liver as the source, the dose to the kidney, S (kidneys←liver) = 6.9×10^{-6} rad per μCi-hr. ICRP Publication 2 lists 10.4 days as the effective half life of ^{203}Hg in the liver, and 8.2 days as the effective half-life of ^{203}Hg in the whole body, Table 6.10.

The total dose to the kidneys is the sum of the doses due to the ^{203}Hg deposited in the kidneys, in the liver, and in the rest of the body. If half the body burden of Hg is in the kidneys and if 20% of the body burden is in the liver, then the accidental exposure resulted in a deposit of 0.2 MBq in the liver and 0.3 MBq distributed throughout the rest of the body.

The cumulated activity in the kidney is given by equation (6.65):

$$\tilde{A}(\text{kid}) = \frac{A_s(0)}{\lambda_E}(\text{kid}) = \frac{5 \times 10^5 \text{ Bq}}{2.6 \times 10^{-2} \text{ day}^{-1} \times 1 \text{ day}/86400 \text{ sec}} = 1.66 \times 10^{12} \text{ Bq-sec.}$$

From Table 6.9 we find S (kidneys←kidneys) to be 8.1×10^{-4} rad/μCi-hr.

To convert rad/(μCi-hr) to Gy/(Bq-sec):

$$\frac{\text{Gy}}{\text{Bq-sec}} = \frac{\text{rad}}{\mu\text{Ci-hr}} \times \frac{1 \text{ Gy}}{100 \text{ rad}} \times \frac{1\ \mu\text{Ci}}{3.7 \times 10^4 \text{ Bq}} \times \frac{1 \text{ hr}}{3.6 \times 10^3 \text{ sec}}$$

$$= \frac{\text{rad}}{\mu\text{Ci-hr}} \times 7.51 \times 10^{-11}. \tag{6.73}$$

Using the conversion factor in equation (6.73) we find that S(kidney ← kidney) = 6.08×10^{-14} Gy/Bq-sec. The dose to the kidney from the mercury within the kidney is calculated from equation (6.71):

$$D(\text{kidney}\leftarrow\text{kidney}) = \tilde{A}(\text{kid}) \times S(\text{kidney}\leftarrow\text{kidney})$$

$$= 1.66 \times 10^{12} \text{ Bq-sec} \times 6.08 \times 10^{-14} \text{ Gy/Bq-sec}$$

$$= 0.1010 \text{ Gy (10.10 rads).}$$

The contribution of the ^{203}Hg in the liver to the kidney dose is calculated from equations (6.65) and (6.71):

$$D(\text{kidney}\leftarrow\text{liver}) = \frac{A_s(0)}{\lambda_E}(\text{liver}) \times S(\text{kidney}\leftarrow\text{liver})$$

$$= \frac{2 \times 10^5 \text{ Bq}}{7.71 \times 10^{-7} \text{ sec}^{-1}} \times 5.18 \times 10^{-16} \text{ Gy/Bq-sec}$$

$$= 1.344 \times 10^{-4} \text{ Gy } (1.344 \times 10^{-2} \text{ rads}).$$

The effective clearance halftime of ^{203}Hg from the liver, 8.2 days, corresponds to the effective clearance constant 7.91×10^{-7} per second, and S is converted from the data in Table 6.9. In a similar manner, the radiation dose to the kidney from the remaining ^{203}Hg

Table 6.10. Data for the Physiological Kinetics of Mercury

Gen. ref.	Element and radio-nuclides	Average daily ingestion I (g/day)	Organ of reference, mass (g) effective radius (cm)	Average concen-tration, C (g/g wet tissues)	Half-life (days)			Fraction from GI tract to blood f_1	Fraction in organ of reference of that in total body f_2		Fraction from blood to organ of Reference f_2'	Fraction reaching organ of reference	
					Physical T_r	Bio-logical T_b	Effective T		Element	Radio-nuclide		By ingestion f_w	By inhalation f_a
$z\backslash$			Ch-1			eq. 44, 45	eq. 49			eq. 41, 42		eq. 47	eq. 46
80	Hg	2×10^{-5} (Sl-1)	Total body 7×10^4 g 30 cm			10 (Ha-85)		0.75 (Sl-1)	1.0 (D)		1.0 (D)	0.75	0.63
	Hg^{197m}				1		0.91			1.0 (D)			
	Hg^{197}				2.7		2.1			1.0 (D)			
	Hg^{203}				45.8		8.2			1.0 (D)			
	Hg		Kidneys 300 g 7 cm	5×10^{-7} (Stk-1)		14.5			0.5 (Ha-85)		0.35 (Ha-85)	0.26	0.22
	Hg^{197m}				1		0.94			0.36			
	Hg^{197}				2.7		2.3			0.38			
	Hg^{203}				45.8		11.0			0.47			
	Hg		Liver 1.7×10^3 g 10 cm	2.0×10^{-7} (Stk-1)		13.5			0.2 (Ha-85)		0.15 (Ha-85)	0.11	0.09
	Hg^{197m}				1		0.93			0.15			
	Hg^{197}				2.7		2.3			0.16			
	Hg^{203}				45.8		10.4			0.19			
	Hg		Spleen 150 g 7 cm	1.3×10^{-7} (Stk-1)		10 (Ha-85)			0.02 (Ha-85)		0.02 (Ha-85)	0.02	0.01
	Hg^{197m}				1		0.9			0.02			
	Hg^{197}				2.7		2.1			0.02			
	Hg^{203}				45.8		8.2			0.02			

From Report of Committee II on *Permissible Dose for Internal Radiation*, Pergamon Press, Oxford, 1960.

that is distributed throughout the rest of the body is calculated from

$$D(\text{kidney}\leftarrow\text{body}) = \frac{A_s(0)}{\lambda_E}(\text{body}) \times S(\text{kidney}\leftarrow\text{body})$$
$$= \frac{3 \times 10^5\ \text{Bq}}{9.78 \times 10^{-7}\ \text{sec}^{-1}} \times 4.58 \times 10^{-16}\ \text{Gy/Bq-sec}$$
$$= 1.406 \times 10^{-4}\ \text{Gy}\ (1.406 \times 10^{-2}\ \text{rads}).$$

The total dose to the kidneys, therefore, according to equation (6.70) is

$$D(\text{kidneys}) = 1.010 \times 10^{-1} + 1.344 \times 10^{-4} + 1.406 \times 10^{-4}$$
$$= 0.1012\ \text{Gy}\ (10.12\ \text{rads}).$$

Neutrons

The absorbed dose from a beam of neutrons may be computed by considering the energy absorbed by each of the tissue elements that react with the neutrons. The type of reaction, of course, depends on the neutron energy. For fast neutrons, up to about 20 MeV, the main mechanism of energy transfer is elastic collision, while thermal neutrons may be captured and initiate nuclear reactions. In cases of elastic scattering, the scattered nuclei dissipate their energy in the immediate vicinity of the primary neutron interaction. The radiation dose absorbed locally in this way is called the first collision dose, and is determined entirely by the primary neutron flux; the scattered neutron is not considered after this primary interaction. For fast neutrons, the first collision dose rate from neutrons of energy E is:

$$\dot{D}_n(E) = \frac{\phi(E)E\sum N_i\sigma_i f_i}{1\ \text{J/kg-Gy}}, \tag{6.74}$$

where $\phi(E)$ = flux of neutrons whose energy is E, in neutrons/cm^2-sec,
E = neutron energy, in joules,
N_i = atoms per kilogram of the ith element,
σ_i = scattering across section of the ith element for neutrons of energy E, in barns $\times 10^{-24}$ cm^2,
f = mean fractional energy transferred from neutron to scattered atom during collision with neutron.

For isotropic scattering, the average fraction of the neutron energy transferred in an elastic collision with a nucleus of atomic mass number M is

$$f = \frac{2M}{(M+1)^2}. \tag{6.75}$$

The composition of soft tissue, for the purpose of radiation dosimetry is given in Table 6.11. The table also lists the average fraction of the neutron energy transferred to each of the tissue elements.

TABLE 6.11. SYNTHETIC TISSUE COMPOSITION

Element	% Mass	N, atoms/kg	f
Oxygen	71.39	2.69×10^{25}	0.111
Carbon	14.89	6.41×10^{24}	0.142
Hydrogen	10.00	5.98×10^{25}	0.500
Nitrogen	3.47	1.49×10^{24}	0.124
Sodium	0.15	3.93×10^{22}	0.080
Chlorine	0.10	1.70×10^{22}	0.053

Adapted from G. L. Brownell, W. H. Ellet, and A. R. Reddy, Absorbed Fractions for Photon Dosimetry, *J. Nuclear Medicine*, Supplement No. 1, MIRD Pamphlet No. 3, February, 1968.

Example 6.11

What is the absorbed dose rate to soft tissue in a beam of 5-MeV neutrons whose intensity is 2000 neutrons per square centimeter per second?

The scattering cross sections of each of the tissue elements for 5 MeV neutrons are listed below:

Element	σ, cm^2	$N_i \sigma_i f_i$
O	1.55×10^{-24}	4.628×10^0
C	1.65×10^{-24}	1.502×10^0
H	1.50×10^{-24}	4.485×10^1
N	1.00×10^{-24}	1.848×10^{-1}
Na	2.3×10^{-24}	7.231×10^{-3}
Cl	2.8×10^{-24}	2.523×10^{-3}
		$\Sigma N_i \sigma_i f_i = 5.117 \times 10^1$ cm^2/kg

Substituting the appropriate values into equation (6.63) yields

$$\dot{D}_n = \frac{2 \times 10^3\ n/\text{cm}^2\text{-sec} \times 5\ \text{MeV}/n \times 1.6 \times 10^{-13}\ \text{J/MeV} \times 51.17\ \text{cm}^2/\text{kg}}{1\ \text{J/kg-Gy}}$$

$$= 8.19 \times 10^{-8}\ \text{Gy/sec}\ (8.19 \times 10^{-6}\ \text{rad/sec}),$$

or

$$8.19 \times 10^{-8}\ \text{Gy/sec} \times 10^6\ \mu\text{Gy/Gy} \times 3.6 \times 10^3\ \text{sec/hr}$$

$$= 295\ \mu\text{Gy/hr}\ (29.5\ \text{mrad/hr}).$$

In the example above, the neutron beam was monenergetic, and thus only one neutron energy was considered. If a beam contains neutrons of several energies, then the calculation must be carried out separately for each energy group.

For thermal neutrons, two reactions are considered, viz, the $^{14}N(n, p)^{14}C$ reaction and the $^1H(n, \gamma)^2H$ reaction. For the former reaction, the dose rate may be calculated from the equation

$$\dot{D}_{np} = \frac{\phi N \sigma Q \times 1.6 \times 10^{-13}\ \text{J/MeV}}{1\ \text{J/kg-Gy}}, \tag{6.76}$$

where ϕ = thermal flux, neutrons per cm^2 per second,
N = number of nitrogen atoms per kg tissue, 1.49×10^{24},
σ = absorption cross section for nitrogen, 1.75×10^{-24} cm^2,
Q = energy released by the reaction = 0.63 MeV.

The latter reaction, $^1H(n, \gamma)^2H$ is equivalent to having a uniformly distributed gamma-emitting isotope throughout the body, and results in an autointegral gamma-ray dose. The specific activity of this distributed gamma emitter, the number of reactions per second per gram, is governed by the neutron flux and is given by equation (6.77):

$$A = \phi N \sigma\ \text{“Bq”/kg}, \tag{6.77}$$

where ϕ = thermal flux, neutrons per cm^2 per second,
N = number of hydrogen atoms per kg tissue = 5.98×10^{25},
σ = absorption cross section for hydrogen = 0.33×10^{-24} cm^2.

Example 6.12

What is the absorbed dose rate to a 70-kg person from a whole body exposure to a mean thermal flux of 10,000 nuetrons per cm^2 per second?

The dose rate due to the n, p reaction is calculated from equation (6.76):

$$\dot{D}_{np} = 1 \times 10^4 \times 1.49 \times 10^{24} \times 1.75 \times 10^{-24} \times 0.63 \times 1.6 \times 10^{-13}$$
$$= 2.628 \times 10^{-9}\ \text{Gy/s} \quad (2.628 \times 10^{-7}\ \text{rad/sec}),$$

or

$$\dot{D}_{np} = 9.461\ \mu\text{Gy/hr} \quad (0.95\ \text{mrad/hr}).$$

The autointegral gamma-ray dose rate is calculated with equations (6.68) and (6.56). The gamma-ray "activity," from equation (6.77) is

$$A = 10^4\ \text{cm}^2\ \text{sec}^{-1} \times 5.98 \times 10^{25}\ \text{atoms/kg} \times 3.3 \times 10^{-25}\ \text{cm}^2/\text{atom}$$
$$= 1.973 \times 10^5\ \text{"Bq"/kg}.$$

The dose rate from this uniformly distributed gamma ray activity is calculated from equation (6.56):

$$\dot{D}_\gamma = \frac{A_s}{m} \times \varphi \times \Delta.$$

The absorbed fraction, φ, for the 2.23-MeV gamma ray is found, by interpolating in Table 6.8 between the 2.000- and 4.000-MeV values, to be 0.278, and Δ, the dose rate in an infinitely large mass whose specific activity is 1 Bq/kg, is calculated from equation (6.55):

$$\Delta = 1.6 \times 10^{-13} \times 2.23 = 3.57 \times 10^{-13}\ \text{Gy sec}^{-1}/\text{Bq kg}^{-1}.$$

The autointegral gamma ray dose rate, therefore, is

$$\dot{D}_\gamma = 1.973 \times 10^5\ \text{Bq/kg} \times 0.278 \times 3.57 \times 10^{-13}\ \text{kg-Gy/Bq-sec}$$
$$= 1.96 \times 10^{-8}\ \text{Gy/sec} \quad (1.96 \times 10^{-6}\ \text{rad/sec}),$$

or

$$70.6\ \mu\text{Gy/hr} \quad (7.06\ \text{mrad/hr}).$$

We cannot, in this case, add the autointegral gamma ray dose to the dose from the n, p reaction because an absorbed dose of 1 Gy of gamma radiation is not biologically equivalent to 1 Gy from proton radiation. This point, which deals with the relative biological effectiveness of the various radiations, is discussed in the next chapter.

Problems

1. A 50-μC/kg ($\sim$200 mR) pocket dosimeter with air equivalent walls has a sensitive volume whose dimensions are 0.5 in. diameter and 2.5 in. long; the volume is filled with air at atmospheric pressure. The capacitance of the dosimeter is 10 pFd. If 200 V are required to charge the chamber, what is the voltage across the chamber when it reads 50 μC/kg ($\sim$200 mR)?

2. An air ionization chamber whose volume is 1 liter is used as an environmental monitor at a temperature of 27°C and a pressure of 700 torrs. What is the exposure rate, in μC/kg per hour and in mR/hr if the saturation current is 10^{-13} amperes?

3. A beam of 1-MeV gamma rays and another of 0.1-MeV gamma rays each produce the same ionization density in air. What is the ratio of 1:0.1 MeV photon flux?

4. Assuming a specific heat of the body of 1 calorie/g, calculate the temperature rise due to a total body dose of 5 grays.

5. Compute the exposure rate, in mGy/hr at a distance of 50 cm from a small vial containing 10 ml of an aqueous solution of

(a) 2 GBq (54.1 mCi) ^{51}Cr,

(b) 2 GBq (54.1 mCi) ^{24}Na, based on the transformation schemes shown below:

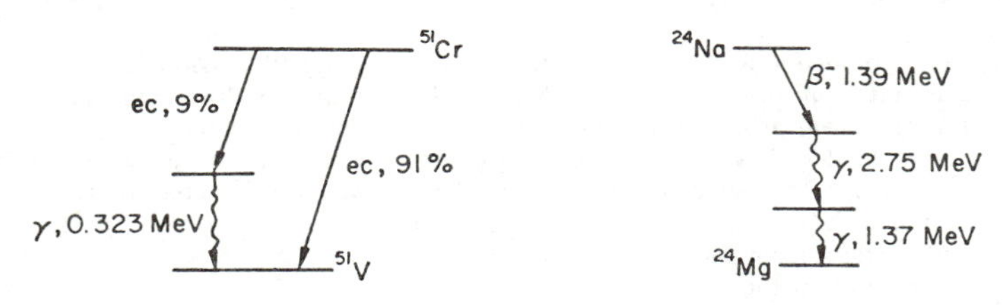

6. What is the dose rate to the flesh during exposure to 25.4 μC/hr (100 mR/hr) of 0.5 MeV gamma radiation?

7. In an experiment, a 250-g rat is injected with 10 μCi ^{203}Hg in the form of $Hg(NO_3)_2$. The rat was counted daily in a total body counter, and the following equation was fitted to the whole body-counting data

$$Y = 0.55e^{-0.0345t} + 0.45e^{-0.0346t},$$

where Y is the fraction of the injected dose retained t after injection. If the long-lived component of the curve represents clearance from the kidneys, while the short-lived component represents clearance from the rest of the body, calculate the radiation absorbed dose to the whole body and the kidneys, if each kidney weighs 0.7 g. Assume the mercury to be uniformly distributed in the whole body and in the kidneys. Base the calculation on the transformation scheme given in Fig. 6.13.

8. Iodine is deposited in the thyroid at a rate of 0.139 per hour. If the radioactive half-life of ^{123}I is 13 hours, what is the effective deposition half-life?

9. A patient with cancer of the thyroid has been found to have a thyroid iodine uptake of 50%. How much ^{131}I must be injected to deliver a dose to the thyroid, which weighs 30 g, of 15 grays (1500 rad) in 3 days?

10. The mean concentration of potassium in seawater is 380 mg/kg. What is the dose rate, in milligrays per year and in millirads per year, in the ocean depths due to the dissolved ^{40}K?

11. Calculate the annual radiation dose to a man from the ^{40}K and from the ^{14}C deposited in his body. The specific activity of carbon is 0.476 Bq (6.9 pCi) per gram. Assume in both instances, that the radioisotopes are uniformly distributed throughout the body.

12. A thin-walled carbon-wall ionization chamber, whose volume is 2 cm^3, is filled with standard air at 0°C and 760 torr and is placed inside a tank of water to make a depth-dose measurement. A 24-MeV betatron beam produces a current of 0.02 μA in the chamber. What was the absorbed dose rate?

13. An aluminum ionization chamber containing 10 cm^3 air at 20°C and 760 torr operates under Bragg–Gray conditions. After a 1-hour exposure to ^{60}Co gamma rays, 3.6×10^{-9} coulomb of charge is collected. If the relative mass stopping power of Al for the electrons generated by the ^{60}Co gammas is 0.875, what was the dose to the aluminum?

14. An ion chamber made of 50 grams copper has a 10-cm^3 cavity filled with air at STP. The temperature of the copper rose 0.002°C after exposure to ^{60}Co gamma rays. If the mass stopping power of Cu is 0.753 relative to air, and if the specific heat of Cu is 0.092 calories per gram per degree C, calculate

(a) The absorbed dose to the copper,

(b) The amount of charge (in coulombs) formed by ionization in the cavity during the exposure.

15. An aqueous suspension of virus is irradiated by X-rays whose half-value layer is 2 mm Cu. If the exposure was 335 C/kg (1.3×10^6 R), and if the depth of the suspension is 5 mm, what was the absorbed dose, and what was the mean ionization density?

16. A child drinks 1 liter of milk per day containing ^{131}I at a mean concentration of 33.3 Bq (900 pCi) per liter over a period of 30 days. Assuming that the child has no other intake of ^{131}I, calculate the dose to the thyroid at the end of the 30-days ingestion period, and the dose commitment.

17. A patient who weighs 50 kg is given an organic compound tagged with 4 MBq (108 μCi) ^{14}C. On the basis of bioassay measurements, the following whole body retention data were inferred:

Day	0	1	2	3	4	5	6	8	10	12	14
MBq	4	2.94	2.32	1.9	1.6	1.4	1.2	0.9	0.8	0.6	0.5

(a) Plot the retention data, and write the equation for the retention curve as a function of time.

(b) Assuming the ^{14}C to be uniformly distributed throughout the body, calculate the absorbed dose to the patient at day 7 and day 14 after administration of the drug.

(c) What is the dose commitment from this procedure?

18. A 2-MeV electron beam is used to irradiate a sample of plastic whose thickness is 0.5 g/cm^2. If a 250-μ amp beam passes through a port 1 cm in diameter to strike the plastic, calculate the absorbed dose rate.

19. Calculate the average power density, in watts per kg, of an aqueous solution of ^{60}Co, at a concentration of 10 MBq per liter, in

(a) An infinitely large medium,

(b) a 6-liter spherical tank.

20. A 20-liter sealed polyethylene cylinder contains 3700 MBq (100 mCi) ^{137}Cs waste uniformly dispersed in concrete. Neglecting absorption by the cover, estimate the dose rate at the top of the container, and at 1 meter over the center of the top.

21. A nuclear bomb is exploded at an altitude of 200 meters. Assuming 10^{18} fissions in the explosion, 6 fission gammas of 1 MeV each and 3 prompt neutrons of 2 MeV each, estimate the dose from the gammas and from the neutrons at 1500 meters from ground zero. Neglect the shileding effect of the air.

22. An unmarked unshielded vial containing 370 MBq (10 mCi) ^{24}Na is left in a hood. A radiochemist not knowing of the presence of the ^{24}Na, spends 8 hours at his bench, which is 2 meters from the ^{24}Na. Based on the ^{24}Na transformation scheme shown in problem 6.5, calculate

(a) The dose rate at 2 meters from the 370 MBq source,

(b) The dose commitment from the 8-hour exposure.

23. Chlormerodrin tagged either with ^{197}Hg or ^{203}Hg is used diagnostically in studies of renal function. Calculate the dose to the kidneys, for the case of normal uptake, from injection of 3.7 MBq (100 μCi) of each of the radioisotopes. Assume very rapid kidney deposition, followed by elimination with a biological half-time of 6.5 hours.

Suggested References

Attix, F. H., Roesch, W. C., and Tochilin, E.., eds. *Radiation Dosimetry*, Vol. I, *Fundamentals*. Academic Press, New York, 1968.

Brodsky, A., ed. *Handbook of Radiation Protection and Measurement*, Vol. I, *Physical Science and Engineering Data*. CRC Press, West Palm Beach, Florida, 1978.

Brodsky, A., ed. *Handbook of Radiation Protection and Measurement*, Vol. II, *Biological and Mathematical Information*. CRC Press, Boca Raton, Florida, 1982.

Fitzgerald, J. J., Brownell, G. L., and Mahoney, F. J. *Mathematical Theory of Radiation Dosimetry*. Gordon and Breach, New York, 1967.

Hendee, W. R. *Medical Radiation Physics*. Yearbook Medical Publishers, Chicago, 1970.

Johns, H. E., and Cunningham, J. R. *The Physics of Radiology*, 3rd ed. Charles C. Thomas, Springfield, Ill, 1973.

Kase, K. R., and Nelson, W. R. *Concepts of Radiation Dosimetry*. Pergamon, New York, 1978.

Morgan, K. Z., and Turner, J. E., eds. *Principles of Radiation Protection*. Krieger, New York, 1973.

Reed, G. W., ed. *Radiation Dosimetry*. Academic Press, New Yrok, 1964.

Spiers, F. W. *Radioisotopes in the Human Body*. Academic Press, New York, 1968.

Wang, Y., ed. *Handbook of Radioactive Nuclides*. Chemical Rubber Co., Cleveland, 1968.

Whyte, G. N. Chap. VI, *Principles of Radiation Dosimetry*. Academic Press, New York, 1959.

The following reports of the National Council on Radiation Protection and Measurements, Washington, D.C.:

No. 25. *Measurement of Absorbed Dose of Neutrons and of Mixtures of Neutrons and Gamma Rays* (1961).

27. *Stopping Powers for Use with Cavity Chambers* (1961).

The following reports of the International Commission on Radiation Units and Measurements, Washington, D.C.:

No. 10b. *Physical Aspects of Irradiation* (1964).

10d. *Clinical Dosimetry* (1963).

13. *Neutron Fluence, Neutron Spectra, and Kerma* (1969).

14. *Radiation Dosimetry: X-Rays and Gamma Rays with Maximum Photon Energies between* 0.6 *and* 50 *MeV* (1969).

16. *Linear Energy Transfer* (1970).

17. *Radiation Dosimetry: X-Rays Generated at Potentials of* 5 *to* 150 *kV* (1970).

21. *Radiation Dosimetry: Electrons with Initial Energies Between* 1 *and* 50 *MeV* (1972).

23. *Measurement of Absorbed Dose in a Phantom Irradiated by a Single Beam of X- or Gamma Rays* (1973).

24. *Determination of Absorbed Dose in a Patient Irradiated by Beams of X- or Gamma Rays in Radiotherapy Procedures* (1976).

26. *Neutron Dosimetry for Biology and Medicine* (1977).

28. *Basic Aspects of High Energy Particle Interactions and Radiation Dosimetry* (1978).

29. *Dose Specification for Reporting External Beam Therapy with Photons and Electrons* (1978).

30. *Quantitative Concepts and Dosimetry in Radiobiology* (1979).

32. *Methods of Assessment of Absorbed Dose in Clinical Use of Radio-nuclides* (1979).

33. *Radiation Quantities and Units* (1980)

34. *The Dosimetry of Pulsed Radiation* (1982).

CHAPTER 7

BIOLOGICAL EFFECTS OF RADIATION

Radiation ranks among the most thoroughly investigated etiologic agents associated with disease. Although much still remains to be learned about the interaction between ionizing radiation and living matter, more is known about the mechanism of radiation damage on the molecular, cellular, and organ system levels than is known for most other environmental stressing agents. Indeed, it is precisely this vast accumulation of quantitative dose-response data that enable health physicists to specify environmental radiation levels so that medical, scientific, and industrial application of nuclear technology may continue at levels of risk no greater than, and frequently less than, the level of risk associated with any other technology.

Dose-Response Characteristics

Observed radiation affects (or effects of other types of noxious agents) may be broadly classified into two categories, viz. *stochastic* and *non-stochastic* effects. Most biological effects fall into the category of non-stochastic effects. Non-stochastic effects are characterized by three qualities: First, a certain minimum dose must be exceeded before the particular effect is observed. Additionally, the magnitude of the effect increases with the size of the dose. Furthermore, there is a clear causal relationship between exposure to the noxious agent and the observed effect. For example, a person must exceed a certain amount of alcohol before he shows signs of drinking. After that, the effect of the alcohol depends on how much he drank. Finally, if he exhibits drunken behavior, there is no doubt that his behavior is the result of his drinking. For such non-stochastic effects, when the magnitude of the effect, or the proportion of individuals who respond at a given dose is plotted as a function of dose in order to obtain a *quantitative relationship* between dose and effect, the dose-response curve A shown in Fig. 7.1 is obtained. Because of the minimum-dose that must be exceeded before an individual shows the effect, non-stochastic effects are also called *threshold effects.*

In an experiment to determine a dose-response curve, the 50% dose, that is, the dose to which 50% of the exposed animals respond, is statistically the most reliable. For this reason, the 50% dose is most frequently used as an index of relative effectiveness of a given agent in eliciting a particular response. When death of the experimental animal is the biological endpoint, the 50% dose is called the LD-50 dose. The time required for the toxic substance to act is also important, and is always specified with the dose. Thus, if 50% of the animals die within 30 days, we refer to the LD-50/30 day dose. This index, the LD-50/30 day dose, is widely used by toxicologists to designate the relative toxicity of a substance.

Stochastic effects, as the name implies, are those effects that occur by chance; and they occur among unexposed people as well as among exposed individuals. Stochastic effects are therefore not unequivocally related to exposure to a noxious agent, as drunkenness is

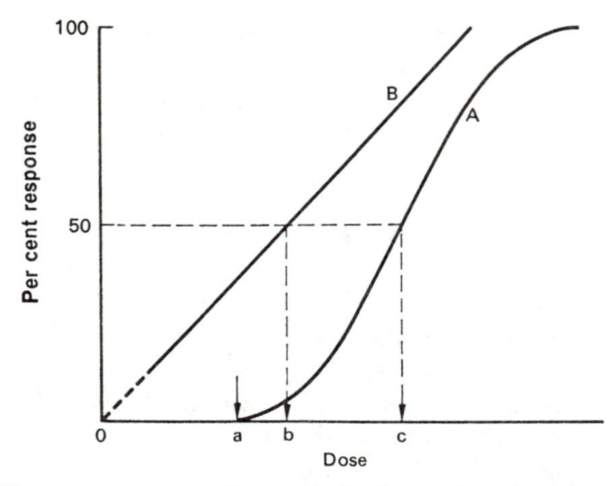

FIG. 7.1. Dose-response curves. Curve *A* is the characteristic shape for a biological effect that exhibits a threshold dose—point *a*. The spread of the curve, from the threshold at *a* until the 100% response is thought to be due to "biological variability" around the mean dose, point *c*, which is called the 50% dose. Curve *B* represents a zero-threshold, or linear response; point *b* represents the 50% dose for the zero-threshold biological effect.

to alcohol ingestion or sunburn is to overexposure to the sun. In the context of radiation protection, the main stochastic effects are cancer and genetic effects. The result of exposure to a carcinogen or to a mutagen is an increase in the probability of occurrence of the effect, with the increase in probability being directly proportional to the size of the dose. Thus, people develop cancer whether or not they are exposed to carcinogenic agents. However, exposure to a carcinogen increases the likelihood of cancer; and the greater the exposure the greater is the increased likelihood. At no time, however, regardless of the size of the exposure, is it certain that cancer will result from exposure to a carcinogen. If cancer does develop after exposure to a carcinogen, we cannot be absolutely certain as we are in the case of the causal relationship between alcohol and drunkenness—that the cancer was caused by the carcinogen. The best that we can do is to estimate the probability that the cancer was caused by the carcinogen. For example, we found lung cancer in a much higher proportion of cigarette smokers than among non-smokers, and among cigarette smokers, lung cancer is seen in a greater proportion of heavy smokers than in light smokers. However, most cigarette smokers do not develop lung cancer. Smoking, even very heavy smoking, does not ensure that the smoker will develop lung cancer, it merely increases the likelihood of developing the disease.

When frequency of occurrence of a stochastic effect is plotted against size of dose, curve *B* in Fig. 7.1, a linear dose-response relationship is observed rather than the *S*-shaped curve that is characteristic of agents associated with a threshold response. The biological model that is compatible with this straight line dose-response relationship and with our knowledge of molecular biology predicts that cancer can be initiated or a genetic change can be wrought by scrembling the genetic information coded in a single molecule. Thus, carcinogenesis and mutagenesis are merely different manifestations of the same basic molecular phenomenon. In this context, a cancer is initiated by damaging the information stored in the chromosomes of a somatic cell, whereas a genetic change is caused by damage to the information stored in the chromosomes of a germ cell (a sperm or an ovum). On the basis of this model, no threshold should exist for a stochastic effect, and even the smallest amount of the carcinogen or mutagen—a single molecule in the case of chemicals or a single photon in the case of X-rays—can produce the effect. For these reasons, stochastic effects are often called *linear, zero-threshold dose-response effects*. According to the linear, zero-threshold model, every increment of radiation, no matter how small, carries with it a corresponding increase in risk of stochastic effect.

Direct Action

The gross biological effects resulting from overexposure to radiation are the sequellae to a long and complex series of events that are initiated by ionization or excitation of relatively few molecules in the organism. For example, the LD-50/30 day dose for man of gamma-rays is about 4 Gy (400 rads). Since 1 gray corresponds to an energy absorption of 1 J/kg, or 6.25×10^{18} eV/g, and since about 34 eV are expended in producing a single ionization, the lethal dose produces, in tissue,

$$\frac{4\ \text{Gy} \times 6.25 \times 10^{18}\ \text{eV kg}^{-1}\ \text{Gy}^{-1}}{34\ \text{eV/ion}} = 7.35 \times 10^{17}$$

ionized atoms per gram tissue. If we estimate that about nine other atoms are excited for each one ionized, we find that about 7.35×10^{18} atoms/kg of tissue are directly affected by a lethal radiation dose. In soft tissue, there are about 9.5×10^{25} atoms/kg. The fraction of directly affected atoms, therefore, is

$$\frac{7.35 \times 10^{18}}{9.5 \times 10^{25}} \approx 1 \times 10^{-7},$$

or about 1 atom in 10 million.

Effects of radiation for which a zero threshold dose is postulated are thought to be the result of a direct insult to a molecule by ionization and excitation and the consequent dissociation of the molecule. Point mutations, in which there is a change in a single gene locus, is an example of such an effect. The dissociation, due to ionization or excitation, of an atom on the DNA molecule prevents the information originally contained in the gene from being transmitted to the next generation. Such point mutations may occur in the germinal cells, in which case the point mutation is passed on to the next individual; or it may occur in somatic cells, which results in a point mutation in the daughter cell. Since these point mutations are thereafter transmitted to succeeding generations of cells (except for the highly improbable instance where one mutated gene may suffer another mutation), it is clear that for those biological effects of radiation that depend on point mutations, the radiation dose is cumulative; every little dose may result in a change in the gene burden which is then continuously transmitted. When dealing quantitatively with such phenomena, however, we must consider the probability of observing a genetic change among the offspring of an irradiated individual. For radiation doses down to about 250 mGy (25 rads), the magnitude of the effect, as measured by frequency of gene mutations, is proportional to the dose. Below doses of about 250 mGy, the mutation probability is so low that enormous numbers of animals must be used in order to detect a mutation that could be ascribed to the radiation. For this reason, no reliable experimental data are available for genetic changes in the range of 0 to about 250 mGy.

Indirect Action

Direct effects of radiation, ionization, and excitation are non-specific, and may occur anywhere in the body. When the directly affected atom is in a protein molecule, or in a molecule of nucleic acid, then certain specific effects due to the damaged molecule may ensue. However, most of the body is water, and most of the direct action of radiation therefore is on water. The result of this energy absorption by water is the production, in the water, of highly reactive free radicals that are chemically toxic (a free radical is a fragment of a compound or an element that contains an unpaired electron) and which may exert their toxicity on other molecules. When pure water is irradiated we have

$$H_2O \rightarrow H_2O^+ + e^-, \tag{7.1}$$

and the positive ion dissociates immediately according to the equation

$$H_2O^+ \rightarrow H^+ + OH, \tag{7.2}$$

while the electron is picked up by a neutral water molecule

$$H_2O + e^- \rightarrow H_2O^-, \tag{7.3}$$

which dissociates immediately

$$H_2O^- \rightarrow H + OH^-. \tag{7.4}$$

The ions H^+ and OH^- are of no consequence, since all body fluids already contain significant concentrations of both these ions. The free radicals H and OH may combine with like radicals, or they react with other molecules in solution. Their most probable fate is determined chiefly by the LET of the radiation. In the case of a high rate of linear energy transfer, such as results from passage of an alpha particle or other particle of high specific ionization, the free OH radicals are formed close enough together to enable them to combine with each other before they can recombine with free H radicals, which leads to the production of hydrogen peroxide

$$OH + OH \rightarrow H_2O_2, \tag{7.5}$$

while the free H radicals combine to form gaseous hydrogen. Whereas the products of the primary reactions of equations (7.1) through (7.4) have very short lifetimes, on the order of a microsecond, the hydrogen peroxide, being a relatively stable compound, persists long enough to diffuse to points quite remote from their point of origin. The hydrogen peroxide, which is a very powerful oxidizing agent, can thus affect molecules or cells that did not suffer radiation damage directly. If the irradiated water contains dissolved oxygen, the free hydrogen radical may combine with oxygen to form the hydroperoxyl radical,

$$H + O_2 \rightarrow HO_2, \tag{7.6}$$

which is not as reactive, and therefore has a longer lifetime, than the free OH radical. This greater stability allows the hydroperoxyl radical to combine with a free hydrogen radical to form hydrogen peroxide, thereby further enhancing the toxicity of the radiation.

Radiation is thus seen to produce biological effects by two mechanisms, viz. directly by dissociating molecules following their excitation and ionization; and indirectly by the production of free radicals and hydrogen peroxide in the water of the body fluids. The greatest gap in our knowledge of radiobiology is the sequence of events between the primary effects described above, and the gross biological effects that may be observed long after irradiation.

Radiation Effects

In health physics, as in other areas of environmental control of harmful agents, we are concerned with two types of exposure: (1) a single accidental exposure to a high dose of radiation during a short period of time, which is commonly called *acute* exposure, and which may produce biological effects within a short time after exposure; (2) long-term, low level overexposure, commonly called *continuous* or *chronic* exposure, where the results of the overexposure may not be apparent for years, and which is likely to be the result of improper or inadequate protective measures.

Acute Effects

Acute whole body radiation overexposure affects all the organs and systems of the body. However, since not all organs and organ systems are equally sensitive to radiation, the pattern of response, or disease syndrome, in an overexposed individual depends on the magnitude of the dose. To simplify classification, the acute radiation syndrome is subdivided into three classes; in order of increasing severity, these are: (1) the hemopoietic syndrome, (2) the gastrointestinal syndrome, and (3) the central nervous system syndrome. Certain effects are common to all categories; these include:

(a) nausea and vomiting,
(b) malaise and fatigue,
(c) increased temperature,
(d) blood changes.

In addition to these effects, numerous other changes are seen.

Blood changes

Of the four common effects listed above, changes in the peripheral blood count are the most sensitive biological indicators of acute overexposure. These changes are seen even in cases of mild overexposure, which results in none of the three syndromes. Although blood changes have been seen in individuals with gamma-ray doses as low as 140 mGy (14 rads), they usually do not appear until doses of 250–500 mGy (25–50 rads) are experienced. Beyond 500 mGy, blood changes are almost certain to appear.

The blood consists of about 55% (by volume) fluid, called the blood plasma, and about 45% of formed elements, including white blood cells, called leucocytes, red blood cells, called erythrocytes, and platelets, or thrombocytes. The white blood cells, which number about 7000/mm^3 of blood in the average adult, function in the body as a major line of defense against bacterial invasion. An infection anywhere in the body stimulates the production of leucocytes in order to combat the infecting organisms. Several major types of leucocytes are found: the granulocytes and the lymphocytes, each with certain specialized functions to aid in the fight against infection. Under normal conditions, the relative proportions of each of these remain approximately constant—the granulocytes form about 70–75% of the white blood cells while the lymphocytes account for about 25–30%. The granulocytes are produced in the bone marrow, and circulate for about 3 days before death and destruction, while the lymphocytes are produced in the lymph nodes and spleen and remain alive in the blood for about 24 hr. The red blood cells are the most abundant of the formed elements; their concentration in the blood is about 5 million per cubic millimeter. The main function of the red blood cells is to transport oxygen from the lungs to the body cells, and to carry the carbon dioxide waste from the cells to the lung. The erythrocytes are formed in the bone marrow, and survive in the circulating blood for about 90–120 days. The platelets, or thrombocytes, which number about 200,000–400,000/mm^3 blood, are concerned with the clotting of the blood. They are manufactured in the marrow, and have a useful lifetime of about 8–12 days.

After an acute radiation exposure in the sub-lethal range, there is a transitory sharp increase in the number of granulocytes followed within a day by a decrease which reaches a minimum several weeks after exposure, and then returns to normal after a period of several weeks to several months. The lymphocytes drop sharply after exposure, and remain depressed for a period of several months. In contrast to the very rapid response of the white cells to radiation overexposure, the red blood count does not reflect an overexposure until about a week after exposure. Depression in the erythrocyte count continues until a minimum is reached between 1 to 2 months after exposure, followed by a slow recovery over a period of weeks. The platelet count falls steadily until a minimum is reached about a month after exposure; recovery is very slow, and may take several months. In all cases, the degree of change in the blood, as well as the rate of change, is a function of the radiation dose. Figure 7.2 shows graphically the trends in rate and degree of blood changes for several different exposure doses.

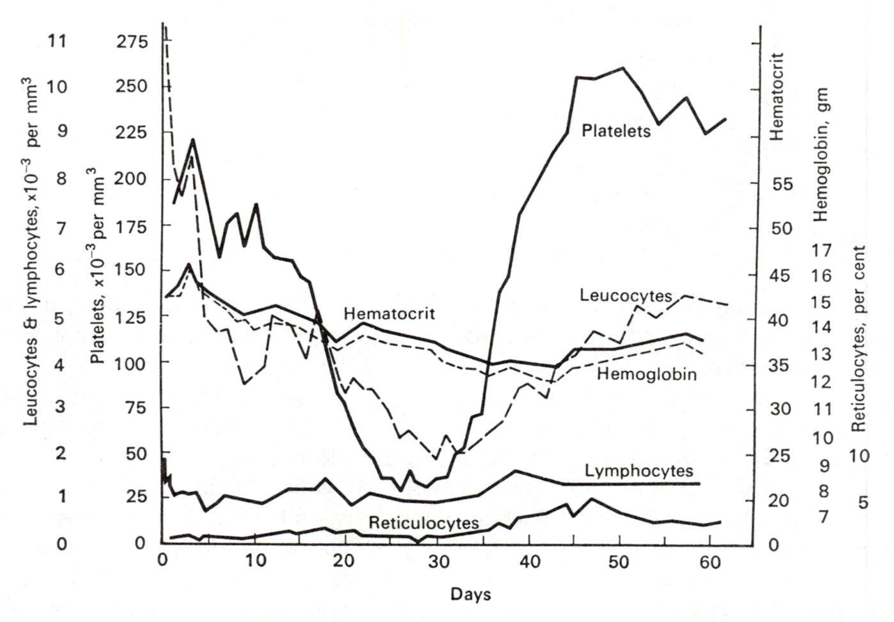

FIG. 7.2. Hematologic effect of radiation overexposure. Average values for five patients who were exposed to 236–365 rad (estimated) during a criticality accident at the Y-12 plant in Oak Ridge on 16 June 1958. (G. H. Andrews, B. W. Sitterson, A. L. Kretchman and M. Brucer, Criticality accident at the Y-12 plant, *Diagnosis and Treatment of Acute Radiation Injury*, pp. 27–48, World Health Organization, Geneva, 1961.)

Hemopoietic syndrome

The hemopoietic syndrome appears after a gamma dose of about 2 Gy (200 rads). This disease is characterized by depression or ablation of the bone marrow, and the physiological consequences of this damage. The onset of the disease is rather sudden, and is heralded by nausea and vomiting within several hours after the overexposure occurred. Malaise and fatigue are felt by the victim, but the degree of malaise does not seem to be correlated with the size of the dose. Epilation, which is almost always seen, appears between the second and third week post exposure. Death may occur within 1 to 2 months after exposure. The chief effects to be noted, of course, are in the bone marrow and in the blood. Marrow depression is seen at 2 Gy; at about 4–6 Gy (400–600 rads), complete ablation of the marrow occurs. In this case, however, spontaneous regrowth of the marrow is possible if the victim survives the physiological effects of the denuding of his marrow. An exposure of about 7 Gy or greater leads to irreversible ablation of the bone marrow. The LD-50/30 day dose for most mammals, including man, falls within this dose range. The blood picture reflects the damage to the marrow; the changes noted above are seen, and the magnitude of the change is roughly correlated with the dose. A very low lymphocyte count, 500/mm^3 or less within the first day or two after exposure suggests that death will probably ensue.

Gastrointestinal syndrome

This disease follows a total body gamma dose of about 10 Gy or greater, and is a consequence of the desquamation of the intestinal epithelium. All the signs and symptoms of the hemopoietic syndrome are seen—with the addition of severe nausea, vomiting, and diarrhea which begins very soon or immediately after exposure. Death within 1 to 2 weeks after exposure is the most likely outcome.

Central nervous system syndrome

A total body gamma dose in excess of about 20 Gy (2000 rads) damages the central nervous system as well as all the other organ systems in the body. Unconsciousness follows within minutes after exposure, and death in a matter of hours to several days; the rapidity of onset of unconsciousness is directly related to dose. In one instance in which a 200-μsec burst of mixed neutrons and gamma-rays delivered a mean total body dose of about 44 Gy (4400 rads), the victim was ataxic and disoriented within 30 sec. In 10 min he was unconscious and in shock. Thirty-five minutes after the accident, analysis of faecal fluid from an explosive water diarrhea showed a copious passage of fluids into the gastro-intestinal tract. Vigorous symptomatic treatment kept the patient alive for $34\frac{1}{2}$ hr after the accident.

Other acute effects

Several other immediate effects of acute overexposure should be noted. Because of its physical location, the skin is subject to more radiation exposure, especially in the case of low energy X-rays and beta rays, than most other tissues. An exposure of about 77 mC/kg (300 R) of low energy (in the diagnostic range) X-rays results in erythema; higher doses may cause changes in pigmentation, epilation, blistering, necrosis, and ulceration. Radiation dermatitis of the hands and face was a relatively common occupational disease among radiologists who practiced during the early years of the twentieth century.

The gonads are particularly radiosensitive. A single dose of only 300 mGy (30 rads) to the testes results in temporary sterility among men; for women, a 3-Gy (300-rad) dose to the ovaries produces temporary sterility. Higher doses increase the period of temporary sterility; one man, whose exposure to the gonads was less than 4.4 Gy (440 rads), was aspermatic for a period of several years. In women, temporary sterility is evidenced by a cessation of menstruation for a period of 1 month or more, depending on the dose. Irregularities in the menstrual cycle, which suggest functional changes in the gonads, may result from local irradiation of the ovaries with doses smaller than that required for temporary sterilization.

The eyes, too, are relatively radiosensitive. A local dose of several grays can result in acute conjunctivitis and keratitis.

Delayed Effects

The delayed effects of radiation may be due either to a single large overexposure or continuing low-level overexposure.

Continuing overexposure can be due to exposure to external radiation fields, or can result from inhalation or ingestion of a radioisotope which then becomes fixed in the body through chemical reaction with the tissue protein or, because of the chemical similarity of the radioisotope with normal metabolites, may be systemically absorbed within certain organs and tissues. In either case, the internally deposited radioisotope may continue to irradiate the tissue for a long time. In this connection, it should be pointed out that the adjectives "acute" and "chronic", as ordinarily used by toxicologists to describe single and continuous exposures respectively, are not directly applicable to inhaled or ingested radioisotopes, since a single, or "acute" exposure may lead to continuous, or "chronic" irradiation of the tissue in which the radioactive material is located. Among the delayed consequences of overexposure which are of concern are cancer, genetic effects, shortening of life span, and cataracts.

Cancer

The carcinogenic effects of doses of 1 Gy (100 rads) or more of gamma radiation delivered at high dose rates are well documented, consistent and definitive. Although any organ or tissue may develop neoplasia after overexposure to radiation, certain organs and tissues seem to be more sensitive in this respect than others. Radiation-induced cancer is observed most frequently in the hemopoietic system, in the thyroid, in the bone, and in the skin. In all these cases, the tumor induction time in man is relatively long—on the order of 5 to 20 years after exposure. Carcinoma of the skin was the first type of malignancy that was associated with exposure to X-rays. Early X-ray workers, including physicists and physicians, had a much higher incidence of skin cancer than could be expected from random occurrence of this disease. Well over 100 cases of radiation induced skin cancer are documented in the literature. As early as 1900, a physician who had been using X-rays in his practice described the irritating effects of X-rays. He recorded that erythema and itching progressed to hyper-pigmentation, ulceration, neoplasia, and finally death from metastatic carcinoma. The entire disease process spanned a period of 9 years. An occupational disease among dentists, before the carcinogenic properties of X-rays were well understood, was cancer of the fingers that were used to hold dental X-ray film in the mouths of patients while X-raying their teeth.

Epidemiologic data on the carcinogenicity of low doses of radiation are contradictory and inconclusive. For purposes of setting radiation safety standards, therefore, it is prudent to estimate the risk from low-level radiation by extrapolation from the high-dose case. The linear zero-threshold model on which this extrapolation is based postulates an upper limit of about 125 excess cancer deaths among a population of 1 million people who received a whole body dose-equivalent of 10 millisieverts (1 rem). (The millisievert and the rem are discussed later in this chapter.) To place these postulated excess cancer deaths into perspective, we note that about 206,000 cancer deaths normally occur among 1 million people.

Leukemia

Leukemia, especially acute myelogenous leukemia, and to a lesser extent chronic myelogenous or acute lymphocytic leukemia, is among the most likely forms of malignancy resulting from overexposure to total body radiation. Chronic lymphocytic leukemia does not appear to be related to radiation exposure. Radiologists and other physicians who used X-rays in their practice before strict health physics practices were common showed a significantly higher rate of leukemia than did their colleagues who did not use radiation. Among American radiologists, the doses associated with the increased rate of leukemia were on the order of 1 Gy (100 rads) per year. With the increased practice of health physics, the difference in leukemia rate between radiologists and other physicians has been continually decreasing. An increased leukemia incidence has also been observed among patients who were treated with X-rays for ankylosing spondilitis (rheumatoid arthritis of the small joints of the spine).

Among the survivors of the nuclear bombings of Japan, there was a significantly greater incidence of leukemia among those who had been within 1500 meters of the hypocenter than among those who had been more than 1500 meters from ground zero at the time of the bombing (Fig. 7.3). The first increase in leukemia incidence among the survivors was seen about 3 years after the bombings, and the leukemia rate continued to increase until it peaked about 4 years later. Since then, it has been steadily decreasing. The leukemia incidence rate in Nagasaki reached the normal Japanese rate in the early 1970's; and in Hiroshima the leukemia incidence rate continues to be slightly higher among the exposed survivors than among the unexposed control population.

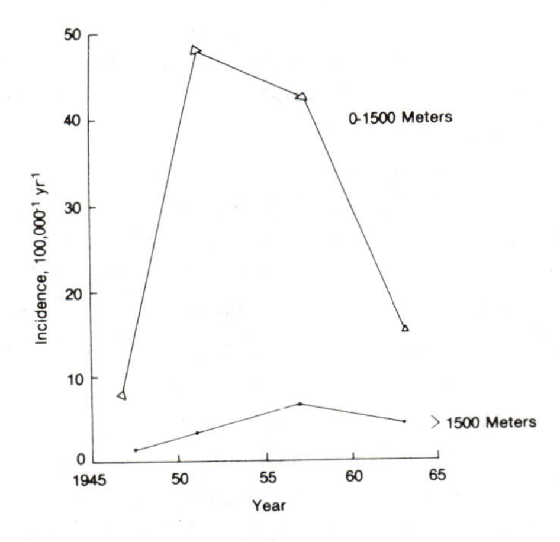

FIG. 7.3. Incidence rate of leukemia among the survivors of the nuclear bombings in Hiroshima and Nagasaki.

The questions regarding the leukemogenicity of low radiation doses and of the existence of a non-zero threshold dose for leukemia induction remains unanswered, and are the subject of controversy. A clear dose-response relationship was seen in Hiroshima and Nagasaki. The lowest leukemogenic dose was seen in Hiroshima in the 0.2–0.4-Gy (20–40 rads) range. In Nagasaki, an increased leukemia incidence was not seen in people who has received less than 1 Gy (100 rads). The marked difference between the findings in the two cities is attributed to the relatively large neutron component of the dose in

Hiroshima and to the virtual absence, less than 2% of the total dose, of neutron radiation in Nagasaki. No radiation induced leukemia was seen among persons who had been exposed *in utero* to radiation from the bombs. However, a study in Great Britain found that children who had been irradiated *in utero* during medical X-ray examinations of the mother had a significantly higher leukemia rate than did children who had not been irradiated *in utero*. On the basis of these studies, it was inferred that as little as 10–50 mGy (1–5) rads of X-rays may lead to leukemia. Since it is estimated that the fetus is about 2 to 10 times more sensitive to radiation damage than post-partum individuals, these findings imply a leukemia threshold dose in the range of 1–25 mGy (0.1–2.5 rads) of X-rays. On the other hand, in other studies involving *in utero* X-ray exposures that were made for routine pelvimetry rather than for some medical indication, no evidence of any radiation effect was seen. The difference between these two divergent observations is thought to be due to the medical factors that required the diagnostic X-rays. This finding is compatible with those of other studies that imply that there is a threshold dose for radiogenic leukemia, and that it lies in the range of 0.1–0.5 Gy (10–50 rads) for *in utero* irradiation. Thus, although all the available data show ionizing radiation to be leukemogenic at high doses, the results of studies on the leukemogenic effects of low doses are inconclusive and controversial. Because the results from low dose studies are conflicting, however, it is reasonable to infer that low level radiation at doses associated with most diagnostic X-ray procedures, with occupational exposure within the recommended limits, and with natural radiation is a very weak leukemogen, and that the attributive risk of leukemia from low level radiation is probably very small.

Bone cancer

Historically, occupational radiation-induced cancer of the bone and of the lung are important in emphasizing the carcinogenicity of internally deposited radioisotopes. Although the cancer-producing properties of radium were known within several years after its discovery by Pierre and Marie Curie in 1902, little or no effort was made to protect people from the harmful effects of radium. One of the first industrial uses to which radium was put was in the manufacture of luminous paint. When powdered radium is mixed with ZnS crystals, the crystals glow owing to absorption of energy from the alpha particles emitted by the radium. Luminous paint made in this way soon was used to paint instrument and clock dials. This application of radium received a great impetus when World War I began. In order to paint fine lines, the girls who were employed as dial painters pointed their brushes between their lips. Minute amounts of radium and mesothorium were swallowed each time that a brush was pointed. In the early 1920's, several girls who had worked as dial painters died from anemia and with degeneration of the jaw bones. A dentist who was treating another dial painter suspected the occupational etiology of the disease. Further investigation revealed more cases of bone damage, including osteogenic sarcoma, and definitely established radium as the etiologic agent. Follow-up studies on radium dial painters and on patients who had received radium injections therapeutically confirmed this finding. It also showed that radium is tenaciously retained in the bones. Significant deposits of radium were found 25–35 years after exposure. Further experimental work with laboratory animals revealed a number of "bone-seeking" radioisotopes which produced the same type of damage as radium. Radio-strontium and barium, as well as radium, are chemically similar to calcium, and are therefore incorporated into the mineral structure of the bone and into the epiphysis. The bone-seeking rare earth fission products such as ^{144}Ce–^{144}Pr also tend to accumulate in the mineral structure. Plutonium, an extremely toxic element, is found to accumulate in the periosteum, endosteum, and trabeculae of the bone. All bone seekers are considered very toxic because they can damage the radiosensitive hemopoietic tissue in the bone marrow; all the bone seekers produce bone cancer when they are injected into laboratory animals in sufficient quantity.

Lung cancer

The susceptibility of the lung to radiation-induced carcinoma has been known for a long time. The mines in Joashimsthal and Schneeberg are rich in pitchblende, a radium-bearing ore (as well as in other minerals which are also suspected of being carcinogens). As a consequence of the radium in the ground, radon gas, the radioactive daughter of radium, is produced and diffuses out of the ground into the air in the mine shafts. Since radon is itself radioactive and gives rise to several radioactive descendants (Table 4.3) a good deal of radioactivity due to the radon daughter products is also present. The atmospheric concentration of radon in the two European mining centers was about 1×10^5 Bq/m^3 (3×10^{-6} μCi/ml) air. To this activity must be added the daughter activities, which are chiefly RaA and RaC. A very high percentage of workers exposed to this atmosphere developed bronchogenic carcinoma within 15 years after beginning to work in the mines. The carcinogenicity of inhaled radon was confirmed in the laboratory using mice who were exposed daily to atmospheric concentrations of 37 MBq/m^3 (10^{-3} μCi/ml). In other laboratory studies, involving the deposition of radioactive materials in the lung by intratracheal injection, by inhalation, by trans-pleural injection of ^{90}Sr-loaded glass beads and by surgical implantation in a bronchus of a small cylinder of ^{106}Ru, radiation was found to be carcinogenic to the lung. From these studies it was found that long-term continuous radiologic insult from radioactive materials residing in the lung was required in order to produce lung cancer. On the order of 100 or more days elapsed, in most instances, between exposure and the observation of lung cancer. However, in one instance, a squamous-cell carcinoma was observed only 48 days after intratracheal injection of 15 μCi ^{144}Ce as CeF_3 particulates. Irradiation of the lungs with X-rays has been found to produce fibrosis and pneumonitis, but no neoplasia.

Genetic Effects

Genetic information necessary for the production and functioning of a new organism is contained in the chromosomes of the germ cells—the sperm and the ovum. The normal human somatic cell contains 46 of these chromosomes; mature sperm and ovum each carry 23 chromosomes. When an ovum is fertilized by a sperm, the resulting cell, called a zygote, contains the full complement of 46 chromosomes. During the 9-month gestation period, the fertilized egg, by successive cellular division and differentiation, develops into a new individual. In the course of the cellular divisions, the chromosomes are exactly duplicated, so that all the cells in the body contain the same genetic information. The units of information in the chromosomes are called the genes. Each gene is an enormously complex macromolecule called deoxyribonucleic acid (DNA), in which the genetic information is coded according to the sequence of certain molecular and sub-assemblies called bases. The DNA molecule, whose molecular weight is on the order of 10^7, consists of two long chains composed of pentose sugars (deoxyribose) and phosphates that wind around each other in a spiral double helix. The two long intertwined strands are held together by the bases, which form cross-links between the long strands in the same manner as the treads in a step-ladder (Fig. 7.4). There are four different bases, 2 purines: adenine (A) and guanine (G), and two pyrimidines: cytosine (C) and thymine (T). The base cross links are formed when two of these bases, one of which is attached to each long strand, join together. The cross-linked bond is specific, with adenine coupling only with thymine (A–T) and cytosine coupling only with guanine (C–G). The bits of information are coded in triplets of various combinations of A–T and C–G cross links, and the sequence of these triplets determine the genetic information contained in a DNA molecule.

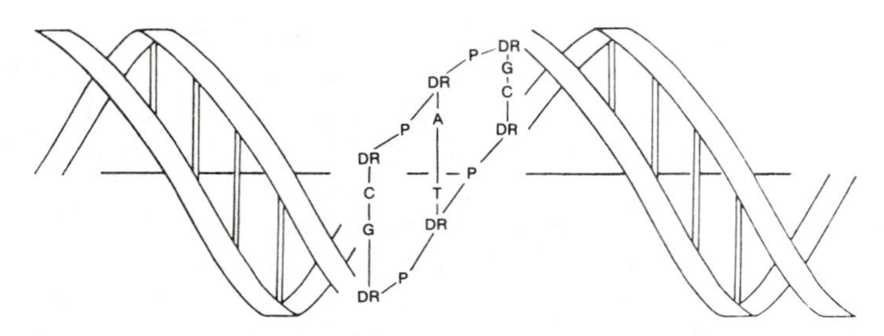

FIG. 7.4. Structure of the DNA molecule. The nucleotides, the strands of the double helix, consist of alternating deoxyribose (DR) and phosphate (p) units. Two pairs of complementary bases, adenine (A) and thymine (T), and guanine (G) and cytosine (C), join the two nucleotide chains.

The genetic information can be altered by many different chemical and physical agents, called mutagens, that disrupt the sequence of bases in a DNA molecule. If this information content of a somatic cell is scrambled, then its descendents may show some sort of an abnormality. If the information that is jumbled is in a germ cell that subsequently is fertilized, then the new individual may carry a genetic defect, or a mutation. Such a mutation is often called a *point mutation*, since it results from damage to one point on a gene. Most geneticists believe that the majority of such mutations in man are undesirable or harmful.

In addition to point mutations, genetic damage can arise through chromosomal aberrations. Certain chemical and physical agents can cause chromosomes to break. In most of these breaks, the fragments reunite, and the only result may be a point mutation at the site of the original break. In a small fraction of breaks, however, the broken pieces do not reunite. When this happens, one of the broken fragments may be lost when the cell divides, and the daughter cell does not receive the genetic information contained in the lost fragment. The other possibility following chromosomal breakage, especially if two or more chromosomes are broken, is the interchange of the fragments among the broken chromosomes, and the production of aberrant chromosomes. Cells with such aberrant chromosomes usually have impaired reproductive capacity as well as other abnormalities.

The mutagenic properties of ionizing radiation were discovered by Hermann J. Muller in 1927 (for which he won the Nobel prize in medicine). He studied the genetic effects of X-rays on fruit flies. Since then, studies on numerous other organisms and with all of the ionizing radiations confirm the mutagenicity of all forms of ionizing radiation. Point mutations are changes on a molecular level. However, many biochemical events that can modify the dose-response relation occur between molecular changes on a gene and the somatic expression of that molecular change. For example, if the genetic damage renders a germ cell or a zygote non-viable, then no living organism that carries that mutation will be born. Empirically, therefore, no detrimental effect is observed. Nevertheless, the studies suggest that the existence of a threshold dose for genetic effects of radiation is unlikely.

However, they also show that the genetic effects of radiation are inversely dependent on dose rate over the dose range of 8 mGy/min (800 mrad/min) to 0.9 Gy/min (90 rads/min). This dose rate dependence clearly implies a repair mechanism that is overwhelmed at the high dose rate.

What is the quantitative relationship between radiation dose and probability of a mutation? The doubling dose, that is, the radiation dose that would eventually lead to a doubling of the mutation rate is estimated by geneticists to probably be in the range of 0.5–2.5 Gy (50–250 rads). Geneticists also believe that about 0.8% of people are born with mutated dominant gene traits, and they estimate that about 4% of these are new mutations that arose in one of the parents. The geneticists thus estimate the proportion of newly mutated dominant genes to be $0.008 \times 0.04 = 0.00032$. If we assume a linear dose-response relationship, and a doubling dose of 0.5 Gy, then the fraction of new dominant mutations due to radiation, if the entire breeding population received a radiation dose of 10 mGy (1 rad) would be $1/50 \times 0.00032$, or about 0.0000064. If only one parent received the radiation dose, then the increased proportion would be one-half of this or 0.0000032. Thus, the probability of having a dominant gene trait mutated by radiation is estimated as about 3 per million per rad in a parent who was irradiated before conception occurred. This increased probability is in addition to the 320 chances per million of a "spontaneous" mutation in a dominant gene trait.

Hazard and toxicity

The experimental work referred to above, in which lung tumors resulted from radioactivity in the lung that was deposited unphysiologically by injection or surgical implantation, clearly cannot serve as a measure of the *hazard* from radioactive dusts. They can serve only to indicate the *toxicity* of radioactive material after the radioactivity is located at the site of its toxic action. The hazard from inhaled radioactive dusts—or indeed the hazard from any toxic material—must include a consideration of the likelihood that the toxic substance will reach the site of its toxic action. In the case of inhaled dusts, this site is assumed to be the bronchial epithelium, the alveolar epithelium, and the pulmonary lymph nodes. The two main factors that influence the degree of hazard from toxic airborne dusts are: (1) the deposition in the lung of the dust particle, and (2) the retention of the particle within the lung.

The deposition of particles within the lung depends mainly on the particle size of the dust, while the retention in the lung depends on the physical and chemical properties of the dust as well as on the physiological status of the lung. Dusts generated by almost any process are found to be randomly distributed in size, or "diameter," around a mean value. This size distribution is found to be "log-normal," that is, the logarithm of the particle size is found to be normally distributed, rather than the size directly. In this case, the mean size, which is called the geometric mean, is defined by

$$m_g = \text{antilog}\left(\frac{\sum \log d_n}{n}\right), \tag{7.7}$$

where d_n is the particle diameter, which most often is expressed in microns ($1\ \mu = 10^{-4}$ cm), of each of the different particles that is measured, and n is the number of particles that are sized. The standard deviation of the size distribution is defined as

$$\sigma_g = \text{antilog}\sqrt{\left(\frac{\sum (\log d_n)^2}{n} - \left(\frac{\sum \log d_n}{n}\right)\right)^2}. \tag{7.8}$$

It corresponds to the inflection point of the Gaussian distribution curve, as shown in Fig. 7.5(a). In a normal distribution, 68.2% of the population falls within the limits bounded by plus and minus one standard deviation from the mean; 95.4% of the population is included within plus and minus two standard deviations of the mean. Because of the logarithmic nature of the distribution, the size distribution must be given as

$$m_g \overset{\times}{\div} \sigma_g, \tag{7.9}$$

rather than, as in the case of an arithmetic mean,

$$m_a \pm \sigma_a. \tag{7.10}$$

For example, if the log-normal size distribution of a dust is given as $1.4 \overset{\times}{\div} 2$, this means that 68% of all the particles lie between 0.7 (or $1.4 \div 2$) and 2.8 (or $1.4 \times 2)\mu$, and that about 96% lie between 0.35 and 5.6 μ.

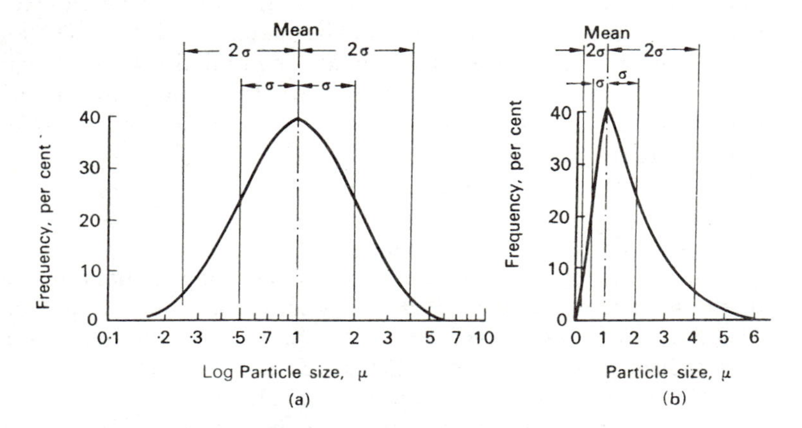

FIG. 7.5. (a). Log-normal distribution of dust particle sizes. The logarithm of the diameter is normally distributed, thus giving a Gaussian distribution when the logarithm of the particle size is plotted against frequency. In this curve, the mean size is 1 μ, and the geometric standard deviation is 2. (b) Size distribution curve for the same data as shown in (a). In this curve, the diameter is plotted against frequency, thus yielding a skewed distribution curve.

The respiratory tree may be roughly divided into two categories on the basis of physiologic function: the upper respiratory tract and the deep respiratory tract. The functional part of the lung, where gaseous exchange between the inspired air and the dissolved gas in the blood occurs, is the deep respiratory tract. The rest of the respiratory tree, or upper respiratory tract, including the nasopharyngeal passages, the trachea, the bronchi, and the terminal bronchioles, are only avenues that permit gas to pass from the atmosphere to and from the alveoli, which form the deep respiratory tract. The volume of the upper respiratory tract, which is really a dead space as far as gas exchange is concerned, is about 150 ml in an average adult; the tidal volume, that is, the volume of air inspired or expired during a single respiratory excursion, is about 500 ml. The functional residual capacity, which is the volume of air remaining in the lung after an ordinary expiration, is about 2.4 liters. It is thus seen that a relatively small fraction, only about 13%, of the air normally present in the lung is exchanged during a single respiratory cycle. The nasal passages are lined with hair, which acts to filter out coarse particles, while the upper respiratory tract from the trachea down to the terminal bronchi are lined with cells containing very fine, hair-like whips, called cilia, which beat rhythmically and thereby carry upward towards the throat those dust particles which are on their surface. The depth of penetration of airborne particulates into the respiratory tract depends on the size of the air-borne particulate. Large dust particles, in excess of about 5 μ, are likely to be filtered out by the nasal hair or to impact on the nasopharyngeal surface. It should be pointed out here that particles must possess a great deal of kinetic energy in order to have motion, except for gravitational settling, independent of the motion of the air. The effect of gravitational settling becomes less pronounced as the particle size decreases. For example, a 20-μ diameter particle of unit density has a settling velocity of 1 cm/sec while a 1-μ particle of this material settles at a rate of only 0.003 cm/sec. For practical purposes, therefore, small particles may be regarded as remaining suspended in the atmosphere, and all but very large particles may be considered to be carried by moving air. In the respiratory tree, because of the relatively small cross-sectional areas of many of the air passages, the inspired air may attain relatively high velocities. Large particles that escape the hair-filter in the nose therefore have high kinetic energies as they pass through the air passages. As a consequence of the momentum of such a heavy particle, it cannot follow the inspired air around sharp curves, and strikes the walls of the upper respiratory tract. As the particle size decreases below 5 μ, this inertial impact decreases, and an increasing number of particles is carried down into the lung. The air in the alveoli is relatively still—since only a small fraction of the air there is exchanged with incoming air during a respiratory excursion. Particles that are carried into the deep respiratory tract, therefore, have the opportunity to settle out under the force of gravity. Gravitational settling, however, decreases with decreasing particle size, and vanishes when the particle size is about $\frac{1}{2}\mu$. As the particle size decreases below about 0.1 μ, the effect of Brownian motion becomes significant. As the particles move randomly about, they may strike the alveolar wall—and be trapped on its moist surface. The combination of these three effects, inertial impact, gravitational settling, and Brownian motion, leads to a maximum likelihood of deposition in the deep respiratory tract for particles in the 1–2 μ size range, and a minimum deposition for particles between 0.1 and 0.5 μ, as shown above in Fig. 7.6.

The retention of particles in the lung depends on the area within the respiratory tract where the particles were deposited, on the physical and chemical properties of the particles,

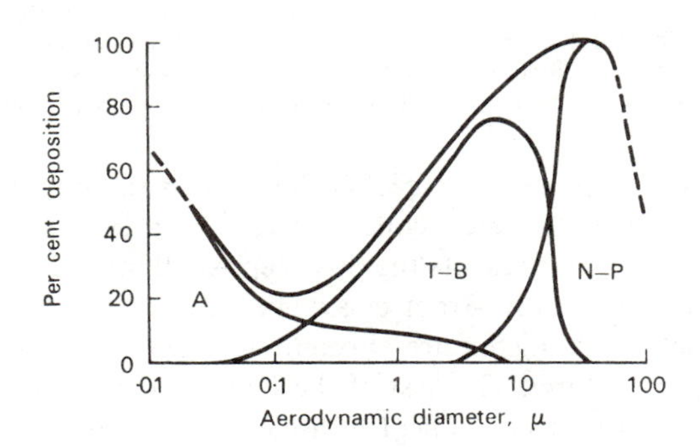

FIG. 7.6. Deposition of dust in the respiratory tract. Region *A* represents principally alveolar deposition, region *T–B* is mainly tracheobronchial deposition, and region *N–P* represents nasopharyngeal deposition. Total deposition is the sum of these three regions. (From P. E. Morrow, Evaluation of inhalation hazards based upon the respirable dust concept and the philosophy and application of selective sampling, *A.I.H.A. Journ.* **25**, 213–36, 1964.)

and on the physiologic properties of the lung. Retention of the inhaled particles, or its inverse—pulmonary clearance—is important in determining the degree of hazard because of its role in tissue exposure time and total dose. Studies of pulmonary retention of various dusts show that the curve of dust remaining in the lung after cessation of exposure is fitted by a complex exponential curve that includes at least two components; one of half-retention time on the order of several hours, and the other on the order of days. Very often, the long-lived component is also complex, and may be resolved into two or three components. Algebraically, this curve is given by the equation

$$D = D_1 e^{-k_1 t} + D_2 e^{-k_2 t} + \cdots. \tag{7.11}$$

Figure 7.7 illustrates a typical two-component retention curve, which is described by the first two terms of equation (7.11).

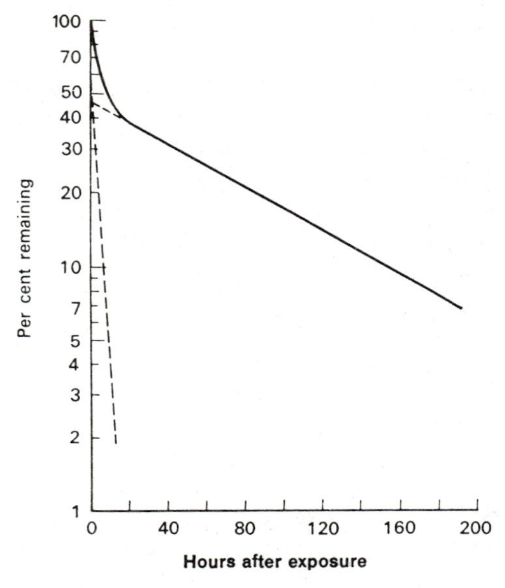

FIG. 7.7. Pulmonary retention curve showing amount of $BaSO_4$ particulate remaining in the lung as a function of time after exposure. According to this curve, 53% of the dust was deposited in the upper respiratory tract; the clearance rate from the URT is 27% per hour. Forty-seven per cent of the dust was deposited in the deep respiratory tract, and was cleared out at a rate of 1% per hour.

In this curve, the first component represents the dust in the upper respiratory tract; D_1 is the amount of dust deposited there, while k_1 gives the rate at which it is cleared from the upper respiratory tract. D_2 represents the dust deposited initially in the deep respiratory tract, and k_2 is the deep respiratory-tract clearance rate.

At least three distinct mechanisms are thought to operate simultaneously to remove foreign particulates from the lung. The first of these mechanisms, ciliary clearance, can act only in the upper respiratory tract. The rhythmic beating of the cilia propel particles upward into the throat—from whence they are swallowed—at very high speeds. Particle velocities ranging from about 2 mm/min in the bronchi to about 3 cm/min in the trachea have been observed. The other two clearance mechanisms deal mainly with particulates in the deep respiratory tract. They include: solubility and absorption into the capillary bed across the alveolar membrane, and removal by phagocytosis. It should be pointed out that the solubility in water of any given substance may not necessarily be a good index of solubility in the lung. For example, mercury from ^{203}HgS, one of the most insoluble compounds known, was found in significant quantities in the kidneys and in the urine of rats exposed to HgS particles of about 1 μ; the tagged mercury could have gotten there only after solution of the particles. On the other hand, ^{144}Ce intratracheally injected as $CeCl_3$ solution was found to be tenaciously retained in the lung for very long periods of time. In this case, the Ce in solution was bound to the tissue protein in the lung; very little of the cerium found its way into the blood. The case of cerium, however, seems to be exceptional. Most inhaled soluble particulates are absorbed into the blood and their chemical constituents translocated to other organs and tissues, where they may be systemically absorbed. The lung can thus be an excellent portal of entry into the body for many different radionuclides.

Phagocytosis, which is the engulfing of foreign particles by alveolar macrophages and their subsequent removal either up the ciliary "escalator" or by entrance into the lymphatic system, is a major pulmonary clearance mechanism. Phagocytes loaded with radioactive particles may be trapped in the sinuses of the tracheobronchial lymph nodes, and may remain there for long periods of time. This accumulation of radioactive dust in the lymph nodes may result in a higher radiation dose to the lymph nodes than to the lungs. The rate of phagocytosis depends to a large degree on the nature of the dust particle. Different particles have been found to be phagocytized at different rates. Furthermore, it has been found that radioactive particles are phagocytized more slowly than non-radioactive particles of the same chemical composition, physical form, and size distribution.

It is clear, from the multiplicity of factors that play a role in determining the biological effects of inhaled radioactive materials, that no simple quantitative relationship between gross atmospheric concentration of a radioisotope and lung effects can be assumed. Very conservative criteria, therefore, must be applied to environmental control of radioactive dust.

Life shortening

Cancer resulting from overexposure to radiation naturally shortens the life span of persons thus overexposed. In addition to this mode of accelerating death, radiation in large doses may shorten the life span by increasing the rate of physiological aging. Such a life shortening effect has been observed among groups of experimental animals. The exact cause of death among these animals cannot be uniquely attributed to radiation; the causes of death are those that are expected among an animal population. Only the age-specific death rate is different from the controls; death occurs earlier among the irradiated animals. Although this life-shortening effect has been clearly demonstrated in animals, no similar clean-cut effect has been seen among man. Comparison of the duration of life among medical specialists who used X-rays in their work with other medical practitioners who did not use radiation in their work gives conflicting implications. While some of the data suggest an increased death rate from non-specific causes among users of X-rays, other data show no statistically significant difference in life expectancy between these two population groups. Quite clearly, therefore, radiation exposure at the levels encountered by radiologists and other X-ray users among physicians is not high enough to accelerate the aging process to the degree that will cause a statistically significant shortening of life span.

Cataracts

Much higher incidences of cataracts among physicists in cyclotron laboratories whose eyes had been exposed intermittently for long periods of time to relative low radiation fields, and among atomic-bomb survivors whose eyes had been exposed to a single high radiation dose, showed that both chronic and acute overexposure of the eyes could lead to cataracts. Radiation may injure the cornea, conjunctiva, iris, and the lens of the eye. In the case of the lens, the principal site of damage is the proliferating cells of the anterior epithelium. This results in abnormal lens fibers, which eventually disintegrate to form an opaque area, or cataract, that prevents light from reaching the retina. The cataractogenic dose to the lens is on the order of 500 rad of beta or gamma radiation. Fast neutrons are more effective in producing cataracts than X- or beta rays; cataracts have been reported after a dose of 200 rad from mixed gamma and neutron irradiation of the lens.

Not all radiations are equally effective in producing cataracts; neutrons are much more efficient than the other radiations. The cataractogenic dose has been found, in laboratory experiments with animals, to be a function of age; young animals are more sensitive than old animals. On the basis of occupational exposure data, it is estimated that the threshold dose for cataracts lies between about 15 and 45 rad of neutrons. No radiogenic cataracts resulting from occupational exposure to X-rays have been reported. From patients who suffered irradiation of the eye in the course of X-ray therapy and developed cataracts as a consequence, the cataractogenic threshold is estimated at about 200 rad. In cases either of occupationally or therapeutically induced radiation cataracts, a long latent period, on the order of several years, usually elapsed between exposure and the appearance of the lens opacity.

Risk Estimates: BEIR III

Public policy in dealing with potential risks from technological innovation depends on the perceived risk as well as on the real risk. In the case of risks from low-level exposure to ionizing radiation, the stochastic nature of the adverse effects of exposure, together with their very low probability, implies that the magnitude of the risk can be measured only by studying large population groups. The latest findings (at this time, 1982) of such a study were published by the (U.S.) National Academy of Sciences Committee on the Biological Effects of Ionizing Radiation in what is known as the BEIR III Report, *The Effects on Populations of Exposure to Low-Levels of Ionizing Radiation*: 1980. In the report, which deals with somatic effects—mainly cancer—and genetic effects the Committee members agreed on the dose response relationships at high doses on the order of 1 Gray (100 rads) or more delivered at high dose rates. They also agreed that any somatic effect that may ensue from the background dose rate, 1 mGy (100 mrads) per year, would be masked by other factors. In the case of genetic effects, the Committee agreed that no radiation-induced transmitted effects have been seen in man, and that it is unlikely that such genetic effects will be seen. Risks of genetic damage, therefore, must be based on animal experiments. Two Committee members disagreed with the conclusions of the majority. One thought the estimated risk factors for low LET radiation were too high, while the other member thought they were too low. The disagreement centered on whether the postulated effects of doses on the order of 0.01 to 0.1 mSv (10 to 100 millirems) per year in addition to the natural background actually will occur in the general population; and whether the postulated effects of doses on the order of 5 to 50 mSv (500 to 5000 millirems) per year will actually occur among the occupationally exposed population. The basis of this minority view is the large degree of uncertainty associated with the exact shape of the dose-response curve at low doses. The functional form of the generalized dose-response curve (Fig. 7.8) is

$$F(D) = (\alpha_0 + \alpha_1 D + \alpha_2 D^2) \exp(-\beta_1 D - \beta_2 D^2), \tag{7.12}$$

where $F(D)$ is the incidence of the effect under consideration (e.g., cancer) at dose D, α_0, α_1, α_2, β_1 and β_2 have positive values and α_0 is the "spontaneous," or "natural" incidence of the effect. β_1 and β_2 are significant only at high doses. Depending on which of these coefficients become insignificant, the generalized curve reduces to the three simple curves shown in Fig. 7.8, namely the linear, the pure quadratic, and the linear quadratic curves.

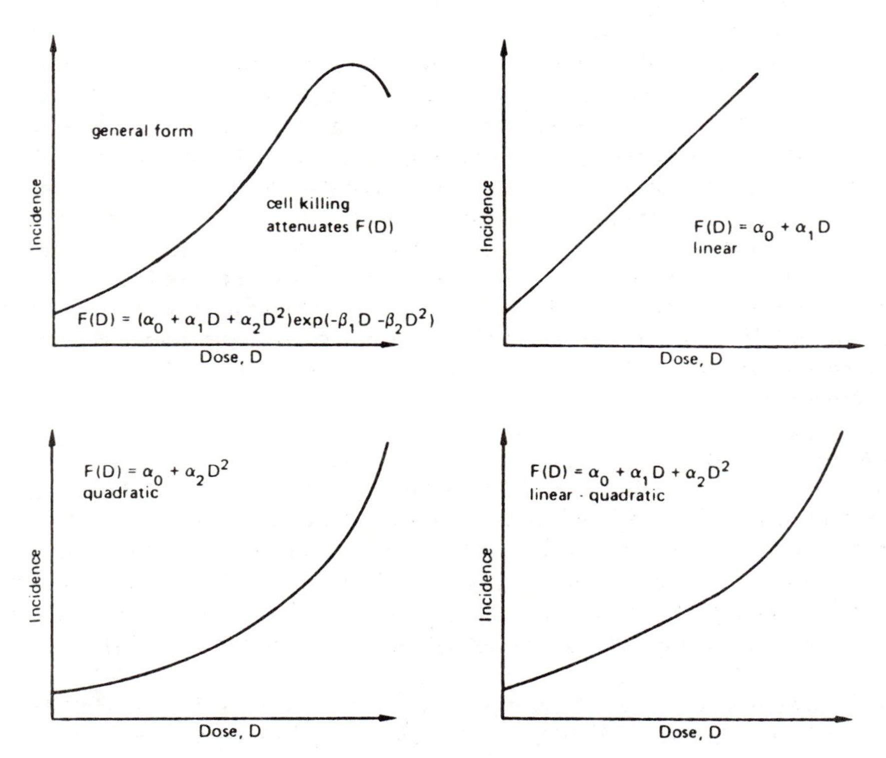

FIG. 7.8. Alternative dose-response curves. (From BEIR III: *The Effects on Populations of Exposure to Low-Levels of Ionizing Radiation*: 1980, National Academy Press, Washington, D.C., 1980.)

The available empirical data do not allow us to choose decisively among the three curves. The majority of the Committee chose to base their risk estimates for low doses on the linear-quadratic curve, the more conservative member chose the linear dose-response, while the other member thought that the pure quadratic model should be used for estimating the risk from low doses.

The chief sources of data on which the risk estimates are based are the Hiroshima and Nagasaki nuclear bomb survivors, patients who were exposed to therapeutic doses of radiation for ankylosing spondylitis and for other diseases, and occupationally exposed populations, such as radium dial painters and uranium miners. The range of risk estimates from BEIR III, as well as from several other sources, are listed in Table 7.1.

TABLE 7.1 RISK ESTIMATES FOR FATAL CANCERS FROM LOW LET RADIATION

Source	Cancer deaths per million man–rem
Beir III (1980)	
Majority	67–226
Minority a. (1 member)	158–501
Minority b. (1 member)	10–28
BEIR I (1972)	115–621
ICRP (1977)	125
UNSCEAR (1977)	100

Estimates of genetic risk were based entirely on the results of animal studies, since no radiation induced genetic effects of low level in man have been observed. Risks are estimated by two methods. In one method, known as the *doubling-dose* method, radiation induced mutation rates observed in experiments with mice are compared with spontaneous rates of genetic disorders in man, and the dose needed to double that spontaneous rate is determined from the experimental studies with mice. The BEIR III Committee estimates the doubling dose to be 50–250 rems (0.5–2.5 Sv), which corresponds to a relative mutation risk of 0.02 to 0.004 per rem. The new equilibrium, at double the present mutation rate, is attained only after many generations, perhaps as many as 10 generations. By this method, BEIR III estimates 60 to 1100 additional genetic disorders when the new equilibrium is attained.

In the other method of estimating genetic risk from ionizing radiation, the dose-response data from experiments with mice are applied directly to humans. By this method, BEIR III postulates that a 0.01-Sv (1-rem) dose to each parent before conception will lead to 5 to 65 additional genetic disorders per million live births during the first generation. Presently, the incidence rate (spontaneous and from causes other than the added radiation) of human genetic disorders is approximately 107,000 cases per million live births.

Relative Biological Effectiveness (RBE) and Quality Factor (QF)

Not only have neutrons been found more effective than X-rays in producing cataracts, but alpha radiation too has been found to be more toxic than beta or gamma radiation. When comparing the relative toxicity, or damage producing potential of various radiations, it is assumed, of course, that the comparison is on the basis of equal amounts of energy absorption. Generally, the higher the rate of linear energy transfer (LET) of the radiation, the more effective it is in damaging an organism. The ratio of the amount of energy of 200-keV X-rays required to produce a given effect to the energy required of any radiation to produce the same effect is called the *relative biological effectiveness* (abbreviated RBE) of that radiation. The RBE of a specific radiation depends on the exact biological effect on a given species of organism under a given set of experimental conditions. The term RBE is thus restricted in application to radiation biology. For health physics purposes, a conservative upper limit of the RBE for the most important effect due to a radiation other than the reference radiation (200-keV X-rays) is used as a normalizing factor in adding doses from different radiations. This normalizing factor, which is called the *quality factor* (abbreviated QF), is related to LET as shown in Table 7.2. For convenience in making health-physics measurements and for administrative purposes, the

TABLE 7.2. RELATIONSHIP BETWEEN QUALITY FACTOR AND LINEAR ENERGY TRANSFER

LET keV per micron in water	QF
3.5 or less	1
3.5–7.0	1–2
7.0–23	2–5
23–53	5–10
53–175	10–20

TABLE 7.3. QUALITY FACTOR VALUES FOR VARIOUS RADIATIONS

Radiation	QF
Gamma-rays from radium in equilibrium with its decay products (filtered by 0.5 mm platinum)	1
X-rays	1
Beta-rays and electrons of energy >0.03 MeV	1
Beta rays and electrons of energy <0.03 MeV	1.7
Thermal neutrons	2
Fast neutrons	10
Protons	10
Alpha-rays	20
Heavy ions	20

various radiations most frequently encountered by health physicists are assumed to have the quality factors listed in Table 7.3.

Dose Equivalent: The Sievert (and the Rem)

The sievert, Sv, is the unit of radiation dose equivalent, H, that is used for radiation protection purposes, for engineering design criteria, and for legal and administrative purposes. The dose equivalent, expressed in sieverts, considers the QF of the radiation as well as the absorbed dose, plus other factors, such as non-uniform distribution DF, which applies only to bone seekers and which may influence the biologic effect of a given absorbed dose. (The distribution factor is discussed in Chapter 8, in connection with the concentration of radioisotopes in drinking water based on comparison with radium. The dose equivalent is defined as

$$H, \text{sieverts} = D, \text{grays} \times \text{QF} \times \text{DF}. \tag{7.13}$$

According to equation (7.13), an absorbed dose of 1 mGy of X- or beta rays corresponds to a dose equivalent of 1 mSv, while a fast neutron dose of 1 mGy corresponds to a dose equivelent of 20 mSv. The gray is based on physical factors only, whereas the sievert considers both physical and biological factors. Dose limits, that is, the maximum allowable radiation dose, are given in units of sieverts or millisieverts. The use of the sievert unit in routine radiation surveying is illustrated by the following example:

> The dose rate outside the shielding of a cyclotron is found to be 5 μGy/hr (0.5 mrad/hr) gamma, 2 μGy/hr (0.2 mrad/hr), thermal neutrons, and 1 μGy/hr (0.1 mrad/hr) fast neutrons. What is the dose equivalent rate of the combined radiations?

$$\begin{aligned} \text{gamma rays: } 5\,\mu\text{Gy/hr} \times 1 &= 5\,\mu\text{Sv/hr} \\ \text{thermal neutrons: } 2\,\mu\text{Gy/hr} \times 2 &= 4\,\mu\text{Sv/hr} \\ \text{fast neutrons: } 1\,\mu\text{Gy/hr} \times 20 &= \underline{20\,\mu\text{Sv/hr}} \\ \text{Dose } H &= 29\,\mu\text{Sv/hr.} \end{aligned}$$

In the cgs system of units, where dose is expressed in rads, the unit dose equivalent is the rem, which is defined as

$$H, \text{rems} = D, \text{rads} \times \text{QF} \times \text{DF}. \tag{7.14}$$

Using cgs units, the dose equivalent in the example above is found to be 2.9 mrem/hour.

The *dose-equivelent commitment* is simply the dose commitment from a particular exposure or practice that has been converted to sieverts through the use of equation (7.13) or to rads through equation (7.14).

For purposes of setting standards, we use the *committed dose equivalent*, H_{50}, which is defined as the dose equivalent that will accumulate during the 50 years following the intake of a radioisotope.

$$H_{50} = \int_0^{50} H(t)\,dt. \tag{7.15}$$

High-Energy Radiation

High-energy radiation, such as that found around high energy accelerators that accelerate particles to the GeV range, poses special problems. The absorbed dose can be measured using a tissue equivalent ionization chamber. The spectral distribution of the radiation is more difficult to measure, but it can be estimated by film track techniques, by pulse height analysis, by threshold detectors, or by absorption methods. The biological effects of these high energy radiations, however, are not sufficiently well known to permit precise determination of numerical values for the quality factor. In this regard, the very high rate of linear energy transfer by very high energy heavy ions is of special interest.

Suggested References

ALEXANDER, P. *Atomic Radiation and Life*. Penguin Books, Harmondsworth, 1957.
ARENA, V. *Ionizing Radiation and Life*. C. V. Mosby, St. Louis, 1971.
BACQ, Z. M., and ALEXANDER, P. *Fundamentals of Radiobiology*. Pergamon Press, Oxford, 1967.
BOND, V. D., FLIEDNER, T. M., and ARCHAMBEAU, J. O. *Mammalian Radiation Lethality*. Academic Press, New York, 1965.
CASARETT, A. P. *Radiation Biology*. Prentice-Hall, Englewood Cliffs, N.J., 1968.
CLAUS, W. D., ed. *Radiation Biology and Medicine*. Addison-Wesley, Reading, 1958.
FABRIKANT, J. I. *Radiobiology*. Medical Publishers, Chicago, 1972.
GROSCH, D. S., and HOPWOOD, L. E. *Biological Effects of Radiations*. Academic Press, New York, 1979.
HOLLAENDER, A., ed. *Radiation Biology*. Mc-Graw-Hill, New York, 1955.

LEA, D. E. *Actions of Radiations on Living Cells*. Cambridge University Press, Cambridge, 1955.
McLEAN, F. C., and BUDY, A. M. *Radiation, Isotopes, and Bone*. Academic Press, New York, 1964.
NORWOOD, W. D. *Health Protection of Radiation Workers*. Charles C. Thomas, Springfield, 1975.
PIZZARELLO, D. J., and WITCOFSKI, R. L. *Basic Radiation Biology*, 2nd ed. Lea and Febiger, Philadelphia, 1975.
PIZZARELLO, D. J., and WITCOFSKI, R. L. *Medical Radiation Biology*. Lea and Febiger, Philadelphia, 1972.
Report of the United Nations Scientific Committee on the Effects of Atomic Radiation, United Nations, New York, 1977.
The Effects on Populations of Exposure to Low Levels of Ionizing Radiation: 1980. Report of the Advisory Committee on the Biological Effects of Ionizing Radiation (BEIR) III Report. National Academy Press, Washington, D.C. 1980.
SCHWARTZ, E. E., ed. *The Biological Basis of Radiation Therapy*. J. B. Lippincott, Philadelphia, 1966.
SAENGER, E. L., ed. *Medical Aspects of Radiation Accidents*. U.S. Atomic Energy Commission, Government Printing Office, Washington, D.C., 1980.
SPEAR, F. G. *Radiations and Living Cells*. John Wiley & Sons, New York, 1953.
SPIERS, F. W. *Radioisotopes in the Human Body*. Academic Press, New York, 1968.
SZIRMAI, E., ed. *Nuclear Hematology*. Academic Press, New York, 1965.
World Health Organization. *Diagnosis and Treatment of Acute Radiation Injury. Geneva*, 1961.
ZIMMER, K. G. *Studies on Quantitative Radiology*. Hafner, New York, 1961.
International Atomic Energy Agency. *Inhalation Risks from Radioactive Contaminants*. Technical Report Series No. 142, IAEA, Vienna, 1973.
The following reports of the National Council on Radiation Protection and Measurements, Washington, D.C.:
No. 42. *Radiological Factors Affecting Decision-Making in a Nuclear Attack*, 1974.
46. *Alpha Emitting Particles in Lungs*, 1975.
53. *Review of the NCRP Radiation Dose Limit for Embryo and Fetus in Occupationally Exposed Women*, 1977.
63. *Tritium and Other Radionuclide Labelled Organic Compounds Incorporated in Genetic Material*, 1979.
64. *Influence of Dose and Its Distribution in Time on Dose-Response Relationships for Low-LET Radiations*, 1980.
66. *Mammography*, 1980.
The following publications of the International Commission on Radiological Protection (ICRP), Pergamon Press, Oxford:
ICRP
Report
No.
8. *The Evaluation of Risks from Radiation*, 1966.
11. *A Review of the Radiosensitivity of the Tissues in Bone*, 1968.
14. *Radiosensitivity and Spatial Distribution of Dose*, 1969.
19. *The Metabolism of Compounds of Plutonium and other Actinides*, 1972.
20. *Alkaline Earth Metabolism in Adult Man*, 1973.

CHAPTER 8

RADIATION PROTECTION GUIDES

Organizations that Set Standards

International Commission on Radiological Protection

The basic responsibility for providing guidance in matters of radiation safety has been assumed by the International Commission on Radiological Protection (ICRP). This organization was established in 1928 by the Second International Congress of Radiology as the International X-ray and Radium Protection Commission. At that time and for many years afterward, its main concern was with the safety aspects of medical radiology. Its interests in radiation protection expanded with the widespread use of radiation outside the sphere of medicine, and in 1950 its name was changed to the ICRP in order to more accurately describe its area of interest. In describing its operating philosophy, the ICRP says that "The policy adopted by the Commission in preparing recommendations is to deal with the basic principles of radiation protection and to leave to the various national protection committees the responsibility of introducing the detailed technical regulations, recommendations, or codes of practice best suited to the needs of their individual countries" (ICRP Publication 6, p. 1, Pergamon Press, Oxford, 1964). The ICRP publishes its reports in the appropriate journals. A listing of such reports is given in the bibliogrphy at the end of this chapter.

International Atomic Energy Agency

The International Atomic Energy Agency (IAEA), a specialized agency of the United Nations that was organized in 1956 in order to promote the peaceful uses of nuclear energy, is concerned with the practical application of the ICRP recommendations. "Under its Statute the International Atomic Energy Agency is empowered to provide for the application of standards of safety for protection against radiation to its own operations and to operations making use of assistance provided by it or with which it is otherwise directly associated. To this end authorities receiving such assistance are required to observe relevant health and safety measures prescribed by the Agency." (From *Safe Handling of Radioisotopes*, Safety Series No. 1, IAEA, Vienna, 1962.) The health and safety measures prescribed by the Agency are published according to subject in the Agency's *Safety Series*.

International Labour Organization

The International Labour Organization (ILO), which was founded in 1919 and then became part of the League of Nations, survived the demise of the League to become the first of the specialized agencies of the United Nations; its concern generally is with the social problems of labor. Included in its work is the specification of international labor standards which deal with the health and safety of workers. These specifications are set forth in the *Model Code of Safety Regulations for Industrial Establishments for the Guidance of Governments and Industries*, in the recommendations of expert committees, and in technical manuals. In regard to radiation, the model code has been amended to incorporate those recommendations of the ICRP that are pertinent to control of occupational radiation hazards, and several manuals dealing with protection of workers against radiation hazards have been published.

International Commission on Radiological Units and Measurements

The International Commission on Radiological Units and Measurements (ICRU), which works closely with the ICRP, has had, since its inception in 1925, as its principle objective the development of internationally acceptable recommendations regarding:

(1) Quantities and units of radiation and radioactivity.
(2) Procedures suitable for the measurement and application of these quantities in clinical radiology and radiobiology.
(3) Physical data needed in the application of these procedures, the use of which tends to assure uniformity in reporting.

In its opearting policy, "The ICRU feels it is the responsibility of national organizations to introduce their own detailed technical procedures for the development and maintenance of standards. However, it urges that all countries adhere as closely as possible to the internationally recommended basic concepts of radiation quantities and units." (ICRU Report 32, 1979.)

National Council on Radiation Protection and Measurements

In accordance with the policy laid down by the ICRP, its recommendations are adapted to the needs and conditions in the various countries by national bodies. In the United States, two such bodies exist. The older one is called the National Council on Radiation Protection and Measurements (NCRP). This committee, which was originally known as the Advisory Committee on X-ray and Radium Protection (founded in 1929), consists of a group of technical experts who are specialists in radiation protection and scientists who are experts in the disciplines which form the basis for radiation protection. The concern of the NCRP is only with the scientific and technical aspects of radiation protection. To accomplish its objectives, the NCRP is organized into a Main Committee and twenty subcommittees. Each of the subcommittees is responsible for preparing specific recommendations in its field of competence. The recommendations of the sub-committees require approval of the Main Committee before publication. Finally, the approved recommendations are published by the Council, such as Report No. 39, *Basic Radiation Protection Criteria*. It should be emphasized that NCRP is not an official government agency, although its recommendations are very often adopted by Federal, state, and local government agencies concerned with the regulation of radiation hazards.

Philosophy of Radiation Protection

Engineering control of the environment by industrial hygienists and by public health personnel is usually based, in the case of non-stochastic effects, on the concept of a threshold dose. If the threshold dose of a toxic substance is not exceeded, then it is assumed that the normally operating physiological mechanisms can cope with the biological insult from that substance. This threshold is usually determined from a combination of experimental animal data and clinical human data; it is then reduced by an appropriate factor of safety, which leads to the maximum allowable concentration (MAC) for the substance. The MAC is then used as the criterion of safety in environmental control. The maximum allowable concentration was defined by the International Association on Occupational Health in 1959 as follows: "The term maximum allowable concentration for any substance shall mean that average concentration in air which causes no signs or symptoms of illness of physical impairment in all but hypersensitive workers during their working day on a continuing basis, as judged by the most sensitive internationally accepted tests."

A different philosophy underlies the control of environmentally based agents, such as ionizing radiation and radioactive isotopes, that lead to increased probability of cancer and genetic effects. Although molecular biologists have found the existance of intra-cellular mechanisms for the repair of damaged DNA in bacteria, and geneticists have observed a dose-rate dependence of radiogenic mutagenesis (both these observations imply the existence of a threshold for stochastic effects), we assume, for the purpose of setting safety standards for radiation as well as for chemical carcinogens and mutagens, that there is no threshold dose for stochastic effects. The dose response curves for carcinogenesis and for mutagenesis is assumed to be linear down to zero dose. The slopes for the dose-response curves for the various stochastic effects are obtained by extrapolation to low doses and to zero dose of the results of high doses.

Furthermore, it is assumed that the effects are independent of the dose rate, and that only the total dose is of biological significance. Since this means that every increment of dose, no matter how small, increases the risks of an adverse effect by a proportional increment, the basis for control of man-made radiation is the limitation of the radiation dose to a level that is compatible with the benefits that accrue to society and to individuals from the use of radiation.

The distinction between those agents that cause non-stochastic effects and those that increase the probability of stochastic effects, which is based on the existence of absence of a threshold dose, is not as clear cut as may first appear. For those substances where a threshold has indeed been established, the threshold is for an *individual*. Different individuals have different thresholds. Thus, although the average threshold value for blood changes due to gamma radiation is taken as 0.25 Gy (25 rads), changes have been observed in people whose dose was as low as 0.14 Gy (14 rads), while others whose doses reached as high as 0.4 Gy (40 rads) showed no blood changes. If a much larger population of exposed people were to be examined for blood changes, it is likely that changes would be seen among those whose dose was even less than 0.14 Gy (14 rads).

It is not unreasonable to expect a distribution of sensitivity to most noxious agents somewhat like that shown in Figure 8.1, in which the sensitivity distribution curve is skewed to the right. The curve should actually intersect the abscissa on the high dose end of the distribution, since we are reasonably certain that there exists some dose that will affect everyone. On the other hand, it is known that there are "hypersensitive" individuals who respond to extremely low doses which would not affect most people. On this basis, it is reasonable to assume that the distribution curve to the left of the mode passes through the origin of the coordinate axes.

In effect, the distribution of susceptibility among the individuals of a population means that the concept of a threshold dose cannot be applied to a very large population. In setting an MAC for a large population group, therefore, a value judgement must be exercised.

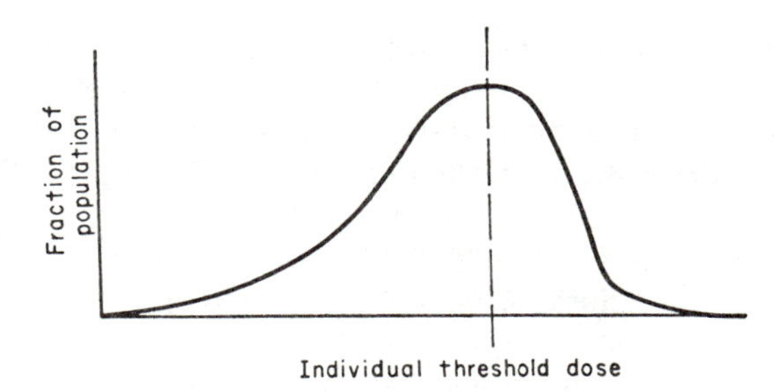

FIG. 8.1. Distribution of individual thresholds among a population.

Someone must decide on what is an acceptable fraction of the population that may be adversely affected by the harmful substance for which the MAC is being set in return for the benefits to be derived from the use of that substance. The MAC is usually set so conservatively that an extremely large number of people would have to be exposed at that level before the hypersensitive person is found.

This same type of reasoning prevails among those who are concerned with recommending radiation dose limits. For occupational exposure, the question of recommending dose limits as a guide to radiation protection is relatively simple. A vast amount of human experience was gained from the promiscuous exposures to radium and X-rays during the first quarter of the twentieth century, and from survivors of the nuclear bombings in Japan; and much more data were obtained from laboratory studies with animals. On the basis of this information, and on the assumption that every additional increment of radiation dose has a correpsonding increment of additional risk, dose limits can be set which, when applied to occupationally exposed radiation workers, will result in a level of risk no greater than that in other occupations that are recognized as having high safety standards and are considered to be "safe." Dose limits for non-occupationally exposed individual members of the general public are set at a level where the resulting postulated radiation risk is very much smaller than the risks that society already accepts in return for other technological benefits to society. From these dose limits, we can derive annual limits of intake and environmental concentrations of the various radionuclides that would result in radiation doses within the prescribed dose limits.

The system of dose limitation recommended by the ICRP is founded on three basic tenets stated in its "Publication 26":

1. No practice shall be adopted unless its introduction produces a net positive benefit;
2. all exposures shall be kept as low as reasonable achievably, economic and social factors being taken into account; and
3. the dose equivalent to individuals shall not exceed the limits recommended for the appropriate circumstances by the Commission.

It should be emphasized that point number 2 above urges that actual operational dose limits for any radiologic activity be more restrictive than the maximum recommended dose limit. This means that processes, equipment (such as shielding, ventilation, etc.), and other operational factors be designed so that workers do not exceed the operational dose limit. This operating philosophy is known as the ALARA (As Low As Reasonably Achievable) concept. To apply the ALARA concept, the ICRP recommends that cost-benefit analyses of alternative lower operational dose limits be made, and then the selection of that level of radiation protection that optimizes the cost of the detrimental effects of the radiation versus the benefits to be derived from the radiation practice. Since economic and social factors must be considered in implementing ALARA, it is clear that widely differing interpretations can be made be equally competent authorities on what is "as low as reasonably achievable." In the United States, the official interpretation is made by the U.S. Nuclear Regulatory Commission, and is published in the *Regulatory Guide* series.

Basic Radiation Safety Criteria

For purposes of radiation safety standards, the ICRP recognizes three categories of exposure:

1. Occupational exposure to adults who are exposed to ionizing radiation in the course of their work. Persons in this category may be called radiation workers. This category contains two subgroups: (a) pregnant women and (b) all other radiation workers.
2. Members of the general public. This category is considered on two levels: (a) individual members of the public and (b) population groups.
3. Medical exposure. This category deals with the intentional exposure of patients for diagnostic and therapeutic purposes by technically qualified medical and paramedical personnel. It does not include exposure to the personnel involved in the administration of radiation to patients.

For occupational exposure, the ICRP recommends the following annual dose equivalent limits

1. To prevent non-stochastic effects:
   (a) 0.5 Sv (50 rem) to all tissues except the lens of the eye.
   (b) 0.15 Sv (15 rem) to the lens of the eye.

These limits apply whether the tissues are exposed singly or together with the organs.

2. To limit stochastic effects, the dose-equivalent limit from uniform whole body irradiation is 50 mSv (5 rem) in one year.

On the principle that the risk of a stochastic effect should be equal whether the whole body is uniformly irradiated or whether the radiation dose is non-uniformly distributed, the ICRP introduced the concept of *effective dose equivalent* in its 1977 review of its radiation safety recommendations (ICRP Publication 26), and recommended that, in order to control stochastic effects, the effective dose equivalent for occupational exposure be limited to 50 mSv (5 rem) in one year. For pregnant radiation workers, the ICRP recommended that the annual exposure not exceed three-tenths of the dose-equivalent limits specified for other radiation workers.

Effective Dose Equivalent

For the purpose of setting radiation safety standards, we assume the probability of a stochastic effect in any tissue to be proportional to the dose equivalent to that tissue. However, because of the differences in sensitivity among the various tissues, the value for the proportionality factors differs among the tissues. The relative sensitivity to harmful stochastic effects, expressed as risk per Sv, of the several organs and tissues that contribute to the overall risk, is shown in Table 8.1. If radiation dose is uniform throughout the body, then the total risk factor is 1. For non-uniform radiation, such as partial body exposure to an external radiation field, or from internal exposure where the isotope concentrates to different degrees in the various organs, weighting factors, listed in Table 8.1, which are based on the relative susceptibility of the organs to stochastic effects, are used to calculate an effective dose equivalent. The effective dose equivalent, H_E, is given by

$$H_E = \sum W_T H_T, \tag{8.1}$$

where W_T is the weighting factor for tissue T, found in Table 8.1, and H_T is the dose equivalent to tissue T.

TABLE 8.1. WEIGHTING FACTORS AND RISK FACTORS FOR TISSUES AT RISK OF STOCHASTIC EFFECTS

Tissue	Risk, Sv^{-1}	Comments	W_T
Gonads	4.0×10^{-3}	Genetic risk to first 2 generations	0.25
Breast	2.5×10^{-3}	Average for all ages and both sexes	0.15
Red bone marrow	2.0×10^{-3}	Leukemia	0.12
Lung	2.0×10^{-3}	Cancer	0.12
Thyroid	5.0×10^{-4}	Fatal cancer	0.03
Bone surface	5.0×10^{-4}	Osteosarcoma	0.03
Remainder	5.0×10^{-3}	Cancer, assuming that no single tissue contributes more than 1/5 of this total	0.30
Total risk	1.65×10^{-2}		

Example 8.1

As a result of a laboratory accident, 370,000 Bq (10 μCi) of ^{131}I were deposited in a radioisotope technician; 74 kBq (2 μCi) were deposited in her thyroid gland, and 296 kBq (8 μCi) were uniformly distributed throughout the rest of her body. Using data from bioassay measurements and body scanning, the health physicist calculated a thyroid dose of 123 mGy (12.3 rad) and a whole body dose of 0.26 mGy (26 mrad).

(a) What was the technician's effective dose equivalent?
(b) Was she overexposed according to the ICRP criteria?

(a) The effective dose equivalent is calculated from equation (8.1), using weighting factors of 0.03 for the thyroid (from Table 8.1) and 0.97 for the rest of the body.

$$\begin{aligned} H_E &= 0.03 \times 123 + 0.97 \times 0.26 \\ &= 3.94 \text{ mSv}. \end{aligned}$$

(b) Since the effective whole body dose equivalent is much less than 50 mSv, and since the dose to the thyroid is much less than 500 mSv, the dose from this accidental exposure did not exceed the ICRP dose limit. Whether her total dose for the year, after suffering this accidental exposure, exceeded the ICRP dose limits depends on her previous exposure history.

Exposure of Individuals in the General Public

For members of the general public, the ICRP recommends a whole body dose-equivalent limit of 5 mSv (500 mrem) in a year. It is believed that the average dose to members of an exposed group will be less than the dose limit. The ICRP points out that the average dose to members of the public would increase if the number of sources increase, even though no single individual exceeds the 5-mSv dose-equivalent limit. For this reason, the Commission recommends that regional or national authorities should maintain surveillance over all the separate sources of exposure in order to control the collective total-dose equivalent.

Exposure of Populations

The ICRP made no specific recommendations for the dose limit to a population. Instead, it emphasized that each man-made contribution to the population dose must be justified by its benefits, and that limits for individual members of the population refer to the total dose equivalent from all sources. The dose limit to a population is thus considered to be the sum of several minimum necessary contributory doses rather than a single permissible total dose equivalent limit that is available for apportionment among several sources.

Medical Exposure

No specific dose limit was recommended by the ICRP for medical exposure. The Commission, however, did recommend that only necessary exposure should be made, that these exposures should be justifiable on the basis of benefits that would not otherwise have been received, and that the doses actually administered should be limited to the minimum dose consistent with the medical benefit to the patient.

Allowable Limit on Intake (ALI)

The current (1981) ICRP recommendations do not specify either maximum permissible body burdens or maximum permissible concentrations. Instead, they specify an annual limit on the intake, ALI, of a radioisotope. The ALI is restricted by the basic requirements for stochastic and for non-stochastic effects, and is defined as the annual intake that would lead to an effective committed dose equivalent (a 50-year dose commitment) not exceeding 50 mSv (5 rem) and an annual dose equivalent to any single organ or tissue not exceeding 500 mSv (50 rem). Expressed symbolically, these requirements are

$$\sum_T W_T H_{50,T} \leq 0.05 \text{ Sv} \tag{8.2}$$

$$H_{50,T} \leq 0.5 \text{ Sv for all } T, \tag{8.3}$$

where W_T is the weighting factor shown in Table 8.1, and $H_{50,T}$ is the total committed dose equivalent in tissue T resulting from intakes of radioactive materials from all sources during the year in question. Equation (8.2) assures that the annual limit on effective whole body dose is not exceeded in order to control stochastic effects, while equation (8.3) assures that the annual limit on the dose to a single tissue or organ is not exceeded in order to remain below the damage threshold for non-stochastic effects. It should be noted that the intake limit is placed on the total intake of radioisotopes in any single year, and that no restrictions are placed by the ICRP on the instantaneous rate of intake. That is, the limit may be met by a single large intake or by continuing intake of small quantities. The principles involved in the application of equations (8.2) and (8.3) may be illustrated by the calculation of the ALI for ingested ^{137}Cs.

Commonly occurring cesium compounds are known to be rapidly and almost completely absorbed from the gastrointestinal tract. After absorption cesium is uniformly distributed throughout the body. In no case is the concentration of cesium in any organ or tissue greater than in the muscle. The retention of cesium in the body over at least the first 1400 days is described by the two compartment equation

$$q(\tau) = 0.1\, q_0 e^{-0.693\tau/2} + 0.9\, q_0 e^{-0.693\tau/110}, \tag{6.33}$$

where q_0 is the quantity of cesium initially taken up by the body. Equation (6.33) tells us that; of the cesium taken up, 10% is transferred to a tissue compartment that is cleared with a half-life of 2 days, while the remaining 90% is transferred to a tissue compartment whose residence half time is 110 days. Since the radiological half life of ^{137}Cs, 30 years, is very much greater than the half-times of each of the two compartments, the effective half-time for each of the compartments is essentially equal to their biological half times. The effective clearance rates, therefore, for the two compartments, as calculated from equation (6.26) are 0.347 and 0.0063 per day, respectively. The total dose from the deposited ^{137}Cs is the sum of the doses for each of the two compartments, as calculated from equation (6.36). To meet the stochastic effects criterion, the dose equivalent must not exceed 0.05 Sv (5 rem); and since ^{137}Cs is a beta-gamma emitted, this corresponds to a dose of 0.05 Gy (5 rad).

$$D = \frac{\dot{D}_{10}}{\lambda_1} + \frac{\dot{D}_{20}}{\lambda_2} = 0.05\ \text{Gy}, \tag{8.4}$$

where $\dot{D}_{10}$ and $\dot{D}_{20}$ are the initial dose rates in compartments 1 and 2, respectively, and λ_1 and λ_2 are their effective clearance rates. These initial dose rates from the uniformly distributed ^{137}Cs may be calculated from equation (6.20) by using the appropriate value for the SEE of the betas plus the gammas. The SEE per transformation may be calculated with the aid of the absorbed fractions, φ, shown in Table 6.8 and the output data for ^{137}Cs

TABLE 8.2. CALCULATION OF SPECIFIC EFFECTIVE ENERGY (SEE) FOR WHOLE-BODY INTERNAL EXPOSURE TO ^{137}Cs

Radiation	Number/ transfrm n_i	Mean energy/ particle E_i, MeV	Absorbed fraction φ_i	Absorbed energy $(\text{MeV/t})_i$
β_1	0.935	0.1749	1	0.1635
β_2	0.065	0.4272	1	0.0278
γ	0.840	0.6616	0.334	0.1856
K int. con. elect.	0.0781	0.6242	1	0.0488
L int. con. elect.	0.0140	0.6560	1	0.0092
M int. con. elect.	0.0031	0.6605	1	0.0020
K $\alpha-1$ X-ray	0.0374	0.0322	0.750	0.0009
K $\alpha-2$ X-ray	0.0194	0.0318	0.754	0.0005
K $\beta-1$ X-ray	0.0105	0.0364	0.703	0.0003
K $\beta-2$ X-ray	0.0022	0.0374	0.690	0.0001
L X-ray	0.0127	0.0045	1	0.0001
KLL Auger elect.	0.0057	0.0263	1	0.0001
KXY Auger elect.	0.0025	0.0308	1	0.0000
KLX Auger elect.	0.0004	0.0353	1	0.0001
LMM Auger elect.	0.0718	0.0034	1	0.0002
MXY Auger elect.	0.173	0.0011	1	0.0002
				$\Sigma = 0.4394$ MeV/t

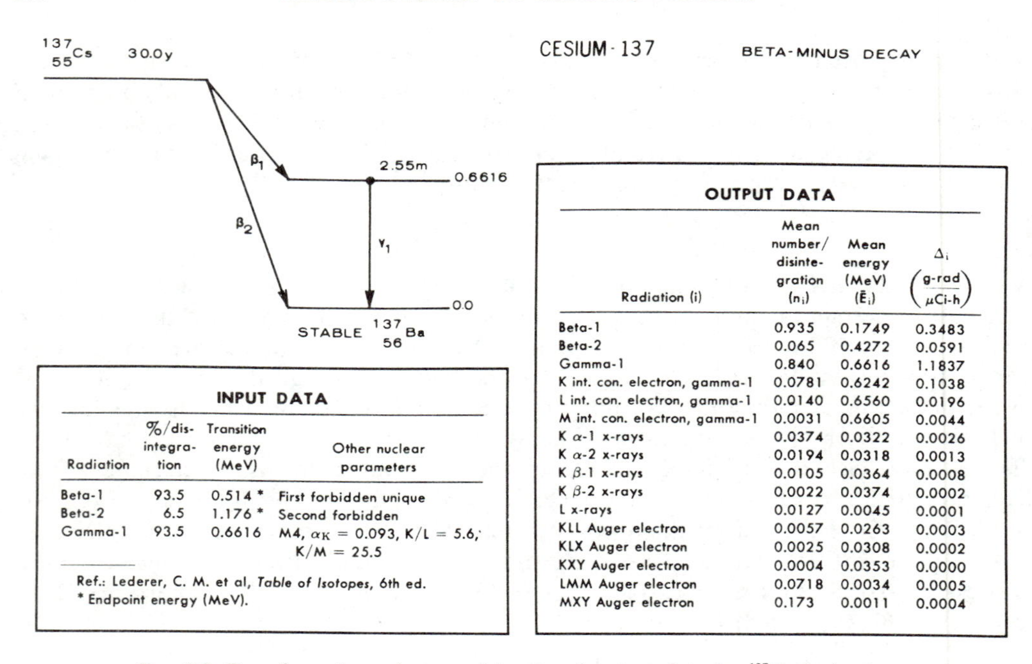

INPUT DATA

Radiation	%/dis-integration	Transition energy (MeV)	Other nuclear parameters
Beta-1	93.5	0.514 *	First forbidden unique
Beta-2	6.5	1.176 *	Second forbidden
Gamma-1	93.5	0.6616	M4, α_K = 0.093, K/L = 5.6, K/M = 25.5

Ref.: Lederer, C. M. et al, *Table of Isotopes*, 6th ed.
* Endpoint energy (MeV).

OUTPUT DATA

Radiation (i)	Mean number/disintegration (n_i)	Mean energy (MeV) ($\bar{E}_i$)	Δ_i $\left(\frac{\text{g-rad}}{\mu\text{Ci-h}}\right)$
Beta-1	0.935	0.1749	0.3483
Beta-2	0.065	0.4272	0.0591
Gamma-1	0.840	0.6616	1.1837
K int. con. electron, gamma-1	0.0781	0.6242	0.1038
L int. con. electron, gamma-1	0.0140	0.6560	0.0196
M int. con. electron, gamma-1	0.0031	0.6605	0.0044
K α-1 x-rays	0.0374	0.0322	0.0026
K α-2 x-rays	0.0194	0.0318	0.0013
K β-1 x-rays	0.0105	0.0364	0.0008
K β-2 x-rays	0.0022	0.0374	0.0002
L x-rays	0.0127	0.0045	0.0001
KLL Auger electron	0.0057	0.0263	0.0003
KLX Auger electron	0.0025	0.0308	0.0002
KXY Auger electron	0.0004	0.0353	0.0000
LMM Auger electron	0.0718	0.0034	0.0005
MXY Auger electron	0.173	0.0011	0.0004

FIG. 8.2. Transformation scheme and input and output data for ^{137}Cs dosimetry. From L. T. Dillman. Radionuclide Decay Schemes and Nuclear Parameters for Use in Radiation Dose Estimation, *J. Nuclear Medicine*, Vol. 10, Suppl. No. 2, MIRD Pamphlet No. 4, 1969.

shown in Fig. 8.2. If we organize the information from Table 6.8 and Fig. 8.2 as shown in Table 8.2, the SEE, in MeV per transformation per kilogram is calculated from

$$\text{SEE} = \frac{\sum n_i E_i \varphi_i}{m}. \tag{8.5}$$

Since the radiation is absorbed uniformly throughout the entire body, m in this case is 70 kg. If we substitute this value for m and 0.4394 MeV/t from Table 8.2 into equation (8.5), we have

$$\text{SSE} = \frac{0.4394\ \text{MeV/t}}{70\ \text{kg}} = 6.28 \times 10^{-3}\, \frac{\text{MeV}}{\text{t}}/\text{kg}.$$

Using this value for the SEE, and an activity $q = 1$ Bq in equation (6.20):

$$\dot{D} = \frac{q\ \text{Bq} \times 1\ \text{tps/Bq} \times \text{SEE}\, \dfrac{\text{MeV}}{\text{t}}\Big/\text{kg} \times 1.6 \times 10^{-13}\ \text{J/MeV} \times 8.64 \times 10^{4}\ \text{sec/day}}{1\, \dfrac{\text{J}}{\text{kg}}\Big/\text{Gy}}, \tag{6.20}$$

we find the dose rate per unit activity of ^{137}Cs to be

$$D = 9 \times 10^{-11}\, \frac{\text{Gy}}{\text{day}}\Big/\text{Bq}.$$

According to equation (6.33), the activity deposited in compartment 2 is nine times greater than the activity deposited in compartment 1. Substituting the appropriate vaues into equation (8.4) allows us to calculate the quantity of ingested cesium 137 that results in a dose of 0.05 Gy (5 rads), corresponding to a dose equivalent of 0.05 Sv (5 rem).

$$0.05\ \text{Gy} = \frac{q\ \text{Bq} \times 9 \times 10^{-11}\ \dfrac{\text{Gy}}{\text{day}}\Big/\text{Bq}}{0.347\ \text{day}} + \frac{9q\ \text{Bq} \times 9 \times 10^{-11}\ \dfrac{\text{Gy}}{\text{day}}\Big/\text{Bq}}{6.3 \times 10^{-3}\ \text{day}^{-1}}$$

$$q = 3.9 \times 10^5\ \text{Bq}$$

$$\therefore\ \text{ALI (stochastic)} = q + 9q = 3.9 \times 10^6\ \text{Bq}.$$

Since cesium is uniformly distributed throughout the body, a similar calculation for the ALI based on non-stochastic effects, where the annual dose-equivalent limit is 0.5 Sv (50 rem), yields a much higher value than the ALI based on stochastic effects. In order to satisfy equations (8.2) and (8.3), we must use the ALI for stochastic effects. Since the ALI is rounded off to the nearest whole number, the ALI for ingested ^{137}Cs is listed in ICRP 30 as 4 MBq.

The calculation of the ALI for ingested ^{137}Cs was simple because of the uniform distribution of the cesium. Generally, however, an internally deposited radioisotope is not uniformly distributed, and doses contributed to the various target organs by each different source organ with its own amount of radioactivity must be calculated. These calculations can be systemized for ease of programming into a computer. For example, the dose equivalent to any organ or tissue, T, at time t after intake of the radioisotope, is given by

$$H(t, T) = \frac{1.6 \times 10^{-13}\ \text{J/MeV} \sum \tilde{A}_{si} \times \text{SEE}_i(T \leftarrow S) \times 1\ t/\text{Bq-sec} \times Q_i}{1\ \dfrac{\text{J}}{\text{kg}}\Big/\text{Gy}}, \tag{8.6}$$

where $\tilde{A}_{si}$ = cumulated activity, in Bq sec, in the various source organs,
SEE_i = specific effective energy, MeV per transformation per kilogram,
Q_i = quality factor of the radiation, Sv/Gy.

Equation (8.6) can be solved for the dose equivalent resulting from an intake of 1 Bq. An ALI based on stochastic effects can be calculated from

$$\text{ALI (stochastic)} = \frac{0.05\ \text{Sv}}{\sum W_T H_{50,T}\ \text{Sv/Bq}}. \tag{8.7}$$

To determine whether the stochastic ALI is limiting, we compare the stochastic ALI to the criterion:

$$\text{Stochastic ALI Bq} \times (H_{50,T})_{\max}\ \text{Sv/Bq} \leq 0.5\ \text{Sv}. \tag{8.8}$$

If the product of the stochastic ALI and the committed dose equivalent to the organ or tissue that receives the greatest dose equivalent from that intake, $(H_{50,T})_{\max}$, exceeds 0.5 Sv (50 rem), then the stochastic ALI is too large, and an ALI based on non-stochastic effects, which is given by

$$\text{ALI (non-stochastic)} = \frac{0.5\ \text{Sv}}{(H_{50,T})_{\max}\ \text{Sv/Bq}}, \tag{8.9}$$

must be calculated.

Example 8.2

Inhalation of 1 Bq of ^{239}Pu in the form of relatively insoluble particulates (Class Y, very long pulmonary retention half-time) leads to the following committed dose equivalents from the ^{239}Pu and its daughters, $H_{50,T}$, in the respective target organs: Lungs, 3.2×10^{-4} Sv; red marrow, 7.6×10^{-5} Sv; bone surfaces 9.5×10^{-4} Sv; and liver, 2.1×10^{-4} Sv. Calculate the ALI for inhalation of this solubility class of ^{239}Pu.

First, we will calculate the weighted committed dose equivalents:

Tissue	$H_{50,T}$, Sv	×	$W_T =$	Weighted $H_{50,T}$
Lungs	3.2×10^{-4}		0.12	3.9×10^{-5}
Red marrow	7.6×10^{-5}		0.12	9.1×10^{-6}
Bone surfaces	9.5×10^{-4}		0.03	2.9×10^{-5}
Liver	2.1×10^{-4}		0.06	1.2×10^{-5}
				$\sum W_T H_{50,T} = 8.9 \times 10^{-5}$ Sv/Bq

The ALI based on stochastic effects is calculated fom equaltion (8.7):

$$\text{ALI (Stochastic)} = \frac{0.05 \text{ Sv}}{8.9 \times 10^{-5} \text{ Sv/Bq}} = 5.6 \times 10^2 \text{ Bq}.$$

Next, let us determine whether the criterion given by equation 8.8 is satisfied. That is, whether the tissue receiving the greatest dose, the bone surfaces in this case, will receive a dose-equivalent exceeding 0.5 Sv from inhaling the stochastic ALI of ^{239}Pu. According to equation (8.8), the product of the stochastic ALI and the greatest committed dose equivalent to any organ or tissue may not exceed 0.5 Sv. In this case, we have

$$5.6 \times 10^2 \text{ Bq} \times 9.5 \times 10^{-4} \text{ Sv/Bq} = 0.53 \text{ Sv}.$$

Since 0.53 Sv exceeds the criterion of equation (8.8), the non-stochastic dose equivalent limit determines the inhalation ALI:

$$\text{ALI (non-stochastic)} = \frac{0.5 \text{ Sv}}{(H_{50,T})_{\max} \text{ Sv/Bq}}$$

$$= \frac{0.5 \text{ Sv}}{9.5 \times 10^{-4} \text{ Sv/Bq}} = 5.3 \times 10^2 \text{ Bq}.$$

Because of the uncertainties in the metabolic models, the ALI's are given to only one significant figure. Thus, the ALI for inhaled "unsoluble" ^{239}Pu is listed in ICRP 30, Part 1, as 5×10^2 Bq.

Inhaled Radioactivity

The ALI for inhaled radioisotopes is based on a multi-compartment lung model. The model accounts for the fact that particulate deposition in the respiratory tract is governed by the size distribution of the inhaled aerosol, and that the clearance rate of deposited particulates is governed by the deposition site as well as by the chemical and physical properties of the particulates. Figure 8.3, a graphical representation of the lung model, shows three regions where inhaled dust may be deposited: the nasopharyngeal region (N–P), the tracheobronchial region (T–B), and the pulmonary region (P), representing the deep respiratory tract where gas exchange occurs. The N–P region is divided into two compartments, a and b. Compartment a represents that part of the dust deposited in the N–P region that dissolves and is absorbed directly into the blood, while compartment b represents the dust that is cleared from the N–P region into the gastrointestinal tract by swallowing. The T–B region is also represented by two compartments, c and d, from which deposited particulates are cleared by the same two mechanisms—dissolution and absorption into the blood and mechanical transfer into the G.I. tract, in this case by way of the

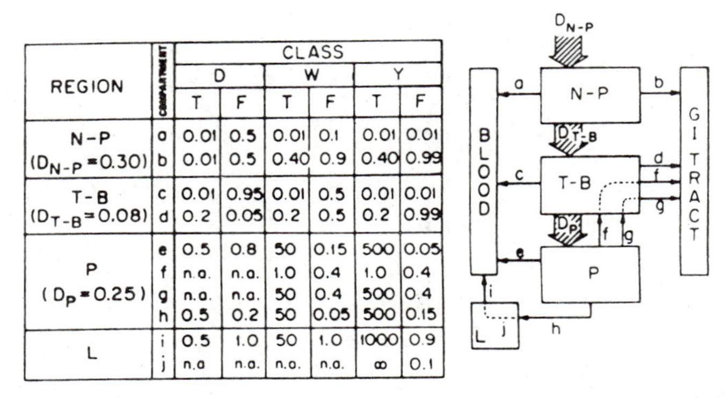

REGION	COMPARTMENT	CLASS D T	D F	W T	W F	Y T	Y F
N-P (D_{N-P}=0.30)	a	0.01	0.5	0.01	0.1	0.01	0.01
	b	0.01	0.5	0.40	0.9	0.40	0.99
T-B (D_{T-B}=0.08)	c	0.01	0.95	0.01	0.5	0.01	0.01
	d	0.2	0.05	0.2	0.5	0.2	0.99
P (D_P=0.25)	e	0.5	0.8	50	0.15	500	0.05
	f	n.a.	n.a.	1.0	0.4	1.0	0.4
	g	n.a.	n.a.	50	0.4	500	0.4
	h	0.5	0.2	50	0.05	500	0.15
L	i	0.5	1.0	50	1.0	1000	0.9
	j	n.a	n.a.	n.a.	n.a.	∞	0.1

FIG. 8.3. Respiratory tract clearance model used in calculating dose from inhaled radioactivity. (From S. B. Watson and M. R. Ford. A User's Manual to the ICRP Code: A Series of Computer Programs to Perform Dosimetric Calculations for the ICRP Committee 2 Report, ORNL/TM-6980, Feb., 1980.) The values for the removal half-times, T_{a-j}, and compartmental fractions, F_{a-j}, are given in the tabular portion of the figure for each of the three classes of retained materials. The values given for D_{N-P}, D_{T-B}, and D_P (left column) are the regional depositions based on an aerosol with an AMAD of 1 μm. The schematic drawing identifies the various clearance pathways in the model, a – j, in relation to the depositions D_{N-P}, D_{T-B}, D_P and the three respiratory regions, N–P, T–B and P. The entry n.a. indicates not applicable.

ciliary escalator to the throat and into the G.I. tract by swallowing. The pulmonary region, P, is modeled by four compartments. One of these compartments e, represents dissolution and absorption into the blood. Compartments f and g represent transfer of undissolved particulates into the G.I. tract via the upper respiratory tract, the T–B region. Compartment f is cleared by mechanical transport, presumably by unbalanced forces during respiratory excursions, and compartment g is cleared by alveolar macrophages that migrate into the T–B region. Compartment h empties into the pulmonary lymph nodes. The pulmonary lymph nodes are represented by two compartments, i and j. Compartment i empties into the blood stream after the particulates have dissolved, while compartment j permanently retains some highly insoluble particulates.

The exact fraction of the deposited aerosol that is cleared by each route and the respective clearance rates are governed by the chemical composition of the aerosol. However, since it is not practical to determine each of these parameters for every compound of every element, the various compounds of all the elements have been assigned, according to the criteria shown in Fig. 8.3, into one of three classes: D, W, and Y. Class D aersols are rapidly cleared from the deep respiratory tract with a clearance half-time on the order of a day or a fraction of a day. Class W aerosols are cleared on the order of weeks, while class Y materials are retained in the lungs on the order of years. Of the various dusts that may be transported to the lymph nodes, only class Y materials are premanently retained in the lymph nodes. For health physics purposes, the lung and the pulmonary lymph nodes are considered as a single organ. That is, the activity in the lung and in the lymph nodes is added together, and the total weight of the lungs and pulmonary lymph nodes is used to calculate dose from inhaled aerosols. ICRP recommendations for inhaled aerosols are based on inhalation and deposition of an aerosol whose activity median diameter is 1 μm and whose geometric standard deviation is 4. This assumed distribution leads to deposition of 30% of the inhaled dust in the N–P region, 8% in the T–B region, and 25% in the P region. The balance, 37%, is exhaled. Deposition for other size distributions is shown in Fig. 8.4.

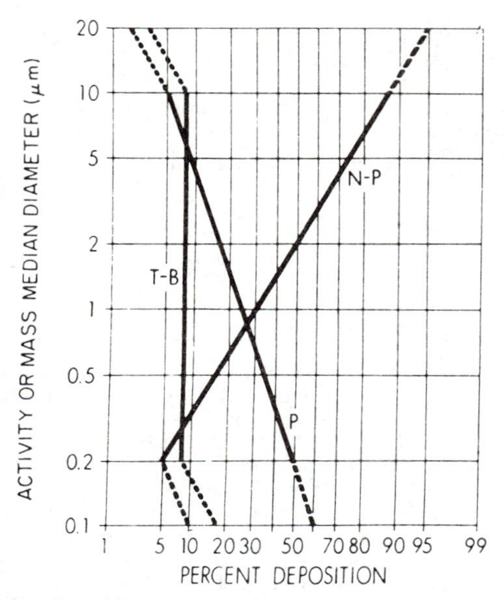

FIG. 8.4. ICRP dust deposition model. The radioactive or mass fraction of an aerosol which is deposited in the N–P, T–B, and P regions is given in relation to the Activity or Mass Median Aerodynamic Diameter (AMAD or MMAD) of the aerosol distribution. The model is intended for use with aerosol distributions having an AMAD or MMAD between 0.2 and 10 μm and whose geometric standard deviations are less than 4.5. Provisional deposition estimates further extending the size range are given by the dashed lines. For the unusual distribution having an AMAD or MMAD greater than 20 μm, complete N–P deposition can be assumed. The model does not apply to aerosols with AMAD or MMAD below 0.1 μm.

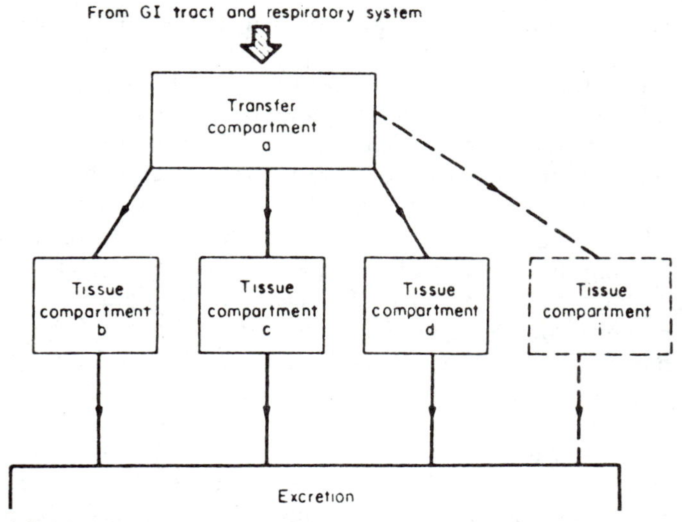

FIG. 8.5. Model usually used to describe the kinetics of radionuclides in the body. (From ICRP Publication 30, Part I, *Limits for Intakes of Radionuclides by Workers*, *Annals of the ICRP*, Vol. 2, No. 3/4, 1979).

Material brought up from the lung and swallowed enters into the G.I. tract, from where it may subsequently be eliminated in the feces, while irradiating the various parts of the G.I. tract and other organs during its passage. It may also undergo dissolution in the G.I. tract, and the dissolved portion may be absorbed into the body fluids and transferred to organs where it may be deposited (Fig. 8.5). Thus, for example, inorganic mercury is deposited mainly in the kidneys, iodine in the thyroid, strontium and radium in the skeleton, and plutonium in the liver and skeleton. The G.I. tract is modeled by four distinct regions, each with its own kinetic parameters, so that organ doses during passage of radionuclides can be calculated. Similarly, the organs where radionuclides are deposited are also modeled by appropriate equations, usually one or more first-order linear differential equations, that allow the organ doses to be calculated.

To illustrate the application of this lung model, let us calculate the ALI, by inhalation, for ^{137}Cs, using particulates of 1-μm activity median aerodynamic diameter and a standard deviation less than 4.5. From Fig. 8.4, we find that 63% of the inhaled dust is deposited in the respiratory tract, distributed as follows:

N–P: 30%,
T–B: 9%
P: 25%,

and 37% is exhaled. For purposes of ALI calculations, particulates of cesium compounds are assigned to clearance category D, since they have been found to be cleared rapidly from the lungs. The amount deposited in each of the compartments, together with the corresponding removal rate, is shown in Table 8.3.

TABLE 8.3. NUMBER OF TRANSFORMATIONS PER SECOND DEPOSITED IN THE SEVERAL COMPARTMENTS IN THE LUNG AFTER AN INHALATION OF 1 Bq ^{137}Cs, AND THE CLEARANCE RATES FOR EACH COMPARTMENT

Compartment	$A_s(0)$, t/s	$T_{1/2}$, day	λ_E, d^{-1}
c (TB→blood)	0.076	0.01	69.3
d (TB→GI)	0.004	0.2	3.47
e (P→blood)	0.200	0.5	1.39
f (P→LN)	0.050	0.5	1.39
LN	0.050	0.5	1.39

We shall calculate the committed dose equivalent from an intake of 1 Bq; then we shall determine how many Bq may be inhaled before reaching the limiting commited dose equivalent. To do this, we shall first calculate the dose to the lung from the ^{137}Cs deposited in the lung, and then from the ^{137}Cs distributed in the rest of the body and which also irradiates the lungs. Next we shall calculate the effective whole body dose from the inhaled activity. In making these calculations, it should benoted that the material deposited in the N–P region does not contribute directly to the intra-pulmonary dose; it contributes to the lung dose only by virtue of its presence in the rest of the body. The intra-pulmonary dose from the 1 Bq intake is due only to the 0.08 Bq ^{137}Cs deposited in the T–B region and the 0.25 Bq deposited in the P region.

The committed dose-equivalent to a target T after deposition of 1 Bq is given by equation (8.6):

$$H_{50,T} = \frac{1.6 \times 10^{-13}\ \text{J/MeV} \sum \tilde{A}_{si} \times \text{SEE}_i(T \leftarrow S) \times 1\ \text{t/Bq-sec} \times Q_i}{1\ \frac{\text{J}}{\text{kg}} \Big/ \text{Gy}}. \tag{8.6}$$

The SEE (lung←lung) is calculated from the output data for ^{137}Cs (Fig. 8.2), together with the specific absorbed fractions listed in Appendix IV. Table 8.4, shows the organization of the data to obtain the SEE for each of the electrons and photons, and their summation to yield 0.283 MeV per kg per transformation.

TABLE 8.4. ORGANIZATION OF DATA FOR CALCULATION OF SEE IN THE LUNG FROM INHALED ^{137}Cs

			Lung←Lung		Lung←Total body	
Radiation	Number/ Transform n_i	Mean energy/ Particle E_i, MeV	Specific absorbed fraction Φ_i, kg^{-1}	SEE_i, $\frac{MeV}{t}$/kg	Specific absorbed fraction Φ_i, kg^{-1}	SEE_i $\frac{MeV}{t}$/kg
β_1	0.935	0.1749	1	1.64E-01	1.43E-02	2.34E-03
β_2	0.065	0.4272	1	2.78E-02	1.43E-02	3.97E-04
γ	0.840	0.6616	4.86E-02	2.70E-02	5.11E-03	2.84E-03
K int. con. elec.	0.0781	0.6242	1	4.88E-02	1.43E-02	6.97E-04
L int. con. elec.	0.0140	0.6560	1	9.18E-03	1.43E-02	1.31E-04
M int. con. elec.	0.0031	0.6605	1	1.52E-03	1.43E-02	2.93E-05
K α-1 X-ray	0.0374	0.0322	2.26E-01	2.72E-04	1.22E-02	1.47E-05
K α-2 X-ray	0.0194	0.0318	2.26E-01	1.39E-04	1.22E-02	7.53E-06
K β-1 X-ray	0.0105	0.0364	2.17E-01	8.29E-05	1.13E-02	4.32E-06
K β-2 X-ray	0.0022	0.0374	2.15E-01	1.77E-05	1.10E-02	9.05E-07
L X-rays	0.0127	0.0045	1	5.72E-05	1.43E-02	8.12E-07
KLL Auger elec.	0.0057	0.0263	1	1.50E-04	1.43E-02	2.14E-06
KXY Auger elec.	0.0025	0.0308	1	7.7E-05	1.43E-02	1.10E-06
KLX Auger elec.	0.0004	0.0353	1	1.41E-05	1.43E-02	2.02E-07
LMM Auger elec.	0.0718	0.0034	1	2.44E-04	1.43E-02	3.49E-06
MXY Auger elec.	0.173	0.0011	1	1.90E-04	1.43E-02	2.72E-06

$$\sum SEE_i \text{ (lung←lung)} = 0.280 \frac{MeV}{t}/kg$$

$$\sum SEE_i \text{ (lung←total body)} = 6.47E\text{-}03 \frac{MeV}{t}/kg$$

Next, we must calculate the cumulated number of transformations in the lung during the 50 years following the 1 Bq intake. To do this, we shall add the number of transformations in each of the compartments of the lung. The cumulated number of transformations in a compartment after time t is given by

$$\tilde{A}(t) = \frac{A_s(0)}{\lambda_E}(1 - e^{-\lambda_E t}), \tag{8.10}$$

where $A_s(0)$ is the activity initially deposited in the compartment and λ_E is the effective clearance constant for the compartment. For a very long time relative to the effective half-time of the isotope, equation (8.10) reduces to equation (6.56):

$$\tilde{A} = \frac{A_s(0)}{\lambda_E}. \tag{6.56}$$

The total cumulated number of transformations in the lung is the sum of the cumulated transformations in each of the compartments:

$$\tilde{A} = \sum \frac{A_{si}}{\lambda_{Ei}}. \tag{8.11}$$

Since 50 years is very long relative to the effective half-time of cesium in every one of the lung compartments, the cumulated number of transformations in the lung is calculated from equation (8.11) with the values listed in Table 8.3.

$$\tilde{A} = 8.64 \times 10^4 \text{ s/d}\left\{\left(\frac{0.076 \text{ t/s}}{69.3 \text{ d}^{-1}}\right) + \left(\frac{0.004 \text{ t/s}}{3.47 \text{ d}^{-1}}\right) + \left(\frac{0.200 \text{ t/s}}{1.39 \text{ d}^{-1}}\right) + \left(\frac{0.050 \text{ t/s}}{1.39 \text{ d}^{-1}}\right) + \left(\frac{0.050 \text{ t/s}}{1.39 \text{ d}^{-1}}\right)\right\}.$$

$$= 1.9 \times 10^4 \text{ transformations.}$$

The committed dose equivalent to the lung from ^{137}Cs in the lung is calculated from equation (8.6).

$$H_{50} = \frac{1.6 \times 10^{-13} \text{ J/MeV} \times 1.9 \times 10^4 \text{ t} \times 0.280 \dfrac{\text{MeV}}{\text{kg}} \Big/ \text{t} \times 1 \text{ Sv/Gy}}{1 \dfrac{\text{J}}{\text{Bq}} / \text{Gy}}$$

$$= 8.6 \times 10^{-10} \text{ Sv.}$$

The next step is to calculate the committed dose equivalent to the lungs from the ^{137}Cs in the rest of the body. The total amount absorbed into the body includes that deposited in the N–P region as well as that deposited in the lung. The total absorbed into the body is

$$0.3 + 0.08 + 0.25 = 0.63 \text{ Bq.}$$

Retention of ^{137}Cs in the body is given by equation (6.33). The equation also tells us that 10% of the absorbed cesium, or 0.063 Bq, is transferred to the compartment whose clearance rate is 0.347 per day, while 90%, or 0.567 Bq, is transferred to the compartment that is cleared at a rate of 0.0063 per day. The total number of transformations in the body from the 0.63 Bq deposited in the respiratory tract after the inhalation of 1 Bq and subsequently transferred to the body is given by equation (8.11):

$$\tilde{A} = 8.64 \times 10^4 \text{ s/d}\left(\frac{0.063 \text{ t/s}}{0.347 \text{ d}^{-1}} + \frac{0.567 \text{ t/s}}{0.0063 \text{ d}^{-1}}\right)$$

$$= 7.8 \times 10^6 \text{ transformations.}$$

Using the gamma and X-ray specific absorbed fractions for (lung←total body) obtained by interpolating between the values listed in Appendix IV, and the beta ray and electron specific absorbed fraction of 1.43×10^{-2}, as calculated from equation (6.58), and then by summing the SEE of each contributing radiation as shown in Table 8.4, we find the SEE (lung←total body) to be 6.47×10^{-3} MeV per transformation per kilogram. The committed dose equivalent to the lung from the ^{137}Cs in the body is calculated from equation (8.6):

$$H_{50} = \frac{1.6 \times 10^{-13} \text{ J/MeV} \times 7.8 \times 10^6 \text{ t} \times 6.74 \times 10^{-3} \dfrac{\text{MeV}}{\text{kg}} \Big/ \text{t} \times 1 \text{ Sv/Gy}}{1 \dfrac{\text{J}}{\text{ky}} / \text{Gy}}$$

$$= 8.4 \times 10^{-9} \text{ Sv}$$

The total committed dose equivalent to the lungs, therefore, from the inhalation of 1 Bq of ^{137}Cs is

$$H_{50,L} = 8.6 \times 10^{-10} \text{ Sv/Bq} + 8.4 \times 10^{-9} \text{ Sv/Bq}$$

$$= 9.3 \times 10^{-9} \text{ Sv/Bq.}$$

In the calculation for the ALI by ingestion, it was found that the committed dose equivalent from ^{137}Cs uniformly distributed throughout the body is

$$\frac{0.05 \text{ Sv}}{3.9 \times 10^{6} \text{ Bq}} = 1.3 \times 10^{-8} \text{ Sv/Bq}.$$

Since only 0.63 Bq is absorbed into the body when 1 Bq is inhaled, using the weighting factors of 0.12 for the lung and 0.88 for the remainder of the body, we find the effective committed dose equivalent from 1 Bq of inhaled ^{137}Cs to be

$$\sum W_T H_{50,T} = 0.12 \times 9.3 \times 10^{-9} + 0.63(0.88 \times 1.30 \times 10^{-8})$$
$$= 8.3 \times 10^{-9} \text{ Sv/Bq}.$$

The ALI for stochastic effects is calculated from equation (8.7)

$$\text{ALI (stochastic)} = \frac{0.05 \text{ Sv}}{8.3 \times 10^{-9} \text{ Sv/Bq}} = 6 \times 10^{6} \text{ Bq}.$$

Now we must determine whether the stochastically based ALI will lead to a committed dose equivalent greater than 0.5 Sv in any organ or tissue. Since the cesium that is transferred out of the lung to the body is uniformly distributed throughout the body, no tissue will receive a committed dose equivalent greater than the 1.3×10^{-8} Sv/Bq delivered to the whole body. The product of the stochastic ALI and the whole body committed dose equivalent from the 0.63 Bq deposited in the body is

$$6 \times 10^{6} \text{ Bq} \times 0.63 \times 1.3 \times 10^{-8} \text{ Sv/Bq} = 0.05 \text{ Sv}.$$

Since the stochastically based ALI leads to a committed dose equivalent less than 0.5 Sv, the stochastically based ALI is applicable, and limits the inhaled intake of ^{137}Cs to 6×10^{6} Bq during one year.

In calculating the ALI's for ^{137}Cs, it was assumed that all the radiations listed in Tables 8.2 and 8.4 came from the cesium as it transformed to barium. Technically this is not true, since the 0.661 MeV gamma ray originates in ^{137m}Ba, whose half life is 2.55 minutes. Because of the very short half life of ^{137m}Ba, however, it is in secular equilibrium with its long-lived (30 years) parent. Furthermore, because of its very short lifetime, the barium does not have the opportunity to follow its own metabolic pathways, but follows the kinetics of metabolism of cesium. For these reasons, the consideration of ^{137}Cs-^{137m}Ba as a single radiation source is valid. In those cases of serial radioactive transformations where the daughters have the opportunity to follow their own metabolic pathways, the contributions of each daughter to the total dose from inhalation or ingestion of the parent is considered separately in calculating the ALI.

Derived Air Concentration (DAC)

The ALI, which is a calculated value based on the dose limits for stochastic and non-stochastic effects, only gives the annual intake limit; it does not deal with the rate of intake or with environmental concentrations of a radionuclide that lead to the intake. For engineering design purposes and for control of routine operations, it is useful to know the environmental concentrations of the radionuclides with which we are dealing. To this end, the *derived air concentration* (DAC) is introduced for airborne contaminants. The DAC is simply that average atmospheric concentration of the radionuclide that would lead to the ALI in a reference person as a consequence of exposure at the DAC for a 2000 hour working year. Since a reference person inhales 20 liters air per minute, or 2400 m^3 during the 2000 hours per year spent at work, the derived air concentration is

$$DAC = \frac{ALI}{2400} \text{ Bq/m}^3. \tag{8.12}$$

Thus, for airborne ^{137}Cs, whose ALI for inhalation is 6×10^6 Bq, the derived air concentration is

$$DAC = \frac{6 \times 10^6 \text{ Bq}}{2.4 \times 10^3 \text{ m}^3} = 2.5 \times 10^3 \text{ Bq/m}^3,$$

which is rounded off to 2×10^3 Bq/m^3.

Gastrointestinal Tract

In cases of ingested radionuclides, and especially for those nuclides that are poorly absorbed from the G.I. tract, the G.I. tract or portions of it may be the tissue or organ receiving the greatest dose. The dose to the G.I. tract is calculated on the basis of the four-compartment model shown in Fig. 8.6. According to this model, the radionuclide enters the stomach, and then passes sequentially through the small intestine, from where most absorption into the body fluids occurs, the upper large intestine, the lower large intestine, and then the remaining activity is excreted in the feces.

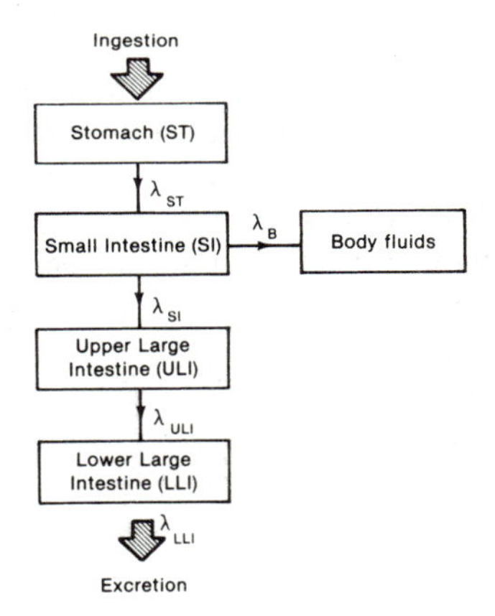

Section of GI tract	Mass of walls* (g)	Mass of contents* (g)	Mean residence time (day)	λ day^{-1}
Stomach (ST)	150	250	1/24	24
Small Intestine (SI)	640	400	4/24	6
Upper Large Intestine (ULI)	210	220	13/24	1.8
Lower Large Intestine (LLI)	160	135	24/24	1

FIG. 8.6. Dosimetric model of the gastrointestinal tract. The clearance rate for transfer from the small intestine into the body fluids is given by

$$\lambda_B = \frac{f_1 \lambda_{si}}{1 - f_1}, \tag{8.13}$$

where f_1 = fraction of the stable element reaching the body fluids after ingestion. (From ICRP 30, Part 1.)

When making dose calculations for the purpose of setting an ALI, we assume the radionuclide to be uniformly distributed throughout the contents of the respective segments of the G.I. tract, and that the quantity of material in each segment be as listed in Fig. 8.6. Furthermore, the movement of the contents between compartments is assumed to follow first order kinetics, with the compartmental clearance rates shown in Fig. 8.6. With this information, we can calculate the dose per Bq of intake to the walls of the G.I. tract (whose masses are given in Fig. 8.6) through the use of the appropriate specific absorbed fractions, and thus to compute an ALI.

Combined Exposure

The ALI's calculated for each isotope assumes exposure to no other isotope and by no other pathway, as well as no external radiation exposure. Generally, a person who may be exposed to several different radioisotopes by inhalation and by ingestion and also to external radiation must meet the following criterion if he is to remain within the recommended dose limit:

$$\sum \frac{\text{intake}_i\text{(inhaled)}}{\text{ALI}_i\text{(inhalation)}} + \sum \frac{\text{Intake}_i\text{(ingested)}}{\text{ALI}_i\text{(ingestion)}} + \sum \frac{\text{External dose}}{\text{Dose limit}} \leq 1. \tag{8.14}$$

Basis for Radiation Safety Regulations

Before the promulgation of the ICRP recommendations on dose limitation that are embodied in ICRP Publication 26 (1977), regulatory agencies of the various countries around the world were guided mainly by the ICRP recommendations that were published in ICRP Publication 2 (1959) and Publication 9 (1965). Publication 2 deals mainly with internally deposited isotopes, while Publication 9 deals generally with radiation protection and dose limitation. These recommendations have been adapted to the needs of the various countries, and regulations based on these recommendations have been promulgated and are currently (1982) in use around the world.

For purposes of radiation safety, the ICRP recognized two main exposure categories:

1. Occupational exposure
   a. Women of reproductive capacity
   b. Pregnant women
   c. All other radiation workers
2. Members of the general public.

The recommended dose limits are summarized in Table 8.5.

The recommendation of 5 rem (50 mSv) per year, when uniformly distributed over a 50-week period, yields a weekly dose rate of 100 mrem (1 mSv). This value of 100 mrem (1 mSv) is frequently, but incorrectly, referred to as the maximum permissible dose rate.

The maximum acceptable dose is the sum of doses due to all types of occupational exposure; it includes external irradiation as well as radiation due to internally deposited radioisotopes. For example, if a radiation worker has deposited in his body a sufficient amount of a radioisotope to irradiate his hemopoietic system at a rate of 25 mrem (0.25 mSv) per week, then his dose due to external radiation is designed to be less than 75 mrem (0.75 mSv) per week. Generally, for radiation safety programs that must consider

TABLE 8.5. SUMMARY OF DOSE LIMITS FOR INDIVIDUALS

Organ or Tissue	Maximum Permissible Doses for occupationally exposed adults	Dose Limits for members of the public
Gonads, red bone marrow	5 rems in a year* (50 mSv in a year)	0.5 rem in a year (5 mSv in a year)
Skin, bone, thyroid	30 rems in a year* (300 mSv in a year)	3 rems in a year** (30 mSv in a year)
Hands and forearms, feet and ankles	75 rems in a year* (750 mSv in a year)	7.5 rems in a year (75 mSv in a year)
Other single organs	15 rems in a year* (150 mSv in a year)	1.5 rems in a year (15 mSv in a year)

*Up to one half the annual dose limit may be accumulated in any period of a quarter of a year. For women of reproductive capacity, the dose limit to the abdomen is 1.3 rems in a quarter, whereas a pregnant woman's abdominal dose should not exceed 1 rem from the time of diagnosis of pregnancy until its termination.

**1.5 rems in a year to the thyroid of children up to 16 years of age.

From ICRP Publication 9.

exposure to more than 1 radioisotope by inhalation and ingestion, as well as external radiation exposure, the working environment is controlled so that

$$\sum \frac{C_i(\text{air})}{\text{MPC}_i(\text{air})} + \sum \frac{C_i(\text{water})}{\text{MPC}_i(\text{water})} + \frac{\text{External dose(rate)}}{\text{MPD(rate)}} \leq 1, \tag{8.15}$$

where C_i(air) and C_i(water) = actual environmental concentrations of the ith isotope,

MPC$_i$(air) and MPC$_i$(water) = maximum permissible environmental concentrations of the ith isotope,

MPD = maximum permissible dose.

For protection against external radiation, the recommended dose limits are biologically meaningful and directly applicable, since it is a simple matter to determine the dose due to exposure for any given time in a radiation field. For protection against internal radiation due to inhaled or ingested radioactive materials, derived limits are used. These derived limits are based on two different criteria, depending on the type of isotope under consideration. For bone seekers, such as ^{90}Sr, ^{227}Ac, ^{232}Th, ^{231}Pa, ^{237}Np, ^{239}Pu, etc., which emit alpha or beta radiation, the derived limits are based on comparison with ^{226}Ra and its daughters. For all other radioisotopes, the amount of radioisotope is computed which, if deposited in the body, will deliver a dose equivalent not greater than 100 mrem (1 mSv) per week if the total body, gonads, or hemopoietic system is the *critical organ* (the critical organ is defined as that part of the body that is most susceptible to radiation damage under the specific conditions under consideration); or 600 mrem (6 mSv) per week if the thyroid is the critical organ; or 300 mrem (3 mSv) per week for any other critical organ or tissue. This quantity of radioisotope was called the *maximum permissible body burden*. Then, with knowledge of the kinetics of metabolism of this radioisotope, we can calculate those concentrations of the isotope in air and water which would lead to a steady-state accumulation in the body equal to the maximum permissible body burden after continuous exposure. These derived concentrations were called the *maximum permissible concentrations* (MPC) in air and water. Maximum permissible concentrations for all known radioisotopes, to which lifetime exposure is assumed, have been calculated and published in ICRP Publication 2. Included in that report are the detailed methods of calculation, together with the biological and physical factors necessary for the calculation. Several of the maximum body burdens and concentrations are given in Table 8.6, which are excerpted from ICRP 2. Since the MPC's were calculated on the basis of continuous lifetime exposure, a short term or accidental exposure to environmental concentrations greater than the derived maxima does not necessarily mean that the person is overexposed.

Calculation of MPC in Drinking Water Based on Dose to Critical Organ

The principles involved in the calculation of the maximum permissible concentration in drinking water may be demonstrated with a simplified example. In making this calculation, it is assumed that the radioisotope is stored in one or more "compartments" or organs in the body. In many cases, deposition in the storage organ may be considered instantaneous with respect to the time that the isotope will be retained by the organ. After deposition, it will be cleared from the organ at the rate determined by its effective half-life. If its passage through subsequent organs or tissues (kidney, liver, etc.) on its way out of the body is rapid compared to the rate of clearance from the storage organ, then this latter rate is the dominant one, and the elimination from the body will appear to follow first order kinetics. In this simplest case, the retention of the isotope in the body is described by equation (6.25)

$$Q = Q_0 e^{-\lambda_E t}, \tag{6.25}$$

where Q is the activity remaining at time t after deposition of Q_0, and λ_E is the effective clearance constant. If intake is continuous, then the accumulation of the isotope is given by

$$Q = \frac{K}{\lambda_E}(1 - e^{-\lambda_E t}), \tag{8.16}$$

where K is the intake of activity per unit time, i.e., Bq/day, for example. Fig. 8.7 is a graphical representation of equation (8.16). As t increases, $e^{-\lambda_E t}$ becomes smaller, and the body burden Q approaches the limiting value

$$Q_s = \frac{K}{\lambda_E}. \tag{8.17}$$

For practical purposes, this limiting value is reached after about six effective half-lives. At this time, we may consider that a steady state has been reached, in which the amount of activity deposited is equal to the amount of activity eliminated.

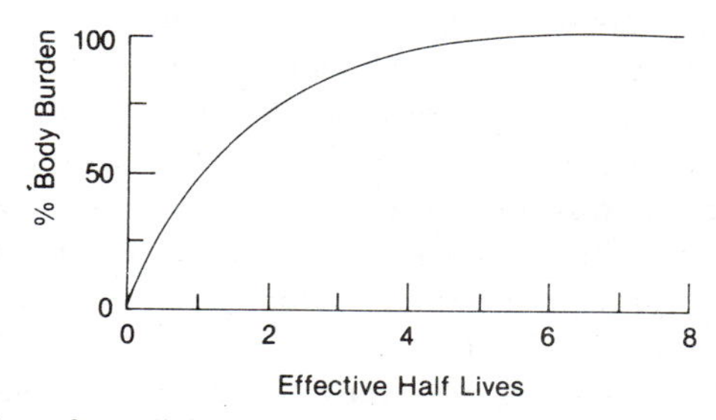

FIG. 8.7. Buildup of a radioisotope in the body resulting from continuous intake.

TABLE 8.6. MAXIMUM PERMISSIBLE BODY BURDENS AND MAXIMUM PERMISSIBLE CONCENTRATIONS OF RADIONUCLIDES IN AIR AND IN WATER FOR OCCUPATIONAL EXPOSURE

Radionuclide and type of decay		Organ of reference[a] (critical organ in **bold** type)	Maximum permissible burden in total body $q(\mu c)$	Maximum permissible concentrations			
				For 40-hr week		For 168-hr week[b]	
				$(MPC)_w$ $\mu Ci/cm^3$	$(MPC)_a$ $\mu Ci/cm^3$	$(MPC)_w$ $\mu Ci/cm^3$	$(MPC)_a$ $\mu Ci/cm^3$
${}^3_1H_2(H^3O)$ (β^-)	(Sol)	**Body tissue**	10^3	0.1	2×10^{-5}	0.03	5×10^{-6}
		Total body	2×10^3	0.2	2×10^{-5}	0.05	7×10^{-6}
(H_2^3)	(Immersion)	**Skin**			2×10^{-3}		4×10^{-4}
${}^{14}_6C(CO_2)$ (β^-)	(Sol)	**Fat**	300	0.02	4×10^{-6}	8×10^{-3}	10^{-6}
		Total body	400	0.03	5×10^{-6}	0.01	2×10^{-6}
		Bone	400	0.04	6×10^{-6}	0.01	2×10^{-6}
	(Immersion)	**Total body**			5×10^{-5}		10^{-3}
${}^{32}_{15}P$ (β^-)	(Sol)	**Bone**	6	5×10^{-4}	7×10^{-8}	2×10^{-4}	2×10^{-8}
		Total body	30	3×10^{-3}	4×10^{-7}	9×10^{-4}	10^{-7}
		GI (LLI)		3×10^{-3}	6×10^{-7}	9×10^{-4}	2×10^{-7}
		Liver	50	5×10^{-3}	6×10^{-7}	2×10^{-3}	2×10^{-7}
		Brain	300	0.02	3×10^{-8}	8×10^{-3}	10^{-6}
	(Insol)	**Lung**			8×10^{-8}		3×10^{-8}
		GI (LLI)		7×10^{-4}	10^{-7}	2×10^{-4}	4×10^{-8}
${}^{41}_{20}Ca$ (β^-)	(Sol)	**Bone**	30	3×10^{-4}	3×10^{-8}	9×10^{-5}	10^{-3}
		Total body	200	2×10^{-3}	3×10^{-7}	7×10^{-4}	9×10^{-5}
		GI (LLI)		0.01	3×10^{-8}	4×10^{-3}	10^{-6}
	(Insol)	**Lung**			10^{-7}		4×10^{-8}
		GI (LLI)		5×10^{-3}	9×10^{-7}	2×10^{-3}	3×10^{-7}
${}^{51}_{24}Cr$ (ec, γ)	(Sol)	**GI (LLI)**		0.05	10^{-5}	0.02	4×10^{-6}
		Total body	800	0.6	10^{-5}	0.2	4×10^{-5}
		Lung	10^3	1	2×10^{-5}	0.4	8×10^{-6}
		Prostate	2×10^2	2	3×10^{-5}	0.5	10^{-5}
		Thyroid	4×10^2	3	6×10^{-5}	1	2×10^{-5}
		Kidney	8×10^2	6	10^{-4}	2	4×10^{-5}

Notes for this table are given on page 213.

TABLE 8.6. (*cont.*)

Radionuclide and type of decay		Organ of reference[a] (critical organ in **bold** type)	Maximum permissible burden in total body $q(\mu c)$	Maximum permissible concentrations: For 40-hr week, $(MPC)_w$ $\mu Ci/cm^3$	For 40-hr week, $(MPC)_a$ $\mu Ci/cm^3$	For 168-hr week[b], $(MPC)_w$ $\mu Ci/cm^3$	For 168-hr week[b], $(MPC)_a$ $\mu Ci/cm^3$
	(Insol)	**Lung**			2×10^{-6}		8×10^{-5}
		GI (LLI)		0.05	8×10^{-6}	0.02	3×10^{-5}
$^{62}_{27}Co$ (β^-, γ)	(Sol)	**GI (LLI)**		10^{-3}	3×10^{-7}	5×10^{-4}	10^{-7}
		Total body	10	4×10^{-3}	4×10^{-7}	10^{-3}	10^{-7}
		Pancreas	70	0.02	2×10^{-5}	7×10^{-3}	6×10^{-7}
		Liver	90	0.03	10^{-6}	9×10^{-3}	5×10^{-7}
		Spleen	200	0.05	4×10^{-6}	0.02	2×10^{-6}
		Kidney	200	0.07	6×10^{-6}	0.03	2×10^{-6}
	(Insol)	**Lung**			9×10^{-9}		3×10^{-9}
		GI (LLI)		10^{-3}	2×10^{-7}	3×10^{-4}	6×10^{-9}
$^{65}_{30}Zn$ (β^+, ϵ, γ)	Sol)	**Total body**	60	3×10^{-3}	10^{-7}	10^{-3}	4×10^{-8}
		Prostate	70	4×10^{-3}	10^{-7}	10^{-3}	4×10^{-8}
		Liver	80	4×10^{-3}	10^{-7}	10^{-3}	5×10^{-8}
		Kidney	100	6×10^{-3}	2×10^{-7}	2×10^{-3}	7×10^{-8}
		GI (LLI)		6×10^{-3}	10^{-6}	2×10^{-3}	4×10^{-7}
		Pancreas	200	7×10^{-3}	3×10^{-7}	3×10^{-3}	9×10^{-8}
		Muscle	200	0.01	4×10^{-7}	4×10^{-3}	10^{-7}
		Ovary	300	0.01	5×10^{-7}	4×10^{-3}	2×10^{-7}
		Testis	400	0.02	6×10^{-7}	6×10^{-3}	2×10^{-7}
		Bone	700	0.04	10^{-4}	0.01	4×10^{-7}
	(Insol)	**Lung**			6×10^{-8}		2×10^{-8}
		GI (LLI)		5×10^{-3}	9×10^{-7}	2×10^{-3}	3×10^{-7}
$^{78}_{33}As$ (β^-, γ)	(Sol)	**GI (LLI)**		6×10^{-4}	10^{-7}	2×10^{-4}	4×10^{-8}
		Total body	20	0.4	5×10^{-6}	0.1	2×10^{-8}
		Kidney	20	0.6	8×10^{-9}	0.2	3×10^{-4}
		Liver	40	1	10^{-5}	0.4	5×10^{-6}
	(Insol)	**GI (LLI)**		6×10^{-4}	10^{-7}	2×10^{-4}	3×10^{-6}
		Lung			6×10^{-7}		2×10^{-7}

TABLE 8.6. (*cont.*)

Radionuclide and type of decay		Organ of reference[a] (critical organ in **bold** type)	Maximum permissible burden in total body $q(\mu c)$	Maximum permissible concentrations			
				For 40-hr week		For 168-hr week[b]	
				$(MPC)_w$ $\mu Ci/cm^3$	$(MPC)_a$ $\mu Ci/cm^3$	$(MPC)_w$ $\mu Ci/cm^3$	$(MPC)_a$ $\mu Ci/cm^3$
$^{84}_{38}Sr$ (β^-)	(Sol)	**Bone**	4	3×10^{-4}	3×10^{-8}	10^{-4}	10^{-8}
		GI (LLI)		10^{-3}	3×10^{-7}	4×10^{-4}	9×10^{-8}
		Total body	40	2×10^{-3}	2×10^{-7}	7×10^{-4}	6×10^{-8}
	(Insol)	**Lung**			4×10^{-8}		10^{-8}
		GI (LLI)		8×10^{-4}	10^{-7}	3×10^{-4}	5×10^{-8}
$^{90}_{38}Sr$ (β^-)	(Sol)	**Bone**	2	4×10^{-6}	3×10^{-10}	10^{-6}	10^{-10}
		Total body	20	10^{-5}	9×10^{-10}	4×10^{-6}	3×10^{-10}
		GI (LLI)		10^{-3}	3×10^{-7}	5×10^{-4}	10^{-7}
	(Insol)	**Lung**			5×10^{-9}		2×10^{-9}
		GI (LLI)		10^{-3}	2×10^{-7}	4×10^{-4}	6×10^{-8}
$^{95}_{40}Zr$ (β^-, γ, e^-).	(Sol)	**GI (LLI)**		2×10^{-3}	4×10^{-7}	6×10^{-4}	10^{-7}
		Total body	20	3	10^{-3}	1	4×10^{-8}
		Bone	30	4	2×10^{-7}	2	6×10^{-8}
		Kidney	30	4	2×10^{-7}	2	6×10^{-8}
		Liver	40	6	3×10^{-7}	2	9×10^{-8}
		Spleen	40	7	3×10^{-7}	2	10^{-7}
	(Insol)	**Lung**			3×10^{-8}		10^{-8}
		GI (LLI)		2×10^{-3}	3×10^{-7}	6×10^{-4}	10^{-7}
$^{95}_{41}Nb$ (β^-, γ)	(Sol)	**GI (LLI)**		3×10^{-3}	6×10^{-7}	10^{-3}	2×10^{-7}
		Total body	40	10	5×10^{-7}	4	2×10^{-7}
		Liver	60	20	7×10^{-7}	6	3×10^{-7}
		Kidney	60	20	8×10^{-7}	6	3×10^{-7}
		Bone	80	20	9×10^{-7}	7	3×10^{-7}
		Spleen	80	20	10^{-6}	7	3×10^{-7}
	(Insol)	**Lung**			10^{-7}		3×10^{-8}
		GI (LLI)		3×10^{-3}	5×10^{-7}	10^{-3}	2×10^{-7}

TABLE 8.6. (*cont.*)

Radionuclide and type of decay		Organ of reference[a] (critical organ in **bold** type)	Maximum permissible burden in total body $q(\mu c)$	Maximum permissible concentrations For 40-hr week (MPC)$_w$ μCi/cm^3	For 40-hr week (MPC)$_a$ μCi/cm^3	For 168-hr week[b] (MPC)$_w$ μCi/cm^3	For 168-hr week[b] (MPC)$_a$ μCi/cm^3
$^{106}_{44}$Ru (β^-, γ)	(Sol)	**GI (LLI)**		4×10^{-4}	8×10^{-8}	10^{-4}	3×10^{-8}
		Kidney	3	0.01	10^{-7}	4×10^{-3}	5×10^{-8}
		Bone	10	0.04	5×10^{-7}	0.01	2×10^{-7}
		Total body	10	0.06	7×10^{-7}	0.02	3×10^{-7}
	(Insol)	**Lung**			6×10^{-9}		2×10^{-9}
		GI (LLI)		3×10^{-4}	6×10^{-8}	10^{-4}	2×10^{-8}
$^{131}_{53}$I (β^-, γ, e$^-$)	(Sol)	**Thyroid**	0.7	6×10^{-5}	9×10^{-8}	2×10^{-5}	3×10^{-9}
		Total body	50	5×10^{-3}	8×10^{-7}	2×10^{-3}	3×10^{-7}
		GI (LLI)		0.03	7×10^{-6}	0.01	2×10^{-6}
	(Insol)	**GI (LLI)**		2×10^{-3}	3×10^{-7}	6×10^{-4}	10^{-7}
		Lung			3×10^{-7}		10^{-7}
$^{137}_{55}$Cs (β, γ, e$^-$)	(Sol)	**Total body**	30	4×10^{-4}	6×10^{-8}	2×10^{-4}	2×10^{-8}
		Liver	40	5×10^{-4}	8×10^{-8}	2×10^{-4}	3×10^{-8}
		Spleen	50	6×10^{-4}	9×10^{-8}	2×10^{-4}	3×10^{-8}
		Muscle	50	7×10^{-4}	10^{-7}	2×10^{-4}	4×10^{-8}
		Bone	100	10^{-3}	2×10^{-7}	5×10^{-4}	7×10^{-8}
		Kidney	100	10^{-3}	2×10^{-7}	5×10^{-4}	8×10^{-8}
		Lung	300	5×10^{-3}	6×10^{-7}	2×10^{-3}	2×10^{-7}
		GI (SI)		0.02	5×10^{-6}	8×10^{-3}	2×10^{-6}
	(Insol)	**Lung**			10^{-8}		5×10^{-9}
		GI (LLI)		10^{-3}	2×10^{-7}	4×10^{-4}	8×10^{-8}
$^{144}_{58}$Ce (α, β^-, γ)	(Sol)	**GI (LLI)**		3×10^{-4}	8×10^{-8}	10^{-4}	3×10^{-8}
		Bone	5	0.2	10^{-8}	0.08	3×10^{-9}
		Liver	6	0.3	10^{-8}	0.1	4×10^{-9}
		Kidney	10	0.5	2×10^{-8}	0.2	7×10^{-9}
		Total body	20	0.7	3×10^{-8}	0.3	10^{-8}

TABLE 8.6. (*cont.*)

Radionuclide and type of decay		Organ of reference[a] (critical organ in **bold** type)	Maximum permissible burden in total body $q(\mu c)$	Maximum permissible concentrations			
				For 40-hr week		For 168-hr week[b]	
				$(MPC)_w$ $\mu Ci/cm^3$	$(MPC)_a$ $\mu Ci/cm^3$	$(MPC)_w$ $\mu Ci/cm^3$	$(MPC)_a$ $\mu Ci/cm^3$
	(Insol)	**Lung**			6×10^{-9}		2×10^{-9}
		GI (LLI)		3×10^{-4}	6×10^{-8}	10^{-4}	2×10^{-8}
$^{147}_{61}Pm$ (α, β^-)	(Sol)	**GI (LLI)**		6×10^{-3}	10^{-6}	2×10^{-3}	5×10^{-7}
		Bone	60	1	6×10^{-8}	0.5	2×10^{-8}
		Kidney	200	4	2×10^{-7}	2	7×10^{-8}
		Total body	300	7	3×10^{-7}	2	10^{-7}
		Liver	300	8	4×10^{-7}	3	10^{-7}
	(Insol)	**Lung**			10^{-7}		3×10^{-8}
		GI (LLI)		6×10^{-3}	10^{-8}	2×10^{-3}	4×10^{-7}
$^{182}_{73}Ta$ (β^-, γ)	(Sol)	**GI (LLI)**		10^{-3}	3×10^{-7}	4×10^{-4}	9×10^{-8}
		Liver	7	0.9	4×10^{-3}	0.3	10^{-8}
		Kidney	20	2	8×10^{-8}	0.7	3×10^{-8}
		Total body	20	2	9×10^{-8}	0.7	3×10^{-8}
		Spleen	30	4	10^{-7}	1	5×10^{-8}
		Bone	50	6	3×10^{-7}	2	9×10^{-8}
	(Insol)	**Lung**			2×10^{-8}		7×10^{-9}
		GI (LLI)		10^{-3}	2×10^{-7}	4×10^{-4}	7×10^{-8}
$^{192}_{77}Ir$ (β^-, γ)	(Sol)	**GI (LLI)**		10^{-3}	3×10^{-7}	4×10^{-4}	9×10^{-8}
		Kidney	6	4×10^{-3}	10^{-7}	10^{-3}	4×10^{-8}
		Spleen	7	4×10^{-3}	10^{-7}	10^{-3}	5×10^{-8}
		Liver	8	5×10^{-3}	2×10^{-7}	2×10^{-3}	6×10^{-8}
		Total body	20	0.01	4×10^{-7}	4×10^{-3}	10^{-7}
	(Insol)	**Lung**			3×10^{-8}		9×10^{-9}
		GI (LLI)		10^{-3}	2×10^{-7}	4×10^{-4}	6×10^{-8}

TABLE 8.6. (*cont.*)

Radionuclide and type of decay	Organ of reference[a] (critical organ in **bold** type)	Maximum permissible burden in total body $q(\mu c)$	Maximum permissible concentrations			
			For 40-hr week		For 168-hr week[b]	
			$(MPC)_w$ $\mu Ci/cm^3$	$(MPC)_a$ $\mu Ci/cm^3$	$(MPC)_w$ $\mu Ci/cm^3$	$(MPC)_a$ $\mu Ci/cm^3$
$^{198}_{79}Au$ (β^-, γ) (Sol)	**GI (LLI)**		2×10^{-3}	3×10^{-7}	5×10^{-4}	10^{-7}
	Kidney	20	0.07	3×10^{-6}	0.02	9×10^{-7}
	Total body	30	0.1	4×10^{-6}	0.04	2×10^{-6}
	Spleen	60	0.2	8×10^{-6}	0.07	3×10^{-6}
	Liver	80	0.3	10^{-5}	0.1	4×10^{-8}
(Insol)	**GI (LLI)**		10^{-3}	2×10^{-7}	5×10^{-4}	8×10^{-8}
	Lung			6×10^{-7}		2×10^{-7}
$^{222}_{86}Rn$[c] (α, β, γ)	**Lung**			3×10^{-8}		10^{-8}
$^{226}_{88}Ra$ $(\alpha, \beta^-, \gamma)$ (Sol)	**Bone**	0.1	4×10^{-7}	3×10^{-11}	10^{-7}	10^{-11}
	Total body	0.2	6×10^{-7}	5×10^{-11}	2×10^{-7}	2×10^{-11}
	GI (LLI)		10^{-3}	3×10^{-7}	5×10^{-4}	10^{-7}
(Insol)	**GI (LLI)**		9×10^{-4}	2×10^{-7}	3×10^{-4}	6×10^{-8}
$^{235}_{92}U$ $(\alpha, \beta^-, \gamma)$ (Sol)	**GI (LLI)**		8×10^{-4}	2×10^{-7}	3×10^{-4}	6×10^{-8}
	Kidney	0.03	0.01	5×10^{-10}	4×10^{-3}	2×10^{-10}
	Bone	0.06	0.01	6×10^{-10}	5×10^{-3}	2×10^{-10}
	Total body	0.4	0.04	2×10^{-9}	0.01	6×10^{-10}
(Insol)	**Lung**			10^{-10}		4×10^{-11}
	GI (LLI)		8×10^{-4}	10^{-7}	3×10^{-4}	5×10^{-8}
$^{238}_{92}U$ (α, γ, e^-) (Sol)	**GI (LLI)**		10^{-3}	2×10^{-7}	4×10^{-4}	8×10^{-8}
	Kidney	5×10^{-3}	2×10^{-3}	7×10^{-11}	6×10^{-4}	3×10^{-11}
	Bone	0.06	0.01	6×10^{-10}	5×10^{-3}	2×10^{-10}
	Total body	0.5	0.04	2×10^{-9}	0.01	6×10^{-10}
(Insol)	**Lung**			10^{-10}		5×10^{-11}
	GI (LLI)		10^{-3}	2×10^{-7}	4×10^{-4}	6×10^{-8}

TABLE 8.6. (*cont.*)

Radionuclide and type of decay	Organ of reference[a] (critical organ in **bold** type)	Maximum permissible burden in total body $q(\mu c)$	Maximum permissible concentrations			
			For 40-hr week		For 168-hr week[b]	
			$(MPC)_w$ $\mu Ci/cm^3$	$(MPC)_a$ $\mu Ci/cm^3$	$(MPC)_w$ $\mu Ci/cm^3$	$(MPC)_a$ $\mu Ci/cm^3$
$^{239}_{94}Pu$ (α, γ) (Sol)	**Bone**	0.04	10^{-4}	2×10^{-12}	5×10^{-5}	6×10^{-13}
	Liver	0.4	5×10^{-4}	7×10^{-12}	2×10^{-4}	2×10^{-12}
	Kidney	0.5	7×10^{-4}	9×10^{-12}	2×10^{-4}	3×10^{-12}
	GI (LLI)		8×10^{-4}	2×10^{-7}	3×10^{-4}	6×10^{-8}
	Total body	0.4	10^{-3}	10^{-11}	3×10^{-4}	5×10^{-12}
(Insol)	**Lung**			4×10^{-11}		10^{-11}
	GI (LLI)		8×10^{-4}	2×10^{-7}	3×10^{-4}	5×10^{-8}

[a] The abbreviations GI, S, SI, ULI, and LLI refer to gastrointestinal tract, stomach, small intestines, upper large intestine, and lower large intestine, respectively.

[b] It will be noted that the MPC values for the 168-hr week are not always precisely the same multiples of the MPC for the 40-hr week. Part of this is caused by rounding off the calculated values to one digit, but in some instances it is due to technical differences discussed in the ICRP report. Because of the uncertainties present in much of the biological data and because of individual variations, the differences are not considered significant. The MPC values for the 40-hr week are to be considered as basic for occupational exposure, and the value for the 168-hr week are basic for continuous exposure as in the case of the population at large.

[c] The daughter isotopes of ^{220}Rn and ^{222}Rn are assumed present to the extent they occur in unfiltered air. For all other isotopes the daughter elements are not considered as part of the intake and if present must be considered on the basis of the rules for mixtures.

In order to calculate the MPC of a radioisotope in drinking water, certain physical and biological data must be available. These include:

A. *Physical*
   1. Type of radiation (α, β, γ)
   2. Energy of the radiation
   3. Radiological half-life

B. *Biological*
   1. Identification of the critical organ
   2. Biological clearance rate from the critical organ
   3. Fraction of the ingested activity that is systemically deposited.

For the case of ^{131}I, for example, ICRP Publication 2 tells us that we have

a beta-gamma emitter,
an effective energy = 0.23 MeV/transformation,
a radiological half-life of 8 days,
critical organ is the thyroid, weight = 0.02 kg,
biological half-life = 138 days,
effective half-life = 7.6 days, therefore,
effective clearance constant = 0.693/7.6 = 0.0913 per day,
30% of ingested iodine is deposited in thyroid,
20% of total body iodine is in thyroid,
daily water intake = 2200 ml,
maximum permissible dose to adult thyroid = 0.3 Gy (30 rads), per year, which corre sponds 0.00086 Gy (0.086 rad) per day.

The quantity of a radioisotope that is uniformly distributed throughout an organ that will deliver the maximum recommended dose rate to that organ is calculated from equation (6.41)

$$\dot{D} = \frac{q \text{ Bq} \times 1 \text{ tps/Bq} \times E_e \text{ MeV/t} \times 1.6 \times 10^{-13} \text{ J/MeV} \times 8.64 \times 10^4 \text{ sec/day}}{m \text{ kg} \times 1 \frac{\text{J}}{\text{kg}} \Big/ \text{Gy}} \tag{6.41}$$

by setting $\dot{D}$ to the appropriate value as shown below,

Daily dose equivalent rate	Critical Organ
1.4×10^{-4} Sv (0.014 rem)	Whole body, hemopoietic system, gonads
8.6×10^{-4} Sv (0.086 rem)	Thyroid
4.3×10^{-4} Sv (0.043 rem)	Other organs or tissues

and then solving for q. For the case of ^{131}I, the maximum dose equivalent rate is 8.6×10^{-4} Sv per day, which corresponds to 8.6×10^{-4} Gy per day (since ^{131}I is a beta-gamma emitter, and therefore has a $QF = 1$). Substituting this and the other values into equation (6.41) gives:

$$8.6 \times 10^{-4} \text{ Gy/day} = \frac{q \text{ Bq} \times 1 \text{ tps/Bq} \times 0.23 \text{ MeV/t} \times 1.6 \times 10^{-13} \text{ J/MeV} \times 8.64 \times 10^{-4} \text{ sec/day}}{0.02 \text{ kg} \times 1 \frac{\text{J}}{\text{kg}} \Big/ \text{Gy}},$$

$$q = 5.4 \times 10^3 \text{ Bq } (0.146 \ \mu\text{Ci}).$$

Since only 20% of the total body iodine is in the thyroid, the maximum permissible body burden is

$$\frac{5.41 \times 10^3 \text{ Bq}}{0.2} = 2.7 \times 10^4 \text{ Bq } (0.73 \text{ } \mu\text{Ci}).$$

Since the ICRP recommended maximum values are rounded off to only 1 significant figure, the maximum permissible body burden for ^{131}I is listed in ICRP-2 as 0.7 μCi (3×10^4 Bq).

To calculate the maximum concentration in drinking water that would load the thyroid with a maximum activity of 5.4×10^3 Bq, we recall that in the steady state condition,

$$\text{activity deposited} = \text{activity eliminated}, \tag{8.18A}$$

$$C \text{ Bq/ml} \times 2.2 \times 10^3 \text{ ml/day} \times f = \lambda_E \text{ day}^{-1} \times q \text{ Bq}, \tag{8.18B}$$

where C = MPC of the isotope in water,
f = fraction of intake that is deposited in critical organ,
λ_E = effective elimination constant,
q = steady-state activity in critical organ.

Substituting the appropriate values into equation (8.18B) and solving for C yields

$$C = \frac{9.13 \times 10^{-2} \text{ day}^{-1} \times 5.4 \times 10^3 \text{ Bq}}{2.2 \times 10^3 \text{ ml/day} \times 0.3},$$

$$= 0.75 \text{ Bq/ml } (2 \times 10^{-5} \text{ } \mu\text{Ci/ml})$$

ICRP Publication 2 lists the MPC in water for continuous ingestion as 2×10^{-5} μCi per ml; for 40 hours per week of occupational exposure, ICRP 2 lists the MPC in water for ^{131}I as 6×10^{-5} μCi per ml.

Concentration in Drinking Water Based on Comparison with Radium

We have a great deal of experience with human exposure to radium. For this reason, and because radium is a "bone seeker," that is, it is deposited in the bone, the maximum permissible body burdens of all bone seekers are established by comparing the dose equivalent of the bone seeker with that delivered to the bone by radium. On the basis of data on humans, 0.1 μg radium, corresponding to 3.7 kBq, in equilibrium with its decay products was recommended as the maximum permissible body burden of ^{226}Ra. Using a quality factor of 10 for alpha particles, the calculated dose equivalent to the bone from 0.1 μgm ^{226}Ra and its daughters was 5.6 mSv (0.56 rem) per week. Radium is deposited relatively uniformly in the bone. Other bone seekers, however, are deposited in the bone in a patchy, non-uniform manner that results in doses to some parts of the bone as much as five times greater than the average bone dose. For this reason, the ICRP introduced the *relative damage factor*, *n*, as a multiplier of the *QF*. This factor has a value of 5 for all corpuscular (alpha or beta) radiation, except for those cases where the corpuscular radiations are due to a chain whose first member is radium. When radium is the first member of the chain, then the relative damage factor is 1, since the distribution of the radioisotopes will be determined by the radium. For example, the relative damage factor for

$$^{228}\text{Th} \xrightarrow{\beta} {}^{224}\text{Ra} \xrightarrow{\alpha}$$

is 5 for each particle, while the same particles are weighted with a relative damage factor of 1 in the chain

$$^{228}\text{Ra} \xrightarrow{\beta} {}^{228}\text{Ac} \xrightarrow{\beta} {}^{228}\text{Th} \xrightarrow{\beta} {}^{224}\text{Ra} \xrightarrow{\alpha}.$$

The energy dissipated in the bone by ^{226}Ra and the daughters that remain in the bone is 11 MeV per transformation. Applying the QF value of 10 brings the effective energy to 110 MeV per transformation. Since 99% of the radium body burden is in the skeleton, we have for the maximum permissible body burden of any other bone seeker

$$q = \frac{3.7 \times 10^3 \text{ Bq} \times 0.99}{f_2} \times \frac{110 \text{ MeV/t}}{E \text{ MeV/t}} = \frac{4 \times 10^5}{f_2 E} \text{ Bq}, \tag{8.19}$$

where E is the effective corpuscular energy per transformation of any other bone seeker, and f_2 is the fraction of the total body burden of the bone seeker that is in the skeleton. For the case of ^{90}Sr, for example, we have

$^{90}Sr - {}^{90}Y$ is a pure beta emitter,

Average energy $= 0.194(^{90}Sr) + 0.93(^{90}Y) = 1.12$ MeV/t,

$QF = 1$,

$n = 5$,

$f_2 = 0.99$.

The effective energy is $5 \times 1.12 = 5.6$ MeV per transformation. From equation (8.19) we find the maximum permissible body burden to be

$$q = \frac{4 \times 10^5}{0.99 \times 1.12} = 7.2 \times 10^4 \text{ Bq } (2\ \mu\text{Ci}).$$

The effective half-life for ^{90}Sr in the skeleton is found in ICRP 2 to be 6400 days, which corresponds to an effective clearance rate, $\lambda_E = 1.08 \times 10^{-4}$ per day. Since 9% of the ingested Sr is deposited in the bone, the maximum permissible concentration in drinking water that will maintain the body burden at 7.2×10^4 Bq (2 μCi) is found to be, with the use of equation (8.18B).

$$C \text{ Bq/ml} \times 2.2 \times 10^3 \text{ ml/day} \times 0.09 = 7.2 \times 10^4 \text{ Bq} \times 1.08 \times 10^{-4} \text{ day}^{-1},$$

$$C = 3.9 \times 10^{-2} \text{ Bq/ml } (1 \times 10^{-6}\ \mu\text{Ci/ml}).$$

Ingestion of water at the rate assumed in the calculation above will result in the maximum permissible body burden *when equilibrium is attained.* Because of the very long effective half-life of ^{90}Sr in the bone, the maximum allowable body burden is not attained during the 50-year occupational exposure time assumed for the purpose of computing values for the radiation protection guide. After 50 years of continuous ingestion at the above rate, the amount of ^{90}Sr in the skeleton will be

$$\begin{aligned} q &= q_m(1 - e^{-\lambda_E t}) \\ &= 7.2 \times 10^4(1 - e^{-1.08 \times 10^{-4} \times 50 \times 365}) \\ &= 6.2 \times 10^4 \text{ Bq } (1.7\ \mu\text{Ci}), \end{aligned}$$

or only 86% of the maximum allowable body burden. It is thus clear that the average body burden, and consequently the average dose rate to the skeleton during a 50-year period of maximum permissible ingestion, will be considerably less than the maximum permissible body burden. The mean body burden during a period of ingestion T starting at time zero when there is no radioisotope of the species in question in the body, and assuming the effective elimination rate for the radioisotope to be λ_E, is given by

$$\bar{q} = \frac{1}{T} \int_0^T q_m(1 - e^{-\lambda_E t})\, dt. \tag{8.20}$$

Integrating equation (8.20), we find that

$$\bar{q} = q_m \left[1 + \frac{1}{\lambda_E T} (e^{-\lambda_E T} - 1) \right]. \qquad (8.21)$$

For ^{90}Sr, whose $\lambda_E = 0.0395$ year^{-1}, we have for a 50-year exposure period

$$\bar{q} = 4.18 \times 10^4 \text{ Bq } (1.13\ \mu\text{Ci}).$$

TABLE 8.7. RADIOISOTOPES THAT DO NOT REACH EQUILIBRIUM WITHIN 50 YEARS

Z	Radioisotope	T_E, years	Percent equilibrium reached in 50 years
38	^{90}Sr	18	86
88	^{226}Ra	44	56
89	^{227}Ac	20	83
90	^{230}Th	200	16
90	^{232}Th	200	16
91	^{231}Pa	200	16
93	^{237}Np	200	16
94	^{238}Pu	62	43
94	^{239}Pu	200	16
94	^{240}Pu	190	16
94	^{241}Pu	12	94
94	^{242}Pu	200	16
95	^{241}Am	140	22
95	^{243}Am	200	16
96	^{243}Cm	30	69
96	^{244}Cm	17	87
96	^{245}Cm	200	16
96	^{246}Cm	190	16
98	^{249}Cf	140	22
98	^{250}Cf	10	97

Several other radioisotopes do not attain their equilibrium values in the body during 50 years of continuous ingestion at the maximum recommended concentrations. These radioisotopes are listed in Table 8.7.

Airborne Radioactivity

In considering airborne radioactivity, the lung is considered from two points of view: first, as a portal of entry for inhaled substances that are systemically deposited, and second, as a critical organ that may suffer radiation damage. For purposes of computing permissible atmospheric concentrations of radioactivity, airborne contaminants may be broadly classified as gaseous and particulate. For gaseous radioactive contaminants, the possible hazards to be considered are immersion and inhalation; in the case of particulate matter, the main hazard is due to deposition of the radioactive particulates in the lung.

The calculation for the permissible concentration of a gas may be illustrated for ^{41}A, a biologically inert gas. Argon 41 decays to ^{41}K by the emission of a 1.2-MeV beta particle and a 1.3-MeV gamma ray. The half-life for ^{41}A is 100 min, or 0.076 days. For the case of immersion, it is assumed that a man is exposed in an infinite hemisphere of the gas. In an infinite medium containing a uniformly distributed isotope, the density of emitted energy is equal to the density of absorbed energy. Using a value of 1.1 for the stopping power of tissue relative to air, and an effective energy of $1.3 + 1/3\ (1.2) = 1.7$ MeV, the concentration of radioactivity that leads to an absorbed dose rate equivalent of 0.000143 Sv (0.0143 rem) per day is computed from

$$1.43 \times 10^{-4}\ \text{Sv/day} = \frac{C\ \text{Bq/m}^3 \times 1\ \text{tps/Bq} \times 1.7\ \text{MeV/t} \times 1.6 \times 10^{-13}\ \text{J/MeV} \times 8.64 \times 10^4\ \text{sec/day} \times 1.1}{1.293\ \text{kg/m}^3 \times 1\ \frac{\text{J}}{\text{kg}}\Big/\text{Gy} \times 1\ \text{Gy/Sv}}$$

$$C = 7.15 \times 10^3\ \text{Bq/m}^3 \quad (1.9 \times 10^{-7}\ \mu\text{Ci/cm}^3).$$

The above calculation was made for a point completely surrounded by radioactive gas. Since people stand on the ground and are thus exposed only through 2π steradians, the permissible concentration is twice that calculated above, or 14.3×10^3 Bq/m^3 (3.8×10^{-7} μCi/cm^3).

When a gas is inhaled, it may dissolve in the body fluids and fat after diffusion across the capillary bed in the lung. In the case of an inert gas, absorption into the body stops after the body fluids and fat are saturated with the dissolved gas. The saturation quantity of dissolved ^{41}A in the body fluids due to inhalation of air contaminated at the maximum level for immersion must now be calculated and ascertained that it does not lead to overexposure. As a first step in this calculation, the molar concentration of ^{41}A that corresponds to 14.3 kBq/m^3 (3.8×10^{-7} μCi/cm^3) is determined. The specific activity of ^{41}A is determined from equation (4.30).

$$\frac{3.7 \times 10^{10} \times 1.6 \times 10^3\ \text{yr} \times 226 \times 365\ \text{d/yr}}{0.076\ \text{day}^{-1} \times 41} = 1.57 \times 10^{18}\ \text{Bq/g},$$

$$\frac{14.3 \times 10^3\ \text{Bq/m}^3}{1.57 \times 10^{18}\ \text{Bq/g}} \times \frac{1\ \text{mole}}{41\ \text{g}} = 2.23 \times 10^{-16}\ \text{mole/m}^3.$$

The molar concentration of air is

$$\frac{1\ \text{mole}}{22.41 \times 10^{-3}\ \text{m}^3/\text{l}} = 44.6\ \frac{\text{mole}}{\text{m}^3}.$$

Since argon constitutes 0.94 volume percent of the air, the molar concentration of naturally occurring argon the air is

$$9.4 \times 10^{-3} \times 44.6\ \text{mole/m}^3 = 0.42\ \text{mole A/m}^3\ \text{air}.$$

The ^{41}A corresponding to the maximum permissible immersion concentration is thus seen to be insignificant relative to the argon already in the air, and the molar concentration of argon in the air may be assumed to be unchanged as a result of adding 14.3 kBq per m^3 (3.8×10^{-7} μCi per cm^3) air. With this amount of ^{41}A in the air, the specific activity of the argon in the air is

$$\frac{14.3 \times 10^3\ \text{Bq/m}^3}{0.42\ \text{mole A/m}^3} = 3.4 \times 10^4\ \text{Bq/mole A} \quad (9.1 \times 10^{-7}\ \text{Ci/mole A}).$$

We shall calculate the concentration of argon in the body fluids when the dissolved argon is in equilibrium with the argon in the air. According to Henry's Law, the amount of a gas dissolved in a liquid is proportional to the partial pressure of the gas above the liquid:

$$P_{\text{gas}} = KN = K\frac{n_g}{n_g + n_s}, \tag{8.22}$$

where P_{gas} = partial pressure of the gas,
K = Henry's constant,
N = mole fraction of the dissolved gas,
n_g = molar concentration of the dissolved gas,
n_s = molar concentration of the solvent.

The solubilities of several gases in water at 38°C, expressed in terms of Henry's constant, are given in Table 8.8. At body temperature, K for argon is 3.41×10^7, and the partial pressure of argon in the atmosphere is

$$P_{gas} = 0.0094 \times 760 = 7.15 \text{ mm Hg}.$$

The total body water in a 70-kg person is 43 liters. The molar concentration of water, the solvent, is

$$n_s = \frac{1000 \text{ g/liter}}{18 \text{ g/mole}} = 55.6 \frac{\text{moles}}{\text{liter}}.$$

Equation (8.22) may now be solved for the concentration of dissolved argon.

$$7.15 = 3.41 \times 10^7 \left(\frac{n_g}{n_g + 55.6}\right),$$

$$n_g = 1.17 \times 10^{-5} \text{ mole/liter}.$$

Since the specific activity of the dissolved argon is 3.4×10^4 Bq/mole (9.1×10^{-7} Ci/mole), the argon activity concentration in the body fluid is 0.4 Bq/liter (1.1×10^{-5} µCi/liter), and the total argon activity in the body fluid is 17.2 Bq (4.6×10^{-4} µCi).

TABLE 8.8. SOLUBILITY OF CERTAIN GASES IN WATER AT 38°C

$$K = \frac{\text{partial pressure of gas in millimeters Hg}}{\text{mole fraction of gas in solution}}$$

Gas	K
H_2	5.72×10^7
He	11.0
N_2	7.51
O_2	4.04
A	3.41
Ne	9.76
Kr	2.13
Xe	1.12
Em	0.651
CO_2	0.168
C_2H_2	0.131
C_2H_4	1.21
N_2O	0.242

Argon is more soluble in fat than in water. At equilibrium, the partition coefficient, which is the concentration ratio of argon in fat to argon in water, is 5.4:1 at body temperature. The amount of ^{41}A in the 10 kg of fat in the reference person is

$$5.4 \times 0.4 \text{ Bq/kg} \times 10 \text{ kg} = 21.6 \text{ Bq} \quad (5.8 \times 10^{-4} \text{ µCi}).$$

The total argon activity in the reference person is the sum of that in the body fluid and in the fat, 38.8 Bq (1.1×10^{-3} μCi). If the argon is assumed to be uniformly distributed throughout the body, then the whole body dose rate from the dissolved ^{41}A may be calculated from equation (6.41)

$$\dot{D} = \frac{q \text{ Bq} \times 1 \text{ tps/Bq} \times E_e \text{ MeV/t} \times 1.6 \times 10^{-13} \text{ J/MeV} \times 8.64 \times 10^4 \text{ sec/day}}{m \text{ kg} \times 1 \dfrac{\text{J}}{\text{kg}} \Big/ \text{Gy}}, \quad (6.41)$$

where the effective absorbed energy per transformation, E_e, is the sum of the average beta ray energy, 0.4 MeV and the energy that is absorbed from the 1.3 MeV gammas. The absorbed fraction of the gamma ray energy, which is found by interpolating between the 1.0 and 1.5 MeV gammas in Table 6.6, is 0.310. The effective energy, therefore, is

$$E_e = 0.4 + 0.31 \times 1.3 = 0.8 \text{ MeV/transformation.}$$

Substituting this value for E_e, together with 38.8 Bq for q and 70 kg for m into equation (6.41) yields a dose rate of 6×10^{-9} Gy/day (6×10^{-4} mrad/day).

According to these calculations, the dose rate due to dissolved argon within the body is negligible in comparison to the immersion dose.

Inhaled particulate matter may be either soluble or insoluble. If soluble, it may be absorbed into the body fluids, or it may form an insoluble precipitate in the lung, or it may interact with the tissue protein. "Insoluble" particulates that are deposited in the deep respiratory tract may remain there for long periods of time, until they are cleared either by phagocytosis, by respiratory excursions that propel them to ciliated surfaces in the bronchioles, or by slow solution. Unless specific biological data are available for specific particulates, recommended protection guides for airborne particulates are based on the following assumptions:

Soluble particles

1. 25% exhaled
2. 50% deposited in the upper respiratory tract and swallowed within 24 hr
3. 25% dissolved and absorbed into the body fluids

Insoluble particles

1. 25% exhaled
2. 50% deposited in the upper respiratory tract and swallowed within 24 hr
3. $12\frac{1}{2}$% deposited in the deep respiratory tract, but cleared into the throat and swallowed within 24 hr
4. $12\frac{1}{2}$% deposited in the deep respiratory tract, and retained with a biological retention half-time of 120 days

The application of these assumptions may be illustrated by calculating the maximum permissibble atmospheric concentration of a dust containing ^{35}S. Sulfur 35 is a pure beta emitter whose mean energy is 0.049 MeV, and whose half-life is 87.2 days. According to the assumptions above, 25% of the inhaled dust is immediately exhaled, 62.5% is deposited in the lung and is cleared out of the lung within 24 hours of deposition, and 12.5% is retained in the deep respiratory tract with an effective half retention time of

$$T_E = \frac{T_B \times T_p}{T_B + T_p}$$

$$= \frac{120 \times 87.2}{120 \times 87.2} = 50.5 \text{ days,}$$

corresponding to a clearance rate, λ_E, of 0.0138 per day. The quantity of activity, Q, in the lung (weight = 1 kg) that will deliver an absorbed dose rate of 0.00043 Gy (0.043 rad) per day is calculated from equation (6.41):

$$0.00043 \text{ Gy/day} = \frac{Q \text{ Bq} \times 1 \text{ tps/Bq} \times 0.049 \text{ MeV/t} \times 1.6 \times 10^{-13} \text{ J/MeV} \times 8.64 \times 10^{4} \text{ sec/day}}{1 \text{ kg} \times 1 \frac{\text{J}}{\text{kg}} \Big/ \text{Gy}},$$

$$Q = 6.4 \times 10^{5} \text{ Bq } (17.2 \text{ µCi}).$$

When the lung burden is in equilibrium with an atmospheric concentration of C Bq/m^3, then

$$\text{amount inhaled} \times \text{fraction deposited} = \text{amount eliminated.} \quad (8.23\text{A})$$

Since a reference person inhales 20 m^3 air per day, equation (8.23A) becomes

$$20 \text{ m}^3\text{/day} \times C \text{ Bq/m}^3 \times 0.75 = \lambda_E \text{ day}^{-1} \times q \text{ Bq} + 0.625 \times 20\, C \text{ Bq/day}, \quad (8.23\text{B})$$

where q is the amount of activity in the deep respiratory tract. However,

$$q + 0.625 \times 20C = Q. \quad (8.24)$$

If we substitute $\lambda_E = 0.0138 \text{ day}^{-1}$ and $Q = 6.4 \times 10^5$ Bq into equations (8.24) and (8.23B), we find that

$$C = 3.3 \times 10^3 \text{ Bq/m}^3 \quad (8.9 \times 10^{-8} \text{ µCi/cm}^3).$$

ICRP-2, in its listing of the MPC in air for insoluble ^{35}S, rounds this off to 9×10^{-8} µCi/cm^3).

The radiation dose to the gastrointestinal tract (weight, including its contents, 2165 grams) due to the fraction of the inhaled activity assumed to be brought up from the lung and swallowed is found, from equation (6.41), to be 1.5×10^{-5} Gy (1.5 mrad) per day.

Maximum Permissible Concentrations for Non-Occupational Exposure

The MPC's in air and water that are tabulated in ICRP-2 apply only to occupational exposure. For non-occupational exposure and for the population at large, the ICRP recommended that the MPC's be reduced by a factor of 100 for those isotopes where either the whole body or the hemopoietic tissue or the gonads is the critical organ, and by a factor of 30 for all other isotopes. In the United States, these recommendations have been incorporated into Title 10, Part 20 of the Code of Federal Regulations, "Standards for Protection Against Radiation." It should be re-emphasized here that the recommended maximum permissible doses and maximum permissible environmental concentrations are upper limits only, and that, in all instances, planning for radiation protection should be based on radiation doses that are as low as readily achievable. Furthermore, the tabulated MPC's that are based on exposure by inhalation or ingestion notwithstanding, radiation protection in any particular case must be based on the most sensitive segment of the exposed population and on the environmental pathway that would lead to the greatest dose to the critical population group. For example, in the case of atmospheric ^{131}I in a region where dairy cattle graze, the critical population group is the milk-drinking infant population, and the critical exposure pathway is air to grass to cow to milk to infant. These considerations lead to a reduction of the tabulated MPC of ^{131}I by a factor of 700.

Suggested References

International Commission on Radiological Protection (ICRP)

X-ray and Radium Protection. Recommendations of the 2nd International Congress of Radiology, 1982. Circular No. 374 of the Bureau of Standards, U.S. Government Printing Office (January 23, 1929). *Br. J. Radiology*, **1**, 359 (1928).

Recommendations of the International X-ray and Radium Protection Commission. Alterations to the 1928 Recommendations of the 2nd International Congress of Radiology. 3rd International Congress of Radiology, 1931. *Br. J. Radiology*, **4**, 485 (1931).

International Recommendations for X-ray and Radium Protection. Revised by the International X-ray and Radium Protection Commission and adopted by the 3rd International Congress of Radiology, Paris, July 1931. *Br. J. Radiology*, **5**, 82 (1932).

International Recommendations for X-ray and Radium Protection. Revised by the International X-ray and Radium Protection Commission and adopted by the 4th International Congress of Radiology, Zurich, July 1934. *Radiology*, **23**, 682–5 (1934). *Br. J. Radiology*, **7**, 695 (1934).

International Recommendations for X-ray and Radium Protection. Revised by the International X-ray and Radium Protection Commission and adopted by the 5th International Congress of Radiology, Chicago, September 1937. British Institute of Radiology (1938).

International Recommendations on Radiological Protection. Revised by the International Commission on Radiological Protection at the 6th International Congress of Radiology, London, 1950. *Radiology*, **56**, 431–9 (March 1951). *Br. J. Radiology*, **24**, 46–53 (1951).

Recommendations of the International Commission on Radiological Protection (Revised December 1, 1954). *Br. J. Radiology*, Supplement 6 (1955).

Report on Amendments during 1956 to the Recommendations of the International Commission on Radiological Protection (ICRP). *Radiation Research*, **8**, 539–42 (June 1958). *Acta Radiol.* **48**, 493–5 (December 1957). *Radiology*, **70**, 261–2 (February 1958). *Fortschritte a.d. Gebiete d Rontgenstrahlen u.d. Nuklearmedizin*, **88**, 500–2 (1958).

Report on Decisions at the 1959 Meeting of the International Commission on Radiological Protection (ICRP). *Radiology*, **74**, 116–19 (1960). *Am. J. Roentg.* **83**, 372–5 (1960). *Strahlentherapie*, Band 112, Heft (3 1960). *Acta Radiol.* **53**, fasc. 2 (February 1960). *Br. J. Radiology*, **33**, 189–92 (1960).

Report of the RBE Committee to the International Commissions on Radiological Protection and on Radiological Units and Measurements. *Health Physics*, **9**, no. 4, 357–84 (1963).

Exposure of Man to Ionizing Radiation Arising from Medical Procedures with Special Reference to Radiation-induced Diseases: An enquiry into methods of evaluation. A report of the International Commissions on Radiological Protection and on Radiological Units and Measurements. *Physics in Medicine and Biology*, **6**, no. 2 (1961).

Exposure of Man to Ionizing Radiation arising from Medical Procedures: An enquiry into methods of evaluation. A report of the International Commissions on Radiological Protection and on Radiological Units and Measurements. *Physics in Medicine and Biology*, **2**, no. 2 (1957).

Radiobiological Aspects of the Supersonic Transport: A report prepared by a Task Group of Committee 1. *Health Physics*, **12**, 209–26 (1966).

Deposition and Retention Models for Internal Dosimetry of the Human Respiratory Tract: A report prepared by a Task Group of Committee 2. *Health Physics*, **12**, 173–207 (1966).

A Review of the Physiology of the Gastro-Intestinal Tract in Relation to Radiation Doses from Radioactive Materials: A report prepared by a consultant to Committee 2. *Health Phyics*, **12**, 131–61 (1966).

Calculation of Radiation Dose from Protons and Neutrons to 400 MeV: A report prepared by a Task Group of Committee 3. *Health Physics*, **12**, 227–37 (1966).

The Following Publications of the ICRP, Pergamon Press, Oxford:

ICRP
Report
No.

1. *Recommendations of the International Commission on Radiological Protection* (adopted September 9, 1958) (1959).
2. *Report of Committee II on Permissible Dose for Internal Radiation*, (1960). Also published in *Health Physics* 3, June, 1960.
3. *Report of Committee III on Protection Against X-rays up to Energies of* 3 *MeV and Beta and Gamma-rays from Sealed Sources* (1960).
4. *Report of Committee IV* (1953–1959) *on Protection Against Electromagnetic Radiation above* 3 *MeV and Electrons, Neutrons, and Protons* (adopted 1962, with revisions adopted 1963) (1964).

5. *Report of Committee V on Handling and Disposal of Radioactive Materials in Hospitals and Medical Research Establishments* (1965).
6. *Recommendations of the International Commission on Radiological Protection* (as Amended 1959 and Revised 1962) (1964).
7. *Principles of Environmental Monitoring Related to the Handling of Radioactive Materials* (1966).
8. *The Evaluation of Risks from Radiation* (1966).
9. *Recommendations of the International Commission on Radiological Protection* (adopted September 17, 1965) (1966).
10. *Report of Committee IV on Evaluation of Radiation Doses to Body Tissues from Internal Contamination Due to Occupational Exposure* (1968).
10a. *The Assessment of Internal Contamination Resulting From Recurrent or Prolonged Uptakes* (1971).
11. *A Review of the Radiosensitivity of the Tissues in Bone* (1968).
12. *General Principles of Monitoring for Radiation Protection of Workers* (1969).
13. *Radiation Protection in Schools for Pupils up to the Age of* 18 *years* (1970).
14. *Radiosensitivity and Spatial Distribution of Dose* (1969).
15. *Protection Against Ionizing Radiation from External Sources* (1970).
16. *Protection of the Patient in X-ray Diagnosis* (1970).
17. *Protection of the Patient in Radionuclide Investigations* (1971).
18. *The RBE for High LET-Radiations with Respect to Mutagenesis.*
19. *The Metabolism of Compounds of Plutonium and Other Actinides* (1972).
20. *Alkaline Earth Metabolism in Adult Man* (1973).
21. Data for Protection Against Ionizing Radiation from External Sources: *Supplement to ICRP Publication* 15 (1973).
22. *Implications of Commission Recommendations that Doses be kept as Low as Readily Achievable* (1974).
23. *Report of the Task Group or Reference Manual* (1975).

The following Reports of the ICRP are Published in the Annals of the ICRP, Pergamon Press, Oxford.

24. Radiation Protection in Uranium and Other Mines, Vol. 1, No. 1, 1977.
25. *The Handling Storage, Use and Disposal of Unsealed Radionuclides in Hospitals and Medical Research Establishments*, Vol. 1, No. 2, 1977.
26. *Recommendations of the International Commission on Radiological Protection*, Vol. 1, No. 3, 1977.
27. *Problems Involved in Developing an Index of Harm*, Vol. 1, No. 4, 1977.
28. *Statement from the* 1978 *Stockholm Meeting of the ICRP*
and
The principles and General Procedures for Handling Emergency and Accidental Exposures of Workers, Vol. 2, No. 1, 1978.
29. *Radionuclide Release into the Environment*: *Assessment of Doses to Man*, Vol. 2, No. 2, 1979.
30-1. *Limits for Intakes of Radionuclides by Workers*, Vol. 2, Nos. 3 and 4, 1979.
30-1. Supplement. *Limits for Intakes of Radionuclides by Workers*, Vol. 3, Nos. 1–4, 1979.
30-2. *Statement and Recommendations of the* 1980 *Brighton Meeting of the ICRP*
and
Limits for Intakes of Radionuclides by Workers, Vol. 4, Nos. 3 and 4, 1980.
30-2. Supplement, *Limits for Intakes of Radionuclides by Workers*, Vol. 5, No. 1–6, 1981.
30-3. *Limits for Intakes of Radionuclides by Workers*, Vol. 6, Nos. 2 and 3, 1981.
30-3. Supplement, *Limits for Intakes of Radionuclides by Workers*, Vol. 7, Nos. 1–3, 1982.
30-3. Supplement B. *Limits for Intakes of Radionuclides by Workers*, Vol. 8, Nos. 1–3, 1982.
31. *Biological Effects of Inhaled Radionuclides*, Vol. 4, Nos. 1–2, 1980.
32. *Limits for Inhalation of Radon Daughters by Workers*, Vol. 6, No. 1, 1981.

International Atomic Energy Agency (IAEA), Vienna

Safety
Series
No.

1. *Safe Handling of Radioisotopes* (1962).
2. *Safe Handling of Radioisotopes—Health Physics Addendum* (1960).
3. *Safe Handling of Radioisotopes—Medical Addendum* (1960).
6. *Regulations for the Safe Transport of Radioactive Materials* (1967).
7. *Regulations for the Safe Transport of Radioactive Materials*: *Notes on Certain Aspects of the Regulations* (1961).
9. *Basic Safety Standards for Radiation Protection* (1967).
14. *The Basic Requirments for Personnel Monitoring* (1980).

20. *Guide to the Safe Handling of Radioisotopes in Hydrology* (1966).
21. *Risk Evaluation for Protection of the Public* (1967).
23. *Radiation Protection Standards for Radioluminous Timepieces* (1967).
37. *Advisory Material for Application of IAEA Transport Regulations* (1973).

International Labour Organization (ILO), Geneva

1. *Manual of Industrial Radiation Protection.* Part I: Convention and recommendation concerning the protection of workers against ionizing radiations (1963).
2. *Manual of Industrial Radiation Protection.* Part II: Model code of safety regulations (ionising radiations) (1959).
3. *Manual of Industrial Radiation Protection.* Part III: General guide on protection against ionising radiations (1963).
4. *Manual of Industrial Radiation Protection.* Part IV: Guide on radiation protection in industrial radiography and fluoroscopy.
5. *Manual of Industrial Radiation Protection.* Part V: Guide on radiation protection in the use of luminous compounds.

Recommendations of the International Commission on Radiological Units and Measurements (ICRU), Washington, D.C.

ICRU
Report
No.
25. *Conceptual Basis for the Determination of Dose Equivalent* (1976).
33. *Radiation Quantities and Units* (1980).

Recommendations of the National Council on Radiation Protection and Measurements (NCRP), Washington, D.C.

NCRP
Report
No.
22. *Maximum Permissible Body Burdens and Maximum Permissible Concentrations of Radionuclides in Air and Water* for Occupational Exposure (1963).
39. Basic Radiation Protection Criteria (1971).
43. Review of the Current State of Radiation Protection Philosophy (1975).
53. Review of the NCRP Radiation Dose Limit for Embryo and Fetus in Occupationally Exposed Women (1977).
54. Medical Radiation Exposure of Pregnant and Potentially Pregnant Women (1977).
56. Radiation Exposure from Consumer Products and Miscellaneous Sources (1977).

CHAPTER 9

HEALTH PHYSICS INSTRUMENTATION

Radiation Detectors

Man possesses no biological sensors of ionizing radiation. As a consequence he must depend entirely on instrumentation for the detection and measurement of radiation. Instruments used in the practice of health physics serve a wide variety of purposes. It is logical, therefore, to find a wide variety of instrument types. We have, for example, instruments such as the Geiger counter that measure particles, film badges and pocket dosimeters that measure accumulated radiation doses, and ionization chamber type instruments that measure dose rate. In each of these categories, one finds instruments that are designed principally for measurement of a certain type of radiation—such as low-energy X-rays, gamma rays, fast neutrons, etc.

Although there are many different instrument types, the operating principles for most radiation-measuring instruments are relatively few. The basic requirement of any radiation-measuring instrument is that the instrument's detector interact with the radiation in such a manner that the magnitude of the instrument's response is proportional to the radiation effect or radiation property under measurement. Some of the physical and chemical radiation effects that are applied to radiation detection and measurement for health physics purposes are listed in Table 9.1.

Particle-Counting Instruments

Particle-counting instruments are frequently used by health physicists to determine the radioactivity of a sample taken from the environment, such as an air sample, or to measure the activity of a biological fluid from someone suspected of being internally contaminated. Another important application of particle-counting instruments is in portable radiation-survey instruments. Particle-counting instruments may be very sensitive—they literally respond to a single ionizing particle. They are, accordingly, widely used for searching for unknown radiation sources, for leaks in shielding, and for areas of contamination. The detector in particle-counting instruments may be either a gas or a solid. In either case, passage of an ionizing particle through the detector results in energy dissipation through a burst of ionization. This burst of ionization is converted into an electrical pulse that actuates a readout device, such as a scaler or a ratemeter, to register a count.

Gas-Filled Particle Counters

Consider a gas detector system such as is shown in Fig. 9.1. This system consists of a variable voltage source, V, a high valued resistor, R, and a gas-filled counting chamber, D, which has two coaxial electrodes that are very well insulated from each other. All the capacitance associated with the circuit is indicated by the capacitor, C. Because of the

TABLE 9.1. RADIATION EFFECTS USED IN THE DETECTION AND MEASUREMENT OF RADIATION

Effect	Type of Instrument	Detector
Electrical	1. Ionization chamber 2. Proportional counter 3. Geiger counter 4. Solid state	1. Gas 2. Gas 3. Gas 4. Semiconductor
Chemical	1. Film 2. Chemical dosimeter	1. Photographic emulsion 2. Solid or liquid
Light	1. Scintillation counter 2. Cerenkov counter	1. Crystal or liquid 2. Crystal or liquid
Thermo-luminescence	Thermoluminescent dosimeter	Crystal
Heat	Calorimeter	Solid or liquid

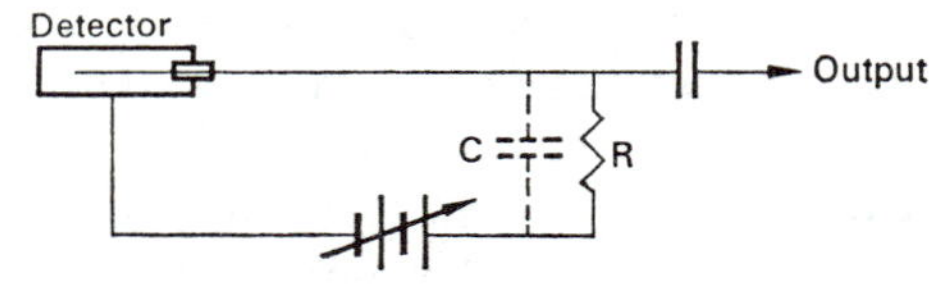

FIG. 9.1. Basic circuit for a gas-filled detector.

production of ions within the detector when it is exposed to radiation, the gas within the detector becomes electrically conducting.

If the time constant RC of the detector circuit is much greater than the time required for the collection of all the ions resulting from the passage of a single particle through the detector, then a voltage pulse of magnitude

$$V = \frac{Q}{C}, \tag{9.1}$$

where Q is the total charge collected and C is the capacitance of the circuit, and of the shape shown by the top curve in Fig. 9.2, appears across the output of the detector circuit.

A broad output pulse would make it difficult to separate successive pulses. However, if the time constant of the detector circuit is made much smaller than the time required to collect all the ions, then the height of the developed voltage pulse is smaller, but the pulse is very much narrower, as shown by the curves in Fig. 9.2. This pulse "clipping," as it is called, allows individual pulses to be separated and counted.

Ionization chamber counter

If a constant flux of radiation is permitted to pass through the detector, and if the voltage, V, is varied, several well-defined regions of importance in radiation measurement may be identified. As the voltage is increased from zero through relatively low voltages, the first region, known as the *ionization chamber region*, is encountered. If the instrument has the electrical polarity shown in Fig. 9.1, then all the positive ions will be collected by the outer cathode, while the negative ions, or electrons, will be collected by the central anode. By "low voltages," in this case, is meant the range of voltage great enough to collect

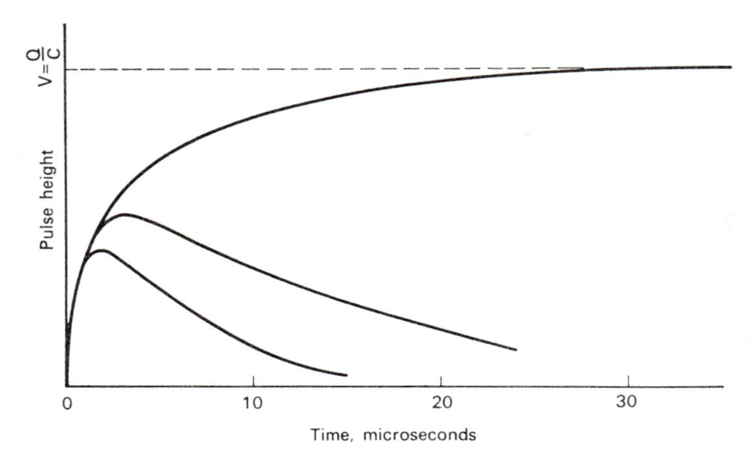

FIG. 9.2. Dependence of pulse shape on the time constant of the detector circuit. The top curve is for the case where $RC = \infty$, the center curve is for the case where RC is less than the ion collection time, while the lowest curve is for the case where RC is much less than the ion collection time.

the ions before a significant fraction of them can recombine, yet not great enough to accelerate the ions sufficiently to produce secondary ionization by collision. The exact value of this voltage is a function of the type of gas, the gas pressure, and the size and geometric arrangement of the electrodes. In this region, the number of electrons collected by the anode will be equal to the number produced by the primary ionizing particle. The pulse size, accordingly, will be independent of the voltage, and will depend only on the number of ions produced by the primary ionizing particle during its passage through the detector. The ionization chamber region may be defined as the range of operating voltages in which there is no multiplication of ions due to secondary ionization; that is, the gas amplification factor is equal to one.

The fact that the pulse size from a counter operating in the ionization chamber region depends on the number of ions produced in the chamber makes it possible to use this instrument to distinguish between radiations of different specific ionization, such as alphas and betas or gammas. For example, an alpha particle that traverses the chamber produces about 10^5 ion pairs, which corresponds to 1.6×10^{-14} C. If the chamber capacitance is $10\ \mu\mu\text{F}$, and if all the charge were collected, then the voltage pulse resulting from the passage of this alpha will be

$$V = \frac{Q}{C} = \frac{1.6 \times 10^{-14}}{10 \times 10^{-12}} = 1.6 \times 10^{-3}\ \text{V}.$$

A beta particle, on the other hand, may produce about 1000 ion pairs within the chamber. The resulting output pulse due to the beta particle will be only 1.6×10^{-5} V. Amplification of these two pulses by a factor of 100 leads to pulses of 0.16 V for the alpha and 0.0016 V for the beta. With the use of a discriminator in the scaler (or other readout device), voltage pulses less than a certain predetermined size can be rejected; only those pulses that exceed this size will be counted. In the case of the example given above, a discriminator setting of 0.1 V would allow the pulses due to the alphas to be counted, but would not pass any of the pulses due to the beta rays. This discriminator setting is often referred to as the input sensitivity of the scaler. Increasing the input sensitivity, in the example above, would allow both alphas and betas to be counted. This ability to distinguish between the two radiations is illustrated in Fig. 9.3, which shows the output pulse height as a function of voltage across the counting chamber.

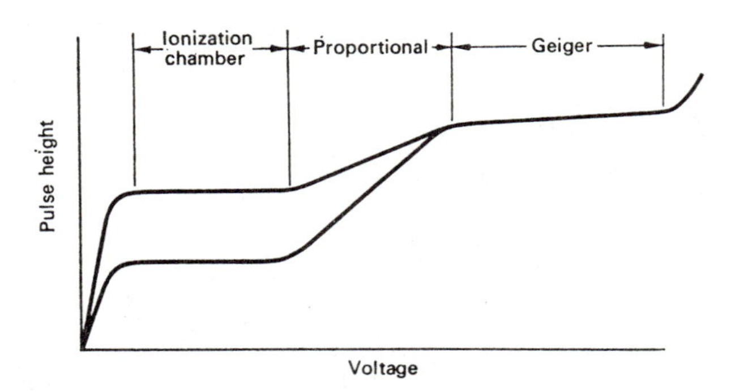

FIG. 9.3. Curve of pulse height versus voltage across a gas-filled pulse counter, illustrating the ionization chamber, proportional, and Geiger regions.

Proportional counter

One of the main disadvantages of operating a counter in the ionization chamber region is the relatively feeble output pulse, which requires either much amplification or a high degree of input sensitivity in the scaler. To overcome this difficulty, and yet to take advantage of pulse size dependence on ionization for the purpose of distinguishing between radiations, the counter may be operated as a proportional counter. As the voltage across the counter is increased beyond the ionization chamber region, a point is reached where secondary electrons are produced by collision. This is the beginning of the proportional region. The voltage drop across resistor R will now be greater than it was in the ionization chamber region because of these additional electrons. The gas amplification factor is greater than 1. This multiplication of ions in the gas, which is called an avalanche, is restricted to the vicinity of the primary ionization. Increasing the voltage causes the avalanche to increase in size by spreading out along the anode. Since the size of the output pulse is determined by the number of electrons collected by the anode, the size of the output voltage pulse from a given detector is proportional to the high voltage across the detector. Besides the high voltage across the tube, the gas amplification depends on the diameter of the collecting electrode (the electric field intensity near the surface of the anode, which is given by equation (2.40), increases as the diameter of the collecting anode decreases) and on the gas pressure. Decreasing gas pressure leads to increasing gas multiplication, as shown in Fig. 9.4. Because of the dependence of gas multiplication, and consequently the size of the output pulse, on the high voltage it is important to use a very stable high-voltage power supply with a proportional counter.

An example of the use of a proportional counter to distinguish between alpha and beta radiation is shown below in Fig. 9.5. At point A, the "threshold" voltage, the pulses produced by the alpha particles that traverse the counter are just great enough to get by the discriminator. A small increase in voltage causes a sharp increase in counting rate because all the output pulses due to alphas now exceed the input sensitivity of the scaler. Further increase in high voltage has little effect on the counting rate, and results in a "plateau," a span of high voltage over which the counting rate is approximately independent of voltage. With the system operating on this alpha plateau, the pulses due

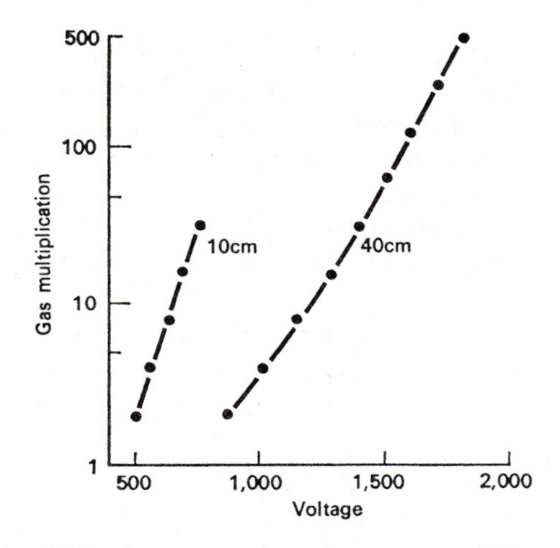

FIG. 9.4. Gas multiplication versus voltage for pressures of 10 and 40 cm Hg tank argon; anode diameter = 0.01 in, cathode diameter = 0.87 in. (From B. B. Rossi and H. H. Staub: *Ionization Chambers and Counters*, McGraw Hill, New York, 1949.)

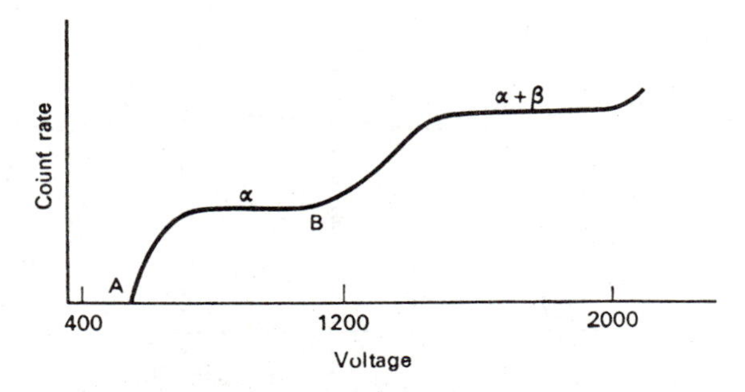

FIG. 9.5. Alpha and alpha–beta counting rates as a function of voltage in a proportional counter.

to beta rays are still too small to get by the discriminator. However, a point, *B*, is reached, as the high voltage is increased, where the gas amplification is great enough to produce output pulses from beta particles that exceed the input sensitivity of the scaler. This leads to another plateau where both alpha and beta particles are counted. By subtracting the alpha count rate from the alpha–beta count rate, the beta-ray activity may be obtained.

Geiger counter

Continuing to increase the high voltage beyond the proportional region will eventually cause the avalanche to extend along the entire length of the anode. When this happens, the end of the proportional region is reached and the Geiger region begins. At this point, the size of all pulses—regardless of the nature of the primary ionizing particle—is the same. When operated in the Geiger region, therefore, a counter cannot distinguish among the several types of radiations. However, the very large output pulses (greater than $\frac{1}{4}$ V) that result from the high gas amplification in a Geiger counter means either the complete elimination of a pulse amplifier or use of an amplifier that does not have to meet the exacting requirements of high pulse amplification.

Figure 9.5 shows the alpha and alpha–beta plateaus of a proportional counter. A Geiger counter too has a wide range of operating voltages over which the counting rate is approximately independent of the operating voltage. This plateau extends approximately from the voltage which results in pulses great enough to be passed by the discriminator to that which causes a rapid increase in counting rate that preceeds an electrical breakdown of the counting gas. In the Geiger region, the avalanche is already extended as far as possible axially along the anode. Increasing the voltage, therefore, causes the avalanche to spread radially, resulting in an increasing counting rate. We therefore have a slight positive slope in the plateau, as shown in Fig. 9.6. Figures of merit for judging the quality of a counter are the length of the plateau, the slope of the plateau, and the resolving time (discussed below). The slope is usually given as percentage increase in counting rate per 100 V:

$$\text{slope} = \frac{(C_2 - C_1)/C_1}{0.01(V_2 - V_1)} \times 100. \tag{9.2}$$

A Geiger counter has a slope of about 3% per hundred volts. The operating voltage for a Geiger counter is about one-third to one-half the distance from the knee of the curve of count rate versus voltage.

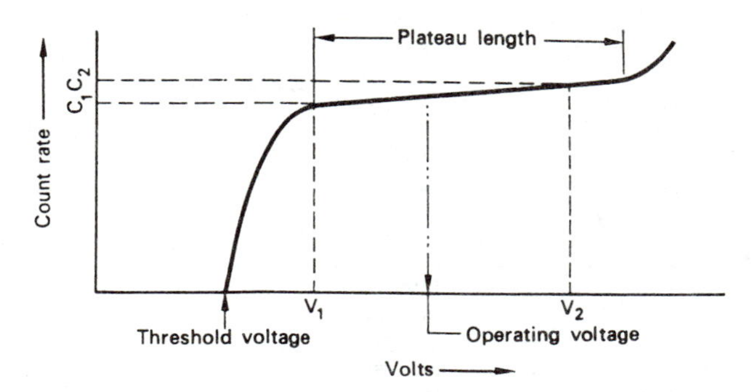

FIG. 9.6. Operating characteristics of a Geiger counter.

Quenching a Geiger counter

When the positive ions are collected after a pulse, they give up their kinetic energy by striking the wall of the tube. Most of this kinetic energy is dissipated as heat. Some of it, however, excites the atoms in the wall. In falling back to the ground state, these atoms may lose their excitation energy by emitting U.V. photons. Since at this time the electric field around the anode is re-established to its full intensity, the interaction of U.V. photon with the gas in the counter may initiate an avalanche, and thereby produce a spurious count. Prevention of such spurious counts is called quenching. Quenching may be accomplished either electronically, by lowering the anode voltage after a pulse until all the positive ions have been collected; or chemically, by using a self-quenching gas. A self-quenching gas is one that can absorb the U.V. photons without becoming ionized. One method of doing this is to introduce a small amount of an organic vapor, such as alcohol or ether, into the tube. The energy from the U.V. photon is then dissipated by dissociating the organic molecule. Such a tube is useful only as long as it has a sufficient number of organic molecules for the quenching action. In practice, an organic vapor Geiger counter has a useful life of about 10^8 counts. Self-quenching also results when the counting gas contains a trace of a halogen. In this case, the halogen molecule does not dissociate after absorbing the energy from the U.V. photon. The useful life of a halogen-quenched counter, therefore, is not limited by the number of pulses that had been produced in it.

Resolving Time

If two particles enter the counter in rapid succession, the avalanche of ions from the first particle paralyzes the counter, and renders it incapable of responding to the second particle. Because the electric-field intensity is greatest near the surface of the anode, the avalanche of ionization starts very close to the anode, and spreads longitudinally along the anode. The negative ions thus formed migrate towards the anode, while the positive ions move towards the cathode. The negative ions, being electrons, move very rapidly, and are soon collected, while the massive positive ions are relatively slow moving, and therefore travel for a relatively long period of time before being collected. The collection time for positive ions formed near the surface of the anode is given by the equation

$$t = \frac{(b^2 - a^2)p \ln b/a}{2V\mu} \text{ sec,} \tag{9.3}$$

where b = radius of cathode, cm,
a = radius of anode, cm,
p = gas pressure in counter, mm Hg,
V = potential difference across counter, volts,
μ = mobility of positive ions (cm/sec)/(V/cm); for air, μ has a value of 1070 and for argon its value is 1040.

Example 9.1

How long will it take to collect all the positive ions in a Geiger counter filled with argon at a pressure of 100 mm Hg, if the operating voltage is 1000 V, and if the cathode and anode have radii of 1 cm and 0.01 cm, respectively?

By substituting the appropriate numbers into equation (9.3), we have

$$t = \frac{(1 - 0.01^2) \times 100 \ln 1/0.01}{2 \times 1000 \times 1040}$$

$$= 221 \times 10^{-6} \text{ sec.}$$

These slow-moving positive ions form a sheath around the positively charged anode, thereby greatly decreasing the electric field intensity around the anode and making it impossible to initiate an avalanche by another ionizing particle. As the positive ion sheath moves towards the cathode, the electric field intensity increases, until a point is reached when another avalanche could be started. The time required to attain this electric field intensity is called the *dead time*. After the end of the dead time, however, when another avalanche can be started, the output pulse from this avalanche is still relatively small, since the electric field intensity is still not great enough to produce a Geiger pulse. As the positive ions continue their outward movement, an output pulse resulting from another ionizing particle would increase in size. When the output pulse is large enough to be passed by the discriminator and be counted, the counter is said to have recovered from the previous ionization, and the time interval between the dead time and the time of full recovery is called the *recovery time*. The sum of the dead time and the recovery time is called the *resolving time*. Alternatively, the resolving time may be defined as the minimum time that must elapse after the detection of an ionizing particle before a second particle can be detected. The relationship between the dead time, recovery time, and resolving time is illustrated in Fig. 9.7. The resolving time of a Geiger counter is on the order of 100 μsec or more. A proportional counter is much faster than a Geiger counter. Since the avalanche in a proportional counter is limited to a short length of the anode, a second avalanche can be started elsewhere along the anode while the region of the first avalanche is completely paralyzed. The resolving time of a proportional counter, therefore, is on the order of several microseconds.

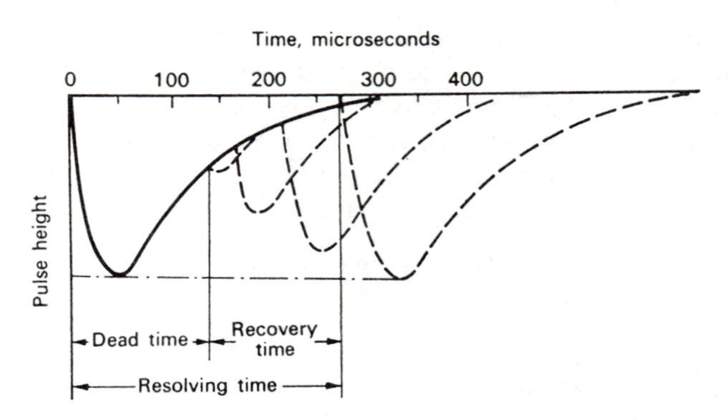

FIG. 9.7. Relationship among dead time, recovery time, and resolving time.

Measurement of resolving time

The resolving time of a counter may be conveniently measured by the two-source method. Two radioactive sources are counted—singly and together. If there were no resolving time loss, the counting rate of the two sources together would be equal to the sum of the two single source counting rates. However, because of the counting losses due to the resolving time of the counting system, the sum of the two single counting rates exceeds that of the two sources together. If R_1 is the counting rate of source 1, R_2 of source 2, $R_{1,2}$ of the two sources together, and R_b the background counting rate, then the resolving time is given by

$$\tau = \frac{R_1 + R_2 - R_{1,2} - R_b}{R_{1,2}^2 - R_1^2 - R_2^2}. \tag{9.4}$$

All the source counting rates above include the background. Because $R_1 + R_2$ is only slightly greater than $R_{1,2}$, all the measurements must be made with a great degree of accuracy when determining the resolving time. For the case where the resolving time is τ and the observed counting rate of a sample is R_0, the counting rate that would have been observed had there been no resolving time loss, that is, the "true" counting rate, R, is

$$R = \frac{R_0}{1 - R_0\tau}. \tag{9.5}$$

Scintillation Counters

A scintillation detector is a transducer that changes the kinetic energy of an ionizing particle into a flash of light. Historically, one of the earliest means of measuring radiation was by scintillation counting. Rutherford, in his classical experiments on scattering of alpha particles, used a zinc sulfide crystal as the primary detector of radiation; he used his eye to see the flickers of light that appeared when alpha particles struck the zinc sulfide. Today, the light is viewed electronically by photomultiplier tubes whose output pulses may be amplified, sorted by size, and counted. The various radiations may be detected with scintillation counters by using the appropriate scintillating material. Table 9.2 lists some of the substances used for this purpose.

TABLE 9.2. SCINTILLATING MATERIALS

Phosphor	Density (g/cm^3)	Wavelength of maximum emission, Å	Relative pulse height	Decay time (μsec)
NaI (Tl)	3.67	4100	210	0.25
CsI (Tl)	4.51	Blue	55	1.1
KI (Tl)	3.13	4100	50	1.0
Anthracene	1.25	4400	100	0.032
Trans-Stilbene	1.16	4100	60	0.0064
plastic		3550–4500	28–48	0.003–0.005
liquid		3550–4500	27–49	0.002–0.008
p-Terphenyl	1.23	4000	40	0.005

From R. Swank, Characteristics of scintillators, *Annual Review of Nuclear Science*, vol. V, 1954.

Scintillation counters are widely used to count gamma rays and low-energy beta rays. The counting efficiency of Geiger or proportional counters for low-energy betas may be very low due to the dissipation of the beta energy within the sample. (This phenomenon is called self-absorption.) This disadvantage can be overcome by dissolving the radioactive sample in a scintillating liquid, such as toluene. Such liquid scintillation counters result in detection efficiencies that approach 100%. They are widely used in research applications, especially in the field of biochemistry, where they are used to measure ^{14}C and ^{3}H. However, liquid scintillation counters find relatively little direct application in operational health physics.

Whereas the inherent detection efficiency of gas-filled counters is close to 100% for those alphas or betas that enter the counter, their detection efficiency for gamma rays is very low—usually less than 1%. Solid scintillating crystals, on the other hand, have high detection efficiencies for gamma rays. Furthermore, since the intensity of the flicker of light in the detector is proportional to the energy of the gamma ray that produces the light, a scintillation detector can, with the aid of the appropriate electronics, be used as a gamma-ray spectrometer. (With a suitable detector, a scintillation counter may also be used as a beta-ray or an alpha-ray spectrometer.)

For gamma-ray measurement, the detector used most frequently is a sodium iodide crystal activated with thallium [NaI(Tl)], that is optically coupled to a photomultiplier tube. The thallium activator, which is present as an "impurity" in the crystal structure to the extent of about 0.2%, converts the energy absorbed in the crystal to light. The high density of the crystal, together with its high effective atomic number, results in a high detection efficiency, Fig. 9.8. Gamma-ray photons, passing through the crystal, interact with the atoms of the crystal by the usual mechanisms of photoelectric absorption,

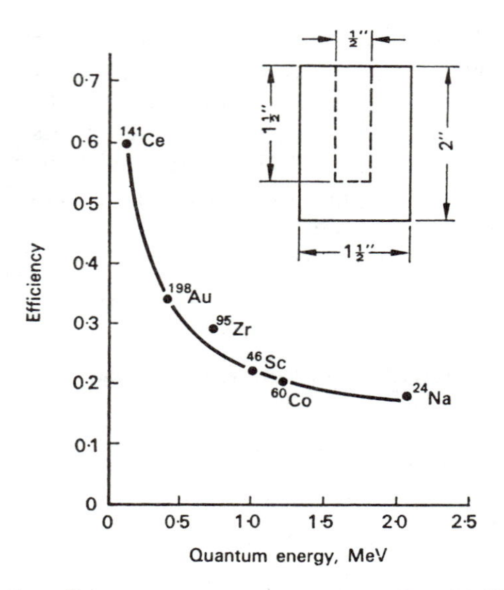

FIG. 9.8. Detection efficiency versus gamma-ray energy for a NaI(Tl) well crystal. (From C. J. Borkowski: O.R.N.L. Progress Report 1160, 1951.)

Compton scattering, and pair production. The primary ionizing particles resulting from the gamma-ray interactions—the photoelectrons, Compton electrons, and positron–electron pairs—dissipate their kinetic energy by exciting and ionizing the atoms in the crystal. The excited atoms return to the ground state by the emission of quanta of light. These light pulses, upon striking the photosensitive cathode of the photomultiplier tube, cause electrons to be ejected from the cathode. These electrons are accelerated to a second electrode, called a dynode, whose potential is about 100 V positive with respect to the photocathode. Each electron that strikes the dynode causes several other electrons to be ejected from the dynode, thereby "multiplying" the original photocurrent. This process is repeated about 10 times before all the electrons thus produced are collected by the plate of the photomultiplier tube. This current pulse, whose magnitude is proportional to the energy of the primary ionizing particle, can then be amplified and counted. Figure 9.9 illustrates schematically the sequence of events in the detection of a photon by a scintillation chamber.

A photoelectric interaction within the crystal produces essentially mono-energetic photoelectrons, which in turn produce light pulses of about the same intensity. These light pulses, being of equal intensity, lead to current output pulses of approximately the same magnitude. In Compton scattering, on the other hand, a continuous spectrum of energy results from the Compton electron—the most energetic electron being that which results from a 180° backscatter of the incident photon. This most energetic Compton electron is called the "Compton edge" in scintillation spectrometry. The scattered photon may pass out of the crystal, or it may interact again, either by photoelectric absorption (which would be the most likely interaction) or by another Compton scattering.

In pair production, a flicker of light representing the original quantum energy minus 1.02 MeV is produced as the positron and negatron simultaneously dissipate their energies in the crystal. After losing its energy, the positron combines with an electron, thus annihilating the two particles and producing two photons of 0.51 MeV. Depending on the time sequence, on the crystal size, and on the geometric location of the initial interaction,

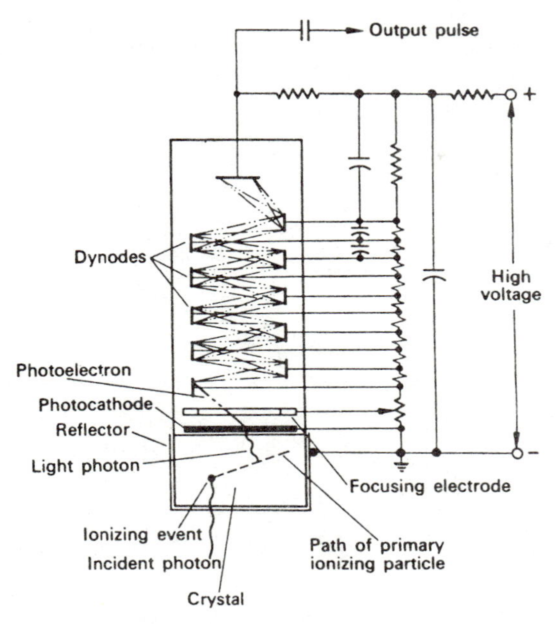

FIG. 9.9. Schematic representation of the sequence of events in the detection of a gamma-ray photon by a scintillation counter. An average of about four electrons are knocked out of a dynode by an incident electron.

we may have two pulses representing 0.51 MeV each, one light pulse representing 1.02 MeV, or one light pulse representing the total energy of the original photon.

Nuclear Spectroscopy

Nuclear spectroscopy is the analysis of radiation sources or radioisotopes by measuring the energy distribution of the source. A spectrometer is an instrument that separates the output pulses from a detector, usually a scintillation detector or a semiconductor detector, according to size. Since the size distribution is proportional to the energy of the detected radiation, the output of the spectrometer provides detailed information that is useful in identifying unknown radioisotopes and in counting one isotope in the presence of others. This technique has found widespread application in X-ray and gamma-ray analysis using NaI(Tl) scintillation detectors and semiconductor detectors, in beta ray analysis using liquid scintillation detectors, and in alpha analysis using semiconductor detectors. Nuclear spectrometers are available in two types, either a single channel instrument or a multi-channel analyzer (MCA). The essentials of a single channel spectrometer consist of the detector, a linear amplifier, a pulse height selector, and a readout device, such as a scaler or a ratemeter (Fig. 9.10). The pulse height selector is an electronic "slit," which may be adjusted to pass pulses whose amplitude lies between any two desired limits of maximum and minimum. The output from the pulse height analyzer is a logic pulse to a scaler or to a count rate meter. The main use of the single channel analyzer is to discriminate between a desired radiation and other radiations that may be considered noise. Thus, the single channel spectrometer is used to measure one radioisotope in the presence of another, or to optimize the signal to noise ratio when measuring a low activity source in the presence of a significant background.

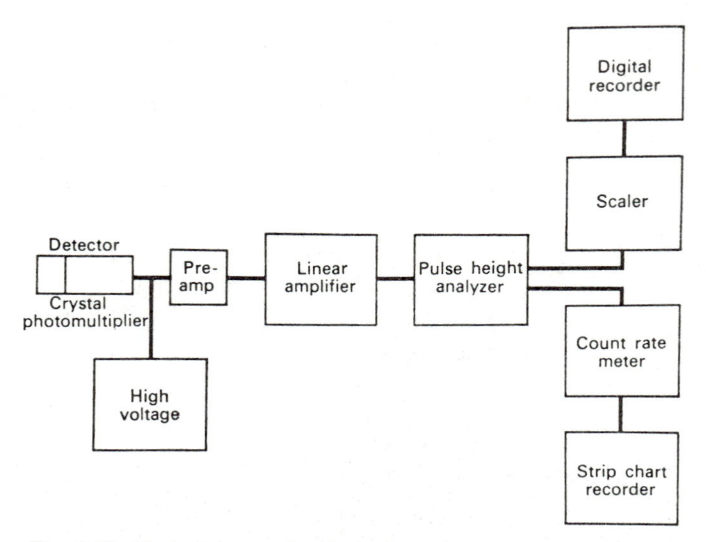

FIG. 9.10. Block diagram of a single-channel gamma-ray spectrometer.

A multi-channel analyzer (Fig. 9.11) has an analog-to-digital converter (ADC) instead of a pulse height selector to sort all the output pulses from the detector according to height. The MCA also has a computer-type memory for storing the information from the ADC or from another source. This feature allows automated data processing operations such as background subtraction and spectrum stripping. Spectrum stripping is a technique for analysis of compound spectra that is based on sequential subtraction of known gamma ray spectra of individual isotopes from the compound spectrum recorded when the sample undergoing nuclear analysis contains several different gamma emitters.

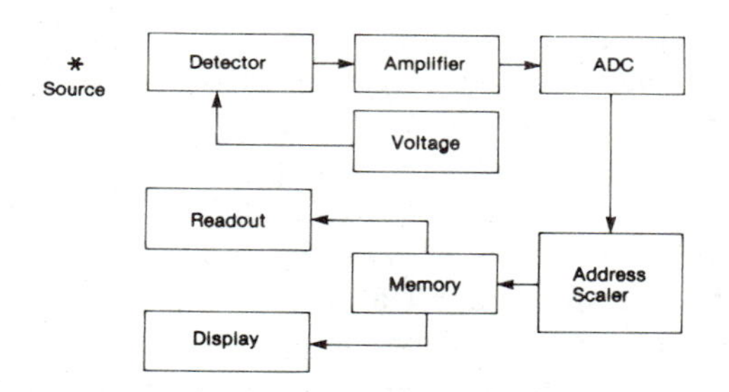

FIG. 9.11. Block diagram of a multi-channel analyzer.

Analog-to-digital conversion is accomplished by charging a capacitor to the peak voltage of the pulse to be analyzed, and then discharging it. While the capacitor is being discharged, "clock" pulses from a high frequency oscillator are counted by a scaler. The number of clock pulses counted during the capacitor discharge is proportional to the time required for the capacitor to discharge, and thus is proportional to the height of the output pulse from the detector. The output from the ADC is a logic pulse that is stored in a channel whose number, or "address" is determined by the number of clock pulses that were counted during the discharge of the capacitor. Most multi-channel analyzers are built with the number of channels varying by a factor of 2 over the range of 128 to 4096, each with a storage capacity of 10^5 to 10^6 counts per channel, and "clock" frequencies ranging from 4 to 100 MHz. The multi-channel spectrometer can print out the counts in each channel as well as visually displaying the gamma-ray spectrum on a cathode ray tube. A typical simple gamma-ray spectrum is shown in Fig. 9.12.

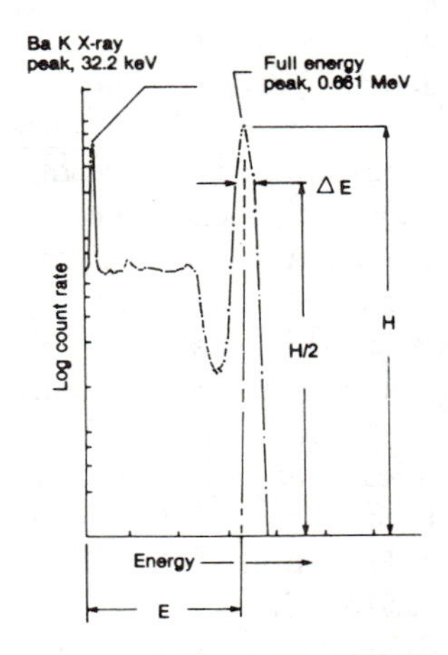

FIG. 9.12. Gamma-ray spectrum of ^{137}Cs. The 0.661-MeV full energy peak is actually due to the isomeric transition of the 2.6-minute ^{137m}Ba daughter of ^{137}Cs. The 32.2-keV barium characteristic X-ray is due to the internal conversion of 8.5% of the 0.661-MeV photons. (From NCRP Report no. 58.)

The basis for nuclear spectroscopy is the location of spectral lines arising from the total absorption of charged particles or photons. For this purpose, the resolution of the detector is important if spectral lines that are close together are to be separated and observed. Resolution is defined as the ratio of the full width at half-maximum (FWHM) of the full energy peak (often called the "photopeak" when dealing with photon detectors) to the energy midpoint of the full energy peak (Fig. 9.12). If the full width at half-maximum is ΔE,

$$\text{percentage resolution} = \frac{\Delta E}{E} \times 100. \tag{9.6}$$

The smaller the energy spread, ΔE, the better is the ability of a detector to separate full energy peaks that are close together. The resolution of a given detector is a function of energy, and improves with increasing energy. For example, for a 100-keV photon, the resolution of a NaI(Tl) detector may be about 14%, whereas for a 1-MeV photon, it may be as good as about 7%.

Cerenkov Detector

Cerenkov radiation is visible light that results from the passage of a charged particle through a medium at a velocity greater than the phase velocity of light in that medium. The emission of Cerenkov radiation is favoured by a high index of refraction of the medium. Accordingly, Cerenkov detectors are made of high-density glass whose index of refraction for the sodium D lines is on the order of 1.6–1.7. Liquids of high refractive index also are used. Cerenkov radiation, after being produced in the detector, is viewed by a photomultiplier tube for measurement. The Cerenkov detector is not usually used in operational health physics. Because of the dependence of Cerenkov radiation on the velocity of charged particles in the detector, the Cerenkov phenomenon is used mainly in high-energy physics research to measure the velocity of very energetic particles.

Semiconductor Detector

A semiconductor detector acts as a solid state ionization chamber. The ionizing particle—beta ray, alpha particle, etc.—interacts with atoms in the sensitive volume of the detector to produce electrons by ionization. The collection of these ions leads to an output pulse. In contrast to the relatively high mean ionization energy of 30–35 eV for most counting gases, a mean energy expenditure of only 3.5 eV is required to produce an ionizing event in a semiconductor detector (silicon).

A semiconductor is a substance that has electrical conducting properties midway between a "good conductor" and an "insulator." Although many substances can be classified as semiconductors, the most commonly used semiconductor materials are silicon and germanium. These elements, each of which has four valence electrons, form crystals that consist of a lattice of atoms that are joined together by covalent bonds. Absorption of energy by the crystal leads to disruption of these bonds—only 1.12 eV are required to knock out one of the valence electrons in silicon—which results in a free electron and a "hole" in the position formerly occupied by the valence electron. This free electron can move about the crystal with ease. The hole too can move about in the crystal; an electron adjacent to the hole can jump into the hole, and thus leave another hole behind. Connecting the semiconductor in a closed electric circuit results in a current through the semiconductor as the electrons flow towards the positive terminal and the holes flow towards the negative terminal.

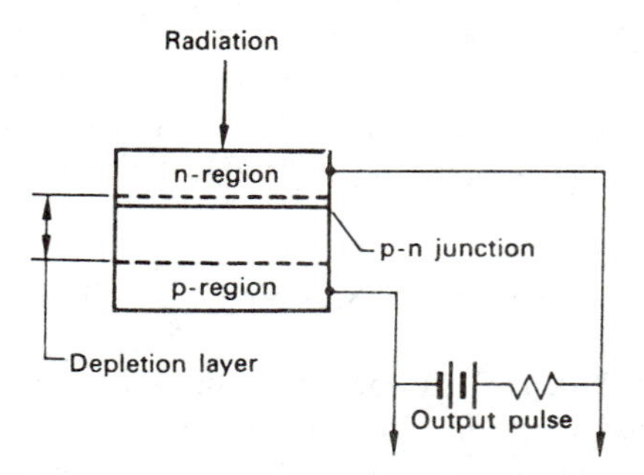

FIG. 9.13. Semiconductor junction detector. This detector is most useful for measuring electrons or other charged particles. For measuring gamma rays, an electron radiator is interposed between the radiation and the detector, and the photo and Compton electrons thus produced are measured by the detector.

The operation of a semiconductor radiation detector depends on its having either an excess of electrons or an excess of holes. A semiconductor with an excess of electrons is called an *n*-type semiconductor, while one with an excess of holes is called a *p*-type semiconductor. Normally a pure silicon crystal will have an equal number of electrons and holes. (These electrons and holes result from the rupture of the covalent bonds by absorption of heat or light energy.) By adding certain impurities to the crystal, either an excess number of electrons (an n region) or an excess number of holes (a p region) can be produced. Germanium and silicon both are in group IV of the periodic table. If atoms from one of the elements in group V, such as phosphorous, arsenic, antimony, or bismuth, each of which has five valence electrons, are added to the pure silicon or germanium, four of the five electrons in each of the added atoms are shared by the silicon or germanium atoms to form a covalent bond. The fifth electron from the impurity is thus an excess electron, and is free to move about in the crystal and to participate in the flow of electric current. Under these conditions, the crystal is of the n type. A p-type semiconductor, having an excess number of holes, can be made by adding an impurity from group III of the periodic table to the semiconductor crystal. Elements from group III, such as boron, aluminum, gadolinium, or indium, have three valence electrons. Incorporation of one of these elements as an impurity in the crystal, therefore, ties up only three of the four valence bonds in the crystal lattice. This deficiency of one electron is a hole, and we have p-type silicon or germanium.

A p region in silicon or germanium that is adjacent to an n region is called an n–p junction. If a forward bias is applied to the junction, that is, if a voltage is applied across the junction such that the p region is connected to the positive terminal and the n region to the negative terminal, the impedance across the junction will be very low, and current will flow across the junction. If the polarity of the applied voltage is reversed, that is, if the n region is connected to the positive terminal and the p region to the negative, we have the condition known as reverse bias. Under this condition, no current (except for a very small current due to thermally generated holes and electrons) flows across the junction. The region around the junction is swept free, by the potential difference, of the holes and electrons in the p and n regions. This region is called the depletion layer, and is the sensitive volume of the solid state detector. When an ionizing particle passes through the depletion layer, electron–hole pairs are produced as a result of ionizing collisions between the ionizing particles and the crystal. The electric field then sweeps the holes and electrons apart, giving rise to a pulse in the load resistor as the electrons flow through the external circuit.

Semiconductor detectors are especially useful for nuclear spectroscopy because of their inherently high energy resolution. For charged particle spectroscopy, surface barrier type semiconductors are used. Fig. 9.14 shows the resolution obtainable with such a detector;

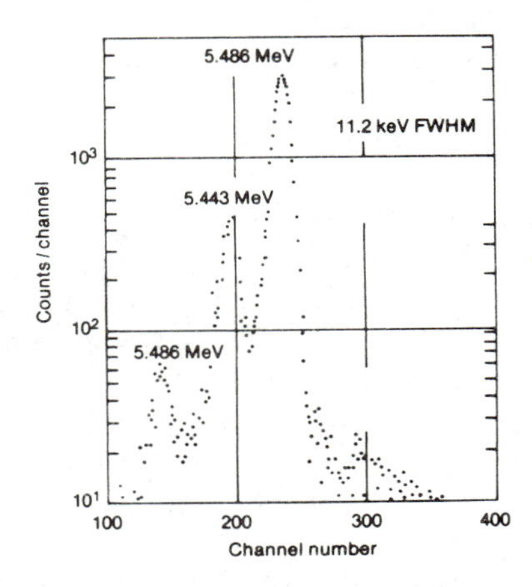

FIG. 9.14. Alpha spectrum of ^{241}Am obtained with a silicon surface barrier detector. (Courtesy of ORTEC, Inc.)

alphas of 5.443 and 5.486 MeV are easily and clearly separated. By using an appropriate neutron-sensitive material such as ^{3}He gas, ^{6}Li, ^{10}B, ^{235}U, or a hydrogenous material together with a surface barrier detector, either slow or fast neutrons can be measured. For photon spectroscopy, lithium-drifted silicon, Si(Li), or lithium-drifted germanium, Ge(Li), is used. These detectors must be operated at cryostatic temperatures. The required low temperature is obtained by mounting the detector on a "cold finger" that is immersed in liquid nitrogen (77°K) in a Dewar flask. Fig. 9.15 illustrates the excellent resolution that is obtainable with these detectors.

Advantages of semiconductor detectors include

(a) High-speed counting due to the very low resolving time—on the order of nanoseconds;

(b) High energy resolution;

(c) Relatively low operating voltage—about 25–300 V.

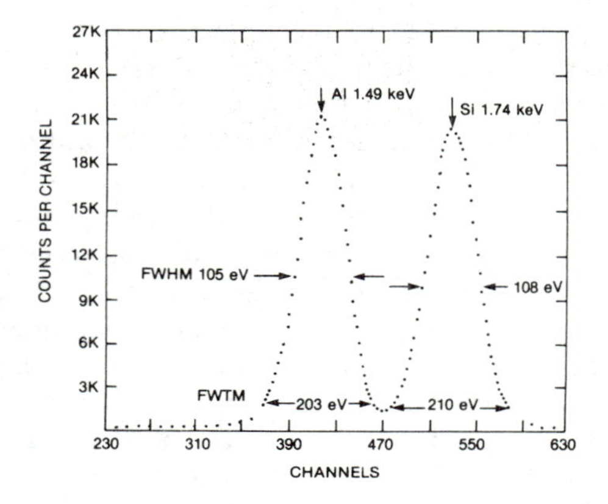

FIG. 9.15. Characteristic X-ray lines from an aluminum-silicate sample. The spectrum shows the excellent resolution capability of a Si(Li) low energy photon detector. (Courtesy ORTEC, Inc.)

Dose-measuring Instruments

Radiation flux is only one of the several factors that determine radiation dose. That a flux measuring instrument does not necessarily measure dose is shown by the following example:

Example 9.2

Consider two radiation fields of equal energy density. In one case, we have a 0.1-MeV photon flux of 2000 photons per cm^2/sec. In the second case, the photon energy is 2 MeV and the flux is 100 photons per cm^2/sec. The energy absorption coefficient for muscle for 0.1-MeV gamma radiation is 0.0252 cm^2/g; for 2-MeV gammas, the energy absorption coefficient is 0.0257 cm^2/g. The dose rates for the two radiation fields are given by:

$$\dot{D} = \frac{\phi \text{ phot/cm}^2\text{/sec} \times E \text{ MeV/phot} \times 1.6 \times 10^{-13} \text{ J/MeV} \times \mu \text{ cm}^2\text{/g}}{10^{-3} \frac{\text{J}}{\text{g}} \Big/ \text{Gy}} \tag{9.7}$$

$= 8.1 \times 10^{-10}$ Gy/sec for the 0.1 MeV radiation, and

7.6×10^{-10} Gy/sec for the 2 MeV photons.

The dose rates for the two radiation fields are about the same. A flux-measuring instrument, however, such as a Geiger counter, would register about 20 times more for the 0.1 MeV radiation than for the higher-energy radiation.

FIG. 9.16. Photograph of a lithium-drifted solid state detector whose front window is $\frac{1}{2}\mu$ gold. This detector can be used to measure charged particle and electromagnetic radiation. Gamma rays are detected with high resolution but with low efficiency, while particles are detected with both high efficiency and high resolution. (Courtesy Technical Measurement Corp.)

Pocket Dosimeters

To measure radiation dose, the response of the instrument must be proportional to absorbed energy. A basic instrument for doing this, the free air ionization chamber, was described in Chapter 6. In that chapter, too, it was shown that an "air wall" ionization chamber could be made on the basis of the operational definition of the exposure unit, or the roentgen, and that such an instrument could be used to measure exposure. Ionization chambers of this type, which are often called "pocket dosimeters," are widely used for personnel monitoring. Two types of pocket dosimeters are in common use. One of these is the condenser type, as illustrated in Fig. 6.2. This type pocket dosimeter is of the indirect reading type; an auxiliary device is necessary in order to read the measured dose. This device, which is in reality an electrostatic voltmeter that is calibrated in roentgens, is called a "charger-reader" (because it is also used to charge the chamber). The term minometer is often used synonymously with charger-reader. Fig. 9.17 shows a photograph of a pocket dosimeter and its charger-reader. Commercially available condenser-type pocket dosimeters measure integrated X- or gamma-ray exposures up to 200 mR with an accuracy of about $\pm 15\%$ for quantum energies between about 0.05 and 2 MeV. For quantum energies outside this range, correction factors, which are supplied by the manufacturer, must be used. These dosimeters also respond to beta rays whose energy exceeds 1 MeV. By coating the inside of the chamber with boron, the pocket dosimeter can also be made sensitive to thermal neutrons. The standard type of pocket dosimeter, however, is designed for measuring X- and gamma radiation only. It is calibrated with radium, ^{60}Co, or ^{137}Cs gamma-rays. Pocket dosimeters discharge slowly even when they are not in a radiation field because of cosmic radiation and because charge leaks across the insulator that separates the central electrode from the outer electrode. A

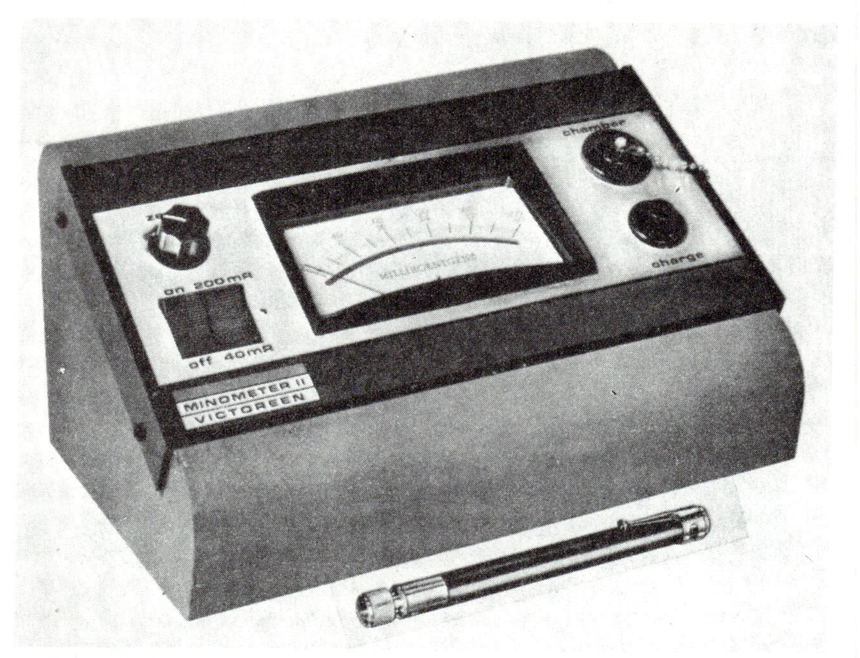

FIG. 9.17. Condenser-type pocket dosimeter and its charger-reader. The dosimeter measures gamma- and X-rays within ±15% from 30 keV to 1.2 MeV in the range 0–200 mR. (Courtesy Victoreen Instrument Co.)

dosimeter that leaks more than 5% of the full-scale reading per day should not be used. Usually, two pocket dosimeters are worn. Since a malfunction will always cause the instrument to read high, the lower of the two readings is considered as more accurate. Because of leakage and possibility of malfunction due to being dropped, pocket dosimeters are usually worn for one day. Reading the instrument erases its information content. It is therefore necessary to recharge the indirect reading pocket dosimeter after each reading.

The second type pocket dosimeter is direct reading, and operates on the principle of the gold-leaf electroscope (Fig. 9.18). A quartz fiber is displaced electrostatically by charging it to a potential of about 200 V. An image of the fiber is focused on a scale and is viewed through a lens at one end of the instrument. Exposure of the dosimeter to radiation discharges the fiber, thereby allowing it to return to its original position. The amount discharged, and consequently the change in position of the fiber, is proportional to the radiation exposure. An advantage of the direct reading dosimeter is that it does not have to be recharged after being read. Commonly used direct reading dosimeters that are commercially available have a range of 0–200 mR, and read within about ±15% of the true exposure for energies from about 50 keV to 2 MeV. An auxiliary charger must be used with this dosimeter.

Pocket dosimeters, as the name implies, are mainly used as personnel monitoring devices, and are worn by persons who may be exposed to X- or gamma radiation in order to measure the actual exposure of the wearer. However, these same instruments may also be used as area monitoring devices by locating them at the points where the exposure dose is to be measured. For this purpose, one or more dosimeters may be left in place for periods up to 1 week. For area monitoring applications, there are available chambers of larger volume (Fig. 9.19) and consequently more sensitive than the pocket dosimeter. Such

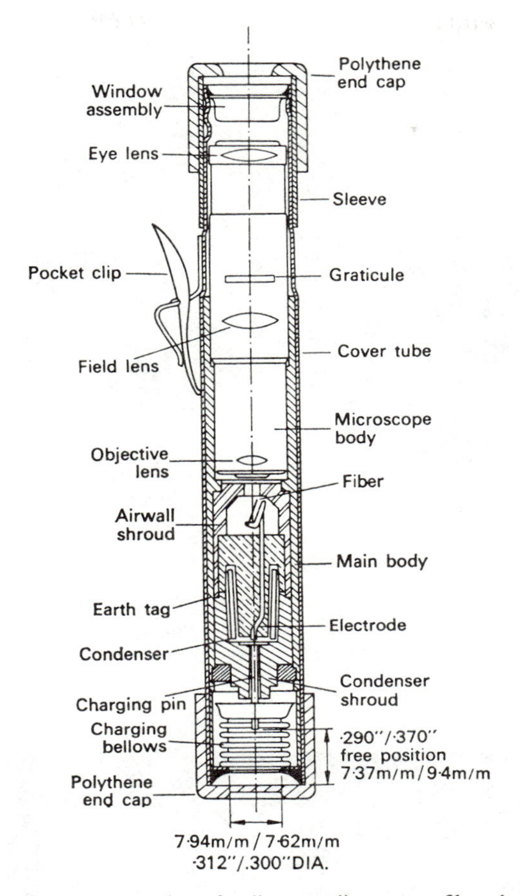

FIG. 9.18. Simplified cross-section of a direct reading quartz fiber electroscope-type pocket dosimeter. The energy dependence characteristics of this dosimeter is shown in Fig. 6.4. (Courtesy R. A. Stephen & Co., Ltd.)

chambers, which are often called stray radiation chambers, are designed to be used with a charger-reader in the same manner as a condenser-type pocket dosimeter. These chambers are especially useful in monitoring scattered radiation from medical and dental X-ray apparatus.

Film Badges

Another very commonly used personnel monitoring device is the film badge (Fig. 9.20), which consists of a packet of two (for X or gamma) or three (for X, gamma, and neutrons) pieces of dental-sized film wrapped in light-tight paper and worn in a suitable plastic or metal container. The two films for X-and gamma radiation include a sensitive emulsion and a relatively insensitive emulsion. Such a film pack is useful over an exposure range of about 10 mR to about 1800 R of radium gamma rays. The film is also sensitive to beta radiation, and may be used to measure beta-ray dose, from betas whose maximum energy exceeds about 400 keV, from about 50 mrad (0.5 mGy) to about 1000 rad (10 Gy). Using appropriate film and techniques, thermal neutron doses of 5 mrads to 500 rads (5 Gy), and fast neutron doses from about 4 mrads (40 μGy) to 10 rads (0.1 Gy) may be measured.

FIG. 9.19. Stray radiation chambers. The smaller chamber has a range of 0–10 mR, while the larger one, which is more sensitive, has a range of 0–1 mR. These chambers are used with the charger-reader shown in Fig. 9.17. (Courtesy Victoreen Instrument Co.)

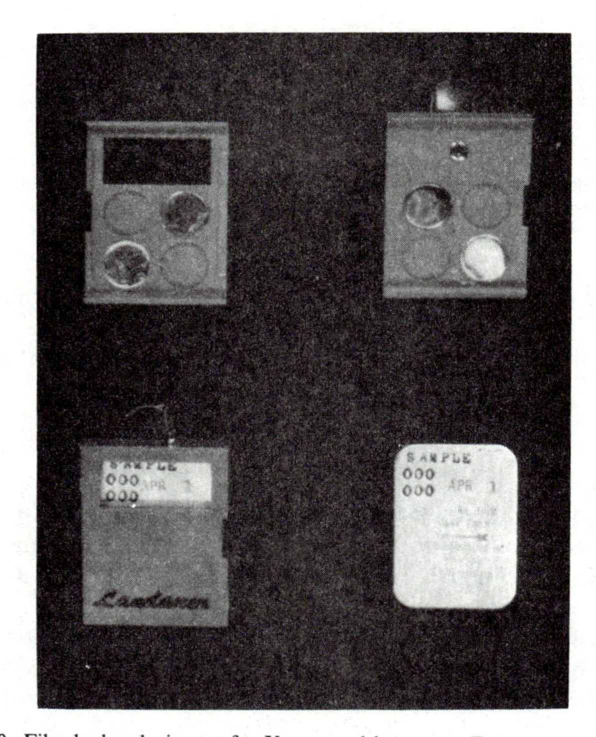

FIG. 9.20. Film badge dosimeter for X-rays and beta rays. For neutron dosimetry, a second film pack sensitive to neutrons, as well as an additional filter, made of cadmium, is added to the badge. (Courtesy R. S. Landauer Co.)

Film badge dosimetry is based on the fact that ionizing radiation exposes the silver halide in the photographic emulsion, which results in a darkening of the film. The degree of darkening, which is called the optical density of the film, can be precisely measured with a photoelectric densitometer whose reading is expressed as the logarithm of the intensity of the light transmitted through the film. The optical density of the exposed film is quantitatively related to the magnitude of the exposure (Fig. 9.21). By comparing the optical density of the film worn by an exposed individual to that of films exposed to known amounts of radiation, the exposure to the individual's film may be determined. Small variations in emulsions greatly affect their quantitative response to radiation. Since the films used in film badges are produced in batches, and since slight variations from batch to batch may be expected, it is necessary to calibrate the film from each batch separately.

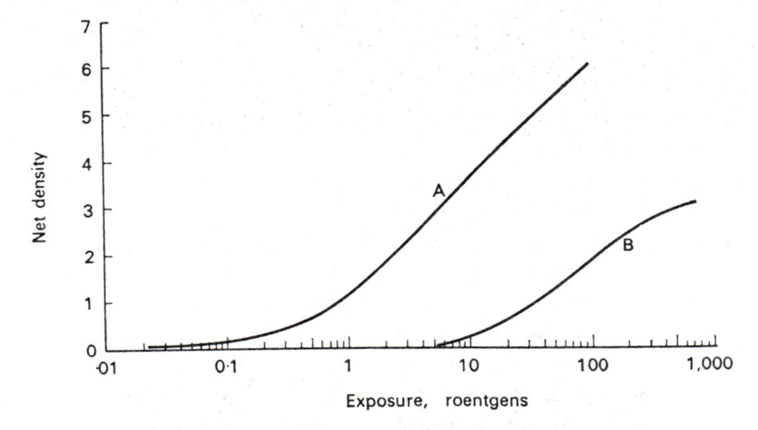

FIG. 9.21. Relationship between radiation exposure and optical density. Curve *A* is the response of duPont type 555 and curve *B* is the response of duPont type 834 dosimeter film to ^{60}Co gamma-rays.

Films used in film badge dosimeters are highly energy dependent in the low energy range, from about 0.2 MeV gamma radiation downward (Fig. 9.22). This energy dependence arises from the fact that the photoelectric cross section for the silver in the emulsion increases much more rapidly than that of air or tissue as the photon energy decreases below about 200 keV. A maximum sensitivity is observed at about 30–40 keV. Below this energy,

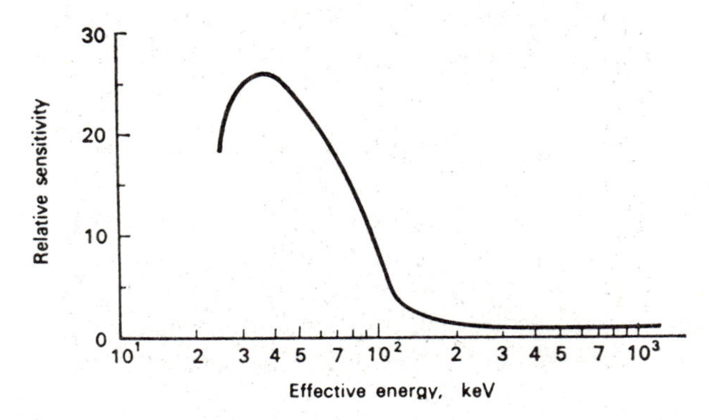

FIG. 9.22. Energy dependence of a film badge dosimeter to X-rays. (From N.B.S. Handbook 50, *Photographic Dosimetry of X- and Gamma-Rays*, 1954.)

the sensitivity of the film decreases because of the attenuation of the radiation by the paper wrapper. As a result of this very strong energy dependence, film dosimetry is useless for X-rays less than 200 keV unless the film was calibrated with radiation of the same energy distribution as the radiation being monitored, or unless the energy dependence of the film is accounted for. This allowance for energy dependence is made by selective filtration. The film badge holder is designed so that radiation may reach the film directly through an open window, or the radiation may be filtered by the film badge holder, or by one of several different filters, such as aluminum, copper, cadmium, tin, silver, and lead. The exact design and choice of filter is governed by the type of radiation to be monitored. The evaluation of the exposure is then made by considering the ratio of the film densities under each of the various filters. Beta-ray dose is determined from the ratio of the open window film reading to that behind the filters. If exposure was to beta radiation only, then film darkening is seen only in the open window area. To help distinguish between low-energy gamma rays and beta rays, for example, comparison is made between the darkening in the open window, and under two thin filters, such as aluminum and silver, which are of the same density thickness, and therefore equivalent beta-ray absorbers. The different atomic numbers, however, result in much greater low-energy X-ray filtration by the silver filter than by the aluminum filter, thereby giving different degrees of darkening under the two filters. Interpretation of mixed beta–gamma radiation with a film badge is difficult because of the greatly different penetrating powers of beta and gamma radiation. For this reason, information from beta-ray monitoring with film badges is used mainly in a qualitative, or in a semi-quantitative manner to evaluate exposure.

Fast neutrons, whose energy exceeds $\frac{1}{2}$ MeV, can be monitored with nuclear track film, such as Eastman Kodak NTA, which is added to the film badge. Irradiation of the film by fast neutrons results in proton recoil tracks due to elastic collisions between hydrogen nuclei in the paper wrapper, in the emulsion, and in the film base. Although the n,p scattering cross section decreases with increasing neutron energy—from 13 barns at 0.1 MeV to 4.5 barns at 1 MeV to 1 barn at 10 MeV—the recoil protons do not have sufficient energy below about $\frac{1}{2}$ MeV to make recognizable tracks, and hence the threshold at $\frac{1}{2}$ MeV. Because the concentration of hydrogen atoms in the film and its paper wrapper is not very much different from that of tissue, the response of the film to fast neutrons is approximately tissue equivalent, and the number of proton tracks per unit area of the film is therefore proportional to the absorbed dose. Fast-neutron exposure is estimated by scanning the developed film with high-powered microscope, and counting the number of proton tracks per square centimeter of film. The maximum recommended dose rate to fast neutrons, 1 mSv (100 mrem)/week, corresponds to a mean proton track density of about $2600/cm^2$ of NTA film for neutrons from a Pu–Be source. Since the area seen by the oil immersion lens is about 2×10^{-4} cm^2, a fast neutron dose of 100 mrem corresponds to a mean track density of about 1 proton recoil track per two microscopic fields.

Thermal neutrons also produce proton recoil tracks in the neutron film as a result of their capture by nitrogen in the film according to the $^{14}N(n,p)\,^{14}C$ reaction. Although the cross section for 2200 m/sec neutrons for this reaction is 1.75 barns, the concentration of nitrogen in the film is much less than that of hydrogen, thus making this reaction less sensitive, on a per-neutron basis, than the n,p scattering reaction for fast neutrons. Nevertheless, because in practice fast neutrons are usually part of a mixed radiation field that includes thermal neutrons (and gamma radiation), and because the permissible flux for thermal neutrons is much higher than for fast neutrons, allowance must be made for the proton tracks due to thermal neutrons. A film badge designed for use in a mixed radiation field that includes neutrons always has at least two metal filters of equal density thickness—one of cadmium and the other usually of tin. Cadmium has a very high cross section, 2500 barns for the $^{113}Cd(n,\gamma)^{114}Cd$ for 0.025 eV neutrons, and 7400 barns for 0.179 eV neutrons. The capture cross section of tin for thermal neutrons is insignificantly small. As a result, a thermal neutron field will show a high track density under the tin filter, but no tracks under the cadmium. Fast neutrons, on the other hand, will produce the same track density under both filters. Furthermore, because of the n,γ reaction in the cadmium, a thermal neutron field will produce a darker area on the gamma-ray film under the cadmium than under the tin. In the absence of any neutrons, gamma radiation would expose the film under each of these filters to the same degree. By counting the tracks and measuring the gamma-ray film density, we determine the thermal neutron flux as well as allowing for the thermal neutron background track density in the determination of the fast neutron flux. It should be reemphasized that the ordinary neutron film badges are not sensitive to neutrons in the energy range between epithermal and $\frac{1}{2}$ MeV. However, if the spectral distribution of the neutron field is known, then allowance for neutrons in the film badge insensitive range can be made.

Thermoluminescent Dosimeter

Many crystals, including CaF_2 containing Mn as an impurity, and LiF, emit light if they are heated after having been exposed to radiation; they are called thermoluminescent crystals. Absorption of energy from the radiation excites the atoms in the crystal, which results in the production of free electrons and holes in the thermoluminescent crystal. These are trapped by impurities or imperfections in the crystalline lattice thus locking the excitation energy into the crystal. Heating the crystal releases the excitation energy as light.

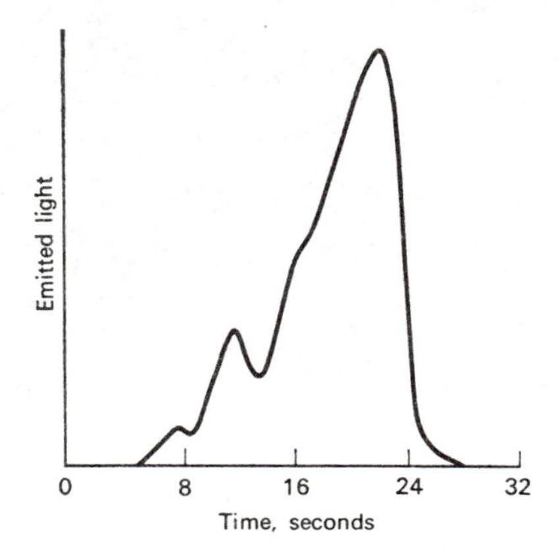

FIG. 9.23. Glow curve for LiF that had been exposed to 100 R (1 Gy). The area under the curve is porportional to the total exposure. (From J. R. Cameron *et al.*, Thermoluminescent dosimetry utilizing LiF, *Health Physics*, **10**, 25–29, 1967.)

Fig. 9.23 shows a characteristic glow curve for LiF, which is obtained by heating the irradiated crystal at a uniform rate, and measuring the light output as the temperature increases. The temperature at which the maximum light output occurs is a measure of the binding energy of the electron on the hole in the trap. More than one peak on a glow curve indicates different trapping sites, each with its own binding energy. The total amount of light is proportional to the number of trapped, excited electrons, which in turn is proportional to the amount of energy absorbed from the radiation. The intensity of the light emitted from the thermoluminescent crystals is thus directly proportional to the radiation dose. In use, a very small quantity, on the order of about 50 mg, is placed into a small capsule (Fig. 9.24), and exposed to radiation. For readout, the phosphor is heated electrically and the intensity of the resulting luminescence is measured by a photomultiplier tube whose output signal, after amplification, is applied to a suitable readout instrument, such as a digital voltmeter. The instrument is calibrated by measuring the intensity of light from phosphors that had been exposed to known doses of radiation. Since the intensity of luminescence is proportional to the quantity of the phosphor as well as to the radiation absorbed dose, the amount of phosphor used in making a measurement must be kept as close as possible to the amount used in calibrating the instrument.

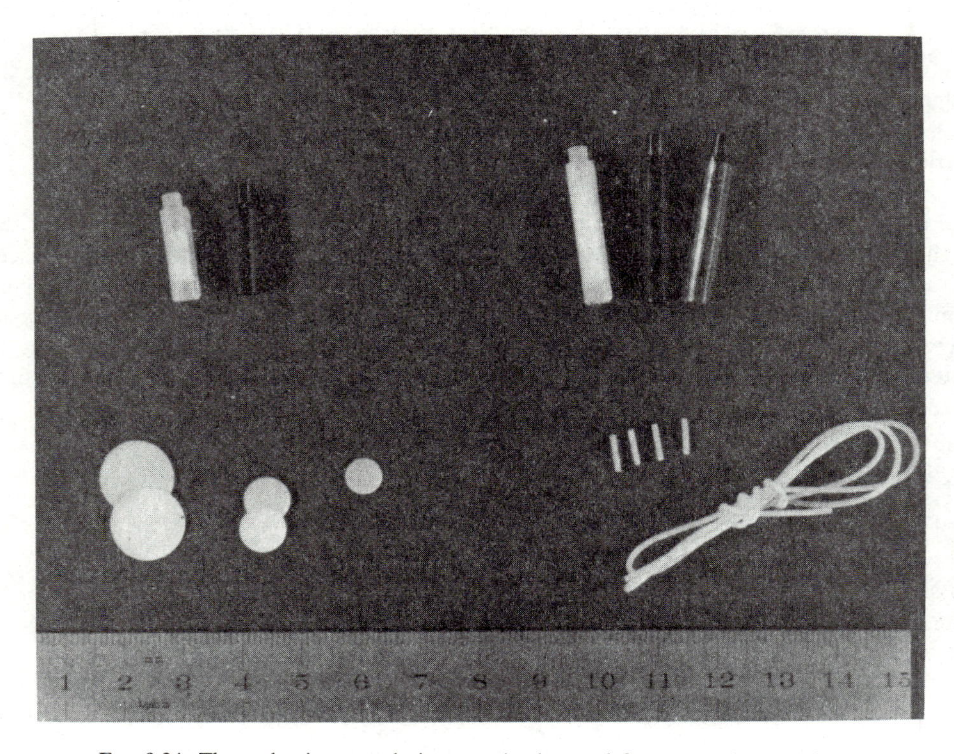

FIG. 9.24. Thermoluminescent dosimeters. At the top left are capsules containing 43 mg LiF powder; the capsules on the top right contain 140 mg LiF, the capsule on the extreme right is surrounded with a low-energy filter. From left to right, on the bottom, are LiF–Teflon discs of two sizes, a CaF_2: Mn–Teflon disc, microrods with LiF, and a long strand of LiF–Teflon from which the microrods are cut. (Courtesy Controls for Radiation Inc.)

Thermoluminescent dosimeters respond quantitatively to X-rays, gamma rays, beta rays, electrons, and protons over a range that extends from about 10 mrad (0.1 mGy) to about 100,000 rad (1000 Gy). LiF thermoluminescent dosimeters are approximately tissue equivalent, since the effective atomic number of the LiF phosphor is 8.1, while the effective atomic number of soft tissue is about 7.4. The response of a LiF thermoluminescent dosimeter is almost energy independent from about 100 keV to 1.3 MeV gamma rays. Below 100 keV, the sensitivity increases somewhat, as shown in Fig. 9.25.

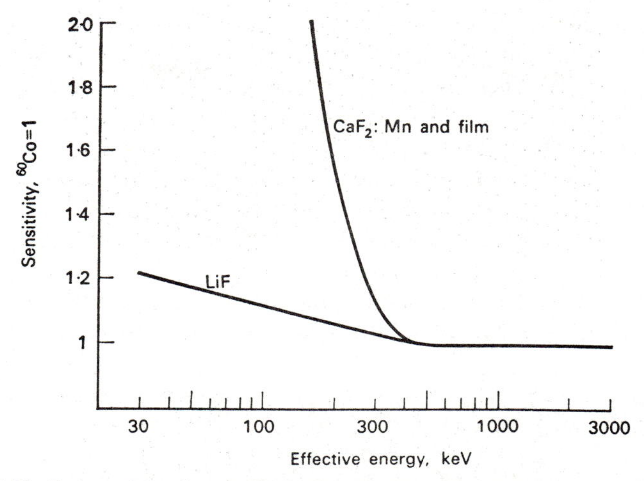

FIG. 9.25. Energy dependence of LiF compared with that of other unshielded dosimeters. (From J. R. Cameron *et al.*, Thermoluminescent dosimetry utilizing LiF, *Health Physics*, **10**, 25–29, 1964.)

Ion Current Chamber

Ion current chambers have a response whose magnitude is proportional to absorbed energy, and hence are widely used by health physicists in making dose measurements. A current ion chamber consists basically of a chamber with two electrodes across which is placed a potential low enough to prevent gas multiplication (Fig. 9.26). The ions that are generated in the chamber by radiation are collected, and flow through an external circuit. The ion chamber thus acts as a current source of infinite internal resistance. Although in principle an ammeter can be placed in the external circuit to read the ion current, in practice this is usually not done because the current is very small. Instead, a high-valued load resistor, R, on the order of 10^{10} ohms is placed in the circuit, and the voltage drop across the resistor is measured with a sensitive electrometer. Because of the capacitance of the counter and the associated circuit, C, the voltage across the load resistor varies with time, t, after closing the circuit according to the equation

$$V(t) = IR(1 - e^{t/RC}). \tag{9.8}$$

The product RC is called the time constant of the detector circuit, and determines the speed with which the detector responds. When t is equal to RC, the exponent in equation (9.8)

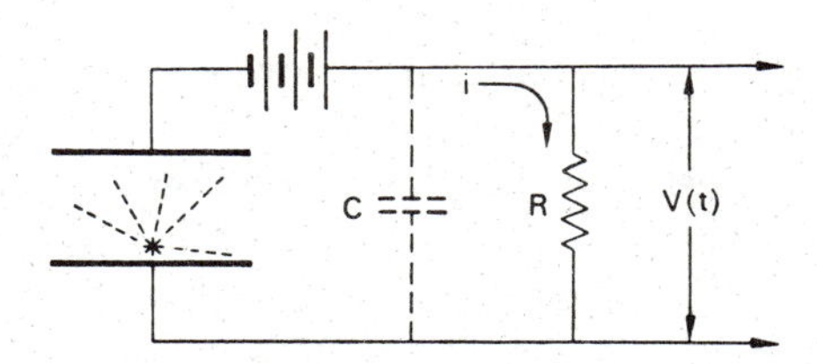

FIG. 9.26. Operating principle of a current ionization chamber. The radiation-produced ions are collected from the chamber, thus causing a current i to flow through the external circuit, resulting in a voltage drop $V(t)$ across the high-valued resistor R. C represents all the capacitance associated with the chamber.

becomes 1, and the voltage attains 63% of its final value. As t increases beyond several time constants the instrument reads the final steady-state voltage. It should be noted that the sensitivity of a detector increases with increasing resistance of the load resistor. Since the capacitance of the detector is fixed, this means that, in an instrument with several ranges—which is accomplished by varying the value of R—the more sensitive ranges have longer time constants, and hence are slower to respond than the less sensitive ranges. The time constants for health physics surveying instruments vary up to about 10 sec. Laboratory instruments, where fast response is not important, may have time constants on the order of 100 sec.

When a current ion chamber is exposed to radiation levels of different intensity, and the voltage across the chamber is varied, a family of curves, as shown in Fig. 9.27, is obtained.

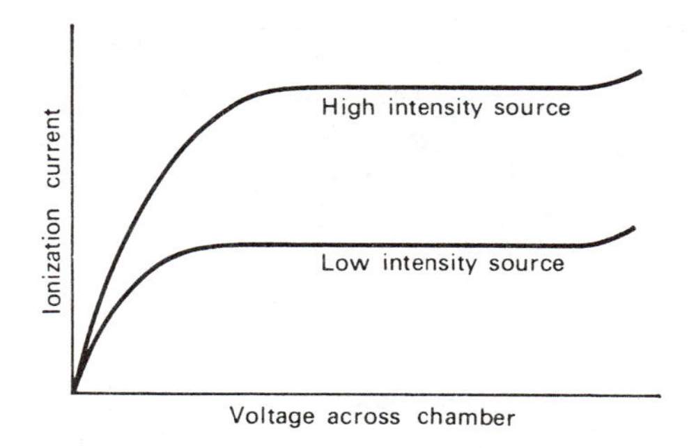

FIG. 9.27. Variation of ionization current with voltage across the ionization chamber for different levels of radiation. The plateau represents the saturation current.

The current plateau is called the *saturation current*. When operated at a voltage that lies on the plateau, all the ions that are produced in the chamber are being collected. The operation of a current ion chamber, and the fact that the magnitude of the response is proportional to absorbed energy, is shown in the following illustrative example.

Example 9.3

A large air-filled ionization chamber has a window whose thickness is 1 mg/cm^2. (a) What ionization current will result if 1200 alpha particles from ^{210}Po enter the chamber per minute? (b) What would be the ionization current if the window thickness were increased to 3 mg/cm^2?

The ionization current within the chamber may be calculated from the equation

$$I = \frac{N \text{ particles/sec} \times \bar{E} \text{ eV/particle} \times 1.6 \times 10^{-19} \text{ C/ion}}{w \text{ eV/ion}}. \tag{9.9}$$

(a) The energy of the alpha particle after it penetrates the window into the chamber, is equal to the difference between its initial kinetic energy, 5.3 MeV, and the energy lost in penetrating the window. Assuming the plastic window to be equivalent to tissue in regard to its stopping power, we calculate, from equations (5.14) and (5.16), that the range of a 5.3-MeV alpha particle in the plastic of which the window is made is 5.1 mg/cm^2. After passing through the 1-mg/cm^2 window, therefore, the remaining kinetic energy of the alpha particle is

$$\frac{5.1 - 1}{5.1} \times 5.3 = 4.25 \text{ MeV},$$

and the resulting ion current is, from equation (9.9),

$$I = \frac{1.2 \times 10^3\ \alpha/\text{min} \times 4.25 \times 10^6\ \text{eV}/\alpha \times 1.6 \times 10^{-19}\ \text{C/ion}}{60\ \text{sec/min} \times 35\ \text{eV/ion}}$$

$$= 3.9 \times 10^{-13}\ \text{amps}.$$

(b) If the window thickness were increased to 3 mg/cm^2, the energy of the alpha particle entering into the ionization chamber would be

$$\frac{5.1 - 3}{5.1} \times 5.3 = 2.18\ \text{MeV},$$

and the ion current would be only 2×10^{-13} A. It should be pointed out that in both instances, alpha particles were entering into the ion chamber at the same rate. If individual pulses had been counted, therefore, the counting rate would have been the same in both cases.

As a result of electronic design considerations, the basic circuit of the ionization chamber survey instrument usually is a Wheatstone bridge (Fig. 9.28), with an electrometer tube being the resistive element in one of the arms of the bridge. The grid of the tube is connected to one of the electrodes of the ion chamber. The bridge is balanced after bypassing the ion chamber current around the bridge. Switching the ion chamber to produce a voltage change on the grid of the electrometer tube as the ionization current flows through the load resistor R_1 unbalances the bridge and results in a deflection of the microammeter A. The amount of deflection of the meter is proportional to the signal voltage applied to the grid of the tube; this in turn is proportional to the ionization current,

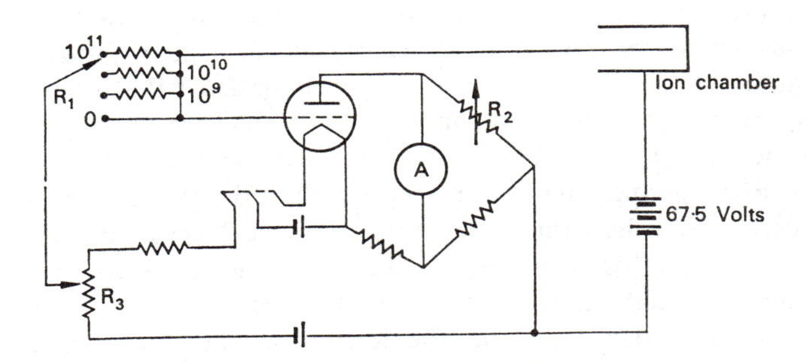

FIG. 9.28. Cutie-pie-type ionization chamber survey instrument circuit diagram. The bridge is balanced, with $R_1 = 0$, by adjusting R_2 and R_3. R_2 is usually inside the case and R_3 is the "zero" knob.

which is proportional to the radiation dose rate. Because of the characteristics of the circuit, an input signal to the bridge of about 1 volt is desired. The current from the ionization chamber is usually very low. The ion current, which depends on the volume of the chamber, the density of the air, ρ, and on the exposure rate $\dot{X}$, is given by

$$i = \frac{V\ \text{cm}^3 \times \rho\ \text{kg/cm}^3 \times \dot{X}\ \mu\text{C/kg-hr} \times 10^{-6}\ \text{A}/\mu\text{C-sec}}{3.6 \times 10^3\ \text{sec/hr}}. \tag{9.10}$$

For a 400-cm^3 chamber filled with standard air and an exposure rate of 6.5 μC/(kg-hr) (25 mR/hr), the ionization current is found, from equation (9.10) to be 9.3×10^{-13} A. To produce a voltage drop of about 1 volt across a resistor with this current requires a resistor of about 10^{11} ohms. If a resistor of this value is used as R_1 on the most sensitive scale of the instrument, then we need resistors of 10^{10} and 10^9 ohms respectively for full-scale dose rate measurements 10 and 100 times greater than that of the most sensitive scale. These very high-valued resistors must be sealed in glass if their resistance is to remain unchanged. Touching with the hands or otherwise dirtying the connections, or in any way lowering the grid to plate or grid to cathode resistance of the electrometer tube, will very greatly affect the sensitivity of the instrument. In making repairs on the instrument, therefore, great care must be taken to assure the maintenance of the proper resistance in the electrometer circuit.

Certain of the ion chamber instruments, such as the Juno type, are designed to respond selectively to alpha, beta, and gamma radiation, while others, such as the cutie pie, can respond to betas and gammas. The span of dose rates on the commonly used ion chamber survey meters is up to several hundred μC/(kg hr) (several thousand mR per hour)—usually in three ranges. Full-scale readings of 6, 60, and 600 μC/(kg hr) (25, 250, and 2500 mR/hr) gamma radiation may be considered typical. However, less sensitive as well as more sensitive ionization-chamber-type survey instruments are commercially available.

Neutron Measurements

Detection Reactions

Neutrons, like gamma rays, are not directly ionizing; they must react with another medium to produce a primary ionizing particle. Because of the strong dependence of neutron reaction rate on the cross section for that particular reaction, we either use different detection media, depending on the energy of the neutrons that we are trying to measure, or we modify the neutron energy distribution in order that it be compatible with the detector. Some of the basic neutron detection reactions used in health physics instrumentation include:

1. $^{10}B(n,\alpha)\,^{7}Li$. Boron, which may be enriched in the ^{10}B isotope, is introduced into the counter either as BF_3 gas or as a thin film on the inside surfaces of the detector tube. The ionization due to the alpha particle and the ^{7}Li recoil nucleus is counted.

2. Elastic scattering of high-energy neutrons by hydrogen atoms. The scattered protons are the primary ionizing particles, and the ionization they produce is detected and measured.

3. Nuclear fission: fissile material (n,f) fission fragments. The fissile material is deposited as a thin film on the inside surface of a counter tube. Capture of a neutron and splitting of the fissile nucleus results in highly ionizing fission fragments, which can easily be detected. Fission reactions are energy dependent, and several fissile isotopes have a threshold energy below which fission cannot occur.

4. Neutron activation: *threshold detectors*. Many neutron reactions produce radioactive isotopes. The degree of neutron-induced radioactivity of any given substance depends on the total neutron irradiation. By measuring the induced activity, and allowing for decay time between exposure and measurement, the integrated neutron exposure can be calculated. Furthermore, since many of these activation reactions have energy thresholds, they can be useful in determining the neutron energy distribution. Tables 9.3 and 9.4 list the materials most commonly used for this purpose. This technique is especially useful in measuring neutron dose at very high dose rates, such as those that would be encountered in a criticality accident. (A criticality accident is an accidental uncontrolled chain reaction

TABLE 9.3. FOIL REACTIONS WITH THERMAL NEUTRONS

Foil material	Target nucleus	Target percentage abundance*	Nuclear reaction	Product nucleus	Product half-life	2200 m/s Activation cross-section (barns)
Ag	Ag-107	51.35	*n*,gamma	Ag-108	2.3 m	44
Ag	Ag-109	48.65	*n*,gamma	Ag-110	25 sec	110
In,In–Al	In-115	95.77	*n*,gamma	In-116 m	54 m	145
U (nat)	U-235	.715	*n*,fission	F.P.	$T^{-1.2}$	582
Dy–Al	Dy-164	28.18	*n*,gamma	Dy-165	2.32 h	2100
V	V-51	99.76	*n*,gamma	V-52	3.77 m	4.5
Mn–Cu	Mn-55	80.00	*n*,gamma	Mn-56	2.56 h	13.4
Cu	Cu-63	69.1	*n*,gamma	Cu-64	12.8 h	4.3
Cu	Cu-65	30.9	*n*,gamma	Cu-66	5.1 m	1.8
Al	Al-27	100	*n*,gamma	Al-28	2.3 m	0.21
Au,Au–Al	Au-197	100	*n*,gamma	Au-198	2.7 d	96
Rh	Rh-103	100	*n*,gamma	Rh-104 m	4.4 m	12
Rh	Rh-103	100	*n*,gamma	Rh-104	44 sec	140
Co,Co–Al	Co-59	100	*n*,gamma	Co-60	5.27 y	36.3
NaCl	Na-23	100	*n*,gamma	Na-24	15 h	0.53
Lu–Al	Lu-175	97.5	*n*,gamma	Lu-176 m	3.7 h	35
B–Al	B-10	19.6	*n*,alpha	Li-7	Stable	4010
Eu–Al	Eu-151	47.77	*n*,gamma	Eu-152	9.2 h	1400
Pb	Pb-208	52.3	*n*,gamma	Pb-209	3.2 h	.06 mb
Nb	Nb-93	100	*n*,gamma	Nb-94 m	6.6 m	1.0

*No adjustment made for composition of alloys or compounds.
This table lists the most useful thermal neutron reactions.
(Courtesy Reactor Experiments, Inc.)

TABLE 9.4. THRESHOLD FOIL REACTIONS

Foil material	Target nucleus	Target percentage abundance*	Nuclear reaction	Product nucleus	Product half-life	Effective activation cross-section (barns)	Effective threshold energy (MeV)
Ni	Ni-58	67.76	*n,p*	Co-58	71.3 d	.42	2.9
Fe	Fe-54	5.84	*n,p*	Mn-54	314 d	.61	~3
Fe	Fe-56	91.68	*n,p*	Mn-56	2.58 h	.110	7.5
Ti	Ti-46	7.95	*n,p*	Sc-46	84.1 d	.23	5.5
S,$(NH_4)_2SO_4$	S-32	95.018	*n,p*	P-32	14.3 d	.30	2.9
Mg	Mg-24	78.60	*n,p*	Na-24	15 h	.060	6.3
Al	Al-27	100.0	*n*,alpha	Na-24	15 h	.130	8.7
Zr	Zr-90	51.46	*n,2n*	Zr-89	3.3 d	1.6	14.0
In,In–Al	In-115	95.77	*n,n'*	In-115 m	4.5 h	0.2	1.0
NH_4I	I-127	100.0	*n,2n*	I-126	13.3 d	.98	11.0
V	V-51	99.76	*n*,alpha	Sc-48	44 h	.08 mb	11.5
Th	Th-232	100.0	*n*,fission	Mo-99	66 h	.060	1.75
U depl (.415%)	U-238	.415	*n*,fission	Mo-99	66 h	0.55	1.45
U depl (378 ppm)	U-238	.0378	*n*,fission	Mo-99	66 h	0.55	1.45
Si–Al	Si-28	92.27	*n,p*	Al-28	2.3 m	.004	6.7

*No adjustment made for composition of alloys or compounds.
This table lists the most useful threshold neutron reactions.
(Courtesy Reactor Experiments, Inc.)

in which a very large amount of energy is liberated during a very brief time.) For measuring high neutron fluxes for health physics purposes, a series of threshold detectors of various threshold energies is packaged in a single unit. Exposure to neutrons activates the detectors. Since the induced activity in the threshold detectors depends on the neutron flux whose energy exceeds the threshold energy, the relative counting rates of the threshold detectors after exposure is used as a measure of the spectral distribution of the neutrons, while the absolute activity of the detector is a measure of exposure. A number of different substances may be used as threshold detectors. One type of pocket criticality dosimeter (Fig. 9.29) uses a combination of indium, cadmium-covered indium, gold, cadmium-covered gold, sulfur, and cadmium-covered copper. By combining the results of the activity in each of these foils, the neutron spectrum may be broken down to the following energy intervals:

Thermal to 0.4 eV
0.4 eV to 2 eV
2 eV to 10 eV
10 eV to 1 MeV
1 MeV to 2.9 MeV
above 2.9 MeV.

The total neutron flux is thus divided into six energy groups, thereby permitting a reasonably accurate means for computing absorbed dose. The threshold detectors are

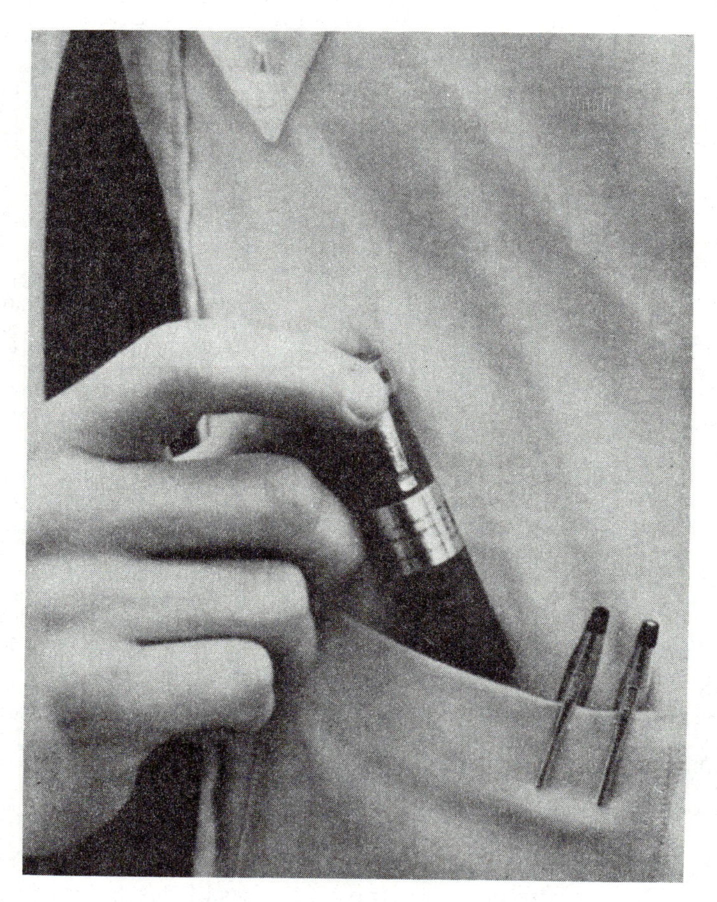

FIG. 9.29. Pocket criticality dosimeter for measuring neutron fluxes. Inside the dosimeter are six different foils whose induced radioactivity, following exposure to neutrons, depends on the neutron energy and fluence. (Courtesy Reactor Experiments, Inc.)

calibrated by exposing the pack to a known beam of neutrons, and then measuring the induced activity. (A thermoluminescent dosimeter may be added to measure the gamma-ray exposure from a criticality accident.)

Neutron Counting with a Proportional Counter

Counting in the proportional region makes it simple to measure neutrons in the presence of gamma radiation. A neutron counter used for this purpose uses BF_3 gas to take advantage of the n, α reaction on ^{10}B:

$$^{10}B(n, \alpha)\, ^7Li.$$

Boron-10 has a high cross section for this reaction, 4010 barns for thermal neutrons, and consequently makes a very sensitive detecting medium. The alpha particles resulting from this reaction are produced inside the detector. Because of the great difference between the output pulses resulting from the alpha particles and Li ions and those due to gamma rays, it is a simple matter to discriminate electronically against all pulses except those due to the alphas and Li ions when a neutron is captured by ^{10}B.

The sensitivity of a BF_3 neutron detector may be increased by using ^{10}B-enriched BF_3 gas. Naturally occurring boron contains about 19.8% ^{10}B. However, it is possible to concentrate the ^{10}B isotope to the extent that BF_3 gas containing 96% ^{10}B is routinely available from commercial suppliers. The BF_3-filled counter is a sensitive and simple thermal neutron detector, and is widely used by health physicists for measuring thermal neutron flux. Furthermore, since 680 thermal neutrons per cm^2/sec for 40 hr corresponds to 100 mrem/week, it is a simple matter to measure dose rate by converting counts per minute of a calibrated counter to neutron flux.

The capture cross section for the $^{10}B\ (n, \alpha)\ ^7Li$ reaction, and hence the counting rate of a BF_3 detector, depends on the neutron energy. Consider the case of a thermal neutron flux of ϕ neutrons per cm^2/sec having a Maxwell–Boltzmann energy distribution and a BF_3 counter containing a total of N atoms of ^{10}B. If there is negligible neutron absorption by the counter wall, and if the intrinsic counting efficiency is 100%, the counting rate is given by

$$CR = N \int \phi(v)\sigma(v)\, dv, \tag{9.11}$$

where $\phi(v)$ is the flux of neutrons of velocity v, and $\sigma(v)$ is the capture cross section for neutrons of velocity v. Substituting $\phi(v) = n(v)v$, where $n(v)$ is the density, in neutrons per cm^3, of neutrons whose velocity is v cm/sec; and, from equation (5.53), substituting $\sigma(v) = \sigma_0(v_0/v)$, into equation (9.11), we have

$$CR = N\sigma_0 v_0 \int n(v)\, dv, \tag{9.12}$$

$$CR = N\sigma_0 v_0 n. \tag{9.13}$$

From equation (9.13), we see that the counting rate of a BF_3 counter is proportional to the total neutron density within the range of energies where the $1/v$ law is valid (up to about 1000 eV). If $\bar{v}$ is the mean neutron velocity (not the velocity corresponding to the mean neutron energy), then

$$\phi = n\bar{v}, \tag{9.14}$$

or $n = \phi/\bar{v}$. Substituting this value for n into equation (9.13), and solving for ϕ, we have

$$\phi = \frac{\bar{v}}{v_0} \times \frac{CR}{N\sigma_0}. \tag{9.15}$$

In the Maxwell–Boltzmann distribution, $\bar{v}/v_0 = 2/\sqrt{\pi} = 1.128$. For thermal neutrons, therefore, the flux is related to the counting rate by

$$\phi = \frac{1.128}{N\sigma_0} \times CR. \tag{9.16}$$

Example 9.4

A thermal neutron counting rate of 600 counts per min is measured with a BF_3 counter whose inside dimensions are 2 cm diameter × 20 cm long, and is filled with 96% enriched $^{10}BF_3$ at a pressure of 20 cm Hg. What is the thermal flux? The volume of the counter is $\pi r^2 l$, or 62.83 cm^3. The molar quantity of BF_3 in the counter is

$$m = \frac{PV}{RT} = \frac{20/76 \text{ atm} \times 0.06283 \text{ liter}}{0.082 \text{ liter atm/mole}^\circ \times 300^\circ} = 6.72 \times 10^{-4} \text{ moles},$$

and the number ^{10}B atoms in the counter is

$$N = 0.96 \frac{^{10}\text{B atoms}}{\text{molecule}} \times 6.03 \times 10^{23} \frac{\text{molecules}}{\text{mole}} \times 6.72 \times 10^{-4} \text{ moles}$$

$$= 3.89 \times 10^{20}\ ^{10}\text{B atoms}.$$

From equation (9.16), we have

$$\phi = \frac{1.128}{3.89 \times 10^{20} \text{ atoms} \times 4.010 \times 10^{-21} \text{ cm}^2/\text{atom}} \times 10 \frac{\text{counts}}{\text{sec}}$$

$$= 7.2 \frac{\text{neutrons}}{\text{cm}^2 \text{ sec}}.$$

The sensitivity of a radiation detector may be defined as

$$\text{sensitivity} = \frac{\text{counting rate}}{\text{flux}}. \tag{9.17}$$

The sensitivity, S, of the BF_3 counter in Example 9.4 is

$$S = \frac{CR}{\phi} = \frac{600 \text{ counts/min}}{7.2 \text{ neutrons/cm}^2 \text{ sec}} = 83.3 \frac{\text{counts/min}}{\text{neutron/cm}^2 \text{ sec}}.$$

Long counter

The BF_3 counter can also be used for fast neutrons if it is surrounded by paraffin or another moderator. Fast neutrons are sufficiently slowed down by the moderator to allow them to be captured by the ^{10}B. The counting rate of a moderated BF_3 counter in a field of fast neutrons increases with increasing moderator thickness until the paraffin is sufficiently thick to absorb a significant fraction of the thermalized neutrons. Beyond this optimum thickness, the counting rate decreases with further increase in thickness. The exact thickness at which this occurs depends on the energy of the neutrons. A paraffin moderator $2\frac{3}{8}$ in. thick results in an approximately flat response over a neutron energy span from about 10 keV to better than 1 MeV. The outside of the paraffin may be covered with a thin sheet of cadmium to absorb thermal neutrons while allowing fast neutrons to pass through into the paraffin. The capture gammas due to absorption of thermal neutrons by the cadmium produce very small pulses in the counter, pulses small enough to be rejected by the discriminator. Such a paraffin surrounded BF_3 counter is called a *long counter* because of its long energy-independent response. In health physics work it is most useful in the energy range of about 10 keV to 500 keV. Measurement of higher energy neutrons with a BF_3 counter requires relatively large amounts of paraffin for slowing down the neutrons, thus making it impractical for many health physics surveying applications. Such a counter, which is reported as having a fairly uniform response from 10 keV to 5 MeV (A. D. Hanson and J. L. McKibben, A neutron detector having uniform sensitivity from 10 keV to 5 MeV, *Phys. Rev.* **72**, 673 (1947), and R. A. Nobles *et al.*, *Rev. Sci. Instr.* **25**, 334 (1954)), is shown below in Fig. 9.30. The counter response is highly directional, and is designed to measure neutrons that are incident only on the front face. The layer of paraffin outside the B_2O_3 is a shield designed to remove neutrons that are incident on the sides. The series of eight holes in the front face permit the lower-energy neutrons to reach the detector tube after being thermalized. The sensitivity of this counter is about 1 count per sec per neutron per square centimeter.

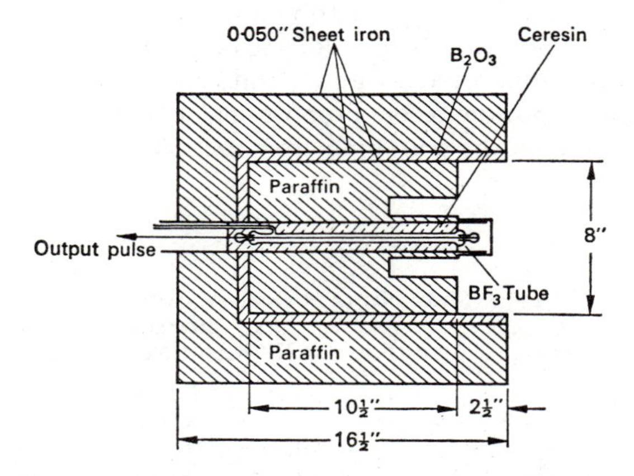

FIG. 9.30. Diagram of a long counter, a neutron counter whose response is approximately uniform from about 10 keV to 5 MeV. (Hanson and McKibben, *Physical Review* **72**, 673, 1947.)

Proton recoil counter

A proportional counter that responds to recoil protons resulting from the collision of fast neutrons with hydrogen atoms may be used for the detection of neutrons whose energy exceeds 500 keV. (Below this energy, the output pulses from the counter are very weak, and high gain amplification would be required to record them.) Such a counter may be made simply by using a hydrogenous gas, such as methane, in the counter. The counter is enclosed in a thin sheet of cadmium to absorb thermal neutrons in order to prevent pulses due to deuteron recoils following absorption of thermal neutrons by the hydrogen. The hydrogenous material may also be a solid, such as paraffin or polyethylene, incorporated into the wall of the counter. The fast neutrons "knock out" protons from these solids, and the protons dissipate their energy in the counter gas. When used in this way as a source of protons, the hydrogenous substance is called a "proton radiator." The sensitivity for fast neutrons of a proton recoil counter is very much less than the sensitivity of a BF_3 counter for thermal neutrons. This is due to two reasons: the cross section of hydrogen for scattering of fast neutrons is very much less than the slow neutron capture cross section of ^{10}B, and also the energy distribution of the scattered protons includes a large fraction of very low-energy protons. For neutron energies up to about 10 MeV the scattering of neutrons is isotropic. This means that the energy of the scattered proton may vary from zero to the energy of the neutron. The pulses that result from protons to which little energy was imparted during the collision are therefore not counted because of the bias against gamma radiation. Above this threshold, the energy response of the recoil proton proportional counter is determined mainly by the energy dependence of the scattering cross section of hydrogen.

Neutron Dosimetry

The dose equivalent (DE) from neutrons depends strongly on the energy of the neutrons as shown in Table 9.5. We therefore cannot simply convert neutron flux into dose equivalent unless we know the energy spectral distribution of the neutrons. For pure thermal neutrons, of course, the energy distribution is known, and a BF_3 counter could therefore be calibrated to read directly in millirems per hour. For higher energy neutrons, neither the moderated BF_3 long counter with its "flat" energy independent response to fast neutrons, nor the simple proton recoil proportional counter, whose response to fast neutrons depends strongly on the scattering cross section, is suitable for dosimetry.

TABLE 9.5. MEAN NEUTRON QUALITY FACTOR AND FLUX CORRESPONDING TO 1 mSv (100 mrem) PER 40-HOUR WEEK (From NCRP Report No. 38)

Neutron energy, MeV	$\overline{QF}$	Neutron flux density, $cm^{-2}\,sec^{-1}$
2.5×10^{-3} (thermal)	2	680
1×10^{-7}	2	680
1×10^{-6}	2	560
1×10^{-5}	2	560
1×10^{-4}	2	580
1×10^{-3}	2	680
1×10^{-2}	2.5	700
1×10^{-1}	7.5	115
5×10^{-1}	11	27
1	11	19
2.5	9	20
5	8	16
7	7	17
10	6.5	17
14	7.5	12
20	8	11
40	7	10
60	5.5	11
1×10^2	4	14
2×10^2	3.5	13
3×10^2	3.5	11
4×10^2	3.5	10

Fast neutrons: Hurst counter

The proton recoil proportional counter can be modified to measure fast neutron dose rate. By using a combination of several different proton radiators and filling gases, as shown schematically in Fig. 9.31, the energy distribution of the recoil protons that enter into the gas-filled cavity of a proportional counter is such that the resulting count rate is proportional to the variation of tissue dose with the neutron energy. Hence, the counting rate meter read out from the proportional counter can be calibrated directly in millirads per hour of fast neutrons. Fig. 9.32 compares the energy dependence of the dose with the response of the count rate dosimeter. Gamma-insensitive fast neutron dosimeters based on this design principle are commercially available. Because of the design, this instrument is limited to measurement of neutrons in the energy range from 0.2 to 14 MeV. The discriminator settings necessary to obtain the correct dose response rejects all counts due to neutrons whose energy is less than 0.2 MeV. Above 14 MeV, the response of this fast neutron dosimeter is not proportional to the absorbed dose rate.

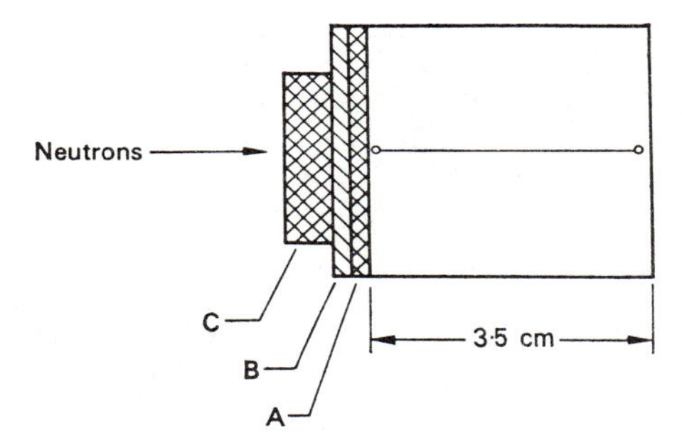

FIG. 9.31. Schematic representation of a count-rate fast-neutron dosimeter. *A*—paraffin (13 mg/cm^2), *B*—aluminum (29 mg/cm^2), *C*—paraffin (100 mg/cm^2). Ratio of the paraffin areas, $a_A/a_C = 2.9$. The gas is methane at 30 cm Hg pressure. The counter has a highly directional response, and must be oriented towards the neutron beam as shown in the diagram. (G. S. Hurst, R. H. Ritchie, and H. N. Wilson, A count rate method of measuring fast neutron dose, *R.S.I.* **22**, 981–6, 1951.)

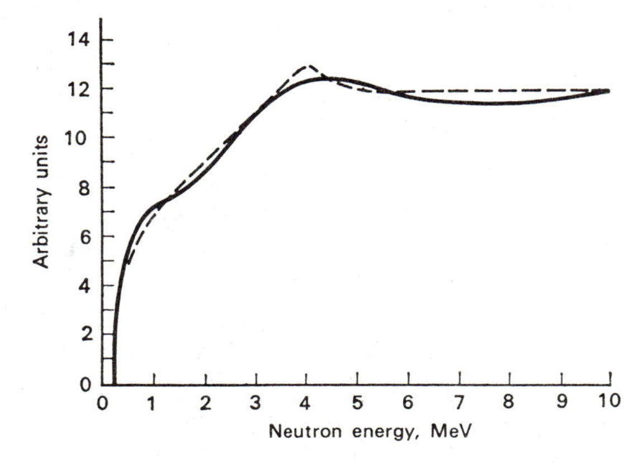

FIG. 9.32. Energy response of the Hurst count-rate fast-neutron dosimeter. The solid curve shows the relationship between counting rate per unit flux and neutron energy; the broken curve relates dose rate per unit flux and neutron energy. (G. S. Hurst, R. H. Ritchie, and H. N. Wilson, A count rate method of measuring fast neutron dose, *R.S.I.* **22**, 981–6, 1951.)

Thermal and fast neutron dose equivalent meter

A neutron counter whose count rate is proportional to the dose equivalent across the energy span of 0.025 eV to 10 MeV can be made by surrounding a BF_3 proportional counter with two cylindrical layers of polyethylene moderator separated by a boron-loaded plastic (Fig. 9.33). Circular discs of the same materials make up the ends of the cylinders.

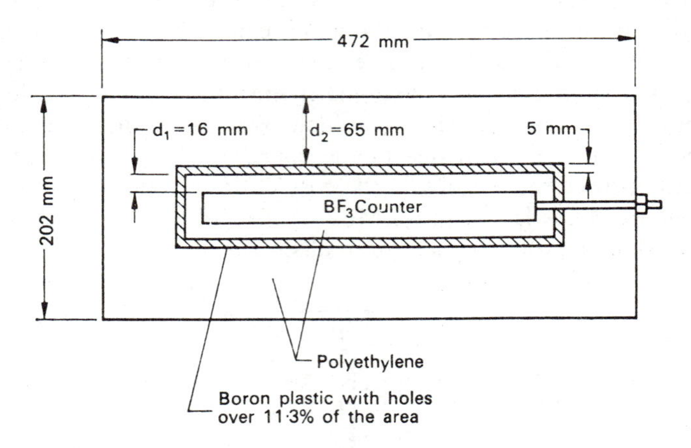

FIG. 9.33. Construction of the neutron rem counter. The BF_3 counter embedded in the polyethylene has an outside diameter of 30 mm, a sensitive length of 200 mm, and a total length of 300 mm; it is filled with 94% ^{10}B-enriched BF_3 at a pressure of 600 mm Hg. (I. O. Anderson and J. Braun, A neutron rem counter with uniform sensitivity from 0.025 eV to 10 MeV, *Neutron Dosimetry*, vol. II, pp. 87–95, I.A.E.A., Vienna, 1963.)

In designing the counter, the simplifying assumptions were made that all neutrons entering the counter and making their first collision in the outer polyethylene layer are absorbed in the boron plastic shield and thus lost for the purpose of detection, while all neutrons making their first collision in the polyethylene inside the boron plastic are considered as having a detection probability k. On the basis of these assumptions, the likelihood of obtaining a pulse from a neutron of energy E that is incident normally to the long axis of the counter is given by

$$P = ke^{-\Sigma(E)d_2}(1 - e^{-2\Sigma(E)d_1}), \tag{9.18}$$

where $\Sigma(E)$ is the macroscopic total scattering cross section for neutron energy E, and d_1 and d_2 are the thicknesses of the inner and outer polyethylene moderators. Varying d_1 and d_2 changes the energy response. With $d_1 = 16$ mm and $d_2 = 32$ mm, the energy response of the counter very closely approximates, within $\pm 15\%$, the curve of rem per neutron per cm^2 versus neutron energy (Fig. 9.34). A commercial survey meter based on this design has a sensitivity of 12 cpm per μSv/hr (120 cpm per mrem/hr) and a range of 1–1000 μSv (0.1–100 mrem) per hour.

Commercially available neutron dose-equivalent meters (Fig. 9.35) that can measure neutron dose equivalent rates from 1 μSv per hour (0.1 mrem per hour) to 100 mSv (10,000 mrem) per hour, utilize a thermal neutron detector surrounded by a spherical or semi-spherical moderator. Either a small (about 4 mm × 4 mm) ^{6}LiI(Eu) scintillating crystal located in the center of the sphere or a $^{10}BF_3$ counter inserted into the moderator can be used as the neutron detector. However, because the $^{10}BF_3$ tube is less sensitive to

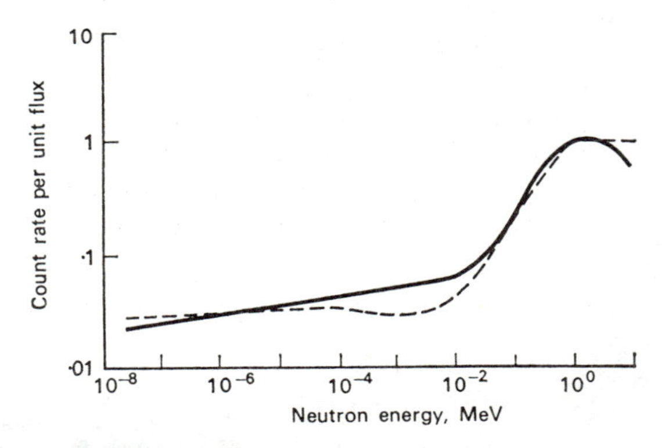

FIG. 9.34. Energy response curve (solid line) of the neutron rem counter shown in Fig. 9.33. The broken line is the dose equivalent per unit flux. Ideally, the response should be 7.6 counts per second per mrem per hour. (I. O. Anderson and J. Braun, A neutron rem counter with uniform sensitivity from 0.025 eV to 10 MeV, *Neutron Dosimetry*, vol. II, pp. 87–95, I.A.E.A., Vienna, 1963.)

gamma rays than is the scintillation crystal, the $^{10}BF_3$ counter is the most widely used neutron detector in neutron dose equivalent meters. The response of a spherical neutron dose equivalent meter, when the sphere is 30 cm in diameter, is approximately proportional to the neutron dose equivalent rate from thermal energies to about 15 MeV. This type of neutron dose-equivalent meter thus may be calibrated with neutrons of any energy within this range. Spheres smaller than 30 m diameter are relatively more sensitive to lower energy neutrons. Since the energy response of the instrument depends on the size of the spherical moderator, it is possible to determine the energy distribution in a neutron field by making a series of measurements with different sized spheres. The set of spheres used for this method of neutron spectroscopy is commonly called "Bonner spheres"; they range in diameter from 5 to 30 cm.

Albedo neutron dosimeter

Personnel monitoring for neutrons can be accomplished through the use of nuclear track film, wherein the neutron knocks a proton out of a molecule in the film emulsion, and the proton leaves a track in the film. A major drawback of this technique, however, is the neutron energy threshold requirement. Unless the neutron energy exceeds about 1/2 MeV, the recoil protons do not have sufficient energy to make recognizable tracks in the film. This disadvantage can be overcome through the use of neutron sensitive thermoluminescent dosimeters whose sensitivity is enhanced by neutrons that are backscattered from the body. Since the human body has many hydrogen atoms, a significant fraction of intermediate energy and fast neutrons can be slowed down to epithermal energies and backscattered, and thus can interact with the neutron-sensitive thermoluminescent material. One neutron monitor of this type is called an *albedo* type neutron dosimeter. Albedo dosimeters of this type are especially useful in the neutron energy range from the Cd cutoff, about 0.2 eV to about 500 keV.

Since the neutron-sensitive TLD also responds to gamma radiation, and since neutrons are almost always accompanied by gamma radiation, and since the neutron sensitive TLD is also sensitive to gamma radiation, a neutron insensitive TLD is used together with the TLD that responds to neutrons. The measured gamma thermoluminescence due to gammas can be separately determined and subtracted from the total thermoluminescence

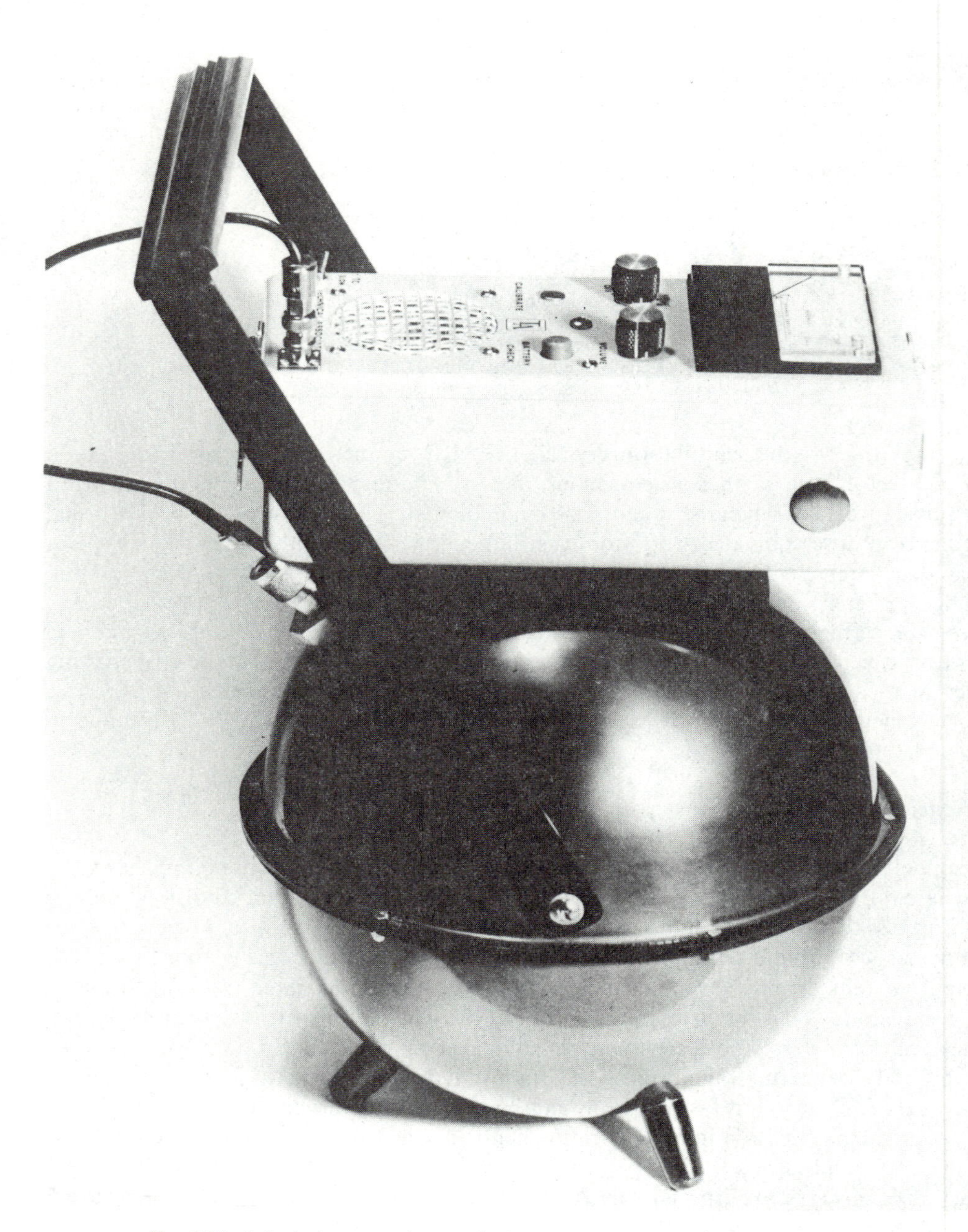

FIG. 9.35. Spherical neutron dose-equivalent meter. (Courtesy Technical Associates.)

of the neutron sensitive detector. To distinguish between the thermoluminescence due to neutrons and gammas in this way, ^{6}LiF TLD material is used as the neutron sensitive detector, while ^{7}LiF is used as the neutron-insensitive dosimeter. Both these TLD materials have about the same response to gamma radiation. For this reason, the gamma thermoluminescence of the ^{7}LiF chip can be subtracted from that of the ^{6}LiF chip to obtain a net luminescence that is due only to the neutrons. This differential thermoluminescence allows a measurement of 0.1 mSv (10 millirems) of neutrons in a gamma field of 2 mSv (200 millirems).

Albedo type neutron dosimeters are highly energy dependent. Their response changes by a factor of about 15 over the neutron energy range of 0.1 to 1.7 MeV. For this reason, an albedo neutron dosimeter must be calibrated with a neutron source whose spectral distribution of energy is as close as possible to the energy distribution of the neutrons to be monitored.

Calibration

Health physics survey instruments are calibrated by exposing the instrument in a known radiation field, and then comparing the meter reading to the known radiation field. These known fields are established with calibrated radiation sources whose calibration can be traced back to a national standards laboratory. Since health physics instruments are often used under many different conditions, a complete calibration should include, in addition to calibration under conditions for which the instrument was designed, an evaluation of the instrument's response to all the radiations and conditions that may reasonably be expected during its actual use in the field. Because of energy dependence, directional dependence, and the possible contribution of scattered radiation to the instrument's reading, the conditions under which the instrument was calibrated should be described in the calibration report.

Gamma Rays

For calibrating gamma ray measuring instruments, one of the most widely used sources is radium (usually a radium salt such as RaBr), in equilibrium with its daughters, sealed into a platinum–iridium capsule. The gamma radiation from such a source is very heterochromatic, as shown in Table 9.6; it originates mainly from the Ra B (^{214}Pb) to Ra C (^{214}Bi) and Ra C′ (^{214}Po) transitions. Since the half-lives of these two transitions are 26.8 and 19.7 minutes, respectively, the equilibrium condition is determined by the equilibrium between radium and its daughter radon. Radium sources are calibrated at the National Bureau of Standards (in the United States) by comparing, under standard conditions of measurement, the gamma radiation output of the unknown with that of a standard source.
measurement, the gamma radiation output of the unknown with that of a standard source.

Knowing the quantity of radium and the physical dimensions of the capsule permits the calculation of the gamma ray exposure rate at various distances from the source. For a radium source containing m mg radium in a capsule whose wall thickness is t mm the exposure rate at a distance d cm from the source is given by

$$\dot{X} = \frac{2.4 \times 10^{-3} m e^{-\mu t}}{d^2} \dot{C}/\text{kg per hr}, \tag{9.19A}$$

where μ is the linear absorption coefficient for the capsule material (for Pt, μ is 0.19 per mm). If the exposure is expressed in roentgens, the corresponding exposure rate is given by

$$\dot{R} = \frac{9.1 me^{-\mu t}}{d^2} \text{ R/hr}. \tag{9.19B}$$

The radiation exposure rate from 1 gm ^{226}Ra in equilibrium with its daughters is 212 μC/kg per hour (825 mR/hr) at a distance of 1 meter. The distance from the source for any other required exposure rate may be calculated with the aid of the inverse square law. Thus, if we had a source containing 100 mg Ra in a capsule of 0.5 mm Pt wall thickness, the exposure rate at 1 m would be 21.2 μC/kg per hour (82.5 mR/hr). To calibrate the 3.2 μC/kg per hour (12.5 mR/hr) midpoint on the scale of an instrument that measures up to 6.4 μC/kg per hour (25 mR/hr), the detector would be placed at a distance from the source calculated as follows:

$$\begin{aligned} \frac{I_1}{I_2} &= \frac{d_2^2}{d_1^2} \\ \frac{21.2}{3.2} &= \frac{d_2^2}{1} \\ d_2 &= 2.6 \text{ meters}. \end{aligned} \tag{9.20}$$

TABLE 9.6. APPROXIMATE GAMMA-RAY SPECTRUM OF RADIUM AND ITS EQUILIBRIUM DECAY PRODUCTS

Transition	Quanta per radium transformation	MeV/photon
Ra→Rn	0.012	0.184
Ra B→Ra C	0.115	0.241
	0.258	0.294
	0.450	0.350
Ra C→Ra C′	0.658	0.607
	0.065	0.766
	0.067	0.933
	0.206	1.120
	0.063	1.238
	0.064	1.379
	0.258	1.761
	0.074	2.198
	Total = 2.3	$\bar{E} = 0.7$

The inverse square law, which assumes the radium to be a point source, is applicable to distances that exceed about 10 times the largest linear dimension of the source. Scatter from the floor, walls, or ceiling of the room in which the calibration is performed will result in a slower decrease in radiation dose rate than that predicted by the inverse square law. This point may be checked by measuring the gamma-ray dose rate at several distances with a calibrated ion chamber or counter. A plot of the logarithm of the dose rate versus the logarithm of distance (or dose rate versus distance on log paper) will result in a straight line of slope -2 if the inverse square law is applicable. A smaller slope would suggest the presence of significant amounts of scattered radiation.

Other sources besides radium may be employed to calibrate gamma-ray measuring instruments. Among the most frequently used sources in this category are ^{60}Co and ^{137}Cs. Such sources must be of high specific activity in order to have a source sufficiently small (in physical dimensions) in order that it could be considered a "point" source. The nominal source strengths for ^{60}Co and ^{137}Cs are 9.2×10^{-3} μC/kg per hr per MBq (1.32 R/hr per Ci) and 2.3×10^{-3} μC/kg per hr per MBq (0.33 R/hr per Ci) at 1 meter. However, since these sources too are required to be encapsulated, the nature of the capsule must be known in order to account for radiation attenuation by the capsule. In using such a source, the radiation output should be measured with a previously calibrated instrument. It should be pointed out that since both these radioisotopes have relatively short half-lives (5.3 years for ^{60}Co and 30 years for ^{137}Cs), appropriate correction factors must be applied to the original source strength measurements or calculations to account for radioactive decay.

Beta Rays

Dose-response characteristics of most portable survey instruments for beta rays are strongly energy dependent. As a consequence, survey instruments are usually used only to detect beta radiation, but not to measure beta-ray dose rate. Such instruments, therefore, are normally not calibrated for beta-ray dose measurements. If beta-ray dose rates are required, an infinitely thick source of known specific activity may be made. The surface dose rate, which is equal to one-half the dose rate in an infinitely large medium of that specific activity, may be easily calculated. The beta-ray source used most frequently for beta dose-rate calibration is an infinitely thick (about 5 mm) slab of metallic uranium; the beta-ray dose rate at the surface is 2.4 mGy/hr (240 mrad/hr).

Alpha Rays

Portable survey instruments used to measure alpha contamination are usually designed to read in units of counting rate. Calibration sources for these instruments are most often metal discs on which are electroplated known amounts of an alpha emitting radioisotope. Such sources are listed in Table 9.7, and are readily available through commercial sources or through national standards laboratories.

TABLE 9.7. CALIBRATION SOURCES FOR ALPHA RAYS

Radionuclide	Alpha energy, MeV	Half-life
^{235}U	4.37–4.6	7.1×10^{8} yr
^{239}Pu	5.11–5.16	2.44×10^{4} yr
^{210}Po	5.3	138.4 days
^{241}Am	5.44–5.49	433 yr

Neutrons

For fast neutrons, commonly used calibration sources include Po–Be, Ra–Be, and Pu–Be. The approximate neutron yield for each of these sources is given in Table 5.3. The exact output of fast neutrons, in neutrons per sec, is usually given by the supplier of the source. For intercomparison of sources, the long counter of Hanson and McKibben may be used with an accuracy of $\pm 5\%$. The neutron flux at any distance, d, from the source, assuming the source to be a "point," is given by

$$\phi \frac{\text{neutrons}}{\text{cm}^2 \text{ sec}} = \frac{N \text{ neutrons/sec}}{4\pi d^2 \text{ cm}^2}. \tag{9.21}$$

All neutron sources supply fast neutrons. For the three α,n sources listed above, the neutron energies span a range from about $\frac{1}{2}$ MeV to about 10 MeV, with a mean energy of about 4–5 MeV (Figs. 5.21, 9.36, and 9.37. In using a fast neutron source for calibration of a dose equivalent neutron survey meter, attention must be paid to the neutron energy spectrum. For example, let us consider the case of a Pu–Be neutron source (Fig. 9.37) and determine the flux that corresponds to 25 μSv/hr (2.5 mrem/hr), or to 1 mSv (100 mrem)

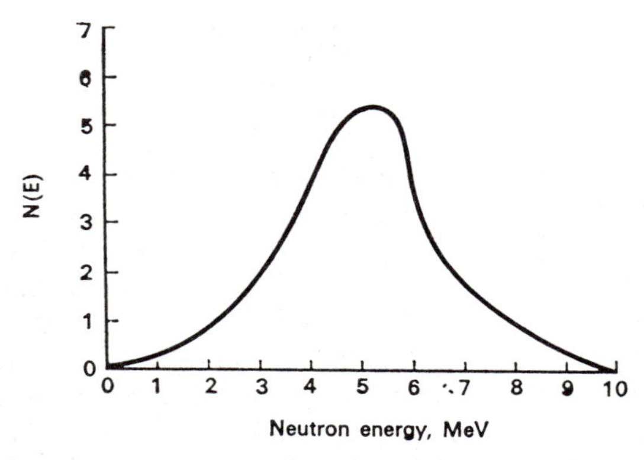

FIG. 9.36. Neutron energy spectrum for Ra–Be (α, n). The neutron spectrum was determined by measuring track lengths in nuclear emulsion. (N.B.S. Handbook 85, *Physical Aspects of Irradiation*, 1964.)

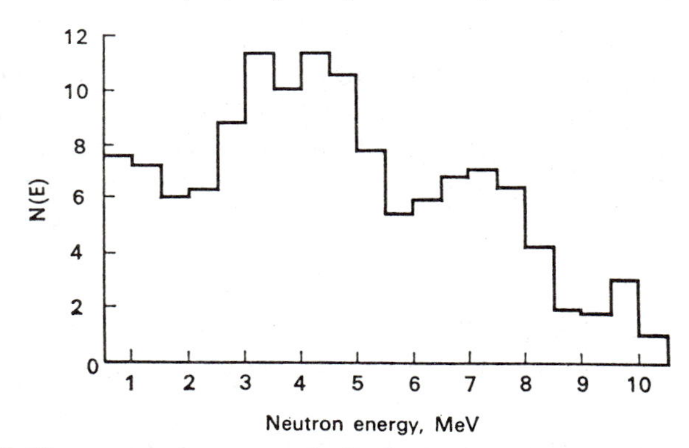

FIG. 9.37. Neutron energy spectrum for Pu–Be (α, n). Nuclear emulsions were used by Stewart (*Phys. Rev.* **98**, 740–3, 1955) for the energy measurements. (From *Neutron Sources and their Characteristics*, Tech. Bulletin NS-1, AECL, Ottawa.)

per 40-hr week. The spectral distribution is given in Table 9.8. From these data, and using the appropriate QF values for the various neutron energies, expressed as flux that will result in a dose equivalent of 25 μSv/hr (2.5 mrem/hr), we calculate that 20 neutrons per square centimeter per second corresponds to 25 μSv/hr (2.5 mrem/hr). If monoenergetic neutrons are required, then one of the γ, n sources listed in Table 5.4 must be used. If thermal neutrons are required for calibration purposes, the fast neutrons emitted from the neutron source must be thermalized. This is easily accomplished by surrounding the source with paraffin. However, it should be emphasized that not all the fast neutrons are thermalized. The beam that emerges from the moderator is heterochromatic, the moderator thus merely extends the lower end of the neutron energy spectrum to thermal energies. If the moderator is too thin, the neutron beam is too rich in fast neutrons; if the moderator is too thick, the ratio of thermal to fast neutrons increases, but the intensity of the neutron beam is decreased. The thickness of paraffin for maximizing the intensity of thermal neutrons from a calibration source is about 4 in. With this thickness, the thermal flux within a distance of 2 m from the source has been found to be about 40% of the fast neutron flux that would have been obtained without the paraffin moderator. If a relatively constant ratio of thermal to fast neutrons is required, the paraffin moderator thickness must be increased beyond four inches.

TABLE 9.8. DATA FOR COMPUTATION OF DOSE EQUIVALENT RATE FROM Pu–Be NEUTRONS

Energy interval E_i	Mean energy $\bar{E}_i$	f_i, fraction of neutrons in E_i	$\bar{E}_i \times f_i$	Flux ϕ_i, for 25 μ Sv/hr	$\phi_i \times f_i$
0–0.5	0.25	0.038	0.0095	82	3.12
0.5–1	0.75	0.049	0.0368	23	1.13
1–1.5	1.25	0.045	0.0563	19	0.86
1.5–2	1.75	0.042	0.0735	19	0.80
2–2.5	2.25	0.046	0.1035	20	0.92
2.5–3	2.75	0.062	0.1705	20	1.24
3–3.5	3.25	0.077	0.2503	18	1.39
3.5–4	3.75	0.083	0.3113	18	1.49
4–4.5	4.25	0.082	0.3485	17	1.39
4.5–5	4.75	0.076	0.3610	17	1.29
5–5.5	5.25	0.057	0.2993	16	0.91
5.5–6	5.75	0.042	0.2415	16	0.67
6–6.5	6.25	0.042	0.2625	17	0.71
6.5–7	6.75	0.052	0.3510	17	0.89
7–7.5	7.25	0.054	0.3915	17	0.93
7.5–8	7.75	0.051	0.3953	17	0.84
8–8.5	8.25	0.038	0.3135	17	0.64
8.5–9	8.75	0.017	0.1488	17	0.29
9–9.5	9.25	0.018	0.1665	17	0.30
9.5–10	9.75	0.022	0.2145	17	0.37
10–10.5	10.25	0.007	0.0718	17	0.11
		1.000	4.5774		$\Sigma = 20.29$

$$\text{Flux corresponding to } 25\frac{\mu\text{Sv}}{\text{hr}}\left(2.5\frac{\text{mrem}}{\text{hr}}\right) = \frac{\Sigma\,\phi_i f_i}{\Sigma f_i} = 20\frac{\text{neutrons}}{\text{cm}^2/\text{sec}},$$

$$\text{mean neutron energy} = \frac{\Sigma\,E_i f_i}{\Sigma f_i} = 4.6\text{ MeV}.$$

The exact ratio of thermal to fast neutrons may be measured by using the neutrons to activate a bare gold foil and a cadmium-covered gold foil. Because of the sharp cadmium absorption cut off at about 0.4 eV, the induced activity in the cadmium shielded foil is due only to neutrons whose energy exceeds 0.4 eV, while the activity of the bare foil is due to thermal neutrons as well as to the higher energy neutrons. The cadmium ratio, which is defined

$$\text{Cd ratio} = \frac{\text{activity of bare foil}}{\text{activity of Cd-covered foil}}, \tag{9.22}$$

is a measure of the purity of a thermal neutron field. The thermal flux may be computed from the gold foil activity measurements with the aid of equation (5.60).

Accuracy

The accuracy required of health physics instruments depends on how close the measured quantity is to a legally prescribed value or to the maximum values implied by ICRP recommendations. When measurements are made at levels on the order of 0.1 or less than the maximum permissible levels, an accuracy of ±200% is usually acceptable. Because of the high degree of conservatism built into the ICRP recommendations, an accuracy of ±30% for levels on the order of maximum permissible is usually acceptable. For levels considerably greater than implied or required maxima, the accuracy should be no worse than ±30%, because medical management of overexposed people is strongly influenced by their estimated dose (as well as by clinical signs and symptoms). In this discussion, health physics instruments must be sharply distinguished from radiation-measuring instruments used in the calibration of radiation producing machines in medical radiology. These instruments must have an accuracy of ±3% or better for the radiations for which their use is intended.

Counting Statistics

Distributions

Radioactive transformations and other nuclear reactions are randomly occurring events, and must therefore be described quantitatively in statistical terms. The sampling distribution of a population of randomly occurring events is called the binomial distribution and is given by the expansion of the binomial

$$(p+q)^n = p^n + np^{n-1}q + \frac{n(n-1)}{2!}p^{n-2}q^2 + \frac{n(n-1)(n-2)}{3!}p^{n-3}q^3 + \ldots, \qquad (9.23)$$

where p is the mean probability of the occurrence of an event, q is the mean probability of non-occurrence of the event, $p + q = 1$, and n is the number of chances of occurrence. The probability of the occurrence of exactly n events is given by the first term of the binomial expansion, the probability of occurrence of $n - 1$ events is given by the second term, and so on. For example, the likelihood of throwing 3 ones in three consecutive throws of a die, in which the mean probability of throwing a one is 1/6, is given by the first term of the expansion, according to equation (9.23), of $(1/6 + 5/6)^3$:

$$\begin{aligned}(1/6 + 5/6)^3 &= (1/6)^3 + 3(1/6)^2(5/6) + \frac{3 \times 2}{2!}(1/6)(5/6)^2 + \frac{3 \times 2 \times 1}{3!}(5/6)^3 \\ &= 1/216 + 15/216 + 75/216 + 125/216 = 1.\end{aligned}$$

The probabilities of 2 ones, 1 one, and no ones are given by the second, third, and fourth terms as 15/216, 75/216, and 125/216, respectively. A plot of these probabilities (Fig. 9.38, curve *A*), shows the distribution to be very asymmetrical. If we were to make similar calculations for the probability of throwing 6, 5, 4, 3, 2, 1 or 0 ones in six throws, that is $(1/6 + 5/6)^6$, we would find the distribution given in Fig. 9.38, curve *B*. By comparing curves *A* and *B* we see that the latter curve is more symmetrical. As n increases, the distribution curve becomes increasingly symmetrical around the center line. For the case where n is infinite, we have the familiar bell-shaped *normal curve*. For cases where $n \geq 30$ the binomial sampling distribution curve is for practical purposes indistinguishable from a normal curve. The normal distribution, which is given by

$$p(n)=\frac{1}{\sigma\sqrt{(2\pi)}}e^{-(n-\bar{n})^2/2\sigma^2}, \tag{9.24}$$

where $p(n)$ = probability of finding exactly n,
$\bar{n}$ = mean value,
σ = standard deviation (also called standard error),
must be fitted by two parameters: the mean and the standard deviation.

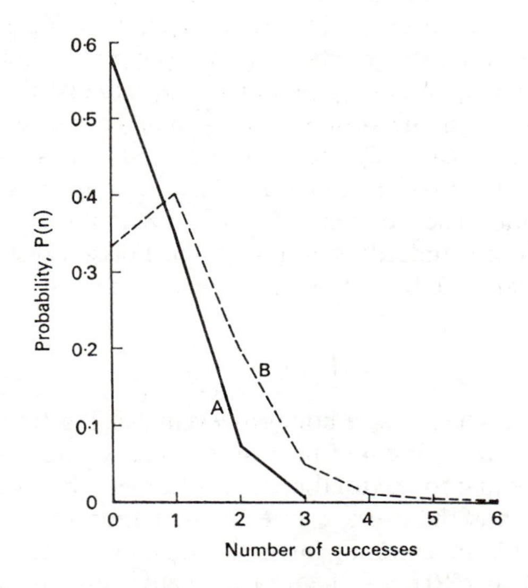

FIG. 9.38. Probability of throwing 0, 1, 2, and 3 ones in three throws of a die, curve *A*; and the probability of throwing 0, 1, 2, 3, 4, 5, and 6 ones in six throws of a die, curve *B*.

For the case where $p \ll 1$, that is where the occurrence of an event is highly improbable, the binomial distribution approaches the *Poisson distribution*. Since radioactive decay of a particular atom is a highly unlikely event (in ^{32}P, for example, whose half-life is 14.3 days, the probability of decay of any particular atom is given by the decay constant as 5.6×10^{-7}/sec), radioactive processes are described by Poisson statistics. According to the Poisson distribution, the probability of the occurrence of exactly n events per unit time, if the true average rate is $\bar{n}$, is given by the $(n + 1)$ term of the expansion of

$$e^{-\bar{n}} \times e^{\bar{n}} = 1, \tag{9.25}$$

which gives, upon series expansion of $e^{\bar{n}}$:

$$e^{\bar{n}}\left(1+\frac{\bar{n}}{1!}+\frac{\bar{n}^2}{2!}+\frac{\bar{n}^3}{3!}+\ldots\right), \tag{9.26}$$

or by the general term for the Poisson distribution

$$p(n)=\frac{(\bar{n})^n \times e^{-\bar{n}}}{n!}. \tag{9.27}$$

For example, if we have 37 Bq (0.001 μCi) of activity, the mean number of transformations per second is 37. The probability of observing exactly 37 transformations in 1 sec is calculated from equation (9.27) as

$$p(37) = \frac{37^{37} \times e^{-37}}{37!}.$$

Using Sterling's approximation

$$n! = (2\pi n)^{1/2} \times \left(\frac{n}{e}\right)^n \tag{9.28}$$

to evaluate 37!, we find $p(37)$ to be equal to 0.066. It thus follows that a wide range of observations around the true mean of 37 would be made in a series of measurements of this activity. This distribution of observations is given by the standard deviation of the distribution of measurements. As in the case of the normal distribution, in a large number of measurements, 68% of all the observations would lie between plus and minus one standard deviation of the mean, 96% between plus and minus two standard deviations, and so on. The normal distribution, equation (9.24), contains, as one of the parameters, σ, the standard deviation. The Poisson distribution, equation (9.27), contains only one parameter, the mean. The standard deviation of the Poisson distribution is equal to the square root of the mean number of observations made during a given measurement interval:

$$\sigma(\text{Poisson}) = \sqrt{n}. \tag{9.29}$$

The "size" of the sample when we are counting events such as radioactive decay or other nuclear events is the arbitrary length of time over which we are making the measurement. In equation (9.29), n is equal to the total number of events that occurred during that time of observation, and is thus the average rate for that time interval. Thus, if we observe 10,000 counts during a 10-min counting interval, the standard deviation of the observation is $\sqrt{10{,}000} = 100$ counts per 10 min. The measurement represents a mean value of 10,000 counts for the 10-min measurement interval. One of the main virtues of the Poisson distribution is that, for practical purposes, when $n \geq 20$, it is indistinguishable from a normal distribution of the same mean and of standard deviation equal to the square root of the mean. Under these conditions, all statistical tests that are valid for normal distribution, such as the t-test, the chi-square criterion, and the variance ratio test (which is called the F-test) may also be used for Poisson distributions. Although all these tests are applicable to radioactivity measurements, a full discussion of them is beyond the scope of this book. Details and applications of these tests may be found in the suggested references at the end of this chapter.

In the example cited above, where 10,000 counts were recorded during 10 min of counting, the mean counting rate, in counts per minute and the standard deviation of the mean rate is given by

$$r \pm \sigma_r = \frac{n}{t} \pm \frac{\sqrt{n}}{t}. \tag{9.30}$$

Since

$$\sigma_r = \frac{\sqrt{n}}{t} = \sqrt{\left(\frac{n}{t} \cdot \frac{1}{t}\right)} = \sqrt{\left(r \cdot \frac{1}{t}\right)},$$

we have

$$r \pm \sigma_r = r \pm \sqrt{\frac{r}{t}}, \tag{9.31}$$

which gives 1000 ± 10 cpm. If the activity had been measured over a 1-min interval, and had given 1000 counts, we would have $1000 \pm \sqrt{1000}$ or 1000 ± 32 cpm. The relative probable error, which is defined as

$$\text{relative probable error} = \frac{\sqrt{n}}{n}, \tag{9.32}$$

is 3.2% in the case of the 1-min count, but only 1% in the case of the 10-min count. The degree of precision of a given counting measurement is thus seen to depend on the total number of counts.

When making radioactivity measurements, we usually must account for background, and are thus interested in the net counting rate; that is, the difference between the gross counting rate of the sample, which includes background, and the background counting rate. Each of these counting rates has its own standard deviation. The standard deviation of the net counting rate is given as

$$\sigma_n = \sqrt{(\sigma_g^2 + \sigma_{bg}^2)} = \sqrt{\left(\frac{r_g}{t_g} + \frac{r_{bg}}{t_{bg}}\right)}, \tag{9.33}$$

where σ_g = standard deviation of gross counting rate,
σ_{bg} = standard deviation of background counting rate,
r_g = gross counting rate,
r_{bg} = background counting rate,
t_g = time during which gross count was made,
t_{bg} = time during which background count was made.

Example 9.5

A 5-min sample count gave 510 counts, while a 1-hr background measurement yielded 2400 counts. What is the net sample counting rate and the standard deviation of the net counting rate?

$$r_n = \frac{510 \text{ counts}}{5 \text{ min}} - \frac{2400 \text{ counts}}{60 \text{ min}} = 102 \text{ cpm} - 40 \text{ cpm}$$

$$= 62 \text{ cpm}.$$

$$\sigma_n = \sqrt{\left(\frac{102}{5} + \frac{40}{60}\right)} = 4.6$$

62 ± 4.6 cpm.

The standard deviation is a measure of the dispersion of randomly occurring events around the mean. If a large number of replicate measurements were made, it was pointed out above that 68% of the observations would fall within ± 1 standard deviation of the mean. For this reason, we say that one standard deviation is the 68% confidence limit. Similarly, two standard deviations represent the 96% confidence limit. If we report data within the limits of two standard deviations, this means that we are 96% certain that the true value lies within the limits given. Several levels of confidence, together with their corresponding number of standard deviations, are given in the table below:

Confidence level	Number of standard deviations
50%	0.6745
68%	1.0
90%	1.645
95%	1.960
96%	2.0
99%	2.575

Example 9.6

A preliminary measurement made during a short counting time suggested a gross counting rate of 55 cpm. The background counting rate, determined by a 1-hr measurement, is 25 cpm. How long should the sample be counted in order to be 96% certain that the measured net counting rate will be within 10% of the true counting rate?

The estimated net counting rate is 30 cpm; 10% of this is 3 cpm. Since we want to be at the 96% confidence level, this allowable error of ± 3 cpm represents two standard deviations; one standard deviation therefore is 1.5 cpm. Using this value in equation (9.33), and substituting the other given values leads to

$$1.5 = \sqrt{\left(\frac{55}{t_g} + \frac{25}{60}\right)},$$

$$t_g = 30.1 \text{ min}.$$

Difference between Means

In making counting measurements, we are almost always interested in the difference between two counting rates—as, for example, the difference between the sample counting rate and the background counting rate, or the difference between the net counting rates of two samples. If this difference is very great, then we know intuitively that there is a real difference between the two samples. However, if the difference is small, then, because of the fact that radioactive decay is a random process in which we know that 96% of the measurements will lie between the true mean and plus or minus two standard deviations, we cannot decide intuitively whether the observed difference is merely due to errors of random sampling, and that the two measurements are in fact two samples of the same population. To help in our decision, we make use of an objective statistical test based on the *Null Hypothesis*. The null hypothesis assumes that there is no difference between the two measurements. On this assumption, we calculate the probability of finding a difference between the two mean counting rates as great as or greater than that actually observed. Furthermore, we arbitrarily set a limit to this probability. If the calculated likelihood of randomly finding the measured difference is greater than this limit, then we say that there is no difference between the two counting rates, and that the two samples come from the same population. If, on the other hand, the probability of finding the measured difference among samples of the same population is less than the calculated value, then we reject the null hypothesis and we say that the difference between the two means is statistically significant; the samples are in fact different. The two arbitrary levels of significance most frequently used in statistical calculations is 1% and 5%. That means, if we choose the 1% level, that if the probability of randomly finding a difference between two samples of the same population as great as that observed in the two sample measurements is greater than 1 in 100, we accept the null hypothesis, and we say that there is no difference between the two samples. If the calculated probability is 1 in 100 or less, then we say that the difference is significant. It should be emphasized that the significance level is purely arbitrary, and is set by the experimenter. The experimenter is not bound to the 1% or 5% level; he may, if he wishes, be more liberal, and use 10%, or any other significance level. However, if he uses a 10% level, he is more likely to accept an apparent difference, which in fact is not real, than he would if he used 5% or 1% criteria. In reading and interpreting experimental data, therefore, it is important to know the probability level that the experimenter used as a criterion for significance.

Determination of the significance of the difference between means is based on the fact that, in a normally distributed population, not only are means of population samples normally distributed about the true means of the population, but differences between means are also normally distributed. If we should draw a very large number of duplicate samples from a population, compute the mean for each sample, then subtract the mean of the second sample, M_2, from the mean of the first sample, M_1, and plot the differences, we will obtain a normal curve about a mean of zero, whose standard deviation is called the standard error of the difference between means. To estimate the standard error of the difference between means from two samples, we use equation (9.33), which can be written in a more general form as

$$\sigma_{\text{diff}} = \sqrt{(\sigma_{M_1}^2 + \sigma_{M_2}^2)}, \tag{9.34}$$

where σ_M^2 is the square of the standard error of the mean. In counting measurements, the standard error of the mean is given in equation (9.31) which gives the standard error of the mean counting rate as

$$\sigma_M = \sqrt{\frac{r}{t}}. \tag{9.35}$$

The t test is used to tell us by how many units of the standard error of the difference between means the difference between two measured means differs from zero:

$$t = \frac{|M_1 - M_2|}{\sigma_{\text{diff}}} = \frac{|M_1 - M_2|}{\sqrt{r_1/t_1 + r_2/t_2}}. \tag{9.36}$$

Example 9.7

A sample of drinking water was counted for 10 min, and gave 530 counts. A 30-min background count gave a background rate of 50 cpm. At the 95% confidence level, was there any activity in the water?

$$M_1 = \frac{530 \text{ counts}}{10 \text{ min}} = 53 \text{ cpm} \qquad t_1 = 10 \text{ min},$$

$$M_2 = 50 \text{ cpm} \qquad t_2 = 30 \text{ min},$$

$$t = \frac{53 - 50}{\sqrt{(53/10 + 50/30)}} = 1.13.$$

This result shows that the measured difference differs from zero by 1.13 standard deviations. From a table of areas under a normal curve, we find that 1.13 standard deviations includes 37% of the area on each side of the mean, or 74% of the area between ± 1.13 standard deviations. We interpret this result as saying that we would expect a difference this great or greater 26 times out of 100 if the two samples came from the same population. Therefore, the difference between the two samples is not significant. To be significant at the 95% confidence level, the difference between the two means would have to be great enough to be observed only 5 times or less out of 100.

Optimization of Counting Time

Optimal use of a given period of time for determining the net counting rate of a sample may be made by dividing the background counting time and the sample counting time in order to minimize the statistical uncertainty of the net counting rate. The standard deviation of the net counting rate is given by equation (9.33) as

$$\sigma_n = \sqrt{\left(\frac{r_g}{t_g} + \frac{r_{bg}}{t_{bg}}\right)}. \tag{9.33}$$

Squaring equation (9.33) gives

$$\sigma_n^2 = \frac{r_g}{t_g} + \frac{r_{bg}}{t_{bg}}, \tag{9.37}$$

and differentiating equation (9.37) with respect to t yields

$$2\sigma_n \, d\sigma_n = -r_g t_g^{-2} \, dt_g - r_{bg} t_{bg}^{-2} \, dt_{bg}. \tag{9.38}$$

To minimize t, set $d\sigma_n = 0$

$$0 = \frac{r_g}{t_g^2} dt_g - \frac{r_{bg}}{t_{bg}^2} dt_{bg}, \tag{9.39}$$

which shows that

$$\frac{r_g}{t_g^2} dt_g = -\frac{r_{bg}}{t_{bg}^2} dt_{bg}. \tag{9.40}$$

Since total counting time, t, is constant

$$t = t_g + t_{bg}, \tag{9.41}$$

and therefore

$$dt_g + dt_{bg} = 0. \tag{9.42}$$

Rearranging equation (9.40) and applying equation (9.42) yields

$$\frac{r_g/t_g^2}{-r_{bg}/t_{bg}^2} = \frac{dt_{bg}}{dt_g} = \frac{dt_{bg}}{-dt_{bg}} = -1, \tag{9.43}$$

which leads to the optimum division of the total counting time between sample counting time and background counting time

$$\frac{t_g}{t - t_g} = \sqrt{\frac{r_g}{r_{bg}}}. \tag{9.44}$$

Example 9.8

A total of 1 hour is available for counting a sample and the background. Preliminary measurements show the background to be about 15 counts per minute, and the sample count rate to be about 22 cpm. How long should the sample and the background be measured in order to minimize the statistical counting error?

From equation (9.44) we have

$$\frac{t_g}{60 - t_g} = \sqrt{\frac{22}{15}},$$

$$t_g = 33 \text{ minutes}$$

$$t_{bg} = 60 - t_g = 27 \text{ minutes}.$$

Weighted Means

Frequently, several different samples are taken from the same batch or universe, and then the value assigned to the universe is the mean of the several samples. For example, the mean concentration of activity in a lake or a river is determined from a number of samples taken from the lake or river. This procedure is statistically valid only if all the measurements are made with the same degree of relative precision. If the precision of the measurements differs, we must weight the more precise measurements more heavily than the less precise measurements in order to arrive at the best estimate of the true mean. The procedure for weighting the means is illustrated by the following example:

Example 9.9

Four samples were taken from a pond for activity determination, and the following counting rates were found, per liter of sample:

$$95 \pm 3$$
$$105 \pm 10$$
$$94 \pm 6$$
$$118 \pm 12.$$

The mean and standard deviation of these data are 103 ± 11.2. The standard error of the mean, which is given by

$$\sigma_M = \frac{\sigma}{\sqrt{(n-1)}}, \tag{9.45}$$

where σ = standard deviation of the distribution,
n = number of samples,

is found to be ± 6.5, and, according to this calculation, the best estimate of the mean counting rate is 103 ± 6.5. However, it should be noted that the precision of the four analytical determinations vary widely, from 3.2% for the first value to 10.2% for the last one. In the calculation above, all the values were given equal consideration in arriving at the mean. Because of the wide variation in precision, the first value should be given greater weight than the last value. This weighting is done in the following manner:

Define a weighting factor

$$w_i = \frac{1}{\sigma_i^2}. \tag{9.46}$$

The weighted mean is then given by

$$M_w = \frac{\sum w_i M_i}{\sum w_i}, \tag{9.47}$$

and the standard error of the weighted mean is given by

$$\sigma_{M_w} = \sqrt{\frac{1}{w_1 + w_2 + \cdots + w_n}}. \tag{9.48}$$

For the data in this example:

$M_i \pm \sigma_i$	σ_i^2	$w_i = 1/\sigma_i^2$	$w_i \times M_i$
95 ± 3	9	0.1111	10.56
105 ± 10	100	0.0100	1.05
94 ± 6	36	0.0278	2.61
118 ± 12	144	0.0070	0.82
		$\Sigma w_i = 0.1559$	$\Sigma w_i M_i = 15.04$

$$M_w = \frac{15.04}{0.156} = 96.5$$

$$\sigma_{M_w} = \sqrt{\frac{1}{0.156}} = 2.54.$$

The best estimate of the true mean, therefore, is 96.5 ± 2.5.

Problems

1. If a certain counting standard has a mean activity of 1000 cpm,
(a) What is the probability of observing exactly 400 counts in 1 min?
(b) What is the probability of measuring 390–410 counts in 1 min?

2. A sample counted 560 counts in 10 min, while the background counted 390 counts in 15 mins.
(a) What is the standard deviation of the gross and background counting rates?
(b) What is the standard deviation of the net counting rate?
(c) What are the 90% and 99% confidence limits for the net counting rate?

3. A background counting rate of 30 cpm was determined by a 60-min count. A sample that was counted for 5 min gave a gross count of 170.
(a) At the 90% confidence level, is there activity in the sample?
(b) Is there activity in the sample at the 95% confidence level?

4. As a test of the operation of a certain counter, two measurements were made on the same long-lived sample. The first gave 10,210 counts in 10 min, and the second gave 4995 counts in 5 min. Is the counter operating satisfactorily?

5. A 1 min count shows a gross activity of 35 counts. If the background is 1560 counts in 60 min, how long must the sample be counted in order to be within $\pm 10\%$ of the true activity at the 95% confidence level?

6. A sample that had been counted for 15 min showed a mean counting rate of 32 cpm. The background, counted for 10 min, was 15 cpm.
(a) What is the net counting rate, at the 95% confidence limit?
(b) What is the probable error (relative error at ± 1 standard deviation of the net counting rate?

7. A sample has an estimated gross counting rate of 35 cpm (based on a 2-min count). The background, determined by a 1 hr count, is 10 counts per minute. How long should the sample be counted if we want to be 95% certain that the net counting rate is within $\pm 5\%$ of the true net counting rate?

8. (a) The gross 1 min count on a sample was 100, and the background, counted during 1 min was 50 counts. What was the net counting rate at the 90% confidence level?
(b) If the sample and background were each counted for 10 min, and gave counting rates of 100 and 50 cpm respectively, what was the net counting rate at the 95% confidence level?

9. A shielded low background counter has an average counting rate of 2 cpm. What is the probability that a 1 min counting period will record?
(a) 2 counts,
(b) 4 counts,
(c) 0 counts.

10. A sample of river water was taken near the waste discharge pipe of an isotope laboratory, and another sample was taken upstream of the discharge point. Each sample was counted for 10 min, and gave 225 and 210 cpm, respectively. At the 99% confidence level, is the downstream water more radioactive than the water upstream?

11. A certain counting standard has a true mean counting rate of 50 cpm.
(a) What is the probability of observing exactly 50 counts in 1 min?
(b) What is the probability of measuring 43–47 counts in 1 min?
(c) What is the probability of finding more than 57 counts in 1 min?

12. An ionization chamber has a window thickness of 2 mg/cm^2. If a 0.01 μCi (370 Bq) ^{210}Po source is located 1 cm in front of the window, so that the counting geometry is 25%, calculate the saturation ionization current.

13. An "air" wall, air-filled ionization chamber, whose volume is 100 cm^3, gives a saturation current of 10^{-12} A when placed in an X-ray field. If the temperature was 27°C, and the atmospheric pressure was 740 mm Hg, what was the radiation exposure rate?

14. (a) What value resistor, to be placed in series with the ion chamber, problem 13, is required to generate a voltage drop of 10 mV?

(b) If the capacity of the chamber is 250 $\mu\mu$F, what is the time constant of detector circuit?

(c) How much time is required before the meter will read 99% of the saturation current?

15. A pocket dosimeter has a capacitance of 5 $\mu\mu$F and a sensitive volume of 1.5 cm^3. What is the charging voltage if it is to be used in the range 0–200 mR (0–51.5 μC/kg) and the voltage across the dosimeter should be one-half the charging voltage when the dosimeter reads 200 mR?

16. A Geiger tube has a capacitance of 25 $\mu\mu$F. The time required to collect all the positive ions is 221×10^{-6} sec. In order to produce sharp output pulses, it is desired to limit the time constant of the detector circuit to 50 μsec.

(a) What is the value of the series resistor?

(b) If 10^8 ion pairs are formed per Geiger pulse, what is the upper limit of the output voltage pulse?

17. A Geiger counter has a resolving time of 250 μsec. What fraction of the counts is lost due to the counter's dead time if the observed counting rate is 30,000 cpm?

18. The fact that the gas multiplication in a proportional counter is very much less than that in a Geiger counter means that a pulse amplifier for use with a proportional counter must have a lower input sensitivity than one used with a Geiger counter. Calculate the input sensitivity for an amplifier to be used with a 2-in. dia hemispherical windowless gas-flow proportional counter whose capacitance is 20 $\mu\mu$F and which is operated to give a gas amplification of 5×10^3. Assume that the output pulse is "clipped" to one-half its maximum height.

19. What is the sensitivity of a thermal neutron detector whose volume is 50 cm^3, and is filled with 96% enriched $^{10}BF_3$ to a total pressure of 70 cm Hg?

20. If the BF_3 tube of problem 13 is used as a current ionization chamber, what saturation current would result from a thermal flux of 10^9 neutrons per cm^2/sec?

21. How long would it take for the sensitivity of the BF_3 detector of problem 19 to decrease by 10%?

22. What is the sensitivity for 1 MeV and for 10 MeV neutrons (amps per neutron per cm^2/sec) of an ion chamber that is filled with CH_4 gas to a pressure of 760 mm Hg, if its volume is 500 cm^3?

23. A 1000 MBq (27 mCi) ^{60}Co source is lost. At what distance can the lost source be detected with a survey meter whose sensitivity is 0.013 μC/kg per hr (0.05 mR/hr) above background?

24. The thermal neutron flux from a moderated ^{252}Cf neutron calibration source is determined by irradiating a gold foil 1 cm diameter × 0.013 cm thick for a period of 7 days at a distance of 100 cm from the source. The foil was counted immediately after the end of the irradiation period, found to have an activity of 100 Bq (2.7 nCi). What was the thermal flux at the point where the foil was irradiated? The activation cross section for gold is 98.5 barns.

25. A thermal neutron counter 1 cm diameter × 10 cm long is filled with BF_3 gas at atmospheric pressure and 20°C. What is the counting rate when the counter is in a thermal neutron flux ($E_{mp} = 0.025$ eV) of 1000 neutrons per square cm per second?

Suggested References

Adams, F., and Dams, R. *Applied Gamma-Ray Spectrometry*, 2nd ed. Pergamon Press, Oxford, 1970.

Attix, F. H., Roesch, W. C., and Tochilin, E., eds. *Radiation Dosimetry*, Vol. II, *Instrumentation*. Academic Press, New York, 1966.

Birks, J. B. *The Theory and Practice of Scintillation Counting*. Pergamon Press, Oxford, 1970.

Cameron, J. R., Suntharalingham, N., and Kenney, G. N. *Thermoluminescent Dosimetry*. University of Wisconsin Press, Madison, 1968.

Eicholz, C. G., and Posten, J. W. *Principles of Nuclear Radiation Detection*. Ann Arbor Science, Ann Arbor, 1979.

Golnick, D. A. *Experimental Radiological Health Physics*. Pergamon Press, Oxford, 1978.

Handloser, J. S. *Health Physics Instrumentation*. Pergamon Press, Oxford, 1959.

Knoll, G. F., *Radiation Detection and Measurement*. Wiley, New York, 1979.

Price, W. S. *Nuclear Radiation Detection*, 2nd ed. McGraw-Hill, New York, 1965.

Sharpe, J. *Nuclear Radiation Detectors*. Methuen, London, 1964.

Siegbahn, K., ed. *Alpha, Beta, and Gamma Ray Spectroscopy*, Vol. 1. North Holland Publishing Co., Amsterdam, 1965.

Snell, A. H., ed. *Nuclear Instruments and Their Uses*. Wiley, New York, 1962.

Tait, W. H. *Radiation Detection*. Butterworths, London, 1980.

U. S. Dept. of Commerce. *Ionizing Radiation Measurement Criteria for Regulatory Purposes*. NBS GCR-79-174,

The following reports of the International Commission on Radiation Units and Measurements, Washington, D.C.:
No.
10b. *Physical Aspects of Irradiation* (1964).
10f. *Methods of Evaluating Radiological Equipment and Materials* (1963).
12. *Certification of Standardized Radioactive Sources* (1968).
20. *Radiation Protection Instrumentation and Its Application* (1971).
22. *Measurement of Low-Level Radioactivity* (1972).
26. *Neutron Dosimetry for Biology and Medicine* (1977).
27. *An International Neutron Dosimetry Intercomparison* (1978).
28. *Basic Aspects of High Energy Particle Interactions and Radiation Dosimetry* (1978).
31. *Average Energy Required to Produce an Ion Pair* (1979).
33. *Radiation Quantities and Units* (1980).
The following reports of the National Council on Radiation Protection and Measurements, Washington, D.C.:
No.
23. *Measurement of Neutron Flux and Spectra* (1960).
25. *Measurement of Absorbed Dose of Neutrons and of Mixtures of Neutrons and Gamma Rays* (1961).
27. *Stopping Powers for Use with Cavity Chambers* (1961).
47. *Tritium Measurement Technics* (1976).
50. *Environmental Radiation Measurements* (1976).
57. *Instrumentation and Monitoring Methods for Radiation Protection* (1978).
58. *A Handbook of Radioactivity Measurements Procedures* (1978).

CHAPTER 10

EXTERNAL RADIATION PROTECTION

Basic Principles

Radiation protection practice is a special aspect of the control of environmental health hazards by engineering means. In the industrial environment, the usual procedure is first to try to eliminate the hazard, as was done when benzene was replaced with carbon tetrachloride, and then the carbon tetrachloride was replaced with trichloroethylene as a solvent for degreasing machined parts. If elimination of the hazard is not feasible, an attempt is made to enclose the hazard, thereby isolating the hazard from man. If neither of these two solutions can be achieved, exposure to the hazard can usually be prevented by isolating the man. The exact manner of application of these general principles to radiation protection depends on the individual situation. It is convenient, in radiation-protection practice, to break down the problem to protection against external radiation and protection against personal contamination resulting from inhaled, ingested, or tactilly transmitted radioactivity.

Technics of External Radiation Protection

External radiation originates in X-ray machines and in other devices specifically designed to produce radiation; in devices in which production of X-rays is a side effect, as in the case of the electron microscope; and in radioisotopes. If it is not feasible to do away with the radiation source, then exposure of personnel to external radiation may be controlled by concurrent application of one or more of the following three technics:

1. minimizing exposure time,
2. maximizing distance from the radiation source,
3. shielding the radiation source.

Time

Although many biological effects of radiation are dependent on dose rate, it may be assumed, for purposes of environmental control, that the reciprocity relationship

$$\text{dose rate} \times \text{exposure time} = \text{total dose}$$

is valid. For total dose within one or two orders of magnitude of the RPG value, we have no data, either clinical or experimental, to contraindicate this assumption. Thus, if work must be performed in a relatively high radiation field, such as the repair of a cyclotron made radioactive by the absorption of neutrons, or manipulation of a radiographic source in a complex casting, restriction of exposure time so that the product of dose rate and exposure time does not exceed the maximum allowable total dose permits the work to be done in accordance with radiation safety criteria. For example, in the case of a radiographer who must make radiograph 5 days per week while working in a radiation field of 0.25 mSv/hr (25 mrem/hr), overexposure can be prevented by limiting his daily working time in the radiation field to 48 min. His total daily dose would then be only 0.2 mSv (20 mrem). If the volume of work requires a longer exposure, then either another radiographer must be used or the operation must be redesigned in order to decrease the intensity of the radiation field in which the radiographer must work.

Distance

Intuitively, it is clear that radiation exposure decreases with increasing distance from a radiation source. When translated to quantitative terms, this fact becomes a powerful tool in radiation safety. We will consider three cases, viz: a point source, a line source, and a surface source.

In the case of a point source, the variation of dose rate with distance is given simply by the inverse square law, equation (9.20). Cobalt 60, for example, which emits one photon of 1.17 MeV and one photon of 1.31 MeV per disintegration, has a source strength which is approximated by equation (6.17):

$$\begin{aligned}\Gamma &= 3.6 \times 10^{-9} \Sigma f_i E_i \frac{(\mathrm{C/kg})\ \mathrm{m}^2}{\mathrm{MBq\text{-}hr}} \\ &= 3.6 \times 10^{-9}(1 \times 1.17 + 1 \times 1.31) \\ &= 9 \times 10^{-9} \frac{(\mathrm{C/kg})\ \mathrm{m}^2}{\mathrm{MBq\text{-}hr}} \left(1.3 \frac{\mathrm{R\text{-}m}^2}{\mathrm{Ci\text{-}hr}}\right).\end{aligned}$$

For a 37-MBq (100-mCi) source, the exposure rate at a distance of 1 meter is about 33.5 μC/kg per hour (130 mR/hr). If a radiographer were to manipulate this source for 1 hour per day, his maximum dose rate should not exceed 0.2 mSv/hr (20 mrem/hr). This restriction could be attained through the use of a remote handling device whose length, as calculated from the inverse square law, equation (9.20) is at least 2.5 meters.

If the radiography is to be done at one end of the shop, which is set aside exclusively for this purpose, then either a barricade must be erected outside of which the dose rate does not exceed the maximum allowable weekly rate, or if this is not possible because of space limitations, a shield must be erected. If the barricade is used, its distance from the source must be such that the dose rate will not exceed

$$\frac{1\ \mathrm{mSv/week}}{40\ \mathrm{hrs/week}} = 0.025\ \mathrm{mSv/hr} \quad (2.5\ \mathrm{mrem/hr}).$$

By the inverse square law, this distance is found to be 7.2 meters.

In the case of a line source of radiation, such as a pipe carrying contaminated liquid waste, the variation of dose rate with distance is somewhat more complex mathematically than in the case of the point source. If the linear concentration of activity in the line is C_l MBq or curies per unit length of a gamma emitter whose source strength is Γ, then the dose rate at point p (Fig. 10.1), at a distance h from the infinitesimal length dl, is given by

$$dD_p = \frac{\Gamma \times C_l \times dl}{l^2 + h^2}, \tag{10.1}$$

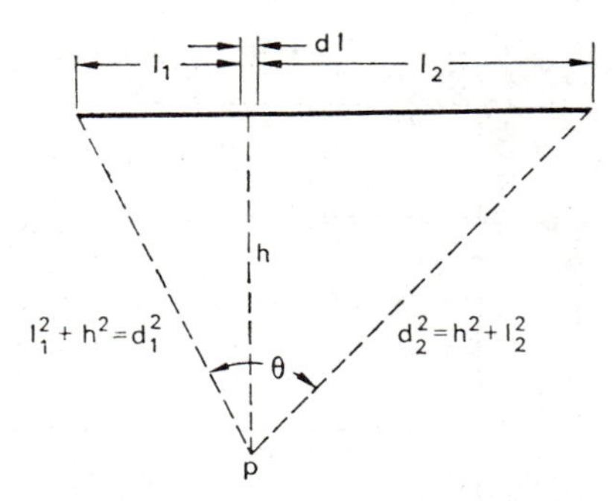

FIG. 10.1. Geometry for computing the gamma-ray dose rate at a finite distance, h, from a line source of uniform activity C_l mCi per unit length.

and for the dose rate due to the activity in the total length of pipe, we have

$$D_p = \Gamma C_l \int_0^{l_1} \frac{dl}{l^2 + h^2} + \Gamma C_l \int_0^{l_2} \frac{dl}{l^2 + h^2}$$

$$= \frac{\Gamma C_l}{h}\left(\tan^{-1}\frac{l_1}{h} + \tan^{-1}\frac{l_2}{h}\right)$$

$$= \frac{\Gamma \times C_l \times \theta}{h}. \qquad (10.2)$$

Example 10.1

Induced ^{24}Na activity in a cooling water line passes through a small diameter pipe in an access room 6 m wide. The door to the room is in the center of the 6 m wall, at a distance of 3 m from the pipe, as shown in Fig. 10.2. If the linear concentration of activity is 100 MBq per meter,

(a) What is the dose equivalent rate in the doorway ($\dot{H}_1$)?

For ^{24}Na, $\Gamma = 1.28 \times 10^{-8}$ x units per hour per MBq at 1 meter (from Table 6.3). The exposure rate at a distance of 3 meters from the pipe is

$$\dot{X}_1 = \frac{2\Gamma C_l}{h}\tan^{-1}\frac{l}{h}$$

$$= \frac{2 \times 1.28 \times 10^{-8} \times 100}{3}\tan^{-1}\frac{3}{3}$$

$$= 8.53 \times 10^{-7} \times \frac{\pi}{4} = 6.7 \times 10^{-7}\,\text{X units per hour.}$$

An exposure of 1 X unit leads to an air dose of 34 grays. Since the quality factor of gamma radiation is one, we have one sievert per gray. The dose equivalent rate, therefore, in the doorway is

$$\dot{H}_1 = 6.7 \times 10^{-7}\,\text{X units/hr} \times 34\,\text{Gy/X-unit} \times 1\,\text{Sv/Gy}$$

$$= 0.023 \times 10^{-3}\,\text{Sv/hr}$$

$$= 0.023\,\text{mSv/hr} \quad (2.3\,\text{mrem/hr}).$$

(b) What is the dose-equivalent rate midway between the pipe and the door, point D_2, at a distance of 1.5 meters from the pipe?

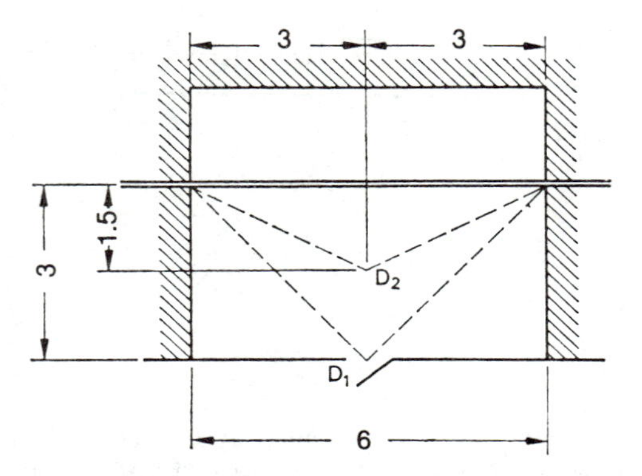

FIG. 10.2. Layout of the room described in Example 10.1.

The ratio of the two dose equivalent rates, H_1 at a distance of 1.5 meters and H_2 at a distance of 3 meters, is

$$\frac{\dot{H}_1}{\dot{H}_2} = \frac{2\Gamma C_l/h_1 \times \theta_1}{2\Gamma C_l/h_2 \times \theta_2}$$

$$\frac{\dot{H}_1}{\dot{H}_2} = \frac{h_2}{h_1} \times \frac{\theta_1}{\theta_2}. \tag{10.3}$$

In this example, $\theta_1 = 2\tan^{-1} 3/3 = \pi/2$ radians, while $\theta_2 = 2\tan^{-1} 3/1.5 = 1.4\pi$ radians. Substituting into equation (10.3) gives

$$\frac{0.023}{\dot{H}_2} = \frac{1.5}{3} \times \frac{\pi/2}{1.4\pi},$$

$$\dot{H}_2 = 0.128 \text{ mSv/hr} \quad (12.8 \text{ mrem/hr}).$$

Frequently the health physicist may find it useful to know the quantitative relationship between dose rate and distance from a plane radiation source. If we have a thin source of radius r meters (Fig. 10.3) and a surface concentration C_a MBq/m^2 of a gamma emitter whose source strength is Γ X units per hour per MBq at 1 meter, then, since an exposure of 1 X unit corresponds to an air dose of 34 grays, and since gamma radiation has a QF of 1, the dose equivalent rate at a point p, a distance h along the central axis is given by

$$\dot{H} = \int_0^R \frac{\Gamma \frac{\text{X-m}^2}{\text{MBq-hr}} \times 34 \text{ Gy/X} \times 1 \text{ Sv/Gy} \times C_a \text{ MBq/m}^2 \times 2\pi r \, dr}{r^2 + h^2}$$

$$= 34\pi \times \Gamma \times C_a \times \ln \frac{R^2 + h^2}{h^2} \text{ Sv/hr}. \tag{10.4}$$

If activity is given in curies and Γ in roentgens per hour per curie at 1 meter, then, since an exposure of 1 R corresponds to a dose equivalent of 1 rem, the dose equivalent rate is given by

$$\dot{H} = \pi \times \Gamma \frac{\text{R-m}^2}{\text{Ci-hr}} \times C_a \text{ Ci/m}^2 \times \ln \frac{R^2 + h^2}{h^2}. \tag{10.5}$$

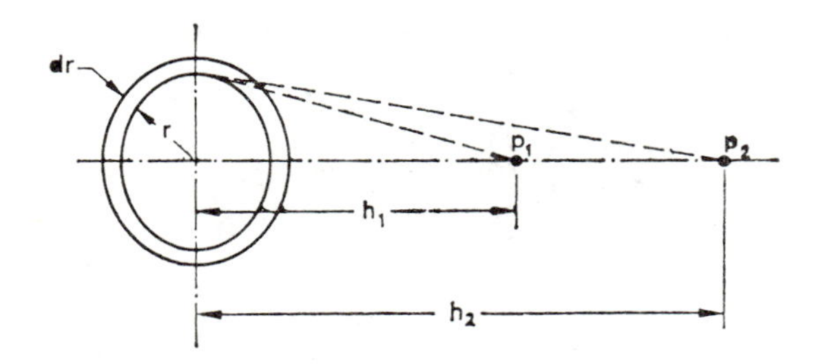

FIG. 10.3. Geometry for calculating the variation of dose rate with distance from a plane source of radiation.

The ratio of the dose equivalent rate at a distance h, to the dose equivalent rate at any other distance is given by

$$\frac{\dot{H}_1}{\dot{H}_2} = \frac{\ln[(R^2 + h_1^2)/h_1^2]}{\ln[(R^2 + h_2^2)/h_2^2]}. \quad (10.6)$$

Example 10.2

Fifty MBq of ^{24}NaCl solution spilled over a circular area 50 cm in diameter. What is the gamma ray dose equivalent rate at heights of

(a) 30 cm?
(b) 1 m?

(a) From Table 6.3, we find the specific gamma ray emission constant for ^{24}Na to be 1.28×10^{-8} X-units per hour per MBq at 1 meter. The areal concentration $C_a = 50\ \text{MBq}/\pi(0.25\ \text{m})^2 = 254.65\ \text{MBq/m}^2$. Substituting the respective values into equation (10.4) yields

$$\dot{H} = 34\pi \times 1.28 \times 10^{-8} \times 254.65 \times \ln\frac{0.25^2 + 0.3^2}{0.3^2}$$

$$= 1.84 \times 10^{-4}\ \text{Sv/hr}$$

$$= 0.184\ \text{mSv/hr} \quad (18.4\ \text{mrem/hr}).$$

(b) By substituting the appropriate values into equation (10.6), we find

$$\frac{0.184}{\dot{H}_2} = \frac{\ln[(0.25^2 + 0.3^2)/0.3^2]}{\ln[(0.25^2 + 1^2)/1^2]}$$

$$\dot{H}_2 = 0.021\ \text{mSv/hr} \quad (2.1\ \text{mrem/hr}).$$

The radiation exposure rate from a thick source containing a uniformly distributed gamma emitting isotope may be estimated from the effective surface activity after allowing for self absorption within the slab. Consider a large slab of thickness t m (Fig. 10.4), containing C_v MBq/m^3 of uniformly distributed radioactivity. The linear absorption coefficient of the slab material is μ per meter. The activity on the surface due to the radioactivity in the layer dx, at a depth of x, is

$$d(C_a) = C_v \cdot dx \cdot e^{-\mu x}. \quad (10.7)$$

Integrating equation (10.7) over the total thickness t yields the effective surface activity:

$$C_a = \int_0^t C_v e^{-\mu x}\, dx = \frac{C_v}{\mu}(1 - e^{-\mu t}). \quad (10.8)$$

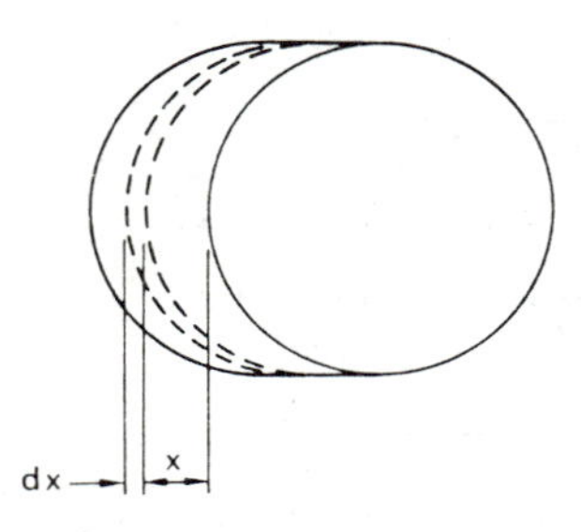

FIG. 10.4. Conditions for setting up equation (10.7).

Substituting equation (10.8) into equation (10.4) yields

$$\dot{H} = 34\pi\Gamma \frac{C_v}{\mu} (1 - e^{-\mu t}) \ln \frac{R^2 + h^2}{h^2} \text{ Sv/hr.} \tag{10.9}$$

The analogous equation for the case where activity is in curies and Γ is in units of roentgens per Ci per hour at 1 meter is

$$\dot{H} = \pi\Gamma \frac{C_v}{\mu} (1 - e^{-\mu t}) \ln \frac{R^2 + h^2}{h^2} \text{ rem/hr.} \tag{10.10}$$

Shielding

In Chapter 5 we saw that, under conditions of good geometry, the attenuation of a beam of gamma radiation is given by:

$$I = I_0 e^{-\mu t}. \tag{5.19}$$

However, under conditions of poor geometry, i.e., for a broad beam or for a very thick shield, equation (5.19) underestimates the required shield thickness because it assumes that every photon that interacts with the shield will be removed from the beam, and thus will not be available for counting by the detector. Under conditions of poor geometry (Fig. 10.5) this assumption is not valid; a significant number of photons may be scattered by the shield into the detector, or photons that had been scattered out of the beam may be scattered back in after a second collision. This effect may be illustrated by Fig. 10.6, which shows the broad beam and narrow beam attenuation of ^{60}Co gamma rays by concrete. To transmit 10% of the incident ^{60}Co radiation, Fig. 10.6 shows that, under conditions of good geometry, about 7 in. of concrete shielding is required. For a broad beam, on the other hand, this thickness of concrete will transmit about 25% of the radiation incident on it. To transmit only 10% of a broad beam requires about 11 in. of concrete. When designing a shield against a broad beam of radiation, experimentally determined shielding

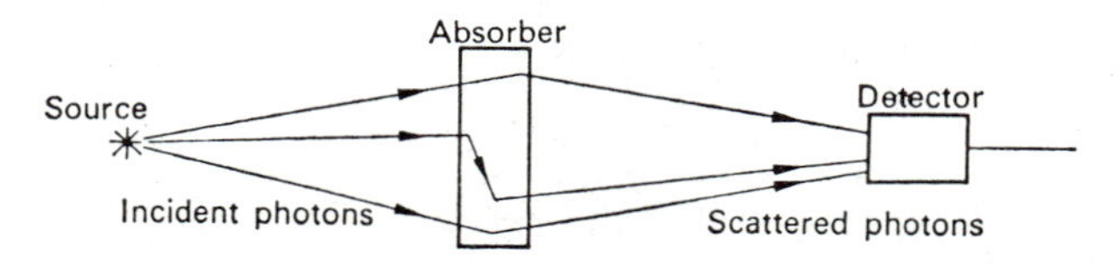

FIG. 10.5. Gamma-ray absorption under conditions of "poor geometry" showing the effect of photons scattered into the detector.

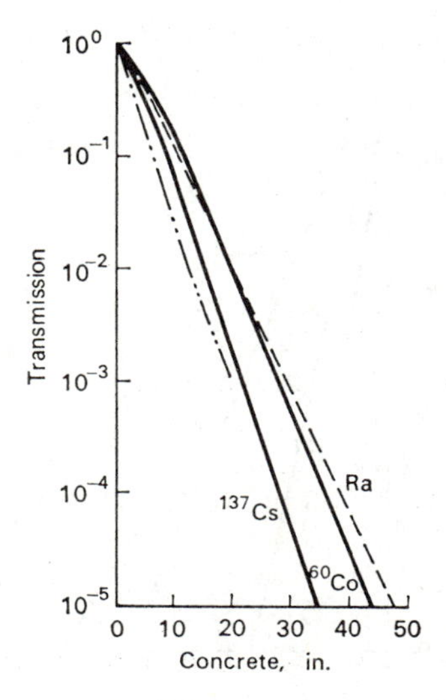

FIG. 10.6. Fractional transmission of gamma rays from ^{137}Cs, ^{60}Co, and Ra (in equilibrium with its decay products) through concrete. The short broken curve represents transmission of ^{60}Co gamma rays under conditions of good geometry. The other curves represent transmission of broad beams.

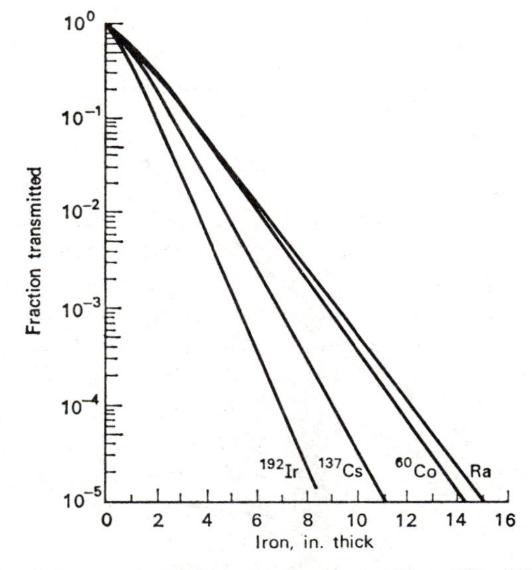

FIG. 10.7. Broad beam attenuation of gamma-rays from ^{192}Ir, ^{137}Cs, ^{60}Co, and radium by iron. (From N.B.S. Handbook 73, *Protection Against Radiations from Sealed Gamma Sources*, 1960.)

data for the radiation in question should be used whenever they are available. (Broad beam attenuation curves for radium, ^{60}Co, and ^{137}Cs for concrete, iron, and lead are given in Figs. 10.6, 10.7 and 10.8). When such data are not available, a shield thickness for conditions of poor geometry may be estimated by modification of equation (5.19) through the use of a *build-up factor*, *B*:

$$I = B \times I_0 e^{-\mu t}. \tag{10.11}$$

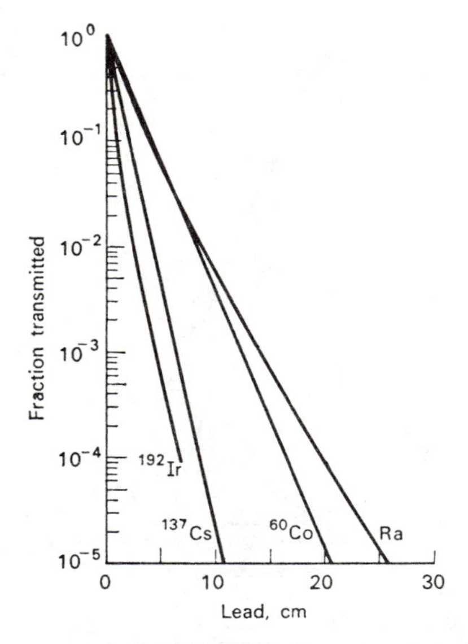

FIG. 10.8. Broad beam attenuation by lead of gamma-rays from ^{192}Ir, ^{137}Cs, ^{60}Co, and radium. (From N.B.S. Handbook 73, *Protection Against Radiations from Sealed Gamma Sources*, 1960.)

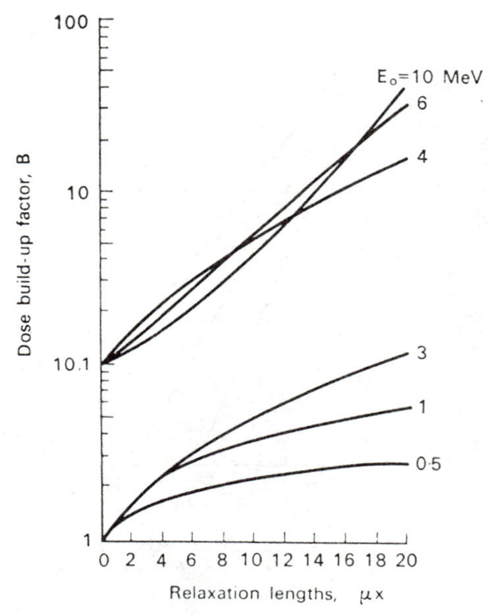

FIG. 10.9. Dose build-up factor in lead for a point isotropic gamma-ray source of energy E_0. (From *Radiological Health Handbook*, 1970.)

The build-up factor, which is always greater than 1, may be defined as the ratio of the intensity of the radiation, including both the primary and scattered radiation, at any point in a beam, to the intensity of the primary radiation only at that point. Build-up factor may apply either to radiation flux or to radiation dose. Build-up factors have been calculated

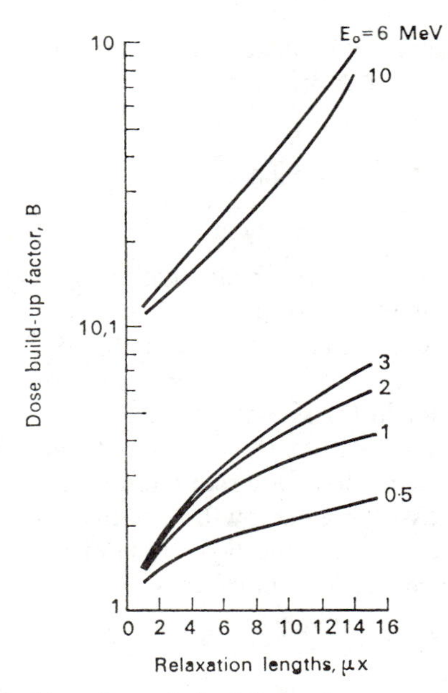

FIG. 10.10. Dose build-up factor in lead for a plane monodirectional gamma-ray source of quantum energy E_0. (From *Radiological Health Handbook*, 1970.)

for various gamma-ray energies and for various absorbers.† Some of these values are given in Figs. 10.9 and 10.10.

In these curves the shield thickness is given in units of *relaxation lengths*. One relaxation length is that thickness of shield that will attenuate a narrow beam to $1/e$ of its original intensity. One relaxation length, therefore, is numerically equal to the reciprocal of the absorption coefficient. The use of the build-up factor in the calculation of a shield thickness may be illustrated with the following examples:

Example 10.3

A 3.7×10^4 MBq (1 Ci) source of ^{137}Cs is to be stored in a spherical lead container when not in use. How thick must the lead be if the dose equivalent rate at a distance of 1 meter from the source is not to exceed 25 μSv/hr (2.5 mrem/hr)? Assume the source to be sufficiently small to be considered a "point."

In Table 6.3, we find the specific gamma ray emission of ^{137}Cs to be 2.3×10^{-9} X-units per hour per MBq at 1 m (0.33 Rhm per Ci). The exposure rate at 1 m from the unshielded source, therefore is

$$\dot{X} = 3.7 \times 10^4 \text{ MBq} \times 2.3 \times 10^{-9} \frac{\text{X-m}^2}{\text{MBq-hr}} \Big/ (1 \text{ m})^2$$

$$= 8.5 \times 10^{-5} \text{ X-units/hr} \quad (0.33 \text{ R/hr}).$$

Since an exposure of 1 X-unit is treated as a dose equivalent of 34 Sv for radiation protection purposes, 8.5×10^{-5} X-units/hr corresponds to 2.89×10^{-3} Sv/hr, or 2890 μSv/hr. If there were no buildup, the required thickness of lead would be calculated

†H. Goldstein and J. E. Wilkins, *Calculation of the Penetration of Gamma Rays*, U.S.A. E.C. report NYO-3075, 1954.

from equation (5.19), using the value for the attenuation coefficient for lead for 0.661 MeV gamma rays from Table 5.2, $\mu = 1.24$ per cm:

$$I = I_0 \times e^{-\mu t},$$

$$25 = 2890 \times e^{-1.24t},$$

$$t = 3.84 \text{ cm}.$$

This is an underestimate of the required shield thickness, since it does not include the additional thickness to account for buildup.

In Figs. 10.9 and 10.10, we see that the buildup factor is a function of the shield thickness. Since the shield thickness is not yet known, equation (10.11) has two unknowns, the buildup factor B and the shield thickness t. To determine the proper shield thickness, we estimate a thickness, then substitute this estimated value into equation (10.11) to determine whether it will satisfy the dose rate reduction requirement. The minimum shield thickness can be estimated by assuming narrow-beam attenuation, and then increasing the thickness thus calculated by one half-value layer (1 HVL) to account for buildup. The half-value layer of lead for 0.661 MeV gamma rays is

$$\text{HVL} = \frac{0.693}{1.24} \text{ cm}^{-1} = 0.59 \text{ cm}.$$

The estimated shield thickness therefore is 3.84 plus 0.59, or 4.43 cm, which corresponds to $1.24 \times 4.43 = 5.5$ relaxation lengths. From Fig. 10.9 we find (by interpolation) the dose build-up factor for 0.661 MeV gamma rays to be 2.12 for a lead shield of this thickness. Substituting these values for B and t into equation (10.9), we have

$$I = 2890 \times 2.12 \times e^{-1.24 \times 4.43}$$

$$= 25.2\ \mu\text{Sv/hr} \quad (2.52 \text{ mrem/hr}).$$

This calculated reduction in gamma-ray dose rate is just slightly less than the desired value of 2.5 mR/hr. The thickness of 4.43, as calculated above, is therefore just about correct. In this example, we found the correct thickness after one attempt in a trial-and-error method. If the calculated reduction in radiation dose rate would not have turned out to be so close to the design value with the estimated shield thickness, we would have continued, by trial and error, to estimate thicknesses until the one that results in the desired reduction of dose rate would have been obtained.

Example 10.4

Design a spherical lead storage container that will attenuate the exposure rate from 1 Ci of ^{24}Na to 10 mR/hr at a distance of 1 m from the source.

In each disintegration of a ^{24}Na atom, two gamma rays are emitted in cascade: one of 2.75 MeV and a second of 1.37 MeV. The dose rate, at a distance of 1 m, due to each of these photons is

$$\Gamma_{2.75} = 0.48 \times 2.75 = 1.32 \text{ Rhm per curie, and}$$

$$\Gamma_{1.37} = 0.48 \times 1.37 = 0.65 \text{ Rhm per curie.}$$

To reduce the exposure rate from the high-energy photon to 10 mR/hr for the condition of good geometry, we have, using a value of 0.475 per cm for the total attenuation coefficient,

$$I/I_0 = 10/1320 = e^{-0.475t}$$

$$t = 10.25 \text{ cm}.$$

The half-value layer of lead for this quantum energy is 1.46 cm. Let us add one half HVL, or 0.73 cm, to the thickness calculated above; and then calculate the attenuation of the 2.75-MeV gamma ray.

$$I = 1320 \times e^{-0.475 \times 10.98}$$
$$= 7.14 \text{ mR per hour.}$$

With this thickness of lead, the 1.37-MeV gamma ray will be attenuated to

$$I = 657 \times e^{-0.621 \times 10.98}$$
$$= 0.72 \text{ mR per hour.}$$

and the total exposure rate at a distance of 1 m will be the sum of the exposure rates due to the two quantum energies, or 7.86 mR/hr.

The calculation above we based on good geometry. Let us now account for build-up in the shield. Consider at this time only the high-energy photon, let us add another half of one HVL to the shield thickness, which gives us 11.71 cm. Since one relaxation length of lead for 2.75-MeV gammas is 2.1 cm, the shield thickness corresponds to 5.6 relaxation lengths. From Fig. 10.7 we find the build-up factor to be 3.13. The attenuation of the shield, therefore, according to equation (10.11), is

$$I = 1320 \times 3.13 \times e^{-0.475 \times 11.71}$$
$$= 15.3 \text{ mR per hour.}$$

This exposure rate is too high; the shield thickness must therefore be increased. If we add another HVL to the shield to give us 13.17 cm, or 6.27 relaxation lengths, and using the corresponding dose build-up factor of 3.42, we find the exposure rate to be 8.5 mR/hr for the high-energy photon. For the lower-energy photon, whose relaxation length in lead is 1.61 cm, this shield thickness corresponds to 8.18 relaxation lengths, and the dose build-up factor is 3.61. With these values in equation (10.11), the exposure rate due to the 1.37-MeV photons is calculated as 0.7 mR/hr. The total exposure rate at a distance of 1 m from the shielded ^{24}Na source is thus 9.2 mR/hr. Since this rate may be considered, for most practical purposes, to be equivalent to the design value of 10 mR/hr, the required shield thickness is 13.17 cm. Since the shield may be interposed anywhere between the source and the point where the desired attenuated dose rate is located, and since the volume of lead, for a given wall thickness in a spherical shield, increases rapidly with increasing outer radius according to the expression

$$\text{Volume} = \frac{4}{3}\pi(r_0^3 - r_1^3),$$

the inner radius of the shield is kept as small as possible consistent with the space requirements set by the physical dimensions of the source. The outside radius then is equal to the sum of the inside radius and the shield thickness.

X-ray shielding

Shielding for protection against X-rays is considered under two categories: source shielding and structural shielding. Source shielding is usually supplied by the manufacturer of the X-ray equipment as a lead shield in which the X-ray tube is housed. The safety regulations recommended by the National Committee on Radiation Protection specify the following two types of protective tube housings for medical X-ray installations†:

†*Structural Shielding Design and Evaluation for Medical Use of X-rays and Gamma Rays of Energies up to 10 MeV*, NCRP Report No. 49, National Council on Radiation Protection and Measurements, Washington, D.C., 1976.

1. *Diagnostic type*: One so built that the leakage radiation at a distance of 1 meter from the target cannot exceed 100 mR (30 μC/kg) in 1 hour when the tube is operated at its maximum continuous rated current for the maximum rated tube potential.
2. *Therapeutic type*:
   (a) For X-ray generators that are incapable of peak voltages of 500 kV or more: A tube housing so built that the leakage radiation at a distance of 1 meter from the target does not exceed 1 R (3,000 μC/kg) in 1 hour when the tube is operated at its maximum continuous rated current for the maximum rated tube potential.
   (b) For X-ray generators that can operate at peak voltages of 500 kV or more: A tube housing so built that the leakage radiation at a distance of 1 meter from the target does not exceed 1 R (3,000 μC/kg) in 1 hour or 0.1% of the useful beam exposure rate at 1 meter from the target, whichever is greater, when the tube is operated at its maximum continuous rated current for the maximum rated accelerating potential.

For non-medical X-rays, a protective tube housing is one which surrounds the X-ray tube itself, or the tube and other parts of the X-ray apparatus (for example, the transformer) and is so constructed that the leakage radiation at a distance of 1 m from the target cannot exceed 1 R in 1 hr when the tube is operated at any of its specified ratings. Leakage radiation, as used in these specifications for tube housings, means all radiation, except the useful beam, coming from the tube housing.

Structural shielding is designed to protect against the useful X-rays, leakage radiation, and scattered radiation. It encloses both the X-ray tube (with its protective tube housing) and the space in which is located the object being irradiated. Structural shielding may vary considerably in form. It may, for example, be either a lead-lined box in the case of an X-ray tube used by a radiobiologist to irradiate small organisms, or it may be the shielding around a room in which a patient is undergoing radiation therapy. In any case, structural shielding is designed to protect people in an occupied area outside an area of high radiation intensity. The structural shielding requirements for a given installation are determined by:

1. The maximum kilovoltage at which the X-ray tube is operated.
2. The maximum milliamperes of beam current.
3. The workload (W), which is a measure, in suitable units, of the amount of use of an X-ray machine. For X-ray shielding design, workload is usually expressed in units of milliampere-minutes per week.
4. The use factor (U), which is the fraction of the workload during which the useful beam is pointed in the direction under consideration.
5. The occupancy factor (T), which is the factor by which the workload should be multiplied to correct for the degree or type of occupancy of the area in question. When adequate occupancy data are not available, the values for T given in Table 10.1 may be used as a guide in planning shielding.

In the discussion that follows, and in the design examples, the roentgen is used as the measure of exposure rather than the newer X-unit, the C/kg. This is done in the interests of practicality, since the necessary design data published in the handbooks currently (1983) in use are based on the old roentgen unit. Since 1 R = 2.58×10^{-4} C/kg, conversion to the new system of units can be done whenever the need arises to use the new system.

According to the recommendations of the ICRP, the maximum permissible dose for occupational exposure is 50 mSv (5,000 millirem) in 1 year. At a uniform rate over 50 weeks, this corresponds to 1 mSv (100 millirem) in 1 week. Since the quality factor for X-rays is 1, and since 1 C/kg, of 1 X-unit is equal to an air dose of 34 Gy, an exposure

TABLE 10.1. OCCUPANCY FACTORS

Full occupancy $T = 1$	Control space, wards, workrooms, darkrooms, corridors large enough to hold desks, waiting rooms, restrooms used by occupationally exposed personnel, children's play areas, living quarters, occupied space in adjacent buildings
Partial occupancy $T = 1/4$	Corridors too narrow for desks, utility rooms, rest rooms not used routinely by occupationally exposed personnel, elevators using operators, and uncontrolled parking lots
Occasional occupancy $T = 1/16$	Stairways, automatic elevators, outside areas used only for pedestrians or vehicular traffic, closets too small for future workrooms, toilets not used routinely by occupationally exposed personnel

of 30 μC/kg corresponds to the maximum permissible occupational dose equivalent. This maximum permissible exposure of 30 μC/kg (100 millirem) in 1 week is used as the design basis for shielding controlled areas. For uncontrolled areas, the basis for shielding design is 3 μC/kg (10 millirem) in 1 week, in accordance with the ICRP recommendation in ICRP Publication 26 that the dose equivalent to an individual who is not a radiation worker shall not exceed 5 mSv (500 millirem) in 1 year.

Some of the physical factors that determine the shielding requirements for protection from X-ray beams are shown in Fig. 10.11. A collimated X-ray beam is directed at patient M (or object to be radiographed in the case of non-medical radiography) from the shielded X-ray generator at A. The beam passes through the patient and is attenuated to an acceptable level by the primary protective barrier before irradiating a person at C. The leakage radiation and the scattered radiations are attenuated to an acceptable level by a secondary protective barrier before reaching point E, where a person may be irradiated.

1. *Primary protective barrier*

The maximum exposure rate at any occupied point at a distance d meters from the target in the X-ray tube is given by

$$\dot{X}_m = \frac{P}{T} \text{ R/week}, \tag{10.12}$$

where P is the maximum permissible weekly exposure (0.1 R/week for controlled areas and 0.01 R/week for uncontrolled areas) and T is the occupancy factor. By applying the inverse square law, we find this radiation field to have an exposure rate at 1 meter from the target that is given by

$$\dot{X}_1 = d^2 \times \dot{X}_m = \frac{d^2 P}{T} \text{ R/week at 1 meter.} \tag{10.13}$$

This exposure is due to the workload WU mA-minutes per week. Let us now define the ratio

$$K = \frac{\dot{X}_1}{WU} = \frac{d^2 P}{WUT} \frac{\text{R/mA} - \text{min}}{\text{week}} \text{ at 1 meter,} \tag{10.14}$$

and measure this value for broad beams of X-rays of various energies that have been

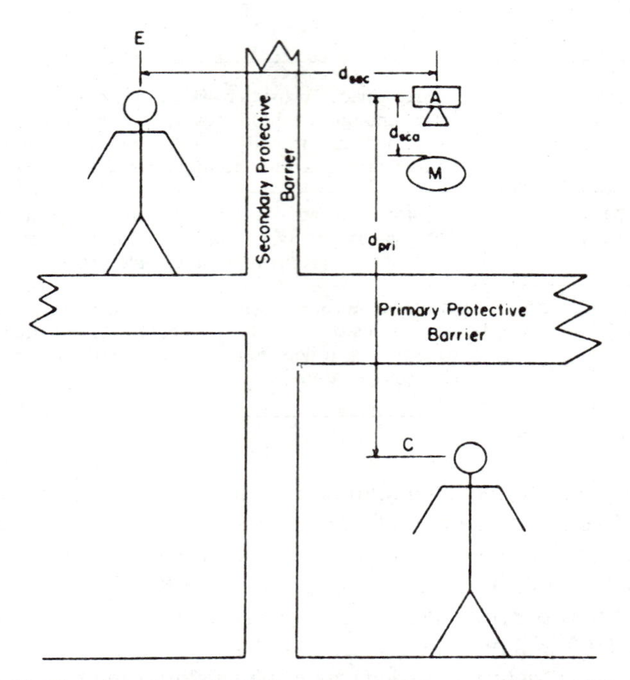

FIG. 10.11. Elevation view of radiation room and its surroundings with indication of distances of interest for radiation shielding calculations. A is the radiation source, M represents the patient, and C and E are positions that may be occupied by persons. (From NCRP Report No. 49: *Structural Shielding Design and Evaluation for Medical Use of X-rays and Gamma Rays of Energies up to* 10 *MeV*, 1976.)

transmitted through lead or concrete shields of varying thicknesses. The results of these measurements are shown in Figs. 10.12 through 10.16. The transmission of X-rays through thick shields has been found experimentally to depend mainly on the highest energy photons in the beam, and, for a beam of any given minimum wavelength, to be influenced relatively little by the quality of the beam (that is, by the half-value layer for that beam). Accordingly, the X-ray transmission data shown in Figs. 10.12 through 10.16 may be used for the design of shielding against any X-rays within the range of the maximum potentials shown in the figures. To design the primary protective barrier, the value of K is computed from equation (10.14) and the required barrier thickness is read from the appropriate figure at the intersection of the ordinate, K, with the curve representing the X-ray energy.

If K is given in units of μC/kg per mA-min, the curves in Figs. 10.12–10.16 can be used if we multiply the value of K by 258. For example, a K value of $3.88 \times 10^{-5}\ \mu$C/kg per mA-min corresponds to $K = 258 \times 3.88 \times 10^{-5} = 1 \times 10^{-2}$ R/mA-min.

Example 10.5

Calculate the thickness of concrete required to protect a controlled area from the useful beam of a 250 kVp X-ray therapy machine whose workload is 20,000 mA min per week. The controlled area is 2.5 meters from the target of the X-ray machine, the occupancy factor is 1, and the use factor is 1/2.

For a controlled area, the maximum weekly exposure may not exceed 0.1 R (30 μC/kg).

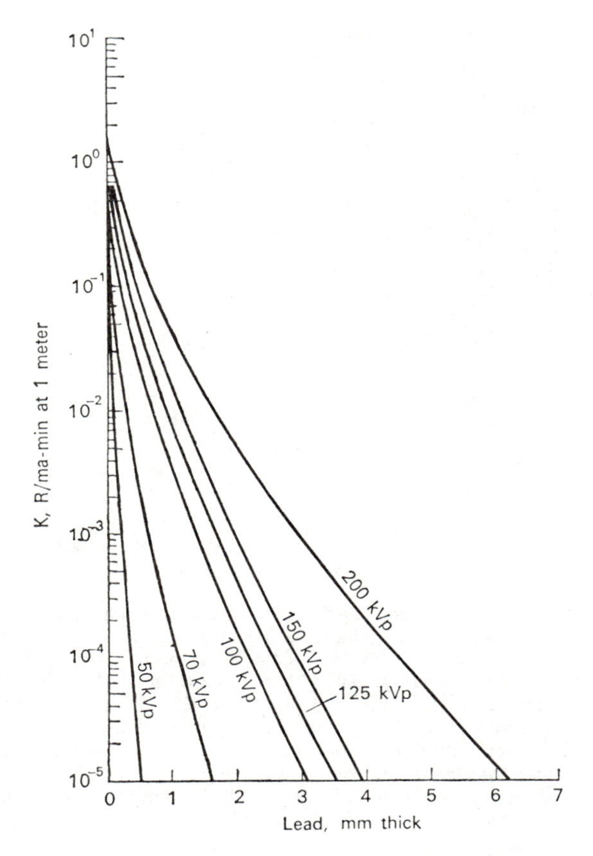

FIG. 10.12. Broad beam attenuation in lead of X-rays produced by potentials of 50–200 kV peak. The measurements were made with a 90° angle between the electron beam and the axis of the pulsed wave form X-ray beam. The 50, 70, 100, and 125 kVp X-rays were filtered with 0.5 mm aluminum; the 150 and 200 kVp X-rays were filtered with 3 mm aluminum. (From *Radiological Health Handbook*, 1970.)

If we substitute this value for P, and the given values for the other factors into equation (10.12), we get

$$K = \frac{(2.5\ \text{m})^2 \times 0.1\ \text{R}}{20{,}000\ \text{mA min} \times 0.5 \times 1} = 6.25 \times 10^{-5}\,\frac{\text{R-m}^2}{\text{mA-min}}.$$

(Note: The dimensions for R/mA-min at 1 meter = R-m^2/mA-min.)

From the curve for 250 kVp in Fig. 10.13, we find the required thickness to be 39 cm concrete. The thickness of lead that will give the same degree of attenuation is found from the 250 kVp curve in Fig. 10.14 to be 8.25 mm.

2. *Secondary protective barrier*

The secondary protective barrier protects against scattered radiation and leakage radiation. To design the secondary protective barrier, we separately calculate the required thickness to protect against each of these components. If the required thicknesses are about the same, we merely add an additional half-value layer, Table 10.2, to the greater thickness.

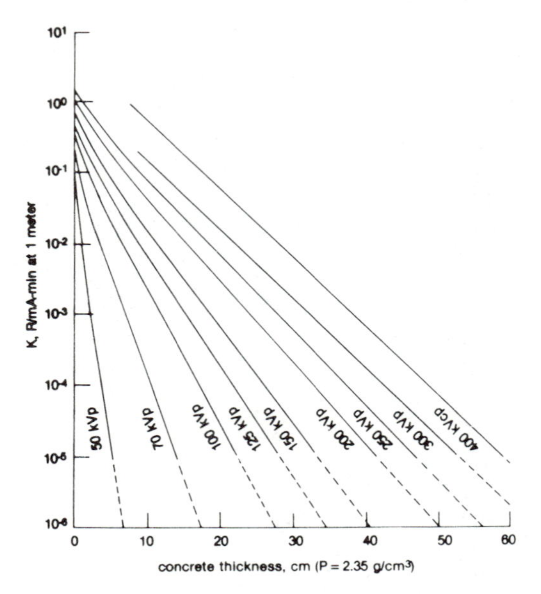

FIG. 10.13. Attenuation in concrete of X-rays produced by potentials of 50 to 300 kV peak; 400 kV constant potential. The measurements were made with a 90° angle between the electron beam and the axis of the X-ray beam. The curves for 50 to 300 kV are for a pulsed waveform. The filtrations were 1 mm Al for 50 kV, 1.5 mm Al for 70 kV, 2 mm Al for 100 kV, and 3 mm Al for 125, 150, 200,250, and 300 kV. The 400-kV curve was interpolated from data obtained with a constant potential generator and inherent filtration of approximately 3 mm Cu. (From NCRP Report No. 49, *Structural Shielding Design and Evaluation for Medical Use of X-Rays and Gamma Rays of Energies up to* 10 *MeV*, 1976. Full size reproductions of the figures giving barrier requirements are available from the NCRP as an adjunct to the report.)

If the difference between the two calculated thicknesses is one tenth-value layer (Table 10.2) or more, the thicker of the two will suffice.

Scattered radiation intensity depends on the scattering angle, on the energy of the primary beam, and on the scattering area (field size). Table 10.3 lists the intensity ratio of the scattered to incident radiation at a distance of 1 meter from the scatterer, for a field size (scattering area) of 400 cm^2. On the assumption that the intensity of the scattered radiation varies inversely with the square of the distance from the scatterer and varies directly with the scattering area, the exposure from the scattered radiation, X_s is given by

$$X_s = \frac{a \times \dot{X}_u}{(d_{sec})^2} \times \frac{F}{400} \times t \tag{10.15}$$

where a = ratio of scattered to incident radiation, Table 10.3,
$\dot{X}_u$ = exposure rate incident on scatterer,
d_{sec} = distance from scatterer to point of interest,
F = scattering field size, cm^2
t = exposure time.

If we call the exposure rate at 1 meter from the target $\dot{X}_n$ when the electron beam current is 1 mA ($\dot{X}_n$ is called the *normalized exposure rate*), if the scatterer is at a distance d_{sca} from the target, and if the barrier attenuates the scattered radiation by a factor B_{sx}, then, for a weekly exposure of P when the electron beam current is I amperes for t seconds, we have

$$P = B_{sx} \times X_s = B_{sx} \times \frac{a \times \dot{X}_n \times I \times t}{(d_{sca})^2 \times (d_{sec})^2} \times \frac{F}{400}. \tag{10.16}$$

TABLE 10.2. HALF-VALUE AND TENTH-VALUE LAYERS.

Peak Voltage (kV)	Attenuation Material: Lead (mm)		Concrete (cm)		Iron (cm)	
	HVL	*TVL*	*HVL*	*TVL*	*HVL*	*TVL*
50	0.06	0.17	0.43	1.5		
70	0.17	0.52	0.84	2.8		
100	0.27	0.88	1.6	5.3		
125	0.28	0.93	2.0	6.6		
150	0.30	0.99	2.24	7.4		
200	0.52	1.7	2.5	8.4		
250	0.88	2.9	2.8	9.4		
300	1.47	4.8	3.1	10.4		
400	2.5	8.3	3.3	10.9		
500	3.6	11.9	3.6	11.7		
1,000	7.9	26	4.4	14.7		
2,000	12.5	42	6.4	21		
3,000	14.5	48.5	7.4	24.5		
4,000	16	53	8.8	29.2	2.7	9.1
6,000	16.9	56	10.4	34.5	3.0	9.9
8,000	16.9	56	11.4	37.8	3.1	10.3
10,000	16.6	55	11.9	39.6	3.2	10.5
Cesium-137	6.5	21.6	4.8	15.7	1.6	5.3
Cobalt-60	12	40	6.2	20.6	2.1	6.9
Radium	16.6	55	6.9	23.4	2.2	7.4

Aproximate values obtained at high attentuation for the indicated peak voltages under broad beam conditions; with low attenuation these values will be significantly less. (From NCRP 49)

Since $It = WT$, and since $U = 1$ for scattered radiation, then, if we solve equation (10.16) for $B_{sx} \times X_n$ and call this product K_{ux}, we have

$$K_{ux} = \frac{P}{aWT} \times (d_{sca})^2 \times (d_{sec})^2 \times \frac{400}{F}. \tag{10.17}$$

When designing secondary protective barriers, we make the following simplifying and conservative assumptions:

1. The energy of the scattered radiation, when the X-rays are generated at 500 kV or less, is equal to the energy of the useful beam.
2. Primary X-ray beams generated at voltages greater than 500 kV are degraded in energy to that of a 500 kV beam after being scattered, and the exposure rate at 1 meter from the scatterer is 0.1% of that in the useful beam at the point of scattering.

These simplifying assumptions are based on the fact that, for quantum energies of 500 keV or less, a photon loses relatively little energy in a scattering interaction, whereas higher energy X-rays that are scattered through 90° are degraded in energy to a quality approximately equivalent in energy distribution to X-rays generated by a potential of 500 kV. For X-ray energies of up to 500 keV, therefore, the transmission curves for the respective kilovoltages are used, whereas for all X-rays of higher energy, the 500 kV transmission curves are used in the design of shielding against scattered radiation. However, the output of an X-ray machine increases as the kilovoltage increases. To account for the increasing X-ray output with increasing voltage, K_{ux} is reduced by a factor

TABLE 10.3. RATIO, a, OF SCATTERED TO INCIDENT EXPOSURE[a] (FROM NCRP 49)

Source	Scattering Angle (from Central Ray)					
	30	45	60	90	120	135
X Rays						
50 kV[b]	0.0005	0.0002	0.00025	0.00035	0.0008	0.0010
70 kV[b]	0.00065	0.00035	0.00035	0.0005	0.0010	0.0013
100 kV[b]	0.0015	0.0012	0.0012	0.0013	0.0020	0.0022
125 kV[b]	0.0018	0.0015	0.0015	0.0015	0.0023	0.0025
150 kV[b]	0.0020	0.0016	0.0016	0.0016	0.0024	0.0026
200 kV[b]	0.0024	0.0020	0.0019	0.0019	0.0027	0.0028
250 kV[b]	0.0025	0.0021	0.0019	0.0019	0.0027	0.0028
300 kV[b]	0.0026	0.0022	0.0020	0.0019	0.0026	0.0028
4 MV[c]	—	0.0027	—	—	—	—
6 MV[d]	0.007	0.0018	0.0011	0.0006	—	0.0004
Gamma Rays						
^{137}Cs[e]	0.0065	0.0050	0.0041	0.0028	—	0.0019
^{60}Co[f]	0.0060	0.0036	0.0023	0.0009	—	0.0006

[a]Scattered radiation measured at one meter from phantom when field area is 400 cm^2 at the phantom surface; incident exposure measured at center of field one meter from the source but without phantom.

[b]From Trout and Kelley (Radiology 104, 161 (1972)). Average scatter for beam centered and beam at edge of typical patient cross-section phantom. Peak pulsating x-ray tube potential.

[c]From Greene and Massey (Brit. J. Radiology **34**, 389 (1961)), cylindrical phantom.

[d]From Karzmark and Capone (Brit. J. Radiology **41**, 222 (1968)), cylindrical phantom.

[e]Interpolated from Frantz and Wyckoff (Radiology **73**, 263 (1959)), these data were obtained from a slab placed obliquely to the central ray. A cylindrical phantom should give smaller values.

[f]From Mooney and Braestrup (AEC Report NYO 2165 (1967)), modified for $F = 400$ cm^2.

f, whose magnitude depends on the kilovoltage. Applying these considerations to equation (10.17) yields

$$K_{ux} = \frac{P \times (d_{sca})^2 \times (d_{sec})^2 \times 400}{a \times W \times T \times F \times f} \text{ R/mA-min per week} \tag{10.18}$$

at a distance of 1 meter from the scatterer. Values assigned to f are:

kV	f
500 or less	1
1000	20
2000	300
2000	700

Example 10.6

Calculate the shield thickness for protection of an uncontrolled area against scattered radiation from a 250-kVp X-ray therapy machine. The weekly workload is 20,000 mA min, the target to skin distance is 50 cm, the maximum treatment field is 20 cm × 20 cm, the distance from the patient to the uncontrolled area is 3 meters, and the occupancy factor in the uncontrolled area is 1.

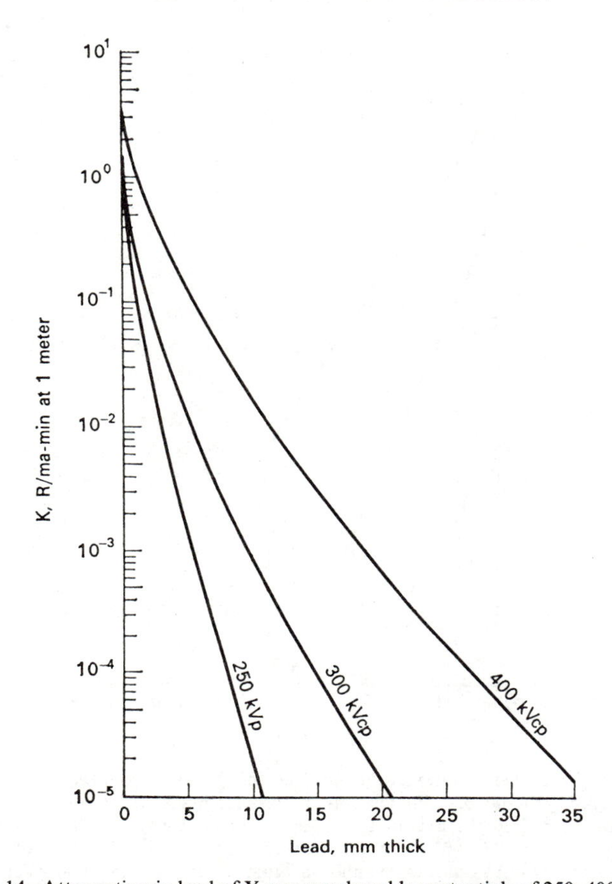

FIG. 10.14. Attenuation in lead of X-rays produced by potentials of 250–400 kV. The measurements were made with an angle of 90° between the axes of the electron beam and the X-ray beam. The 250-kVp curve is for a pulsed wave form and a filtration of 3 mm Al. The 300 and 400 kVcp curves were interpolated from data obtained with a constant potential generator and inherent filtration of approximately 3 mm Cu. (From NCRP Report No. 49, *Structural Design and Evaluation for Medical Use of X-Rays and Gamma Rays of Energies up to* 10 *MeV*, 1976.)

Since we are protecting an uncontrolled area, the maximum weekly exposure is $P = 0.01$ R (3 μC/kg). The other values for use in equation (10.18) are:

$$d_{sca} = 0.5 \text{ m},$$

$$d_{sec} = 3 \text{ m},$$

$$a = 0.0019 \text{ (from Table 10.5)},$$

$$W = 20{,}000 \text{ mA min},$$

$$T = 1,$$

$$F = 400 \text{ cm}^2,$$

$$f = 1.$$

After substituting these values into equation (10.18) and solving for K_{ux}, we will find the thickness of concrete that corresponds to this K_{ux} value in Fig. 10.13; if we wish to use

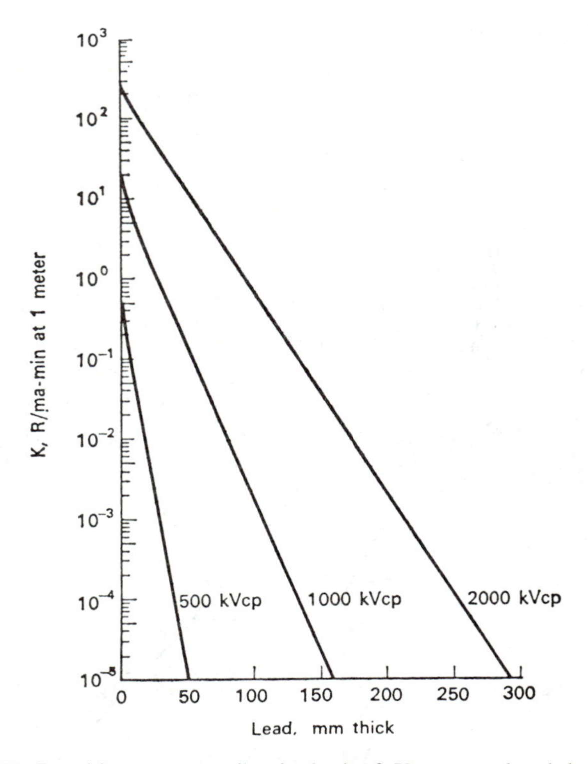

FIG. 10.15. Broad-beam attenuation in lead of X-rays produced by constant potentials of 500 to 2000 kV. The measurements were made with a 0° angle between the electron beam and the axis of the X-ray beam, and with a constant potential generator. The 500 and 1000 kVcp curves were obtained with filtration of 2.8 mm tungsten, 2.8 mm Cu, 2.1 mm brass, and 18.7 mm water. For the 2000-kVcp curve, the inherent filtration was equivalent to 6.8 mm lead. (From NCRP Report No. 49, *Structural Design and Evaluation for Medical Use of X-Rays and Gamma Rays of Energies up to* 10 *MeV*, 1976.)

lead as the shielding material, the thickness will be found in Fig. 10.14.

$$K_{ux} = \frac{0.01 \times (0.5)^2 \times (3)^2 \times 400}{0.0019 \times 20{,}000 \times 400 \times 1} = 5.92 \times 10^{-4}.$$

From the curve for 250 kVp in Fig. 10.13, the required barrier thickness is found to be 30 cm concrete; the 250 kVp curve in Fig. 10.14 shows that 6 mm Pb is required if we wish to use lead as the shielding material.

Since the protective tube housing limits the leakage radiation to a known and fixed value at a distance of one meter, it is a relatively simple matter to determine the required barrier thickness at any other distance if the energy of the leakage radiation is known. The leakage radiation, having been filtered and hardened by passing through the tube housing, is essentially monochromatic; its half-value layer therefore depends only on the kilovoltage across the tube.

For the case of an X-ray tube with a diagnostic type protective tube housing, the maximum leakage is restricted to 0.1 R (30 μC/kg) in 1 hour at a distance of 1 meter from the target when the tube is operating at its maximum rated voltage and current. If the

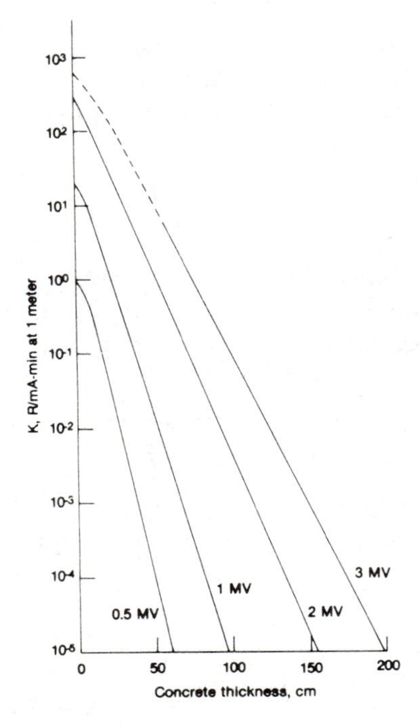

FIG. 10.16. Attenuation in concrete of X-rays produced by potentials of 0.5 to 3 MV constant potential. The measurements were made with a 0° angle between the electron beam and the axis of the X-ray beam and with a constant potential generator. The 0.5- and 1-MV curves were obtained with filtration of 2.8 mm Cu, 2.1 mm brass, and 18.7 mm water. The inherent filtration for the 2- and 3-MV curves was equivalent to 6.8 mm Pb. (From NCRP Report No. 49, *Structural Shielding Design and Evaluation for Medical Use of X-Rays and Gamma Rays of Energies up to 10 MeV*, 1976.)

X-ray machine is operated for t minutes in a week, the weekly leakage exposure at a distance d meters from the target is

$$X_L = \frac{0.1}{d^2} \times \frac{t}{60}. \tag{10.19}$$

For design purposes, WUT, the workload, is equal to the product of the electron beam current and the weekly "on" time, It, and since $U = 1$ for leakage radiation, equation (10.19) can be written as

$$X_L = \frac{0.1}{d^2} \times \frac{WT}{60I}. \tag{10.20}$$

If the maximum weekly exposure is P, and if the barrier against leakage radiation attenuates the leakage radiation by a factor B_{Lx}, the shielding requirement is given by

$$P = B_{Lx} \times X_L = B_{Lx} \times \frac{WT}{d^2 \times 600I}. \tag{10.21}$$

Solving equation (10.21) for the required amount of attenuation leads to

$$B_{Lx} = \frac{P \times d^2 \times 600I}{WT}. \tag{10.22}$$

For the case of an X-ray tube with a therapeutic protective tube housing, and an operating voltage not over 500 kV, the maximum leakage is limited to 1 R in 1 hour at 1 meter from the target. The barrier attenuation equation, therefore, that is analogous to equation (10.22) is

$$B_{Lx} = \frac{P \times d^2 \times 60I}{WT}. \tag{10.23}$$

For X-rays generated at potentials greater than 500 kV, the therapeutic protective tube housing limits the leakage radiation at 1 meter to 0.1% of the intensity of the useful beam at 1 meter. The attenuation required of the secondary protective barrier is therefore given by

$$B_{Lx} = \frac{1{,}000P \times d^2}{\dot{X}_n \times WT}, \tag{10.24}$$

where $\dot{X}_n$ is the normalized exposure rate.

The number of half-value layers needed to give the required degree of attenuation is calculated from

$$B_{Lx} = 1/2^n, \tag{10.25}$$

and the thickness of the barrier is calculated with the aid of Table 10.2, which lists the half-value layers of concrete and lead for various X-ray energies.

Example 10.7

Calculate the required secondary barrier thickness for the X-ray therapy installation given in example 10.6. The machine operates at 250 kVp and 20 mA, and has a weekly workload of 20,000 mA min. The distance from the X-ray machine to the uncontrolled area is 3 meters, and the occupancy factor is 1.

For an uncontrolled area, $P = 0.01$ roentgen in a week. Substituting all the factors into equation (10.23) for a therapeutic protective tube housing operating at 500 kV or less, we have

$$B_{Lx} = \frac{0.01 \times 3^2 \times 60 \times 20}{20{,}000} = 5.4 \times 10^{-3}.$$

The number of half-value layers that leads to this degree of attenuation is given by equation (10.25)

$$5.4 \times 10^{-3} = 1/2^n$$

$$n = 7.53$$

In Table 10.2, we find the HVL of concrete for 250 kV X-rays to be 2.8 cm. The required thickness for protection against the leakage radiation is

$$7.53 \text{ HVL} \times 2.8 \text{ cm/HVL} = 21.1 \text{ cm}.$$

The barrier thickness for protection against the scattered radiation was found in example 10.6 to be 30 cm concrete. The difference between these two thicknesses is 8.9 cm. In Table 10.2, we see that the tenth-value layer for 250-kV X-rays is 9.4 cm concrete. Since the difference between the required thickness for scattered and leakage radiation is less than one tenth-value layer, the final secondary barrier thickness is one half-value layer greater than the thicker barrier:

$$30 \text{ cm} + 2.8 \text{ cm} = 32.8 \text{ cm}.$$

Example 10.8

The room shown in Fig. 10.17 will be used for diagnostic radiology. From the floor to the ceiling is 9 ft, 7 in (292 cm). The room above is another office that is not controlled by the radiologist. The X-ray room is on the ground floor of the building; there is no occupied space below. The floor and ceiling are made of concrete 5 in. (12.7 cm) thick, wall A is made of concrete 4 in. (10.2 cm) thick. Walls B, C, and D are made of hollow tile and thin plaster, as shown in section A-A. The maximum machine ratings are 125 kVp and200 mA. As a fluoroscope, the machine will be operated 5 hours a week at 3.5 mA, and for radiography the machine will be operated at 200 mA for 2 minutes in a week. The average target to skin distance if 0.5 meter. Compute the required shielding for the ceiling and for wall B.

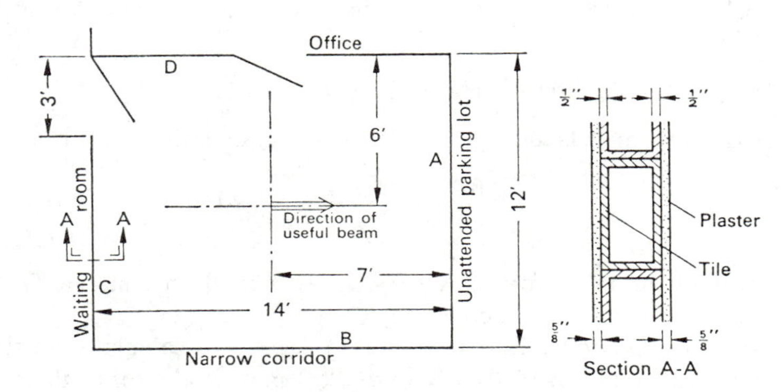

FIG. 10.17. Layout of diagnostic X-ray room of Example 10.8.

To compute the primary barrier:
For radiography, the X-rays will be directed only at wall A. This wall, therefore, must be a primary barrier, and all the other walls are secondary barriers. During fluoroscopy, the useful beam always is intercepted by the viewing screen, which is shielded by leaded glass. No other primary barrier is needed. Considering the radiographic operation only, therefore, the workload is

$$W = 2\text{ min/week} \times 200\text{ mA} = 400\text{ mA-min/week},$$

the use factor, U, is equal to 1, and the occupancy factor, T, is equal to 1/4 (for a parking lot). The distance from the target in the X-ray tube to the wall is 7 ft (2.13 m), and the maximum permissible weekly exposure to an uncontrolled area is 0.01 R (3 μC/kg). Substituting these values into equation (10.14) gives

$$K = \frac{(2.13)^2 \times 0.01}{400 \times 1 \times 1/4} = 4.54 \times 10^{-4}\text{ R/mA-min at 1 meter.}$$

From Fig. 10.13, using the 125 kVp curve, we find that 17.7 cm (7 in.) of concrete are required. The wall is only 10.2 cm thick, therefore lead of thickness equivalent to 5.5 cm concrete must be added to the wall. From Table 10.2, we find the HVL for 125 kV X-rays to be 2.0 cm concrete or 0.28 mm lead. Since 5.5 cm concrete is equal to 2.75 half-value layers, the thickness of the additional lead is 2.75 times 0.28, or 0.77 mm.

To compute the secondary barrier:

(a) *Ceiling*: *leakage radiation.* Assume the X-ray tube to be midway between the floor and the floor above, at a height of 5 ft (1.5 m) above the floor. For design purposes, when the radiation originates below the floor, the dose rate at a height of 3 ft above the floor is considered as the criterion of safety. In this case, therefore, we have a total of 8 ft (2.5 m) between the target in the X-ray machine and the point we wish to protect. The required degree of attenuation of the leakage radiation is calculated from

$$P = B_{Lx} \times X_L. \tag{10.26}$$

Substituting equation (10.19) for X_L, the weekly leakage exposure, and then solving for the attenuation factor, B_{Lx} gives

$$B_{Lx} = P/X_L = 0.01 \times 600d^2/t. \tag{10.27}$$

Since the X-ray machine is "on" 302 minutes in a week, the attenuation factor required to reduce the weekly exposure from leakage radiation to 0.01 R (3 μC/kg) at 2.5 meters is

$$B_{Lx} = 0.01 \times 600 \times (2.5)^2/302 = 0.124,$$

and the number of half-value layers that leads to this attenuation factor is

$$0.124 = 1/2^n$$

$$n = 3.$$

In Table 10.2, we find the HVL for 125 kV X-rays to be 2.0 cm concrete. The required thickness, therefore is 3 × 2.0 = 6 cm concrete.

(b) *Ceiling: scattered radiation.* Equation (10.17) is used to calculate the thickness to attenuate the scattered radiation to 0.01 R (3 μC/kg) in a week. First, the workload is calculated:

$$W = 5\ \text{hr/wk} \times 60\ \text{min/hr} \times 3.5\ \text{mA} = 2\ \text{min} \times 200\ \text{mA}$$
$$= 1450\ \text{mA-min/week}.$$

The other values for use in equation (10.17) are:

$$a = 0.0015\ \text{(from Table 10.3)},$$

$$T = 1,$$

$$d_{sca} = 0.5\ \text{m},$$

$$d_{sec} = 2.5\ \text{m},$$

$$F = 400\ \text{cm}^2,$$

$$K_{ux} = \frac{0.01}{0.0015 \times 1450} \times (0.5)^2 \times (2.5)^2 \times \frac{400}{400} = 7.2 \times 10^{-3}.$$

In Fig. 10.13, on the 125-kVp curve, we find that 9 cm concrete corresponds to $K = 7.2 \times 10^{-3}$ R/mA-min at 1 meter.

The difference between the required barrier thicknesses for leakage and for scatter is 3 cm concrete. Since this difference is less than one tenth-value layer (6.6 cm concrete, Table 10.2), we add one half-value layer, 2.0 cm, to the thicker barrier to obtain a required barrier thickness of 11 cm concrete. No additional shielding is necessary because the ceiling thickness is already 12.7 cm concrete.

(c) *Wall B: leakage radiation.* The attenuation factor for wall B can be calculated from equation (10.27), using a value of 6/3.28 = 1.8 meters for d, and 302 minutes for t.

However, since the occupancy factor for a narrow corridor is 1/4, we must multiply the value of the attenuation factor by 4. This yields

$$B_{Lx} = 4 \times 0.01 \times 600 \times (1.8)^2/302 = 0.257.$$

The number of half-value layers required to attain this degree of attenuation is

$$0.257 = 1/2^n$$
$$n = 2$$

Since the half-value layer for 125-kV X-rays is 2.0 cm concrete, the required barrier thickness is $2 \times 2.0 = 4$ cm concrete.

(*d*) *Wall B*: *scattered radiation.*

If we substitute the following values into equation (10.15):

$$P = 0.01 \text{ R},$$
$$W = 1450 \text{ mA-min/wk at 1 m},$$
$$a = 0.0015 \text{ (from Table 10.3)},$$
$$T = 1/4,$$
$$d_{sca} = 0.5 \text{ m},$$
$$d_{sec} = 1.8 \text{ m},$$
$$F = 400 \text{ cm}^2,$$

we obtain

$$K_{ux} = \frac{0.01}{0.0015 \times 1450 \times 1/4} \times (0.5)^2 \times (1.8)^2 \times \frac{400}{400}$$
$$= 1.5 \times 10^{-2} \text{ R/mA-min at 1 meter.}$$

In Fig. 10.13, on the 125-kVp curve, we find that 7.3 cm concrete corresponds to $K = 1.5 \times 10^{-2}$ R/mA min at 1 m.

Since the difference between the leakage and scatter barrier requirements, 3.3 cm concrete, is less than one tenth-value layer (6.6 cm concrete, Table 10.2), we add one half-value layer, or 2.0 cm concrete to the thicker barrier to give a total secondary barrier thickness of 2.0 cm + 7.3 cm = 9.3 cm concrete.

The wall is made of hollow tile and plaster, as shown in Fig. 10.17. The concrete equivalent of the wall is calculated as follows, using the densities of commercial building materials listed in Table 10.4:

$$(\text{density} \times \text{thickness})_{\text{wall}} = (\text{density} \times \text{thickness})_{\text{concrete}},$$
$$1.54 \text{ g/cm}^2 \times 2 \times 5/8 \text{ in} + 1.9 \text{ g/cm}^3 \times 2 \times 1/2 \text{ in} = 2.35 \text{ g/cm}^3 \times t \text{ in},$$
$$t = 1.63 \text{ in} = 4.1 \text{ cm concrete.}$$

The required barrier thickness is 9.3 cm concrete or its equivalent thickness for other materials. Since we already have the equivalent of 4.1 cm concrete, we need an additional 5.2 cm concrete, or the equivalent of 2.6 half-value layers. The half-value layer for lead for 125 kV X-rays is 0.28 mm (Table 10.2). The total additional lead that must be added to the plastered wall is

$$2.6 \text{ HVL} \times 0.28 \text{ mm Pb/HVL} = 0.73 \text{ mm Pb.}$$

TABLE 10.4. DENSITIES OF COMMERCIAL BUILDING MATERIALS

Material	Density range g/cm^3	Density of average sample g/cm^3
Brick	1.6–2.5	1.9
Granite	2.60–2.70	2.63
Limestone	1.87–2.69	2.30
Marble	2.47–2.86	2.70
Sandplaster	—	1.54
Sandstone	1.90–2.69	2.20
Siliceous concrete	2.25–2.40	2.35
Tile	1.6–2.5	1.9

Commercially available lead sheet is listed in Table 10.5. In the example above, the commercial sheet of the next thickness greater than the calculated value would be used in the actual installation. When shielding the walls of an X-ray room, the lead should extend to a height of at least 7 ft (213 cm).

TABLE 10.5. COMMERCIAL LEAD SHEETS

Thickness		Nominal weight	
mm	in	#/ft^2	kg/m^2
0.79	1/32	2	10
1.00	5/128	$2\frac{1}{2}$	12
1.19	3/64	3	15
1.58	1/16	4	20
1.98	5/64	5	24
2.38	3/32	6	29
3.17	1/8	8	39
4.76	3/16	12	59
6.35	1/4	16	78
8.50	1/3	20	98
10.1	2/5	24	117
12.7	1/2	30	146
16.9	2/3	40	195
25.4	1	60	293

Beta-ray shielding

Two factors must be considered when designing a shield against high-intensity radiation, viz. the beta rays and the bremsstrahlung that are generated due to absorption in the source itself and in the shield. Because of these factors, the beta shield consists of a low-atomic-numbered substance (to minimize the production of bremsstrahlung) sufficiently thick to stop all the beta rays, followed by a high-atomic-numbered material thick enough to attenuate the bremsstrahlung intensity to an acceptable level.

Example 10.9

Fifty milliliters of aqueous solution containing 37×10^4 MBq (10 Ci) carrier-free ^{90}Sr in equilibrium with ^{90}Y, is to be stored in a laboratory. The health physicist requires the dose-equivalent rate at a distance of 1 m from the source to be no greater than 0.1 mSv/hr (10 mrem/hr). Design a source holder and shield to meet the specifications for radiation safety during storage.

The maximum and mean beta-ray energies of ^{90}Sr and ^{90}Y are:

	E_{max}, MeV	$\bar{E}$, MeV
^{90}Sr	0.54	0.19
^{90}Y	2.27	0.93

The beta shield must be sufficiently thick to stop the 2.27-MeV ^{90}Y betas. From Fig. 5.4, the range of a 2.27-MeV beta is found to be 1.1 g/cm^2. Let us use a bottle made of polyethylene, specific gravity 0.95, as the container for the radioactive solution. The wall thickness, therefore, must be

$$E = \frac{1.1 \text{ g/cm}^2}{0.95 \text{ g/cm}^3} = 1.06 \text{ cm}.$$

The bremsstrahlung are especially important, since ^{90}Sr and ^{90}Y are pure beta emitters, and thus there are ordinarily no gammas to be shielded. The radiation dose rate due to the bremsstrahlung at a distance of 1 m from the source may be estimated with the aid of equation (5.11),

$$f = 3.5 \times 10^{-4} Z \times E_{max}.$$

In this case, most of the beta-ray energy will be absorbed in the water. If we use for Z, in equation (5.11), the effective atomic number for water,

$$Z_w = \frac{2}{18} \times 1 + \frac{16}{18} \times 8 = 7.22,$$

and 2.27 MeV for E_{max}, we find f, the fraction of the ^{90}Y beta-ray energy that is converted into bremsstrahlung to be 5.73×10^{-3}. The rate at which energy is carried off by the ^{90}Y betas is

$$E_\beta = 3.7 \times 10^{11} \text{ Bq} \times 1 \frac{\text{tps}}{\text{Bq}} \times 0.93 \frac{\text{MeV}}{d} = 3.44 \times 10^{11} \frac{\text{MeV}}{\text{sec}}.$$

If the bremsstrahlung are considered to radiate from a virtual point in the center of the source, then the dose rate at a distance of 1 m from this point is given by:

$$\dot{D} = \frac{f \times E \text{ Mev/s} \times 1.6 \times 10^{-13} \text{ J/MeV} \times \mu_e \text{ m}^{-1} \times 3.6 \times 10^3 \text{ sec/hr}}{1.293 \text{ kg/m}^3 \times 4\pi(1 \text{ m})^2 \times 1 \frac{\text{J}}{\text{kg}} \Big/ \text{Gy} \times 10^{-3} \text{ Gy/mSv}}, \tag{10.28}$$

where μ_e is the energy absorption coefficient corresponding to the quantum energy of the bremsstrahlung. In calculating bremsstrahlung dose rate for pusposes of radiation protection, the quantum energy is assumed to be equal to that of the maximum energy beta particle. Using the value $\mu = 3 \times 10^{-3} \text{ m}^{-1}$ (from Fig. 5.18), the bremsstrahlung dose rate due to the ^{90}Y beta rays is 0.21 mSv/hr (21 mrem/hr) at 1 m. Similarly, the dose rate due to ^{90}Sr betas is 0.013 mSv/hr (1.3 mrem/hr).

If we have a shield of thickness t cm, then the dose rate attenuation from the bremsstrahling can be calculated from equation (5.19), using a value of 0.51 cm^{-1} for the total linear absorption coefficient for lead for the 2.27-MeV X-rays from ^{90}Y and 1.5 cm^{-1} for the 0.54-MeV X-rays from ^{90}Sr. Furthermore, the sum of the ^{90}Y and ^{90}Sr dose rates must not exceed 0.1 mSv/hr (10 mrem/hr). These considerations give the following set of simultaneous equations

$$I_Y = 0.21\,e^{-0.51t},$$

$$I_{Sr} = 0.013\,e^{-1.5t},$$

$$I_Y + I_{Sr} = 0.1,$$

which can be solved to give $t = 1.75$ cm.

Consideration of a build-up factor in this case is not necessary because it was assumed, in the calculation of the thickness of lead, that all the bremsstrahlung had a quantum energy equal to the maximum energy of the beta particle that gave rise to the X-ray. Since this quantum energy is in fact the upper energy limit of the bremsstrahlung, and since most of the bremsstrahlung are much lower in energy than this upper limit, the thickness calculated above overestimates the required thickness. This overestimate just about compensates for the build-up factor, thus making consideration of the build-up factor unnecessary. The shipping container, therefore, consists of a polyethylene bottle, whose wall thickness is 1.06 cm, placed into a lead container whose sides, top, and bottom are 1.75 cm thick. However, a shield of this construction would be unnecessarily heavy. The total amount of lead in the shield increases with the square of the diameter; thus, even a small decrease in shield diameter may result in a significant weight reduction. If weight were a problem, such a reduction could be achieved by decreasing the thickness of the polyethylene bottle. If, instead of 1.06 cm, the thickness were reduced to $\frac{1}{2}$ cm, all of the ^{90}Sr betas, plus most of the ^{90}Y betas, would be stopped. Furthermore, those ^{90}Y betas that would get through would be greatly reduced in energy. The resulting bremsstrahlung from these betas, therefore, would contribute very little to the radiation dose rate outside the lead shield.

Neutron shielding

Shielding against neutrons is based on slowing down fast neutrons and absorbing thermal neutrons. In Chapter 5 it was seen that attenuation and absorption of neutrons is a complex series of events. Despite the complexity, however, the required shielding around a neutron source can be estimated by the use of removal cross sections. (For neutron energies up to 30 MeV, the removal cross section is about three-quarters of the total cross section.) In designing shielding against neutrons, it must be borne in mind that absorption of neutrons can lead to induced radioactivity and to the production of gamma radiation.

Example 10.10

Design a shield for an 18.5×10^4 MBq (5 Ci) Pu–Be neutron source that emits 5×10^6 neutrons per sec, such that the dose rate at the outside surface of the shield will not exceed 15 μSv/hr (1.5 mrem/hr). The mean energy of the neutrons produced in this source is 4 MeV.

Let us make the shield of water, and compute the minimum radius for the case of a spherical shield. Since we know that the capture of a neutron by hydrogen produces a 2.26-MeV gamma ray, let us allow for the gamma-ray dose by designing the shield to give a maximum fast-neutron dose rate of 10 μSv/hr (1 mrem/hr), which corresponds to a fast flux of 6 neutrons per cm^2/sec (Table 9.5). The total cross section for 4 MeV neutrons for hydrogen and oxygen are 1.9 and 1.7 barns, respectively. Since water contains 6.7×10^{22}

hydrogen atoms and 3.35×10^{22} oxygen atoms per cm^3, the linear absorption coefficient of water is

$$\Sigma = 3/4(1.9 \times 10^{-24}\ cm^2/atom \times 6.7 \times 10^{22}\ atom/cm^3 + 1.7 \times 10^{-24}\ cm^2/atom \times 3.5 \times 10^{22}\ atoms/cm^3) = 0.138\ cm^{-1},$$

which corresponds to a half-value layer of 5.01 cm. The Pu–Be may be considered as a point source of neutrons, with the neutron flux decreasing with increasing distance as a result of both inverse square dispersion and attenuation by the water. If S is the source strength in neutrons per second, T is the half-value layer in cm, n is the number of half-value layers, and B is the build-up factor, the fast-neutron flux, after passing through a thickness of nT cm, is

$$\phi = \frac{BS}{4\pi(nT)^2} \times \frac{1}{2^n} \frac{\text{neutrons}}{cm^2\ sec}. \tag{10.29}$$

For radioactive neutron sources on the order of several curies, the shield thickness is relatively thick, and a significant dose build-up due to scattered neutrons results. For a hydrogenous shield at least 20 cm thick, the dose build-up factor is approximately 5. Using a value of 6 neutrons per cm^2/sec for ϕ, 5 cm for T, and 5 for B, equation (10.29) may be solved for n to give about 7.8 half-value layers, which corresponds to a thickness of 39 cm water.

The thermal neutrons that would escape from the surface of a spherical water shield may be estimated with the aid of equation (5.52):

$$\phi_{th} = \frac{n_0}{\pi RD} e^{-R/L}.$$

Since the shield radius calculated above is much greater than the fast diffusion length (which is equal to 5.75 cm), we may assume, for the purpose of this calculation, that essentially all the fast neutrons are thermalized, and that the thermal neutrons are diffusing outward from the center. Substituting the appropriate numbers into equation (5.52), we have

$$\phi_{th} = \frac{5 \times 10^6\ \text{neutrons/sec}}{4\pi \times 39\ cm \times 0.16\ cm} e^{-39/2.88} = 0.08$$

thermal neutrons per cm^2/sec. This thermal neutron flux is so small relative to the maximum permissible thermal flux of 680 neutrons per cm^2/sec, that it may, for most practical purposes, be ignored.

Capture of a thermal neutron by a hydrogen atom results in the emission of a 2.26-MeV gamma ray. The water shield, therefore, acts as a distributed source of gamma radiation. Since 6.08 neutrons per cm^2/sec escape from the surface, the total number that escape from a sphere of radius 39 cm is 1.2×10^5 neutrons per sec, or approximately 2.4% of the source neutrons. The remaining 97.6% are absorbed in the water, thus giving a mean "specific activity" for 2.26 MeV photons of

$$\frac{4.9 \times 10^6\ \text{photons/sec}}{\frac{4}{3}\pi(39\ cm)^3} = 19.7\ \text{"Bq"}/cm^3 \quad (0.53\ \text{"pCi"}/cm^3).$$

The dose rate at the surface of a sphere containing a uniformly distributed gamma emitter is, from equations (6.40) and (6.45),

$$\dot{D} = \tfrac{1}{2} \times C\Gamma \times \frac{4\pi}{\mu}(1 - e^{-\mu r}). \tag{10.30}$$

Using a value 2.7 mGy/hr per MBq (10^4 mrad/hr per mCi) at 1 cm for Γ, 0.046 cm^{-1} for μ for 2.26 MeV photons in water, and 39 cm for the radius gives

$$\dot{D} = \tfrac{1}{2} \times 19.7 \times 10^{-6} \times 2.7 \times \frac{4\pi}{4.6 \times 10^{-2}}(1 - e^{-0.046 \times 39})$$

$$= 6 \times 10^{-3}\ \text{mGy/hr} \quad (0.6\ \text{mrad/hr})$$

The dose rate at the surface of the shield due to both neutrons and gamma rays is 16 μSv/hr (1.6 mrem/hr), which is very close to the desired figure of 15 μSv/hr (1.5 mrem/hr). The gamma-ray dose rate could be reduced either by increasing the gamma-ray absorption coefficient of the water shield by dissolving a high-atomic-numbered substance, such as $BaCl_2$; or by reducing the rate of production of the gamma radiation. Of these possible alternatives, the simplest one is the reduction in the production of gamma radiation. This is easily accomplished merely by dissolving a boron compound in the water. Boron captures thermal neutrons with a capture cross section of 755 barns, according to the reaction $^{10}B + n^1 \rightarrow {}^7Li + \gamma$ (0.48 MeV). The 0.48 MeV gamma is emitted in 93% of the captures. Either sodium tetraborate (borax), $Na_2B_4O_7 \cdot 10\ H_2O$, or boric acid, H_3BO_3, both of which are highly soluble in water and very inexpensive, may be considered for this application. If suppression of gamma radiation is the objective, boric acid may be preferred over borax, since the sodium in the borax has a relatively high cross-section, 505 millibarns, for the $^{23}Na(n, \gamma)\ ^{24}Na$ reaction. As a consequence of this reaction, a 6.96-MeV capture gamma is emitted and radioactive ^{24}Na, which emits one 1.39-MeV beta, one 1.37-MeV gamma, and one 2.75-MeV gamma per disintegration is produced.

The solubility of boric acid in water at room temperature is 63.2 g/l. The formula weight of H_3BO_3 is 61.84. The concentration of boron atoms in the saturated solution is

$$\frac{63.2\ \text{g/l} \times 10^{-3}\ \text{l/ml} \times 6.03 \times 10^{23}\ \text{molecules/mole} \times 1\ \text{atom B/molecule}}{61.84\ \text{g/mole}}$$

$$= 6.17 \times 10^{20}\ \frac{\text{atoms B}}{\text{ml}}.$$

If we consider the macroscopic cross sections for thermal neutron capture of the dissolved boron and of the hydrogen, we find that

$$\frac{\Sigma H}{\Sigma B} = \frac{0.13\ \text{cm}^{-1}}{0.42\ \text{cm}^{-1}} = 0.31.$$

The flux of 2.26 MeV hydrogen gamma rays, and consequently the dose rate, will be reduced by this factor to $0.31 \times 6 \times 10^{-3} = 1.2 \times 10^{-3}$ mGy/hr (0.12 mrad/hr). The dose rate due to the ^{10}B capture gammas, which is calculated from equation (10.16) using a photon specific activity of $0.69 \times 0.93 \times 19.7 \times 10^{-6}$ "MBq"/cm^3 (3.4×10^{-7} "mCi"/cm^3), an absorption coefficient for 0.48-MeV photons in water of 0.097 cm^{-1}, and a value for Γ of 0.62 mGy-cm^2/MBq-hr (2300 mrad-cm^2/mCi-hr), is found to be 7.7×10^{-4} mGy/hr (0.077 mrad/hr). The total dose equivalent rate at the shield surface, therefore is 10 μSv/hr (1 mrem/hr) fast neutrons plus 1.2 μSv/hr (0.12 mrem/hr) due to the hydrogen capture gammas, plus 0.8 μSv/hr (0.08 mrem/hr) due to the boron capture gammas, or 12 μSv/hr (1.2 mrem/hr) if we saturate the water with boric acid.

Problems

1. A Po–Be neutron source emits 10^7 neutrons per second, of average energy 4 MeV. The source is to be stored in a paraffin shield of sufficient thickness to reduce the fast flux at the surface to 10 neutrons per cm^2/sec. Consider paraffin to be essentially CH_2 (for the purpose of this problem) and to have a density of 0.89 gm/cm^3.

(a) What is the minimum thickness of the paraffin shield?

(b) If the slowing down length is 6 cm, the thermal diffusion length is 3 cm, and the diffusion coefficient is 0.381 cm, what will be the thermal neutron leakage flux at the surface of the shield?

(c) What is the gamma-ray dose rate, due to the hydrogen capture gammas, at the surface of the paraffin shield?

2. An X-ray therapy machine operates at 250 kVp and 20 mA. At a target to skin distance of 100 cm, the exposure rate is 20 R/min. The work load is 10,000 mA min/week. The X-ray tube is constrained to point vertically downward. At a distance of 4 m from the target is an uncontrolled waiting room. Calculate the thickness of lead to be added to the wall if the total thickness of the wall (which is made of hollow tile and plaster, density 2.35 g/cm^3) is 2 in.

3. A 7.4×10^{13} Bq (2000 Ci) ^{60}Co teletherapy unit is to be installed in an existing concrete room in the basement of a hospital so that the source is 4 m from the north and west walls—which are 30 in. thick. Beyond the north wall is a fully occupied controlled room. Beyond the west wall is a public parking lot. The useful beam is to be directed towards the north wall for a maximum of 5 hr per week during radiation therapy. The beam will be directed at the west wall 1 hr per week. Considering only the radiation from the primary beam, how much additional shielding, if any, is required for each of the walls?

4. A radiochemist wants to carry a small vial containing 2×10^9 Bq ($\sim$ 50 mCi) ^{60}Co solution from one hood to another. If the estimated carrying time is 3 min, what would be the minimum length of the tongs used to carry the vial in order that his dose not exceed 60 μGy (6 mrad) during the operation?

5. A viewing window for use with an isotope that emits 1-MeV gamma rays is to be made from a saturated aqueous solution of KI in a rectangular battery jar. What will be the attenuation factor, assuming conditions of good geometry, if the solution thickness is 10 cm, and if the glass walls are equivalent in their attenuation property to 1 mm lead? A saturated solution of KI may be made by adding 30 g KI to 21 ml water to give 30 ml solution at 25°C. Total attenuation cross sections for 1 MeV gamma-rays for the elements in the solution are:

K = 4 barns	H = 0.2 barns
I = 12 barns	O = 1.7 barns

6. Lead foil consists of an alloy containing 87% Pb and 12% Sn, and 1% Cu. Its specific gravity is 10.4. If the mass attenuation coefficients for these three elements are 3.50, 1.17, and 0.325 cm^2/g respectively for X-rays whose wavelength is 0.098 Å, and if the specific gravities of the three elements are 11.3, 7.3, and 8.9 respectively,

(a) calculate the mass and linear absorption coefficients for lead foil,

(b) what thickness of lead foil would be required to attenuate the intensity of ^{57}Co gamma-rays by a factor of 25?

7. A hypodermic syringe that will be used in an experiment in which ^{90}Sr solution will be injected has a glass barrel whose wall is 1.5 mm thick. If the density of the glass is 2.5 g/cm^3, how many mm thick must we make a Lucite sleeve that will fit around the syringe if no beta particles are to come through the Lucite? The density of the Lucite is 1.2 g/cm^3.

8. A room in which a 7.4×10^{13} Bq (2000 Ci) ^{137}Cs source will be exposed has the following layout

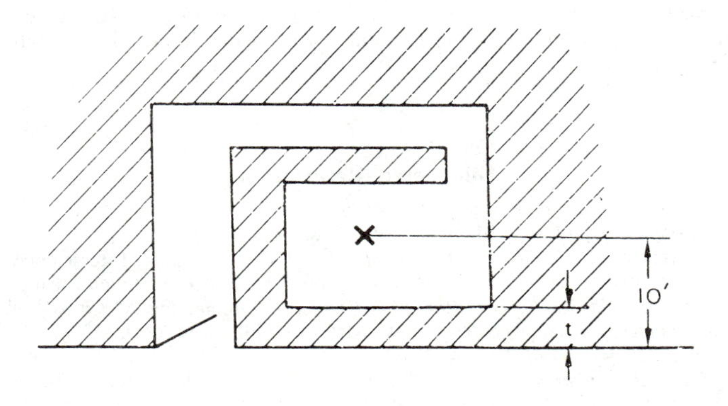

Calculate the thickness t of concrete so that the exposure rate at the outside surface of the wall does not exceed 0.64 μC/kg (2.5 mR) per hour.

9. What minimum density thickness must a pair of gloves have to protect the hands from ^{32}P radiation?

10. When a radium source containing 50 mg Ra encapsulated in 0.5-mm-thick Pt is placed into a Pb storage container, the measured exposure rate at a distance of 1 m from the source is 10.4 μC/kg (21 mR) per hr. If this same container is used for storing ^{137}Cs, how many MBq may be kept in it for a period of 4 hr without exceeding an exposure of 20.6 μC/kg (40 mR) at a distance of 50 cm from the source?

11. What is the maximum working time in a mixed radiation field consisting of 6 μC/kg (20 mR) per hour gamma, 40 μGy (4 mrad) per hour fast neutrons, and 50 μGy (5 mrad) per hour thermal neutrons, if a maximum dose equivalent of 3 mSv (300 mrem) has been specified for the job?

12. Maintenance work must be done on a piece of equipment that is 2 meters from an internally contaminated (with ^{137}Cs) valve. The exposure rate at 30 cm from the valve is 500 R/hr (0.13 C/kg-hr). If 4 hours is the estimated repair time, what thickness of lead shielding is required to limit the dose equivalent of the maintenance men to 1 mSv (100 millirem)?

13. Calculate the exposure rate from a 100,000-MBq (2.7-Ci) ^{60}Co "point" source, at a distance of 1.25 meters, if the source is shielded with 10 cm Pb.

14. A stainless steel bolt came loose from a reactor vessel. It is planned to pick up the bolt with a remotely operated set of tongs and transport it for inspection and study. The bolt had been in a mean thermal neutron flux of 2×10^{12} n/cm^2-sec for a period of 900 days, and will be picked up 21 days after reactor shut-down. Calculate the gamma-ray dose rate at a distance of 1 meter from the bolt if the bolt weighs 200 grams and has the following composition by weight:

Fe	80%
Ni	19%
Mn	0.5%
C	0.5%

15. A circular area 1 meter in diameter is accidentally contaminated with 10 MBq (270 μCi) ^{131}I. What is the maximum dose-equivalent rate at a distance of 1 meter above the contaminated area?

16. Design a spherical lead storage container that will attenuate the radiation dose rate from 5×10^{10} Bq (1.35 Ci) ^{24}Na to 100 μGy/hr (10 mrad/hr) at a distance of 1 meter from the source. (The source is physically small enough to be considered a "point.")

17. What thickness of standard concrete is needed to reduce the intensity of a collimated beam of 10-MeV X-rays from 10^4 mW/cm^2 to an intensity corresponding to 2.5×10^{-3} centisieverts per hour?

18. A Ra–Be neutron source emits about 1.2×10^7 fast neutrons (average energy = 4 MeV) per gram Ra. What fraction of the dose equivalent from an unshielded source is due to the neutrons?

19. Transport regulations for shipping a radioactive package specify a maximum surface dose rate equivalent of 2 mSv/hr (200 mrem/hr) and a maximum of 0.1 mSv/hr (10 mrem/hr) at 1 meter from the surface. If aqueous ^{137}Cs waste is to be mixed with cement for disposal, what is the maximum specific activity of the concrete if it is to be cast in 20-liter cylindrical polyethylene containers 30 cm diameter for shipment to the waste burial site?

20. A technician's job in a radiopharmaceutical laboratory involves simultaneous handling of 5000 MBq (135 mCi) ^{125}I, 4000 MBq (108 mCi) ^{198}Au and 2000 MBq (54 mCi) ^{24}Na for 1 hour per day, 5 days per week, for an indefinitely long period of time. Her average dose equivalent during the other 7 hours will be 0.01 mSv (1 millirem). Her body will be 75 cm from the sources while she works with them, and manipulators will be provided so that her hands will not be exposed inside the shield.

(a) What is the source strength for each of the sources?

(b) What thickness of lead shielding is required if her weekly dose equivalent is to be within ALARA guidelines, that is, at 1/10 of the maximum permissible dose?

21. Design a spherical shield for a 1×10^{11} Bq (2.7 Ci) ^{90}Sr "point" source so that the dose-equivalent rate at the surface will not exceed 2 mSv (200 millirem) per hour. What is the dose-equivalent rate at a distance of 1 meter from the shielded source?

Suggested References

BLATZ, H. *Radiation Hygiene Handbook*. McGraw-Hill, New York, 1959.

BLIZZARD, E. P., and ABBOTT, L. S. *Reactor Handbook*, Vol. III B, *Shielding*. Interscience, New York, 1962.

BRAESTRUP, C. B., and WYCKOFF, H. O. *Radiation Protection*. Charles G. Thomas, Springfield, 1958.

DRESNER, L., translator. *Jaeger's Principles of Radiation Protection Engineering*. McGraw-Hill, New York, 1965.

ETHERINGTON, H., ed. *Nuclear Engineering Handbook*. McGraw-Hill, New York, 1958.

GOLDSTEIN, H. *Fundamental Aspects of Reactor Shielding*. Addison-Wesley, Reading, Mass., 1959.

JAEGER, R. G., ed. *Engineering Compendium on Radiation Shielding*. Springer Verlag, Berlin, 1968.

PRICE, B. T., HORTON, C. C., and SPINNEY, K. T. *Radiation Shielding*. Pergamon, New York, 1957.

PROFIO, E. *Radiation Shielding and Dosimetry*. Wiley, New York, 1979.

Radiological Health Handbook, revised edition. U.S. Dept. of Health, Education, and Welfare, Rockville, Maryland, 1970.

ROCKWELL, T., ed. *Reactor Shielding Design Manual*. McGraw-Hill, New York, 1956.

SCHAEFER, N. M., ed. *Reactor Shielding for Nuclear Engineers.* TID-25951, U.S. Atomic Energy Commission, Washington, D.C., 1973.

The following publications of the International Commission on Radiological Protection (ICRP), Pergamon Press, Oxford:

ICRP
Pub.
No.

3. *Report of Committee III on Protection Against X-rays up to Energies of* 3 *MeV and Beta and Gamma-Rays from Sealed Sounds* (1960).
15. *Protection Against Ionizing Radiation from External Sources* (1970).
16. *Protection of the Patient in X-Ray Diagnosis* (1970).
21. *Data for Protection Against Ionizing Radiation from External Sources: Supplement to ICRP Publication* 15 (1973).

The following Reports of the National Council on Radiation Protection and Measurements (NCRP), Washington, D.C.

NCRP
Report
No.

32. *Radiation Protection in Educational Institutions* (1966).
33. *Medical X-ray and Gamma Ray Protection for Energies Up to* 10 *MeV: Equipment Design and Use* (1968).
35. *Dental X-ray Protection* (1970).
36. *Radiation Protection in Veterinary Medicine* (1970).
38. *Protection against Neutron Radiation* (1971).
39. *Basic Radiation Protection Criteria* (1971).
48. *Radiation Protection for Medical and Allied Health Personnel* (1976).
49. *Structural Shielding Design and Evaluation for Medical Use of X-Rays and Gamma Rays of Energies Up to* 10 *MeV* (1976).
51. *Radiation Protection Guidelines for* 0.1–100 *MeV Particle Accelerator Facilities* (1977).

CHAPTER 11

INTERNAL RADIATION PROTECTION

Internal Radiation Hazard

Internal radiation results when the body is contaminated—either internally or externally—with a radioisotope. Accordingly, internal radiation protection is concerned with preventing or minimizing the deposition of radioactive substances on or in personnel. This is accomplished by a program designed to keep the contamination of the environment within acceptable limits, and at levels as low as practical. The last point—keeping environmental levels as low as practical—is especially important in the context of internal radiation protection. External radiation exposure is due to radiation originating in sources outside the body; there is no physical contact with the radiation source, and exposure ceases when one leaves the radiation area or the source is removed. Since external radiation may be measured with relative ease and accuracy, the potential or actual hazards may be estimated with a good deal of confidence. In the case of contamination, on the other hand, the radioactive material is deposited on or within the body. As a consequence, the contaminated person continues to suffer irradiation even after he leaves the area where he became contaminated. Furthermore, radioisotopes within the body may become systemically fixed. Their elimination then can be hastened—in those cases where possible—only with relative difficulty. Biological turnover rates, which may exhibit a marked variability, then become significant in determining the radiation absorbed dose from the internally deposited isotopes. For these reasons, it is relatively difficult to assess the hazard from internal emitters, and great emphasis is therefore placed on the prevention of contamination of personnel with radioactive materials.

Principle of Control

Radioactive substances, like other toxic agents, may gain entry into the body through three portals:

1. Inhalation—by breathing radioactive dust and gas.
2. Ingestion—by drinking contaminated water, eating contaminated food, or by tactilly transferring radioactivity to the mouth.
3. Absorption through the intact skin or through wounds.

Basically, therefore, protective measures to counter internal radiation are designed either to block the portals of entry into the body, or to interrupt the transmission of radioactivity from the source to man. This interruption can be effected either at the source—by enclosing and confining it, or by controlling the environment—by ventilation and good housekeeping, or at the man—by providing him with protective clothing and with respiratory protective devices. It should be noted that these control measures do not differ from those employed by the industrial hygienist in the protection of workers from the effects of non-radioactive noxious substances. However, the degree of control required for radiological safety almost always greatly exceeds the requirements for chemical safety. This point is made clear by the figures in Table 11.1, which compare the maximum allowable atmospheric concentrations of several non-radioactive toxic substances to the maximum concentrations recommended by the ICRP for radioactive forms of the same element.

TABLE 11.1. MAXIMUM ALLOWABLE CONCENTRATIONS OF SEVERAL SUBSTANCES BASED ON CHEMICAL AND RADIOLOGICAL TOXICITY

	Milligrams per meter^3		
	Non-radioactive	Radioactive	
Beryllium	0.002	^{7}Be	1.7×10^{-8}
Mercury	0.1	^{203}Hg	5×10^{-9}
Lead	0.05	^{210}Pb	1×10^{-9}
Arsenic	0.5	^{74}As	3×10^{-9}
Cadmium	0.1	^{115}Cd	4×10^{-10}
Zinc	5	^{65}Zn	1.2×10^{-8}

Control of the Source: Confinement

The simplest type of confinement and enclosure may be accomplished by limiting the handling of radioactive materials to well-defined, separated areas within a laboratory, and by the use of sub-isolating units, such as trays. For low-level work where there is no likelihood of atmospheric contamination with more than one maximum recommended body burden, this may be sufficient. If the possibility exists of the release to the atmosphere, either as a gas or an aerosol, of amounts of activity between 1 and 10 times the maximum recommended body burden, the usual practice is to use a ventilated hood. The purpose of the ventilated hood is to dilute and to sweep out the released radioactivity with the air that flows through the hood. It is thus essential to have a sufficient amount of air flowing through the hood at all times in order to accomplish this purpose. Constant air-flow may be maintained by a by-pass that opens as the face is closed. Openings along the bottom of the front-face frame facilitate the flow of air when the face is closed. Figure 11.1 shows a typical radiochemistry fume hood. The face velocity must be great enough to prevent contaminated air from flowing out of the face into the laboratory, but not great enough to produce turbulence around the edges that would allow contaminated air from the hood to spill out into the laboratory. It has been found that face velocities of 0.6–1.4 m/sec (125–275 ft/min) are required. To minimize the probability of contaminating the working environment with the exhaust from the hood, all the ductwork must be kept under a negative pressure. Any leakage in the ductwork will then be *into* the duct. This is most easily accomplished by locating the exhaust fan at the discharge end of the exhaust line, as shown in Fig. 11.1.

If the hood is designed to remove only gases, vapors, or fumes, an air velocity in the ductwork of about 10 m/sec (2000 ft/min) is sufficient. However, since particulate matter tends to settle out, the air transport velocity must be on the order of 18–22 m/sec (3500–4500 ft/min) if particulate matter fallout is to be minimized. If the exhaust from the hood is of such a nature that it may create a radioactive pollution problem, then the effluent from the hood should be decontaminated by an appropriate air cleaning device. For this purpose, if the pollutant is a dust, either a rough filter alone or a rough filter, followed by a fire-resistant, high-efficiency filter is commonly employed. As used in this

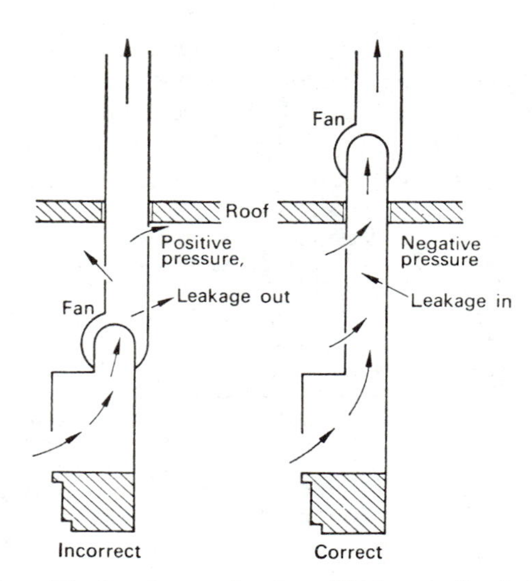

FIG. 11.1. Effect of fan location on direction of leakage in the ductwork. The fan should be close to the discharge end, thereby creating a negative pressure in the ductwork and causing leakage in the ductwork to be into the duct.

context, a high-efficiency filter is one that removes at least 99.995% of 0.3-μ-diameter homogeneous particles of dioctyl phthalate (DOP). The filter should not offer a resistance greater than one inch water when air at 70°F and 29.9 in Hg flows through it at its rated capacity. A manometer, or other device, should be used to indicate when the filter is loaded and ready to be changed. A filter loaded with radioactive dust can easily become a source of contamination if adequate precautions to prevent the dust from falling off the filter during the changing operation are not taken. A simple way to minimize dispersal of loose dust when removing the filter is to spray the filter faces with an aerosol laquer before removing the filter, thereby trapping the radioactive dust in the filter. For this purpose, access ports upstream and downstream of the filter should be provided in the ductwork.

If the nature of an operation involving radioactivity is such that it must be completely enclosed, that is, if the operation is potentially capable of contaminating the working environment with more than 10 times the recommended maximum body burden, or when the large quantities of air required by a hood are not avialable, then a glove box (Fig. 11.2) is used. It should be re-emphasized here that, whereas the main function of a fume hood is to dilute and remove atmospheric contaminants, the main function of a glove box is to isolate the contaminant from the environment by confining it to the enclosed volume. This is especially true if the contaminant is a particulate. Accordingly, the air flow through the glove box may be very small—on the order of 0.01–0.02 m^3/sec (25–50 ft^3/min). Air is usually admitted into the glove box through a high efficiency fiber-glass filter (to prevent discharge of radioactive dust into the room in case of an accidental positive pressure inside the glove box) and is exhausted through a series of fire-resistant rough and high efficiency filters. Air-borne particles small enough to be carried by this flow of air are thus transferred out of the glove box into the filter; larger particles fall out inside the glove box and remain there until cleaned out. A negative pressure of at least 13 mm (0.5 in.) water inside the glove box assures that any air leakage will be into the box. Despite the negative pressure, however, it may be assumed that a small fraction, about 10^{-8}, of the activity inside the glove box will leak out during the course of normal use of the glove box. The laboratory should be prepared to handle such contamination, and the health physicist should be prepared to account for this activity in the design and operation of his surveillance program. For maximum safety, transfer of materials and apparatus into or out of the glove box always is done through an air lock. The viewing panel may be heat-resistant safety plate glass. Glove boxes are unshielded when used for handling radioisotopes that do not create high-level radiation levels. For radioisotopes that do create such high levels of radiation, shielding must be added. When handling a high energy beta-emitting radioisotope, it may be necessary to use extra-thick gloves.

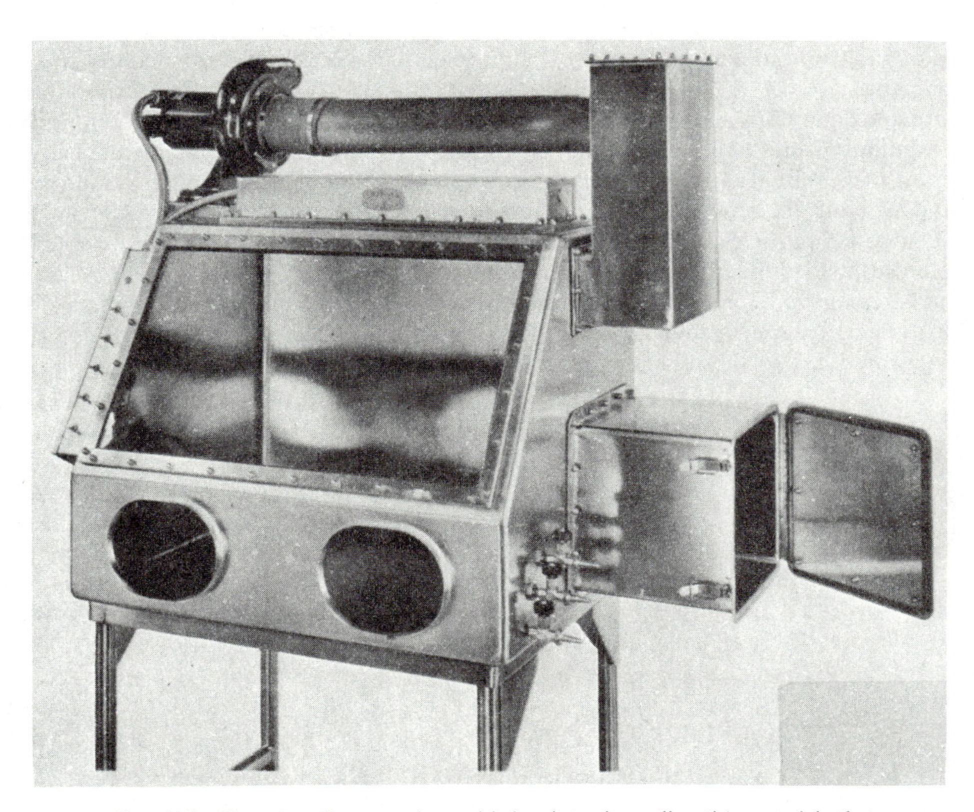

FIG. 11.2. Glove box for operations with low-intensity radioactive materials that might accidentlly become dispersed into the environment if not handled in an enclosed volume. In use, long rubber gloves fit over the port-flanges; material is transferred into and out of the glove box through the air lock. (Courtesy S. Blickman, Inc.)

Environmental Control

Environmental control of hazards from radioactive contamination begins with the proper design of the buildings, rooms, or physical facilities in which radioisotopes will be used; and continues with the proper design of the procedures and processes in which radioactivity will be employed. Since a finite probability exists that an accidental breakdown of a mechanical device or a human failure will occur despite the best efforts to prevent such a breakdown, the course of action to be taken in the event of an emergency must be known before the emergency occurs. In the design of the physical facilities, attention must be paid to the decontaminability of working surfaces, floors, and walls; to plumbing and means for monitoring or storing radioactive waste, both liquid and solid; to means for incineration of radioactive waste; to isotope storage facilities; to change rooms and showers; and to ventilation and the direction of air flow: office to corridor to area where radioisotopes are handled to exhaust through an air-cleaning system that will assure radiological safety outside the building. Strict control, including monitoring of all persons, materials, and equipment leaving the radiation area, must be maintained over the area where radioisotopes are being used or stored in order to prevent the spread of contamination outside the radiation area. The degree to which each of these control measures is implemented depends, of course, on the types and amounts of isotopes handled, and on the consequences of an accidental release of radioactivity to the environment.

In order to maintain control over internal radiation hazards, good housekeeping and good ventilation must be practiced. In regard to ventilation requirements, several important facts should be emphasized. The first is that fine particles under the influence of gravity do not, for practical purposes, move independently of the air in which they are suspended. Such particles behave effectively as if they are weightless, and can be assumed to remain suspended indefinitely in the air. *Control of airborne dust particles thus is reduced to a matter of air flow control.* A particle released into the atmosphere (with no horizontal motion) is acted upon by two forces—the downward force of gravity and the upward retarding force due to the resistance to free fall offered by the air (Fig. 11.3). When the

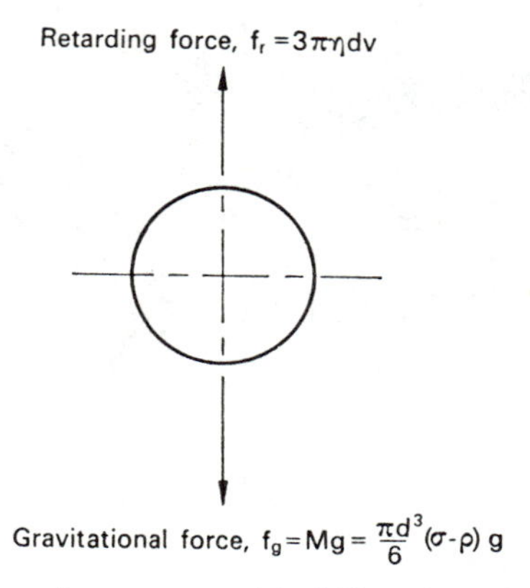

FIG. 11.3. The forces acting on a particle falling through air. η = viscosity of air = 185 μpoise at room temperature (1 poise = 1 g/cm-sec; d = particle diameter, cm; v = velocity of fall, cm/sec; M = particle mass, g; g = acceleration due to gravity, 980 cm/sec^2; σ = particle density, g/cm^3; ρ = air density, g/cm^3.

retarding force is equal to the gravitational force, there is no unbalanced force to accelerate the falling particle, and a constant velocity, called the *terminal velocity*, is attained by the falling particle. Equating f_r to f_g and solving for the terminal velocity, we have

$$f_r = 3\pi\eta \text{ d}v = \frac{\pi d^3}{6}(\sigma - \rho)g = f_g, \tag{11.1A}$$

$$v_t = v = \frac{d^2(\sigma - \rho)g}{18\eta}. \tag{11.1B}$$

Example 11.1

Calculate the terminal velocities of spherical particles of U_3O_8 whose diameters are 1 μ and 20 μ. The density of U_3O_8 is 8.30 g/cm^3.

Substituting into equation (11.1B), we have for the 1-μ particle,

$$v_t = \frac{(10^{-4}\,\text{cm}^2)\,(8.30 - 1.29 \times 10^{-3})\,\text{g/cm}^3 \times 980\,\text{cm/sec}^2}{18 \times 1.85 \times 10^{-4}\,\text{g/cm-sec}}$$

$$= 0.00244\,\frac{\text{cm}}{\text{sec}} \quad (0.0047\,\text{ft/min}),$$

and for the 20-μ particle, the terminal velocity is calculated as

$$v_t = 0.976\,\frac{\text{cm}}{\text{sec}} \quad (1.9\,\text{ft/min}).$$

The difference between the terminal settling velocities of the 1-μ and 20-μ-diameter particles is striking. However, even the settling velocity of the 20-μ particle is very much less than the ambient air velocities of about 0.13 m/sec (25 ft/min) in occupied space. A 20-μ particle could thus be carried a relatively long distance by air currents before falling out.

Control of Man: Protective Clothing

The philosophy of radiation protection advocates the restriction of radiation exposure to levels as far below the maximum recommended doses as practically possible. Since it is extremely difficult to maintain absolute radiological asepsis when working with unsealed sources, and since the possibility of an accidental spill or release to the environment of radioactivity always exists, it is customary to require isotope workers to wear protective clothing whose use is restricted to the radioactivity area. Such protective clothing may include laboratory coats, coveralls, caps, gloves, and shoes or shoe covers. Protective clothing is always assumed to be contaminated and, therefore, is removed when the worker leaves the radioactivity area. To be most effective, the protective clothing should be so designed that the worker can remove it easily and without transferring contamination from the clothing either to himself or to his environment. He should always be monitored before leaving the radioactivity area.

Protective clothing must, by its very nature, become contaminated; its main function is to intercept radioactivity that would otherwise contaminate the worker's skin or the clothing that he wears outside the radioactivity area. The degree of allowable contamination in the protective clothing varies with the type of work that the wearer does. For this reason, the maximum degree of contamination permitted on protective clothing is determined by the individual installation.

For most isotope laboratories, laundering protective clothing presents no special problem. In most instances, ordinary laundering procedures, repeated more than once if necessary, suffice. Sodium hexametaphosphate, or a complexing agent, such as sodium ethylene–diamine–tetraacetic acid (Na EDTA), may be added to the pre-wash rinse to facilitate the removal of radionuclides. After laundering, the protective clothing should be monitored to ascertain that it has, in fact, been decontaminated to some previously determined limit. If a piece of protective clothing is unusually or very severely contaminated, it may be simpler and cheaper to dispose of the item rather than to try to decontaminate it. In most cases of industrial or medical use of radioisotopes, the concentration of radioactivity in the wash water is low enough to be discharged directly into the sanitary sewer system. However, before being discharged, the wash water must be checked to verify that it can be safely discharged. If the activity level is too high, then the wash water must be treated as radioactive waste.

Control of Man: Respiratory Protection

When a person is likely to be exposed to a high concentration of air-borne radioactivity, respiratory protection is required. The exact type of respiratory protection depends on the nature of the air-borne contaminant. It must be emphasized that *respiratory protective devices may be used only for those hazards for which they are designed.*

Respiratory protective devices for radiological applications may be classified into two main categories:

1. Filter type respirators that are suitable only for dusts. *These respirators are not designed* to provide protection against radioactive gases. When using the half-face respirator, the wearer must try to eliminate possible leaks around the face-piece. Half-face respirators are considered suitable for air-borne dust concentrations up to ten (10) times the recommended maximum atmospheric concentration; respirators with full face masks are considered suitable up to fifty (50) times the recommended maximum atmospheric concentration.
2. Supplied air masks that may be used either against dusts or gases or both. In this category we have two subdivisions: (a) air line hoods, which utilize uncontaminated air, under positive (with respect to the atmosphere) pressure supplied from a remote source, and (b) self-contained breathing apparatus, in which breathing air is supplied either from a bottle carried by the man, or from a cannister containing oxygen generating chemicals. The advantage of the supplied air device is that the pressure in the breathing zone is higher than atmospheric pressure. As a consequence, all leakage is from the inside out. When using a supplied air device, it is imperative to know the time limitation on the supply of air.

A third type of respiratory protective device is the gas-mask. In this device, contaminated air is cleared by chemicals in a cannister through which the air passes. Because of the specific action of the chemical agents on the contaminant, different cannisters must be used for different gases. For this reason, as well as for the fact that air may leak into the face-piece of a gas-mask, gas-masks are not recommended for use against radioactive gas.

Surface Contamination Limits

Contamination of personnel and/or equipment may occur either from normal operations or as a result of the breakdown of protective measures. An exact quantitative definition of contamination that would be applicable in all situations cannot be given. Generally, contamination means the presence of undesirable radioactivity—undesirable either in the context of health or for technical reasons, such as increased background, interference with tracer studies, etc. In this discussion, only the health aspects of contamination are considered.

Surface contamination falls into two categories, fixed and loose. In the case of fixed contamination, the radioactivity cannot be transmitted to personnel, and the hazard, consequently, is that of external radiation. For fixed contamination, therefore, the degree of acceptable contamination is directly related to the external radiation dose rate. Setting a maximum limit for fixed surface contamination thus becomes a relatively simple matter. The hazard from loose surface contamination arises mainly from the possibility of tactile transmission of the radioactive contaminant to the mouth or to the skin, or of resuspending the contaminant and then inhaling it. It follows that the degree of hazard from surface contamination is strongly dependent on the degree to which the contaminant is fixed to the surface.

Dealing with loose surface contamination limits is not as straightforward as dealing with contamination of air and water. In the case of air and water contamination, safety standards can be easily set—at least in theory—on the basis of recommended dose limits. Using these criteria, we can calculate maximum permissible body burdens for each of the radioisotopes. From the calculated body burden, we go one step further from the basic radiation safety criteria, and compute maximum concentrations in air and water which, if continuously inhaled or ingested, would result in a body burden less than the calculated maximum. For the case of surface contamination, we go one more step away from the basic criteria; we try to estimate the surface contamination which, if it should be dispersed into the environment, would result in concentrations that may lead to an excessive body burden. Thus, specification of limits for loose surface contamination is three steps removed from the basic safety requirements.

From the foregoing discussion, it is clear that limits for surface contamination cannot be fixed in the same sense as limits for the concentration of radionuclides in air and water. Nevertheless, it is useful to compute a number that may serve as a guide in the evaluation of the hazard to workers from surface contamination, and to assist the health physicist in deciding whether or not to require the use of special protective measures for workers in contaminated areas.

On the basis of per-unit quantity of radioactivity, inhalation is considered the most serious route of exposure. Surface contamination, therefore, is usually limited by the inhalation hazard that may arise from resuspension of the contaminant. The quantitative relationship between the concentration of loose surface contamination and consequent atmospheric concentration above the contaminated surface due to stirring up the surface is called the *resuspension factor*, f_r, and is defined by

$$f_r = \frac{\text{atmospheric concentration, Bq/m}^3}{\text{surface concentration Bq/m}^2}. \tag{11.2}$$

Experimental investigation of the resuspension of loose surface contamination shows that the resuspension factor varies from about 10^{-4} to 10^{-8}, depending on the conditions under which the studies were conducted. A value of 10^{-6} is reasonable for the purpose of estimating the hazard from surface contamination.

Example 11.2

Estimate the maximum surface contamination of "insoluble" ^{90}Sr dust that may be allowed before taking special safety measures to protect personnel against a contamination hazard.

The derived atmospheric concentration of ^{90}Sr recommended by the ICRP is 60 Bq/m^3 (2×10^{-9} μCi/cm^3). Using a value of 10^{-6} m^{-1} for the resuspension factor in equation (11.2), we have

$$10^{-6}\,\text{m}^{-1} = \frac{60\ \text{Bq/m}^3}{\text{surface concentration}}$$

$$\therefore \text{ surface concentration} = 60\ \text{MBq/m}^2\ (2 \times 10^{-3}\,\mu\text{Ci/cm}^2).$$

It should be emphasized that a figure for loose surface contamination calculated by the method of Example 11.2 is intended only as a guide. In any particular case, the health physicist may, at his discretion, and depending on the nature of the operation, the degree of ventilation, and other relevant factors, insist on more or less stringent requirements for surface contamination before requiring the use of protective devices for the worker.

Various laboratories and nuclear installations have set their own limits for contamination of personnel, equipment, and protective clothing. Tables 11.2 and 11.3 are given to illustrate some of the contamination standards maintained by several large users of radioisotopes.

TABLE 11.2. U.S.S.R SURFACE CONTAMINATION LIMITS

Object of contamination	Contamination from 150 cm² in 1 min			
	Alpha particles		Beta particles	
	Before cleaning	After cleaning	Before cleaning	After cleaning
Hands	75	bg	5000	bg
Special linens and towels	75	bg	5000	bg
Cotton special work clothes	500	100	25,000	5000
"Pellicular" clothing	500	200	25,000	10,000
Gloves, outside	500	100	25,000	5000
Special shoes, outside	500	200	25,000	5000
Work surfaces and equipment	500	200	25,000	5000

Note. No contamination of the body is permitted.

(From *Sanitary Regulations for Work with Radioactive Substances and Sources of Ionizing Radiation*, Ministry of Health, U.S.S.R., Moscow, 1960.)

Waste Management

Proper collection and management of radioactive waste is an inherent part of contamination control and internal radiation protection. In one sense, we cannot "dispose" of radioactive waste. All other types (non-radioactive) of hazardous wastes can be treated either chemically, physically, or biologically in order to reduce their toxicity. In the case of radioactive wastes, on the other hand, nothing can be done to decrease their radioactivity, and hence the inherent toxicity of the waste. The only means of ultimate disposal is time—to allow the radioactivity to decay.

Radioactive wastes originate from any operation in which radioisotopes are used or produced. For purposes of management and treatment, wastes may be classified as high, intermediate, or low level. Low-level wastes may be defined as those which would have to be diluted by a factor no greater than 10^3, if they were to be discharged to the environment; for intermediate levels, $10^3 < DF < 10^5$; and for high level wastes, the dilution factor would have to exceed 10^5 if they were to be released to the environment. (These figures are for illustrative purposes only, and are not intended to imply that dilution is sufficient treatment for release of these wastes to the environment.) For the case of liquid

TABLE 11.3. SURFACE CONTAMINATION LIMITS IN THE UNITED KINGDOM

Category	Surface	Levels of contamination that should not be exceeded ($Bq\ cm^{-2}$)						
A	Surfaces of the interiors the contents of glove boxes and fume cupboards	The minimum reasonably achievable						
		Extent of contamination (m^2)	Class I	Class II	Class III		Class IV	Class V
B	Surface of active areas and of plant, apparatus, equipment (including personal protective clothing materials and articles within active areas other than in Category A	< 1	3	3×10^1	3×10^1		3×10^2	3×10^3
		> 1	3×10^{-1}	3				
					α emitters	others		
C	Surface of the body		3×10^{-1}	3×10^{-1}	3×10^{-1}	3	3×10^1	3×10^2
D	Inactive areas, personal clothing		3×10^{-1}	3×10^{-1}	3		3×10^1	3×10^2

Notes:
1. In relation to Category C (surfaces of the body) the values for all radionuclides may be increased by a factor of 10 when the skin is monitored with a small area probe.
2. For ^{231}Pa, use a tenth of the values in Class I for surfaces in categories B, C, and D. For ^{237}Np use a tenth of the values in Class I for surfaces in category C.
3. These values are not applicable to volatile compounds and radionuclides in forms that can readily penetrate the skin.
4. Direct monitoring should be employed wherever practicable. If wipe testing is employed, the assumption should be made that 10% of the contamination has been removed.

RADIONUCLIDES BY CLASS

Class	Radionuclide
I	^{227}Ac, ^{228}Th, ^{230}Th, ^{232}Th, Th-nat, ^{231}Pa, ^{232}U, ^{233}U, ^{234}U, ^{236}U, alpha emitters with $Z > 92$
II	^{147}Sm, ^{210}Pb, ^{227}Th, ^{235}U, ^{238}U, U-depl, U-nat, U-enr, ^{241}Pu
III	Other nuclides except those in Classes IV and V
IV	^{14}C, ^{35}S, ^{54}Mn, ^{57}Co, ^{65}Zn, ^{67}Ga, ^{75}Se, ^{77}Br, ^{85}Sr, ^{99m}Tc, ^{109}Cd, ^{123}I, ^{125}I, ^{129}Ca, ^{197}Hg
V	^{3}H, ^{51}Cr, ^{55}Fe, ^{63}Ni, ^{131}Cs

From NRPB-DL2, *Derived Limits for Surface Contamination*, by A. D. Wrixon, G. S. Linsley, K. C. Binns, and D. F. White, National Radiological Protection Board, Harwell, 1979.

wastes, low level generally means concentrations in the range of 10 kBq per liter (about 1 μCi per gallon), while high level means activity concentrations on the order of 10^{12} Bq per liter (about 100 Ci per gallon) or more. High level wastes are associated with the inventory of fission products in the burned-up fuel or nuclear reactors, Table 12.5, and with the chemical and metallurgical processes involved in the separation of the fission products from the unspent uranium or plutonium in the burned-up fuel. Intermediate and low level wastes are derived from numerous sources, such as hospitals where radioisotopes are used for diagnosis and therapy, research laboratories, and from the various parts of the nuclear fuel cycle, including operation of nuclear reactors and fuel reprocessing plants (Table 11.4). Because of the wide range of radioactivity in nuclear wastes, several basically different methods are used in the management of radioactive waste. For large amounts of activity, the general principle is to concentrate and confine the waste, whereas for very small amounts of activity, the waste may be diluted and dispersed. The exact manner of waste management depends on the activity level of the waste and on whether the waste is in a liquid, solid, or gaseous state.

High Level Liquid Wastes

High-level liquid wastes originate mainly as highly acidic solutions from the chemical processing of burned-up fuel, at specific activities of the order of 10^{12} Bq per liter (about hundreds of curies per gallon). In the early years of the nuclear industry, these liquids were reduced in volume by evaporation, and then were stored in underground tanks. The problem of storing high-level liquid waste was complicated by the fact that the rate of heat production due to radioactive transformation is high. If we assume the mean decay energy to be 1 MeV per transformation, then 10^{13} Bq (270 Ci) generates 1 watt of power as heat. Provision must therefore be made to remove this decay heat. Storage tanks for high level waste were designed for strength, corrosion resistance, heat removal, and monitoring for leakage. Typical tanks, on the order of 3800 m^3 (10^6 gallons) are steel-lined, reinforced concrete, with an outer steel shell to serve as a backup in case of a leak, and with an integrated monitoring system for leak detection (Fig. 11.4).

Although storage in tanks has so far been practical, the potential hazards from leakage during the long-term retention time, which is on the order of hundreds of years, are relatively serious. The potential hazards from high level radioactive waste are twofold: The greatest threat is from indiscriminate release of the fission product waste to the biosphere. If this were to happen, successive bioconcentrations of radionuclides by plants and animals in the food web could lead to unacceptably high levels of radioactivity in man's food, which in turn could lead to an unacceptably high internal dose. The second threat, of smaller magnitude than the first, is external radiation from the radioactive waste. Results of extensive research that was directed towards new waste treatment methods that would remove these threats show that the first of these threats can be eliminated by transforming the radionuclides into a form that makes them unavailable to the biosphere. One way to do this is to incorporate the radioactive atoms into glass beads. Tests have shown the leaching rate of the radionuclides from the glass to be much slower than the rate of radioactive decay. Thus, even if the radioactive beads were to escape into the environment, the radioactivity would remain locked within the glass, and would not be available for uptake by the flora or fauna. Should an animal swallow one of these glass beads, the bead would pass through its gastrointestinal tract and be eliminated. Of course, the GI tract would be irradiated during the passage of the bead, but the animal would not absorb radioactivity from the bead. Experience with obsidian, a naturally occurring volcanic glass,

TABLE 11.4. RADIOACTIVE WASTES FROM THE FUEL CYCLE

	Types of wastes and principal constituents	Approximate radioactivity level (Ci/ton U)	(Bq/ton U)
Mining and milling	Gaseous: ^{222}Rn, ^{218}Po, ^{214}Bi, ^{214}Po	10^{-4}–10^{-3}	4×10^6–4×10^7
	Liquid		
	Solid: U, ^{226}Ra, ^{230}Th, ^{210}Pb	0.5–1	2×10^{10}–4×10^{10}
Refining	Liquid: ^{238}U, ^{234}Th, ^{234}Pa, ^{226}Ra	10^{-4}–10^{-3}	4×10^6–4×10^7
Fuel fabrication	Liquid		
	Solid: U, Pu, Th	10^{-4}–10^{-3}	4×10^6–4×10^7
Reactor operation	Gaseous: ^{13}N, ^{41}A, ^{89}Kr, ^{87}Kr, ^{138}Xe, ^{135}Xe	10–100[a]	4×10^{11}–4×10^{12}
	Liquid		
	Solid: ^{58}Co, ^{60}Co, ^{59}Fe, ^{51}Cr, ^{3}H	50–100[a]	2×10^{12}–4×10^{12}
Chemical processing	Gaseous: ^{85}Kr, ^{133}Xe, ^{131}I, ^{129}I, ^{3}H	7000[b]	26×10^{13}
	Liquid		
	Solid: Fission products, Pu, Am, Cm	6,000,000[b]	22×10^{16}

[a]At time of waste discharge or shipment based on fuel exposure of 20,000 Mwd/ton of U.
[b]Waste from fuel at 20,000 Mwd/ton, 120 days cooled.
From ORNL Drawing 69–83 R2.

proves that glass is stable over geologic periods of time. Having thus assured that the high-level waste will not enter into the biosphere, the second task is to isolate the radioactive beads so that they do not irradiate people or other living creatures. This can be accomplished by putting the glass beads into a suitable container, and then isolating the container deep underground, in a geologically acceptable formation such as a salt dome, or in the deep tunnels that were used for underground nuclear bomb tests, and thus are already radioactive and useless for most purposes. By these methods, which effectively cut off exposure pathways, inherently toxic high level radioactive wastes can be rendered non-hazardous.

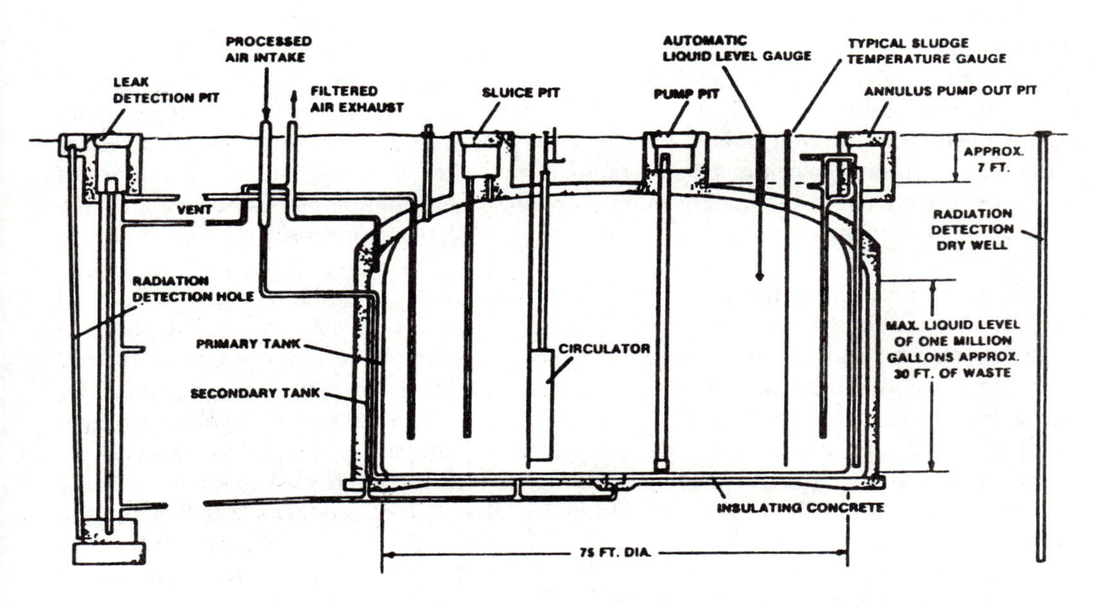

FIG. 11.4. Tank for storing high level liquid radioactive waste. (From *Radioactive Waste Management at Hanford*, 4th revision, 1979.)

Intermediate- and Low-level Liquid Wastes

Intermediate- and low-level wastes can, under certain conditions, be discharged into the sea or into the ground; or they may be treated either chemically, physically, or biologically in order to separate the radioactive solutes from the non-radioactive solvent.

The tremendous volume of the ocean seems to make it an ideal medium for the dilute and disperse technic for management of low and intermediate levels of waste. Furthermore, since sea water already contains a significant amount of radioactivity, mostly in the form of ^{40}K (it is estimated that the total activity content of the oceans is on the order of 2×10^{22} Bq, or about one-half million megacuries) addition of relatively small quantities of activity in the form of low and intermediate levels of waste would thus seem to add very little to the total activity of the oceans. However, because of the uncertainties regarding the diverse physical, chemical, and biological processes that govern the distribution of radioisotopes in the sea and the transmission of the radioisotopes through the food chain to man, it is very difficult to specify maximum amounts of radioactivity that may be discharged into the seas. According to the recommendations of the International Atomic Energy Agency (*Radioactive Waste Disposal into the Sea*, 1961), wastes of low and intermediate activity may be disposed of into the sea under controlled and specified conditions; providing that accurate records of all waste disposal activities are kept, and that the oceans be under continuous radiological surveillance in order to ascertain that radiologic safety is being maintained.

Disposal of low- and intermediate-level wastes into the ground may be practiced if the hydrologic factors, the ion-exchange properties of the soil, and the population density are favorable. This method of disposal can be called the "delay and decay" method because the slow movement of the radioactivity through the ground affords the activity sufficient time to decay by a significant degree. Characteristics favorable to ground disposal include a deep water table, good ion-exchange properties of the soil in order to extract relatively large fractions of radionuclides from the waste as the waste percolates through the ground, few bodies of surface water in order to maximize the time of underground flow, a large volume of underground water flow in order to maximize dilution, and a very low population density in the area of the ground disposal site. In the practice of ground disposal, a wood-lined pit, called a "crib," of appropriate capacity, is built in the ground, and is filled with gravel. The liquid waste is pumped into the crib, and slowly percolates out into the ground. When dispersing radioactive wastes into the ground, it is important to follow the course of the underground activity and to monitor the surface waters. In a suitable environment, such as is found at Hanford, ground disposal may be practised safely.

Chemical processes for the decontamination of intermediate- and low-level liquid wastes include the standard methods of water treatment and ion-exchange methods. Hydroxide flocs, which are produced by adding alum or ferric salts to the liquid wastes, and then increasing the pH until aluminum or ferric hydroxide is precipitated, are useful for removing cations other than those of the alkali metals and alkaline earths. This treatment is especially effective for removing alpha emitters; it is not very effective for removing ^{90}Sr. Removal of about 95% of ^{90}Sr may be effected with a calcium phosphate floc under highly basic conditions (pH $\sim$ 11.5). Radiostrontium can also be effectively removed by lime-soda softening the water. The degree of removal of ^{90}Sr is proportional to the degree of softening, since the $^{90}SrCO_3$ is precipitated with $CaCO_3$. Under certain conditions, liquid wastes may be decontaminated by ion-exchange methods. However, since non-radioactive ions are also absorbed on the ion exchanger, the effectiveness of this method depends on the relative concentrations of radioactive and non-radioactive ions. Better than 99% reduction in radioactivity can be achieved under optimum conditions.

Decontamination of water by biological means may also be used. However, biological removal of radionuclides is less effective than chemical treatment. Its main use, consequently, is for those cases where organic matter must be destroyed, as in sewage treatment or where high concentrations of organic complexing agents makes ordinary chemical treatment difficult.

For non-volatile radioactivity, evaporation is an effective means for decontaminating water. However, because evaporation requires removal of the solvent or the suspending medium, and since this component of the liquid waste usually accounts for more than 95% of the total pre-treated volume, evaporation is a relatively expensive method for the treatment of liquid waste. Evaporation is usually reserved for those cases where a very high degree of decontamination is required. By means of evaporation, decontamination factors on the order of 10^4–10^6 may be obtained at vapor mass velocities ranging from about 20–3000 $kg/m^2/hr$. The separated radioactivity, now in a relatively small volume, is processed further, as described in the next section, for disposal.

After removal of the bulk of the radioactivity from the suspending liquid, the decontaminated water can often be discharged into the storm sewer. Very low-level wastes, such as those produced in a laboratory handling tracer amounts of radioactivity, also are usually discharged into the sewer. Such discharges cannot be done indiscriminately, but are subject to regulatory control. The rule that is usually adopted for discharge into the public sewer system is that the quantity released per day, when diluted by the average daily quantity of flow into the sewer from the institution, does not cause the mean concentration of the sewage to exceed a fraction of the maximum permissible concentration for drinking water.

A convenient way to immobilize the radioactivity in aqueous waste is to convert the liquid into concrete, then to package the solidified waste in an impermeable container and either to bury it on land or dump it into the sea. Several methods are used in packaging the radioactive concrete. The simplest is to pour a concrete bottom in a 208-liter drum several cm thick, then, after it hardens, to pour the radioactive mix to within several cm of the top. After the radioactive concrete hardens, it is topped with more concrete, then the drum is sealed with its cover. For higher-level waste, the radioactive concrete is poured into a metal container (whose maximum size is a 114-liter, or 30-gal drum). This sealed container is then placed into a concrete-bottomed 208-liter drum, concrete is poured around and on top of the smaller radioactive container, and the drum is sealed. If the drum is to be transported elsewhere for burial (on land or sea), it must be properly marked, according to the regulations; and the radiation level must not exceed 2 mGy/hr (200 mR/hr) on the surface or 0.1 mGy/hr (10 mR/hr) at 1 m from the surface.

Airborne Wastes

Airborne radioactivity may be either gaseous or particulate. Gases may arise from neutron activation of cooling air in a reactor and from gaseous fission products, and from radiochemical reactions in which a gaseous product is produced. Particulates may be due to a large variety of processes, ranging from condensate droplets formed during the treatment of high-level liquid wastes to dusts from incinerators in which inflammable solids are burned. Hazards from air-borne wastes are best controlled at the source of the waste by limiting the production of air-borne wastes. If air-borne wastes are produced, the air must be sufficiently decontaminated that it may be safely diluted and discharged into the atmosphere. If the levels of the air-borne radioactivity are sufficiently low, the waste may be diluted and dispersed into the environment without further treatment.

Gases are usually difficult to remove. For small quantities of iodine and the noble gases, adsorption on activated charcoal may be used. Most of the radioactive gaseous wastes of the atomic energy industry are very short lived. Accordingly, these gases may be compressed and stored in tanks until they decay. Some of the methods used against radioactive gases are summarized in Table 11.5. In many instances, the most expedient method for dealing with radioactive gas is to discharge it to the atmosphere from a high stack, and thus to dilute the radioactivity to an acceptable level.

Particulate matter may be removed from gases by a variety of different devices, listed in Table 11.6, whose operating principles may be based on gravitational, inertial, electrostatic, thermal, or sonic forces; on physicochemical effects, or on filtration or barrier effects. The collection efficiencies of the different devices vary over a wide range. In considering an air-cleaning device for radioactive dusts, it should be borne in mind that the collection efficiency given by the manufacturer of air-cleaning devices for non-radioactive dusts is usually based on *mass* collection. Since the mass of a particle is proportional to the cube of its diameter, a single 10-micron particle is equivalent to 1000 one-micron particles. Reference to Table 11.1 shows that the maximum allowable concentration of non-radioactive particles is on the order of a million times or more greater than the allowable concentration for radioactive particles. Air-cleaning devices that are designed to remove much mass from the air, and are thus designated as high-efficiency collectors, may nevertheless be inadequate for respirable radioactive dusts. When this is the case, the final air-cleaning device usually is a high-efficiency filter that is designed for radioactive dusts. The performance of some high-efficiency filters is given in Table 11.7.

The extremely rigorous filtration requirements for radioactive dusts makes it desirable to specify the performance of a filter in a more meaningful way than "collection efficiency." Rather than designate the effectiveness of filters by filtration efficiency, in which there appears to be only a small difference between 99.99% and 99.995% (the former passes twice as many particles as the latter, 10 per 100,000 versus 5 per 100,000), we often used the *decontamination factor* as the figure of merit for a filter. The decontamination factor, d*f*, for a filter whose efficiency is E percent, is defined as

$$\mathrm{d}f = \frac{100}{100 - E}. \tag{11.3}$$

The filter whose efficiency is 99.99% and, thus, passes 10 particles per 100,000 has a decontamination factor of 10,000, while the filter of 99.995% efficiency, which passes 5 particles per 100,000, has a decontamination factor of 20,000.

After filtration, the remaining radioactive particulates are discharged into the atmosphere for dispersion of the non-filterable low levels of activity. If the particles are small, i.e. $< 1\ \mu$, the particulate terminal settling velocity is very low, and the particles may be considered as part of the gas in regard to their diffusion into the atmosphere and transport with the gases that issue forth from the exhaust stack.

Meteorological considerations

When a contaminant is discharged from a chimney, it is assumed that the contaminant will be carried downwind, while at the same time it diffuses laterally and vertically. The two main consequences of this dispersion in the atmosphere are dilution of the contaminant, and its eventual return to the breathing zone at ground level. Of particular

TABLE 11.5. TREATMENT METHODS FOR RADIOACTIVE GAS

Treatment	Gas	Efficiency %	Velocity fpm	Pressure drop, in. of water	Comments
Detention chamber	Noble gases	100	0	0	Use to hold up relatively small volumes
Spray tower	Halogens, HF	70–99	50	0.1–1.0	Precleaning or final cleaning for iodine removal
Packed tower	Radioiodine	95–99	50–200	1–10	Heated Berl Saddles coated with silver nitrate
Adsorbent beds	Iodine and noble gases	99.95	168	2.8	Activated charcoal or molecular sieves: may be used to decay xenon. May be refrigerated
Limestone beds	Halogens, HF	94–99.9	30	1–3	Experimental only. Some hood applications
Liquifaction column	Noble gases	99.9	—	—	Used to recover small amounts
Stripping column		90–95	—	—	Pilot studies only
Refrigerated carbon catalyst and carbon pellets	Xenon, krypton	99.9	—	—	Liquid nitrogen used for refrigerant. Gases recovered by desorption

From L. Silverman, Economic aspects of air and gas cleaning for nuclear energy processes, *Disposal of Radioactive Wastes*, Vol. 1, p. 147, I.A.E.A., Vienna, 1960.

TABLE 11.6. BASIC CHARACTERISTICS OF AIR CLEANING EQUIPMENT

Type of equipment	Particle size range mass median μ	Efficiency for size in col. 2 %	Velocity fpm	Pressure loss, in. of water	Current application in U.S. atomic energy programs
Simple settling chambers	> 50	60–80	25–75	0.2–0.5	Rarely used except for chips and recovery operations
Cyclones, large diameter	> 5	40–85	2000–3500 (entry)	0.5–2.5	Precleaners in mining, ore handling and machining operations
Cyclones, small diameter	> 5	40–95	2500–3500 (entry)	2–4.5	Same as above
Mechanical centrifugal collectors	> 5	20–85	2500–4000	—	Same as large cyclone application
Baffle chambers	> 5	10–40	1000–1500	0.5–1.0	Incorporated in chip traps for metal turning
Spray washers	> 5	20–40	200–500	0.1–0.2	Rarely used, occasionally as cooling for hot gases
Wet filters	Gases and 0.1–25 μ mists	90–99	100	1–5	Used in laboratory hoods and chemical separation operations
Packed towers	Gases and soluble particles > 5	90	200–500	1–10	Gas absorption and precleaning for acid mists
Cyclone scrubber	> 5	40–85	2000–3500 (entry)	1–5	Pyrophoric materials in machining and casting operations, mining, and ore handling. Roughing for incinerators
Inertial scrubbers, power-driven	8–10	90–95	—	3 to 5 HP/1000 cfm	Pyrophoric materials in machining and casting operations, mining and ore handling
Venturi scrubber	> 1	99 for H_2SO_4 mist. SiO_2, oil smoke, etc. 60–70	12,000 24,000 at throat	6–30	Incorporated in air cleaning train of incinerators
Viscous air conditioning filters	10–25	70–85	300–500	0.03–0.15	General ventilation air

Dry spun-glass filters	5	85–90	30–35	0.1–0.3	General ventilation air. Precleaning from chemical and metallurgical hoods
Packed beds of graded glass fibres 1–20 μ. 40 in. deep	< 1	99.90–99.99	20	10–30	Dissolver off-gas cleaning
High-efficiency cellulose-asbestos filters	< 1	99.95–99.98	5 through media. 250 at face	1.0–2.0	Final cleaning for hoods, glove boxes, reactor air and incinerators
All-glass web filters	< 1	99.95–99.99	5 through media. 250 at face	1.0–2.0	Same as above
Conventional fabric filters	> 1	90–99.9	3–5	5–7	Dust and fumes in feed materials production
Reverse-jet fabric	> 1	90–99.9	15–50	2–5	Same as above
Single-stage electrostatic precipitator	< 1	90–99	200–400	0.25–0.75	Final clean-up for chemical and metallurgical hoods. Uranium machining
Two-stage electrostatic precipitator	< 1–5	85–99	200–400	0.25–0.50	Not widely used for decontamination

From L. Silverman, Economic aspects of air and gas cleaning for nuclear energy processes, *Disposal of Radioactive Wastes*, Vol. 1, pp. 139–79, I.A.E.A., Vienna, 1960.

TABLE 11.7. PERFORMANCE OF HIGH-EFFICIENCY FILTERS
(AT NORMAL AIR TEMPERATURES AND STANDARD DENSITY AIR)

Medium	Test aerosol		Air velocity		Resistance, in. of water	Efficiency %	Method	Remarks
	Name	Size μ (homogeneous except*)	fpm	cm/sec				
CC-6 Cellulose-asbestos paper	Methylene blue	—	4	2	0.8	99.9871	Discoloration	
	Dioctyl phthalate (DOP)	0.3	5	2.5	0.67	99.85	Penetrometer	
	Atmospheric dust	0.5*	5	2.5	0.67	99.9+	Count	
	Duralumin	0.18	500	250	100	97.7	Count	Note excessive velocity causes greater penetration of fine size
	Duralumin	0.18	2	1	0.28	99.7	Count	Reduced velocity improves performance
	Potassium permanganate ($KMnO_4$)	0.02	20	10	2.7	93.0	Count	Size for maximum penetration
AEC No. 1 Cellulose-asbestos	DOP	0.3	5	2.6	0.7	99.78	Penetrometer	
	Duralumin	0.18	2	1	0.28	92.9	Count	
	Duralumin	0.18	40	20	5.6	99.6	Count	
	Atmospheric dust	0.58	5	2.5	0.7	99.98	Count	
	$KMnO_4$	0.01 0.02	4	2	0.56	91.0	Count	Size for maximum penetration

All-glass superfine fibres—Hurlbut-MSA1106B	DOP	0.3	5	2.5	1.05	99.999	Penetrometer	
	Atmospheric dust	0.5*	5	2.5	1.05	99.9+	Count	
	$KMnO_4$	0.015	20	10	4.4	93.0	Count	Size for maximum penetration
Sand	Cell ventilation gases	1	3–5	1.5–2.5	4.5–5.5	99.5–99.8	Radioactivity	69 in. deep: graded sizes from 2¼–50 mesh. Will not withstand high moisture conditions
Composite glass wool	Process off-gases	1	20	10	4.0	99.0		Composition given in reference
	Methylene blue	0.6 MMD	20	10	4.0	99.99	Gravimetric	Same
Compressed glass fibres	Atmospheric dust	0.5	5.25	2.6	0.69	99.997	Count	0.02 in. thick 50% 1.3 μ and 50% 3.0 μ fibers
Resin wool	Atmospheric dust	0.5	14	7	0.3	99.6	Discoloration	These filters are known to decrease in performance when exposed to ionizing radiation
Glass	Uranium oxide	0.12	2.3–7.8	1.2–3.9	0.30–1.23	95.5–99.5	Gravimetric	Special glass formulation developed by A.D. Little. Aluminum separators and furnace cement seals

From L. Silverman, Economic aspects of air and gas cleaning for nuclear energy processes, *Disposal of Radioactive Wastes*, Vol. 1, pp. 139–79, I.A.E.A., Vienna, 1960.

interest in evaluating the safety of discharge into the air is the relationships between the rate of discharge and the ground level concentrations—both in the breathing zone and on the ground (as fallout)—of the discharged radioactivity. The ground-level distribution of the discharged radioactivity depends on a number of factors, including atmospheric stability, wind velocity, type of terrain and the nature of the boundary layer of air (the air layer immediately over the ground for a distance of several hundred feet) and height of the chimney. It is thus very difficult to predict precisely the pattern of ground-level distribution, although reasonable estimates may be made from one of several different sets of atomspheric diffusion equations.

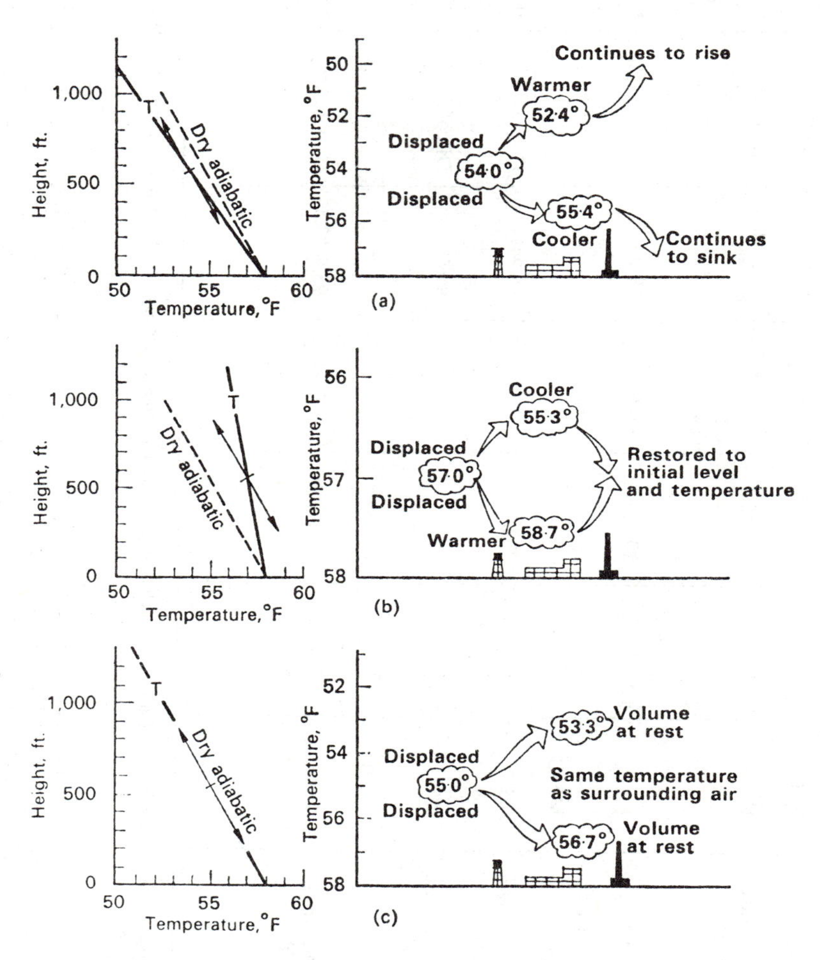

FIG. 11.5. Effect of atmospheric temperature gradient—or lapse rate—on a displaced volume of air. (a) Unstable lapse rate. (b) Stable lapse rate. (c) Neutral lapse rate. (From *Meteorology and Atomic Energy*, U.S.A.E.C., Washington, D.C., 1955.)

Atmospheric stability depends on the temperature gradient of the air (Fig. 11.5). Meteorologists refer to the temperature gradient of the atmosphere as the *lapse rate*. A parcel of air that is rising expands as a result of the decreasing atmospheric pressure. If no heat is gained or lost by this parcel of air, the expansion will be adiabatic, and the temperature of the air parcel will drop. For dry air, this adiabatic cooling results in a temperature decrease of 1°C per 100 m (5.4°F per 1000 ft) of ascent; for average moist air, the lapse rate is 0.65°C per 100 m (3.5°F per 1000 ft). If the temperature gradient of the atmosphere is less than adiabatic, but still negative, we have a *stable lapse rate*. In this case, a rising parcel of air cools faster than the surrounding atmosphere. It, therefore, is denser than the air in which it is immersed, and tends to sink. A sinking parcel of air is warmer than the surrounding air, and thus is less dense, which results in a tendency to rise. A stable lapse rate, therefore, tends to restrict the width of the plume in the vertical direction, thereby decreasing the dilution effect of the atmosphere.

If the lapse rate is positive, that is, if the air temperature increases with increasing height, then the super-stable condition known as an *inversion* occurs (since the temperature gradient is "inverted"). The rising effluent from the chimney becomes much denser than the surrounding air as it cools adiabatically, and thus sinks. The overall effect of an inversion is to trap the effluent from the chimney, and to prevent its ascent to higher altitudes.

A super adiabatic lapse rate, one in which the rate of decrease of temperature with increasing height is greater than 1°C per hundred meters (5.4°F per thousand feet), produces an unstable condition that helps to promote vertical dispersion of the contaminated effluent from the chimney. Under the conditions of such an unstable lapse rate, a rising parcel of air does not cool fast enough, because of its adiabatic expansion, and therefore remains warmer and less dense than the surrounding air, and thus continues to rise. By the same reasoning, a falling parcel of air continues to fall.

Dispersion of gas from a continuous source

Although we often speak of atmospheric diffusion, the fact is that very little atmospheric dispersion of gases is due to diffusion. The effects of turbulence usually are so great that molecular diffusion is completely masked. For this reason, estimates of the dispersion of gases in the atmosphere are based on mathematical models that consider the meteorological state of the atmosphere rather than on classical diffusion theory. One of the more commonly used models for estimating the ground level concentration of a gaseous effluent from a point source, such as a chimney, is the Gaussian plume, straight-line trajectory model (Fig. 11.6). In this model, the contaminant is assumed to be normally distributed around the central axis of the plume, and that atmospheric stability and wind speed determine the atmospheric dispersion characteristics of the contaminant in the downwind direction. This model is described by the Pasquill–Gifford equation:

$$\chi(x, y) = \frac{Q}{\pi \sigma_y \sigma_z \mu} \exp\left[-\frac{1}{2}\left(\frac{y^2}{\sigma_y^2} + \frac{H^2}{\sigma_z^2}\right)\right], \tag{11.4}$$

where $\chi(x, y)$ = ground level concentration in Bq (or Ci) per cubic meter at point x, y,
x = downwind distance on plume center line, meters,
y = cross-wind distance, meters,
Q = emission rate, Bq (or Ci) per second,
σ_y, σ_z = horizontal and vertical standard deviations of contaminant concentration in the plume, meters,
μ = mean wind speed at level of plume center line, meters per second,
H = effective chimney height, meters.

If the effluent gas has a significant exit velocity, or if it is at a high temperature, it will rise to a level higher than the chimney. The effective chimney height, therefore, is the sum of the actual chimney height, plus a factor that accounts for the exit velocity and the temperature of the effluent gas:

$$H = h + d\left(\frac{v}{\mu}\right)^{1.4}\left(1 + \frac{\Delta T}{T}\right), \tag{11.5}$$

where h = actual chimney height, meters,
d = chimney outlet diameter, meters,
v = exit velocity of gas, meters per second,
μ = mean wind speed, meters per second,
ΔT = difference between ambient and effluent gas temperatures,
T = absolute temperature of effluent gas.

The maximum ground level concentration occurs on the plume center line, at the downwind distance where

$$\sigma_z = H\sqrt{2}. \tag{11.6}$$

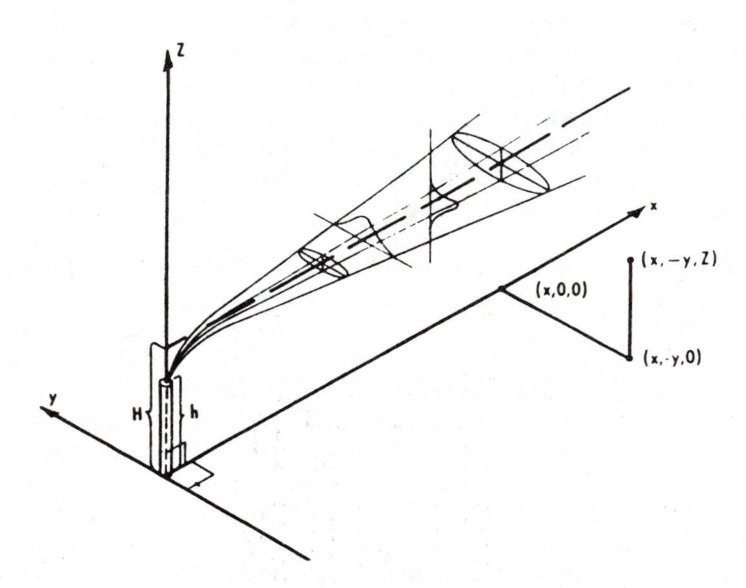

FIG. 11.6. Gaussian plume dispersion model for a continuous point source.

Although SI units are shown in equations (11.4) and (11.5), any consistent set of units may be used.

The spread of the plume at any downwind distance is determined by the atmospheric stability, wind speed, and the downwind distance. For purposes of calculating ground level concentrations with the use of equation (11.4), Pasquill proposed the stability categories listed in Table 11.8. For each of the stability categories, the values of the standard deviations in the horizontal and vertical planes through the plume center line, σ_y and σ_z, as a function of downwind distance, are given in Figs. 11.7 and 11.8.

The many uncertainties implicit in equation (11.4), such as type of terrain, fluctuations in meteorological conditions, etc., lead to a degree of imprecision in the calculated ground

TABLE 11.8. PASQUILL'S CATEGORIES OF ATMOSPHERIC STABILITY

A: Extremely unstable conditions
B: Moderately unstable conditions
C: Slightly unstable conditions
D: Neutral conditions*
E: Slightly stable conditions
F: Moderately stable conditions

Surface sind speed, m/sec	Daytime insolation			Thin overcast or ≥ 4/8 cloudiness†	≤ 3/8 cloudiness
	Strong	Moderate	Slight		
<2	A	A–B	B		
2	A–B	B	C	E	F
4	B	B–C	C	D	E
6	C	C–D	D	D	D
>6	C	D	D	D	D

*Applicable to heavy overcast, day or night.
†The degree of cloudiness is defined as that fraction of the sky above the local apparent horizon which is covered by clouds. (Manual of Surface Observations [WBAN], Circular N [7th ed.], paragraph 1210, U.S. Government Printing Office, Washington, July 1960.) (From W. F. Hilsmeier and F. A. Gifford, Jr. *Graphs for Estimating Atmospheric Dispersion*. Report ORO-545, Oak Ridge National Laboratory.

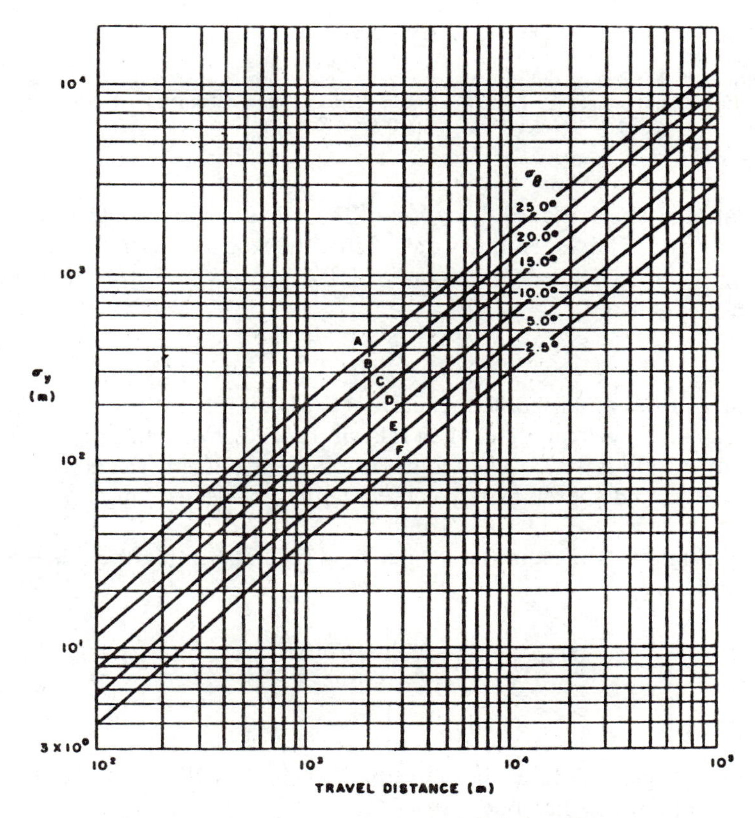

FIG. 11.7. Horizontal diffusion, σ_v, *versus* downwind distance from a point source for Pasquill's atmospheric stability categories. (From D. H. Slade, ed. *Meteorology and Atomic Energy*. Tech. Inform. Div., U.S.A.E.C., Washington, D.C., 1968.)

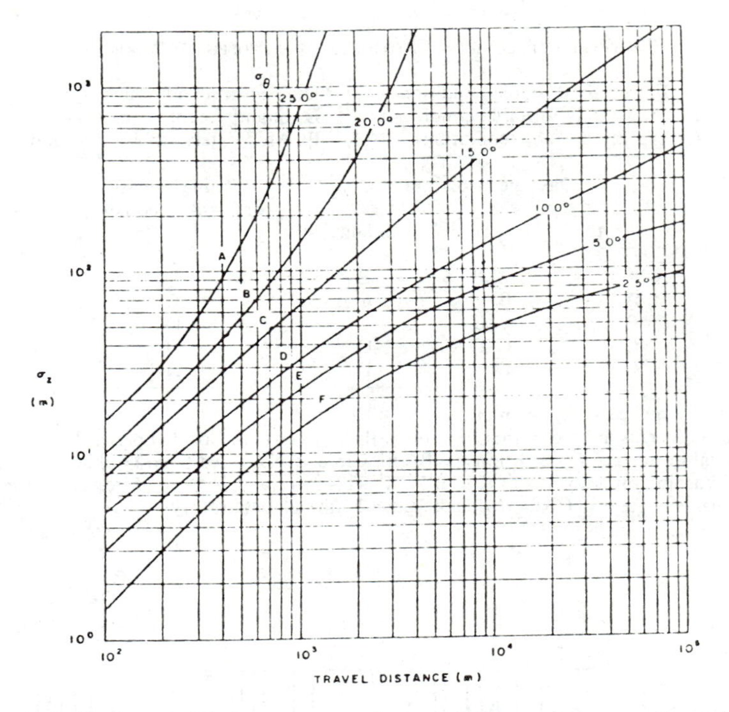

FIG. 11.8. Horizontal diffusion, σ_z, *versus* downwind distance from a continuously emitting point source for Pasquill's atmospheric stability categories. (From Slade's *Meteorology and Atomic Energy*, 1968.)

level concentrations. The standard deviation of the calculated concentration is thought to be about a factor of 3. That is, 68 times out of 100, the true concentration can be expected to lie between $\chi/3$ and 3χ, while 96 times out of 100, the true concentration can be expected to lie between $\chi/6$ and 6χ.

Example 11.3

The ^{41}A effluent from an air cooled reactor is 40 MBq (1.08 mCi) per second on a clear night, through a chimney 75 m high, when the wind speed at the chimney height is 4 m/sec and the air temperature is 17°C. The temperature of the effluent gas is 87°C, the effluent velocity is 10 m/sec, and the diameter of the chimney orifice is 2 m.

(a) Calculate the effective chimney height.

The effective height of the chimney is calculated by substituting the appropriate values into equation (11.5):

$$H = 75 + 2\left(\frac{10}{4}\right)^{1.4}\left(1 + \frac{70}{273 + 87}\right)$$

$$= 83.6 \text{ m}.$$

(b) Calculate the ground level concentration at 500, 1000, 2000, 4000, and 8000 m downwind on the plume center line.

For the given atmospheric conditions, the Pasquill stability category is found from Table 11.8 to be Category E. If we substitute the respective values into equation (11.4), including the values of σ_y (Fig. 11.7) and σ_z (Fig. 11.8) for $x = 500$ m, we have

$$\chi(500, 0) = \frac{40 \times 10^6 \text{ Bq}}{\pi \times 29 \text{ m} \times 14 \text{ m} \times 4 \text{ m/sec}} \exp\left[-\frac{1}{2}\left(\frac{83.6 \text{ m}}{14 \text{ m}}\right)^2\right]$$
$$= 1.4 \times 10^{-4} \text{ Bq/m}^3 \quad (3.8 \times 10^{-15} \ \mu\text{Ci/cm}^3).$$

Similar calculations for the other distances yields

χ	σ_y	σ_z	Bq/m³	μCi/cm³
500 m	29 m	14 m	1.5×10^{-4}	4.2×10^{-15}
1000	50	23	3.7	1.0×10^{-10}
2000	100	37	67	1.8×10^{-9}
4000	190	54	94	2.5×15^{-9}
8000	340	78	68	1.8×15^{-9}

(c) How far downwind is the maximum ground level concentration?

Equation (11.6) tells us that the maximum ground level concentration occurs where the vertical standard deviation is

$$\sigma_z = 83.4/\sqrt{2} = 59 \text{ m}.$$

From Fig. 11.8, we find this distance to be 4,500 m downwind from the source.

(d) What is the ground level concentration 50 m crosswind from the point of maximum concentration? Using equation (11.4) and the values for σ_y and σ_z at 4500 m, we have

$$\chi(4500, 50) = \frac{40 \times 10^6}{\pi \times 205 \times 60 \times 4} \exp\left[-\frac{1}{2}\left(\frac{50}{205}\right)^2 - \frac{1}{2}\left(\frac{83.4}{60}\right)^2\right]$$
$$= 95.6 \text{ Bq/m}^3 \quad (2.6 \times 10^{-9} \ \mu\text{Ci/cm}^3).$$

In this calculation, no allowance was made for the fact ^{41}A has a half-life of 110 minutes, and thus will undergo a significant decrease during the transit times required for travel of the effluent to the point in question. Generally, if this travel time is significant relative to the half-life of the radioisotope, the emission rate, Q_1 in equation (11.4), is multiplied by the decay factor, $e^{-\lambda t}$, where t is the transit time.

Particulates

Equation (11.4), which gives the ground level concentration of a gas that is continuously emitted from a point source, is based on *total reflection* of the gas by the ground. If the pollutant in the plume were retained on the ground, however, as would be true in the case of particulates, then the ground level concentration would be only one-half that given by equation (11.4):

$$\chi(x, y) = \frac{Q}{2\pi\sigma_y\sigma_z u} \exp\left[-\frac{1}{2}\left(\frac{y^2}{\sigma_y^2} + \frac{H^2}{\sigma_z^2}\right)\right]. \tag{11.7}$$

Furthermore, if a chimney emits particulates, then depletion of the radioactivity in the plume due to gravitational settling, impaction on surfaces protruding from the ground, and by precipitation scavenging must be considered in estimating downwind concentrations. Gravitational settling is important for large particles, about 15 microns or larger; impaction and wet deposition are important mechanisms of plume depletion mainly for small particulates. Generally, because of the great number of factors that determine depletion of particulates, such as particle size distribution, wetability, solubility, humidity, etc., we can only at best estimate depletion of particulates from a plume. The effect of gravitational settling is to tilt the axis of the plume downward through an angle $\theta = \tan^{-1} v_t/\mu$ from the horizontal, where v_t is the terminal settling velocity and μ is the mean wind speed. The effective height of the plume centerline line, H', at any downwind distance x becomes

$$H' = H - x \tan \theta, \tag{11.8}$$

or

$$H' = H - xv_t/\mu. \tag{11.9}$$

Equation (11.7) may be modified to estimate the concentration of particulates at or near ground level by substituting H', from equation (11.9), for H in equation (11.7):

$$\chi(x, y) = \frac{Q}{2\pi\sigma_y\sigma_z\mu} \exp\left[-\frac{1}{2}\left(\frac{y^2}{\sigma_y^2} + \frac{(H - xv_t/\mu)^2}{\sigma_z^2}\right)\right]. \tag{11.10}$$

The rate of ground deposition of particles at point (x, y) is found by multiplying the ground level concentration, equation (11.10) by the deposition velocity, v_g, of the particulate matter:

$$w \text{ Bq/m}^2\text{-sec} = \chi \text{ Bq/m}^3 \times v_g \text{ m/sec}. \tag{11.11}$$

Deposition velocity is determined mainly by micrometeorological conditions near the surface, and thus cannot be calculated with any reasonable degree of accuracy. Experimentally, values ranging from about 0.001 to 0.10 meter per second, with an average of about 0.01 meter per second, have been obtained.

Solid Wastes

Not very much can be done to solid radioactive waste in order to reduce its radioactivity content. The chief type of processing is reduction of volume prior to storage, or burial in land, or dumping into the sea. Long-term storage is usually restricted to high-level solids, and may be accomplished by use of an underground concrete vault. Properly packaged waste may be buried, that is, placed directly into the ground and covered with earth in special areas that are designated by the appropriate regulative agencies as radioactive waste-disposal burial grounds. The exact type of packaging depends on the quantity and the half-life of the activity. High-level wastes may be encased in concrete filled steel drums, whereas for low-level or short-lived wastes it may be sufficient to place the activity into polyethylene bags, and then to seal the bags into steel cans or drums. A good-quality, painted (inside and outside) steel drum may be expected to maintain its integrity for at least 10 years; concrete casing may be expected to retain its integrity for 100 years.

When dumping solid radioactive waste into the sea, consideration is given to the density of the package (in order to assure sinking to the ocean bottom), corrosion resistance, and sufficient strength to resist the great hydrostatic pressures at the bottom of the sea. Concrete encased waste in a steel drum is considered to be satisfactory. A good drum has a lifetime in sea water of about 10 years; concrete may be expected to have a lifetime in the ocean of about 30 years. Clearly, the concrete serves more than a weighting function; it is, in fact, the chief containment vessel. Accordingly, it should be of good quality and low porosity. A package for sea disposal should be free of voids for two reasons: (1) an empty space inside the package may make the package subject to collapse from the hydrostatic pressure of the ocean, and (2) it decreases the overall density of the package. For disposal into the deep sea, the specific gravity should not be less than 1.2 (75 lb/ft^2, or 10 lb/gal). If the waste is to be dumped on the continental shelf, the specific gravity of the container and its contents should not be less than 1.5 (12.5 lb/gal). Furthermore, the loaded container should contain no voids if collapse of the container due to the tremendous pressure of the water is to be prevented.

Relatively large quantities of combustible low-level radioactive waste, such as paper towels, protective clothing, rags, animal excreta and carcasses, etc., as well as non-combustible low-level wastes, such as broken glassware, hypodermic needles, etc., are produced in many radioisotope laboratories and nuclear medicine clinics. Often, the non-combustible and combustible waste is accumulated in different containers. The non-burnable waste is packaged for disposal into the sea or into the ground, while the combustible waste is burned. Incineration may either concentrate the activity by burning away the substrate in which activity is held, if the activity is non-volatile, or it may disperse the activity with the effluent from the chimney if the activity is volatile or if the contaminated waste is transformed physically into particulates. In the instance where the activity is concentrated in the ashes, we have a case of volume reduction; the ashes still must be picked up and packaged for disposal. (The collection of ashes from an incinerator in which radioactivity had been burned should be done under the supervision of a health physicist. Appropriate respiratory protective equipment should be available if necessary.) If the activity may go up the stack to be dispersed in the atmosphere, the rate of incineration of radioactive waste should be controlled in order to limit the activity in the stack effluent to acceptable levels.

Example 11.4

Dead rats that had been injected with a ^{14}C-tagged compound are to be burned in an incinerator. The carbon is expected to go up the stack as CO_2, but some of it may also be discharged as carbon particulates. About 125 kg of dry non-radioactive waste are burned per 8-hr day in this incinerator. How much activity may be incinerated if the concentration at the top of the chimney is not to exceed 10 times the maximum allowable concentration of 37 kBq/m^3 (10^{-6} μCi/cm^3)? (It should be noted that this does not violate the ICRP recommendation, since the concentration referred to is at a point where it cannot be inhaled. The higher discharge concentration in this case is based on dilution by the atmosphere to a safe level before the activity can be inhaled.)

At least 3.5 kg of air per kg waste must be supplied to the incinerator. Since 1 cubic meter of air weighs 1.2 kg at 22°C and standard atmospheric pressure, the amount of air used during the day is

$$125\,\frac{\text{kg waste}}{\text{day}} \times \frac{3.5\text{ kg air/kg waste}}{1.2\text{ kg air/m}^3} = 365\text{ m}^3\text{/day}.$$

To attain a maximum concentration of 370 kBq/m^3 (10^{-5} μCi/cm^2), we cannot incinerate more than

$$3.7 \times 10^5\text{ Bq/m}^3 \times 365\text{ m}^3\text{/day} = 1.3 \times 10^8\text{ Bq/day} \quad (3650\ \mu\text{Ci/day}).$$

Furthermore, since N.B.S. handbook 53 recommends a maximum specific activity of 4 μCi (148 kBq) per gram carbon for particulates discharged from a chimney, we shall adhere to this limit. If the incinerated waste is assumed to be 25% carbon by weight, then the amount of C incinerated is

$$125\text{ kg waste/day} \times 0.25\text{ kg C/kg waste} = 31.3\text{ kg C/day}.$$

The mean specific activity, therefore, if 1.3×10^8 Bq (3650 μCi) are burned per day is

$$\frac{1.3 \times 10^8 \text{ Bq/day}}{31.3 \text{ kg C/day}} = 4.2 \times 10^6 \text{ Bq/kg C} \quad (0.1\ \mu\text{Ci/g C}),$$

which is very much less than the recommended specific activity. Let us now consider the rat. It consists of 18% carbon by weight. The maximum activity, A_m, in a rat weighing W grams that could be incinerated, if there were no isotopic dilution of the carbon is

$$A_m, \text{ Bq/rat} = 148 \text{ kBq/g C} \times 0.18 \text{ g C/g} \times W \text{ g}.$$

For a 300-gram animal, this corresponds to 8 MBq (216 μCi) per rat. However, because of the very large isotopic dilution of the carbon by the carbon in other waste, we may allow the daily maximum incinerated activity of 1.3×10^8 Bq (3650 μCi) to be distributed over any number of rats.

In certain other instances, the radioactivity could be converted into a gas, and then discharged to the atmosphere. An illustration of how this may be accomplished within limits prescribed by radiation safety regulations is shown below.

Example 11.5

Three hundred and seventy MBq (10 mCi) ^{14}C waste, in the form of 1 g $BaCo_3$, will be disposed of by changing the chemical form of the carbon to $^{14}CO_2$, and then discharging the radioactive gas to the atmosphere. The chemical manipulations will be carried out in a fume hood whose face opening is 2 m wide and 0.8 m high, and whose face velocity is 0.5 m/sec (100 feet/min). The $^{14}CO_2$ will be vented to the atmosphere through an exhaust stack from the hood. The chemical conversion from the carbonate to the gas will be accomplished by the addition of 1 NHCl. What is the maximum rate at which acid may be added to the $BaCO_3$ if the maximum permissible atmospheric concentration of 3.7×10^4 Bq/m^3 ($10^{-6}\ \mu$Ci/cm^3) is not to be exceeded at the discharge end of the exhaust stack?

The conversion of the carbonate to CO_2 proceeds according to the reaction

$$BaCO_3 + 2HCl \rightarrow BaCl_2 + CO_2.$$

Since the formula weight of $BaCO_3$ is 197.4 (the additional weight due to the ^{14}C is very small, and may be neglected), 1 g $BaCO_3$ is

$$\frac{1 \text{ g}}{197.4 \text{ g/mole}} = 0.00506 \text{ mole}.$$

To convert all the $BaCO_3$ to CO_2, $2 \times 0.00506 = 0.01012$ mole HCl is needed. Since 1 NHCl contains 1 mole acid per liter, the required amount of acid will be contained in 0.01012 liter, or 10.12 ml HCl. According to the chemical equation, 1 mole $BaCO_3$ reacts with 2 moles acid to yield 1 mole CO_2. Therefore, if the reaction goes to completion, 0.00506 moles $^{14}CO_2$ will be produced. The gas will occupy a volume, under standard conditions of temperature and pressure of

$$5.06 \times 10^{-3} \text{ mole} \times 22.4 \text{ l/mole} = 0.1155 \text{ liter}.$$

The specific activity of the $^{14}CO_2$ produced in this reaction is

$$\frac{3.7 \times 10^8 \text{ Bq}}{1.155 \times 10^{-1} \text{ l}} \times 10^3 \text{ l/m}^3 = 3.2 \times 10^{12} \text{ Bq/m}^3.$$

The maximum permissible concentration for atmospheric ^{14}C is 3.7×10^4 Bq/m^3

(10^{-6} μCi/cm^3). The amount of air in which the generated $^{14}CO_2$ must be mixed in order to meet the required concentration at the point of emission is

$$\frac{3.7 \times 10^8 \text{ Bq}}{X \text{ m}^3} = \frac{3.7 \times 10^4 \text{ Bq}}{1 \text{ m}^3},$$

$$X = 10^4 \text{ m}^3.$$

The volume of air that flows out of the discharge stock is

$$\begin{aligned} Q &= \text{face area} \times \text{velocity} \\ &= 2 \text{ m} \times 0.8 \text{ m} \times 30 \text{ m/min} \\ &= 48 \text{ m}^3/\text{min}. \end{aligned}$$

If the conversion of the $BaCO_3$ to CO_2 proceeds at a uniform rate of speed, it must take at least

$$\frac{10^4 \text{ m}^3}{48 \text{ m}^3/\text{min}} = 208 \text{ minutes}.$$

The 1 NHC, therefore, must flow into the gas generator at a rate not exceeding

$$\frac{10.12 \text{ ml}}{208 \text{ min}} = 4.87 \times 10^{-2} \text{ ml/min},$$

or about 20.6 minutes per ml.

Assessment of Hazard

A realistic assessment of a hazard from an internally deposited radioisotope requires more consideration than merely comparing an environmental concentration with a legally prescribed maximum allowable concentration. The MPC's, it should be emphasized, are based on continuous exposure for 50 years. In recognition of this fact, the Federal regulations in the United States limit inhalation of radioactivity in any calendar quarter to that which would result from inhalation for 40 hours per week for 13 weeks of air containing the maximum permissible concentration of radioactivity. The I.C.R.P. does not even recommedn a maximum permissible environmental concentration of radioactivity. Instead, it recommends annual limits on the intake (ALI) of the various radioisotopes. From these recommended ALI's, the I.C.R.P. derives atmospheric concentrations (DAC) which, if inhaled for 2000 hours, would lead to the ALI. Furthermore, the recommended maxima do not, in most instances, consider the chemical form of the radioisotope, nor the influence of the chemical form on the metabolic properties of the isotope, nor the consequent effect on the absorbed dose from exposure to the isotope. If the metabolic properties are known, they may be used to assess the hazard from an internal emitter, as shown below in the following example:

Example 11.6

^{14}CO will be produced in a pilot study in which excess H_2SO_4 will react with H–^{14}CO–ONa, whose specific activity is 7400 MBq (200 mCi) per millimole, to produce ^{14}CO. The threshold limit value (TLV) for CO gas, based on its chemical effects, is 50 parts per million for occupational exposure. The maximum atmospheric concentration, based on radiological considerations, for occupational exposure to ^{14}C is listed in the U.S. Code of Federal Regulations (10CFR-20) as 5×10^{-5} μCi/ml (1.85 MBq/m^3).

(a) Will the industrial hygiene control that limits CO to 50 ppm be sufficient to meet the handbook requirements of radiological safety?

To find the molar concentration of CO in the atmosphere that corresponds to 50 ppm:

$$50 \text{ ppm} = \frac{50 \text{ moles CO}}{10^6 \text{ moles atmosphere}} = 5 \times 10^{-5} \frac{\text{moles CO}}{\text{mole atmosphere}}.$$

Since there is one carbon atom per molecule of sodium formate, and also one carbon atom per CO molecule, the specific activity of the ^{14}CO will also be 7400 MBq (200 mCi) per millimole. The radioactivity concentration corresponding to 50 ppm is

$$\frac{5 \times 10^{-5} \text{ mole CO/mole atm} \times 7.4 \times 10^{12} \text{ Bq/mole CO}}{0.0224 \text{ m}^3\text{/mole atm}}$$
$$= 1.65 \times 10^{10} \text{ Bq/m}^3 \quad (0.446\ \mu\text{Ci/cm}^3).$$

Use of industrial hygiene criteria would, in this case, lead to an atmospheric concentration of ^{14}C of $4.46 \times 10^{-1} \div 5 \times 10^{-5} = 8920$ times the regulatory limit for continuous exposure.

(b) The industrial hygienist, believing that control of the ^{14}CO according to the chemical TLV is sufficient for the radiological hazard, allows a chemical engineer to be exposed to 50 ppm of the ^{14}CO for a period of 2 hr. What is the chemical engineer's dose commitment as a result of his exposure?

In order to calculate the absorbed dose, certain facts must be known about the physiological behavior of CO. When CO is inhaled, it diffuses across the capillary bed in the lungs, and dissolved in the blood. It then is absorbed by the erythrocytes and combines with the hemoglobin to form carboxyhemoglobin. Since carboxyhemoglobin is incapable of transporting oxygen, the inhalation of CO leads to cellular anoxia, which in turn may lead to unconsciousness or to death—depending on the amount of CO that is absorbed into the blood. The maximum amount of an inhaled gas that can be absorbed, which is called the *saturation value*, depends on the partial pressure of the gas in the atmosphere. The saturation value for CO, S_∞, as percent hemoglobin tied up as carboxyhemoglobin, is given by

$$S_\infty = \frac{210 \times \text{p CO}}{210 \times \text{p CO} + \text{p O}_2} \times 100, \tag{11.12}$$

where p CO is the percent CO in the air, and p O_2 is the percent oxygen in the alveolar air (p O_2 is usually equal to 15). One hundred percent saturation corresponds to 20 ml CO per 100 ml blood. Rate of absorption, in most cases, follows first-order kinetics; that is, the fractional approach to saturation per unit time remains constant. Thus, if 1% of the saturation value is absorbed in 1 min after beginning inhaling the gas, 1% of the remaining 99% will be absorbed during the second minute, then 1% of the 98.01% left and so on. Since saturation is approached assymptotically, we usually refer to the *half saturation time* to designate the rate of absorption of an inhaled gas. The numerical value for the half saturation time is independent of the atmospheric concentration of the gas (except for very high concentrations). For CO, the half saturation time is about 47 min. The absorption of CO is analogous to the buildup of a radioactive daughter as it approaches secular equilibrium, and is described by a similar equation:

$$S = S_\infty(1 - e^{-(0.693/T) \times t_i}), \tag{11.13}$$

where S_∞ is the saturation value corresponding to a particular atmospheric concentration of the CO, S is the percent of the hemoglobin bound with CO, T is the half saturation time, and t_i is the inhalation time.

For an atmospheric concentration of 50 ppm (50×10^{-6} parts CO per part air), which corresponds to 50×10^{-4}%, or 0.005%, the hemoglobin saturation value is calculated from equation (11.12) to be

$$S_\infty = \frac{210 \times 0.005}{210 \times 0.005 + 15} \times 100 = 6.55\%.$$

After the 2-hr exposure, the percentage of the worker's hemoglobin that is bound with CO is calculated from equation (11.13) to be

$$S = 6.55(1 - e^{-0.693/47 \times 120}) = 5.45\%.$$

The blood volume of the standard man is 7.7% of his weight, or 5.4 liters for a 70-kg man; it, therefore, can hold 1/5 of 5.4 liters, or 1080 ml CO or oxygen. Since 5.45% of this capacity is tied up with CO, the quantity of CO in the man's body is

$$0.0545 \times 1080 \text{ ml} = 59 \text{ ml } ^{14}\text{CO at NTP}.$$

Since the specific activity of the ^{14}CO is 7.4×10^9 Bq (200 mCi) per millimole, the body burden following two hours of inhalation is

$$\frac{59 \text{ ml}}{22.4 \text{ ml/mmole}} \times 7.4 \times 10^9 \text{ Bq/mmole} = 1.95 \times 10^{10} \text{ Bq (527 mCi)}.$$

Assuming the blood, and hence the ^{14}C, to be uniformly distributed throughout the body of a 70 kg man, the dose rate due to this body burden of ^{14}C is calculated from equation (6.41):

$$\dot{D} = \frac{1.95 \times 10^{10} \text{ Bq} \times 1 \text{ tps/Bq} \times 5 \times 10^{-2} \text{ MeV/t} \times 1.6 \times 10^{-13} \text{ J/MeV} \times 3.6 \times 10^3 \text{ sec/hr}}{70 \text{ kg} \times 1 \dfrac{\text{J/kg}}{\text{Gy}}}$$

$$= 8.03 \times 10^{-3} \text{ Gy/hr} \quad (8.03 \times 10^{-1} \text{ rad/hr}).$$

If inhalation had continued until the hemoglobin saturation value was attained, the body burden would have reached

$$\frac{6.55}{5.45} \times 1.95 \times 10^{10} \text{ Bq} = 2.34 \times 10^{10} \text{ Bq} \quad (633 \text{ mCi}),$$

and the dose rate would have proportionately increased to $\dot{D}_\infty$, the maximum possible value under the conditions of exposure, of 9.65×10^{-3} Gy/hr (0.965 rad/hr). In this case, the body burden, and hence the dose rate, varied with time, as shown in (Fig. 11.9). The instantaneous dose rate during the period of inhalation (period I in Fig. 11.9) is given by

$$\dot{D} = \dot{D}_\infty(1 - e^{-kt_i}), \tag{11.14}$$

where k is the carboxyhemoglobin dissociation constant, $0.693/T$, and t_i is the inhalation time. The total dose during the period of inhalation is

$$D_i = \dot{D}_\infty \int_0^{t_i} (1 - e^{-kt_i})\, dt, \tag{11.15}$$

which, when integrated, yields

$$D = \dot{D}_{\infty}\left[t_i + \frac{1}{k}(e^{-kt_i} - 1)\right]. \qquad (11.16)$$

The dose absorbed during the time period t_e (period II in Fig. 11.9) after termination of inhalation, under conditions of decreasing body burden, is given by equation (6.31) as

$$D_e = \frac{\dot{D}_0}{\lambda_E}(1 - e^{-\lambda_E t_e}), \qquad (6.31)$$

where $\dot{D}_0$ is the instantaneous dose rate at time $t = 0$, that is, the time when inhalation ceased, and λ_E has the same meaning as k in equation (11.14). Substituting 9.65×10^{-3} Gy/hr (0.965 rad/hr) for $\dot{D}_{\infty}$, 2 hr for t_i, and 0.885 per hour for k into equation (11.16), the radiation dose absorbed during build-up is found to be 0.01 Gy (1.0 rad). When 8.03×10^{-3} Gy/hr (0.803 rad/hr) is substituted for $\dot{D}_0$, 0.885 per hour for λ_E, and ∞ for t in equation (6.31), the radiation dose absorbed from the ^{14}C during the period of elimination is found to be 9.07×10^{-3} Gy (9.07×10^{-1} rad). The dose commitment from the 2-hour inhalation of the ^{14}CO is the sum of these two doses, 1.94×10^{-2} Gy (1.94 rad). More than 99% of this dose commitment is absorbed within 7 hours after beginning inhalation of the radioactive carbon monoxide.

Although the worker was exposed to a concentration very much higher than the maximum permissible concentration, the consequent radiation absorbed dose—which is the basic criterion for radiation safety—is very much less than the 30 mSv (3000 mrem) maximum that is allowed by the U.S. regulations during a period of 13 consecutive weeks.

The concept of a maximum credible accident is useful in advance planning for the purpose of minimizing radiation dose in the event of an accident, or for designing safety limitations into an experiment. Consider the following example.

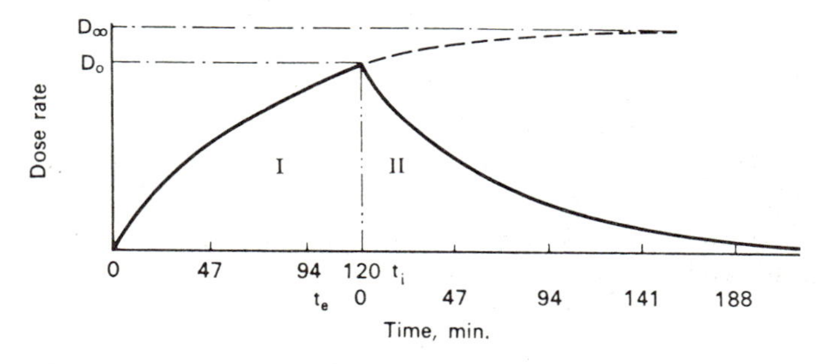

FIG. 11.9. Variation of dose rate with time after start of ^{14}CO inhalation. Region I under the curve represents the period of inhalation, and region II represents the period of exhalation. $\dot{D}_0$ is the dose rate at the end of the inhalation period, and $\dot{D}_{\infty}$ is the dose rate due to the saturation amount of radioactive CO.

Example 11.7

An engineer wishes to use tritiated water in an experimental study of a closed pressurized system. The system's capacity is 3 liters water, which will be kept at a temperature of 150°C (302°F). The experiment will be done in a ventilated laboratory whose dimensions are 3 m × 3 m × 3 m. The maximum credible accident is one in which the system will rupture, and the entire 3 liters of tritiated water will be sprayed into the room. The laboratory ventilation rate is 7 m^3/min (250 cfm). The laboratory has its own exhaust line and stack, so there is no possibility of spreading the tritium, in the event of an accident, to other laboratories in the building.

What is the maximum amount of tritium, as triated water, ^3HOH, that may be in the system, assuming a maximum credible accident, if the engineer is not to inhale more tritium than that which would deliver a dose of 30 mGy (3 rad) over a period of 13 weeks. In the event of such an accident, it is estimated that the engineer might remain in the laboratory for as long as 2 minutes.

For tritium:

(a) the critical organ is the total body, weight = 70 kg;
(b) the biological half-time is 12 days;
(c) the radiological half-life is 12.3 years;
  (i) the effective half-life, from equation (6.28), is 12 days;
  (ii) the effective elimination rate, from equation (6.26), is 0.0578 per day;
(d) pure beta emitter, average beta energy = 0.006 MeV;
(e) all the inhaled tritium is assumed to be absorbed.

The initial dose rate, $\dot{D}_0$, that will result in a total dose of 30 mGy (3000 mrad) over a period of 13 weeks (91 days) is calculated from equation (6.31):

$$30\ \text{mGy} = \frac{\dot{D}_0}{0.0578\ \text{day}^{-1}}(1 - e^{-0.0578 \times 91}),$$

$$\dot{D}_0 = 1.74\ \text{mGy (174 mrad)/day}.$$

The body burden, Q Bq, that will deliver this initial dose rate is calculated from equation (6.21), and is found to be

$$1.74 \times 10^{-3}\ \text{Gy/day} = \frac{Q\ \text{Bq} \times 1\ \text{tps/Bq} \times 6 \times 10^{-3}\ \text{MeV/t} \times 1.6 \times 10^{-13}\ \text{J/MeV} \times 8.64 \times 10^4\ \text{sec/day}}{70\ \text{kg} \times 1\ \dfrac{\text{J/kg}}{\text{Gy}}}$$

$$Q = 1.47 \times 10^9\ \text{Bq} \quad (39.7\ \text{mCi}).$$

If the entire 3 liters of water were vaporised, the density of the steam would be 3 kg/1000 m^3 = 0.003 kg/m^3. For the density of water vapor to be this high, the temperature must be 56°C (132°F). This is an unreasonably high ambient temperature. If we assume an ambient temperature of 38°C (100°F), then the saturated water vapor density is 1.275×10^{-3} kg/m^3, and the amount of water in the air is

$$\frac{1.275}{3.00} \times 3\ \text{kg} = 1.275\ \text{kg}.$$

The tritium activity in this amount of water must be restricted to that quantity that would lead to an inhalation of no more than 1.47×10^9 Bq (39.7 mCi) during two minutes of breathing air saturated with the tritiated water vapor at a temperature as high as 38°C.

Assume that the worker's breathing rate is 20 respirations per minute, and that the tidal volume is 0.5 liters. If there were no ventilation, and the concentration of the tritium had remained constant, the atmospheric concentration C Bq/liter that would lead to the required body burden is

$$C\frac{\text{Bq}}{\text{liter}} \times \text{Respiration rate, } \frac{\text{liter}}{\text{min}} \times \text{Exposure time, min} = Q\ \text{Bq}.$$

However, the atmospheric concentration does not remain constant; the ventilation system changes one-half the air of the laboratory in 2 min, which corresponds to a rate of 0.3465 per minute. The atmospheric concentration of tritium in the laboratory, C, at any time t after release, assuming instantaneous release and distribution to give an initial concentration of C_0 is given by

$$C = C_0 e^{-kt}, \tag{11.17}$$

where k is the turnover rate of the air. The total amount of inhaled tritium, during any exposure at a mean respiration rate of RR, is

$$Q = RR \times C_0 \int_0^t e^{-kt} \, dt, \tag{11.18}$$

which yields, upon integration,

$$Q = RR \times \frac{C_0}{k} (1 - e^{-kt}). \tag{11.19}$$

In the case under consideration, the respiration rate is 0.5 liter per inspiration × 20 inspirations per minute = 10 liters per minute. Substituting into equation (11.19) to solve for C_0, we have

$$1.47 \times 10^9 \text{ Bq} = 10 \text{ l/min} \times \frac{C_0 \text{ Bq/l}}{0.3465 \text{ min}^{-1}} (1 - e^{-0.3645 \times 2}),$$

$$C_0 = 1.02 \times 10^8 \text{ Bq/l} \quad (2.75 \text{ mCi/l}).$$

Since the room volume is 27 m^3, or 27,000 liters, and since only 1.275 kg of the 3 kg of water will be in the vapor state, the maximum amount of tritium that may be contained in the tritiated water is:

$$\frac{3}{1.275} \times 1.02 \times 10^8 \text{ Bq/l vapor} \times 2.7 \times 10^4 \text{ l vapor} = 6.48 \times 10^{12} \text{ Bq (175 Ci)}.$$

According to these calculations the use of no more than 6.5×10^{12} Bq (175 Ci) tritium in the 3 liters of tritiated water would insure against overexposure in the event of the maximum credible accident. If this is not enough activity for the purpose of the experiment, then additional precautions, such as enclosure of the process or increased ventilation, would have to be employed.

Problems

1. A health physicist finds that a radiochemist was inhaling $Ba^{35}SO_4$ particulates that were leaking out of a faulty glove box. The radiochemist had been inhaling the dust, whose mean radioactivity concentration was 3.3 MBq/m^3 (9×10^{-5} μCi/cm^3), for a period of 2 hours. Using the two-compartment ICRP lung model, calculate the absorbed dose to the lung during the 13-week period and during the 1-year period immediately following inhalation.

2. A tank, of volume 100 liters, contained ^{85}Kr gas at a pressure of 10.0 kg/cm^2. The specific activity of the krypton is 20 Ci/g. The tank is in an unventilated storage room, at a temperature of 27°C, whose dimensions are 3 × 3 × 2 m. As a result of a very small leak, the gas leaked out until the pressure in the tank was 9.9 kg/cm^2. A man unknowingly then spent 1 hr in the storage room. Assume the half-saturation time for krypton solution in the body fluids to be 3 min. Henry's law constant for Kr in water at body temperature is 2.13×10^7. Calculate (a) the immersion dose, (b) the internal dose due to the inhaled krypton. The partition ratio of Kr in water to Kr in fat is 1:10.

3. If the man in problem 2 turned on a small ventilation fan of capacity 100 ft^3/min as he entered the room, calculate his immersion and inhalation doses.

4. An accidental discharge of ^{89}Sr into a reservoir resulted in a contamination level of 37 Bq (10^{-3} μCi) per cm^3 of water.

(a) Using the *basic* radiological health criterion of the ICRP, would this water be acceptable for drinking purposes for the general public if the turn-over half time of the water in the reservoir is 30 days?

(b) If the water were ingested continuously; what maximum body burden would be reached?

(c) How long after ingestion started would this maximum occur?

(d) What would be the absorbed dose during the first 13 weeks of ingestion?

(e) What would be the absorbed dose during the first year?

(f) What would be the absorbed dose during 50 years following the start of ingestion?

5. Nickel carbonyl $Ni(CO)_4$ has a maximum permissible atmospheric concentration of 1 part per billion (ppb) based on its chemical toxicity. A chemist is going to use this compound tagged with ^{63}Ni. The specific activity of the nickel is 2.5×10^8 Bq/g (6.75 mCi/g). The industrial hygienist is planning to limit the atmospheric concentration of $Ni(CO)_4$ in the lab to 0.5 ppb. Will this restriction meet the requirement for the radioactivity maximum allowable concentration of 6×10^{-8} μCi/ml?

6. Chlorine-36 tagged chloroform, $CHCl_3$, whose specific activity is 100 μCi/mole, is to be used under such conditions that 100 mg/hr may be lost by evaporation. The experiment is to be done in a laboratory of dimensions $15 \times 10 \times 8$ ft. The lab is ventilated at a rate of 100 ft^3/min.

(a) Do any special measures have to be taken in order to control the atmospheric concentration of the ^{36}Cl to 10% of its MPC (MPC $= 4 \times 10^{-7}$ μCi/cm^3)?

(b) To what concentration of chloroform, in parts per million, does the radiological MPC correspond for this compound? Compare this concentration to the chemical MPC for chloroform.

7. For the purpose of estimating hazards from toxic vapors or gases of high molecular weight, it is sometimes *incorrectly* assumed that settling of the vapor is determined by the specific gravity of the pure vapor, which is defined as

$$\frac{\text{Molecular weight of the pure vapor}}{\text{"Molecular weight" of air}}$$

instead of the *correct* specific gravity given by

$$\frac{\text{"Molecular weight" of air and vapor mixture}}{\text{"Molecular weight" of air}}.$$

(a) If the vapor pressure of benzene (benzol), C_6H_6, is 160 mm Hg at 20°C, calculate the correct specific gravity of a saturated air mixture of benzene vapors, and compare it to the specific gravity of the pure vapor.

(b) If the chemical MPC for benzene is 10 ppm by volume, calculate the specific gravity of an air–benzene mixture of this concentration.

(c) What is the maximum specific activity of ^{14}C-tagged benzene in order that one-half the radiological MPC for ^{14}C (MPC $= 4 \times 10^{-6}$ μCi/cm^3) not be exceeded by a benzene concentration of 10 ppm?

8. Iodine-131 is to be continuously released to the environment through a chimney whose effective height is 100 m, and whose discharge rate is 100 m^3/min. The average wind speed is 2 m.p.s. and the lapse rate is stable.

(a) At what maximum rate may the radioiodine be discharged if the maximum downwind ground level concentration is not to exceed 10% of the ICRP's DAC of 700 Bq/m^3 (2×10^{-8} μCi/cm^3).

(b) How far from the chimney will this maximum occur?

9. Inhalation exposure is often described as the product of atmospheric concentration and time, as in units of Bq-sec/m^3. Using the ICRP assumptions that 23% of inhaled iodine is deposited in the thyroid, and that the thyroid weighs 20 g, calculate the dose corresponding to an acute exposure of 1 Bq-sec/m^3 of (a) ^{131}I, (b) ^{133}I. (c) Assuming that the other 73% of the inhaled iodine is absorbed into the blood and is bound to the protein, calculate the total body doses due to the protein-bound iodine.

10. A possible way for a small isotope user to treat his radioactive waste prior to disposal into the sanitary sewer is to isotopically dilute the radioisotope to a specific activity such that continuous ingestion of the contaminated water would result in a dose to the critical organ that does not exceed the maximum permissible. For the case of ^{75}Se, calculate the maximum specific activity to meet this criterion. According to the ICRP, the critical organ for Se is the kidneys, weight 300 g. The maximum permissible amount in the kidneys is 3.33 MBq (90 μCi), and this activity is 4% of the total body burden. Four per cent of the ingested Se is deposited in the kidney.

11. A graphite-moderated reactor is cooled by passing 680,000 kg air per hour through the core. The mean temperature in the core is 300°C, and the thermal neutron flux is 5×10^{13} neutrons/cm^2/sec. If the air spends an average of 10 sec in the reactor core, what is the rate of production of ^{41}A? If the chimney through which the air is discharged is 100 meters high and has an orifice diameter of 2 meters; and the temperature of the effluent air is 170°C, while the ambient temperature is 30°C on a sunny day and if the mean wind velocity is 2 m/sec, at what distance from the chimney will the ground level concentration of ^{41}A be a maximum? What will be the value of this maximum concentration (in Bq/m^3). How does this figure compare to the DAC for ^{41}A?

12. About 10^{13} Bq (270 Ci) of ^{14}C waste is generated per year from biomedical sources in the United States. If this waste will continue to be generated at the same annual rate,

(a) what will be the resultant steady-state quantity of ^{14}C waste?

(b) How long will it take until 99% of the steady-state inventory is reached?

13. Analysis of albacore in the Pacific Ocean for ^{137}Cs from nuclear bomb fallout showed the mean concentration to be 2.74 Bq/kg (7.4 pCi/kg) wet weight during the period 1965–1971. Calculate the committed dose equivalent due to the consumption of 1 kg albacore per week for one year.

14. Krypton gas, tagged with ^{85}Kr to a specific activity of 1.3×10^{11} Bq/mole, (3.5 Ci/mole), will be transferred from a tank into another vessel at a rate of 0.1 cm^3 per minute (at 25°C and 760 torr) through plastic tubing. There is a remote possibility that the tubing connection will break, and the gas will escape into the laboratory. If the laboratory dimensions are 3 m × 4 m × 3 m, what must be the minimum ventilation rate if the steady-state concentration is not to exceed 1/10 of the 10 CFR 20 limit of 3.7×10^5 Bq/m^3 (1×10^{-5} μCi/cm^3)?

Suggested References

American Conference of Governmental Industrial Hygienists. *Industrial Ventilation, A Manual of Recommended Practice*, 16th ed. ACGIH Committee on Industrial Ventilation, Lansing, MI, 1981.

Ayers, J. A., ed. *Decontamination of Nuclear Reactors and Equipment*. Ronald Press, New York, 1970.

Burchsted, C. A., Kahn, J. E., and Fuller, A. B. *Nuclear Air Cleaning Handbook*. ERDA 76-21, NSTIS, Springfield, VA, 1976.

Clayton, G. D., and Clayton, F. E., eds. *Patty's Industrial Hygiene and Toxicology*, Vol. I, *General Principles*. Wiley Interscience, New York, 1978.

Dennis, R., ed. *Handbook on Aerosols*. TID-26608, NTIS, Springfield, VA, 1976.

Design Guide for a Radioiostope Laboratory. American Institute of Chemical Engineers, New York, 1964.

Dresner, L., Translator. *Jaeger's Principles of Radiation Protection*. McGraw Hill, New York, 1965.

Drinker, P., and Hatch, T. F. *Industrial Dust*. McGraw Hill, New York, 1954.

Eicholz, G. G. *Environmental Aspects of Nuclear Power*. Ann Arbor Science Publishers, Ann Arbor, 1976.

Eisenbud, M. *Environmental Radioactivity*, 2nd ed. Academic Press, New York, 1973.

Green, H. L., and Lane, W. R. *Particulate Clouds: Dusts, Smokes, and Mists*. D. Van Nostrand, Princeton, 1964.

Hawkins, M. B. The design of laboratories for the safe handling of radioisotopes. In *Laboratory Design*, edited by H. S. Coleman. Reinhold Publishing Co., New York, 1951.

International Atomic Energy Agency. *Manual on Early Medical Treatment of Possible Radiation Injury*, IAEA, Vienna, 1978.

Lewis, E. E. *Nuclear Power Reactor Safety*. Wiley Interscience, New York, 1977.

Olishifski, J. B., ed. *Fundamentals of Industrial Hygiene*, 2nd ed. National Safety Council, Chicago, 1979.

Pritchard, J. A. *A Guide to Industrial Respiratory Protection*. LA-6671-M, NSTIS, Springfield, VA, 1976.

Sagan, L. A., ed. *Human and Ecologic Effects of Nuclear Power Plants*. Charles C. Thomas, Springfield, IL, 1974.

Slade, D. H., ed. *Meteorology and Atomic Energy*. TID-24190, NSTIS, Springfield, VA, 1968.

Straub, C. P. *Public Health Implications of Radioactive Waste Releases*. World Health Organization, Geneva, 1970.

Wrixon, A. D., Linsley, G. S., Binns, K. C., and White, D. F. *Derived Limits for Surface Contamination*. NRPB-DL2, National Radiological Protection Board, Harwell, 1979.

The following publications in the Safety Series of the International Atomic Energy Agency (IAEA), Vienna:

Series
No.

5. *Radioactive Waste Disposal into the Sea* (1961).
10. *Disposal of Radioactive Wastes into Fresh Water* (1963).
12. *The Management of Radioactive Wastes Produced by Radioisotope Users* (1963).
15. *Radioactive Waste Disposal into the Ground* (1965).
22. *Respirators and Protective Clothing* (1967).
24. *Basic Factors for the Treatment and Disposal of Radioactive Wastes* (1967).
28. *Management of Radioactive Wastes at Nuclear Power Plants* (1968).
29. *Application of Meteorology to Safety at Nuclear Plants* (1968).
30. *Manual on Safety Aspects of the Design and Equipment of Hot Laboratories* (1981).
36. *Disposal of Radioactive Wastes into Rivers, Lakes, and Estuaries* (1971).
39. *Safe Handling of Plutonium* (1974).
44. *Management of Wastes from the Mining and Milling of Uranium and Thorium Ores* (1976).
45. *Principles for Establishing Limits for the Release of Radioactive Materials into the Environment* (1978).
46. *Monitoring of Airborne and Liquid Radioactive Releases from Nuclear Facilities to the Environment* (1978).
47. *Manual on Early Medical Treatment of Possible Radiation Injury* (1978).
48. *Manual on Decontamination of Surfaces* (1979).
49. *Radiological Surveillance of Airborne Contaminants in the Working Environment* (1979).
50. *Atmospheric Dispersion in Nuclear Power Plant Siting* (1980).
53. *Shallow Ground Disposal of Radioactive Wastes* (1981).
54. *Underground Disposal of Radioactive Wastes* (1981).

The following publications in the Technical Report Series of the International Atomic Energy Agency (IAEA), Vienna:

101. *Standardization of Radioactive Waste Categories* (1970).
122. *Air Filters for Use at Nuclear Facilities* (1970).

The following publications in the Symposium Series of the International Atomic Energy Agency (IAEA), Vienna:

a. *Environmental Contamination by Radioactive Materials* (1969).
b. *Management of Low- and Internediate-Level Radioactive Wastes* (1970).
c. *Environmental Aspects of Nuclear Power* (1971).

The following reports of the National Council on Radiation Protection and Measurement (NCRP), Washington, D.C.:

8. *Control and Removal of Radioactive Contaminants in Laboratories* (1951).
9. *Recommendations for Waste Disposal of Phosphorus*-32 *and Iodine*-131 *for Medical Users* (1951).
12. *Recommendations for Waste Disposal in the Ocean* (1954).
16. *Radioactive Waste Disposal in the Ocean* (1954).
44. *Krypton*-85 *in the Atmosphere—Accumulation, Biological Significance, and Control Technology* (1975).
55. *Protection of the Thyroid Gland in the Event of the Release of Radioiodine* (1977).
60. *Physical, Chemical, and Biological Properties of Radiocerium Relevant to Radiation Protection Guidelines* (1978).
62. *Tritium in the Environment* (1979).
65. *Management of Persons Accidently Contaminated with Radionuclides* (1979).

CHAPTER 12

CRITICALITY

Criticality Hazard

Of all the potential radiation hazards with which the health physicist deals, that of an accidental criticality is among the most serious. Criticality may be defined as the attainment of physical conditions such that a fissile material will sustain a chain reaction. During this chain reaction, the nuclei of the fissile material (material whose atoms can be made to fission) splits, thereby liberating tremendous amounts of energy in the form of radiation and producing large quantities of radioactive fission products. In a nuclear reactor, criticality is attained under conditions that are very rigorously controlled in regard to safety and power level. If criticality is accidentally attained outside a reactor during the processing or handling of nuclear fuel, the consequences to personnel and equipment are very grave. The utmost in care and controls, both technical and administrative, must be exercised in the handling, use, or transport of fissile materials if death or serious injury or property damage due to a criticality accident is to be avoided. Generally, such efforts to prevent criticality accidents are called *criticality control* or *nuclear safety*.

Nuclear Fission

The liquid drop model pictures the nucleus as a sphere inside of which the nucleons are in constant motion. As a consequence of this motion, the sphere may become distorted, and, under certain conditions, may become highly deformed and split into several parts: two smaller nuclei, which are called fission fragments, and several neutrons.

For fission to be possible, the following mass–energy relationship must hold:

$$E_f = (M - m_1 - m_2 - m_n)c^2, \tag{12.1}$$

where E_f is the energy released during fission, M is the mass of the fissioned nucleus, m_1 and m_2 are the masses of the fission fragments, and m_n is the mass of the neutrons. This condition can be met only by those isotopes whose atomic number and atomic mass number are such that $Z^2/A \geq 15$. However, although many isotopes at the upper end of the periodic table meet this requirement, spontaneous nuclear fission is an extremely unlikely event. Table 12.1 lists the spontaneous fission rates for several isotopes that are theoretically fissionable. Although the likelihood of spontaneous fission is almost infinitesimally small, spontaneous fission nevertheless is extremely important in criticality control, since it can, under the proper conditions, initiate an *accidental criticality*, or an uncontrolled nuclear chain reaction. For an isotope where $Z^2/A \geq 49$, the nucleus is unstable towards fission, and the isotope would undergo instantaneous spontaneous fission if it should be produced. All the naturally occurring fissionable isotopes are highly stable against spontaneous fission.

TABLE 12.1.
SPONTANEOUS FISSION RATES

Isotope	Fissions/g/sec
^{232}Th	4.1×10^{-5}
^{233}U	$<1.9 \times 10^{-4}$
^{234}U	3.5×10^{-3}
^{235}U	3.1×10^{-4}
^{236}U	2.8×10^{-3}
^{238}U	7.0×10^{-3}

From Studier and Huizenga, *Phys. Rev.* **96**, 546 (1954).

Fissionable isotopes require a certain amount of activation energy in order to cause them to fission. This is due to a potential barrier that must be exceeded before the nucleus splits. For example, let us consider the case where a nucleus of atomic number Z splits into two equal fission fragments of atomic number $Z/2$ and nuclear radius r. In order to part into two distinct fragments as a result of coloumbic repulsion, these two fission fragments must be separated by a minimum distance of $2r$. At this separation, the potential energy in the system is at its maximum value of

$$E_m = \frac{(Ze/2)^2}{2r},$$

and decreases as the distance between the fission fragments increases. At distances less than $2r$, nuclear forces become operative, and the potential energy in the system again decreases, as shown in Fig. 12.1. In order for fission to occur, E_f must be equal to or greater than E_m. Thus, if sufficient energy is added to the system to exceed the height of the potential barrier, the fission can be initiated. From Fig. 12.1 it can be seen that an isotope whose nuclear potential well is represented by the dotted line requires less activation energy than one whose nuclear potential well is described by the solid line. This difference in activation energy explains why some isotopes, such as ^{235}U and ^{239}Pu are more easily fissionable than others, such as ^{238}U and ^{240}Pu.

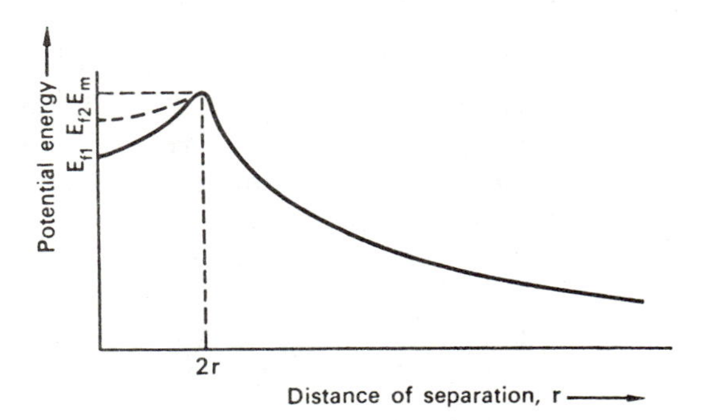

FIG. 12.1. Potential energy in a system of two equal fission fragments, each of atomic number $Z/2$ and radius r.

Activation energy for nuclear fission may be obtained from the binding energy released when a neutron is absorbed by a fissile nucleus. It was pointed out in Chapter 4 that even–even nuclei are most stable; that is, they have more binding energy per nucleon than isotopes with an odd number of nucleons. Because of this fact, the addition of a neutron to a nucleus containing an odd number of nucleons, thus producing a nucleus with an even number of nucleons, releases more energy than the addition of a neutron to a nucleus containing an even number of nucleons. Nuclei with odd numbers of nucleons are consequently more easily fissioned than those with an even number of nucleons. For example, ^{235}U, which fissions after capturing a thermal neutron, is first transformed into an even–even nucleus by the captured thermal neutron, and then fissions:

$$^{235}_{92}\mathrm{U} + {}^{1}_{0}n \rightarrow {}^{236}_{92}\mathrm{U} \rightarrow \text{fission},$$

whereas ^{238}U, which can also capture a thermal neutron, is transformed into an even–odd nucleus, and rids itself of its excitation energy by emitting a gamma-ray:

$$^{238}_{92}\mathrm{U} + {}^{1}_{0}n \rightarrow {}^{239}_{92}\mathrm{U} \rightarrow \gamma.$$

The reason for the two different modes of de-excitation of the compound nucleus, nuclear fission and gamma emission, is shown in the following calculation, in which the energy of binding a neutron to ^{235}U and to ^{238}U is calculated. From equation (3.20), the nuclear mass for ^{235}U is calculated, then the mass of a neutron is added. We once more use equation (3.20), this time to calculate the mass of ^{236}U, then find the binding energy by subtracting the ^{236}U mass from the sum of the masses of ^{235}U and a neutron. This procedure gives

$$\begin{array}{ll} \text{mass of } ^{235}\mathrm{U} & = 235.11240 \\ \text{mass of neutron} & = \underline{1.00893} \\ \text{sum of masses} & 236.12133 \\ \text{—mass of } ^{236}\mathrm{U} & = 236.11401 \\ \text{mass defect} & = 0.00732 \text{ amu,} \end{array}$$

or

$$0.00732 \text{ amu} \times 931 \frac{\text{MeV}}{\text{amu}} = 6.81 \text{ MeV}.$$

In a similar manner, the binding energy of a neutron captured by a ^{238}U nucleus is calculated as 5.31 MeV. Thus, we see that more energy, in the amount of 1.5 MeV, is liberated when a ^{235}U nucleus binds a thermal neutron then when ^{238}U binds such a neutron. In the case of ^{235}U, this additional energy is sufficient to cause fission. Uranium-238 can also be made to fission. This occurs if the ^{238}U nucleus captures a fast neutron. The kinetic energy of the neutron plus the binding energy is sufficient to cause nuclear fission. Experimentally, it has been found that a neutron must have at least 1.1 MeV of kinetic energy in order to induce fission in ^{238}U. Although ^{238}U can undergo "fast fission", the probability of such a reaction is very low in comparison to the probability of thermal fission in ^{235}U. The cross section for fast fission of ^{238}U is 0.29 barns, while for thermal fission of ^{235}U, the cross section is 588 barns. This great difference in fission cross section is one of the chief reasons for the popularity of ^{235}U fueled thermal neutron reactors.

When an atom fissions, it splits into two fission fragments plus several neutrons (the mean number of neutrons per fission of ^{235}U is 2.5) plus gamma rays—according to the conservation equation

$$^{236}_{92}\mathrm{U} \rightarrow {}^{A_1}_{Z_1}F + {}^{A_2}_{Z_2}F + \nu\, {}^{1}_{0}n + Q. \tag{12.2}$$

The value of Q in equation (12.2), which may be calculated from the mass balance for any particular pair of fission fragments, is about 200 MeV. An approximate distribution of this energy is

Fission fragments, kinetic energy	167 MeV
Neutron kinetic energy	6
Fission gamma rays	6
Radioactive decay	
Beta particle	5
Gamma rays	5
Neutrinos	11
	200

Most of this energy is dissipated as heat within the critical assembly. In a power reactor, this heat energy is converted into electrical energy. Using a mean value of 190 MeV heat energy per fission, the rate of fission to generate one watt of power is calculated as follows:

$$1\ \mathrm{W} = X\frac{\text{fiss}}{\text{sec}} \times 190\frac{\text{MeV}}{\text{fiss}} \times 1.6 \times 10^{-13}\frac{\text{J}}{\text{MeV}} \times 1\frac{\text{W}}{\text{J/sec}}$$

$$X = 3.3 \times 10^{10}\frac{\text{fissions}}{\text{second}}.$$

The amount of ^{235}U burned up to produce 1 megawatt day of heat energy is

$$\frac{3.3 \times 10^{10}\ \text{fiss/sec-W} \times 10^{6}\ \text{W/MW} \times 8.64 \times 10^{4}\ \text{sec/day}}{\dfrac{6.03 \times 10^{23}\ \text{atoms/mole}}{235\ \text{g/mole}}} = 1.125\frac{\text{g fissioned}}{\text{MWd}}.$$

Fission Products

The atomic numbers of the fission fragments range from 30 (^{72}Zn) to 64 (^{158}Gd). All fission fragments are radioactive, and decay, usually in chains of several members in length, to form *fission products*. Two such chains of fission products that are of special interest to health physicists are

$$\text{(1)}\quad {}^{90}_{36}\text{Kr} \xrightarrow[33\,\text{s}]{\beta} {}^{90}_{37}\text{Rb} \xrightarrow[2.74\,\text{m}]{\beta} {}^{90}_{38}\text{Sr} \xrightarrow[28\,\text{y}]{\beta} {}^{90}_{39}\text{Y} \xrightarrow[64.2\,\text{h}]{\beta} {}^{90}_{40}\text{Zr (stable)},$$

and

$$\text{(2)}\quad {}^{137}_{53}\text{I} \xrightarrow[22\,\text{s}]{\beta} {}^{137}_{54}\text{Xe} \xrightarrow[3.9\,\text{m}]{\beta} {}^{137}_{55}\text{Cs} \xrightarrow[30\,\text{y}]{\beta} {}^{137}_{56}\text{Ba (stable)}.$$

The fission yield for the various fission products are shown in Fig. 12.2. After production, each species of fission product decays according to its unique disintegration rate. Although the total activity at any time after production, therefore, is the sum of the exponential decay curves for each fission product, the collective activity, between about 10 sec and 1000 hr after fission, decreases as (time)$^{-1.2}$. The quantity of radioactivity at time T days after fission is given by

$$A = 3.81 \times 10^{-6}T^{-1.2}\ \text{Bq/fission}, \tag{12.3A}$$

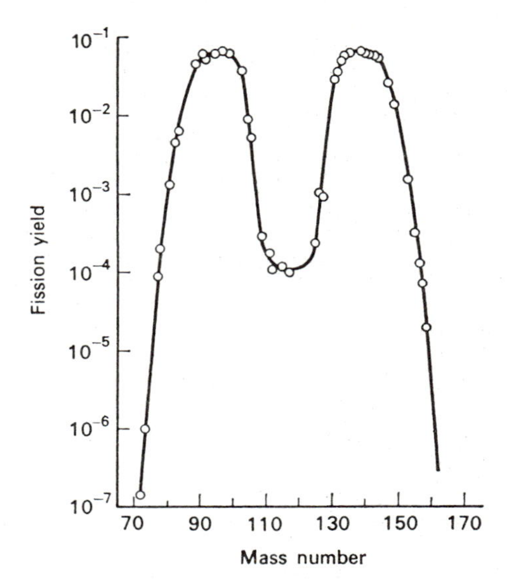

FIG. 12.2. Fission product yield from thermal fission of ^{235}U.

or

$$A = 1.03 \times 10^{-16} T^{-1.2} \frac{\text{curies}}{\text{fission}}, \tag{12.3B}$$

where T is the time in days after fission.

Example 12.1

A certain criticality accident was due to the accumulation, in a 225-gal tank, of 3 kg ^{239}Pu dissolved in an organic solvent (tributyl phosphate). When a stirrer began to operate, changes in the density and geometry of the system caused it to become critical, and, in a single burst of about 10 msec, about 10^{17} nuclei fissioned.

(a) Now much energy, in kilowatt hours, was released?

$$E = 10^{17}\,\text{fiss} \times 190\,\frac{\text{MeV}}{\text{fiss}} \times 1.6 \times 10^{-6}\,\frac{\text{erg}}{\text{MeV}} \times 1\,\frac{\text{J}}{10^7\,\text{erg}} \times 1\,\frac{\text{W sec}}{\text{J}}$$

$$\times 1\,\frac{\text{kWh}}{3.6 \times 10^6\,\text{W sec}} = 0.845\,\text{kWh}.$$

(b) What was the mean power level during criticality?

$$\frac{0.845\,\text{kWh} \times 3.6 \times 10^3\,\text{sec/hr}}{1 \times 10^{-3}\,\text{sec}} = 304\,\text{megawatts}.$$

(c) What was the fission product inventory 1 min, 1 hr, 1 day after the accident? From equation (12.3), we find the activity at 1 min ($1/1440 = 6.95 \times 10^{-4}$ days) to be

$$A = 3.81 \times 10^{-6} (6.95 \times 10^{-4})^{-1.2}\,\frac{\text{Bq}}{\text{fiss}} \times 10^{17}\,\text{fiss} = 2.35 \times 10^{15}\,\text{Bq},$$

or

$$6.34 \times 10^4\,\text{Ci}$$

after 1 hr, the activity is related to the 1 min activity by

$$\frac{A_1}{A_2} = \frac{(T_1)^{-1.2}}{(T_2)^{-1.2}},$$

$$A_2 = A_1\left(\frac{T_1}{T_2}\right)^{1.2} \tag{12.4}$$

$$= 2.35 \times 10^{15} \times 10^4\left(\frac{1}{60}\right)^{1.2}$$

$$= 1.73 \times 10^{13}\ \text{Bq} \qquad (465\ \text{Ci}),$$

and after 1 day, we have left

$$A_2 = 2.35 \times 10^{15}\left(\frac{1}{1440}\right)^{1.2}$$

$$= 3.81 \times 10^{11}\ \text{Bq} \qquad (10.3\ \text{Ci}).$$

Criticality

Criticality is attained when at least one of the several neutrons that are emitted in a fission process causes a second nucleus to fission. However, although a chain reaction occurs whenever fissionable material is irradiated with neutrons, not all systems of fissionable material can go critical. If more neutrons are lost by escape from the system or by non-fission absorption in impurities or "poisons" than are produced in fission, then the chain reaction is not self sustaining, and dies out. In this case, the assembly of fissionable material is called *sub-critical.* If we have a sustained chain reaction, and if the rate of fission neutron production exceeds the rate of loss, the assembly is called *super-critical.* When exactly one neutron per fission is available for initiating another fission, the system is called *critical.*

Multiplication Factor: The Four-factor Formula

The measure of criticality in a given system is expressed by the effective multiplication factor, k_{eff}, which is defined as

$$k_{\text{eff}} = \frac{N_{f+1}}{N_f}, \tag{12.5}$$

where N_{f+1} is the number of neutrons produced in generation $f+1$ by the number of neutrons N_f of the previous generation. Neutrons of a given generation remain in that generation until they either cause fission or are lost through non-fission process. When k_{eff} is less than 1.0000, the system is sub-critical; when k_{eff} exceeds 1.0000, the system is super-critical, and when k_{eff} is exactly equal to 1, the system is critical.

Criticality, or the value of k_{eff}, depends on the supply of neutrons of proper energy to initiate fission, and on the availability of fissile atoms. The numerical value of k_{eff}, in turn, depends on two conditions: the composition and the physical arrangement of the fissile assembly, and the size of the fissile assembly. Neutrons are lost from the assembly by several non-fission processes, including leakage from the fissile assembly. If the assembly is infinitely large, then no neutrons are lost through leakage, and the multiplication factor is known as the infinite multiplication factor, k_∞. The relationship between k_{eff} and k_∞ is

$$k_{eff} = L \times k_\infty, \tag{12.6}$$

where L is the non-leakage probability.

The value of k_∞ is determined by four factors whose mutual interrelationship can be demonstrated by following a group of n fission neutrons through a life-cycle. If η is the mean number of neutrons emitted per *absorption in uranium*, then capture of n thermal neutrons will result in $n\eta$ fission neutrons. The mean number of neutrons emitted per fission, v, depends on the fuel. For ^{235}U, $v = 2.5$; and for ^{239}Pu, $v = 3.0$. However, since both ^{235}U and ^{239}Pu also absorb thermal neutrons without fissioning, not every absorption leads to fission, and hence the mean number of fission neutrons per absorption by the fissile material (the fuel) must be less than v. For pure ^{235}U, $\eta = 2.1$, while for any other degree of enrichment, $\eta < 2.1$ because of non-fission absorption by ^{238}U. The exact value of η depends on the degree of enrichment of the uranium.

Example 12.2

Calculate η for natural uranium

$$\eta = \frac{\Sigma_f}{\Sigma_a} \times v = \frac{N_5\,\sigma_{f5}}{N_5\,\sigma_{a5} + N_8\,\sigma_{a8}} \times v = \frac{\sigma_{f5}}{\sigma_{a5} + (N_8/N_5)\sigma_{a8}} \times v \tag{12.7}$$

where Σ_f = macroscopic fission cross section,
Σ_a = macroscopic absorption cross section,
N_5 = ^{235}U atoms per cm^3,
N_8 = ^{238}U atoms per cm^3,
σ_{f5} = fission cross section for ^{235}U = 549 barns,
σ_{a5} = absorption cross section for ^{235}U = 650 barns,
σ_{a8} = absorption cross section for ^{238}U = 2.8 barns,
v = average number of neutrons per fission of ^{235}U = 2.5.

For natural uranium,

$$\frac{N_8}{N_5} = 139,$$

and

$$\eta = \frac{549}{650 + (139 \times 2.8)} \times 2.5 = 1.32.$$

Uranium 238 has a small cross section (0.29 barn) for fission by fast neutrons. Before becoming thermalized, therefore, some of the fast neutrons will be captured by ^{238}U, and will cause "fast fission." If we define the *fast fission factor*, ϵ, as

$$\epsilon = \frac{\text{Total number of fission neutrons}}{\text{Number of thermal fission neutrons}}, \tag{12.8}$$

then the capture of n thermal neutrons in the fuel will produce $n\,\eta\epsilon$ fission neutrons. The value for ϵ depends on the ratio of moderator to fuel, on the ratio of inelastic scattering cross section to fission cross section, and on the geometrical relationship between uranium and moderator. The maximum value of ϵ is 1.29 in the case of unmoderated pure uranium metal. In the case of a homogeneous fuel assembly, such as a solution of fuel, ϵ is very close to 1.

While the fast neutrons are being slowed down, they may be captured by ^{238}U without producing fission; resonances for such capture occur between 200 and 5 eV. The probability that a neutron will escape this resonance capture is called the *resonance escape probability*, p, and is defined as the fraction of the fast, fission-produced neutrons that finally become thermalized. The value of p depends on the ratio of moderator to fuel. For a very high ratio, p approaches 1, whereas for a very low ratio, p is very small. For pure unmoderated natural uranium p is 0, which means that natural uranium cannot become critical under any conditions if it is not moderated. The resonance escape probability is given by

$$p = \exp\left[-\frac{N_8}{\xi \Sigma_s} \int (\sigma_a)_{\text{eff}} \frac{\mathrm{d}E}{E} \right], \tag{12.9}$$

where N_8 = number of ^{238}U atoms per cm^3,
ξ = average logarithmic energy decrement, as defined by equation (5.48),
Σ_s = macroscopic scattering cross section for the moderator–uranium mixture,
$(\sigma_a)_{\text{eff}}$ = effective absorption cross section for the moderator–uranium mixture.

Experimentally, it has been found that, for a homogeneous mixture of uranium in a moderator, the effective resonance integral can be approximated by

$$\int (\sigma_a)_{\text{eff}} \frac{\mathrm{d}E}{E} = 3.9 \left(\frac{\Sigma_s}{N_8} \right)^{0.415} \text{barns}, \tag{12.10}$$

for cases where $\Sigma_s/N_8 \leq 1000$ barns. When Σ_s/N_8 increases beyond 1000 barns, the value of the effective resonance integral increases to a limit of 240 barns. Some values for the integral are given in Table 12.2.

TABLE 12.2 EFFECTIVE RESONANCE INTEGRAL FOR SEVERAL VALUES OF Σ_s/N_8

Σ_s/N_o, barns	$\int (\sigma_a)_{\text{eff}} \frac{\mathrm{d}E}{E}$, barns
8.2	9.3
50	20
100	26
300	42
500	51
1,000	69
2,000	90
3,000	101
10,000	125
∞	240

For a heterogeneous assembly consisting of natural uranium fuel rods,

$$\int (\sigma_a)_{\text{eff}} \frac{\mathrm{d}E}{E} = 9.25 + 24.7 \frac{S}{M} \text{ barns}, \tag{12.11}$$

S = surface area of the uranium, cm^2, M = weight of uranium, grams.

From the original n thermal neutrons, we thus have $n\eta\epsilon p$ thermal neutrons. Not all of these thermal neutrons produce fission; some are absorbed by non-fuel atoms, and some are absorbed by ^{235}U without producing fission. (Only 84% of the thermal neutrons absorbed by ^{235}U cause fission.) The fraction of the total number of thermalized neutrons absorbed by the fuel (including all the uranium) is called the *thermal utilization factor*, f. The total number of new neutrons thus produced by the original n thermal neutrons is $n\eta\epsilon pf$. From the definition of k in equation (12.5), we obtain the *four-factor formula* for criticality in the infinitely large system:

$$k_{\infty} = \frac{N_{f+1}}{N_f} = \frac{n\eta\epsilon pf}{n} = n\epsilon pf. \tag{12.12}$$

One of the four factors, η, depends only on the fuel. The other factors, ϵ, p, and f, depend on the composition and physical arrangement of the fuel: ϵ varies from a maximum of 1.29 for unmoderated uranium to almost 1 for a homogeneous dispersion of fuel in a moderator; p is on the order of 0.8 to almost 1 (for pure ^{235}U fuel, $p = 1$; for high degrees of enrichment, $p \approx 1$). The thermal utilization factor can be calculated from its definition

$$f = \frac{\Sigma_{aU}}{\Sigma_{aU} + \Sigma_{aM} + \Sigma_{ap}} = \frac{\sigma_{a5}\ N_5 + \sigma_{a8}\ N_8}{\sigma_{a5}\ N_5 + \sigma_{a8}\ N_8 + \sigma_{aM}\ N_M + (\Sigma\ \sigma_{ai}\ N_i)_p}, \tag{12.13}$$

Σ_{aU} = macroscopic absorption cross section of uranium,
Σ_{aM} = macroscopic absorption cross section of the moderator,
Σ_{ap} = macroscopic absorption cross section of other substances in the fuel assembly,
σ_{ai} = absorption cross section of the ith element in the "other substances",
N_i = atoms/cm^3 of the ith element in the other substances,
σ_{a5} = absorption cross section of ^{235}U = 650 barns,
σ_{a8} = absorption cross section of ^{238}U = 2.8 barns,
N_5, N_8 = atoms/cm^3 of ^{235}U and ^{238}U respectively.

Example 12.3

Calculate f for a solution of 925 g uranium as uranyl sulfate, UO_2SO_4, in 14 liters water. The uranium is enriched to 93%.

The problem may be solved by application of equation (12.13). However, it can be slightly simplified since, in a homogeneous mixture, the number of atoms per cm^3 is directly proportional to the molar concentration of the various substances. Therefore, equation (12.13) can be rewritten as

$$f = \frac{\sigma_{a5}M_5 + \sigma_{a8}M_8}{\sigma_{a5}M_5 + \sigma_{a8}M_8 + \sigma_{aH_2O}M_{H_2O} + \sigma_{aO_2SO_4}M_{O_2SO_4}}, \tag{12.14}$$

where M represents the number of moles of the respective substances in the solution. In 925 g of 93% enriched uranium, we have

$$\frac{0.94\ \text{g}^{235}\text{U/g U} \times 925\ \text{g U}}{235\ \text{g}^{235}\text{U/mole}} = 3.66\ \text{moles}\ ^{235}\text{U}$$

and

$$\frac{0.07 \times 925}{238} = 0.27\ \text{moles}\ ^{238}\text{U}$$

for a total of 3.93 moles uranium as UO_2SO_4.

From a table of absorption cross sections, we find that

$$\sigma_{aH} = 0.332 \text{ barns} \qquad \sigma_{a5} = 650$$
$$\sigma_{aO} = 0.0002 \qquad \sigma_{a8} = 2.8$$
$$\sigma_{aS} = 0.49$$

This leads to the following:

Material	σ_a, barns	Moles	$\sigma_a \times$ moles
Uranium-235	650	3.66	2379.00
Uranium-238	2.8	0.27	0.76
Water	0.664	778	516.59
O_2SO_4	0.491	3.93	1.93

Substituting into equation (12.14), we have

$$f = \frac{2379 + 0.76}{2379 + 0.76 + 516.59 + 1.93} = 0.82.$$

Note that in this case we used absorption cross sections for 2200 m/sec neutrons, which leads to a conservative result when making criticality calculations for nuclear safety.

The effective multiplication constant is useful for estimating, for purposes of nuclear safety, the minimum concentration of a fissile material in a solution of moderator that could become critical.

Example 12.4

Estimate the minimum concentration of 93% enriched uranyl sulfate, UO_2SO_4, in water, in order that the solution form a critical mass.

The minimum concentration will occur when the size (or volume) of the soution is infinitely large. In that case, in order to attain criticality,

$$k_{\text{eff}} = \eta \epsilon p f = 1.$$

The value for η is calculated from equation (12.7), using the molar ratio of 0.27 to 3.66 for $^{238}N/^{235}N$:

$$\eta = \frac{549}{650 + 0.27/3.66 \times 2.8} \times 2.5 = 2.11$$

To calculate f, using equation (12.14), let us arrange the data as shown in the following table, letting M represent the number of moles of UO_2SO_4 that is to be added to 1 mole of water, to give the minimum concentration that will result in $k_\infty = 1$. In Example 12.3, we calculated an atomic ratio of $^{235}U/U$ of 3.66/3.93. Using this value gives:

Material	σ_a, barns	Moles	$\sigma_a \times$ moles
Uranium-235	650	$\frac{3.66}{3.93}M$	605 M
Uranium-238	2.8	$\frac{0.27}{3.93}M$	0.192 M
Water	0.664	1	0.664
O_2SO_4	0.491	M	0.491 M

From the definition of f, equation (12.14), we have:

$$f = \frac{605\ M + 0.192\ M}{605\ M + 0.192\ M + 0.664 + 0.491\ M} = \frac{605.2\ M}{605.7\ M + 0.664}.$$

Making the very reasonable assumption that ϵ and p both are equal to 1, the four factor formula, equation (12.12) becomes:

$$k_\infty = \eta\epsilon p f = 1,$$

$$1 = 2.11 \times 1 \times 1 \times \frac{605.2\ M}{605.7\ M + 0.664},$$

$$M = 9.85 \times 10^{-4}\ \text{moles},$$

and, since the molecular weight of the 93%-enriched uranium is $0.93 \times 235 + 0.07 \times 238 = 235.2$, we need

$$235.2 \frac{\text{g U}}{\text{mole}} \times 9.85 \times 10^{-4}\ \text{moles} = 0.232 \frac{\text{g U}}{\text{mole water}}.$$

Since 1 mole water = 18 g ≈ 18 ml, 0.232 g U/mole water corresponds to:

$$\frac{0.232\ \text{g U}}{18\ \text{ml}\ H_2O} = \frac{x\ \text{g U}}{1000\ \text{ml}\ H_2O}$$

$$x = 12.9\ \text{g U/liter water},$$

or

$$\frac{363\ \text{g}\ UO_2SO_4}{235.2\ \text{g U}} \times 12.9 \frac{\text{g U}}{\text{liter water}} = \frac{20\ \text{g}\ UO_2SO_4}{\text{liter water}}.$$

Since the above calculation was based on an infinitely large volume, a concentration less than 12.9 g 93%-enriched uranium as UO_2SO_4 can never go critical.

Nuclear Reactor

In a nuclear reactor, these various factors are combined to produce a controlled, sustained chain reaction. The core of a nuclear reactor consists of fuel (^{235}U or ^{239}Pu), a moderator to thermalize the neutrons, a coolant to remove the heat, and control rods to control the chain reaction. In the case of a uranium-fueled reactor, the uranium is usually enriched in the ^{235}U isotope, since natural uranium contains only about 0.7% ^{235}U. Generally, the greater the degree of enrichment, the smaller is the size of the reactor. The control rod is made of a metal, such as cadmium, hafnium, or boron steel, that has a very high thermal neutron capture cross section. When the control rod is fully inserted into the reactor core, the multiplication constant is less than one because of the loss of neutrons to the absorber. As the control rod is withdrawn, neutrons that would have been captured by the control rod are now free to initiate a fission reaction. At a certain point of withdrawal, the multiplication factor becomes equal to 1 and the reactor is critical. Should the control rod be kept at that point, then there would be no further increase in the power level of the reactor. To increase the power level of the reactor, the rod is withdrawn, so that the multiplication factor k is greater than 1. The power level then increases, and when the desired power level is attained, the control rod is *reinserted* until

k is decreased to exactly 1.00000. The reactor then continues to operate at that power level. To decrease the power level, the contol rod is inserted into the core, causing k to become less than 1.0000, and the rate of nuclear fission to decrease. When the desired new power level is attained, the control rod is withdrawn to make k exactly equal to 1.0000 once more, and the reactor continues to operate at the reduced power level.

Reactivity and Reactor Control

The increase in the multiplication factor above 1 is called excess reactivity, and is defined by

$$\Delta k = k - 1. \tag{12.15}$$

For every n neutrons in one generation, we have $n\Delta k$ additional neutrons in the succeeding generation. If the lifetime of a neutron generation is l sec, the time rate of change of neutrons is

$$\frac{dn}{dt} = \frac{n\Delta k}{l}, \tag{12.16}$$

which yields, when integrated from n_0 to n,

$$\frac{n}{n_0} = e^{\Delta k/l \cdot t}. \tag{12.17}$$

If we define the *reactor period*, T, as the time during which the neutrons (and consequently the power level) would increase by a factor of e (an e-fold increase), then in equation (12.17),

$$\frac{\Delta k}{l} = \frac{1}{T},$$

and the reactor period, or e-folding time, is

$$T = \frac{l}{\Delta k}. \tag{12.18}$$

The mean lifetime, from birth of a neutron until its absorption in pure ^{235}U, is about 0.001 sec. Consider the case where the excess reactivity is 0.1%, that is, $\Delta k = 0.001$. The reactor period is

$$T = \frac{0.001}{0.001} = 1 \text{ sec},$$

and the power level would increase by a factor of e, or 2.718 each second. If Δk were increased to 0.5%, then

$$T = \frac{0.001}{0.005} = 0.2 \text{ sec},$$

and the power level increase in 1 sec would be

$$\frac{n}{n_0} = e^{t/T} = e^{1/0.2} = 150.$$

Such rapid increases in the power level as calculated in the example above would make it extremely difficult, if not impossible, to control a reactor. Fortunately, the calculation above is not applicable to a real reactor, because although the mean lifetime of a single neutron, from birth until absorption, is about 0.001 sec, the mean lifetime of a whole generation of a large number of neutrons is much greater than 0.001 sec. This increased mean generation time is due to the fact that 0.6407% of fission neutrons are delayed, that is, they are emitted as long as 80.39 sec after fission. Six distinct groups of delayed neutrons are observed, each group having its own mean delay time. Table 12.3 lists these groups, together with their mean delay (or generation) time.

TABLE 12.3. DELAYED NEUTRONS FRON THE FISSION OF ^{235}U

Group i	Yield, % n_i	Mean generation time, sec T_i	Yield × mean time $n_i \times T_i$
1	0.0267	0.33	0.009
2	0.0737	0.88	0.065
3	0.2526	3.31	0.836
4	0.1255	8.97	1.125
5	0.1401	32.78	4.592
6	0.0211	80.39	1.688
	$\Sigma n_i = 0.6407$		$\Sigma n_i T_i = 8.315$

The mean generation time for all the fission neutrons of a given generation is

$$\bar{T} = \frac{\sum_{i=0}^{6} n_i T_i}{\sum_{i=0}^{6} n_i} = \frac{8.315 + 99.359 \times 10^{-3}}{100} = 0.084 \text{ sec.}$$

The group $i = 0$ is the group of *prompt neutrons*, whose yield is 99.359%, and whose mean generation time is 0.001 sec.

If Δk is equal to or greater than 0.006407, the reactor is said to be in the *prompt critical* condition, since the chain reaction can be sustained by the prompt neutrons alone. If Δk is less than 0.006407, the reactor is in the *delayed critical* condition, because the delayed neutrons are essential to sustaining the chain reaction. In the delayed critical condition, the reactor period is sufficiently long to allow the power level to be easily controlled.

Example 12.5

Compute the reactor period and the increase in power level in 1 sec, for the case where the mean neutron generation time is 0.084 sec, for excess reactivity of 0.1% and 0.5%.

For $\Delta k = 0.001$,

$$T = \frac{l}{\Delta k} = \frac{0.084}{0.001} = 84 \text{ sec,}$$

and

$$\frac{n}{n_0} = e^{t/T} = e^{1/84} = 1.012;$$

for

$$\Delta k = 0.005,$$

$$T = \frac{l}{\Delta k} = \frac{0.084}{0.005} = 16.8 \text{ sec},$$

and

$$\frac{n}{n_0} = e^{t/T} = e^{1/16.8} = 1.06.$$

Excess reactivity is measured in units of *dollars* and *cents* ($1 = 100¢) and in *inhours*. One dollar's worth of reactivity is that amount of excess reactivity that will cause the reactor to go prompt critical. One inhour (*in*verse hour) is that amount of excess reactivity that results in a reactor period of one hour. Two inhours of reactivity give a reactor period of one-half hour, etc.

Fission Product Inventory

As a reactor countinues to operate, fission products are produced at a rate proportional to the power level, and the activity per fission is given by equation (12.3). If a nuclear reactor operates at a power level of P watts for a time dt, then the fission product activity at a time T days after shut down is

$$dA = 3.81 \times 10^{-6} T^{-1.2} \frac{\text{Bq}}{\text{fiss}} \times (3.3 \times 10^{10} \times 8.64 \times 10^{4}) \frac{\text{fiss}}{\text{day}} /\text{watt} \times \text{P watts} \times dt \text{ days} \tag{12.19}$$

or, combining the numerical constants,

$$dA = 1.09 \times 10^{10} PT^{-1.2} dt. \tag{12.20}$$

Now referring to Fig. 12.3,

if t = reactor operating time, days,
T = cooling time after reactor shut down, days,
τ = total time = $t + T$,

then $T = \tau - t$, and equation (12.20) may be written as

$$dA = 1.09 \times 10^{10} P(\tau - t)^{-1.2} dt, \tag{12.21}$$

$$A = 1.09 \times 10^{10} P \int_0^t (\tau - t)^{-1.2} dt, \tag{12.22}$$

$$A = 5.45 \times 10^{10} P[(\tau - t)^{-0.2} - \tau^{-0.2}] \text{ Bq}, \tag{12.23A}$$

or , if activity is expressed in Ci,

$$A = 1.46 P[(\tau - t)^{-0.2} - \tau^{-0.2}] \text{ Ci} \tag{12.23B}$$

Example 12.6

A power reactor operates at a thermal power level of 500 megawatts for 200 days. It is then shut down for refuelling. What is the fission product inventory (a) a day after shutdown and (b) 10 days after shut down?

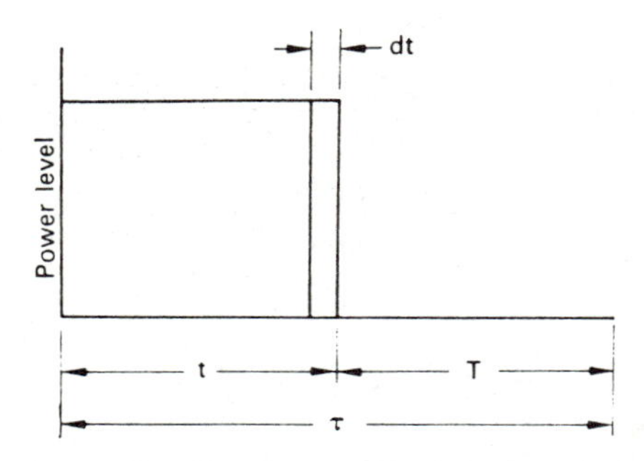

FIG. 12.3. Curve showing relationship between reactor operating time at a constant power level and post operation cooling time, that is used in the derivation of equation (12.23).

Substituting into equation (12.23B),

(a) $A = 1.46 \times 5 \times 10^8[(201 - 200)^{-0.2} - (201)^{-0.2}]$
$= 1.46 \times 5 \times 10^8[1 - 0.346] = 4.8 \times 10^8$ Ci $\quad (1.78 \times 10^{19}$ Bq).

(b) $A = 1.46 \times 5 \times 10^8[(210 - 200)^{-0.2} - (210)^{-0.2}]$
$= 1.46 \times 5 \times 10^8[0.631 - 0.343] = 2.1 \times 10^8$ Ci $\quad (7.78 \times 10^{18}$ Bq).

Tables 12.4 and 12.5 list the fission product activity resulting from operation at a power level of 1 megawatt; Fig. 12.4 graphically illustrates the fission product activity–time relationship.

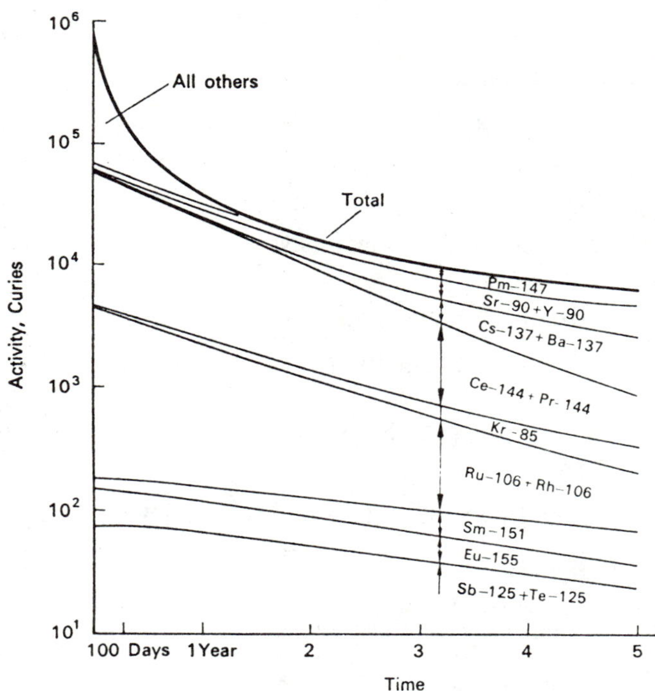

FIG. 12.4. Activity-time reationship of certain fission products after removal from a reactor that had been operating at a power level of 1 Mw for 1 year. (From *Radiological Health Handbook*, 1970.)

TABLE 12.4. PRODUCTION OF IMPORTANT FISSION PRODUCTS IN A REACTOR

Fission product	Activity (curies)[a] after selected periods of continuous operation of a reactor at a power level of 1000 kW		
	100 days	1 year	5 years
Kr-85	53	191	818
Rb-86	0.25	0.26	0.26
Sr-89	28,200	38,200	38,500
Sr-90	402	1,430	6,700
Y-90[b]	402	1,430	6,700
Y-91	34,800	48,900	49,500
Zr-95	32,900	49,200	50,300
Nb-95(90 H)[b]	446	687	704
Nb-95(35 D)[b]	20,900	48,200	50,500
Ru-103	25,100	30,900	31,000
Rh-103[b]	25,100	30,900	31,000
Ru-106	753	2,180	4,220
Rh-106[b]	753	2,180	4,220
Ag-111	151	151	151
Cd-115	4.8	5.9	5.9
Sn-117	83	84	84
Sn-119	<24	<24	<100
Sn-123	4	9	10
Sn-125	100	101	101
Sb-125[b]	12	43	139
Te-125[b]	5	34	136
Sb-127	787	787	787
Te-127(90 D)[b]	146	260	277
Te-127(9.3 H)[b]	808	922	939
Te-129(32 D)	1,410	1,590	1,590
Te-129(70 M)[b]	1,410	1,590	1,590
I-131	25,200	25,200	25,200
Xe-131[b]	250	252	252
Te-132	36,900	36,900	36,900
I-132[b]	36,900	36,900	36,900
Xe-133	55,300	55,300	55,300
Cs-136	52	52	52
Cs-137	300	1,080	5,170
Ba-137[b]	285	1,030	4,910
Ba-140	51,500	51,700	51,700
La-140[b]	51,300	51,700	51,700
Ce-141	43,000	47,800	47,800
Pr-143	45,000	45,300	45,300
Ce-144	9,860	26,700	44,000
Pr-144[b]	9,860	26,700	44,000
Nd-147	21,800	21,800	21,800
Pm-147[b]	1,290	4,900	16,000
Sm-151	9	37	175
Eu-155	23	74	207
Eu-156	108	109	109
Total	563,691	693,573	767,547

[a] Calculated using fission product yields. [b] Daughter product.
From *Radiological Health Handbook* 1970.

TABLE 12.5. ACTIVITY OF FISSION PRODUCTS IN CURIES AT SPECIFIED TIMES (T) AFTER REMOVAL FROM A REACTOR THAT HAS OPERATED AT 1000 kW FOR 1 YEAR

Fission product	$T = 0$[a]	$T = 100$ days	$T = 1$ year	$T = 5$ years
Kr-85	191	187	177	132
Rb-86	0.26	—	—	—
Sr-89	38,200	10,300	321	—
Sr-90	1,430	1,420	1,380	1,200
Y-90[b]	1,430	1,420	1,380	1,200
Y-91	48,900	14,500	577	—
Zr-95	49,200	17,000	1,000	—
Nb-95(90 H)[b]	687	152	15	—
Nb-95(35 D)[b]	48,200	28,700	2,140	—
Ru-103	30,900	5,920	74	—
Rh-103[b]	30,900	5,920	74	—
Ru-106	2,180	1,800	1,090	68
Rh-106[b]	2,180	1,800	1,090	68
Ag-111	151	—	—	—
Cd-115	5.9	1.2	—	—
Sn-117	84	0.7	—	—
Sn-119	< 64	< 48	< 23	< 0.4
Sn-123	9	5	1	—
Sn-125	101	0.1	—	—
Sb-125[b]	43	41	34	12
Te-125[b]	34	39	36	13
Sb-127	787	—	—	—
Te-127(90 D)[b]	260	123	16	—
Te-127(9.3 H)[b]	922	124	16	—
Te-129(32 D)	1,590	182	0.6	—
Te-129(70 M)[b]	1,590	182	0.6	—
I-131	25,200	4	—	—
Xe-131[b]	252	1.5	—	—
Te-132	36,900	—	—	—
I-132[b]	36,900	—	—	—
Xe-133	55,300	—	—	—
Cs-136	52	0.3	—	—
Cs-137	1,080	1,070	1,060	970
Ba-137[b]	1,030	1,020	1,010	920
Ba-140	51,700	230	—	—
La-140[b]	51,700	265	—	—
Ce-141	47,800	4,740	10	—
Pr-143	45,300	288	—	—
Ce-144	26,700	20,800	10,700	268
Pr-144[b]	26,700	20,800	10,700	268
Nd-147	21,800	40	—	—
Pm-147[b]	4,900	4,800	3,950	1,360
Sm-151	37	37	36.6	35
Eu-155	74	67	52	13
Eu-156	109	1.2	—	—
Total	693,573	144,130	36,964	6,527

[a]Calculated using fission product yields. [b]Daughter product.

From *Radiological Health Handbook*, 1970.

Criticality Control

The occurrence of an accidental criticality depends on the following factors:

1. Quantity of fissile material.
2. Geometry of the fissile assembly.
3. Presence or absence of a moderator.
4. Presence or absence of a neutron reflector.
5. Presence or absence of a strong neutron absorber (Poison).
6. Concentration of fissile material, if the fissile material is in solution.
7. Interaction between two or more assemblies or arrays of fissile materials,each one of which is sub-critical by itself. Consideration of this possibility is important in the transport and storage of fissile materials.

The influence of some of these factors is shown in Table 12.6, which lists the minimum critical mass of size of 93.5% enriched ^{235}U for several different conditions.

TABLE 12.6. MINIMUM CRITICAL MASS OR SIZE OF 93.5% ENRICHED ^{235}U

Condition	Critical mass or size of ^{235}U
Aqueous solution containing not more than 11.94 g ^{235}U per liter	∞
Bare sphere of metallic uranium	48.6 kg
Water reflected sphere of metallic uranium	22.8 kg
Unreflected sphere containing an aqueous solution of 75 g ^{235}U per liter	1.44 kg
Water reflected sphere containing an aqueous solution of 75 g ^{235}U per liter	0.83 kg
Infinitely long unreflected cylinder containing an aqueous solution of 75 g ^{235}U per liter	<8.7 in. diameter
Infinitely long water reflected cylinder containing an aqueous solution of 75 gr ^{235}U per liter	<6.3 in. diameter

Generally, nuclear safety can be assured by limiting at least one of the factors that determine criticality in such a manner that it becomes physically impossible to initiate a sustained chain reaction. The basic control methods for assuring nuclear safety include:

1. *Mass control*: limiting the mass of fissile material to less than the critical mass under any conceivable condition.

2. *Geometry control*: having a geometric configuration that is "always safe," that is, it can never become critical because the surface to volume ratio is such that excessive neutron leakage makes it impossible to attain a multiplication factor as great as one.

3. *Concentration control*: The size of a critical mass depends strongly on the ratio of fissile atoms to moderating atoms. Thus, if this ratio is kept sufficiently small, that is, if the solution of fissile material is sufficiently dilute, absorption of neutrons by the hydrogen atoms makes a sustained chain reaction impossible. The degree of enrichment of ^{235}U is especially important in concentration control. In this connection it should be pointed out that a homogeneous mixture or a solution of natural uranium in water can never become

TABLE 12.7. VALUES OF BASIC NUCLEAR SAFETY PARAMETERS

Isotope	Parameter	Value				Minimum safety factor
		Moderated		Unmoderated		
		Recommended maximum	Minimum critical	Recommended maximum	Minimum critical	
^{235}U	Mass	350 g	820 g	10.0 kg	22.8 kg	2.3
	Diameter of infinite cylinder	5.0 in.	5.4 in.	2.7 in.	3.1 in.	1.1
	Thickness of infinite slab	1.4 in.	1.7 in.	0.5 in.	0.6 in.	1.2
	Volume of solution	4.8 liters	6.3 liters	—	—	1.33
	Concentration (aqueous)	10.8 g/liter	12.1 g/liter	—	—	1.12
	Enrichment of ^{235}U	0.95	1.0	—	—	
^{233}U	Mass	250 g	590 g	3.2 kg	7.5 kg	2.3
	Diameter of infinite cylinder	3.7 in.	4.4 in.	1.7 in.	1.9 in.	1.1
	Thickness of infinite slab	0.8 in.	1.2 in.	0.2 in.	0.3 in.	1.2
	Volume of solution	2.3 liters	3.3 liters	—	—	1.33
	Concentration (aqueous)	10.0 g/liter	11.2 g/liter	—	—	1.12
^{239}Pu	Mass	220 g	510 g	2.6 kg	5.6 kg	2.3
	Diameter of infinite cylinder	4.2 in.	4.9 in.	1.4 in.	1.7 in.	1.1
	Thickness of infinite slab	0.9 in.	1.3 in.	0.18 in.	0.24 in.	1.2
	Volume of solution	3.4 liters	3.5 liters	—	—	1.33
	Concentration (aqueous)	6.9 g/liter	7.8 g/liter	—	—	1.12

From *Nuclear Safety Guide*, U.S. Atomic Energy Commission Report, TID 7016, Rev-1, 1961, U.S.A.E.C.

Notes: 1. Moderation by H_2O is assumed.
2. All values, moderated and unmoderated, assume water reflection.

critical, regardless of the concentration. For an aqueous mixture or solution of uranium, the ^{235}U must be enriched to about 1% before it can be made to attain criticality.

Nuclear safety by one of these three basic methods of control can be maintained by restricting any one of the control parameters to the maximum values listed in Table 12.7. It should be emphasized that the values listed in the table are independent of each other, that is, restriction of any one of the parameters to the value listed in the table allows other restrictions to be relaxed. It is considered good practice, however, to design processes and equipment in such a way that at least two highly unlikely, independent events occur simultaneously before a criticality accident can occur. In this connection, it should be pointed out that it is considered better practice to rely primarily on safety designed into the equipment rather than to rely principally on safety designed into a process. Thus, for example, a process may be designed to meet the concentration or enrichment criteria of Table 12.7. However, it would not be too difficult to cause a criticality inadvertently by using the wrong material. On the other hand, if the equipment were designed according to "always safe" geometry, such as limiting the diameter of a pipe or the capacity of a vessel, then an accidental criticality could not possibly occur under any conceivable operating condition.

When fissile material is being either stored or transported, it is especially important to prevent an interaction between two sub-critical units. This is accomplished by packing the fissile materials (in subcritical quantities) in containers of such design that the fissile materials are properly spaced. Such containers consist of two parts: a unit container in which the fissile material is placed, and a spacing container, whose function is to insure the physical separation of the unit containers. One common type of shipping and storage container, which is called a "bird-cage", is a weldment consisting of a steel pipe with endcaps (the unit container), in the center of a large angle iron frame (the spacing container). A variation of this type of container may be made by welding the unit container coaxially, using appropriate spacers, in a 55-gal. steel drum. The drum, in this case, is the spacing container. The spacing containers should be of such size that the closest distance (surface to surface) between unit containers cannot be less than 8 in. With this much distance between the unit containers, criticality due to interaction between two or more units is impossible even if the containers are immersed in water. No criticality restrictions on transportation or storage are necessary for:

(a) Uranium enrichment to 0.95% or less as a homogeneous aqueous mixture.
(b) Uranium metal enriched to 5% or less, provided that there is no hydrogeneous material within the container.
(c) Aqueous solutions of ^{235}U at concentrations that do not exceed 10.8 g/liter, of ^{233}U at concentrations that do not exceed 10.0 g ^{233}U per iiter, or of ^{239}Pu at concentrations that do not exceed 6.9 g/liter.

Problems

1. Cooling water circulates through a water boiler reactor core at a rate of 4 liters/min through a coiled stainless-steel tube 6.4 mm. inside diameter and 213 cm long. The concentration of Na and Cl in the water is 5 atoms each per million molecules H_2O. What is the concentration of induced Na and Cl radioactivity in the cooling water after a single passage through the reactor core, if the mean thermal flux is 10^{11} neutrons per cm^2/sec and the mean temperature in the core is 80°C?

2. If the cooling water in problem 1 circulates through a heat-exchange reservoir containing 400 liters (including the water in the pipes between the core and the reservoir), what will be the concentration of induced activity in the reservoir after 7 days operation of the reactor?

3. If the tank of problem 2 is spherical, what will be the surface dose rate due to the induced radioactivity?

4. A research reactor, after going critical for the first time, operates at a power level of 100 watts for 4 hr. How many curies of fission product activity does that core contain?

5. An accidental criticality occurred in an aqueous solution in a half-filled mixing tank 25 cm diameter × 100 cm high. The energy released during the burst was estimated as 1800 joules. Assuming that, on the average, each disintegration of a fission product is accompanied by a 1-MeV gamma-ray, estimate the gamma-ray dose rate at the surface of the tank (which maintained its integrity during the criticality) and at a distance of 25 ft from the tank at 1 min, 1 hr, 1 day, and 1 week after the criticality accident.

6. A slab of pure natural uranium metal weighing 1 kg is irradiated in a thermal flux of 10^{12} neutrons per cm^2/sec for 24 days at a temperature of 150°F. If the fission yield for ^{131}I is 2.8%, how many millicuries of ^{131}I will be extracted 5 days after the end of the irradiation?

7. What is the uranium concentration of uranyl sulfate UO_2SO_4 aqueous solution that can go critical if the uranium is enriched to (a) 10% and (b) 90%?

8. Calculate η for ^{239}Pu, given that the fission cross section is 664 barns and the non-fission absorption cross section is 361 barns.

9. The blood plasma from a worker who was overexposed during a criticality accident had a ^{24}Na specific activity of 37 Bq (0.001 μCi) per ml 15 hr after the accident. The accidental excursion lasted 10 msec. What was the absorbed dose due to (a) the ^{14}N (n, p) ^{14}C reaction and (b) the autointegral gamma-ray dose due to the n, γ reaction on hydrogen. All the Na in nature is ^{23}Na. The thermal neutron activation cross section at 20° is 0.53 barns.

10. (a) For the case where $k = 1.0025$, and an initial number of 1000 neutrons, how many neutrons will be present after 10 generations?

(b) After how many generations will the neutron flux be doubled?

11. Twenty minutes after a criticality accident the dose rate in a laboratory from the fission products was 15 Gy/hr (1500 rad/hr). If the laboratory ventilation system was shut down at the time of the criticality, how long would it take before a person could enter the laboratory if his dose equivalent during a projected 15-minute exposure time is not to exceed 50 mGy (5 rad)?

12. A transient burst of 1×10^{15} fissions in an unshielded accumulation of fissile materials causes a total dose equivalent of 0.25 Sv (25 rem) at a distance of 2 meters. If the neutron to gamma dose-equivalent ratio is 9, what were the absorbed doses from the gammas and from the neutrons?

13. The composition, by weight percent, of a concrete mix used in reactor shielding consists of oxygen: 52.17%, Si: 34.0%, Ca: 4.4%, Al: 3.5%, Na: 1.6%, Fe: 1.5%, K: 1.3%, H: 1.0%. The density of the concrete is 2.35 g/cm^3.

(a) Find and tabulate the thermal (2200 m/s) absorption cross section for each element.

(b) Calculate the linear absorption coefficient (macroscopic cross section) of the concrete.

Suggested References

AYERS, J. A., ed. *Decontamination of Nuclear Reactors and Equipment.* Ronald Press, New York, 1970.

CLARK, H. K. *Handbook of Nuclear Safety.* Report No. DP-J32, TID-4500, Office of Technical Services, Washington, 1961.

ETHERINGTON, H., ed. *Nuclear Engineering Handbook.* McGraw Hill, New York 1958.

FARMER, F. R., ed. *Nuclear Reactor Safety.* Academic Press, New York, 1977.

FOSTER, A. R., and WRIGHT, R. L., JR. *Basic Nuclear Engineering*, 3rd ed. Allyn and Bacon, Boston, 1977.

GLASSTONE, S., and EDMUND, M. C. *The Elements of Nuclear Reactor Theory.* D. Van Nostrand, Princeton, 1952.

HENRY, H. F. *Guide to Shipment of ^{235}U Enriched Uranium Materials.* TID-7019, Office of Technical Services, Washington, D.C., 1959.

International Atomic Energy Agency. *Criticality Control of Fissile Materials.* IAEA, Vienna, 1966.

LEWIS, E. E. *Nuclear Power Reactor Safety.* Wiley Interscience, New York, 1977.

MCCULLOUGH, C. R. *Safety Aspects of Nuclear Reactors.* D. Van Nostrand, Princeton, 1952.

Organization for Economic Cooperation and Development. *Criticality Control.* OECD, Karlsruhe, 1961.

PENDLEBURY, E. D., WOODCOCK, E. R., THOMAS, A. F., JOHNSON, K. D. B., and NICHOLS, C. M. *Lectures on Criticality.* UKAEA Report AHSB(S) R4, H.M. Stationery Office, London, 1961.

RUST, J. H., and WEAVER, L. E. eds. *Nuclear Power Safety.* Pergamon, New York, 1976.

THOMAS, A. F., and ABBEY, F. *Calculational Methods for Interacting Arrays of Fissile Materials.* Pergamon, Oxford, 1977.

U.S. Nuclear Regulatory Commission. *Nuclear Safety Guide.* TID-7016, Office of Technical Services, Washington.

U.S. Nuclear Regulatory Commission. *Reactor Safety Study* (Rasmussen Report). WASH-1400, U.S.N.R.C., Washington, D.C., 1975.

CHAPTER 13

EVALUATION OF PROTECTIVE MEASURES

The effectiveness of protective measures against radiation hazards is evaluated by a surveillance program that includes observations on both man and his environment. Such a surveillance program may employ one or more of a variety of techniques, depending on the nature of the hazard and the consequences of a breakdown in the system of controls. These techniques may include pre-employment physical examinations, periodic physical examinations, estimation of internally deposited radioactivity by bioassay and total body counting, personnel monitoring, radiation and contamination surveys, and continuous environmental monitoring.

Medical Surveillance

The great degree of overexposure required before clinical signs or symptoms of overexposure appear precludes the use of medical surveillance of radiation workers as a routine monitoring device. Nevertheless, medical supervision may play an important role in protecting radiation workers against possible radiation damage. Among the main tasks of medical supervision is the proper placement of radiation workers according to their medical histories and physical condition, and history of previous radiation exposure. Dermatitis, cataracts, and blood dyscrasias, including leukemia are associated with radiation exposure. A pre-employment physical examination, therefore, should be given if the nature of the work, including consideration to possible accidental overexposures, warrants it—in which special attention is paid to physical conditions that may lead to, or be suggestive of, susceptibility to any of these effects. Possible indirect effects from working with radioisotopes also are considered by the examining physician. For example, sensitivity or allergy may contraindicate work that requires the wearing of rubber gloves or that may require washing the hands or body with strong detergents or harsh chemicals in order to decontaminate the skin. In addition to the pre-employment examination, the radiation worker may be routinely examined at periodic intervals to ascertain that he continues to be free of signs that would contraindicate further occupational exposure to radiation. The physician is thus instrumental in preventing damage or injury that could otherwise have arisen, either directly or indirectly, as a consequence of working with radioisotopes or exposure to radiation. Medical supervision of radiation workers may also be necessary to evaluate overexposure, to treat radiation injuries, and to decontaminate personnel. These activities of the physician are, of course, in addition to the routine health services that he provides which are not connected to radiation hazards. It should be pointed out that medical surveillance of workers is not unique to the field of radiation health. All good occupational health programs include pre-employment examinations, consideration of medical findings in job placement, and continuing medical surveillance to help maximize the protection of workers against the harmful effects of toxic substances.

Estimation of Internally Deposited Radioactivity

One of the technics for evaluating a contamination control program is the determination of the body burdens of personnel who are at risk. This determination is done indirectly by bioassay methods, and directly by total body counting in the case of gamma-emitting radionuclides or beta emitters that give rise to suitable bremsstrahlung. The underlying rationale for bioassay is that a quantitative relationship exists among inhalation or ingestion of a radionuclide, the resulting body burden, and the rate at which the radionuclide is eliminated either in the urine or in the feces. From measurements of activity in the urine and feces, therefore, we should be able to infer the body burden. Unfortunately, the kinetics of metabolism of most substances is influenced by a large number of factors, and, as a consequence, the desirable quantitative relationships between body burden and elimination rates are known for relatively few cases. In most instances, therefore, bioassay data give only a very approximate estimate of the degree of internal deposition of radioactivity. Although both urine and feces are available for bioassay measurements in case of an accidental inhalation or ingestion of a large amount of radioactivity, routine bioassay monitoring is usually done only with urine samples, because of the ease of sample collection and also for esthetic reasons.

For purposes of bioassay, we distinguish between readily soluble and difficultly soluble compounds. This distinction is especially important in the case of inhaled particulates that are relatively insoluble, in which the difficultly soluble material is brought up from the lung and swallowed, while at the same time some of the inhaled particulate matter is dissolving and being absorbed into the body fluids. The resulting complexities due to varying pulmonary deposition and clearance rates and the varying urinary to fecal ratios of the radionuclide makes it difficult to quantitatively estimate the lung dose from difficultly soluble air-borne contamination. Nevertheless, an estimate of the minimum amount of difficulty soluble radioactive particulates that was deposited in the upper respiratory tract following a single accidental inhalation could be made from the cumulative fecal activity during the first few days after the inhalation. With this information, and using the ICRP model for lung clearance, a less reliable estimate can then be made of the activity remaining in the deep respiratory tract.

Readily soluble radionuclides may be grouped into three categories according to their distribution and metabolic pathways: (1) those that are uniformly distributed throughout the body, such as ^3H in tritiated water or radiosodium ions, (2) those that concentrate mainly either in specific organs, such as iodine in the thyroid gland, mercury in the kidney, or in the intracellular fluid, such as potsasium or cesium, and (3) those which are deposited in the skeleton. Bioassay data are most reliable in the case of the first category, the widely distributed radionuclides. In this case, the radioactivity is excreted exponentially at a rate given by the effective elimination constant, λ_e and the body burden, $A(t)$, at any time t after an intake of $A(0)$ is given by

$$A(t) = A(0)\, e^{-\lambda_e t} \tag{13.1}$$

If a constant fraction of the isotope, f_U, is eliminated in the urine, then the activity in the urine, $U(t)$, at time t after ingestion is given by

$$U(t) = f_U \frac{\mathrm{d}A(t)}{\mathrm{d}t}, \tag{13.2}$$

$$U(t) = f_U A(0) \lambda_e e^{-\lambda_e t}. \tag{13.3}$$

For tritium as tritiated water, for example, the ICRP (in Publication 10, *Evaluation of Radiation Doses to Body Tissues from Internal Contamination due to Occupational Exposure*) recommends the following equation for estimating the fractional retention of tritium at a time t days after a single contamination event:

$$R(t) = A(t)/A(0) = e^{-0.069t}, \tag{13.4}$$

and the fractional excretion by all routes on day t after the deposition of the tritium is given as

$$Y(t) = 0.07\, e^{-0.069t}. \tag{13.5}$$

According to the ICRP water balance for reference man (ICRP Publication 23, *Report of the Task Group on Reference Man*), 47% of the body's daily water loss is via the urinary tract. Accordingly, the amount of tritium eliminated in the urine on day t following the accidental exposure is related to the amount initially deposited by

$$U(t) = 0.032A(0)\, e^{-0.069t}. \tag{13.6}$$

Example 13.1

One week after a maintenance technician at a heavy water reactor worked on a certain repair job, it was learned that the working environment had been contaminated with tritiated water vapor at an unknown atmospheric concentration. To estimate the technician's dose, a urine assaying program was started. A 24-hour urine sample from the seventh day post exposure was analyzed, and was found to contain 20×10^4 Bq (5.4 μCi) ^{3}H activity. Is there cause for alarm about harmful medical consequences of this exposure?

First we estimate the initial ^{3}H deposition, and then calculate the resulting dose. Using equation (13.6), we find the initial ^{3}H deposit to have been

$$0.2 \text{ MBq} = 0.032A(0)\, e^{-0.069 \times 7}$$

$$A(0) = 10.13 \text{ MBq}.$$

The average energy of a tritium beta is 0.006 MeV. Assuming the tritiated water to be distributed uniformly throughout the body, and assuming a body weight of 70 kg, we calculate initial dose rate from equation (6.21).:

$$\dot{D} = \frac{10.13 \times 10^6 \text{ Bq} \times 1 \text{ tps/Bq} \times 0.006 \text{ MeV/t} \times 1.6 \times 10^{-13} \text{ J/MeV} \times 8.64 \times 10^4 \text{ sec/day}}{70 \text{ kg} \times 1 \frac{\text{J}}{\text{kg}} \Big/ \text{Gy}}$$

$$= 1.2 \times 10^{-5} \text{ Gy/day} \quad (1.2 \times 10^{-3} \text{ rad/day}).$$

The total dose is calculated from equation (6.32):

$$D = \frac{1.2 \times 10^{-5} \text{ Gy/day}}{0.069 \text{ day}^{-1}} = 1.8 \times 10^{-4} \text{ Gy} \quad (1.8 \times 10^{-2} \text{ rad}).$$

Since the effective half-life of the ^{3}H is 10 days, essentially all of this dose would be absorbed during the two-month period following the accidental exposure. This dose is very much less than that which would result in an observable biological effect, and much less than the maximum permissible annual dose equivalent of 0.05 Sv (5 rem). The worker, therefore, was not even overexposed (in the regulatory context), and thus no adverse medical consequences are expected.

The second category, those radionuclides that are concentrated in one or more organs, are not as amenable to quantitative monitoring by bioassay as the widely distributed radionuclides. These radionuclides are absorbed, after ingestion or inhalation, into the body fluids and the blood plasma. From these fluids, they pass into the organs in which they concentrate; a dynamic equilibrium eventually results between the concentration of the nuclide in the organ and in the body fluids. While the isotope is equilibrating between the body fluids and the organ of concentration, it is also being filtered by the kidney into the urine. This leads to a clearance curve, Fig. 13.1, that is the sum of at least two exponential components. The first component, which falls steeply, represents the clearance of the isotope from the body fluids, while the second component represents the clearance of the isotope from the organ of concentration. The slope and magnitude of the first component may be influenced by a number of factors, such as the amount of non-radioactive form of the same element that was inhaled or ingested, the amount of water intake, the physiologic state of the kidney, etc., which makes it extremely difficult to relate the intake of the radionuclide to the urinary excretion data during the first few days after a single intake. The component which represents the clearance from the organ of concentration is much less influenced by these factors. An approximate estimate, therefore, of the radionuclide in the organ of concentration following a single exposure often can be made from the urinary excretion after sufficient data are available to establish the second component of the curve.

The third category, which comprises the elements absorbed into the bone, is a special case of the category of isotopes concentrated in an organ or tissue. Bone seekers differ from

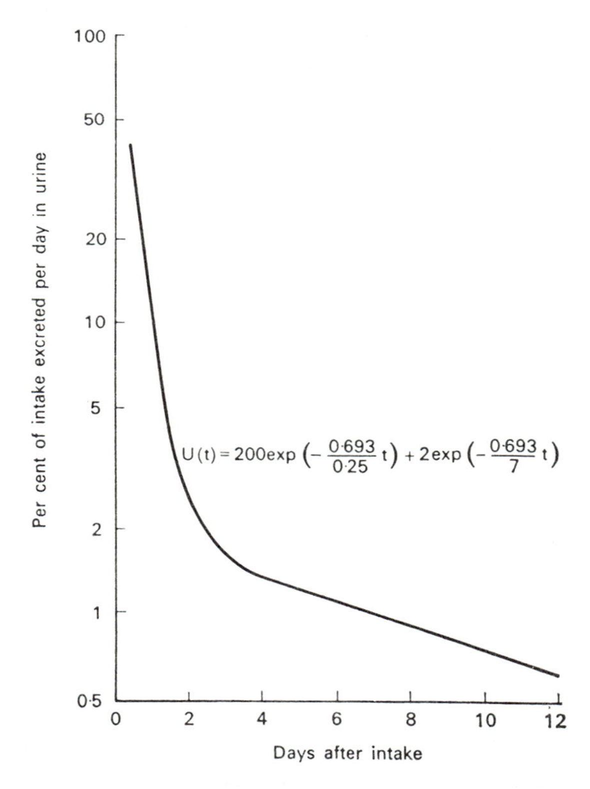

FIG. 13.1. Urinary excretion curve of ^{35}S after a single intake of a soluble, inorganic sulfate. The general shape of the curve is typical for a readily soluble compound that concentrates in a single organ. (From S. Jackson and G. W. Dolphin, Report AHSB (RP) R 51, U.K.A.E.A., 1965.)

other radionuclides mainly in the rate of elimination of the isotope. Whereas clearance half-times for the non-bone seekers are measured in days or weeks, clearance half-times for the bone seekers are measured in years. Furthermore, the rate of clearance from the skeleton is not constant, as is usually the case for the non-bone seekers, but decreases with increasing time. This is due to the fact that the skeleton is not a single "compartment", but rather a number of different "compartments", each of which has its own clearance rate. Over a long period of time ($t \gg 1$), therefore, the sum of all the exponentials representing these different compartments is described mathematically by a power function of the form

$$R(t) = At^{-n}, \tag{13.7}$$

where $R(t)$ = fractional retention t days after intake,
A = normalized fraction of the dose retained at the end of 1 day,
n = a constant.

The fraction of the intake that is eliminated in the urine, per day, if f_U is the fraction of the eliminated isotope that leaves in the urine, is

$$U_f(t) = -f_U \frac{\mathrm{d}R(t)}{\mathrm{d}t}, \tag{13.8}$$

$$U_f(t) = f_U Ant^{-(n+1)}. \tag{13.9}$$

For the case of radium, $A = 0.54$, $n = 0.52$, and $f_U = 0.02$. The fraction of the intake that may be found in a 24-hr urine sample, therefore is, from equation (13.9),

$$U_f(t) = 0.0056t^{-1.52}. \tag{13.10}$$

The body burden of radium may also be inferred from measurements of the concentration of radon in the breath. Radium decays directly to radon; some of the radon dissolves in the body fluids and in the adipose tissue, and the balance is exhaled. For body burdens on the order of the maximum permissible, i.e. 3700 Bq (0.1 μCi), 65% of the radon is exhaled. The exhaled activity, A_e, is related to the body burden q by the equation

$$A_e \frac{\text{Bq}}{\text{min}} = 0.65 \times q \text{ Bq} \times \lambda \text{ min}^{-1} \tag{13.11}$$

where λ, the decay constant for radium, is 8.1×10^{-10} per min. The concentration of radon in the breath is given by

$$C \frac{\text{Bq}}{\text{liter}} = \frac{A_e \text{ Bq/min}}{V \text{ liter/min}}. \tag{13.12}$$

Under resting conditions, the respiration rate is about 20 per min and the tidal volume is about $\frac{1}{2}$ liter; the ventilation rate, V, therefore is about 10 liters/min. Breath radon can be conveniently determined by the method of Hursch (*Nucleonics* **12**, No. 1, 63, 1954). Radon from a measured volume of exhaled breath is adsorbed on activated charcoal. The radon is then desorbed by heating, and transferred into an ionization chamber for measurement.

Direct determination, by whole-body counting, of body burdens of gamma-emitting nuclides provides a more accurate estimate of the body burden than does excreta analysis (Fig. 13.2). However, because of the high cost of installation and operation of a total-body counting facility, and also because of its limitation to the determination of gamma emitters or suitable bremsstrahlung generating beta emitters, total-body counters are not generally used for routine monitoring. Their main use is in research studies or in the assessment of internal contamination following an accidental exposure.

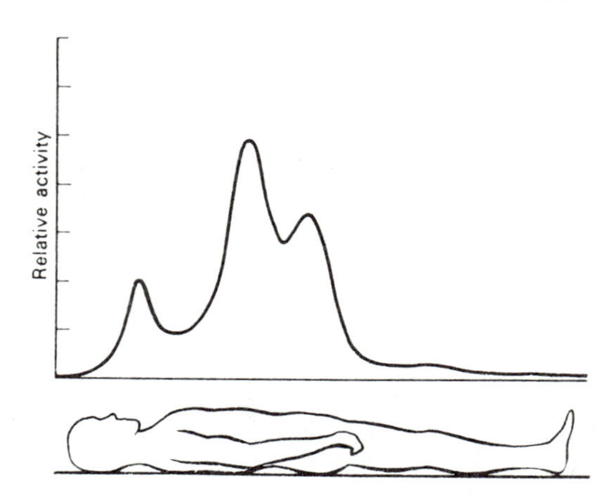

FIG. 13.2. Whole-body scan, with a crystal gamma-ray detector, of a man 2 hr after ingestion of Tc. (From T. M. Beasley, H. E. Palmer, and W. B. Nelp, Distribution and excretion of technetium in humans, *Health Physics*, **12**, 1425, 1966.)

Personnel Monitoring

Personnel monitoring is the continuous measurement of an individual's exposure dose by means of one or more types of suitable instruments, such as pocket meters, film badges, and thermoluminescent dosimeters (Chapter 8), which are carried by the individual at all times. The choice of personnel monitoring instrument must be compatible with the type and energy of the radiation being measured. For example, a worker who is exposed only to ^{14}C would wear no personnel monitoring instrument, since these isotopes emit only beta rays of such low energy that they are not recorded by any of the commercially available personnel monitoring devices. Bioassay procedures would be indicated if personnel monitoring were necessary.

The film badge has a number of advantages that makes it widely used for monitoring personnel who may be exposed to X-rays, gamma rays, high-energy beta rays, and neutrons. The chief advantage claimed for the film badge is the fact that it provides a permanent exposure record. In addition, it can measure dose over a very large range, from about 0.1 mGy (10 mrad) (X-rays) to several tens of grays (several thousand rads), it can be made to record the dose of a wide range of energies on the same film and, conversely the type and energy of the radiation may be inferred from the film exposure, and it is mechanically rugged. The main disadvantages of film badges are the long delay time between exposure and development and interpretation of the film and the relatively inaccurate exposures that have sometimes been reported by commercial film badge suppliers in several tests in which the films had been exposed to known amounts of radiation (Fig. 13.3). An average accuracy of about $\pm 25\%$ of the true exposure to the film may be expected from a good film badge service. Despite these large inaccuracies reported in tests of routine film, Brodsky *et al.* reported (*Health Physics* **11**, 1071–82, 1965) an accuracy of 10–20% is possible for film badge measurements of annual doses of 100 mR or more.

Many of the alleged disadvantages of the film badge dosimeter (particularly that of its inaccuracy) can be overcome by the use of a thermoluminescent dosimeter (TLD). The accuracy of a LiF TLD in the dose range of 0.5 mGy to 10 Gy (50 mR to 1000 R) is about $\pm 9\%$. Furthermore, the TLD is sensitive to doses as low as about 50 μGy (5 mrad), and

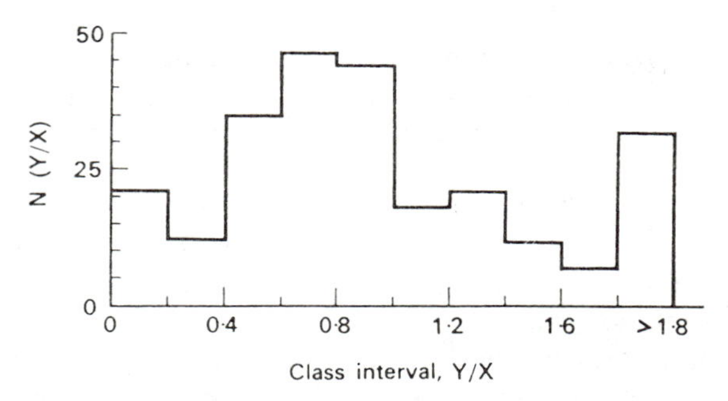

FIG. 13.3. Distribution of errors in reading radiation exposure from film badges. Y/X is defined as the ratio of reported exposure to delivered exposure. (From D. W. Barber, Film badge service performance, *Rad. Health Data and Reports* 7, 623, 1966.)

is reliable for measuring a dose of 0.1 mGy (10 mrad). With a TLD the waiting period before the exposure dose is known is less than that with film badge dosimeter because of the chemical processing required for the film. The exposure information stored in the TLD can be read out in several seconds, thus making it useful for emergency dosimetry. However, for routine batch processing of large numbers of film, the total processing time per dosimeter of film badges may be about the same as that of TLD. For these reasons, we might expect the TLD (or other solid state dosimeter), to gradually displace the film badge for personnel monitoring purposes. However, it should be pointed out that TLD readout is destructive, while developed films can be stored for possible future reference.

The main purpose of personnel monitoring is to obtain information on the exposure of an individual. In addition to this main purpose, personnel monitoring is also used to observe trends or changes (in time) in the working habits of a single individual or of a department, and thus to measure the effectiveness of a radiation control program. Whereas the distribution of personnel monitoring data might all appear to lie within a normal range when viewed as individual readings, statistical analysis of the grouped data may reveal small but significant differences among different control measures, or different operating procedures or work habits that might otherwise have escaped the attention of the radiation safety officer.

Radiation and Contamination Surveys

A survey is a systematic set of measurements made by a health physics surveyor in order to determine one or more of the following:

(a) an unknown radiation source,
(b) dose rate,
(c) surface contamination,
(d) atmospheric contamination.

In order to make these determinations, the health physics surveyor must choose the appropriate instruments, and must use it properly.

Choosing a Health Physics Instrument

The choice of a surveying instrument for a specific application depends on a number of factors. Some general requirements include portability, mechanical ruggedness, ease of use and reading, ease of servicing, ease of decontamination, and reliability. In addition to these general requirements, health physics survey instruments must be calibrated for the radiation that they are designed to measure, and they must have certain other characteristics:

1. *Ability to respond to the radiation being measured.* This point can be clarified with a practical example: a commonly used side window beta-gamma probe has a window thickness of 30 mg/cm^2. This probe would be worse than useless if one wished to survey for low-energy beta radiation, such as ^{14}C or ^{35}S, or for an alpha contaminant such as ^{210}Po. Each of these radioisotopes emits only radiation whose range is less than 30 mg/cm^2—radiation not sufficiently penetrating to pass through the window of the probe. Incorrect use of this probe, therefore, may falsely indicate safe conditions when, in fact, there may be severe contamination. Similarly, incorrect inferences may be drawn if a neutron monitor is used to measure gamma radiation or if an instrument designed to measure gamma-rays is used for neutrons. It is essential that radiation survey instruments be used only for the radiations for which they are designed.

2. *Sensitivity.* The instrument must be sufficiently sensitive to measure radiation at the desired level. Thus, an instrument to be used in a search for a lost radium needle should be more sensitive than a survey meter used to measure the radiation levels inside the shielding of an accelerator. In the latter case, where the radiation levels may reach hundreds of mGy per hr (thousands of mrad per hr), an ionization chamber, such as a Juno, whose sensitivity is about 0.01 mGy/hr (1 mrad/hr), is suitable. In searching for the lost radium needle, on the other hand, a sensitivity of 0.01 mGy/hr (1 mrad/hr) would greatly limit the area that could be covered in the search; a Geiger counter survey meter that has a sensitivity of about 0.05 mGy/hr (1 mrad/hr) is much more useful. For example, if a 1-mg radium needle were lost, the distance within which it could be detected with the Juno is about 90 cm, while the Geiger counter will respond to the lost radium at a distance of 412 cm. The Geiger counter can thus cover an area of $53\frac{1}{2}$ m^2, while the Juno can cover only about $2\frac{1}{2}$ m^2. Too great a sensitivity, on the other hand, may be equally undesirable. The range of radiation levels over which the instrument is to be used should be matched by the range of radiation levels for which the instrument is designed. Sensitivity is determined mainly by the value of the input resistor across the detector, R in Fig. 9.1 and R_1 in Fig. 9.28. The sensitivity of the detector is directly proportional to the size of the input resistance.

3. *Response time.* The response time of a survey instrument may be defined as the time required for the instrument to attain 63% of its final reading in any radiation field. This time is determined by the product of the input capacity (in farads) of the detector and the shunting resistance (in ohms) across the detector, RC in Fig. 9.1. The time constant is usually expressed in seconds. A low value for the time constant means an instrument that responds to rapid changes in radiation level—such as would be experienced when passing the probe rapidly over a small area of contamination on a bench top or over a small crack in a radiation shield. A fast response time, however, may mean a decrease in sensitivity due to a smaller value of R. Furthermore, a fast response time may result in rapid fluctuations of the meter reading, thus making it difficult to obtain an average level. In practice, the response time of a survey instrument is designed to optimize these divergent factors. Many instruments offer a range of response times, the appropriate one being selected by the surveyor turning the time constant selector switch to the desired value.

4. *Energy dependence.* Most radiation-measuring instruments have a limited span of energy over which the radiation dose is accurately measured. One of the figures of merit of a radiation dosimeter is the energy range over which the instrument is useful. This

information must be known by the health physicist in order to choose a proper instrument for a particular application or if he is to properly interpret his measurements. The energy dependence is usually specified by the manufacturer as "Accurate to $\pm 10\%$ of the true value from 80 keV to 2 MeV", or by means of an energy dependence curve (Fig. 13.4). The magnitude of the errors that can arise when the energy dependence factor is overlooked is shown in Table 13.1.

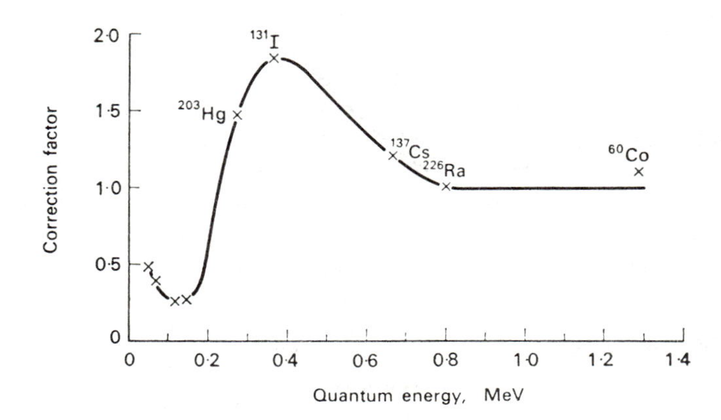

FIG. 13.4. Energy dependence of a Geiger counter survey meter. The meter reading is multiplied by the correction factor appropriate to the quantum energy in order to obtain the true exposure rate. (Courtesy of Electronica Lombarda S.P.A.)

TABLE 13.1. ENERGY DEPENDENCE OF DOSE-RATE RESPONSE OF G.M. AND SCINTILLATION COUNTERS

Meter reading for a true exposure rate of 1 mR/hr

Isotope	Gamma-ray energy, MeV	G.M. counter	Scintillation counter
^{60}Co	1.25	1.15 mR/hr	0.6 mR/hr
^{226}Ra	0.84	1.0	0.96
^{137}Cs	0.661	0.92	1.39
^{198}An	0.411	0.82	2.65
^{203}Hg	0.279	1.29	7.5
^{141}Ce	0.145	2.4	14.1
^{241}Am	0.06	6.0	9.8

From D. H. Peirson, *Physics in Medicine and Biology* **7**, 450 (1963).

Surface Contamination

Surface contamination can be located by scanning with a sensitive detector, such as a thin-end window Geiger counter. After finding a contaminated spot or area, a dose-measuring instrument may be employed to measure the dose rate at some appropriate distance from the surface. The main hazard from surface contamination is transmission of the contamination from the surface into the body via inhalation or ingestion. To estimate this hazard, a *smear test* is performed to determine whether the surface contamination is fixed or loose, and therefore transmissible. A smear test consists of wiping the suspected area with a piece of filter paper several centimeters in diameter and then measuring the activity in the paper. The area to be smeared varies according to the extent of the suspected contamination and the physical conditions under which the survey is made; a wipe-area of 100 cm^2 is not uncommon. A smear survey, which is a systematic series of smears without first using a scanning instrument to detect the contamination, is often done in a work area that is subject to contamination, and where the background due to radiation sources is high enough to mask the activity due to contamination. It should be emphasized that a smear test is a qualitative, or at best a semiquantitative determination whose chief purpose is to allow an estimate to be made of the degree to which surface contamination is fixed. If significant transmissible contamination is found, and, if in the opinion of the health physicist this contamination may be hazardous, then prompt decontamination procedures are instituted.

Leak Testing Sealed Sources

Sealed gamma-ray, beta-ray, bremsstrahlung and neutron sources are used in a wide variety of applications in medicine and in industry. In all cases, the radioactive material is permanently enclosed either in a capsule or another suitable container. Before being shipped from the supplier, all such sources must pass inspection for freedom from surface contamination and leakage. Either during transport from the supplier or in the course of time, however, the capsule may develop faults through which the radioactive source material may escape into the environment. Because of the serious consequences of such an escape, a sealed source must be tested before being put into use and periodically thereafter for surface contamination and leakage. The testing cycle depends on the nature of the source and on the kind of use to which it is put. However, it is usually recommended that such tests be performed at least once every six months. The following technics may be employed to perform these tests: to test for surface contamination, wipe all exposed external surfaces of the source thoroughly with a piece of filter paper or a cotton swab moistened with an appropriate solvent, then measure the activity on the paper or the swab. The source is considered free of surface contamination if less than 200 Bq (0.005 μCi) alpha or less than 2000 Bq (0.05 μCi) beta activity was wiped off. To test for leakage, one of the following tests may be performed:

1. Wipe the source with either a piece of wet filter paper or a cotton swab. Repeat at least 7 days later. If less than 200 Bq (0.005 μCi alpha or less than 2000 Bq (0.05 μCi) beta activity was wiped off each time, then the source is considered free of leaks.
2. For high activity sources such as those used in teletherapy, where wiping the source might be hazardous, accessible surfaces of the housing port or collimator may be wiped while the source is in the "off" position.
3. Immerse the source in ethanediol, and reduce the pressure on the liquid to 100 mm Hg for a period of 30 sec. A leak is indicated if a stream of fine bubbles issues forth from the source. This method is reliable only for such sources where enough gas would be trapped to produce a stream of fine bubbles.

Air Sampling

Air sampling is considered an important part of a survey where there is a possibility of significant atmospheric contamination. Allowable working levels of contaminated air involve quantities of radioactivity very much less than those which would be considered hazardous if the activity were in a sealed source and if the hazard were limited only to external radiation. Furthermore, even if only sealed sources are used, a program of air sampling is recommended if the nature of the source is such that radioactive gaseous or particulate matter could escape in the event that the source capsule develops a flaw. An air sample, in such a case, might detect the contamination and the leaky source before a significant amount of radioactivity escaped.

An air sampling system consists of three basic elements: (1) a source of suction (a vacuum pump) for drawing the air to be sampled through (2) a collecting device, which usually separates the contaminant from the air, and (3) a metering device for measuring the quantity of air sampled. After collection, the sample of the contaminant is counted to determine the radioactivity content, and then, when this information is combined with the size of the air sample, the concentration of atmospheric radioactivity is calculated. The exact type of the sampling system depends on the nature of the radioactive contaminant—mainly whether the contaminant is gaseous or particulate. Regardless of the nature of the contaminant, however, there exist several problems that are common to all types of contaminants (including non-radioactive contaminants). These common problems include:

1. Obtaining a sample of air *that is representative of the situation under investigation*. In most cases, we are interested in the radioactivity that a person might inhale. For this purpose, air samples are usually taken in the "breathing zone," that is, at a height of about 6 ft above the ground. If the contaminant is a dust, then the collector should be oriented so that the collection orifice is vertical, in order that it collect respirable dust particles suspended in the air, rather than non-respirable particles that might fall down on the collector. There is a good deal of evidence showing that concentrations of air-borne contamination varies significantly in time and location. Obtaining a "representative sample" of what a person might inhale is thus a fairly difficult task. To simplify this task, a worker may wear a personal air sampler whose collector is as near to his nose as practicable. Measurements made, under actual working conditions, with a personal air sampler and a fixed air sampler have shown that there is little correlation between the activity on the fixed air sampler and that of the personal air samples.

A special problem in obtaining a representative sample arises in sampling dusts that are moving at a high velocity, as in the case of an exhaust duct or a chimney. Air-borne particulates are carried by an airstream, and they tend to follow the streamlines of flow. If the streamlines bend or curve, then the path of an air-borne particle is determined by the ratio of the viscous forces (which tend to keep the particle in the streamlines) to the inertial force (which tends to cause the particle to cut across streamlines). Consider the case where a sampling device is oriented at right angles to the direction of flow in a duct, that is, the gas is blowing directly into the sampling device. If the gas is drawn through the sampler at a velocity less than that of the gas in the duct, then some of the gas must flow around the sampling device. Large particles will, because of their inertia, tend to continue in a straight line, and thus cut across the streamlines and enter into the sampler. This results in an excessive number of particles in the collector, which results in an overestimate of the particulate concentration in the gas. If the gas is drawn through the sampler at a faster velocity than that of the gas, then more of the heavy particles will be undeflected than will the lighter particles, thus leading to a smaller deposition of particles and consequently to an underestimate of the true concentration of particulate matter in the stream of gas. When the velocity of the gas through the sampler is equal to the velocity of gas in the duct, the streamlines of the gas are not disturbed by the sampling orifice, and no sampling error due to the inertia of the suspended particles occurs. This condition is called *isokinetic sampling*, and must be met if the dust sample is to be representative of the dusts in the airstream. The magnitude of the error due to *anisokinetic sampling* conditions increases as the mass of the particles increases and as the difference between the sampling, velocity and the gas velocity increases.

2. Obtaining a sample that is large enough to give a reasonably accurate estimate of the mean concentration of dust particles in the air, and also large enough to meet the sensitivity requirements of the radioactivity detector.

Example 13.2

A $^{90}SrCl_2$ aerosol will be produced in an inhalation experiment in which the particulates are cubic, have a mean edge length of 1 micron, and a mean activity of 0.04 Bq (10^{-6} μCi) per particle. The recommended DAC for soluble ^{90}Sr for a 40-hour week is listed by the ICRP (ICRP Publication 30, Supplement to Part 1) as 300 Bq/m^3 (8×10^{-9} μCi/cm^3). In keeping with the ALARA philosophy, we wish to limit the atmospheric concentration to 30 Bq/m^3 (8×10^{-10} μCi/cm^3). To meet the ALARA requirement in this case means an average particle concentration of

$$\frac{30\ \text{Bq/m}^3}{0.04\ \text{Bq/particle}} = 750\ \text{particles/m}^3.$$

For health physics monitoring purposes, we wish to detect 10% of this concentration, or 75 particles/m^3, within $\pm 20\%$ at the 96% confidence level. The air will be sampled with a membrane filter, 0.45-micron pore size, that is 100% efficient for the collection of these particles.

(i) How large must the air sample be?

(ii) The radiation detector, a windowless proportional counter whose geometry is 50%, requires a net activity of 50 transformations per minute in order to make a statistically significant measurement. Will the air sample size calculated from (i) meet the counting requirements?

(a) We wish to determine a mean particle concentration of $75 \pm 20\%$, or 75 ± 15 particles per cubic meter at the 96% confidence level. Since 96% includes two standard deviations, the standard deviation of the mean, σ_c in this case, is $15 \div 2$ or 7.5 particles per cubic meter.

Since the capture of a particular particle on the filter is a highly unlikely event, Poisson statistics are applicable.

Following the development of equation (9.31), we have, in this example, the mean particle concentration and the standard error of the mean concentration:

$$c \pm \sigma_c = \frac{n}{V} \pm \frac{\sqrt{n}}{V}, \tag{13.13}$$

where n = the number of particles collected in the sample, and V = the volume of air in the sample.

Since

$$\sigma_c = \frac{\sqrt{n}}{V} = \sqrt{\left(\frac{n}{V} \cdot \frac{1}{V}\right)} = \sqrt{\frac{c}{V}},$$

we have

$$c \pm \sigma_c = c \pm \sqrt{\frac{c}{V}}. \tag{13.14}$$

Substituting the appropriate numerical values for σ_c and c into equation (13.14), we find that

$$\sigma_c = \sqrt{\frac{c}{V}},$$

$$7.5 = \sqrt{\frac{75}{V}},$$

$$V = 1.3 \text{ cubic meters.}$$

(b) The expected activity of the sample is

$$2\ \text{m}^3 \times 50 \frac{\text{particles}}{\text{m}^3} \times 10^{-6} \frac{\mu\text{Ci}}{\text{particle}} \times 2.22 \times 10^6 \frac{\text{tpm}}{\mu\text{Ci}} = 222\ \text{tpm}.$$

Since there are only two chances in 100 that the sample will be less active than this, and since we require only 50 tpm for counting purposes, this sample is sufficiently large for the radioactivity determination.

3. Choice of collecting device. The main factor in choosing a collecting device is the nature of the contaminant—whether gaseous or particulate. Gas may be collected by a number of different technics including:

(a) Adsorption, in which a monomolecular layer of the gas binds to the surface of certain particulate substances. The capacity of such a substance depends on the specific surface area of the adsorbant (square meters per gram), the partial paressure of the gas, and the temperature. This binding capacity is represented by a curve called an *isotherm* (Fig. 13.5), in which the moles of gas adsorbed per gram of adsorbant is plotted against the equilibrium pressure at constant temperature. Commonly used adsorbants are activated carbon, activated alumina, and silica gel. In use, the adsorbant is packed into a suitable container, and the air to be sampled is drawn through the adsorbant. If desired, the adsorbed gas can be driven off in the laboratory by the application of heat.

(b) Absorption, in which the contaminated air is bubbled through a liquid with which the contaminant will interact. For optimum operation, the gas must be brought into intimate contact with the solution, and it must remain in contact long enough to allow the desired chemical reaction to take place. A commonly used absorption collector is the Greenburg–Smith impringer with a fritted glass bubbler. This collector can hold up to 500 ml absorbing solution, and, at a sampling rate of 1–5 liters/min, has a collection efficiency (for the gas appropriate to the absorbing solution) that approaches 100%. A smaller version of this instrument is called a midget impringer (Fig. 13.6).

(c) Grab sample in which an evacuated container is opened in the atmosphere to be sampled, thus permitting contaminated air to enter into the container. The activity of this contaminated air may be determined by transferring the gas to an ionization chamber, and then measuring the ionization current due to the gaseous activity. A vibrating-reed electrometer is useful for this purpose.

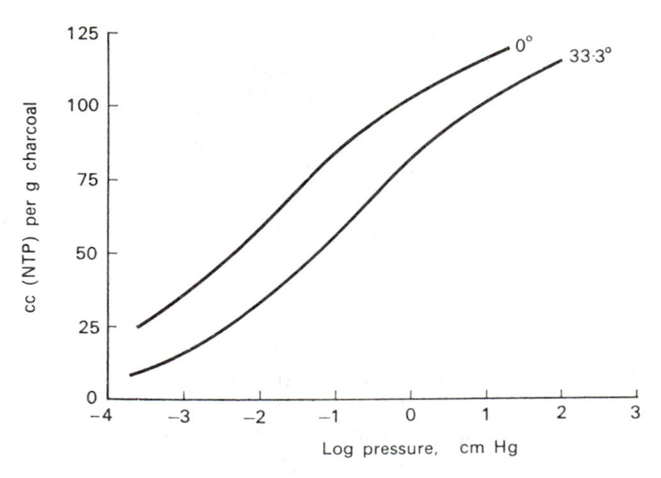

FIG. 13.5. Adsorption isotherm of benzene by charcoal. (From J. W. McBain, *The Sorption of Gases and Vapors by Solids*, George Routledge & Sons, London, 1932.)

FIG. 13.6. Midget impringer. A self-contained unit, including a hand-operated pump that samples air at a rate of 0.1 cfm. For sampling gases and vapors, a fritted gas bubbler is used; for sampling dusts, the air is drawn into the collecting liquid at a relatively high velocity through a glass nozzle, causing the dust particles to impinge on the bottom of the flask and to be retained in the liquid. (Courtesy Mine Safety Appliances Co.)

Airborne Particulates

Measurement of the concentration of airborne particulates requires the separation of the particles from a known volume of air and their collection in a manner appropriate to the parameter under measurement (mass, radioactivity, particle size, particle number, etc.). Basically, particulates may be sampled by five methods: sedimentation, filtration, electrostatic precipitation, impaction, and thermal precipitation. The first method, sedimentation, is widely used for fallout measurements. A piece of adhesive-bearing paper of known area is exposed for a predetermined length of time. The radioactive fallout per unit area per unit time is then calculated from radiometric measurements of the paper. The last-named method, thermal precipitation, is not widely used in the field by health physicists. It samples the air at a very low rate—on the order of several milliliters per minute—and is thus not very useful for general purpose air sampling. Its main function is to collect dust for accurate determination of size distribution—and this it does extraordinarily efficiently. The thermal precipitator is 100% efficient for particles 0.1 μ and larger. The operating principle in the thermal precipitator is that airborne particles in a stream that passes through a thermal gradiant (which in a thermal precipitator, Fig. 13.7, is maintained between a hot wire and a cold collecting surface made of a glass microscope slide or cover slip) are deflected by the hot wire with a force that is proportional to the temperature gradient, and are deposited on the cold surface.

FIG. 13.7. Thermal precipitator. The part on the right has an electrically heated wire across the channel through which the dust-laden air is drawn. The cold collecting plate, a square glass cover slip used by microscopists, is placed into the part on the left. The two parts are then assembled, with the cover slip sandwiched between them. The cover slip is rotated at 1 r.p.m. in order to obtain a uniform size distribution of the collected particles.

Filter

For radioactive particulates, filtration is the most commonly used collection method. Several different types of filter media are available: paper, glass fiber, and membrane (cellulose ester). Each of these has its own flow rate and filtering characteristics. Filters trap particulates mainly by two mechanisms: (1) sieving, which captures particles that are larger than the pore size, and (2) impaction, which captures particles smaller than the pore size. In impaction, the inertia of the particles in the air stream cause the particles to tend to move in straight lines. As the air bends on its tortuous path through the filter's pores, the particles continue in straight lines, and thus strike the filter matrix, and are captured. For this reason, the filtration efficiency of a filter increases as the flow rate through the filter increases. Filters made of glass fiber and paper trap particles within the matrix; membrane filters trap particles on the filter surface. This point is important when sampling for alpha or for low-energy beta particles because of the corrections for self-absorption that must be used. For this reason, as well as its very high retention efficiency for particulates of respirable size, membrane filters are very widely used for sampling of radioactive aerosols. Membrane filters may be made transparent with immersion oil, thus allowing direct microscopic observation of the particles for sizing or for particle concentration measurements. Membrane filters, however, retain a strong electrostatic charge. If the filter is placed into a windowless counter, this electrostatic charge would distort the electric field around the anode, and thus would introduce a counting error. To prevent this, the membrane filter is treated with a mixture of dioxane and petroleum ether, which eliminates the static charges from the filter without interfering with the dust particles on the filter.

The filtration efficiencies, at a sampling velocity of 0.61 m/sec (2 ft/sec), are shown in Table 13.2.

TABLE 13.2. COLLECTING EFFICIENCY OF CERTAIN FILTERS, PERCENT

	<0.4	0.4–0.6	0.6–0.8	0.8–1.0	1–2	>2
Whatman 41	23	28	64	74	80	100
Whatman 4	23	32	38	79	84	100
MSA S	48	47	77	92	94	100
H-70	99.3	99.3				
Glass fiber	99.9	99.9				
Membrane	99.9	99.9				

Electrostatic precipitators

Electrostatic precipitators, which are used by industrial hygienists for dust sampling, are also used, though less frequently, by health physicists. The electrostatic precipitator (Fig. 13.8) consists of two coaxial electrodes, a central wire cathode about 1 mm in diameter, and an aluminum anode, several centimeters in diameter, which serves as the dust collector. When a potential difference of about 12,000 volts is applied across these two electrodes (the collecting anode is kept at ground potential) a corona discharge around the central electrode charges the dust particles negatively as they are carried by in the air stream. These negatively charged dust particles are collected by the outer cylinder, which is electrically positive with respect to the charged dust particles. After collection, the collecting cylinder is removed and the particles are washed off the inside of the cylinder for counting. Alternatively, the cylinder may be incorporated into a gas-flow counter (Fig. 13.9), for counting with no further sample preparation. The electrostatic precipitator samples air at a rate of 3 ft^3/min; it is about 99% efficient for particles of 0.2–5 μ. Above 5 μ, the collecting efficiency decreases rapidly.

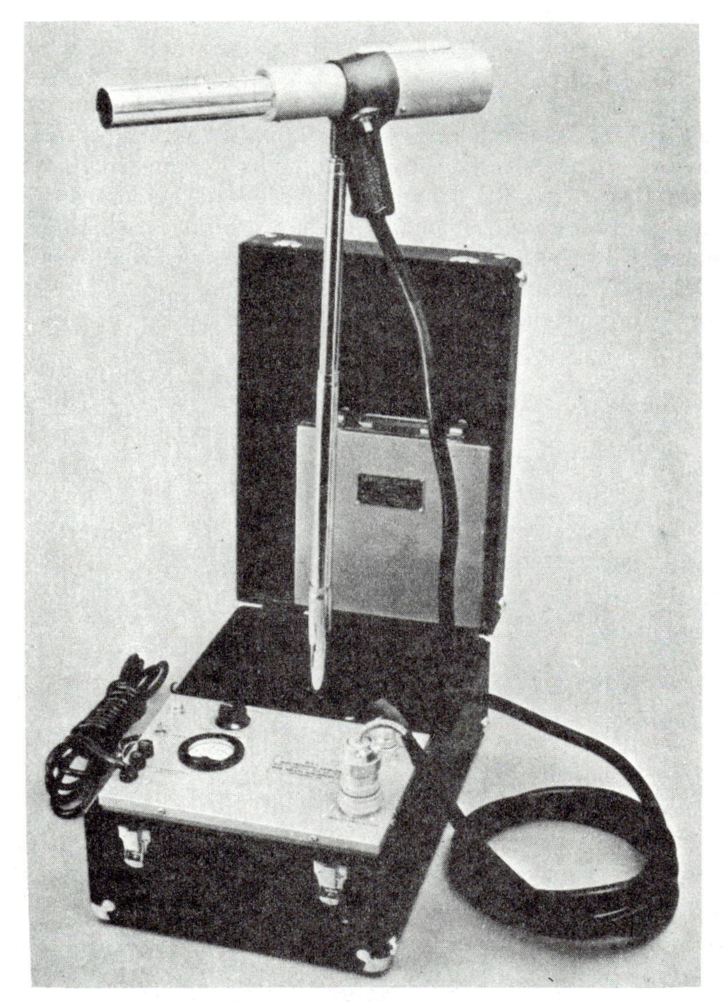

FIG. 13.8. Electrostatic precipitator for collecting samples of atmospheric dust, fumes, and smoke at a rate of 3 ft^3/min. (Courtesy Mine Safety Appliances Co.)

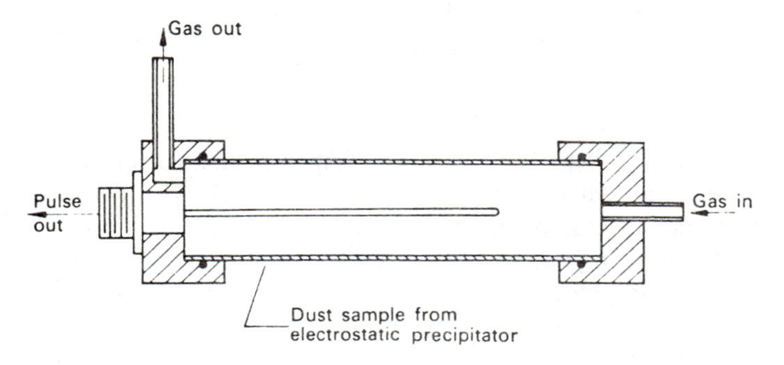

FIG. 13.9. Arrangement for counting the radioactivity in an electrostatic precipitator dust sample.

Cascade impactor

The cascade impactor, as its name implies, collects particles by impaction as a jet of high-velocity air strikes a surface perpendicular to its direction of travel and is thus abruptly deflected. Particulate matter, by virtue of its inertia, tends to continue in the original path, and thus strikes the deflecting surface—which is also a collecting surface. The main advantage of the cascade impactor is that it separates particles according to their size (or their mass). How this is accomplished may be seen in Fig. 13.10. The air is drawn through a series of openings of successively decreasing width, thus resulting in an increasing velocity through the successive sizes. Because of their greater inertia, massive

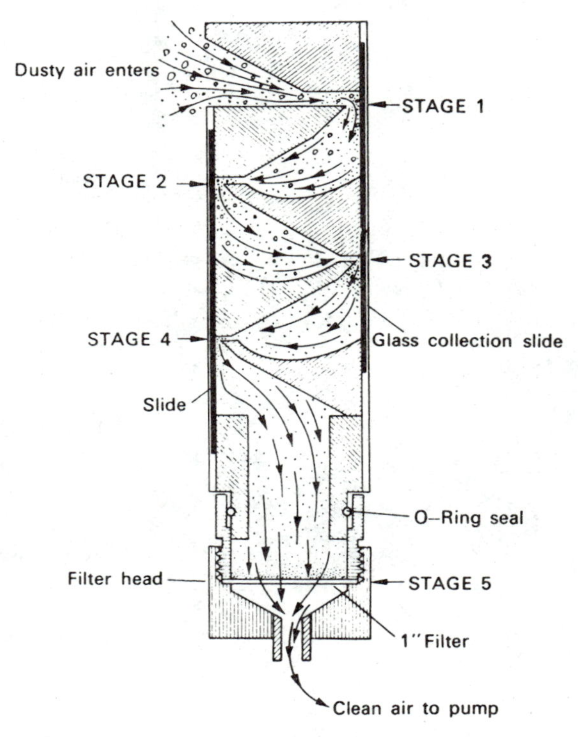

FIG. 13.10. Cascade impactor. The dust particles are collected on two standard 1 × 3 in glass microscope slides (stages 1, 2, 3, and 4) and on a membrane filter (stage 5). (Courtesy Union Industrial Equipment Co.)

particles do not get past the first stage, while lighter particles can be carried by the air stream to the next stage. There the air stream passes through the opening at a higher velocity than through the first opening; thus imparting sufficient momentum to certain particles to cause them to impact on the second stage. In commercially available cascade impactors, this process is repeated four or more times, and then the air passes through a membrane filter that removes those particles small enough to have escaped impaction on the last stage. The various stages are glass slides, coated with a very thin layer of adhesive material. The collected particles can either be examined microscopically, or can be determined either gravimetrically (for non-radioactive particles) or radiometrically.

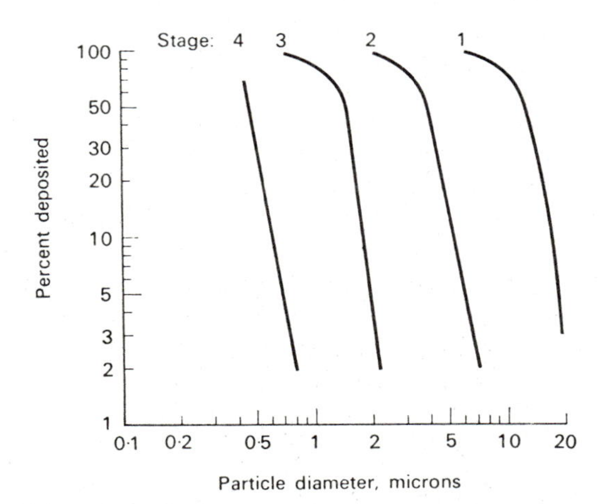

FIG. 13.11. Typical particle size distributions on the first four stages of a cascade impactor for particles of unit density and a sampling rate of 17.5 liters/min.

Before use, the cascade impactor stages are calibrated with particles of known density to determine the size distributions collected by each stage (Fig. 13.11). Each stage is thus uniquely associated with a certain mass median diameter (MMD), that is, the particle size that is collected with 50% efficiency by that stage.

To determine the size distribution of an aerosol with a calibrated cascade impactor, the dust is sampled, and the total amount of collected material on each stage is determined. Table 13.3 shows a typical set of data for U_3O_8 dust (density = 8.3 g/cm^3) that may be obtained in this manner. The median stage diameter is related to the density of the particle by

$$\frac{d_1}{d_2} = \sqrt{\frac{\rho_2}{\rho_1}}, \tag{13.15}$$

where d_1 and d_2, and ρ_1 and ρ_2 are the unit size and density and sample size and density respectively.

TABLE 13.3. RELATIONSHIP BETWEEN MASS AND SIZE DISTRIBUTION IN AN AEROSOL SAMPLED WITH A CASCADE IMPACTOR

Stage	MMD	% of total mass on stage	Cumulative %
1	4.26	11.2	94.4
2	1.35	12.4	71.0
3	0.52	49.6	40.0
4	0.16	12.2	9.1
5	(filter)	3.0	1.5

The stage MMD's (obtained from the calibration curves and corrected for density by equation (13.15) are plotted on log probability paper (Fig. 13.12) against the cumulative percentage up to the respective stage—that is, all the weight on the preceding stages plus one-half the weight on the stage in question. When plotting the data, greater emphasis is

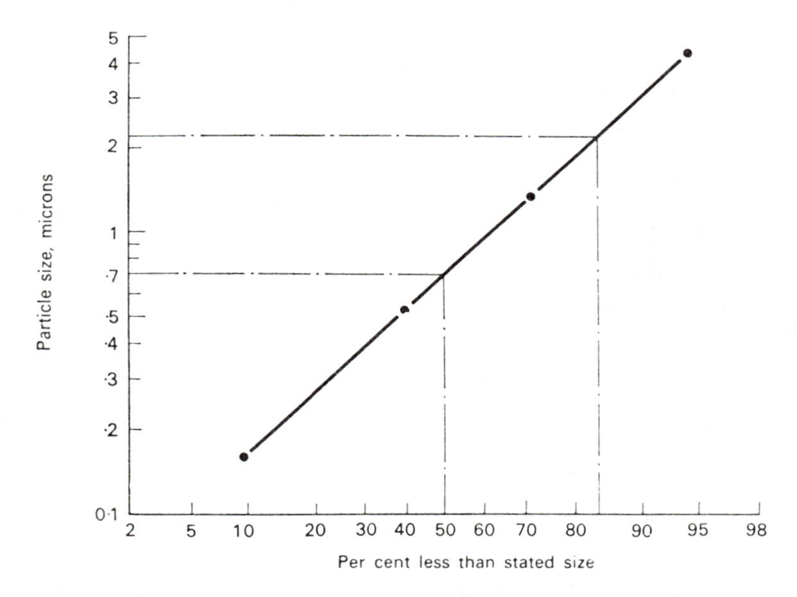

FIG. 13.12. Data of Table 13.3 plotted on log-probability paper. The particle size corresponding to the 50% point is the geometric mean, 0.7 μ, and the ratio of the 84% size to the 50% size gives the geometric standard deviation as 3.

given to the weights on stages 2, 3, and 4, since the upper limit on stage 1 can be very large and the lower limit on stage 5 (the membrane filter) can be very small. From the curve of Fig. 13.12, the mass median diameter is found to be 0.7 μ, and the geometric standard deviation, σ_g, is 3.

The particle size determined with the cascade impactor is the mass median diameter, that is, the diameter such that 50% of the mass (or volume) of the aerosol is in particles less than this diameter. The linear median diameter, D, that is, the diameter such that 50% of the particles are smaller than this diameter, is related to the mass-median diameter by the equation

$$\log D = \log \text{MMD} - 6.9 \log^2 \sigma_g. \tag{13.16}$$

The standard deviation of the distribution of linear diameters is exactly the same as that of the mass-diameters. For the particles in Table 13.3, the linear median diameter is, from equation (13.16), 0.02 microns, and the geometric standard deviation is 3.

Since the size separation of particles in a cascade impactor depends on the velocity of the air in the jets, the sampling rate which for routine use of a cascade impactor is 17 liters/min, is very important. To avoid errors due to variations in the jet velocity, a critical orifice is used to keep the sampling rate constant.

Natural Airborne Radioactivity

Determination of an airborne contaminant is complicated by the existence of naturally occurring airborne radioactivity. This arises from the gaseous isotopes ^{222}Rn (Radon) and ^{220}Rn (Thoron) that seep out of the ground from the ubiquitous uranium and thorium series (the average concentration of uranium and thorium, in the first 0.3 m of soil, is 1 ton and 3 tons respectively per square mile) and from their decay products. These decay products attach themselves to air-borne particulates with which they come in contact, thus contaminating the dust particles with radioactivity. The limiting activity in the radon daughter chain is ^{214}Pb (RaB), a beta emitter whose half-life is 26.8 min. In the thoron daughter chain, the limiting activity is ^{212}Pb (ThB), a 10.6 hr half-lived beta-emitting isotope.

For routine monitoring of long-lived contaminants, allowance is made for these activities by: (1) counting the air sample several hours after collection, thereby allowing the 26.8 min activity to decay away, and (2) counting the air sample about two ThB half-lives later. From these two measurements, the activity of the long-lived contaminant can be calculated. The counting rate, C_1, of the first measurement is due to the natural activity and the long-lived contaminant, C_{LL}:

$$C_1 = C_{n_1} + C_{LL}; \tag{13.17}$$

when the second count is made, the natural activity is reduced to

$$C_{n_2} = C_{n_1} e^{-\lambda \Delta t}, \tag{13.18}$$

where λ, the decay constant of ThB, is 0.0655 per hr, and Δt is the time interval, in hours, between the two counts. The second measurement, C_2, includes the following

$$C_2 = C_{n_2} + C_{LL}, \tag{13.19}$$

$$C_2 = C_{n1} e^{-\lambda \Delta t} + C_{LL}. \tag{13.20}$$

Solving equations (13.19) and (13.20) simultaneously for the long-lived activity gives

$$C_{LL} = \frac{C_2 - C_1 e^{-\lambda \Delta t}}{1 - e^{-\lambda \Delta t}}. \tag{13.21}$$

Example 13.3

In monitoring for air-borne ^{90}Sr, a 10 m^3 air sample is taken with a membrane filter. The first measurement, taken 4 hr after sampling, gives a net counting rate of 100 counts/min. The second count, 20 hr later, gives a net counting rate of 50 counts/min. If the filter (after treatment to prevent electrostatic charge accumulation) was counted in a windowless 2π gas-flow counter, what was the mean concentration of ^{90}Sr in the air?

Substituting the respective values for C_1 and C_2 into equation (13.21) gives

$$C_{LL} = \frac{50 - 100\, e^{-0.0655 \times 20}}{1 - e^{-0.0655 \times 20}}$$

$$= 31.5 \text{ counts per minute}$$

for the long-lived (^{90}Sr) activity. Since the counter has an efficiency of 50% and the volume of air sampled was 10 m^3, the mean concentration of ^{90}Sr is calculated as follows:

$$C = \frac{31.5 \text{ cpm} \times 2 \text{ tpm/cpm}}{60 \text{ tpm/Bq} \times 10 \text{ m}^3}$$

$$= 0.11 \text{ Bq/m}^3 \quad (2.84 \times 10^{-12} \,\mu\text{Ci/cm}^3).$$

If the contaminant is of such half-life that a significant amount will decay during the waiting period, then due allowance must be made for this fact. In this case, the activity in the sample at the times of the first and second counts is:

$$C_1 = C_{n1} + C_{c1}, \tag{13.22}$$

$$C_2 = C_{n2} + C_{c2}, \tag{13.23}$$

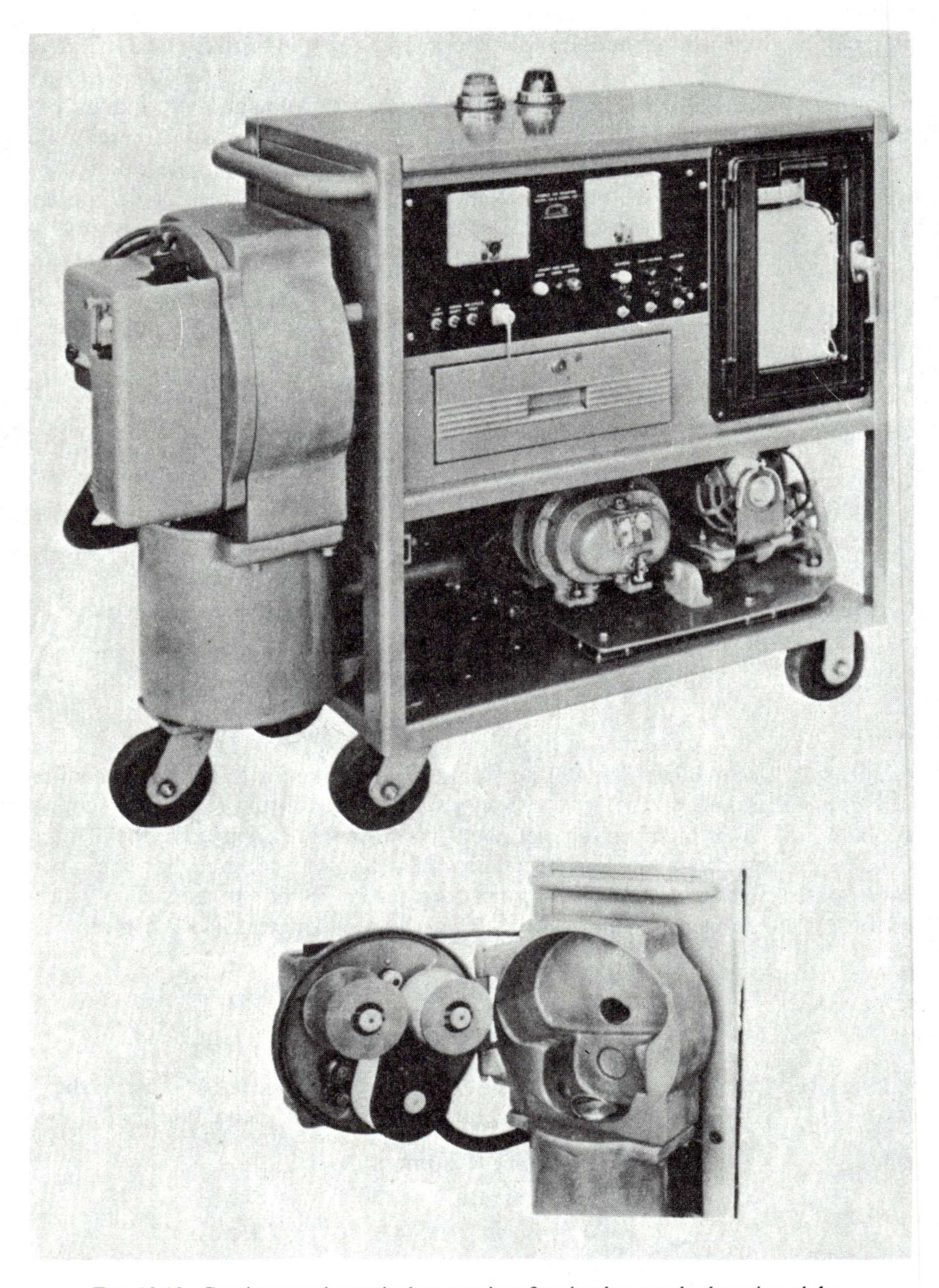

FIG. 13.13. Continuous air particulate monitor for simultaneously detecting alpha, beta, and gamma activity at a rate of 1 ft^3 air per min. The sensitivity at this sampling rate is 10^{-12} μCi/cm^3 of beta–gamma and of alpha activity. (Courtesy Nuclear Measurements Corp.)

where C_n is the naturally occurring activity and C_c is the contaminant. At the time of the second count the natural and contaminating activities will have decayed to

$$C_{n2} = C_{n1}\, e^{-\lambda_n \Delta t}, \tag{13.24}$$

$$C_{c2} = C_{c1}\, e^{-\lambda_c \Delta t}. \tag{13.25}$$

Solving these equations simultaneously gives

$$C_{c1} = \frac{C_2 - C_1\, e^{-\lambda_n t}}{e^{-\lambda_c t} - e^{-\lambda_n t}}. \tag{13.26}$$

Continuous Environmental Monitoring

Continuous monitoring of the environment is usually done if a breakdown of control measures could lead to a serious hazard. For example, if a threat to life could result from a source inadvertently left unshielded, then a continuous radiation monitor coupled to an alarm would be indicated. Continuous monitoring may also be required in a laboratory where low-level activities are measured, and where precise knowledge of fluctuations in the background is therefore necessary. Another application of continuous monitoring is where the amount of radioactivity discharged into the environment must be known—as in the case of gaseous and particulate effluent from an incinerator used to burn radioactive waste or the case of a sewage line from a building in which much liquid radioactive waste is disposed of via the sink.

Continuous environmental monitors fall into three classes, those used to measure radiation levels (these are often called area monitors), those used to measure atmospheric radioactivity, and those used to measure liquid radioactivity.

Area monitoring systems usually consist of an appropriate detector, a ratemeter, a recorder, and, if necessary, an alarm that is actuated when a preset radiation level is exceeded. Liquids or gases may be monitored by letting them flow around or through a suitable detector. Basically, a liquid monitoring system is the same as the area monitoring system—except that the read-out is calibrated to read in activity units, such as microcuries per cm^3 rather than units of radiation dose. For air-borne particulates, air is sucked through a filter at a known rate. The filter is placed in close proximity to an appropriate detector which responds to the radioactive dust caught by the filter, and which is shielded against environmental radiation (Fig. 13.13). The pulses from the detector are measured by a rate meter whose output is recorded on a strip chart. Since radioactivity continues to accumulate on the filter as more air is sampled, the rate-meter reading continues to increase. The index of the degree of atmospheric contamination, therefore, is the rate of increase of the count rate rather than the value of the count rate. Monitors for airborne dust are therefore equipped with a derivative alarm which is actuated when the rate of increase of the counting rate exceeds a preset value.

Problems

1. A series of measurements with threshold detectors showed the following spectral distribution of neutrons:

Energy	Per cent neutrons
Thermal	40
1000 eV	20
10,000 eV	10
0.1 MeV	10
1 MeV	10
10 MeV	10

500 mg ^{32}S was irradiated for 2 hr in this field, and when counted in a 2π counter 24 hr after the end of irradiation gave 500 counts/min. What is the dose rate in this neutron field?

2. A sealed ^{90}Sr source is leak-tested. The wipe, counted in a 2π gas-flow counter, gave 155 counts in 5 mm. The background was 130 counts in 5 min. At the 95% confidence level, is the source contaminated?

3. An air sample on a filter paper was counted in a 2π gas-flow counter, and gave 800 counts in 5 min. A background count gave 260 counts in 10 min. What was the standard deviation of the net counting rate?

4. A radioisotope worker weighing 70 kg inadvertantly drinks water containing 3.7 MBq (100 μCi) ^{22}Na. Following this accidental exposure, his body burden was measured by whole body counts made over a period of 2 months. The following retention function was fitted to the whole body counting data:

$$Q(t) = 1.8 \exp(-0.082t) + 1.9 \exp(-0.052t) \text{ MBq}.$$

Calculate:

(a) The cumulative activity, in Bq days.

(b) The initial dose rate, assuming the ^{22}Na to be uniformly distributed throughout the body (see Fig. 4.8 for ^{22}Na transformation scheme).

5. The maximum permissible skeletal burden of ^{90}Sr is 74 kBq (2 μCi). Calculate the number of transformations per minute per 24-hr urine sample that may be expected from one-fourth of this skeletal burden if 0.05% per day is eliminated in the urine.

6. Using the ICRP two-compartment lung model and the data for the reference person, calculate the ratio of concentration of soluble uranium particulates in the air to uranium in the urine, Bq/m^3 air per Bq/liter urine, for the case where a steady state has been attained through continuous inhalation of the uranium. Assume that all the uranium that is brought up from the lung and is swallowed is eliminated in the feces.

7. The body burden of ^{137}Cs at time t days following a single intake $Q(0)$ is given by

$$Q(t) = Q(0)(0.1\, e^{-0.693t} + 0.9\, e^{-0.011t}).$$

If the ratio of urinary to fecal excretion is 9:1, calculate the activity per 24-hr urine sample 1 day and 10 days after ingestion of 50,000 Bq (1.35 μCi) ^{137}Cs.

8. A chemist accidentally inhaled a ^{14}C-tagged organic solvent that is readily absorbed from the lungs. The solvent is known to concentrate in the liver. That part of the solvent that is eliminated before deposition in the liver leaves in the urine; the detoxification products are eliminated from the liver into the G.I. tract and into the urinary tract; 25% is eliminated in the urine, and 75% in the feces. Following the inhalation, 24-hr urine samples were collected over a 2-week period, and the following data were obtained:

Days after inhalation	1	2	3	4	5	6	8	10	12	14
kBq/sample	98	57	39	26	20	18	12	10	7.4	5.9

(a) How much activity was absorbed into the body?

(b) What was the total dose to the body during the 13 weeks after inhalation?

(c) What was the total dose to the liver during the 13 weeks after inhalation?

9. A health physicist samples waste water to ascertain that the water may be safely discharged into the environment. The water analysis is made by chemically separating the ^{90}Sr, allowing the ^{90}Y daughter to accumulate, then extracting and counting the ^{90}Y activity. The volume of the sample was 1 liter, the ^{90}Y in growth time was 7 days, and the ^{90}Y activity was determined 15 hr after extraction in an internal gas-flow counter having an overall efficiency of 50%. The background counting rate, determined by a 60 minute count was 35 counts/min. The sample (including background) gave 2766 counts in 60 min. What was the ^{90}Sr concentration, at the 90% confidence level?

10. An air sample that was counted 4 hr after collection gave 1450 counts in 10 min. The background was counted for 30 min, and gave a rate of 45 counts/min. The sample was counted again 20 hr later, and gave 990 counts in 10 min; a 60-min background count gave 2940 counts. If the volume of the air sample was 1.0 m^3, and if the counting geometry was 50%, calculate the atmospheric concentration of the long-lived contaminant, Bq/m^3 and μCi/cm^3, and the 95% confidence limits.

11. A film badge worn by a worker in a fast neutron field showed the following distribution of proton recoil tracks among 100 random microscopic fields of 2×10^{-4} cm^2 each:

Observed track per field	Frequency
0	40
1	40
2	18
3	2

(a) If 2600 tracks per cm^2 correspond to 1 mSv (100 mrem), what was the fast neutron dose?

(b) What is the 95% confidence limit of this measurement?

12. Using ICRP lung model and the physiologic data for the standard man, compute dose to the lungs and to the bone following a single acute exposure of 1 Bq-sec (2.7×10^{-5} μCi-sec) per cubic meter of a respirable aerosol, MMAD = 2 μ, of (a) strontium titinate, (b) strontium chloride.

13. The following size distribution was obtained on a sample of an aerosol.

Per cent by number	Class interval, μ
10	0.5–1.0
15	1.0–1.5
15	1.5–2.0
10	2.0–2.5
10	2.5–3.0
10	3.0–3.5
10	3.5–4.5
10	4.5–6.0
5	6.0–8.0
5	8.0–10.0

(a) Plot the cumulative frequency distributions on linear graph paper, on linear probability paper, and on log probability paper, by number, surface area (assume the particles to be spherical), and by mass (assume the particles to have a density of 2.7 g/cm^3).

(b) Are the size distributions normally or log-normally distributed?

(c) Compute the geometric mean and standard deviations for each of the three types of distributions.

14. An instrument repairmen suffered an accidental exposure to ^{131}I while working in a customer's laboratory. Two days later his thyroid gland was found to contain 2×10^4 Bq (0.54 μCi) ^{131}I. Assuming he is a normal healthy man who weighs 70 kg, calculate

(a) the amount of ^{131}I activity originally deposited in the thyroid,

(b) the dose commitment to the thyroid as a result of the accident.

Note: The following thyroid retention function is given in ICRP 10:

$$R(t) = 0.7 \exp\left(-\frac{0.693}{0.35}t\right) + 0.3 \exp\left(-\frac{0.693}{100}t\right).$$

15. A 20-liter breath sample was collected over 2 minutes. Analysis for ^{222}Rn showed the radon concentration to be 1×10^{-7} Bq per liter. Estimate the body burden of ^{222}Ra from these data.

16. A lab worker accidentally ingested ^{210}Po by using a contaminated cup for his coffee. Twenty-four hour urine samples were taken over a 60-day period and analyzed. The following data were obtained:

Days after ingestion	1	5	10	15	20	25	30	40	50	60
Bq per sample	25	23	21	19	18	16	15	12	11	9

(a) Plot the data on semi-log paper and fit an equation to the elimination data.

(b) If 10% of ingested Po is known to be eliminated in the urine, and 90% is eliminated in the feces, how much ^{210}Po was ingested?

(c) If 13% of the ^{210}Po was deposited in the kidneys, what was the committed dose equivalent to the kidneys from this accidental ingestion?

17. What is the dose commitment to the skeleton due to the ingestion of 100 Bq/day, for 1 year, of ^{90}Sr dissolved in drinking water?

18. An accidental release of $^{210}PoO_2$ from a glove box leads to an atmospheric concentration of 1500 Bq/m^3 (4.05×10^{-8} μCi/cm^3). From a recording air monitor whose alarm had failed it was later learned that a worker had been exposed to the airborne $^{210}PoO_2$ for 1 hour. Measurements made with a cascade impactor showed the mass-median aerodynamic particle size (MMAD) to be 0.5 micron. Using the data for the reference man, calculate

(a) the amount of activity deposited in the lung,

(b) the dose commitment to the lung from this accidental exposure.

19. A demineralizer 20-cm-diameter × 20-cm-high processes 200 liters per minute contaminated water, and removes the following long lived isotopes:

Isotope	Bq/liter	μCi/liter
^{60}Co	1.48×10^4	0.4
^{137}Cs	1.11×10^5	3.0
^{144}Ce	1.85×10^6	50.0

The demineralizer operates for 180 days. Thirty days later,

(a) what is the activity of each of these isotopes in the demineralizer?

(b) If the demineralizer approximates a point source at 4 meters, estimate the gamma ray dose rate there.

(c) Estimate the gamma-ray dose rate at the surface of the demineralizer.

20. In accidental releases to the air in a fuel reprocessing plant, the following mixture of isotopes is usually found. Using the MPC values for the air given in 10CFR-20, calculate the atmospheric MPC for the total activity that must be applied during cleanup of the contamination.

Isotope	% of total activity
^{89}Sr	7
^{90}Sr	1
^{91}Y	10
^{95}Zr	15
^{95}Nb	25
^{144}Ce	13
^{147}Pm	2

21. Tritiated water vapor was unknowingly released in a laboratory. An air sample was taken using a freeze-out technic (100% freeze out) when the leak was discovered. Further investigation revealed that the system had been leaking for 24 hours prior to the discovery. Five hundred liters air were drawn through the cold trap, and the collected moisture was diluted to 50 ml. One ml of the dilution was counted for tritium betas in a liquid scintillation counter whose background was 12 cpm and whose counting efficiency was 30%. The 1-ml sample gave 3200 cpm.

(a) What was the tritium concentration in the air?

(b) A technician who had been working in the lab for 8 hours left for a vacation without leaving a urine sample. If the principal route of intake was inhalation, and if all the inhaled tritium was taken up by the technician, estimate her dose commitment. (Use the biological data given for the reference person.)

(c) The technician submitted a urine sample 21 days later. What concentration of tritium would be expected in the urine?

Suggested References

BLATZ, H, ed. *Radiation Hygiene Handbook*. McGraw-Hill, New York, 1959.

BRODSKY, A., ed. *Handbook of Radiation Measurements and Protection*, Vol. II: *Biological and Mathematical Information*. CRC Press, Boca Raton, FL, 1982.

BRODSKY, A., ed. *Handbook of Radiation Measurements and Protection*, Vol. I: Physical Science and Engineering Data. CRC Press, West Palm Beach, Fla., 1978.

CLAYTON, G. D., and CLAYTON, F. E., eds. *Patty's Industrial Hygiene and Toxicology*, Vol. I: *General Principles*. Wiley Interscience, New York, 1978.

CADLE, R. D. *Particle Size*. Reinhold, New York, 1965.

CRALLEY, L. J., and CRALLEY, L. V. *Patty's Industrial Hygiene and Toxicology*, Vol. III: *Theory and Rationale of Industrial Hygiene Practice*. Wiley Interscience, New York, 1979.

DRINKER, P., and HATCH, T. F. *Industrial Dust*. McGraw-Hill, New York, 1954.

EISENBUD, M. *Environmental Radioactivity*, 2nd ed. McGraw-Hill, New York, 1973.

GODBOLD, B. C., and JONES, J. K., eds. *Radiological Monitoring of the Environment*. Pergamon, Oxford, 1965.

GREEN, H. L., and LANE, W. R. *Particlate Clouds: Dusts, Smokes, and Mists*. D. Van Nostrand, Princeton, 1964.

HOBSON, W. *The Theory and Practice of Public Health*. Oxford University Press, London, 1965.

KAMATH, P. R. *The Environmental Surveillance Laboratory*. W.H.O., Geneva, 1970.

LANZL, L. H., PINGEL, J. H., and RUST, J. H. *Radiation Accidents and Emergencies in Medicine, Research, and Industry*. Charles Thomas, Springfield, Ill., 1965.

MORGAN, K. Z., and TURNER, J. E., eds. *Principles of Radiation Protection*. John Wiley, New York, 1967.

NORWOOD, W. D. *Health Protection of Radiation Workers*. Charles Thomas, Springfield, Ill., 1975.

OLISHIFSKI, J. *Fundamentals of Industrial Hygiene*, 2nd ed. National Safety Council, Chicago, 1979.

SAENGER, E. L., ed. *Medical Aspects of Radiation Accidents*. U.S.A.E.C., Washington, D.C., 1963.

SARTWELL, P. E., ed. *Preventive Medicine and Public Health*, 10th ed. Appleton-Century-Crofts, New York, 1973.

SHAPIRO, J. *Radiation Protection*. Harvard University Press, Cambridge, Mass., 1972.

STRAUB, C. P. *Public Health Implications of Radioactive Waste Releases*. W.H.O., Geneva, 1970.

U.S. Department of Energy. *EML Procedures Manual*. HASL-300, Environmental Measurements Laboratory, New York.

U.S. Public Health Service. *Radioassay Procedures for Environmental Samples*. U.S.P.H.S. Publication 999-RH-27, Washington, D.C., 1967.

World Health Organization. *Protection of the Public in the Event of Radiation Accidents*, W.H.O., Geneva, 1965.

World Health Organization. *Diagnosis and Treatment of Acute Radiation Injury*. W.H.O., Geneva, 1961.

ZENC, C. *Occupational Medicine: Principles and Practical Applications*. Year Book Medical Publishers, Chicago, 1975.

The following publications in the Safety Series, International Atomic Energy Agency (IAEA), Vienna:

Safety
Series
No.

1. *Safe Handling of Radioisotopes* (1962).
2. *Safe Handling of Radioisotopes: Health Physics Addendum* (1960).

3. *Safe Handling of Radioisotopes: Medical Addendum* (1960).
6. *Regulations for the Safe Transport of Radioactive Materials* (1979).
8. *The Use of Film Badges for Personnel Monitoring* (1962).
14. *Basic Requirements for Personnel Monitoring* (1980).
16. *Manual on Environmental Monitoring in Normal Operation* (1966).
18. *Environmental Monitoring in Emergency Situations* (1966).
21. *Risk Evaluation for Protection of the Public in Radiation Accidents* (1967).
25. *Medical Supervision of Radiation Workers* (1968).
32. *Planning for the Handling of Radiation Accidents* (1969).
37. *Advisory Material for the Application of the IAEA Transport Regulations* (1982).
38. *Radiation Protection Procedures* (1973).
41. *Objectives and Design of Environmental Monitoring Programmes for Radioactive Contaminants* (1975).
42. *Radiological Safety Aspects of the Operation of Neutron Generators* (1976).
43. *Manual on Radiological Safety in Uranium and Thorium Mines and Mills* (1976).
45. *Principles for Establishing Limits for the Release of Radioactive Materials into the Environment* (1978).
46. *Monitoring of Airborne and Liquid Radioactive Releases from Nuclear Facilities to the Environment* (1978).
47. *Manual on Early Medical Treatment of Possible Radiation Injury* (1978).
49. *Radiological Surveillance of Airborne Contaminants in the Working Environment* (1979).

The following reports of the International Commission on Radiation Units and Measurements (ICRU), Washington, D.C.:

ICRU
Report
No.

20. *Radiation Instrumentation and its Application* (1971).
22. *Measurement of Low-Level Radioactivity* (1972).

The following reports of the National Council on Radiation Protection and Measurements (NCRP), Washington, D.C.:

NCRP
Report
No.

32. *Radiation Protection in Educational Institutions* (1966).
35. *Dental X-ray Protection* (1970).
36. *Radiation Protection in Veterinary Medicine* (1970).
38. *Protection against Neutron Radiation* (1971).
39. *Basic Radiation Protection Criteria* (1971).
40. *Protection Against Radiation from Brachytherapy Sources* (1972).
47. *Tritium Measurement Techniques* (1976).
48. *Radiation Protection for Medical and Allied Health Personnel* (1976).
50. *Environmental Radiation Measurements* (1976).
57. *Instrumentations and Monitoring Methods for Radiation Protection* (1978).
58. *A Handbook of Radioactivity Measurements Procedures* (1978).
59. *Operational Radiation Safety Program* (1978).
62. *Tritium in the Environment* (1979).
65. *Management of Persons Accidentally Contaminated with Radionuclides* (1979).

The following publications of the International Commission on Radiological Protection (ICRP), Pergamon Press, Oxford:

ICRP
Publication
No.

7. *Principles of Environmental Monitoring Related to the Handling of Radioactive Materials* (1965).
10. *Report of Committee IV on Evaluation of Radiation Dose to Body Tissues from Internal Contamination due to Occupational Exposure* (1968).
10A. *The Assessment of Internal Contamination Resulting from Recurrent or Prolonged Uptakes* (1971).
12. *General Principles of Monitoring for Radiation Protection of Workers* (1969).
13. *Radiation Protection in Schools for Pupils up to the Age of* 18 *years* (1970).
24. *Radiation Protection in Uranium and Other Mines* (1977).
28. *The Principles and General Procedures for Handling Emergency and Accidental Exposures of Workers* (1978).
29. *Radionuclide Release to the Environment: Assessment of Doses to Man* (1979).

CHAPTER 14

NON-IONIZING RADIATION

The hazards from non-ionizing radiation received little attention before the end of World War II. At that time, we already has a good deal of experience with damage to the eye from observing solar eclipses, from exposure to ultraviolet light among welders, and from exposure to infrared energy among glass blowers and steel workers. We also had evidence of damage to the skin from exposure to ultraviolet and infrared radiation. However, it was the post-war boom in electronics and communications based on the microwave portion of the electromagnetic spectrum, followed by the mushrooming use of lasers, that focused attention on the possible public health aspects of non-ionizing radiation, especially from these two sources of radiant energy. In 1968, the Radiation Control for Health and Safety Act (Public Law 90–602) was passed by the Congress of the United States for the purpose of regulating the hazards from consumer electronic produces, and in 1970 the Occupational Safety and Health Act (PL 91-596) was passed to protect workers from hazards, including ionizing and non-ionizing radiation hazards, associated with their occupations. These legislative acts, and acts in other countries, have led to the promulgation of safety regulations for microwaves and for lasers.

Units

Illumination is measured in the *photometric* system units. Photometric units relate the light to the response of the human eye to light. *Radiometric* units on the other hand, are absolute physical units that are defined for the entire electromagnetic spectrum (Table 14.1). Safety standards and criteria for laser energy and for microwave energy are specified in the radiometric system of units.

TABLE 14.1. RADIOMETRIC UNITS

Quantity	Symbol	Description	Units
Radiant energy	U	Energy emitted from the source, per pulse	joule, J
Radiant power	P	Power emitted from the energy source	watt, W
Radiant intensity	I	Radiant power emitted per unit solid angle	W/sr
Radiance	L	Power emitted from the source per unit solid angle per unit area	$\frac{W/sr}{cm^2}$
Radiant emittance	W	Power emitted per unit area of the source	W/cm^2
Radiant exposure	H	Areal density of total radiant energy incident on a surface	J/cm^2
Irradiance*	E	Areal density of power incident on a surface	W/cm^2

*In the context of lasers, power per unit area is called "irradiance," in the context of microwaves, power per unit area is called power density."

Lasers

The word laser is an acronym for Light Amplification by Stimulated Emission of Radiation. The laser is a device for producing a beam of monochromatic "light," in the ultraviolet, visible, or infrared regions of the electromagnetic spectrum in which the waves are all in phase. That is, the light beam is said to be *coherent* in both space (since the waves are all in phase) and time (since the waves are all of the same frequency). As a result of this coherency, we have a beam that has relatively little divergence together with a high concentration of energy per unit area of the beam, both at the laser end and at the far end of the beam. This almost constant power density at both ends of the beam (which differs sharply from the inverse-square fall-off of the intensity from a point source of incoherent light or of ionizing radiation, but see problem 3) is a significant factor in the hazard potential of a laser.

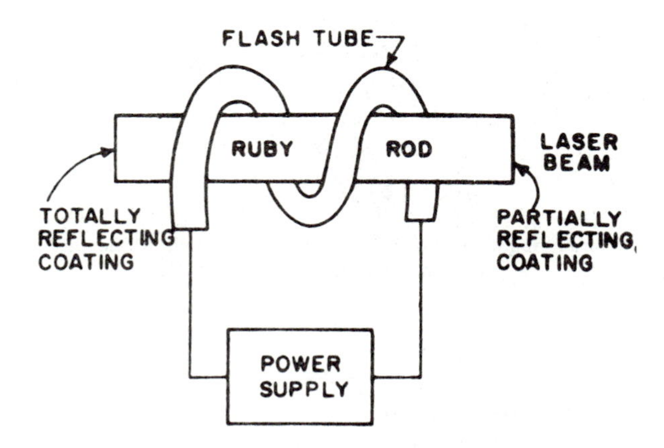

FIG. 14.1. Components of a laser system, as illustrated by a schematic of a ruby laser with optical pumping, which is characteristic of solid state lasers.

All lasers include the three basic components (Fig. 14.1):

1. A high-Q optical cavity with one end completely mirrored and the other end partially mirrored. Q, as used here, has the same physical significance as the Q used to describe the quality of certain circuit configurations in electronics. Q is a figure of merit that gives the ratio of the energy stored in a particular device or circuit configuration to the energy dissipated per unit time interval:

$$Q = \frac{\text{energy stored}}{\text{energy dissipated/unit time interval}} = \frac{\text{energy stored}}{\text{dissipated power}}. \tag{14.1}$$

2. Lasing medium. The lasing medium is a substance that can be excited to a metastable state through the addition of energy that is "pumped" into the lasing medium. The lasing medium may be either in solid, liquid, or gaseous form. Commonly used lasing media include ruby rods (a ruby consists of Al_2O_3 crystals in which about 0.5% of the Al has been replaced by Cr; the chromium atoms are the atoms which are excited to the metastable state, and thus are responsible for the lasing action), neodymium in glass, gallium-arsenide, helium-neon, argon, and carbon dioxide.

3. Energy pump. The source of energy needed to excite the atoms of the lasing medium may be either an intense source of light that emits a wide range of photon energies, and

necessarily includes photons of exactly the right energy to excite the lasing atoms (Fig. 14.2); or in the case of gas laser, a radio frequency voltage generator of about 1000 V accelerates ions, which in turn excites the lasing atoms by colliding with them. Semiconductor (or diode) lasers, such as galium-arsenide, are pumped by passing an electric current of very high density, on the order of hundreds to thousands of amperes per square centimeter, across the P–N junction of the semi-conductor.

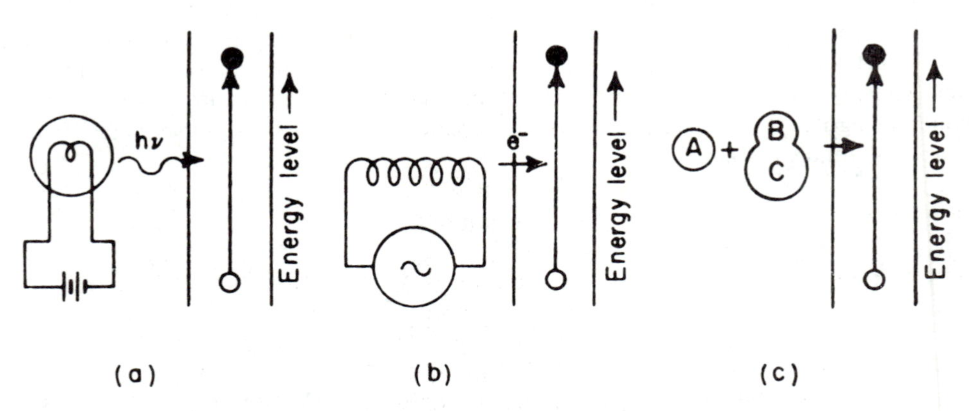

FIG. 14.2. Energy pumping systems for lasers. (a) Light (optical), (b) Electron collision (radiofrequency), (c) Chemical.

Laser Operation

According to the Bohr atomic model (Chapter 3), photons of ordinary light are produced when excited electrons fall to a lower energy level. These electrons may be excited to the higher unstable energy levels through the addition of energy by one of several ways:

1. By absorbing energy from photons, as in the case of fluorescence,
2. By absorbing energy from charged particles, as in the case of luminescent paint, or cathode ray tube phosphors,
3. By heating, as in the case of an incandescent light bulb, or a piece of metal or glass heated to a high temperature,
4. By collisions with other electrons, as in the case of a fluorescent lamp or a "neon" sign,
5. By exothermic chemical reactions, as in the case of a flame.

The electronic transition, in the case of ordinary light, occur randomly, and the photons, as a consequence, are unrelated to each other. In a laser on the other hand, the electrons are excited by an energy "pump" into a relatively long-lived metastable state, where they remain until a passing photon of exactly the correct energy "stimulates" a transition to the lower energy level, and all the excited atoms emit photons of the same energy at the same time. Einstein, in the development of the theory underlying the photoelectric effect (for which he was awarded the Nobel Prize in Physics in 1921), showed that a photon whose energy is exactly equal to that of an electron in an excited state, can stimulate the excited electron to fall to the ground state and thus to emit a photon whose frequency corresponds to the excitation energy. Not only are the emitted and the stimulating photons of the same frequency, they are also in phase.

Lasing Action

Under normal conditions, most of the atoms in any medium are in the ground state. Brownian motion leads to collisions among the atoms in which sufficient energy to raise an atom to an excited level may be transferred. Thus, although most of the atoms are in the ground state, some atoms may be in one of the several possible excited states. The relationship between the number of atoms in any two energy levels is given by the Boltzmann equation

$$N_2 = N_1 \, e^{(E_2 - E_1)/kT}, \tag{14.2}$$

where N_1 and N_2 are the numbers of atoms in energy levels E_1 and E_2, respectively; k is Boltzmann's constant, 1.38×10^{-23} J/degree K, and T is the absolute temperature, degrees K.

In materials where the atoms can be excited to a metastable state, it is possible, by "pumping" large amounts of energy into them, to attain a population inversion, in which most of the atoms are in an excited state.

After a population inversion has been obtained, lasing action is initiated by a photon that is emitted from an excited atom whose electron spontaneously falls to the ground state. This photon then stimulates another excited atom to emit a photon by falling to a lower energy level. Most of these stimulated photons strike the walls of the optical cavity and are lost. Those photons, however, which are released in a direction parallel to the long axis of the optical cavity continue to stimulate emission and to combine coherently with the emitted photons until they strike one of the mirrored ends of the optical cavity. Thus, as the photons progress within the optical cavity, the beam intensity continuously increases, while the phase relationships remain constant, thereby maintaining the coherence of the beam. When the beam strikes the totally reflecting end, its direction is reversed, and it continues to stimulate emission of photons and to increase in intensity until it reaches the partially reflecting end. There, some of the beam escapes, and the remainder is reflected back to continue the process of stimulated photon emission. This lasing action continues as long as energy is supplied to the lasing medium in order to excite the atoms and thus to maintain a population inversion. The wavelength of this light depends on the difference between the energy level of the metastable state and of the lower energy level of the lasing medium. For a ruby laser, the wavelength is 694 nm, for a He–Ne laser the

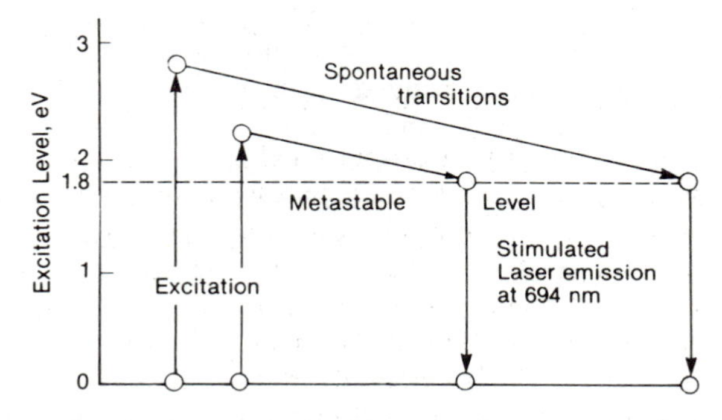

FIG. 14.3. Energy level diagram for a ruby laser. Electrons are excited to a high energy level by the pumping system. They then fall down spontaneously to the 1.8-eV metastable level, from which they are stimulated, and they all cascade simultaneously to the ground state.

wavelength is 633 nm, and for a CO_2 laser the emitted infrared energy has a wavelength of 1060 nm. Fig. 14.3 shows schematically these energy levels for the ruby laser; similar energy level diagrams may be constructed for other lasing materials.

The laser is an extremely inefficient device. In most types of lasers, less than 0.1% of the energy pumped into the system is converted to useful coherent radiation.

Manner of Operation

Lasers operate in one of three different manners:

1. Continuous wave (cw),
2. Pulsed (long pulse or normal pulse),
3. Q-switched (or Q-spoiled).

The optical cavity has one end that is completely reflecting and one end that is partially reflecting and partially transmitting. If the partially transmitting end allows a fraction of the light energy that strikes it to escape, and if energy can be pumped into the lasing medium at such a rate that the laser output can be maintained uninterruptedly, then we have a *continuous wave* laser. Most cw lasers employ gas as the lasing medium, although solid state lasers can also be made to operate in the cw mode. The first cw laser used a mixture of helium and neon (He–Ne), where the He atoms were excited by an applied radio frequency high voltage. The excited He atoms, in turn, raised the Ne atoms to a metastable state by colliding with them. He–Ne cw lasers operate at power levels ranging from a fraction of a milliwatt to about 50 milliwatts. At the high end of the cw power spectrum are CO_2 lasers, which operate at power levels up to about 10,000 watts.

Normally pulsed lasers deliver their output in bursts of light whose duration is on the order of about 0.1 to 10 milliseconds. The ruby laser is an example of this operating mode. A very intense flash of light, produced by discharging large capacitors across the flash-lamp, pumps the lasing medium and creates a population inversion. The resultant stimulated radiation builds in intensity and emerges as a long pulse of coherent laser radiation. The energy content of a normal pulse from ruby lasers varies from a fraction of a joule to about 30 joules, while the pulse repetition frequency (PRF) ranges from about 1/30 to 10 pulses per second.

A pulsed laser may also be operated in another manner, called *Q-switched* (or Q-spoiled). A Q-switch is an acousto-optical or an electro-optical device within the optical cavity that is analogous to a shutter; it prevents laser emission until it is opened. In Q-switching, the Q of the optical cavity is suddenly increased from a low-value, when lasing does not occur despite a large population inversion, to a high Q, when lasing can occur. The very large population inversion built-up during the low-Q part of the operation suddenly falls to the ground state in a very short time, on the order of nanoseconds, to produce a very intense pulse. Because of the combination of high energy and narrow pulse width, very high powers, on the order of megawatts, are readily attainable with Q-switched lasers.

TEM Modes

The distribution of light energy across the laser beam is determined by diffraction effects within the optical cavity. It is described by the *mode pattern* of the transverse electromagnetic waves, which is designated as TEM_{pq}, where p and q are integers. The TEM_{00} mode, which corresponds to a circular beam of laser light, has the least amount of diffraction, and is the main mode of oscillation within the optical cavity. The light intensity across a TEM_{00} laser beam approximates a Gaussian distribution. Because of this variation in intensity across the beam, and because the edge of the beam is not sharply defined as a result of diffraction effects at the edges of the mirrors, an arbitrary definition of the beam diameter is frequently used. One such value for describing the beam diameter when the laser is operating in the TEM_{00} mode is the $1/e^2$ power point. This is defined as the diameter of the circle that intercepts $1\text{-}1/e^2 = 0.865$ of the energy in the laser beam. Beam cross sections for several different TEM modes are shown in Fig. 14.4.

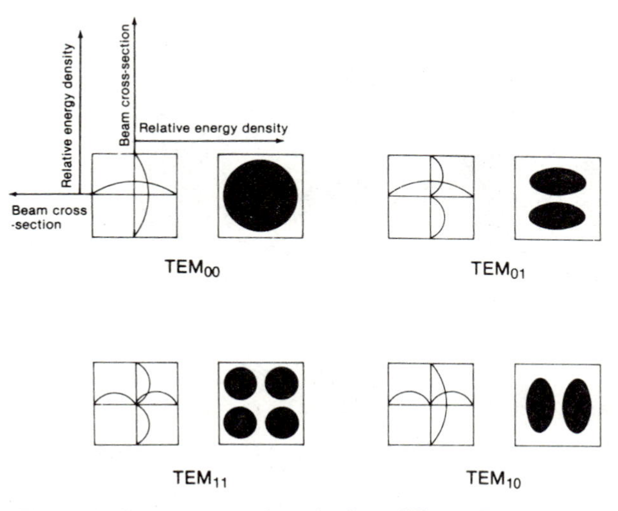

FIG. 14.4. Beam cross sections for four different TEM modes.

Biological Effects

Light, both visible and ultraviolet, have long been known to be biophysically active. Tanning and sunburn from exposure to sunlight are common experiences, and the use of ultraviolet lamps for bacteriocidal purposes is widespread. In an early 20th-century study on the lethal effects of filtered (by window glass of various colors) sunlight on the tubercle bacillus, the following killing times were reported:

TABLE 14.2. LETHAL EFFECTIVENESS OF COLORED LIGHT AGAINST THE TUBERCLE BACILLUS

Color	Killing time
Clear	5–10 minutes
Blue	10–20
Red	20–30
Green	45

For killing bacteria and bacterial spores, U.V. in the wavelength range of 250–270 nm has been found most effective. For *E. coli*, for example, the most efficient wavelength is 265 nm; at this wavelength, the absorption of 14-MeV total energy is required to kill the bacterium.

Mechanisms of biological damage from light include both temperature effects due to absorbed energy and photochemical reactions. The chief mode of damage depends on the wavelength of the light and on the tissue being exposed. For control of hazards from lasers, the principal mode of damage is believed to be due to temperature effects, and the critical organs are the eye and the skin.

The Eye

The structure of the eye is shown in Fig. 14.5. The optical system of the eye includes

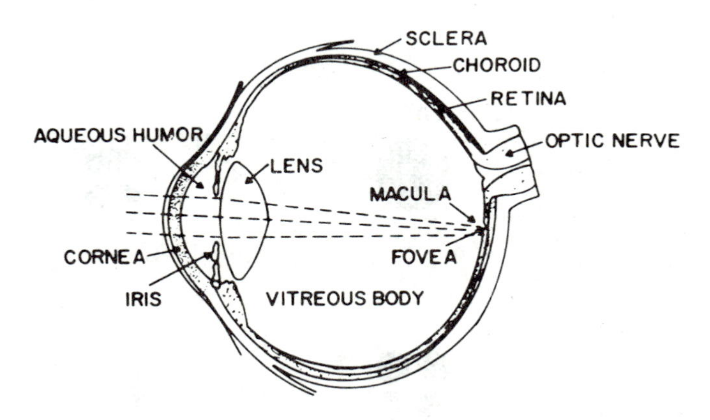

FIG. 14.5. The gross structures of the human eye.

those tissues which act together to focus a real image of the object on the retina. The system includes:

1. The cornea, a transparent layer about 0.5 mm thick with a mean index of refraction of 1.376. The anterior surface has a radius of curvature of about 7.7 mm, while the radius of curvature of the posterior surface is about 6.8 mm.
2. The aqueous humor, a clear, transparent, dilute solution ($>99\%$ water) of albumium, globulin, and sugar; its index of refraction is 1.336.
3. The lens, a biconvex, clear transporent semi-solid tissue encased in a transparent membrane called the *capsule*; its index of refraction is 1.413. Focusing of the eye is accomplished through a process called *accommodation* by thickening or elongating the lens, thus charging the radii of curvature.
4. The vitreous humor, a soft, jellylike, clear, transparent substance that fills the eye between the crystalline lens and the retina. It consists of about 99% water and has an index of refraction of 1.336.

When the eye is illuminated, the light incident on the cornea is concentrated by the eye's optical system to form an image on the retina of very much greater light intensity than that on the cornea. The light enters through the pupil, whose diameter is varied by the iris diaphragm according to the intensity of the light and the age of the person. This dependency is shown in Table 14.3. After passing through the pupil the light is focused by the lens on the retina, where the light is transduced to nerve impulses that are transmitted to the brain via the optic nerve (second cranial nerve) to give the sensation of sight.

The retina is a complex structure which is, in fact, an extension of the *optic disc*, its point of entry into the interior of the eyeball. On the visual axis, the retina is formed into a slightly elevated yellow spot, about 0.6 mm in diameter, called the *macula lutea*; in the center of the macula is a small depressed area called the *fovea centralis*.

TABLE 14.3. MEAN PUPIL DIAMETER

Age	Daylight	Night-time	Difference
20	4.7 mm	8.0 mm	3.3 mm
30	4.3	7.0	2.7
40	3.9	6.0	2.1
50	3.5	5.0	1.5
60	3.1	4.1	1.0
70	2.7	3.2	0.5
80	2.3	2.5	0.2

*From M. Luckiesh and F. K. Moss, *The Science of Seeing*, D. Van Nostrand and Co., New York (1937).

Sharp vision is dependent on the formation of a real image on the macula. Two types of photoreceptor nerve endings are found in the retina: the *rods* and the *cones*. The cones are concentrated mainly in the fovea, and serve to resolve fine details and to discriminate among the various hues of color. However, they are relatively insensitive, and function only under conditions of good illumination. The rods, on the other hand, are much more light-sensitive than the cones, and thus are useful in dim illumination and for "night vision." However, the rods neither resolve fine details nor discriminate among hues. At low levels of illumination, therefore, things tend to appear fuzzy and grayish regardless of their color. The retina is transparent through the layers of photoreceptors; the last layer consists of opaque pigmented epithelium. Behind the pigmented epithelium is a layer of highly vascularized (rich in blood vessels) tissue called the *choroid*, which serves as the source of nutrition for the retina and to carry away the cytometabolic wastes from the retinal cells. The retina itself is not vascular. Heat energy absorbed in the retina, must therefore by conducted to the choroid before it can be transferred to the heat-exchange medium, the blood, for removal.

Eye damage

Laser irradiation of the eye may cause damage to the cornea, the lens, or the retina, depending on the wavelength of the light and the energy absorption characteristics of the ocular tissues. Fig. 14.6 shows the percent transmission of light through the occular media as a function of wavelength. The figure shows that most of the visible part of the electromagnetic spectrum is transmitted, and that the transmission of the near ultraviolet and the near infrared drops very sharply. Fig. 14.7 shows that the visible light that is transmitted by the ocular media is strongly absorbed in the retina. Because of these transmission and absorption characteristics, we may infer that visible light is much less likely to damage the cornea, per unit of incident light energy, than to damage the retina. This expectation has been experimentally confirmed. Although functional changes in vision have been reported following prolonged exposure of Rhesus monkeys to levels of laser light as low as 3.1×10^{-5} W/cm^2 on the retina for three hours per day over a seven-day period, chorioretinal burns have been found to be the chief eye hazard from lasers operating in the visible region. For purposes of setting laser safety standards, the criterion

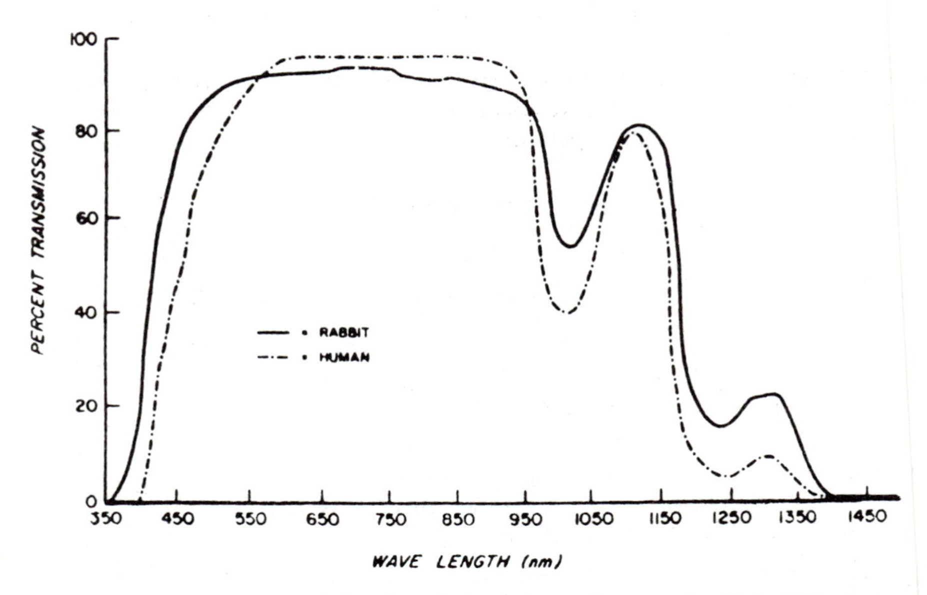

FIG. 14.6. Percent transmission through the ocular media vs wavelength for light of uniform intensity incident on the cornea; comparison between man and rabbit. (From Geeraets, W. J. and Berry, E. R. Ocular spectral characteristics as related to hazards from lasers and other light sources, *Am. J. Ophth.*, **66**:15, 1968.)

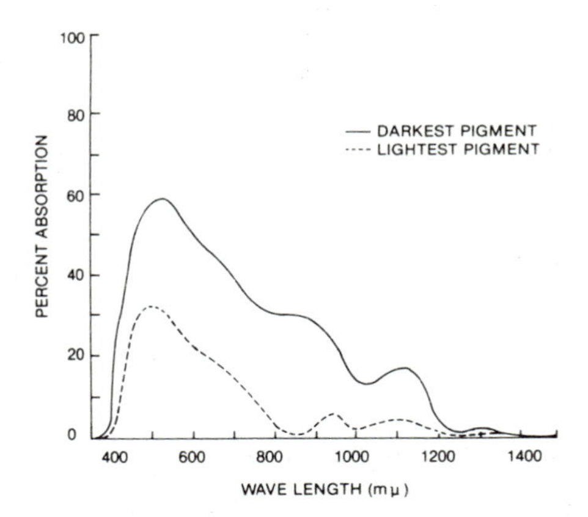

FIG. 14.7. Graph of percent absorption in human pigment epithelium vs wavelength for light of uniform intensity incident on the cornea; plots are for the lightest and darkest pigmented eyes studied. (From Geeraetes, W. J. Berry, E. R. Ocular spectral characteristics as related to hazards from lasers and other light sources, *Am. J. Ophth.*, **66**:15, 1968.)

for damage is a retinal lesion visible through an ophthalmoscope. The magnitude of the damage and the latent period before a visible lesion is seen are functions of the irradiated area, exposure rate, and total exposure. The basic factor in retinal damage is the rate at which heat energy can be removed from the irridiated tissue and the consequent temperature change. A temperature increase of only several degrees higher than that experienced during fevers is believed to be capable of producing permanent retinal damage.

Threshold values for retinal damage from visible light are listed in Table 14.4. Although the values listed are for specific wavelengths, it is reasonable to assume little or no difference over the entire visible spectrum.

TABLE 14.4. RETINAL DAMAGE THRESHOLDS

Laser type	Wavelength	Pulse	Level
Continuous wave	white light	—	6 W/cm^2
Normal pulse	694 nm	200 microsec	0.85 J/cm^2
Q-switched pulse	694 nm	30 nanosec	0.07 J/cm^2

From D. H. Sliney and W. A. Palmisano. The Evalualtion of laser hazards, *AIHA Jour.*, **20**:425, 1968.

Skin damage

Acute exposure of the skin to sufficiently high amounts of energy from laser may lead to burns that do not differ from ordinary thermal or solar burns. The overexposed skin undergoes non-specific coagulation necrosis whose extent depends on the degree of overexposure. The incident radiant energy is converted to heat which is not rapidly dissipated because of the poor thermal conductivity of the tissue. The resulting local temperature rise leads to denaturation of the tissue proteins. If enough energy is absorbed, the water in the tissue may be vaporized, and the tissue itself may be heated to incandensence and carbonized. The response of the skin to laser light increases as the degree of pigmentation increases. In addition to the inherent optical properties of the skin (reflection, transmission, and absorption), injury to the skin depends on the wavelength of the laser light and the exposure time Table 14.5 lists the minimal reactive doses, MRD, to the flexor surface forearm of a Caucasian adult from laser light under several different conditions of exposure.

TABLE 14.5. SKIN DAMAGE THRESHOLDS

Laser	λ, nm	Exposure time	Area, cm^2	MRD, J/cm^2
Ruby, normal pulse	694	0.2 millisec	$2.4\text{–}3.4 \times 10^{-3}$	14–20
Argon	500	6 sec	95×10^{-3}	13–17
CO_2	1060	4–6 sec	1	4–6
Ruby, Q-switched	694	10–12 nanosec	0.33–1.0	0.5–1.5

From L. Goldman. The Skin, *Arch. Environmental Health*, **18**:435 (1969).

Scars may develop when severe lesions from acute overexposure heal. Chronic low-level exposure to laser light generally does not lead to injury. Neither acute overexposure or chronic low-level exposure has yet been found to be associated with increased incidence of cancer. It should be pointed out, however, that ultraviolet light can lead to skin cancer. The exact portion of the UV spectrum for photocarcinogenesis has not yet been determined. However, experimental evidence suggests that in the absence of photo-sensitizing substances, UV in the wavelength range of 290 to 320 nm is carcinogenic. Epidemiological studies show that exposure to sunlight, which includes this band of wavelengths, is associated with an increased incidence of skin cancer, including melanoma, a highly malignant form of skin cancer. These studies show the incidence of all types of skin cancer to double about every 8–10° latitude as the equator is approached.

Protection Guides and Standards

In setting ocular safety standards the focusing action of the crystalline lens must be considered. The total amount of light energy entering into the eye is determined by the area of the pupillary opening. The transmitted light energy that reaches the retina is absorbed by the pigmented epithelium, where most of it is converted into heat. Because of the focusing action of the lens, the image of the limiting aperture (the pupillary opening) formed on the retina is very much smaller than the pupillary opening. For light of wavelength λ cm, and an eye whose pupillary diameter is d_p and whose lens has a focal length f cm, the diameter of the image on the retina, d_r, is given by

$$d_r = \frac{2.44\,\lambda f}{d_p}. \tag{14.3}$$

Since radiant exposure (H) or irradiance (E) is related to the illuminated area by

$$H\ (\text{or } E) = \frac{\text{energy}}{\text{area}}, \tag{14.4}$$

the ratio of H or E of the cornea to that of the retina varies inversely with the square of the ratio of the pupillary diameter to the diameter of the image on the retina:

$$\frac{H \text{ or } E(\text{retina})}{H \text{ or } E(\text{cornea})} = \left(\frac{d_p}{d_r}\right)^2. \tag{14.5}$$

Example 14.1

A Q-switched ruby laser, $\lambda = 694.3$ nm, emits 15 J per pulse. If a pulse of this radiation, in a beam 1.6 cm in diameter, were accidentally to fall on an eye whose iris was opened to 7 mm diameter, calculate the radiant exposure at (a) the cornea and (b) the retina.

(a)

$$H = \frac{\text{energy}}{\text{area}} = \frac{15\text{ J}}{(\pi/4)(1.6\text{ cm})^2} = 7.46\text{ J/cm}^2.$$

(b) The diameter of the image on the retina, if the focal length of the lens is 1.7 cm is, from equation (14.4),

$$d_r = \frac{2.44 \times 6.943 \times 10^{-5}\text{ cm} \times 1.7\text{ cm}}{0.7\text{ cm}} = 4.11 \times 10^{-4}\text{ cm},$$

and the radiant exposure at the retina is, from equation (14.3)

$$H = 7.46\frac{\text{J}}{\text{cm}^2}\left(\frac{0.7\text{ cm}}{4.11 \times 10^{-4}\text{ cm}}\right)^2 = 2.16 \times 10^7\text{ J/cm}^2$$

The maximum permissible exposure of the cornea must allow for this enormous concentration of light energy due to the focusing action of the lens. On the basis of retinal damage thresholds and concentration of light by the crystalline lens, maximum permissible exposure limits have been recommended by the American Conference of Government industrial Hygienists (A.C.G.I.H.) and by the American National Standards Institute (A.N.S.I.). These two sets of recommended limits are essentially equivalent. Portions of

TABLE 14.6. MAXIMUM PERMISSIBLE EXPOSURE (MPE) FOR DIRECT OCULAR EXPOSURES, INTRABEAM VIEWING, FROM A LASER BEAM

λ, μm	Exposure time, t, seconds	MPE
0.400–0.700	10^{-9}–1.8×10^{-5}	5×10^{-7} J/cm^2
0.400–0.700	1.8×10^{-5}–10	$1.8t^{\frac{3}{4}} \times 10^{-3}$ J/cm^2
0.400–0.550	10–10^4	10×10^{-3} J/cm^2
0.550–0.700	10–T_1	$1.8t^{\frac{3}{4}} \times 10^{-3}$ J/cm^2
0.550–0.700	T_1–10^4	$10\, C_B \times 10^{-3}$ J/cm^2
0.400–0.700	10^4–3×10^4	$C_B \times 10^{-6}$ W/cm^2

Notes: $C_B = 1$ for $\lambda = 0.400$–0.550 μm; $C_B = 10^{[15(\lambda - 0.550)]}$ for $\lambda = 0.550$–0.700 μm; $T_1 = 10 \times 10^{[20(\lambda - 0.550)]}$ sec for $\lambda = 0.550$–0.700 μm. The values listed in the table are based on a 7-mm maximum aperture diameter for averaging the radiant exposure. From ANSI Z136.1 (1976).

TABLE 14.7. MAXIMUM PERMISSIBLE EXPOSURE (MPE) FOR VIEWING A DIFFUSE REFLECTION OF A LASER BEAM OR AN EXTENDED SOURCE LASER

λ, μm	Exposure time, t_1, seconds	MPE
0.315–0.400	10^{-9}–10	$0.56\, t^{\frac{1}{4}}$ J/cm^2
0.315–0.400	10–10^3	1 J/cm^2
0.315–0.400	10^3–3×10^4	0.001 W/cm^2
0.400–0.700	10^{-9}–10	$10\, t^{\frac{1}{3}}$ J/cm^2/sr
0.400–0.550	10–10^4	21 J/cm^2/sr
0.550–0.700	10–T_1	$3.83\, t^{\frac{3}{4}}$ J/cm^2/sr
0.550–0.700	T_1–10^4	$21\, C_B$ J/cm^2/sr
0.400–0.700	10^4–3×10^4	$2.1\, C_B\, t \times 10^{-3}$ W/cm^2/sr

Notes: $C_B = 1$ for $\lambda = 0.400$–0.550 μm; $C_B = 10^{[15(\lambda - 0.550)]}$ for $\lambda = 0.550$–0.700 μm; $T_1 = 10 \times 10^{[20(\lambda - 0.550)]}$ for $\lambda = 0.500$–0.700 μm. From ANSI Z136.1 (1976).

TABLE 14.8. MAXIMUM PERMISSIBLE EXPOSURE (MPE) FOR SKIN EXPOSURE TO A LASER BEAM

λ, μm	Exposure time, t, seconds	MPE
0.315–0.400	10^{-9}–10	$0.56\, t^{\frac{1}{4}}$ J/cm^2
0.315–0.400	10–1000	1 J/cm^2
0.315–0.400	10^{-3}–3×10^4	0.001 W/cm^2
0.4 –1.4	10^{-9}–10^{-7}	0.02 J/cm^2
0.4 –1.4	10^{-7}–10	$1.1\, t^{\frac{1}{4}}$ J/cm^2
0.4 –1.4	10–3×10^4	0.2 W/cm^2
1.4 –1000	10^{-9}–10^{-7}	0.02 J/cm^2
1.4 –1000	10^{-7}–10	$0.56\, t^{\frac{1}{4}}$ J/cm^2
1.4 –1000	> 10	0.1 W/cm^2

From ANSI Z136.1 (1976).

the A.N.S.I. exposure limits for direct intrabeam exposure and for exposure to diffuse reflected light are listed in Tables 14.6 and 14.7. Table 14.8 lists some of the frequency and time dependent limits for exposure of the skin. Because the size of the image on the retina varies inversely with the pupil diameter, the degree of concentration of light on the retina increases as the diameter of the pupil increases. For this reason, to be conservative, the MPE values listed in Table 14.6 for visible light are based on a pupil diameter of 7 mm, which is considered to be the maximum opening of the iris diaphragm of the eye. For other wavelengths, where retinal damage is not the limiting harmful effects, the incident laser energy is averaged over a 1 mm diameter circle. Tables 14.6, 14.7, and 14.8 are shown here for purposes of illustration only. For practical application in any real situation, the complete tables in the latest edition of the A.N.S.I. or A.C.G.I.H. recommendations should be consulted.

Regulatory requirements

In the United States, laser safety regulations are promulgated by two agencies: the Department of Health and Human Services through the Bureau of Radiological Health (BRH), and the Department of Labor (DOL), through the Occupational Safety and Health Administration (OSHA). The BRH regulates manufacturers only, not users, through requirements for performance specifications. All lasers are classified into one of four different classes, depending on the level of risk from the laser. Class I is the least hazardous and Class IV is the most hazardous. Table 14.9 lists the potentially hazardous capabilities of the lasers in each of the four classes:

TABLE 14.9. SUMMARY OF HAZARDOUS CAPABILITIES OF LASER CLASSES

Class	Hazardous capabilities
I	Cannot produce hazardous radiation
II	Continuous intrabeam exposure damages the eye. Momentary intrabeam exposure (<0.25 second) is not damaging to the eye.
III	Can damage the eye during momentary intrabeam viewing.
IV	Can damage the skin as well as the eye during momentary intrabeam exposure or exposure to diffuse reflection.

According to the laser's class, certain engineering and labeling requirements are specified in the regulations. The engineering requirements include:

1. A protective housing, which prevents exposure to laser radiation not necessary for the performance of the intended function of the laser (leakage radiation).
2. Safety interlocks, designed to prevent human access to laser radiation upon removal or displacement of the protective housing.
3. A remote control connector to allow additional interlocks and remote on–off controls.
4. Key control to prevent unauthorized use of the laser. The key must be removable and the laser must be inoperable unless the key control is turned on by the key.

The labeling requirements include information to be prominently displayed on the appropriate signs, warning of a laser hazard, and information about the laser and its output radiation which must be prominently affixed to the laser. The engineered features and labeling requirements for each laser class are listed in Table 14.10.

TABLE 14.10. SUMMARY OF LASER-ENGINEERED SAFETY FEATURES AND LABELING REQUIREMENTS

Safety feature	Class I	Class II	Class III	Class IV
Protective housing	X	X	X	X
Safety interlock	X	X	X	X
Remote connector			X	X
Key control			X	X
Emission indicator		X	X	X
Beam attenuator		X	X	X
Label				
Certification and Manufacturer	X	X	X	X
Class designation and Warning logotype		X	X	X
Aperture label		X	X	X
Radiation output				
Non-interlocked protective housing		X	X	X

From Federal Register, Vol. 39, No. 172, Sept. 4, 1974, pp. 32105–32108.

OSHA regulates the use of lasers in industry and specifies user qualifications, posting and labeling requirements, maximum permisssible exposure levels, and suitable laser safety goggles. According to OSHA regulations (29CFR, 1926.54), employees shall not be exposed to light intensities above:

1. 1 microwatt per cm^2 for direct staring,
2. 1 milliwatt per cm^2 for incidental observing,
3. $2\frac{1}{2}$ watts per cm^2 for diffuse reflected light.

Direct viewing

Intrabeam exposures to laser energy (Fig. 14.8) are specified in units of *radiant exposure*, H joules per cm^2, or in units of *irradiance*, E watts per cm^2. Irradiance is related to radiant exposure by

$$H = E \times t, \tag{14.6}$$

where t is the exposure time in seconds. The corresponding quantities at the laser aperture are called respectively emergent radiant exposures, H_0 and emergent irradiance E_0, and are specified by the energy output, J, per pulse per unit area of the laser aperture or the laser power output P per unit area of laser aperture.

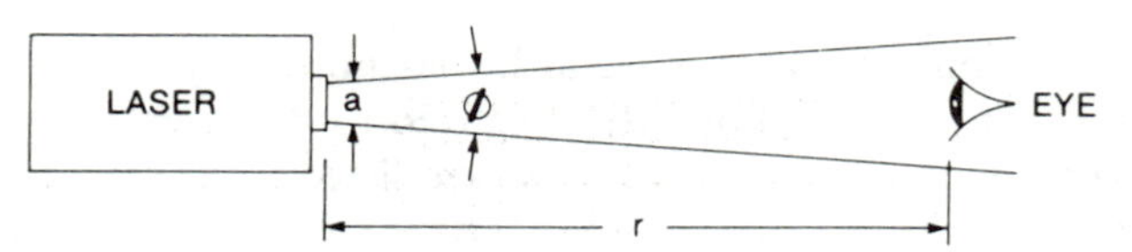

FIG. 14.8. Intrabeam viewing of the primary—or direct—beam. (From *A Guide for Control of Laser Hazards*, American Conference of Governmental Industrial Hygienists, 1973.)

Example 14.2

What is the MPE value for direct occular exposure for 5 seconds to the 442-nm light from a He–Cd laser operating at a power level of 20 mW cw?

From Table 14.6, the MPE for this exposure condition is

$$\text{MPE} = 1.8\, t^{3/4} \times 10^{-3} \frac{\text{J}}{\text{cm}^2}.$$

for $t = 5$ seconds, we have

$$\begin{aligned}\text{MPE} &= 1.8(5)^{3/4} \times 10^{-3}\\ &= 6.02 \times 10^{-3} \frac{\text{J}}{\text{cm}^2} = 6.02 \frac{\text{mJ}}{\text{cm}^2},\end{aligned}$$

since

$$H \frac{\text{mJ}}{\text{cm}^2} = E \frac{\text{mW}}{\text{cm}^2} \times t \text{ sec},$$

$$\text{MPE}(E) = \frac{6.02 \text{ mJ/cm}^2}{5 \text{ sec}} = 1.2 \frac{\text{mW}}{\text{cm}^2}.$$

Less is known about the biological effects of pulsed radiation, delivered in repetitive narrow pulses, than about the same average amount of energy delivered at a uniform rate. Accordingly, for pulses that are less than 10 microseconds in duration it is recommended that the maximum permissible exposure limits listed in Tables 14.6 and 14.7 be reduced by a factor, given in Fig. 14.9, that depends on the pulse repition frequency (PRF). In the case of skin exposed to repetitively pulsed laser light, it is recommended that the MPE should conform to that for a single pulse, and that the mean irradiance of the pulse train should be limited to the MPE applicable for the total exposure time of the pulse train. The smaller of these two values is used as the actual MPE. In the context of laser safety, a laser may be considered to be "repetitively pulsed" if it generates a continuous train of pulses lasting 0.25 seconds or more, with the individual pulses being equally spaced within the train.

Example 14.3

A ruby laser, $\lambda = 694.3$ nm, is pulsed at a rate of 2 per minute, and each pulse is 10 microseconds wide. What is the MPE for (a) direct occular exposure, and (b) skin exposure.

(a) Since for safety analysis this is not considered a repetitively pulsed laser, we find, in Table 14.6, that the maximum permissible radiant exposure for a pulse width (or exposure time) of 10 microseconds is

$$H(\text{MPE}) = 5 \times 10^{-7} \text{ J/cm}^2.$$

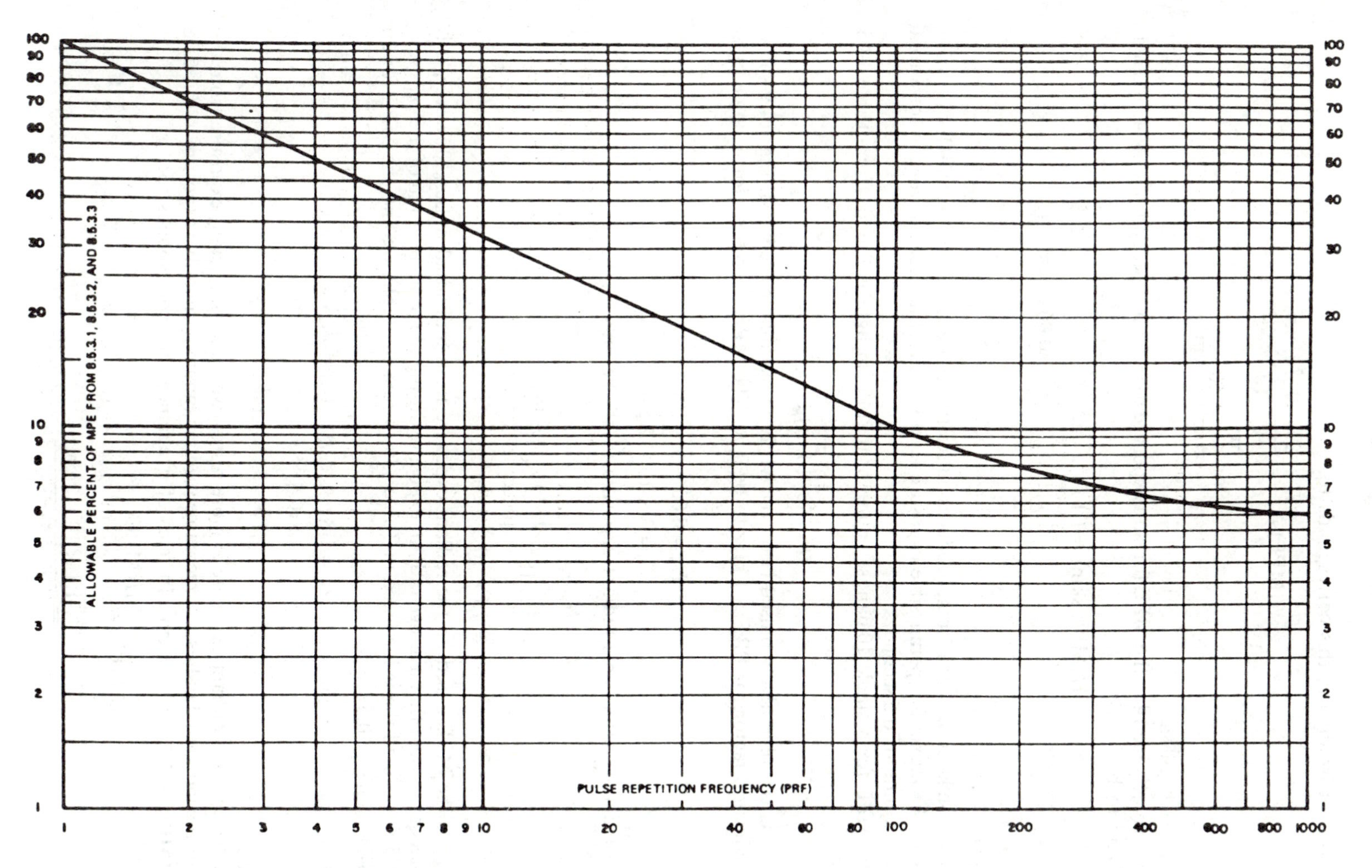

FIG. 14.9. Reduction of MPE for Repetitively Pulsed or Multiple Exposures from Scanning Lasers, Individual Pulse or Exposure Less than 10 μs. Note: For repetition rates greater than 1000 pps, the percent MPE = 6. (From ANSI Z136.1 1976.)

In terms of irradiance, this corresponds to

$$E(\text{MPE}) = \frac{H}{t} = \frac{5 \times 10^{-7}\ \text{J/cm}^2}{10 \times 10^{-6}\ \text{sec}}$$
$$= 0.05 \frac{\text{W}}{\text{cm}^2} = 50 \frac{\text{mW}}{\text{cm}^2}$$

during each pulse.

(b) From Table 14.8, the MPE for skin exposure to a 10 microseconds of 0.6943 μm light is found to be

$$H(\text{MPE skin}) = 1.1\, t^{1/4}\ \text{J/cm}^2$$
$$= 1.1\,(1 \times 10^{-5})^{1/4} = 0.062\ \text{J/cm}^2.$$

Example 14.4

An argon laser, $\lambda = 514.5$ nm, produces pulses 10 nanoseconds wide at a PRF of 1 MHz. What is the maximum permissible exposure level for direct intrabeam viewing for a total exposure time of 0.5 second.

The MPE for a single 10 nanosec pulse of 0.5145 μm light is found, from Table 14.6, to be

$$H = 5 \times 10^{-7}\ \text{J/cm}^2.$$

Since this is a repetitively pulsed laser, the MPE above must be reduced. Fig. 14.9 shows the reduction factor for PRF > 1 kHz to be 0.06. The MPE, therefore, for a single pulse is

$$H_p = 0.06 \times 5 \times 10^{-7}$$
$$= 3 \times 10^{-8}\ \text{J/cm}^2$$

The maximum permissible radiant exposure to all the pulses is

$$H = H_p \times f \times t$$
$$= 3 \times 10^{-8} \frac{\text{J/cm}^2}{\text{pulse}} \times 10^6 \frac{\text{pulse}}{\text{sec}} \times 0.5\ \text{sec}$$
$$= 15 \times 10^{-3}\ \text{J/cm}^2.$$

This is the MPE based on the radiant exposure from individual pulses. However, it is also recommended that average radiant exposure during the total duration of the exposure should not exceed the value listed in Table 14.6. From the table, we find she MPE for an exposure time of 0.5 sec to be

$$H = 1.8\, t^{3/4} \times 10^{-3}\ \text{J/cm}^2$$
$$= 1.8\,(0.5)^{3/4} \times 10^{-3}$$
$$= 1.1 \times 10^{-3}\ \text{J/cm}^2.$$

Since the average radiant exposure is more restrictive than the radiant exposure from the sum of the individual pulses, the recommended maximum permissible exposure in this case is $H(\text{MPE}) = 1.1 \times 10^{-3}\ \text{J/cm}^2$. In tems of irradiance, this corresponds to

$$E(\text{MPE}) = \frac{H(\text{MPE})}{t}$$
$$= \frac{1.1 \times 10^{-3}\ \text{J/cm}^2}{0.5\ \text{sec}} = 2.2 \times 10^{-3}\ \text{W/cm}^2.$$

Example 14.5

A small He–Ne laser, of the type often used in the classrooms, has the following characteristics:

wavelength = 632.8 nm,
power output = 0.5 mW cw radiation,
aperture diameter = 2 mm,
beam divergence = 0.2 millirad.

Calculate (a) The emergent irradiance
(b) The irradiance at a distance of 1 meter
(c) The irradiance at a distance of 10 meters
(d) The hazardous intrabeam viewing distance, that is the distance at which the irradiance exceeds $2.5\,mW/cm^2$.

(a) The emergent irradiance is

$$E_0 = \frac{\text{power output}}{\text{aperture area}} = \frac{0.5\,\text{mW}}{\frac{\pi}{4}(0.2\,\text{cm})^2} = 15.9\,\text{mW/cm}^2. \tag{14.7}$$

(b) Irradiance at a distance from the aperture will be less than the emergent irradiance because of beam divergence. Absorption of the laser light by air is negligible, unless the path is very long. For visible laser light, the attenuation coefficient varies from about 10^{-2} per meter in thick fog to about 10^{-5} per meter is clean air.

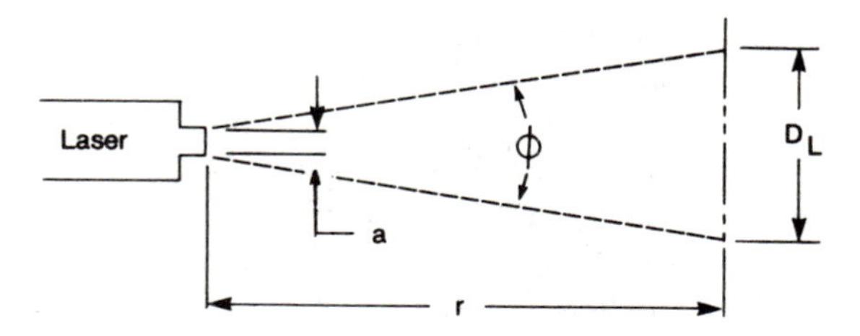

FIG. 14.10. Geometry for evaluating irradiance at a distance from the laser aperture.

From Fig. 14.10, we see that the irradiance at a distance r cm is given by

$$E = \frac{Pe^{-\mu r}}{A_L} = \frac{Pe^{-\mu r}}{\pi\left(\frac{1}{2}D_L\right)^2} = \frac{Pe^{-\mu r}}{\pi\left[\frac{1}{2}(a + r\varphi)\right]^2}, \tag{14.8}$$

where E = irradiance mW/cm^2,
P = power output of the laser, mW,
A_L = area of laser beam at distance r cm,
a = laser aperture, cm,
D_L = diameter of laser beam at distance r cm,
φ = beam divergence, radians,
μ = attentuation coefficient, cm^{-1}.

At a distance of 1 m, $e^{-\mu r} \approx 1$, and the irradiance from this laser, whose emergent beam irradiance, E_0, is 15.9 mW/cm², is

$$E = \frac{0.5\ \text{mW}}{\pi\left[\frac{1}{2}(0.2\ \text{cm} + 100\ \text{cm} \times 0.0002\ \text{rad}\right]^2}$$

$$E = 13.15\ \frac{\text{mW}}{\text{cm}^2}.$$

(c) At a distance of 10 m, the irradiance decreases to 3.98 mW/cm². It should be noted that the light intensity from a laser does not follow the inverse square law when distances are measured from the laser aperature.

(d) The hazardous intra-beam viewing distance may be calculated by neglecting attenuation by air, rearranging equation (14.8) and sugstituting MPE = 2.5 mW for E to give

$$r\ (\text{hazardous}) = \frac{\left(\frac{4}{\pi} \times \frac{P}{\text{MPE}}\right)^{1/2} - a}{\varphi\ \text{radians}} \tag{14.9}$$

$$= \frac{\left(\frac{4}{\pi} \times \frac{0.5\ \text{mW}}{2.5\ \text{mW/cm}^2}\right)^{1/2} - 0.2\ \text{cm}}{2 \times 10^{-4}\ \text{rad}} = 1523\ \text{cm}.$$

For higher-power lasers, the "safe" viewing distance can become very great. If the power of the laser in this example is increased to 100 mW, then the "safe" viewing distance is found to be 5.3×10^6 cm, or 5.3 km!

Reflections

When a laser beam falls on a surface, some fraction of the incident light, depending on the nature of the surface and on the wavelength of the light, will be reflected. Where the reflecting surface is polished and mirror like, and where the angle of incidence of the beam is equal to the angle of reflection, we have specular reflection. If the reflecting surface is rough or dull, and if the illuminated surface appears equally bright at all viewing angles, the surface is matte, and we have diffuse reflection. We have mixed reflections if some of the incident light is diffusely reflected and some undergoes specular reflection. The reflection coefficient of a surface gives the fraction of the incident light which that surface reflects. Reflection coefficients for some materials are listed in Table 14.11.

An *extended source* is a source whose dimensions are large relative to the viewing distance. In this case, the source can be resolved into a geometrical image on the retina (a point source cannot be resolved into a geometrical image on the retina). For purposes of laser hazard control we have an extended source when the source subtends a viewing angle greater than α_{min},

$$\alpha_{\text{min}} = \frac{d_r}{f}\ \text{radian}, \tag{14.10}$$

where d_r, the diameter of the smallest possible retinal spot is calculated from equation (14.4), and f is the effective focal length of the eye, 0.7 cm. In the case of a ruby laser, for example, where $\lambda = 6943$ Å, the minimum-sized retinal spot may occur when the pupillary diameter is 7 mm. From equation (14.3), we find this diameter to be 4.11×10^{-4} cm. The minimum

TABLE 14.11. REFLECTING MATERIALS

Type	Material	Reflection coefficient
Specular	Mirrored glass	0.8–0.9
	Aluminum foil	0.84–0.87
	Rhodium	0.7–0.9
	Aluminum, polished	0.6–0.7
	Chromium	0.60–0.65
	Stainless steel	0.55–0.65
	Black structural glass	0.04–0.05
Mixed	Aluminum, oxidized	0.70–0.85
	Aluminum, brushed	0.54–0.58
	Aluminum paint	0.6–0.7
	Stainless steel, satin	0.51–0.56
Diffuse	White plaster	0.90–0.92
	White paint, flat	0.75–0.90
	Limestone	0.35–0.65
	Sandstone	0.30–0.42

viewing angle for an extended source is, from equation (14.10),

$$\alpha_{min} = \frac{4.11 \times 10^{-4}\,\text{cm}}{1.7\,\text{cm}} = 2.4 \times 10^{-4}\,\text{rad}.$$

As long as the eye subtends an angle greater than α_{min} when viewing a reflection, the intensity of the light energy on the retina is independent of the viewing distance, since the light falloff with distance is exactly compensated by the increased area that is viewed. Under these conditions, the maximum safe brightness of the reflecting area can be calculated. Then, if we know the reflection coefficient for that surface, we can calculate the maximum irradiance at the reflecting surface (Fig. 14.11).

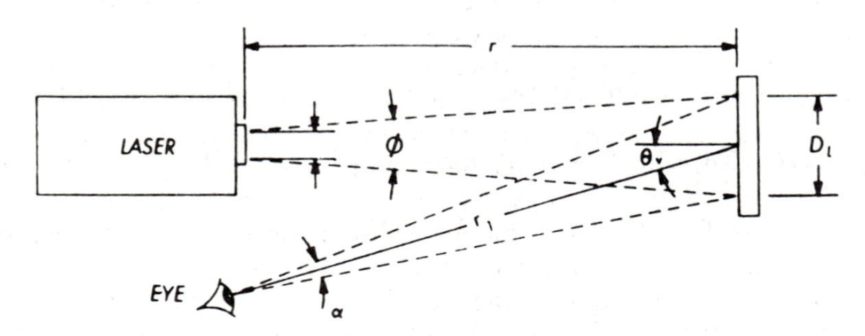

FIG. 14.11. Extended source viewing of a diffuse reflection (From *A Guide for Control of Laser Hazards*, American Conference of Governmental Industrial Hygienists, 1973.)

Example 14.6

A cw He–Ne laser, $\lambda = 633$ nm, having an output of 1 mW, a 3-mm-diameter aperture and a beam divergence of 1 milliradian is used in a darkened laboratory. The beam illuminates a white painted matte wall, whose reflection coefficient is 0.8, at a distance of 10 meters. May the reflected spot be viewed for 60 seconds from a point that is 30° off the central axis ($\theta_v = 30°$) and at a distance of 1 meter from the wall without eye protection?

The diameter of the reflection is, from Fig. 14.10,

$$D_L = a + r\varphi \tag{14.11}$$
$$= 0.3 \text{ cm} + 1000 \text{ cm} \times 10^{-3} \text{ rad.}$$
$$= 1.3 \text{ cm},$$

and the viewing angle α at a distance r_1 of 1 meter is calculated from

$$\alpha = \frac{D_L \cos \theta_V}{r_1} \tag{14.12}$$
$$= \frac{1.3 \text{ cm} \times \cos 30°}{100 \text{ cm}} = 1.13 \times 10^{-2} \text{ rad.}$$

To distinguish between a point source and an extended source, compare the viewing angle to $\alpha_{\min}$. Combining equations (14.4) and (14.10), we have

$$\alpha_{\min} = \frac{2.44\,\lambda}{d_p} \tag{14.13}$$
$$= \frac{2.44 \times 6.328 \times 10^{-5} \text{ cm}}{0.7 \text{ cm}}$$
$$= 2.2 \times 10^{-4} \text{ rad.}$$

Since the viewing angle exceeds $\alpha_{\min}$, we have an extended source. The brightness of the reflection is given by the product of the irradiance and the reflection coefficient:

$$B = E \times \rho, \tag{14.14}$$
$$B = \frac{1 \text{ mW}}{\pi\left(\frac{1.3}{2} \text{ cm}\right)^2} \times 0.8$$
$$= 0.6 \text{ mW/cm}^2,$$

which corresponds to a radiant exposure, during a 60-second exposure period, of

$$H = 0.6 \times 10^{-3} \frac{\text{W}}{\text{cm}^2} \times 60 \text{ sec} \times 1 \frac{\text{J}}{\text{W-sec}} = 3.6 \times 10^{-2} \text{ J/cm}^2.$$

From Table 14.7, we find that the MPE for 6328 nm light, for an exposure time of 60 seconds, is

$$H(\text{MPE}) = 3.83\, t^{3/4} \text{ J/cm}^2/\text{sr}$$
$$= 3.83(60)^{3/4} = 82.57 \text{ J/cm}^2/\text{sr}.$$

The solid angle β included in the view is given by

$$\beta = \frac{A \cos \theta_v}{r_1^2} \text{ steradian,} \tag{14.15}$$

where A = area of reflection,
r_1 = viewing distance,
θ_v = viewing angle.

From the equation above, we have

$$\beta = \frac{(\pi/4)(1.3\ \text{cm})^2(\cos 30°)}{(100\ \text{cm})^2} = 1.15 \times 10^{-4}\ \text{sr}.$$

The reflected radiant exposure per unit solid angle is

$$H_r = \frac{H}{\beta} \tag{14.16}$$

$$= \frac{3.6 \times 10^{-2}\ \text{J/cm}^2}{1.15 \times 10^{-4}\ \text{sr}} = 313\ \text{J/cm}^2/\text{sr}.$$

Since the reflected radiant exposure is greater than the MPE, the reflection may not be viewed for 60 seconds without using goggles.

Protective Eyewear

The eye may be protected against laser radiation by means of protective goggles that attenuate the intensity of the laser light while transmitting enough ambient light for safe visibility.

The attenuation of light by a protective goggle is given by its *optical density*, OD, which is defined by

$$\text{OD} = \log \frac{E\ (\text{or}\ H)}{\text{MPE}}, \tag{14.17}$$

where MPE is the maximum allowable exposure on the cornea, E is the radiance, and H is the radiant exposure. The determination of the minimum required density is illustrated in the following example:

Example 14.7

A He–Ne laser has a power output of 15 mW cw. The $1/e^2$ beam diameter at the aperture is 1.7 mm, and the beam divergence is 1.0 mrad. What is the minimum optical density of goggles in order to comply with the MPE recommended in the ANSI standard for intrabeam exposure to the 6328 Å radiation from this laser at a distance of 1 meter from the aperture?

Since the $1/e^2$ diameter encompasses 86.5% of the energy in the laser beam, the irradiance at a distance of 100 cm is

$$E = \frac{15 \times 10^{-3}\ \text{W} \times 0.865}{(\pi/4)(0.17 + 100 \times 10^{-3}\ \text{cm})^2} = 2.27 \times 10^{-1}\ \text{W/cm}^2.$$

In Table 14.6, we find that the MPE for the eye is

$$E(\text{MPE}) = C_B \times 10^{-6}\ \text{W/cm}^2,$$

where

$$C_B = 10^{15(\lambda - 0.550)}$$

$$= 10^{15(0.6328 - 0.5500)} = 17.5.$$

$$\therefore\ E(\text{MPE}) = 17.5 \times 10^{-6}\ \text{W/cm}^2$$

The minimum required optical density is

$$\text{OD} = \log \frac{0.227\ \text{W/cm}^2}{17.5 \times 10^{-6}\ \text{W/cm}^2} = 4.1\ \text{at 633 nm}.$$

Example 14.8

A ruby laser, $\lambda = 694$ nm, is pulsed at a rate of 2 pulses per minute. Each pulse is 20 nanosec wide, and contains 10 J of energy. The exit beam diameter is 18 mm, and the beam divergence is 7 mrad. What is the minimum optical density required to comply with the recommendations in ANSI Z136.1- 1976?

The radiant exposure per pulse is

$$H = \frac{10\ \text{J}}{(\pi/4)(1.8\ \text{cm})^2} = 3.93\ \text{J/cm}^2$$

From Table 14.6, we find that the MPE value for a pulse 2×10^{-8} second wide is

$$H\ (\text{MPE}) = 5 \times 10^{-7}\ \text{J/cm}^2,$$

and Fig. 14.9 shows that there is no reduction in MPE value if the pulse repetition frequency is one per second or less. The minimum required optical density, therefore, is:

$$\text{OD} = \log \frac{3.93\ \text{J/cm}^2}{5 \times 10^{-7}\ \text{J/cm}^2} = 6.9 \text{ at } 694\ \text{nm}.$$

When choosing goggles, the *luminous transmission* must be considered as well as the optical density. The optimum goggle is the one that provides maximum attenuation of the laser light while transmitting the maximum amount of ambient light. Luminous transmission is given as the per cent transmission of light from a standard source, whose spectral distribution is equivalent to that of a black body at 6500 Å. This source is known as the C.I.E. standard source C, and is approximately equivalent to average daylight. Fig. 14.12 shows the light absorbing properties of one of the commercially available goggles. The curve shows that although these goggles provide an optical density of 7 for the 694-nm wavelength light from a ruby laser, they are almost useless for protection against the 455–515 nm wavelengths from He–Ne lasers.

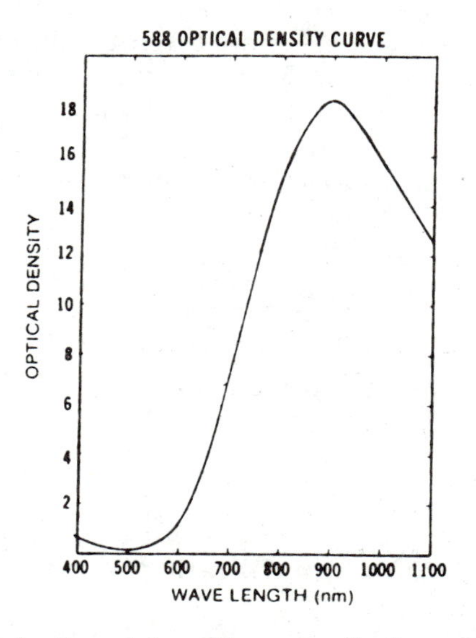

FIG. 14.12. Optical density variation with wavelength of one of the commercially goggles for protection against laser light. The luminous transmission of this glass for a standard C source is 35%. (Courtesy of American Optical Co.)

When choosing goggles or safety glasses for protection against laser light, the resistance of the lenses to damage from the laser beam must be considered. The absorption of laser energy by the lenses of the protective eyewear leads to an increase in temperature of the lens. The lens must be capable of absorbing the amount of energy under the expected operating conditions without suffering changes in light transmission, without softening or shattering, and without suffering other surface damage that might impair its usefulness or create a hazard to the user. It must be understood that no single lens material is useful for all wavelengths and for all radiant exposures. In choosing a protective eyewear, therefore, careful consideration must be given to the laser operating parameters, the maximum permissible exposure levels, the conditions of use, and the response of the protective eyewear to the insult from the laser beam under the most severe conditions of use. The protective eyewear that is finally chosen must meet the safety criteria in each of these categories.

Safety Measurements

Evaluation of the safety aspects of a laser includes measurement of power or energy and beam divergence. Two general types of power and energy instruments are used: one based on the photoelectric effect, which is very fast, and is suitable for Q-switched measurements; and one based on a thermal effect, which is very stable and wavelength independent, but is slower and less sensitive than photo-electric devices.

Power and energy

Photoelectric devices. These instruments employ either photomultipliers or photodiodes as the detection elements. In either case, the output signal is directly proportional to the irradiance, and the meter usually is calibrated to give average power levels for cw or high pulse repetition frequencies; connectors are provided to display pulses on an oscilloscope, and thus to determine the energy per pulse, pulse width, pulse frequency and peak power. Like almost quantum electronic devices, photo-electric detectors are highly energy dependent, and each instrument must be calibrated. A typical calibration curve, as supplied by the manufacturer, is shown in Fig. 14.13.

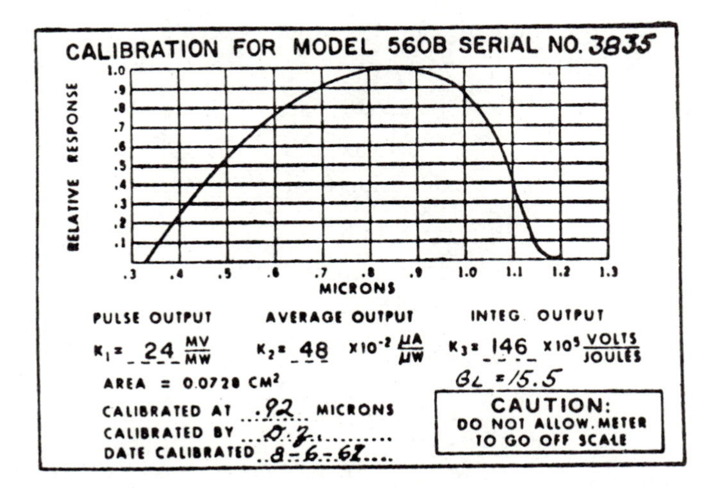

FIG. 14.13. Calibration curve for a photoelectric detector (Courtesy E.G.& G.)

Example 14.9

The output from a pulsed ruby laser, $\lambda = 694.3$ nm, is measured with a photodiode type meter whose calibration curve is shown above. With an aid of an oscilloscope, we find rectangular shaped pulses as shown below:

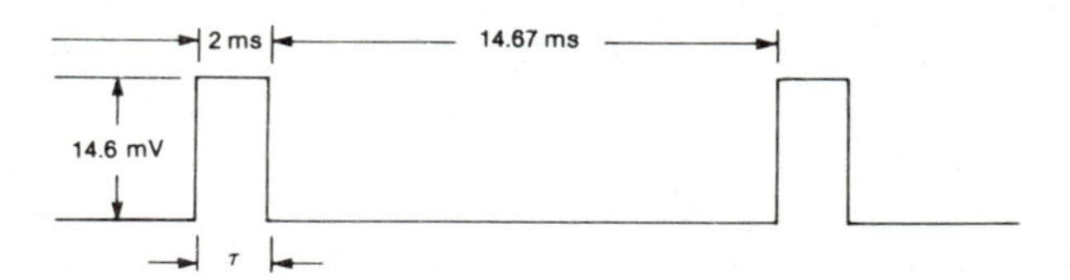

The meter reading, which is proportional to the average power, is 35 μA. What is the

(a) duty cycle (fraction of time that lasing actions occurs),
(b) average power, in μ watts,
(c) peak power,
(d) energy per pulse,

(a) duty cycle $= \dfrac{2}{2 + 14.67} = 0.120$

(b) The meter was calibrated at 920 nm (0.92 microns), where the calibration curve shows the relative response to be 0.98. At 694.3 nm, the relative response is 0.9; the meter's average current output is 0.48 μA/μW.

A meter reading of 35 μA therefore corresponds to

$$\frac{\dfrac{0.98}{0.9} \times 35\ \mu\text{A}}{0.48\ \dfrac{\mu\text{A}}{\mu\text{W}}} = 79.4\ \mu\text{W}.$$

(c) Peak power P_0, may be calculated by any of the following methods

$$P_p = \frac{\text{average power}}{\text{duty cycle}} = \frac{79.4\ \mu\text{W}}{0.12} = 662\ \mu\text{W}, \qquad (14.18\text{A})$$

$$P_p = \frac{P}{n\tau} = \frac{79.4\ \mu\text{W}}{60\ \text{sec}^{-1} \times 2 \times 10^{-3}\ \text{sec}} = 662\ \mu\text{W}, \qquad (14.18\text{B})$$

$$P_p = \frac{\text{pulse height}}{\text{calibration factor}} \qquad (14.19)$$

$$= \frac{\dfrac{0.98}{0.9} \times 14.6\ \text{mV}}{24\ \dfrac{\text{mV}}{\text{mW}}} = 0.662\ \text{mW} = 662\ \mu\text{W}.$$

(d) Energy per pulse, J, is given by

$$J\frac{\text{joule}}{\text{pulse}} = P_0 \times \tau$$

$$= 662 \times 10^{-6}\ \text{W} \times 2 \times 10^{-3}\ \text{sec}$$

$$= 1.32 \times 10^{-6}\ \text{J/pulse}$$

Thermal devices. Light energy can also be measured by transforming it into heat energy and then measuring the heat. Thermocouples and bolometers are commonly used transducers for this purpose. A thermocouple consists of two wires of different metals joined together at their ends. If the two junctions are at different temperatures, a current whose magnitude depends on the temperature difference flows through the circuit. This current can be measured, and its magnitude related to the intensity of the light incident on the detector. In practice, in order to increase the sensitivity of the instrument, a number of thermocouples are joined together to form a thermopile. Bismuth-silver, copper-constantan, and manganim-constantan are the metals used in thermopiles. The surfaces that receive the light energy are coated either with lampblack, gold black, or parson's black, all of which have very high absorbances ($\geq 95\%$) for the wavelengths associated with most lasers.

A ballistic thermopile is a frequently used transducing element for measuring the energy in a laser pulse. This device employs two absorbers, one of which absorbs energy from the laser light to which it is exposed, and thus experiences a temperature rise, while the other is a reference absorber that is exposed only to the ambient light. The temperature difference between them generates a voltage difference that is quantitatively related to the energy in the pulse of laser light. The ballistic thermopile is useful for measuring pulses in the range of about 1 mJ to 100 J.

A bolometer is a resistive element whose electrical resistance is a function of temperature. One special form of such a circuit element is the *thermistor*, which is a semiconductor with a high negative temperature coefficient of resistivity. The resistive element, enclosed in a light-receiving case that is treated with lampblack to maximize energy absorption, forms one of the arms of a bridge that is balanced in the absence of the laser light. The incident light raises the temperature of the bolometer, thus changing its resistance and unbalancing the bridge. The degree of unbalance is quantitatively related to the incident light energy.

Detectors based on thermal effects generally are less sensitive than quantum devices. However, they are wavelength independent, very stable, easy to calibrate, and retain their calibration for long periods of time.

Beam divergence

The beam divergence is determined by measuring the beam diameter at two different distances from the laser, and then calculating the divergence. Because of the wave nature of light, however, there is no sharp, measurable beam diameter. Accordingly, the diameter may be defined as the circle that encloses some arbitrary fraction of the total light energy. The most frequently used arbitrary fraction is $1/e^2$, or 86.5%. To measure the beam diameter, we require a power meter whose aperture exceeds the circle that would enclose 100% of the laser beam and a coaxial iris diaphragm. A measurement is first made with the iris wide open; the iris is then slowly closed until the meter reading is down to 86.5% of the first reading. The diameter of the iris opening at that point is defined as the $1/e^2$ diameter.

Example 14.10

The $1/e^2$ diameters of a laser beam at 2 m and 6 m from the laser were found to be 2.82 mm and 5.30 mm respectively. What was the beam divergance?

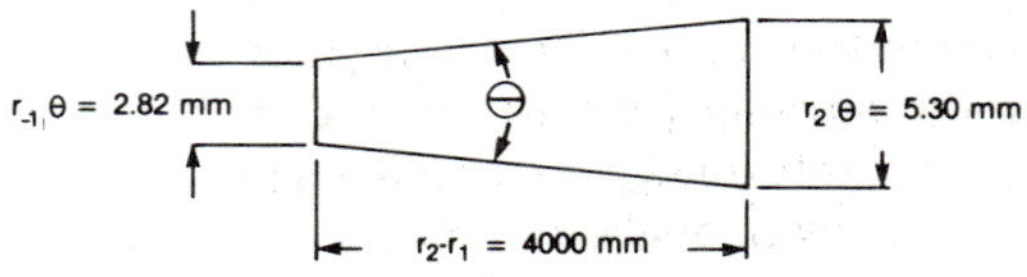

$r_1\theta = 2.82$ mm $\quad$ $r_2\theta = 5.30$ mm $\quad$ $r_2 - r_1 = 4000$ mm

$$r_2\theta - r_1\theta = (r_2 - r_1)\theta,$$

$$\theta = \frac{5.30 - 2.82}{4000} = 6.2 \times 10^{-4} \text{ rad} = 0.62 \text{ mrad}.$$

Radiofrequency (RF) and Microwaves

Radiofrequency radiation is defined arbitrarily as electromagnetic radiation in the frequency range of 0.3–30 MHz, while the arbitrary definition of microwaves includes electromagnetic radiation whose frequencies range from 30 MHz to 300 GHz. Frequency bands assigned to radar systems are listed in Table 14.12, while industrial, scientific, and medical applications of RF and microwaves have been assigned the frequency bands listed in Table 14.13.

TABLE 14.12. MICROWAVE FREQUENCY BANDS

Band	Frequency, MHz	λ, cm
L	1,000– 1,400	27.3–21.4
S	2,600– 3,950	11.5– 7.6
C	3,950– 5,850	7.6 – 5.13
X	8,200–12,400	3.66– 2.42
K_u	12,400–18,000	2.42– 1.67
K	18,000–26,000	1.67– 1.16
K_a	26,000–40,000	1.16– 0.75

TABLE 14.13. INDUSTRIAL, SCIENTIFIC, AND MEDICAL (ISM) FREQUENCY BANDS

Frequency, MHz	λ
13.56	66.37 m
27.12	33.19 m
40.68	22.12 m
915	32.8 cm
2,450	12.2 cm
5,800	5.2 cm
22,125	1.4 cm

Communications

The short wave lengths in the Microwave Frequency Bands, on the order of millimeters to centimeters, contrast sharply with the very much longer wavelengths, on the order of tens to hundreds of meters, in the radiofrequency portion of the electromagnetic spectrum. This wavelength difference is the basis for the widespread application of microwaves in communications and in radar. The short wavelengths allow electromagnetic energy to be transmitted through the air or through free space in relatively sharply focussed, intense beams similar to a beam of light. Furthermore, these focussed, high frequency beams can carry more information than can be transmitted by the more conventional radiofrequency waves. This increased information carrying capacity of microwaves is due to the bandwidths associated with microwaves. Informaton is transmitted in a band of frequencies. For example, speech requires a bandwith of about 3,000 Hz, music is transmitted in a bandwidth of about 20,000 Hz, and television requires a bandwidth of about $4\frac{1}{2}$ MHz. In the case of microwaves, a 200-MHz bandwidth centered on 2,000 MHz, that is, a 10% bandwidth, has more frequency bands available for transmission of speech and music than all the AM and FM freqeuncies combined.

In radar applications, a short, intense highly focused pulse of microwave radiation is emitted from an antenna. If this emitted radiation pulse strikes an object, some of the incident energy is reflected back to the radar antenna which is connected to a receiver as well as a transmitter, and is detected as an incoming signal by the reciver. The distance between the object and the radar is determined from the time between the emitted pulse and the arrival of the reflected signal. The direction of the object is determined from the direction in which the antenna was pointed. From these two measurements, the location of the object, in polar coordinates, can thus be determined at any moment. By continuous measurements, the object, if it is moving, can be tracked, and its speed and directions can thus be determined.

Antennas

The focussing of energy into relatively narrow, intense beams is possible because the size of microwave antennas greatly exceeds the wavelength of the emitted radiation. A microwave antenna usually consists of a microwave feeding device, such as a dipole or a horn that is connected to the microwave generator, and a parabolic reflector, as shown in Fig. 14.14.

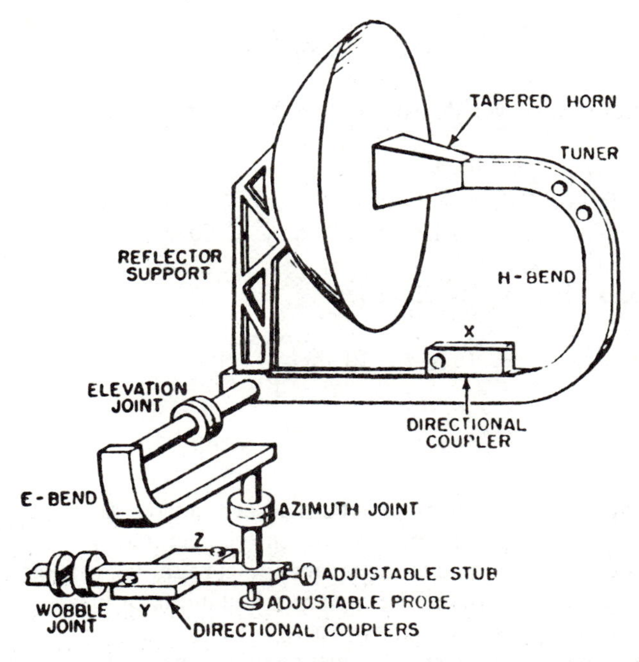

FIG. 14.14. Microwave antenna, showing the feeder (tapered horn) and the parabolic reflector.

In a dipole, electromagnetic energy is radiated from solid conductors that are coupled to a transmitter. The horn is an opening in a wave guide from which electromagnetic radiation emerges in the direction in which the aperture is facing. The microwave feed is mounted at the focus of the parabolic reflector, and emits radiation into the reflector. The radiation is then reflected in a (theoretically) parallel beam. However, the beam does spread, with the amount of spread being inversely proportional to the size of the reflector. The measure of beam spread is called the *beamwidth.* The beamwidth is defined by the width of the beam at the half-power points, and is measured in degrees. The ability of an antenna to concentrate the radiated electromagnetic energy into a beam is described by a figure of merit called the *antenna gain*. The gain of an antenna is defined as the ratio of the intensty of the radiated signal in the far field beam to the intensity at the same point if the same amount of transmitter power were radiated by an isotropic antenna.

$$\text{Antenna gain, } G_a = \frac{\text{power from antenna}}{\text{power from isotropic radiator}} = \frac{P}{P_0}. \tag{14.20}$$

The gain of an antenna is related to its area and to the wavelength of the radiation. For a parabolic antenna, this relationship is given by

$$G = \frac{P}{P_i} = \frac{4\pi A}{\lambda^2}, \tag{14.21}$$

where A is the area of the antenna aperture and λ is the wave length of the radiation.

All microwave antennas can be used either for transmission or receiving of microwave energy, depending on whether the dipole or horn is connected to a transmitter or to a receiver. In either case, the gain of the antenna is the same in both applications. Antenna gain is usually expressed in dB (decibels) relative to a reference power density, rather than as a power density ratio, as in equation (14.21). Gain in dB is given by

$$G = 10 \log \frac{P_2}{P_1} \, \text{d}B, \tag{14.22}$$

where P_1 is a reference power density and P_2 is the power density of interest.

Example 14.11

A "standard gain horn" designed for use in the 1.7–2.6 GHz frequency range, has a gain of 16.5 dB with reference to an isotropic radiator. To what power ratio does this gain correspond?

Substituting into equation (14.22) gives:

$$16.5 = 10 \log \frac{P_2}{P_1},$$

$$\frac{P_2}{P_1} = \log^{-1} \frac{16.5}{10} = 44.7.$$

Not all the energy in the electromagnetic field around the antenna is radiated. Part of the energy, called the *reactive* energy, is stored in the field and is recovered and re-emitted during successive oscillations, in a manner analogous to the stored electrical and magnetic energy that is interchanged between the capacitor and inductor in a low frequency resonant circuit. The region where we find reactive energy is called the *near-field.* In the near-field, the electric and magnetic fields are not perpendicular to each other, their exact orientation varying from point to point. Thus, the simple relationship between power density and electric or magnetic field strength illustrated in Example 2.17 and equation (2.71) does not apply to near-field conditions. True near-field conditions exist only close to the radiator. At a distance less than $\lambda/2\pi$, the reactive energy exceeds the radiated energy; the reactive energy is equal to the radiated energy at this distance. At distances from the antenna aperture greater than $\lambda/2\pi$, the proportion of reactive energy rapidly decreases to zero. Thus, the near-field extends for a distance considerably less than one wavelength from the radiator.

In the far-field (also called the Fraunhofer region), the electric and magnetic fields are perpendicular to each other, thus making possible meaningful measurements of power density. The distance to the start of the far-field region from the radiator depends on the size of the antenna and on the wavelength of the radiation. Although there is no sharp boundary marking the beginning of the far-field region, *for measurement and hazard estimation purposes*, its distance from the antenna is approximated by

$$R_{ff} = \frac{2D^2}{\lambda}, \tag{14.23}$$

where D = longest linear dimension of the antenna aperture
λ = wavelength

Example 14.12

At about what distance from a two-meter-diameter antenna that is transmitting a beam of 3,000-MHz (10-cm) microwaves can we expect to find far-field conditions?

From equation (14.23), the estimated distance is

$$R_{ff} = \frac{2D^2}{\lambda} = \frac{2(2\text{m})^2}{0.1\text{ m}} = 80\text{ m}.$$

Until this distance, the beam undergoes very little divergence; at this distance, the beam starts to spread and to become conical in shape.

Between the end of the near-field and the start of the far-field is the intermediate region (also called the Fresnel region), a transition between two regions. In the near-field and in the intermediate region, differences in path length of the radiation due to different points of origin on the antenna lead to differences in phase relations of the electromagnetic waves, and consequently to alternating maxima and minima in radiation intensity. The average power density at a uniformly illuminated antenna aperture is simply

$$W_0 = \frac{P}{A}, \tag{14.24}$$

where W_0 = average power density,
P = microwave power output,
A = area of aperture.

In the near-field, the maxima are four times greater than the average power density. The maximum power density in the near field, therefore is

$$W_{nf} = \frac{4P}{A} = 4W_0. \tag{14.25}$$

For purposes of safety assessment, the power density throughout the near field is considered to be $4W_0$. For the case of a *circular* antenna aperture, this power density is assumed to extend for a distance

$$R_{nf} = \frac{D^2}{5.66\lambda}, \tag{14.26}$$

where D = diameter of aperture,
λ = wavelength.

The intermediate zone is assumed, for purposes of safety assessment, to be as long as the near-field region. That is, the distance from the antenna until the end of the intermediate zone and the start of the far-field, is assumed to be

$$R_{\mathrm{ff}} = \frac{D^2}{2.83\lambda} = 2R_{\mathrm{nf}}. \tag{14.27}$$

Within the intermediate zone, the power level is assumed to decrease linearly from $4W_0$ at the start of the intermediate zone to $2W_0$ at the end of the intermediate zone. The power density within the intermediate field at a distance R from the antenna, may therefore be calculated from

$$W_{\mathrm{if}} = 4W_0 \frac{R_{\mathrm{nf}}}{R}. \tag{14.28}$$

For distances R from the antenna greater than $D^2/2.83\lambda$, the power density falls off inversely with the square of the distance, according to

$$W_{\mathrm{ff}} = 2W_0\left(\frac{R_{\mathrm{ff}}}{R}\right)^2. \tag{12.29}$$

These radiation zones are illustrated in Fig. 14.15.

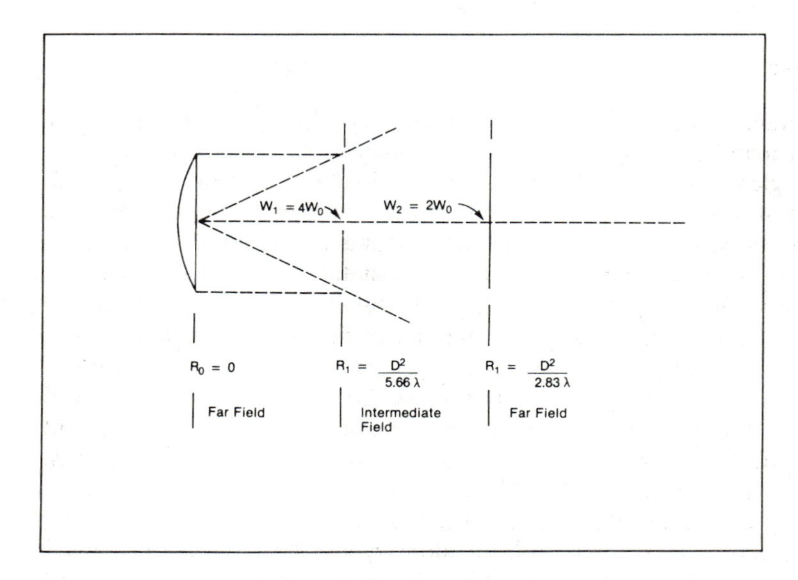

FIG. 14.15. Radiation zones. Distances from antenna for approximating the several radiation zones are given as a function of antenna area and radiation wavelength.

Example 14.13

To estimate the hazard from a dipole antenna that radiates 1,000 watts of 3,000-MHz microwaves from a circular parabolic reflector 0.75 m in diameter, calculate:

(a) the mean power density at the aperture,

(b) the maximum power density in the near field,
(c) the distance to the far-field,
(d) the power density at a distance of 250 m,
(e) the distance at which the power density will be 10 mW/cm^2 (100 W/m^2),

$$\text{(a)}\ W_0 = \frac{P}{A} = \frac{1000\ \text{W}}{(\pi/4)(0.75\ \text{m})^2} = 2263\,\frac{\text{W}}{\text{m}^2} = 226.3\,\frac{\text{mW}}{\text{cm}^2}.$$

$$\text{(b)}\ W_{nf} = 4W_0 = 4 \times 226.3\,\frac{\text{mW}}{\text{cm}^2} = 905\,\frac{\text{mW}}{\text{cm}^2}.$$

$$\text{(c)}\ R_{ff} = \frac{D^2}{2.83\lambda} = \frac{(0.75\ \text{m})^2}{2.83 \times 0.01\ \text{m}} = 20\ \text{m}.$$

$$\text{(d)}\ W(250\ \text{m}) = 2W_0\left(\frac{R_{ff}}{R}\right)^2 = 2 \times 226.3\,\frac{\text{mW}}{\text{cm}^2}\left(\frac{20\ \text{m}}{250\ \text{m}}\right)^2$$

$$= 2.9\,\frac{\text{mW}}{\text{cm}^2}.$$

$$\text{(e)}\ R\left(\frac{10\ \text{mW}}{\text{cm}^2}\right) = \sqrt{\frac{2 \times W_0 \times (R_{ff})^2}{W_{ff}}}$$

$$= \sqrt{\frac{2 \times 226.3\ \text{mW/cm}^2(20\ \text{m})^2}{10\ \text{mW/cm}^2}} = 135\ \text{m}.$$

Heating Effect

In its interaction with matter, microwave energy may either be reflected, as in the case of metals, it may be transmitted with little energy loss to the transmitting medium, as in the case of glass, or it may be absorbed by the irradiated matter, and thus raise the temperature of the absorber. This heating is attributed to two effects: the main mechanism is believed to be joule heating due to ionic currents induced by the electric fields that are set up within the absorbing medium by the radiation. The second mechanism is due to the interaction between polar molecules in the absorber and the applied high-frequency electric field. The alternating electric field causes these polar molecules to oscillate forth and back in an attempt to maintain the proper alignment in the electric field. These oscillations are resisted by other intermolecular forces, and the work done by the alternating electric field in overcoming these resistive forces is converted into heat.

The 27-MHz band is widely used in industry in *dielectric heating*, that is, for heating materials that are not electrically conducting, such as wood being dried or glued, plastics being heat sealed, textiles undergoing heat treatment, etc., while the 915- and 2450-MHz bands are used for *microwave heating* in industry and in the home. Heating in both these modes is very rapid. In dielectric heating, the object to be heated is placed between two electrodes across which a high frequency voltage, on the order of 100,000 V/m, is applied. Effectively, we have an electrical capacitor, in which the material to be heated is the dielectric. Most dielectric materials, such as ceramics, wood, paper, plastics, etc., absorb energy when placed in an alternating electric field because these "non-conductors" exhibit some degree of conductivity as polar molecules rotate under the influence of the alternating electric field. The loss tangent, or the degree of lossiness of a dielectric material is a measure of energy absorption by this mechanism. The degree of lossiness for any given material, as shown in equation (2.61), is a function of the frequency of the alternating electric field.

Generally, the efficiency of dielectric heating is 50–60%. An application of industrial dielectric heating is given in the following example:

Example 14.14

500 kg of resin-bonded sand cores per hour are to be baked in a dielectric core oven. The sand mix contains 3% water, and resin that cures at 107°C. If the specific heat of the resin-sand mixture is 1250 J/kg-°C (0.3 Btu/lb-°F), the ambient temperature is 18°C, and the overall efficiency is 50%, calculate the required power.

1. To heat the water in the mixture from 18°C to 100°C, using specific heat of water = 4178 J/kg-°C:

$$0.03 \times 500 \frac{\text{kg}}{\text{hr}} (100°\text{C} - 18°\text{C}) \times 4178 \frac{\text{J}}{\text{kg-°C}} \times \frac{1 \text{ hr}}{3600 \text{ sec}} = 1428 \frac{\text{J}}{\text{sec}}.$$

2. To vaporize the water using latent heat of vaporization of water = 2.25×10^6 J/kg:

$$0.03 \times 500 \frac{\text{kg}}{\text{hr}} \times 2.25 \times 10^6 \frac{\text{J}}{\text{kg}} \times \frac{1 \text{ hr}}{3600 \text{ sec}} = 9381 \frac{\text{J}}{\text{sec}}.$$

3. To heat the sand-resin mixture from 18°C to 107°C:

$$0.97 \times 500 \frac{\text{kg}}{\text{hr}} (107°\text{C} - 18°\text{C}) \times 1250 \frac{\text{J}}{\text{kg-°C}} \times \frac{1 \text{ hr}}{3600 \text{ sec}} = 14{,}988 \frac{\text{J}}{\text{sec}}.$$

Total energy dissipated in the cores

$$= 1428 + 9381 + 14{,}988 = 25{,}797 \frac{\text{J}}{\text{sec}},$$

and, at 50% efficiency, the required power

$$\frac{25{,}797 \frac{\text{J}}{\text{sec}}}{0.5} \times \frac{1 \text{ W}}{\text{J/sec}} \times \frac{1 \text{ kW}}{1{,}000 \text{ W}} = 51.6 \text{ kW}.$$

Microwave cooking is widely used because of the speed with which food can be heated. In a conventional oven, the oven's interior, including the walls and the air must be heated. Heat then is transferred from the hot air to the surface of the food, and then flows by conduction, into the material being cooked. The overall process is relatively inefficient and slow. In microwave cooking, on the other hand, the oven and its interior environment are not heated. All the microwave energy is absorbed by the food. Furthermore, because of the penetration of the microwaves to a depth of 1–2 cm below the surface of the food, direct deep heating of the food occurs. The volume of food that must be heated by conduction from the outer layers is thus greatly diminished. This combination of efficient energy utilization and deep heating leads to rapid cooking, as shown in Example 14.15. The deep heating effect of microwave irradiation is the basis of medical-microwave diathermy. In medical microwave diathermy, beneficial amounts of heat can be applied to inflamed or injured joints and tissues inside the body without overheating the skin.

Example 14.15

How long will it take to heat 0.45 kg fresh peas in 0.05 liter water to a temperature of 75°C from a room temperature of 20°C, in a microwave oven whose microwave power output is 700 watts, if the specific heat of the pea-water mixture is 3760 J/kg-°C.?

The required amount of heat energy is

$$Q = 0.5 \text{ kg} \times 3760 \frac{\text{J}}{\text{kg-}^\circ\text{C}} (75^\circ\text{C} - 20^\circ\text{C}) = 103{,}400 \text{ J}.$$

Since there is 1 J/sec in a watt,

$$700 \text{ W} \times 1 \frac{\text{J}}{\text{sec}} / \text{W} \times t \text{ sec} = 103{,}400 \text{ J}.$$

$$\therefore t = 148 \text{ sec} = 2\tfrac{1}{2} \text{ min}.$$

A microwave oven consists of a microwave generator, usually a magnetron whose maximum power output is 600 to 1000 W, a wave guide for conducting the microwaves from the generator to the oven cavity, the oven cavity, and a metallic rotating "stirrer" which produces a relatively uniform radiation field inside the oven by preventing the establishment of standing waves within the oven cavity. Additionally for safety reasons, all microwave ovens have doors that are interlocked with the microwave generators, and which stop the generation of microwaves if the door should be opened while the oven is operating. If the door is not properly closed, the interlocks prevent the activation of the microwave generator. If the oven door were not properly closed, then microwave radiation would leak through the gap between the oven cavity and the door, thus creating a possible health hazard. Verification of the proper operation of the door interlock is thus an important part of the safety evaluation of a microwave oven.

Penetration Depth

As microwaves pass through a medium, they lose energy to the medium through joule heating from ionic currents induced by the electric field and through the vibration of polar molecules, such as those in water, under the influence of the changing electric of the changing electric field. This resulting continuous decrease in the intensity of the electromagnetic field is related exponentially to the depth of penetration into the absorbing medium by

$$E_2 = E_1 e^{-2\alpha t}, \tag{14.30}$$

where E_1 = initial power density,
t = absorber thickness,
E_2 = power density at depth t,
α = absorption coefficient.

The absorption coefficient α, is dependent on the frequency of the radiation and on the conductivity, permittivity, and permeability of the absorber:

$$\alpha = \omega \sqrt{\frac{\mu\epsilon}{2}} \left\{ \sqrt{\left[1 + \left(\frac{\sigma}{\omega\epsilon}\right)^2\right]} - 1 \right\}^{\frac{1}{2}}, \tag{14.31}$$

where ω = angular frequency = 2π f,
σ = conductivity, $(\text{ohm-meter})^{-1}$,
ϵ = permittivity = relative dielectric constant $\times \epsilon_0$,
μ = permeability. Since the permeability of biological substances is very close to that of free space, we may use $\mu = \mu_0 = 4\pi \times 10^{-7}$ N/A^2.

Example 14.16

(a) Calculate the absorption coefficient of muscle tissue for 2450-MHz radiation, given that $\sigma = 2.21$ per ohm-meter, and the relative dielectric coefficient is 47.

(b) By what depth will 95% of the radiation be absorbed?

(a) Substituting the appropriate values into equation (14.31) gives

$$\alpha = \omega\left(\frac{\mu\epsilon}{2}\right)^{\frac{1}{2}}\left\{\sqrt{\left[1+\left(\frac{\sigma}{\omega\epsilon}\right)^2\right]}-1\right\}^{\frac{1}{2}}$$

$$= 2\pi \times 2.45 \times 10^9\ \text{s}^{-1}[\tfrac{1}{2}(4\pi \times 10^{-7}\ \text{N/A}^2)(47 \times 8.85 \times 10^{-12}\ \text{C}^2/\text{N-m}^2)]^{\frac{1}{2}}$$

$$\times \left\{\left[1+\left(\frac{2.21\ (\text{ohm-m})^{-1}}{2\pi \times 2.45 \times 10^9\ \text{s}^{-1} \times 47 \times 8.85 \times 10^{-12}\ \text{C}^2/\text{N-m}^2}\right)^2\right]^{\frac{1}{2}} - 1\right\}^{\frac{1}{2}}$$

$$= 59.87\ \text{m}^{-1} = 0.6\ \text{cm}^{-1}.$$

(b) At the depth that will include 95% of the absorbed energy, the power density will be 5% of the initial power density. Therefore, from equation (14.30), we have

$$\frac{E_2}{E_1} = 0.05 = e^{-2 \times 0.6\ \text{cm}^{-1} \times t\ \text{cm}},$$

$$\ln 0.05 = -2 \times 0.6 \times t,$$

$$t = 2.5\ \text{cm}.$$

The effectiveness of an absorber for radiation of any given wavelength is measured by the *penetration depth* in the absorber. At an absorber thickness $t = 1/\alpha$, the power density is reduced by absorption to

$$\frac{E_2}{E_1} = e^{-2} = 0.135$$

of its original value. Since the power density is reduced to 13.5%, this means that 86.5% of the energy is absorbed in a thickness equal to $1/\alpha$. This thickness is called the penetration depth (δ), and its exact value depends on the frequency of the radiation and on the electrical and magnetic properties (μ, σ, and ϵ) of the absorber. In the case of the 2450-MHz radiation of Example 14.16, the penetration depth in muscle tissue is

$$\delta = \frac{1}{\alpha} \tag{14.32}$$

$$= \frac{1}{0.6\ \text{cm}^{-1}} = 1.67\ \text{cm}.$$

The rate of heat generation in any absorber, such as biological tissue, is inversely proportional to the square of the penetration depth. Thus, a tissue with a relatively small penetration depth because of high water content, such as muscle, will heat much faster, under microwave radiation, than will a tissue such as fat, whose penetration depth is relatively large because of its very low water content. At 2450 MHz, for example, the penetration depths in muscle and fat are 1.67 cm and 8.1 cm, respectively. The rate of heating in the muscle, therefore from a given power density, will be about $(8.1/1.67)^2 = 23.5$ times greater than in the fat.

Biological Effects

Organs and tissues are made of a structural matrix that is bathed in biological fluids. The structural matrix is built of fixed molecules that often are electrically polarized, while the biological fluids contain ions of dissolved electrolytes and macromolecules. Under the influence of the electric fields from high frequency electromagnetic radiation, these polar molecules and ions experience electric forces whose magnitude is proportional to the product of the electric field intensity and the charge on the ion or on the polar molecule.

$$f = q\mathscr{E}. \tag{2.34}$$

These induced forces lead to current flow in the case of the dissolved ions and consequently to joule heating of the biological material. The rapidly alternating electrical forces on the immobile structural molecules may cause them to vibrate or to rotate, which in turn leads to heat production. Additionally, the electric field-induced forces may change the spatial distribution of polar molecules from a random orientation to an orientation alligned with the electric field. In all cases of microwave irradiation of living material (single cells to complex organisms) both of these induced force effects occur simultaneously. When biological effects are due mainly to heating, we say that we have *thermal effects*; when a biological effect cannot be attributed to heating, we say that we have a *non-thermal effect*. Thermal effects are associated with exposures greater than 10 mW per cm^2, while non-thermal effects generally are associated with exposures less than 10 mW per cm^2.

Animal studies show that microwave radiation, in frequencies ranging from 200 to 24,000 MHz, is lethal if the product of exposure intensity and time is sufficiently great to increase the body temperature beyond the body's homeostatic capabilities (> ~ 5°C). For example, rats exposed to 3,000 MHz radiation at a power density of 300 mW/cm^2 suffered a temperature increase of 8–10°C and died after 15 minutes of exposure; at a power density of 100 mW/cm^2, rats died after 25 minutes of exposure, when their body temperature had risen 6–7°C.

The temperature rise in any absorbing medium is related to the absorbed power density P_a, by

$$P_a = \frac{\rho \dfrac{\text{kg}}{\text{m}^3} \times c \dfrac{\text{J}}{\text{kg-}^\circ\text{C}} \times \Delta T\,^\circ\text{C}}{t \text{ sec}} \frac{\text{W}}{\text{m}^3}, \tag{14.33}$$

where ρ = density of the absorbing medium,
c = specific heat of the absorbing medium,
ΔT = temperature increase,
t = exposure time.

If the absorbing medium is living tissue, the temperature increase will differ from that predicted by equation (14.33) because of heat loss due to evaporative cooling, and convective and conductive heat loss or gain. The net amount of heat stored in the body is given by the heat balance equation

$$S = M \pm R \pm C - E, \tag{14.34}$$

where M = metabolic heat rate,
R = radiative heat gain or loss,
C = conductive and convective heat gain or loss,
E = heat loss due to evaporative cooling,
S = rate of heat storage.

While resting, a person generates metabolic heat at a rate of about 75 W; while engaged in moderate work or exercise, the metabolic heat output increases to about 300 W. This metabolic heat is dissipated in the environment when the temperature and humidity do not exceed the comfort range. If the temperature and humidity are too high to dissipate the metabolic heat, the person's body temperature increases. The additional heat load due to absorption of microwave energy must be dissipated in exactly the same manner as any other heat load. An unacceptable heat stress results when the combination of heat load and environmental conditions lead to an increased body temperature of 1°C or more. The "comfort range" is measured by the *temperature-humidity index* (THI), which is defined by

$$\text{THI} = 0.72\ (T_d + T_w) + 40.6, \tag{14.35}$$

where T_d = dry bulb temperature, °C,
T_w = wet bulb temperature, °C.

In terms of percent relative humidity, RH, equation (14.35) may be approximated by

$$\text{THI} = 1.44\ T_d + 0.1\ \text{RH} + 30.6. \tag{14.36}$$

THI values of 65 to 80 are usually considered comfortable. As the THI value increases above 80, the heat strain on a person becomes increasingly difficult, and additional heat loads may lead to overheating.

Now let us consider the thermal stress from absorpton of microwave radiation. If we assume the projected area of a person to be 0.9 m^2, an incident radiation beam whose power density is 10 mW/cm^2 will deliver energy to a person at a rate of

$$0.9\ \text{m}^2 \times 10 \frac{\text{mW}}{\text{cm}^2} \times 10^4 \frac{\text{cm}^2}{\text{m}^2} \times 10^{-3} \frac{\text{W}}{\text{mW}} = 90\ \text{W},$$

if all the incident energy is absorbed. This additional heat load is about the same as that which a person generates in the course of his normal activities, and usually can be dissipated without undue thermal strain if the THI does not exceed about 70. As the THI increases above 80, a radiation exposure at 10 mW/cm^2 could lead to unacceptable thermal strain. Thus, we see that the biological thermal stress from whole body exposure at power densities on the order of 1–10 mW/cm^2 depends strongly on the temperature-humidity index.

Most of the documented harmful biological effects in man from microwaves are attributed to hyperthermia. These include damage mainly to the eyes and to the testicles. Both these tissues are relatively ischemic, and thus are unable to efficiently dissipate energy that is absorbed at a rate greater than 10 to 15 milliwatts per square cm. The lens of the eye is avascular and encapsulated, and thus is particularly vulnerable to heat buildup and temperature rise from high radiation intensities. Fig. 14.16 shows the relationship between energy absorption in the lens of the eye and the consequent temperature rise. Microwave radiation through heating, and possibly through a non-thermal mechanism, initiates a chain of events that ultimately may lead to cataracts. The site of the initially observed lesions from microwave radiation is not in the lens substance, but rather on the posterior surface of the lens capsule. The pathogenesis of cataracts from ionizing radiation is similar to that of microwave induced cataracts. "Senile" cataracts (those due to aging), on the other hand, originate in the anterior surface of the lens.

Although the time-intensity relationship for cataractogenesis is not precisely known, one of the principal factors in microwave cataratogenesis is the increased temperature of the

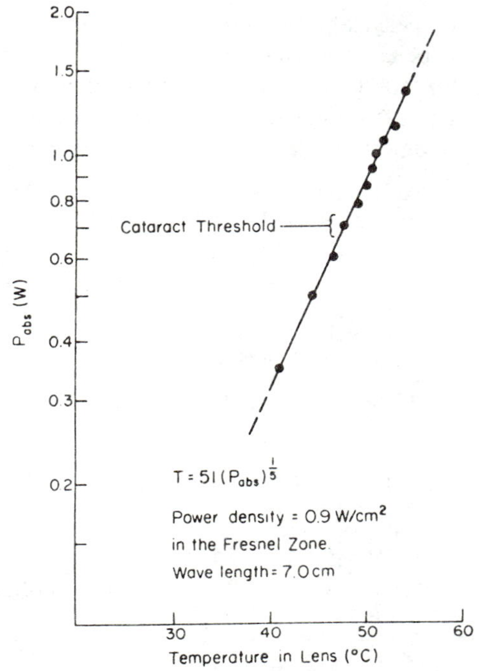

FIG. 14.16. A log–log plot of power absorption versus temperature rise in the lens. (From F. G. Hirsch, Microwave cataracts: A case report reevaluated. In *Electronic Product Radiation and the Health Physicist*, BRH/DEP 70-26, Bureau of Radiological Health, Rockville, Md, 1970.)

lens. Time-intensity exposure conditions that lead to an intraocular temperature of 45°C or higher are believed to be cataractogenic. When translated to practical conditions, a high risk of developing a cataract is associated with exposures on the order of a hundred or more mW per square cm.

Testicular function is strongly influnced by temperature. Normally the temperature of the testicles, which are outside the core of the body, is about 2°C less than the core temperature of 37°C. Elevation of the testicular temperature even to 37°C depresses spermatogenesis. This effect is reversible and spermatogenesis again proceeds normally when the temperature of the testes is lowered. Animal experiments have shown the threshold for minimal reversible testicular damage, consisting of depressed spermatogenesis and degeneration of the epithelial lining of the seminiferous tubules, to be 10 mW per square cm.

Although the biological mechanisms for non-thermal effects are not clearly understood, many such effects have been observed in the laboratory. The main impetus for experimental study of non-thermal effects was the persistent reports from the East-European countries of physiological effects on workers who had prolonged histories of exposure to low-level ($< 10\ mW/cm^2$) microwave exposure. Table 14.14 lists the findings in one of these reports, while Table 14.15 summarizes the various subjective effects mentioned in these reports. Because of the lack of dosimetric quantification of the radiation fields and also because all the symptoms described in these reports are normally associated with a population that is simply growing older, it is difficult to evaluate the public health significance of these reports. Laboratory studies with animals at relatively high exposure levels confirm changes in electroencephalograms, alteration of the blood-brain barrier, alteration of cell membrane permeability, hematologic effects, teratogenic effects, central nervous system effects, and behavioral changes. Until dosimetry problems are solved and uncertainties in dose-response relationships for repeated and for continuous low-level exposure are eliminated, a prudent degree of conservatism must continue to be exercised in the control of hazards from microwave and radio-frequency radiation.

TABLE 14.14. CLINICAL FINDINGS AMONG 525 WORKERS CHRONICALLY EXPOSED TO MICROWAVES

Symptoms
1. Increased fatigability
2. Periodic or constant headaches
3. Extreme irritability
4. Sleepiness during work
5. Decrease in olfactory sensitivity
Signs
1. Bradycardia
2. Hypotension
3. Hyperthyroid
4. Increase in blood histamine level

From Letavet, A. A., and Z. V. Gordon, Biological action of ultra high frequencies. U.S.S.R. Academy of Medical Science, Translation 12 471, U.S. Joint Pub. Res. Service, 1962.

TABLE 14.15. SUBJECTIVE EFFECTS ON WORKERS IN MICROWAVE FIELDS

Headaches
Eyestrain
Fatigue
Dizziness
Disturbed sleep
Sleepiness during daytime
Moodiness
Irritability
Unsociability
Hypochondriac reactions
Feelings of fear
Nervous tension
Mental depression
Impaired memory
Pulling sensation in scalp and brow
Pain in muscles
Pain region of heart
Breathing difficulties
Increased sweating of extremities
Difficulties in sex life

From Marha, K., J. Musil, and H. Tuha. *Electromagnetic Fields and the Life Environment*. San Franscisco Press, San Francisco, 1970.

Microwave Measurements

At low frequencies, we measure current through a circuit element and the voltage across the circuit element, and thus determine the power dissipated in the circuit element. In electromagnetic fields, we may measure the electric field intensity or the magnetic field.

When the relationship between the electric and magnetic field intensities is known, as in the case of the far-field, then the power density can be calculated from measured values of $\mathscr{E}$ or H, as illustrated in example 2.17. In any case, the electromagnetic field is probed by an antenna that interacts with the electromagnetic field. In the case of a dipole antenna, for example, a voltage is induced across the antenna. This voltage is proportional to the electric field intensity.

$$V = l\frac{\mathscr{E}_0}{\sqrt{2}}, \tag{14.37}$$

where l, the constant of proportionality is the effective length of the antenna and $\mathscr{E}_0$ is the maximum electric field strength. Thus, the electrical response of any given dipole probe is quantitatively related to the intensity of the electric field. By using a calibrated probe, therefore, we can measure the electric field strength. For far-field conditions, the power density can be readily calculated from electric field strength using equation (2.70). Radiation survey meters usually are calibrated to read the far field power density corresponding to the measured electric field intensity.

The devices most often used in microwave probes to convert the microwave energy into electric current or voltage are crystal diodes and bolometers. The crystal diode is a non-linear device that rectifies the received signal; and has an output voltage that is proportional to the antenna power input (typically, the sensitivity is on the order of about 0.5 millivolts per microwatt). Since microwave power density is proportional to the square of the electric field strength, the crystal detector is called a "square-law" detector. This square-law response is seen only for small power inputs; when the input power exceeds about 10 microwatts, the square-law response starts to change into a linear response, and at input powers in excess of 100 microwatts, the departure from the square-law response is significant, and may lead to serious measurement errors.

The bolometer often used in microwave measurements is the thermistor. Since the resistance of a thermistor decreases with increasing temperature, and since the temperature increase is proportional to the intensity of the microwave radiation, the radiation field strength may be determined by measuring changes in thermistor resistance. Thermistors are characterized by long time constants, on the order of 0.1 to 1 second, and therefore are useful for measuring average power. A *barretter* is a resistive element whose electrical resistance increases with temperature. Barretters are characterized by short time constants, on the order of 250–500 microseconds, and thus are suitable for measuring peak power. Thermistors and barretters can be used over a power range of one microwatt to several milliwatts.

A thermocouple is another temperature sensitive device. Heating one junction to establish a temperature difference between the heated junction and the cold junction generates a voltage whose magnitude depends on this temperature difference. In principle, a thermocouple detector simply requires an antenna and a meter for measuring the voltage.

A power meter may be a bridge in which one arm of the bridge is the transducing element. The bridge is balanced in the absence of microwave radiation, and the voltage drop across the transducer due to the absorption of microwave energy unbalances the bridge, with the degree of unbalance being proportional to the absorbed microwave power. Because the response of the detector follows the square-law only for a low power, it is common to use an attenuator between the receiving antenna and the detector. The degree of attenuation is given in decibels. For example, a 20-dB attenuator will attenuate the input power to output power by a factor P_i/P_0, which can be calculated as follows:

$$dB = 10 \log \frac{\text{input power}}{\text{output power}} = 10 \log \frac{P_i}{P_0},$$

$$20 = 10 \log \frac{P_i}{P_0},$$

$$\log^{-1} \frac{20}{10} = \frac{P_i}{P_0} = 100.$$

A 10-dB attenuator will reduce the power by a factor of 10, while a 30-dB attenuator will reduce the power by a factor of 1000.

Microwave power measuring instruments usually include four basic components:

1. calibrated antenna;
2. attenuator, which reduces the microwave power received by the antenna by a known factor;
3. transducer;
4. meter, a microammeter or millivoltmeter for measuring current or voltage changes due to absorption of microwave energy.

These components may be combined together into a single direct reading instrument that gives power density directly, Fig. 14.17, or they may be joined together as individual units

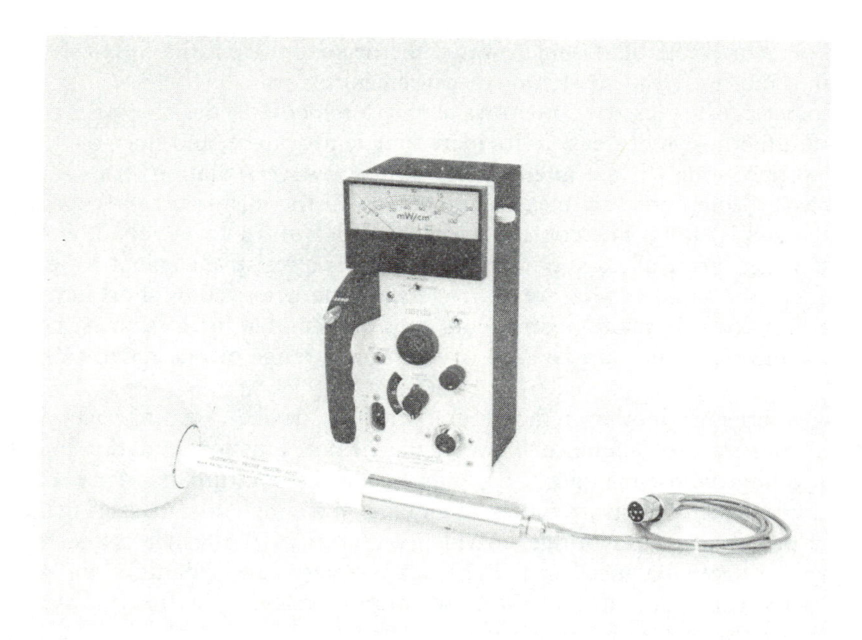

FIG. 14.17. Broad-band isotropic electromagnetic radiation survey meter. The instrument pictured above resonds over a frequency range of 0.3 to 26 GHz, and can measure a power density as low as 0.02 mW/cm². Perturbation of the radiation field by the probe is minimized through the use of resistive dipoles that minimize coupling into the field and current flow. (Courtesy Narda Microwave Corp.)

to measure the power density received by the antenna. In this case, the power density is related to the measured power by the relationship

$$W = \frac{4\pi\alpha P}{\lambda^2 G_a}, \tag{14.38}$$

where W = power density,
P = absorbed power,
λ = wavelength,
G_a = antenna gain, absolute power gain, not dB,
α = actual attenuation factor, not dB.

Example 14.17

A 200-ohm thermistor whose sensitivity is 25 ohms per mW is used with an antenna whose gain is 16 dB at 10,000 MHz and a 30-dB attenuator to measure the power density in a 10,000 MHz field. From the resistance change, the absorbed power is found to be 0.4 mW. What is the power density in the microwave field?

Antenna gain, G_a;

$$16\text{ dB} = 10 \log G_a,$$

$$G_a = \log^{-1}\frac{16}{10} = 40.$$

Wave length, λ:

$$f \times \lambda = 3 \times 10^{10}\text{ cm/sec},$$

$$\lambda = \frac{3 \times 10^{10}\text{ cm/sec}}{1 \times 10^{10}\text{ sec}^{-1}} = 3\text{ cm}.$$

Attenuation factor, α:

$$30\text{ dB} = 10 \log \alpha,$$

$$\alpha = \log^{-1}\frac{30}{10} = 1000.$$

Power density

$$W = \frac{4\pi\alpha P}{\lambda^2 G_a}$$

$$= \frac{4\pi \times 1000 \times 0.4\text{ mW}}{(3\text{ cm})^2 \times 40}$$

$$= 14\text{ mW/cm}^2.$$

Several factors must be considered when choosing or using a microwave radiation monitoring instrument. First, introduction of the detector probe causes some distortion of the field, and the power density that is measured while the probe is in the field is not necessarily the power density at that point in the absence of the probe. This source of error can be minimized by using a small probe. A second factor is polarization. If we have a dipole receiving antenna, and if the dipole is oriented perpendicularly to the plane of polarization of the electric field, then no voltage will be induced in the dipole and no microwave energy will be detected. On the other hand, aligning the dipole parallel to the plane of polarization results in an induced voltage, and thus to detection of the microwave field. One method of accounting for polarization is to use a probe that contains three mutually perpendicular dipoles. The sum of the induced voltage in each of the three dipoles will be the same regardless of their orientation. Thirdly, microwave detectors are frequency dependent, and the frequency response of the probe must be matched to the frequency of the field being measured. Finally, detector probes will burn out if they absorb too much power. There is an upper limit, therefore, on the power density that can be measured with any particular probe.

A commonly used instrument for microwave radiation monitoring employs a dipole probe that has a thermocouple as the thermo-electric element for measuring the radiation (Fig. 14.17). The thermocouple generates an output voltage whose magnitude is proportional to the power absorbed in the element, thereby allowing the voltmeter used with this probe to be calibrated in units of power density.

Protection Guides and Standards

The mechanisms of interaction of ionizing radiation with living systems are reasonably well understood, and therefore energy absorption patterns can be easily calculated and measured. Dosimetry of ionizing radiation is thus on sound theoretical and practical bases, and radiation dose units that are based on energy absorption, such as the gray or the rad, are practically useful and biologically meaningful. In the case of microwave radiation, however, the mechanisms of its interaction with living systems are not completely understood. Irradiation of people and animals leads to complex, frequency specific, non-uniform energy absorption patterns that are difficult to calculate theoretically and to measure experimentally. Equation (14.31) shows the absorption coefficient for microwaves to depend on frequency, on the dielectric properties of the absorber, and on the electrical conductivity of the absorbing medium. Fig. 14.18 and Table 14.16 show the strong frequency dependence of electrical conductivity and dielectric coefficient for muscle. These electrical properties are strongly dependent on the amount of water in the absorber, and thus energy absorption by different organs and tissues in living systems is strongly dependent

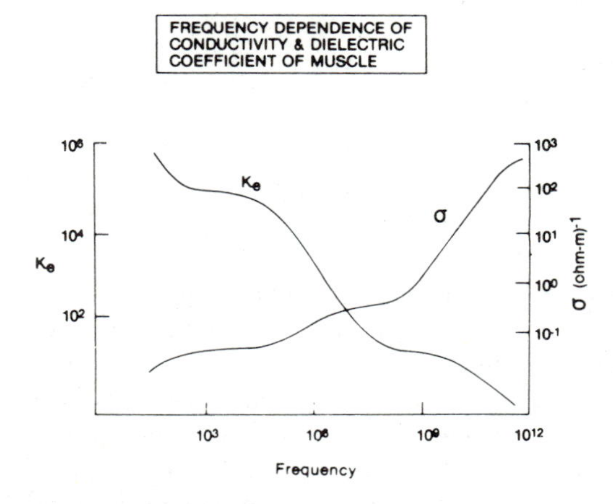

FIG. 14.18. Curve showing the dependence of conductivity and dielectric coefficient of muscle tissue on frequency of the irradiating electromagnetic field.

TABLE 14.16. ELECTRICAL PROPERTIES OF HUMAN TISSUES

	Muscle					Fat			
Frequency (MHz)	Wavelength in air (cm)	Dielectric constant ϵ_m	Conductivity Q_m (mho/m)	Depth of penetration (cm)	Wavelength in tissue (cm)	Dielectric constant ϵ_f	Conductivity Q_l (millimho/m)	Depth of penetration (cm)	Wavelength in tissue (cm)
27.12	1006	113	0.612	14.3	68.1	20	10.9–43.2	159	241
40.68	738	97.3	0.693	11.2	51.3	14.6	12.6–52.8	118	187
100	300	71.7	0.889	6.66	27	7.45	19.1–75.9	60.4	106
433	69.3	53	1.43	3.57	8.76	5.6	37.9–118	26.2	28.8
750	40	52	1.54	3.18	5.34	5.6	49.8–138	23	16.8
915	32.8	51	1.60	3.04	4.46	5.6	55.8–147	17.7	13.7
1500	20	49	1.77	2.42	2.81	5.6	70.8–171	13.9	3.41
2450	12.2	47	2.21	1.70	1.76	5.5	96.4–213	11.2	5.21

Adapted from Guy, A. W., Lehmann, J. F., and Stoneridge, J. B. Therapeutic Applications of Electromagnetic Power, *Proc. I.E.E.E.*, **62**:61, 1974.

on the degree of hydration of the organs or tissues. For example, skin, muscle, and internal organs all have large water contents, and therefore high absorption coefficients; fat, bone, and yellow bone marrow have small water contents, and thus low absorption coefficients. Furthermore, the situation is complicated by reflection at interfaces between different types of tissue, such as the interface between skin and fat, fat and muscle, and muscle and bone. This effect is implicit in the influence of physical size and geometry on energy absorption, as illustrated in Fig. 14.19, which shows the relative absorption cross section of spheres as a function of their radii and wavelength of the radiation. Because of the complexities, uncertainties, and unknowns in our knowledge of microwave radition absorption, a dose unit analogous to the gray or the rad, such as the *specific absorption rate* (SAR) which is specified in joules per second per kilogram, is impractical at this time as the basis for setting protection standards. Instead, radiation protection standards are specified in terms of radiation exposure, such as the environmental power density, which is expressed in milliwatts per square centimeter, or electric field strength in units of volts per meter, or magnetic field strength in units of amperes per meter. Power density is applicable to far-field conditions, while electric or magnetic field intensity is applicable to near-field conditions.

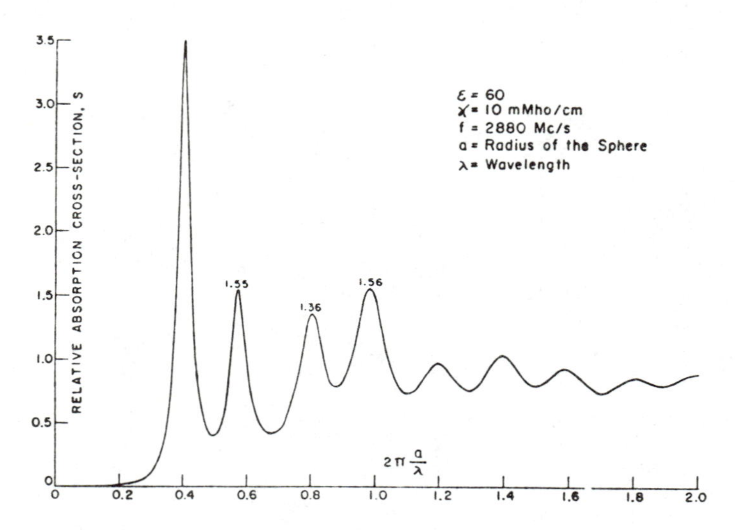

FIG. 14.19. Curve showing the dependence of energy absorption cross section (for 28.8-GHz radiation) of a sphere of muscle equivalent material on the ratio of size to wavelength. (From Sher, L. D. Interaction of microwave and RF energy on biological material, *Electronic Product Radiation and the Health Physicist*, BRH/DEP 70-26, Bureau of Radiological Health, Rockville, MD, 1970.)

Regulations for the protection of workers and members of the public from the harmful effects of microwave radiation have been promulgated in numerous countries around the world. In the United States, the Occupational Safety and Health Administration (OSHA) established a maximum level of 10 mW/cm^2 incident electromagnetic power density for frequencies of 10 MHz to 100 GHz, inclusive, averaged over a 6-minute period. No distinction is made between pulsed and continuous radiation in the standard. The Bureau of Radiological Health in the United States is charged with the responsibility of protecting the public from the harmful radiation effects from consumer products. For the case of radiation from microwave ovens, the Bureau set a standard of 1 mW/cm^2 leakage at a distance of 5 cm from the oven when the oven leaves the factory; furthermore, the oven must be so designed and built that the maximum leakage that may be expected during its lifetime is 5 mW/cm^2 at 5 cm. Many other countries have set the same occupational control standards as those in the United States. Notable exceptions, however, include the Soviet Union and her European neighbors. In the U.S.S.R., the maximum occupational exposure limits are: 10 microwatts/cm^2 for continuous exposure, 100 microwatts/cm^2 for 2 hours exposure per day, and 1000 microwatts/cm^2 for exposures not exceeding 20 minutes per day. In Poland, the maximum level for continuous exposure during a 10-hour working day is 200 microwatts/cm^2. Exposure at higher power densities, up to a maximum of 10 mW/cm^2, is permissible for shorter periods of time. The maximum occupational exposure level of 10 mW/cm^2 in the United States was set on the basis of thermal considerations only. It was believed to be the maximum sustained heatload which the homeostatic thermo-regulating system of a healthy adult worker can accommodate. The Soviet Union's 10-microwatt/cm^2 maximum level for continuous occupational exposure, on the other hand, is based on the central nervous system effects; it is believed to be an exposure level that will elicit no biological response, not even those responses that are included within a person's usual range of biological reactions to the ordinary stresses of living. For non-occupational exposure, and despite the fact that 10 microwatts/cm^2 is believed to be below the threshold of detectable biological effects, the Soviet Union has set a maximum exposure level of one microwatt/cm^2 for the general population. In this context, it should be noted that measurements made in several American cities, whose combined population was 8,300,000, showed the median population exposure level to man-made and natural background to be 0.014 microwatts of cm^2, while 99% of the population was found to be exposed to power densities less than one microwatt/cm^2. Both occupational standards, 10 mW/cm^2 and 10 microwatts/cm^2, are controversial, the former believed by many critics to be too high and the latter too low. On the basis of bioeffects studies, especially non-thermal bioeffects, new protection guides and standards are being considered in the United States. In contrast to the current (and soon to be replaced) standard which deals only with frequencies of 10 MHz to 100 GHz, and treats all these frequencies equally, the proposed new standards span the frequency range of 0.3 MHz to 300,000 Mhz (300 GHz). Furthermore, the proposed new standards consider the relative biological effectiveness, which is based on the frequency dependent penetration depth and on the electrical and thermal properties of the tissues that consequently absorb energy from the electromagnetic field, of the various frequencies. The American National Standards Institute (ANSI) is considering the levels tabulated in Table 14.16, while The National Institute of Occupational Safety and Health (NIOSH) is considering levels shown in Fig. 14.20 and 14.21 as safety criteria for exposure to microwaves. It should be noted that the two sets of standards under consideration do not differ very much. For example, for 2 MHz, ANSI allows 100 mW/cm^2, while NIOSH allows 375 V/m, which corresponds to 37.3 mW/cm^2. For 100 MHz, ANSI allows 1 mW/cm^2, while NIOSH allows 60 V/m, which corresponds to 1 mW/cm^2, while at a frequency of 3000 MHz, ANSI allows 5 mW/cm^2 and NIOSH allows 5.2 mW/cm^2. Whole-body exposure at any of the levels given in Table 14.16 is believed to result in a whole body average specific absorption rate (SAR) of about 0.4 W/kg. If a worker is exposed simultaneously to radiation of several different frequencies, then, in order to stay within the recommended exposure limits,

$$\frac{W_1}{L_1} + \frac{W_2}{L_2} + \cdots + \frac{W_n}{L_n} \leq 1, \tag{14.39}$$

where W_1, W_2, W_n = power density for frequency 1, 2, n,
L_1, L_2, L_n = limiting power density for frequency 1, 2, n.

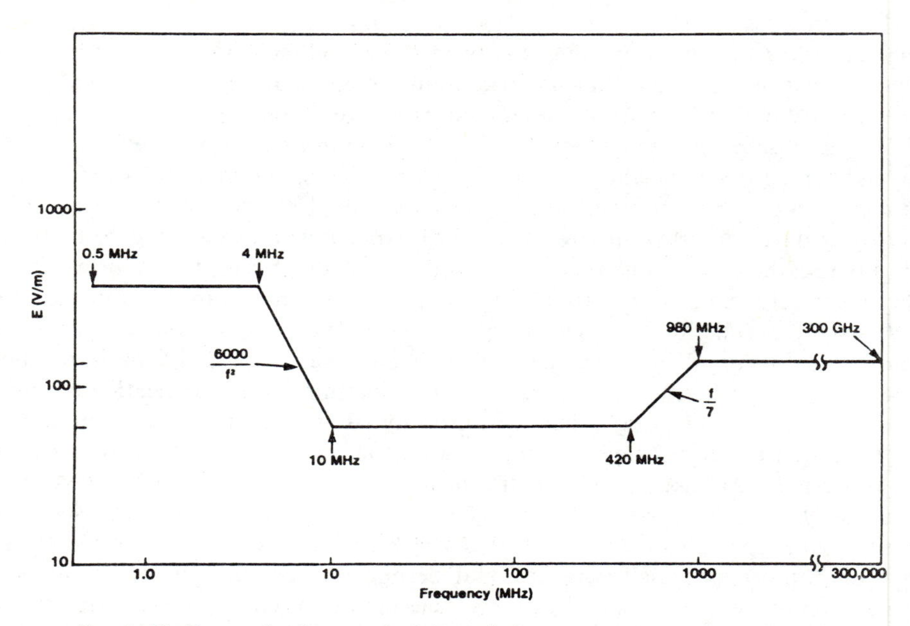

FIG. 14.20. Proposed safety standard for radiofrequency and microwave radiation based on electric field intensity and frequency. This standard is proposed by the National Institute of Occupational Safety and Health.

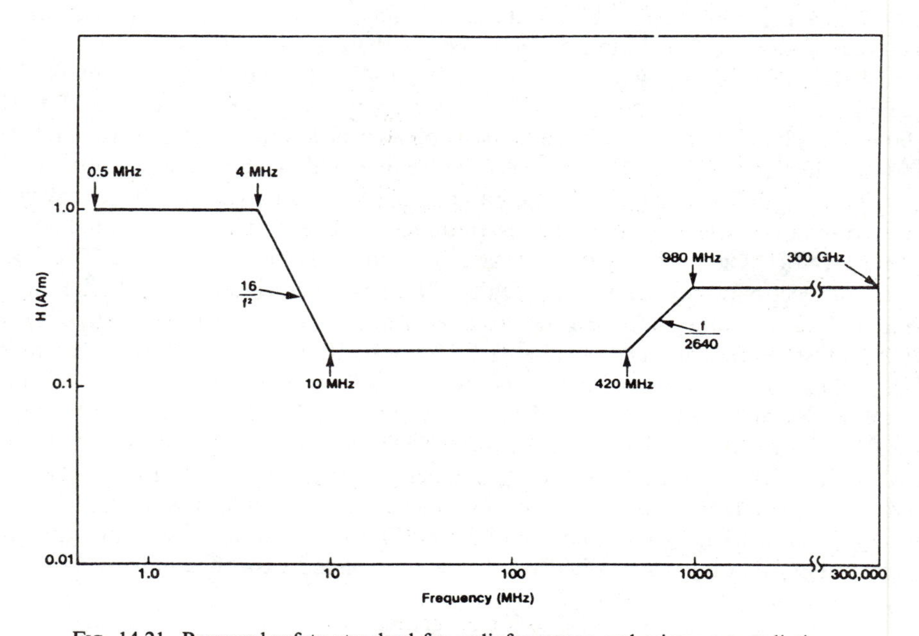

FIG. 14.21. Proposed safety standard for radiofrequency and microwave radiation based on magnetic field intensity and frequency. This standard is proposed by the National Institute of Occupational Safety and Health.

TABLE 14.17. WHOLE-BODY RADIOFREQUENCY PROTECTION GUIDES

Frequency range (MHz)	Power density (mW/cm²)	E^2 (V/m)²	H^2 (A/m)²
0.3–3	100	400,000	2.5
3–30	$900/f^2$	4,000 $(900/f^2)$	0.025 $(900/f^2)$
30–300	1.0	4,000	0.025
300–1500	$f/300$	4,000 $(f/300)$	0.025 $(f/300)$
1500–300,000	5	20,000	0.125

Notes: (1) f is the frequency in MHz. (2) Power density is applicable only to far field-exposure. (3) Electric field and magnetic field strengths only are applicable to near-field exposure.

Problems

1. The lethal absorbed dose of 265 nm ultraviolet light for *E. coli* bacteria is 14 MeV. To how many photons of this UV radiation does the lethal absorbed dose correspond?

2. An yttrium–alunimum–garnet (YAG) laser emits near-infrared radiation 1060 nm in wavelength at a power level of 10 watts cw. The exit aperture is 3 mm diameter, and the beam divergence is 5 mrad. Calculate
(a) the $1/e^2$ diameter at the aperture,
(b) the $1/e^2$ diameter at a distance of 10 meters,
(c) the irradiance at a distance of 10 meters.

3. A 0.1-J ruby laser has an aperture of 7 mm and a beam divergence of 1 mrad.
(a) What is the radiant exposure at distances of 5 and 10 meters from the aperture?
(b) How far behind the laser aperture is the virtual focal point from where the laser light seems to originate?
(c) Is the inverse square law applicable if distances are measured from the virtual focal point of the laser beam?

4. The output of a cw pumped Nd:YAG laser is Q-switched at a pulse repetition frequency of 10 kHz. If each pulse is 50 nanoseconds wide, and if the mean power output is 10 W, calculate
(a) the duty cycle,
(b) the peak power per pulse,
(c) the energy per pulse.

5. What is the recommended MPE value, in mW/cm², for direct ocular exposure for 1 second to a beam from a He–Ne laser operating at a power level of 5 mW cw?

6. What is the protection standard, i.e., the maximum permissible irradiance for a direct intrabeam exposure to a 0.8-millisecond-wide pulse from a 694.3-nm ruby laser?

7. A pulsed ruby laser, $\lambda = 694.3$ nm, is operated in a laboratory at a level of 5×10^{-4} J/pulse, 100-μsec pulse width, and a PRF of 60 pulses per second. The beam exit aperture is 1 mm diameter, and the beam divergence is 0.1 mrad. What optical density is required for protective goggles for
(a) incidental exposure for as long as 2 seconds?
(b) continuous exposure?

8. A He–Ne laser, $\lambda = 632.8$ mm, is operated at a power level of 3 W cw. The beam aperture is 0.9 mm diameter, and the beam divergence is 0.9 mrad. If the possibility exists for momentary accidental intrabeam ocular exposure not exceeding 0.25 second, calculate the minimum required optical density of protective goggles for exposure,
(a) at the laser,
(b) at a distance of 100 meters,
(c) What is the minimum required OD for continuous intrabeam viewing for up to 100 seconds at a distance of 100 meters?

9. A scanning He–Ne laser that scans at a rate of 10 per second emits 5 mW through a aperture of 0.7 cm. If the beam divergence is 5 mrad, then, for an intra-veam viewing distance of 200 cm, calculate
(a) the time during each scan that the pupil of the eye can be exposed,
(b) the radiant exposure per scan,
(c) the average irradiance at the cornea.
(d) What hazard class should be assigned to this laser?

10. A microwave beam is pulsed 100 times per second, the pulse width is 1 μsec, and the peak power is 1 MW.
(a) What is the duty cycle?
(b) What is the average power?

11. A radar operates at an average power level of 20 W; its pulse width is 1 μsec, and the PRF is 1000 per second. Calculate
(a) the duty cycle,
(b) the peak power.

12. Calculate the gain of a 30-cm-diameter parabolic antenna used to transmit at a frequency of 3,000 MHz. Express the answer as
(a) power ratio,
(b) dB.

13. A far-field measurement shows the power density in a 1,000 MHz radiation field to be 4 mW/cm^2. Calculate
(a) the electric field strength, in V/m, and the square of the electric field strength,
(b) the magnetic field strength, in A/m, and the square of the magnetic field strength,
(c) the energy density, in pJ/cm^3.

14. How many dB attenuation of power density are required to reduce a 1 W/cm^2 field to 1 mW/cm^2?

15. What is the maximum magnetic intensity in a plane electromagnetic wave whose maximum electric intensity is 100 V/m?

16. A dipole antenna with a 50 cm parabolic dish radiates 100 W of power at a frequency of 2,400 MHz. Calculate the
(a) mean power density at the aperture,
(b) maximum power density in the near field,
(c) distance fo the far-field,
(d) power density at 1 meter.
(e) distance at which the power density is down to 1 mW/cm^2.

17. A rectangular horn antenna, 17 × 24 cm and operating at a frequency of 2400 MHz has an effective area of 200 cm^2. If the radiated power is 100 W, calculate the
(a) mean power density at the aperture,
(b) the maximum power density in the near field,
(c) the distance to the start of the far field,
(d) the power density in the far field, at a distance of 5 meters,
(e) distance at which the power density is down to 1 mW/cm^2.

18. A radar installation, using a parabolic dish antenna 1.2 m in diameter, has a peak power output of 2 MW of 10 GHz radiation. If the pulse repetition frequency is 200 per second and the pulse width is 5 μsec, calculate the
(a) duty cycle,
(b) average power output,
(c) antenna gain in dB,
(d) downrange distance to an average power density of 1 mW/cm^2.

19. A radar that rotates at 3 rpm operates at a frequency of 10 GHz and peak power of 2 MW. The PRF = 400 per second, the pulse width is 3 microseconds, the beam width is $4\frac{1}{2}^\circ$, and the parabolic dish is 0.5 meter in diameter. Calculate
(a) the duty cycle,
(b) the average power,
(c) the power density at a distance of 50 meters,
(d) the energy density per 6-minute exposure at a distance of 50 meters. Is this calculated exposure within the OSHA limit?

20. Show by dimensional analysis that the unit for α in equation (14.31) is per meter.

21. Calculate the wavelength of 100-MHz microwaves in muscle and in fat.

22. Calculate the penetration depth of 100 MHz radiation in fat and in muscle.

23. How many dB attenuation will reduce a field of 800 mW/cm^2 to the power density of 1 mW/cm^2 recommended by ANSI for occupational exposure?

24. (a) How far from a 10-GHz radar transmitter would the far-field be expected to begin if the antenna diameter is 0.6 m?

(b) A power meter, whose calibration at 10 GHz gives -10 dB as the power level that corresponds to 1 mW/cm^2, is used to measure the power density at the distance calculated in part (a). The meter reading is $+3.5$ dB. What is the power density at this point?

At what distance from the antenna will the power density be down to -7 dB?

25. A microwave survey meter that reads in dB is calibrated to read -8 dB in a field whose power density is 0.01 mW/cm^2. It is then used in a radiation survey, and gives a reading of $+2.5$ dB at a certain point in a microwave field. What is the power density at the point?

26. A power meter reads 5 mW in a 10 GHz microwave field. If the standard gain horn has a gain of 16 dB at this frequency, and if a 30 dB attenuator was used, what is the power density in the microwave field?

27. A radar whose beam width is 12° rotates at 3 rpm. What is the irradiation time for a person who remains in the scanning field for 0.1 hour?

28. A radar has the following characteristics:

$f = 10$ GHz	Beam width = 2.5°
Peak power = 2 MW	Rotational frequency = 4 rpm
PRF = 200 pps	Dish diameter = 1.22 m
Pulse width = 5 μsec	

(a) What is the duty cycle?
(b) What is the average power?
(c) What is the power density at 100 meters?
(d) A person spends 1 hour at this distance. According to the Occupational Safety and Health Administration (OSHA) standards, is his exposure within acceptable limits?

Suggested References

American Conference of Governmental Industrial Hygienists. *A Guide for Control of Laser Hazards*. ACGIH, 1976.

American National Standards Institute. *Techniques and Instrumentation for the Measurement of Potentially Hazardous Electromagnetic Radiation at Microwave Frequencies*. ANSI C953-1973, Institute of Electrical and Electronic Engineers, New York, 1973.

American National Standards Institute. *The Safe Use of Lasers*. ANSI Z-136-1-1976, American National Standards Institute, New York, 1976.

BARANSKI, S., and CZERSKI, P. *Biological Effects of Microwaves*. Dowden, Hutchinson, and Ross, Stroudsberg, Pa., 1976.

CARLSON, F. P. *Introduction to Applied Optics for Engineers*. Academic Press. New York, 1971.

CHARSCHAN, S. S. *Lasers in Industry*. Van Nostrand Reinhold, New York, 1972.

CLEARY, S. Biological effects of microwaves and radiofrequency radiation. *CRC Critical Reviews in Environmental Control*, **7**:121, 1977.

Department of the air Force. *Laser Health Hazards Control*. AFM 161–8, Department of the Air Force, Washington, D.C., 1969.

Department of the Army. *Hazards from Microwave ovens and Inspection Guidelines*. USAEHA-RL, Environmental Hygiene Agency, Edgewood Arsenal, MD, 1974.

Department of the Army. *Control of Hazards to Health from Laser Radiation*. TB-MED 279, Department of the Army, Washington, D.C., 1975.

Department of the Army and the Air Force. *Control of Hazards to Health from Microwave Radiation*. TB-MED 270/AFM161, Department of the Army, Washington, D.C., 1965.

DURNEY, C. H., JOHNSON, C. C., BARBER, P. W., MASSOUDI, H. ISKANDER, M. F., LORDS, J. L., RYSER, D. K., ALLEN, S. J., and MITCHELL, I. C. *Radiofrequency Radiation Dosimetry Hanbook*, 2nd ed. SAM-TR-78-22, School of Aerospace Medicine, Brooks Air Force Base, TX, 1978.

Electronic Engineering Association. *A General Guide to the Safe Use of Lasers*. EEA, London, 1966.

GOLDMAN, L., ROCKWELL, R. J. Jr., and HORNBY, P. Laser laboratory design and personnel protection from high energy lasers. In *Handbook of Laboratory Safety* 2nd ed., N.C. Steere, ed. Chemical Rubber Publishing Company, Cleveland, 1971.

GORDON, Z. V. Occupational health aspects of radiofrequency electromagnetic radiation. In *Ergonomics and Physical Environmental Factors*. Occupational Safety and Health Series, No. 21, International Labor Office, Geneva, 1970.

HALLIDAY, D., and RESNICK, R. *Physics*, 3rd ed. John Wiley and Sons, New York, 1977.

HAZZARD, D. G. ed. *Symposium on Biological Effects and Measurement of Radiofrequency/Microwaves*. HEW Publication (FDA) 77-8026, Bureau of Radiological Health, 1977.

JENKINS, F. A., and WHITE, H. E. *Fundamentals of Optics*, 3rd ed. McGraw Hill, New York, 1957.

JOHNSON, C. C., DURNEY, C. H., BARBER, P. W., MASSOUDI, H., ALLEN, S. J., and MITCHELL, J. C. *Radiofrequency Radiation Dosimetry Handbook*. SAM TR-76-35, USAF School of Aerospace Medicine, Brooks Airforce Base, Tx, 1976.

LARGENT, E. J., and OLISHIFSKI, J. B. Non-Ionizing Radiation. in *Fundamentals of Industrial Hygiene*, J. B. Olishifski, ed. National Safety Council, Chicago, 1979.

Laser Institute of America. *Laser Safety Guide*. LIA, Cincinnati, 1976.

LENGYEL, B. A. *Introduction to Laser Physics*. John Wiley and Sons, New York, 1966.

MAGID, L. M. *Electromagnetic Fields, Energy and Waves*. John Wiley and Sons, New York, 1972.

MOORADIAN, A., JAEGER, T., and STOKSETH, P., eds. *Tunable Lasers and Applications*. Springer–Verlag, New York, 1976.

National Council on Radiation Protection and Measurement. *Radiofrequency Electromagnetic Fields*. NCRP Report No. 67, NCRP, Washington, D.C., 1981.

New York Academy of Medicine. Symposium on health effects on non-ionizing radiation. *Bulletin of the New York Academy of Medicine*, Vol. 55, No. 11, December, 1979.

PRESSLEY, R. J. *Handbook of Lasers*. Chemical Rubber Publishing Company, Cleveland, 1971.

SEARS, F. W., and ZEMANSKY, M. W. *University Physics*, 4th ed. Addison-Wesley, Reading, MA, 1970.

SIEGMAN, A. E. *Introduction to Lasers and Masers*. McGraw Hill, New York, 1971.

SLINEY, D., and WOLBARSHT, M. *Safety with Lasers and Other Optical Sources*. Plenum Press, New York, 1980.

SVELTO, O. *Principles of Lasers*. Plenum Press, 1976.

TAYLOR, L. S., and CHENG, A. Y., eds. *The Physical Basis of Electromagnetic Interactions with Biological Systems*. HEW Publication (FDA) 78-8055, Bureau of Radiological Health, Rockville, MD, 1978.

TYLER, P. E., ed. Biologic effects of non-ionizing radiation. *Annals of the New York Academy of Sciences*, vol. 247, 1975.

APPENDIX I

VALUES OF SOME USEFUL CONSTANTS

Quantity	Symbol	Value SI	Value cgs
Electron charge	e	1.6 E-19 C	4.8 E-10 SC
Electron mass	m_0	9.0186 E-31 kg; 0.0000548 amu	9.0186 E-28 g; 0.000548 amu
Proton mass		1.6726 E-27 kg; 1.007277 amu	1.6726 E-24 g; 1.007277 amu
Neutron mass		1.67492 E-27 kg; 1.008665 amu	1.67492 E-24 g; 1.008665 amu
Atomic mass unit	amu	1.660531 E-27 kg; 931 MeV	1.660531 E-24 g; 931 MeV
Speed of light	c	2.997928 E8 m/s	2.997928 E10 cm/s
Avogadro's number	N	6.0247 E23 mol^{-1}	6.0247 E23 mol^{-1}
Planck's constant	h	6.6262 E-34 $J \cdot s$	6.6262 E-27 $erg \cdot s$
Gas constant	R	8.3144 $J \cdot mol^{-1} \cdot K^{-1}$	8.3144 E7 $ergs \cdot mol^{-1} \cdot {}^{\circ}K^{-1}$; 0.0805 $liter \cdot atm \cdot mol^{-1} \cdot {}^{\circ}K^{-1}$
Boltzman's constant	k	1.38062 E-23 $J \cdot K^{-1}$	1.38062 E-16 $erg \cdot {}^{\circ}K^{-1}$
Acceleration of gravity	g	9.807 $m \cdot s^{-2}$	980.665 $cm \cdot s^{-2}$
Gravitational constant	γ	6.673 E-11 $N \cdot m^2 \cdot kg^{-2}$	6.673 E-8 $dyne \cdot cm^2 \cdot g^{-2}$

APPENDIX II

TABLE OF THE ELEMENTS

Name	Symbol	At. No.	Atomic weight
Actinium	Ac	89	227
Aluminum	Al	13	26.98
Americium	Am	95	(243)
Antimony	Sb	51	121.76
Argon	A	18	39.944
Arsenic	As	33	74.92
Astatine	At	85	(210)
Barium	Ba	56	137.36
Berkelium	Bk	97	(249)
Beryllium	Be	4	9.013
Bismuth	Bi	83	208.99
Boron	B	5	10.82
Bromine	Br	35	79.916
Cadmium	Cd	48	112.41
Calcium	Ca	20	40.08
Californium	Cf	98	(251)
Carbon	C	6	12.011
Cerium	Ce	58	140.13
Cesium	Cs	55	132.91
Chlorine	Cl	17	35.457
Chromium	Cr	24	52.01
Cobalt	Co	27	58.94
Columbium, *see* Niobium			
Copper	Cu	29	63.54
Curium	Cm	96	(247)
Dysprosium	Dy	66	162.51
Einsteinium	E	99	(254)
Erbium	Er	68	167.27
Europium	Eu	63	152.0
Fermium	Fm	100	(253)
Fluorine	F	9	19.00
Francium	Fr	87	(223)
Gadolinium	Gd	64	157.26
Gallium	Ga	31	69.72
Germanium	Ge	32	72.60
Gold	Au	79	197.0
Hafnium	Hf	72	178.50
Hahnium	Ha	105	(262)
Helium	He	2	4.003
Holmium	Ho	67	164.94
Hydrogen	H	1	1.0080
Indium	In	49	114.82
Iodine	I	53	126.91
Iridium	Ir	77	192.2
Iron	Fe	26	55.85
Kurchatovium	Ku	104	(261)
Krypton	Kr	36	83.80
Lanthanum	La	57	138.92
Lawrencium	Lw	103	(260)
Lead	Pb	82	207.21
Lithium	Li	3	6.940
Lutetium	Lu	71	174.99
Magnesium	Mg	12	24.32

Name	Symbol	At. No.	Atomic weight
Manganese	Mn	25	54.94
Mendelevium	Mv	101	(256)
Mercury	Hg	80	200.61
Molybdenum	Mo	42	95.95
Neodymium	Nd	60	144.27
Neon	Ne	10	20.183
Neptunium	Np	93	(237)
Nickel	Ni	28	58.71
Niobium (columbium)	Nb	41	92.91
Nitrogen	N	7	14.008
Nobelium	No	102	(254)
Osmium	Os	76	190.2
Oxygen	O	8	16.000
Palladium	Pd	46	106.4
Phosphorus	P	15	30.975
Platinum	Pt	78	195.09
Plutonium	Pu	94	(242)
Polonium	Po	84	210
Potassium	K	19	39.100
Praseodymium	Pr	59	140.92
Promethium	Pm	61	(147)
Protactinium	Pa	91	231
Radium	Ra	88	226
Radon	Rn	86	222
Rhenium	Re	75	186.22
Rhodium	Rh	45	102.91
Rubidium	Rb	37	85.48
Ruthenium	Ru	44	101.1
Samarium	Sm, Sa	62	150.35
Scandium	Sc	21	44.96
Selenium	Se	34	78.96
Silicon	Si	14	28.09
Silver	Ag	47	107.873
Sodium	Na	11	22.991
Strontium	Sr	38	87.63
Sulfur	S	16	32.066
Tantalum	Ta	73	180.95
Technetium	Tc	43	(99)
Tellurium	Te	52	127.61
Terbium	Tb	65	158 93
Thallium	Tl	81	204.39
Thorium	Th	90	232
Thulium	Tm	69	168.94
Tin	Sn	50	118.70
Titanium	Ti	22	47.90
Tungsten (wolfram)	W	74	183.86
Uranium	U	92	238.07
Vanadium	V	23	50.95
Xenon	Xe	54	131.30
Ytterbium	Yb	70	173.04
Yttrium	Y	39	88.91
Zinc	Zn	30	65.38
Zirconium	Zr	40	91.22

APPENDIX III

THE STANDARD MAN[a]

1. MASS AND EFFECTIVE RADIUS OF ORGANS OF THE ADULT HUMAN BODY[b]

	Mass in grams	Per cent of total body	Effective radius in cm
Total body[c]	70,000	100	30
Muscle	30,000	43	30
Skin and subcutaneous tissue[d]	6100	8.7	0.1
Fat	10,000	14	20
Skeleton:			
Without bone marrow	7000	10	5
Red marrow[e]	1500	2.1	—
Yellow marrow	1500	2.1	—
Blood	5400	7.7	—
Gastrointestinal tract[c]	2000	2.9	30
Contents of GI tract:			
Lower large intestine	150	—	5
Stomach	250	—	—
Small intestine	1100	—	—
Upper large intestine	135	—	—
Liver	1700	2.4	10
Brain	1500	2.1	—
Lungs (2)	1000	1.4	30
Lymphoid tissue	700	1.0	—
Kidneys (2)	300	0.43	7
Heart	300	0.43	—
Spleen	150	0.21	7
Urinary bladder	150	0.21	—
Pancreas	70	0.10	—
Salivary glands (6)	50	0.071	—
Testes (2)	40	0.057	—
Spinal cord	30	0.043	—
Eyes (2)	30	0.043	—
Thyroid gland	20	0.029	3
Teeth	20	0.029	—
Prostate gland	20	0.029	—
Adrenal glands of suprarenal (2)	20	0.029	—
Thymus	10	0.014	—
Miscellaneous (blood vessels, cartilage, nerves, etc.)	390	0.56	—

[a] *Radiological Health Handbook.*

[b] The reports, *Standard Man*, by Hermann Lisco, ANL–4253, Nov. 1948–Feb. 1949, p. 96, and *A Survey Report of the Characteristics of the Standard Man*, Sept. 1948, by M. J. Cook were used as the principal sources of reference in the original selection of values given in this table. (Data taken from report of the International Sub-Committee II on Permissible Dose for External Radiations, K. Z. Morgan, Chairman, ICRP/54/4.)

[c] Does not include contents of gastrointestinal tract.

[d] The mass of the skin alone is taken as 2000 g. The minimum thickness of the epidermis is 0.07 mm.

[e] The average depth of the blood-forming organs is assumed to be at 5 cm.

2. Chemical Composition

Element	Proportion %	Approximate mass in the body in grams
Oxygen	65.0	45,500
Carbon	18.0	12,600
Hydrogen	10.0	7000
Nitrogen	3.0	2100
Calcium	1.5	1050
Phosphorus	1.0	700
Sulfur	0.25	175
Potassium	0.2	140
Sodium	0.15	105
Chlorine	0.15	105
Magnesium	0.05	35
Iron	0.006	4
Manganese	0.00003	0.02
Copper	0.0002	0.1
Iodine	0.00004	0.03

The figures for a given organ may differ considerably from these averages for the whole body. For example, the nitrogen content of the dividing cells of the basal layer of skin is probably nearer 6% than 3%.

3. Applied Physiology

Average data for normal activity in a temperate zone:

1. *Water balance*

Daily water intake

Water of oxidation	0.3 liters
In food	0.7 liters
As fluids	1.5 liters
Total	2.5 liters

Calculations of maximum permissible levels for radioactive isotopes in water have been based on the total intake figure of 2.5 liters a day.

Daily water output

Sweat	0.5 liters
From lungs	0.4 liters
In feces	0.1 liters
Urine	1.5 liters
Total	2.5 liters

(The total water content of the body is 42 liters.)

2. *Respiration*[a]

Area of Respiratory Tract

Respiratory interchange area	50 m^2
Nonrespiratory area (upper tract and trachea to bronchioles)	20 m^2
Total	70 m^2

RESPIRATORY EXCHANGE

Physical activity	Hours per day	Tidal air (liters)	Respiration per min	Volume per 8 hr (m^3)	Volume per day (m^3)
At work	8	1.0	20	10	20
Not at work	16	0.5	20	5	20

Carbon dioxide content (by volume) of air

Inhaled air (dry, at sea level)	0.03%
Alveolar air	5.5%
Exhaled air	4.0%

3. *Retention of particulate matter in the respiratory tract*

Retention of particulate matter in the lungs depends on many factors, such as the size, shape and density of the particles, the chemical form and whether or not the person is a mouth breather; however, when specific data are lacking it is assumed the distribution is as follows:

Distribution	Readily soluble compounds %	Other compounds %
Exhaled	25	25
Deposited in upper respiratory passages and subsequently swallowed	50	50
Deposited in the lungs (lower respiratory passages)	25 (this is taken up into the body)	25[b]

[a] As stated in U.S. Department of Commerce, *Bureau of Standards Handbook* 47, 1950, Appendix I.

[b] Of this, half is eliminated from the lungs and swallowed in the first 24 hr making a total of 62½% swallowed. The remaining 12½% is retained in the lungs with a half-life of 120 days, it being assumed that this portion is taken up into body fluids.

4. Duration of Exposure

1. *Duration of occupational exposure*

The following figures have been adopted in calculations pertaining to occupational exposure:

8 hr per day
40 hr per week
50 weeks per year
50 years continuous work period.

2. *Duration of "lifetime for nonoccupational exposure"*

A conventional figure of 70 years has been adopted.

5. Blood Count

Relative and Absolute Values for Leukocyte Counts in Normal Adults per Cubic Millimeter of Blood[a]

Type of cell	%	Absolute number		
		Average	Minimum	Maximum
Total leukocytes	—	7000	5000	10,000
Myelocytes	0	0	0	0
Juvenile neutrophils	3–5	300	150	400
Segmented neutrophils	54–62	4000	3000	5800
Eosinophils	1–3	200	50	250
Basophils	0–0.75	25	15	50
Lymphocytes	25–33	2100	1500	3000
Monocytes	3–7	375	285	500

Normal Range of Values—Human Adults[a]

	Male	Female
Red cell count (millions per mm^3)	5.4 ± 0.8	4.8 ± 0.6
Hemoglobin (g per 100 cm^3)	16.0 ± 2.0	14.0 ± 2.0
Hematocrite or vol.-packed R.B.C. (cm^3 per 100 cm^3)	47.0 ± 7.0	42.0 ± 5.0

[a] M. M. Wintrobe, *Clinical Hematology*, Lea & Febiger, Philadelphia, 3rd edition.

APPENDIX IV

Specific Absorbed Fraction of Photon Energy

† Build-up factor method. ⊕ Adjustment on density or composition. * Extrapolation from higher energy. ⊗ Calculated by difference.

Source in Bladder Contents

Target	Energy (MeV) 0.010	0.015	0.020	0.030	0.050	0.100
Bladder wall	8.49E—04	1.40E—03	1.43E—03	9.83E—04	4.49E—04	2.56E—04
Stomach wall	2.83E—18*	4.25E—18†	1.20E—11†	3.28E—08†	4.53E—07	1.05E—06
Small intestine plus contents	1.52E—10*	2.28E—10†	1.76E—07	5.11E—06	1.24E—05	1.22E—05
Upper large intestine wall	1.71E—10*	2.57E—10†	1.02E—07†	3.04E—06	8.59E—06	8.31E—06
Lower large intestine wall	3.42E—07*	5.12E—07	7.80E—06	3.51E—05	4.16E—05	3.01E—05
Kidneys	1.52E—19*	2.28E—19†	3.95E—12†	2.46E—08†	4.05E—07	9.98E—07
Liver	2.61E—19*	3.91E—19†	2.37E—12†	2.05E—08	2.72E—07	5.55E—07
Lungs	4.25E—29*	6.39E—29†	5.55E—17†	7.65E—11†	1.31E—08	8.89E—08
"Other tissues" (suggested for muscle)	2.00E—07⊗	2.02E—06⊗	5.42E—06⊗	9.69E—06⊗	9.26E—06⊗	6.86E—06⊗
Ovaries	6.23E—09*	9.35E—09†	1.62E—06†	2.78E—05	3.58E—05	2.82E—05
Pancreas	3.63E—22*	5.45E—22†	2.11E—13†	6.51E—09†	2.92E—07	7.24E—07
Skeleton (suggested for total endosteal cells)	2.56E—11*	3.84E—11⊕	1.60E—08	1.16E—06	4.30E—06	4.06E—06
Red marrow	2.84E—08*	4.25E—08*	5.67E—08	3.88E—06	1.23E—05	1.00E—05
Skin	2.66E—08*	3.99E—08	3.75E—07	1.70E—06	2.26E—06	2.12E—06
Spleen	2.63E—22*	3.94E—22†	1.52E—13†	4.78E—09†	1.96E—07	3.87E—06
Testes	1.79E—09*	2.69E—09†	1.39E—06	1.54E—05	2.75E—05	1.88E—05
Thymus	5.91E—37*	8.87E—37†	1.32E—20†	2.05E—12†	2.06E—09†	2.67E—08†
Thyroid	6.20E—25*	9.31E—25*	1.24E—24†	2.86E—14†	1.55E—10†	4.05E—09†
Uterus	3.88E—06*	5.83E—06	4.28E—05	1.09E—04	1.02E—04	6.21E—05
Total body	1.43E—05	1.43E—05	1.42E—05	1.31E—05	1.00E—05	7.28E—06

Target	Energy (MeV) 0.200	0.500	1.000	1.500	2.000	4.000
Bladder wall	2.47E—04	2.56E—04	2.22E—04	2.07E—04	1.97E—04	1.57E—04
Stomach wall	1.02E—06	1.10E—06	1.86E—06	1.30E—06	1.83E—06	1.53E—06
Small intestine plus contents	1.03E—05	8.97E—06	8.84E—06	8.17E—06	8.42E—06	6.65E—06
Upper large intestine wall	8.37E—06	8.18E—06	6.19E—06	6.15E—06	5.80E—06	5.29E—06
Lower large intestine wall	2.51E—05	2.44E—05	2.09E—05	2.16E—05	2.02E—05	1.36E—05
Kidneys	1.00E—06	1.33E—06	1.17E—06	1.69E—06	1.26E—06	1.37E—06
Liver	7.70E—07	9.76E—07	1.02E—06	9.60E—07	1.19E—06	1.04E—06
Lungs	9.33E—08	1.65E—07	2.83E—07	3.07E—07	3.31E—07	4.61E—07
"Other tissues" (suggested for muscle)	6.28E—06⊗	6.13E—06⊗	5.71E—06⊗	5.42E—06⊗	5.11E—06⊗	4.27E—06⊗
Ovaries	2.72E—05	2.17E—05	1.35E—05	2.31E—05	1.66E—05	1.48E—05
Pancreas	1.10E—06	8.92E—07	1.00E—06	8.00E—07	1.40E—06	8.26E—07
Skeleton (suggested for total endosteal cells)	2.70E—06	2.03E—06	1.78E—06	1.72E—06	1.71E—06	1.55E—06
Red marrow	6.10E—06	4.33E—06	3.72E—06	3.47E—06	3.36E—06	3.07E—06
Skin	2.01E—06	2.24E—06	2.35E—06	2.22E—06	2.12E—06	1.89E—06
Spleen	5.29E—07	7.80E—07	8.59E—07	6.34E—07	1.12E—06	4.63E—07
Testes	1.59E—05	1.75E—05	1.57E—05	1.28E—05	1.40E—05	1.21E—05
Thymus	6.22E—08†	1.10E—07†	1.54E—07†	1.78E—07†	1.94E—07†	2.10E—07†
Thyroid	1.33E—08†	3.27E—08†	5.68E—08†	7.33E—08†	8.55E—08†	1.05E—07†
Uterus	5.51E—05	5.11E—05	4.68E—05	3.98E—05	4.04E—05	3.30E—05
Total body	6.58E—06	6.44E—06	5.98E—06	5.63E—06	5.35E—06	4.44E—06

SPECIFIC ABSORBED FRACTION OF PHOTON ENERGY (*contd.*)

† Build-up factor method. ⊕ Adjustment on density or composition. * Extrapolation from higher energy. ⊗ Calculated by difference.

Source in Stomach Contents

Target	Energy (MeV) 0.010	0.015	0.020	0.030	0.050	0.100
Bladder wall	1.61E—18*	2.41E—18†	8.64E—12†	2.82E—08†	5.32E—07	1.17E—06
Stomach wall	3.21E—04	7.61E—04	8.96E—04	6.87E—04	3.28E—04	1.88E—04
Small intestine plus contents	1.99E—07*	2.99E—07	1.23E—06	6.72E—06	1.22E—05	1.08E—05
Upper large intestine wall	8.09E—07*	1.21E—06	5.67E—06	1.62E—05	1.90E—05	1.40E—05
Lower large intestine wall	1.71E—09*	2.56E—09†	2.80E—07	1.83E—06	4.78E—06	4.44E—06
Kidneys	3.83E—10*	5.74E—10†	4.17E—07	8.18E—06	1.65E—05	1.36E—05
Liver	2.14E—07*	3.22E—07*	4.29E—07	3.99E—06	8.27E—06	8.06E—06
Lungs	1.76E—07*	2.64E—07	2.60E—06	7.91E—06	9.61E—06	6.92E—06
"Other tissues" (suggested for muscle)	3.98E—08⊗	1.15E—06⊗	3.92E—06⊗	7.64E—06⊗	7.31E—06⊗	5.37E—06⊗
Ovaries	1.99E—14*	2.98E—14†	1.05E—09†	2.89E—07†	2.36E—06†	2.13E—06
Pancreas	8.30E—06*	1.24E—05	6.95E—05	1.40E—04	1.12E—04	7.10E—05
Skeleton (suggested for total endosteal cells)	3.86E—08*	5.79E—08	6.31E—07	3.05E—06	5.52E—06	4.00E—06
Red marrow	4.00E—08*	6.00E—08	6.67E—07	3.65E—06	8.53E—06	7.12E—06
Skin	1.38E—08*	2.07E—08	3.62E—07	1.53E—06	1.94E—06	1.66E—06
Spleen	1.12E—07*	1.69E—07†	1.06E—05	5.19E—05	5.98E—05	3.96E—05
Testes	3.42E—26*	5.13E—26†	1.89E—15†	5.20E—10†	5.50E—08†	1.58E—07
Thymus	2.19E—18*	3.29E—18†	1.18E—11†	3.41E—08†	1.58E—06	1.07E—06
Thyroid	5.67E—26*	8.51E—26†	2.27E—15†	5.70E—10†	5.82E—08†	2.59E—07
Uterus	7.51E—16*	1.13E—15†	2.81E—10†	1.90E—07†	1.45E—06	2.90E—06
Total body	1.43E—05	1.43E—05	1.42E—05	1.31E—05	9.75E—06	6.87E—06

Target	Energy (MeV) 0.200	0.500	1.000	1.500	2.000	4.000
Bladder wall	7.91E—07	1.58E—06	1.65E—06	1.48E—06	1.03E—06	1.44E—06
Stomach wall	1.78E—04	1.79E—04	1.65E—04	1.48E—04	1.41E—04	1.15E—04
Small intestine plus contents	9.27E—06	9.08E—06	8.16E—06	7.37E—06	7.66E—06	6.38E—06
Upper large intestine wall	1.26E—05	1.13E—05	9.62E—06	1.17E—05	8.40E—06	7.91E—06
Lower large intestine wall	4.65E—06	4.35E—06	4.53E—06	4.51E—06	3.98E—06	2.76E—06
Kidneys	1.26E—05	1.13E—05	1.06E—05	9.84E—06	8.97E—06	7.42E—06
Liver	7.00E—06	7.02E—06	6.34E—06	6.15E—06	6.20E—06	4.89E—06
Lungs	5.97E—06	6.40E—06	5.72E—06	5.51E—06	4.91E—06	4.28E—06
"Other tissues" (suggested for muscle)	4.96E—06⊗	4.84E—06⊗	4.60E—06⊗	4.38E—06⊗	4.10E—06⊗	3.45E—06⊗
Ovaries	1.53E—06	1.70E—06	5.09E—06	3.02E—06†	2.86E—06†	2.42E—06†
Pancreas	6.37E—05	6.02E—05	5.23E—05	5.06E—05	4.40E—05	3.39E—05
Skeleton (suggested for total endosteal cells)	2.59E—06	2.08E—06	1.89E—06	1.79E—06	1.73E—06	1.61E—06
Red marrow	4.64E—06	3.61E—06	3.18E—06	2.90E—06	2.78E—06	2.66E—06
Skin	1.67E—06	2.04E—06	2.07E—06	1.96E—06	2.09E—06	1.70E—06
Spleen	3.56E—05	3.26E—05	3.04E—05	2.80E—05	2.49E—05	1.84E—05
Testes	2.41E—07	8.06E—08	5.72E—07†	5.91E—07†	5.95E—07†	5.61E—07†
Thymus	8.91E—07	7.81E—07	2.87E—06	2.26E—06	1.65E—06	1.42E—06†
Thyroid	4.35E—07†	5.26E—07†	5.82E—07†	5.96E—07†	6.02E—07†	5.67E—07†
Uterus	2.97E—06	3.08E—06	3.22E—06	3.28E—06	2.54E—06	4.00E—06
Total body	6.21E—06	6.11E—06	5.74E—06	5.39E—06	5.07E—06	4.22E—06

SPECIFIC ABSORBED FRACTION OF PHOTON ENERGY (*cont.*)

† Build-up factor method. ⊕ Adjustment on density or composition. * Extrapolation from higher energy. ⊗ Calculated by difference.

Source in Small Intestine Plus Wall

Target	Energy (MeV)					
	0.010	0.015	0.020	0.030	0.050	0.100
Bladder wall	4.21E—10*	6.31E—10†	1.74E—07†	5.86E—06	1.02E—05	1.02E—05
Stomach wall	5.69E—07	1.80E—06	8.17E—06	1.42E—05	1.77E—05	1.45E—05
Small intestine plus contents	7.40E—04	6.61E—04	5.60E—04	3.71E—04	1.87E—04	1.13E—04
Upper large intestine wall	4.50E—04	4.48E—04	3.95E—04	2.74E—04	1.42E—04	8.66E—05
Lower large intestine wall	2.81E—05	5.12E—05	6.84E—05	6.33E—05	4.28E—05	2.82E—05
Kidneys	4.94E—09*	7.41E—09†	8.43E—07	6.92E—06	1.28E—05	1.30E—05
Liver	7.60E—08	3.52E—07	1.33E—06	4.59E—06	7.67E—06	7.17E—06
Lungs	4.44E—18*	6.67E—18†	9.65E—12†	6.31E—08	4.61E—07	7.93E—07
"Other tissues" (suggested for muscle)	6.96E—07⊗	2.14E—06⊗	4.24E—06⊗	7.31E—06⊗	7.49E—06⊗	5.95E—06⊗
Ovaries	1.38E—05*	2.07E—05	4.43E—05	6.26E—05	6.51E—05	4.23E—03
Pancreas	2.64E—11*	3.96E—11†	4.11E—08†	2.45E—06	7.92E—06	8.13E—06
Skeleton (suggested for total endosteal cells)	2.15E—07	7.10E—07	1.79E—06	5.23E—06	8.26E—06	5.61E—06
Red marrow	9.04E—07	2.97E—06	7.39E—06	2.08E—05	3.06E—05	1.92E—05
Skin	5.31E—08*	7.96E—08*	1.06E—07	7.67E—07	1.49E—06	1.52E—06
Spleen	2.12E—10*	3.18E—10†	6.00E—08†	2.67E—06	5.11E—06	5.95E—06
Testes	2.76E—18*	4.13E—18†	1.68E—11†	4.80E—08†	4.88E—07	1.10E—06
Thymus	1.05E—26*	1.58E—26†	1.07E—15†	4.01E—10†	4.76E—08†	2.52E—07†
Thyroid	4.34E—35*	6.50E—35†	1.11E—19†	5.17E—12†	3.48E—09†	3.84E—08†
Uterus	1.36E—06*	2.05E—06	1.97E—05	5.49E—05	5.74E—05	3.83E—05
Total body	1.43E—05*	1.43E—05	1.43E—05	1.37E—05	1.08E—05	7.82E—06

Target	Energy (MeV) 0.200	0.500	1.000	1.500	2.000	4.000
Bladder wall	9.28E—06	1.18E—05	6.79E—06	1.07E—05	7.97E—06	5.56E—06
Stomach wall	1.31E—05	1.16E—05	1.14E—05	1.08E—05	9.88E—06	7.84E—06
Small intestine plus contents	1.07E—04	1.07E—04	9.57E—05	9.07E—05	8.23E—05	6.68E—05
Upper large intestine wall	8.23E—05	7.83E—05	7.21E—05	6.69E—05	6.31E—05	4.98E—05
Lower large intestine wall	2.52E—05	2.32E—05	2.19E—05	2.03E—05	2.05E—05	1.63E—05
Kidneys	1.08E—05	1.09E—05	1.03E—05	8.67E—06	8.59E—06	6.26E—06
Liver	6.64E—06	6.20E—06	6.11E—06	5.67E—06	5.48E—06	4.39E—06
Lungs	9.02E—07	1.16E—06	1.17E—06	1.07E—06	1.17E—06	1.13E—06
"Other tissues" (suggested for muscle)	5.51E—06⊗	5.33E—06⊗	5.00E—06⊗	4.73E—06⊗	4.43E—06⊗	3.77E—06⊗
Ovaries	3.75E—05	3.00E—05	3.39E—05	2.31E—05	2.13E—05	2.81E—05
Pancreas	7.30E—06	7.17E—06	6.55E—06	5.08E—06	5.74E—06	5.21E—06
Skeleton (suggested for total endosteal cells)	3.54E—06	2.69E—06	2.53E—06	2.34E—06	2.26E—06	1.95E—06
Red marrow	1.16E—05	8.57E—06	7.78E—06	7.07E—06	6.77E—06	5.52E—06
Skin	1.63E—06	1.83E—06	1.93E—06	1.97E—06	1.75E—06	1.54E—06
Spleen	5.55E—06	5.41E—06	5.05E—06	5.28E—06	5.01E—06†	3.59E—06
Testes	1.28E—06	1.24E—06	1.79E—06	1.92E—06	2.12E—06	1.18E—06
Thymus	3.89E—07†	4.81E—07†	5.38E—07†	5.54E—07†	5.62E—07†	5.34E—07†
Thyroid	8.31E—08†	1.38E—07†	1.85E—07†	2.10E—07†	2.26E—07†	2.40E—07†
Uterus	3.31E—05	2.99E—05	2.71E—05	2.42E—05	2.21E—05	1.85E—05
Total body	7.02E—06	6.74E—06	6.27E—06	5.92E—06	5.51E—06	4.60E—06

SPECIFIC ABSORBED FRACTION OF PHOTON ENERGY (*cont.*)

† Build-up factor method. ⊕ Adjustment on density or composition. * Extrapolation from higher energy. ⊗ Calculated by difference.

Source in Upper Large Intestine Contents

Target	Energy (MeV) 0.010	0.015	0.020	0.030	0.050	0.100
Bladder wall	3.32E—10*	4.97E—10†	1.25E—07†	3.87E—06	7.40E—06	8.92E—06
Stomach wall	2.11E—06*	3.16E—06	1.14E—05	2.22E—05	2.05E—05	1.49E—05
Small intestine plus contents	5.46E—06	5.06E—05	1.19E—04	1.50E—04	1.03E—04	6.49E—05
Upper large intestine wall	3.63E—04	8.01E—04	8.62E—04	5.57E—04	2.61E—04	1.52E—04
Lower large intestine wall	2.47E—06*	3.70E—06	1.14E—05	1.81E—05	1.51E—05	1.30E—05
Kidneys	3.18E—10*	4.77E—10†	3.55E—07	5.76E—06	1.25E—05	1.14E—05
Liver	1.58E—07*	2.37E—07	1.93E—06	7.43E—06	1.18E—05	1.00E—05
Lungs	3.97E—18*	5.95E—18†	1.18E—11†	7.16E—08	6.14E—07	9.96E—07
"Other tissues" (suggested for muscle)	7.42E—08⊗	8.77E—07⊗	3.19E—06⊗	6.95E—06⊗	7.28E—06⊗	5.74E—06⊗
Ovaries	2.01E—05*	3.02E—05	5.33E—05	7.41E—05	5.59E—05	4.53E—05
Pancreas	2.25E—11*	3.38E—11†	5.01E—08†	2.71E—06	8.10E—06	9.01E—06
Skeleton (suggested for total endosteal cells)	1.64E—07*	2.47E—07	1.37E—06	4.87E—06	7.35E—06	5.08E—06
Red marrow	6.91E—07*	1.04E—06	5.73E—06	1.93E—05	2.62E—05	1.66E—05
Skin	8.49E—10*	1.34E—09†	9.59E—08	9.69E—07	1.64E—06	1.52E—06
Spleen	5.77E—11*	8.66E—11†	5.09E—08†	1.57E—06	6.08E—06	5.82E—06
Testes	1.23E—17*	1.85E—17†	2.55E—11†	4.81E—08†	5.08E—07	9.33E—07
Thymus	1.34E—26*	2.02E—26†	1.58E—15†	5.43E—10†	5.78E—08†	3.51E—07
Thyroid	5.27E—35*	7.90E—35†	1.59E—19†	6.86E—12†	4.15E—09†	4.30E—08†
Uterus	4.28E—09*	6.41E—09†	1.21E—06	1.39E—05	2.81E—05	2.27E—05
Total body	1.43E—05	1.43E—05	1.43E—05	1.35E—05	1.05E—05	7.61E—06

Target	Energy (MeV) 0.200	0.500	1.000	1.500	2.000	4.000
Bladder wall	7.67E—06	6.53E—06	7.06E—06	4.83E—06	7.28E—06	4.26E—06
Stomach wall	1.32E—05	1.24E—05	1.16E—05	9.48E—06	1.01E—05	7.76E—06
Small intestine plus contents	5.89E—05	5.61E—05	5.05E—05	4.54E—05	4.21E—05	3.29E—05
Upper large intestine wall	1.47E—04	1.47E—04	1.39E—04	1.24E—04	1.18E—04	9.33E—05
Lower large intestine wall	1.06E—05	1.03E—05	1.08E—05	8.63E—06	8.14E—06	5.99E—06
Kidneys	9.84E—06	9.32E—06	9.40E—06	8.59E—06	7.31E—06	6.25E—06
Liver	9.31E—06	8.45E—06	8.11E—06	7.67E—06	7.24E—06	6.32E—06
Lungs	9.83E—07	1.17E—06	1.38E—06	1.43E—06	1.28E—06	1.13E—06
"Other tissues" (suggested for muscle)	5.22E—06⊗	5.10E—06⊗	4.80E—06⊗	4.55E—06⊗	4.36E—06⊗	3.60E—06⊗
Ovaries	4.10E—05	4.39E—05	2.31E—05	1.87E—05	3.33E—05	2.08E—05
Pancreas	8.74E—06	6.11E—06	6.85E—06	6.31E—06	7.00E—06	7.77E—06
Skeleton (suggested for total endosteal cells)	3.18E—06	2.49E—06	2.21E—06	2.13E—06	2.04E—06	1.75E—06
Red marrow	9.96E—06	7.45E—06	6.40E—06	6.05E—06	5.86E—06	4.85E—06
Skin	1.63E—06	1.92E—06	1.87E—06	1.86E—06	1.82E—06	1.76E—06
Spleen	5.01E—06	4.21E—06	4.55E—06	4.73E—06	3.66E—06	3.79E—06
Testes	1.15E—06	1.78E—06	1.92E—06	1.30E—06	1.74E—06†	1.51E—06†
Thymus	4.33E—07†	5.27E—07†	5.80E—07†	5.94E—07†	6.01E—07†	5.66E—07†
Thyroid	9.11E—08†	1.48E—07†	1.95E—07†	2.21E—07†	2.37E—07†	2.49E—07†
Uterus	1.68E—05	1.53E—05	1.53E—05	1.55E—05	1.44E—05	1.16E—05
Total body	6.79E—06	6.57E—06	6.17E—06	5.77E—06	5.46E—06	4.50E—06

Specific Absorbed Fraction of Photon Energy (*cont.*)

† Build-up factor method. ⊕ Adjustment on density or composition. * Extrapolation from higher energy. ⊗ Calculated by difference.

Source in Lower Large Intestine Contents

Target	Energy (MeV) 0.010	0.015	0.020	0.030	0.050	0.100
Bladder wall	1.38E—07*	2.08E—07†	6.01E—06	2.70E—05	3.34E—05	2.80E—05
Stomach wall	1.25E—08*	1.88E—08†	8.62E—07	5.03E—06	7.97E—06	6.96E—06
Small intestine plus contents	2.42E—06	2.78E—05	6.14E—05	7.81E—05	5.42E—05	3.65E—05
Upper large intestine wall	1.03E—05	2.22E—05	3.16E—05	3.07E—05	2.21E—05	1.61E—05
Lower large intestine wall	5.53E—04	1.13E—03	1.16E—03	7.02E—04	2.97E—04	1.76E—04
Kidneys	4.24E—12*	6.36E—12†	1.31E—08†	8.91E—07	2.59E—06	3.12E—06
Liver	2.92E—17*	4.38E—17†	2.34E—11†	3.25E—08	4.60E—07	8.95E—07
Lungs	2.61E—17*	3.92E—20†	4.15E—13†	2.91E—09†	1.15E—07	2.87E—07
"Other tissues" (suggested for muscle)	6.79E—08⊗	1.35E—06⊗	4.12E—06⊗	7.92E—06⊗	8.26E—06⊗	6.51E—06⊗
Ovaries	4.10E—05*	6.16E—05	1.62E—04	1.77E—04	1.06E—04	6.25E—05
Pancreas	1.71E—14*	2.57E—14†	1.18E—09†	5.69E—07	1.69E—06	3.05E—06
Skeleton (suggested for total endosteal cells)	4.30E—08	1.36E—06	5.69E—06	1.31E—05	1.30E—05	7.25E—06
Red marrow	1.81E—07	5.70E—06	2.37E—05	5.24E—05	4.68E—05	2.28E—05
Skin	8.10E—08	2.36E—07	4.75E—07	1.14E—06	1.88E—06	1.79E—06
Spleen	1.52E—12*	2.29E—12†	7.72E—09†	5.72E—07	3.20E—06	3.22E—06
Testes	4.67E—10*	7.01E—10†	2.21E—07†	4.41E—06	8.59E—06	6.27E—06
Thymus	1.02E—29*	1.53E—29†	2.27E—17†	4.46E—11†	1.03E—08†	7.60E—08†
Thyroid	1.05E—37*	1.58E—37†	3.76E—21†	7.26E—13†	8.56E—10†	1.26E—08†
Uterus	5.13E—09*	7.69E—09†	2.33E—06	2.25E—05	3.65E—05	2.79E—05
Total body	1.43E—05	1.43E—05	1.43E—05	1.37E—05	1.08E—05	7.62E—06

Target	Energy (MeV) 0.200	0.500	1.000	1.500	2.000	4.000
Bladder wall	2.34E—05	1.94E—05	1.62E—05	2.02E—05	1.67E—05	1.04E—05
Stomach wall	6.34E—06	6.23E—06	4.98E—06	4.73E—06	5.49E—06	4.92E—06
Small intestine plus contents	3.32E—05	3.07E—05	2.84E—05	2.64E—05	2.51E—05	1.99E—05
Upper large intestine wall	1.50E—05	1.45E—05	1.25E—05	1.35E—05	1.08E—05	9.87E—06
Lower large intestine wall	i.74E—04	1.77E—04	1.62E—04	1.52E—04	1.39E—04	1.08E—04
Kidneys	3.47E—06	2.84E—06	3.08E—06	2.98E—06	3.54E—06	2.50E—06
Liver	1.02E—06	1.19E—06	1.29E—06	1.50E—06	1.24E—06	1.18E—06
Lungs	3.14E—07	3.49E—07	5.49E—07	5.49E—07	7.81E—07	5.78E—07
"Other tissues" (suggested for muscle)	6.01E—06⊗	5.84E—06⊗	5.48E—06⊗	5.18E—06⊗	4.89E—06⊗	4.09E—06⊗
Ovaries	7.13E—05	5.42E—05	5.50E—05	5.02E—05	4.63E—05	3.51E—05
Pancreas	2.49E—06	2.58E—06	2.43E—06	2.00E—06	2.48E—06	1.45E—06
Skeleton (suggested for total endosteal cells)	4.49E—06	3.59E—06	3.25E—06	3.02E—06	2.97E—06	2.43E—06
Red marrow	1.36E—05	1.07E—05	9.65E—06	8.77E—06	8.46E—06	6.82E—06
Skin	1.89E—06	2.03E—06	2.05E—06	2.23E—06	2.03E—06	1.85E—06
Spleen	2.76E—06	3.24E—06	3.56E—06	3.05E—06	3.20E—06	2.22E—06
Testes	7.25E—06	7.14E—06	4.86E—06	4.82E—06	6.10E—06	3.92E—06
Thymus	1.39E—07†	2.01E—07†	2.51E—07†	2.74E—07†	2.89E—07†	2.93E—07†
Thyroid	3.19E—08†	6.26E—08†	9.34E—08†	1.13E—07†	1.27E—07†	1.45E—07†
Uterus	2.53E—05	1.91E—05	1.79E—05	1.78E—05	1.52E—05	1.29E—05
Total body	6.77E—06	6.53E—06	6.10E—06	5.76E—06	5.44E—06	4.48E—06

SPECIFIC ABSORBED FRACTION OF PHOTON ENERGY (*cont.*)

† Build-up factor method. ⊕ Adjustment on density or composition. * Extrapolation from higher energy. ⊗ Calculated by difference.

Source in Kidneys

Target	Energy (MeV) 0.010	0.015	0.020	0.030	0.050	0.100
Bladder wall	4.59E—19*	6.88E—19†	5.48E—12†	2.72E—08†	3.14E—07	1.04E—06
Stomach wall	1.35E—09*	2.03E—09†	5.45E—07	7.99E—06	1.59E—05	1.47E—05
Small intestine plus contents	6.51E—09*	9.77E—09†	9.63E—07	6.48E—06	1.23E—05	1.14E—05
Upper large intestine wall	7.46E—10*	1.12E—09†	3.70E—07	6.25E—06	1.17E—05	1.13E—05
Lower large intestine wall	8.29E—12*	1.24E—11†	1.16E—08†	6.63E—07	1.81E—06	2.49E—06
Kidneys	3.28E—03	2.74E—03	2.04E—03	1.03E—03	3.93E—04	2.35E—04
Liver	4.43E—07*	6.65E—07	4.68E—06	1.54E—05	1.95E—05	1.53E—05
Lungs	3.57E—08*	5.36E—08*	7.15E—08	1.18E—06	3.27E—06	3.37E—06
"Other tissues" (suggested for muscle)	1.42E—06⊗	4.42E—06⊗	7.39E—06⊗	8.82E—06⊗	6.78E—06⊗	5.12E—06⊗
Ovaries	3.68E—14*	5.53E—14†	1.76E—09†	4.81E—07†	3.71E—06	4.13E—06
Pancreas	2.34E—08*	3.51E—08†	4.45E—06	2.44E—05	3.22E—05	2.63E—05
Skeleton (suggested for total endosteal cells)	2.24E—07*	3.36E—07	2.18E—06	7.54E—06	1.03E—05	6.40E—06
Red marrow	5.99E—07*	8.99E—07	5.72E—06	2.01E—05	2.76E—05	1.69E—05
Skin	1.66E—07*	2.49E—07	1.16E—06	2.50E—06	2.23E—06	1.93E—06
Spleen	6.19E—06*	9.28E—06	3.63E—05	6.69E—05	5.32E—05	3.30E—05
Testes	4.39E—27*	6.58E—27†	8.52E—16†	4.33E—10†	5.25E—08†	2.76E—07†
Thymus	7.07E—22*	1.06E—21†	2.52E—13†	6.17E—09†	3.51E—07	8.06E—07
Thyroid	8.15E—30*	1.22E—29†	3.87E—17†	9.30E—11†	2.07E—08†	1.40E—07†
Uterus	1.46E—15*	2.19E—15†	4.32E—10†	3.26E—07	1.85E—06	3.31E—06
Total body	1.43E—05	1.43E—05	1.40E—05	1.23E—05	9.01E—06	6.45E—06

Target	Energy (MeV) 0.200	0.500	1.000	1.500	2.000	4.000
Bladder wall	1.12E—06	1.32E—06	1.59E—06	1.47E—06	1.13E—06	1.92E—06
Stomach wall	1.20E—05	1.16E—05	1.07E—05	8.44E—06	1.00E—05	8.14E—06
Small intestine plus contents	1.02E—05	9.42E—06	9.02E—06	8.91E—06	8.05E—06	6.66E—06
Upper large intestine wall	1.03E—05	9.90E—06	9.20E—06	9.23E—06	7.29E—06	6.25E—06
Lower large intestine wall	3.05E—06	3.03E—06	2.84E—06	3.39E—06	2.93E—06	2.52E—06
Kidneys	2.39E—04	2.52E—04	2.26E—04	2.14E—04	1.93E—04	1.63E—04
Liver	1.36E—05	1.38E—05	1.22E—05	1.15E—05	1.10E—05	8.79E—06
Lungs	2.99E—06	3.28E—06	3.30E—06	2.96E—06	2.80E—06	2.72E—06
"Other tissues" (suggested for muscle)	4.95E—06⊗	4.93E—06⊗	4.71E—06⊗	4.47E—06⊗	4.22E—06⊗	3.55E—06⊗
Ovaries	4.44E—06	3.96E—06	5.43E—06	3.83E—06†	3.59E—06†	1.77E—08
Pancreas	2.28E—05	2.25E—05	2.24E—05	1.96E—05	1.50E—05	1.47E—05
Skeleton (suggested for total endosteal cells)	4.06E—06	3.39E—06	3.10E—06	2.99E—06	2.76E—06	2.46E—06
Red marrow	1.05E—05	8.59E—06	7.77E—06	7.43E—06	6.81E—06	5.85E—06
Skin	2.06E—06	2.29E—06	2.24E—06	2.13E—06	2.15E—06	1.89E—06
Spleen	3.11E—05	2.85E—05	2.70E—05	2.41E—05	2.20E—05	1.73E—05
Testes	4.23E—07†	5.18E—07†	5.77E—07†	5.93E—07†	5.98E—07†	5.65E—07†
Thymus	6.92E—07	1.41E—06	2.72E—07	1.13E—06†	4.67E—07	9.90E—07†
Thyroid	2.42E—07†	3.27E—07†	3.86E—07†	4.11E—07†	4.24E—07†	4.15E—07†
Uterus	3.91E—06	2.87E—06	2.77E—06	3.66E—06	1.95E—06	3.98E—06
Total body	5.90E—06	5.84E—06	5.48E—06	5.21E—06	4.87E—06	4.12E—06

SPECIFIC ABSORBED FRACTION OF PHOTON ENERGY (*cont.*)

† Build-up factor method. ⊕ Adjustment on density or composition. * Extrapolation from higher energy. ⊗ Calculated by difference.

Source in Liver

Target	Energy (MeV) 0.010	0.015	0.020	0.030	0.050	0.100
Bladder wall	5.75E—19*	8.62E—19†	3.52E—12†	1.33E—08†	2.04E—07	6.16E—07
Stomach wall	3.85E—07*	5.78E—07*	7.71E—07	4.08E—06	8.90E—06	7.07E—06
Small intestine plus contents	1.62E—07*	2.43E—07	7.70E—07	2.95E—06	6.39E—06	6.32E—06
Upper large intestine wall	1.89E—07*	2.84E—07	1.71E—06	8.56E—06	1.18E—05	1.01E—05
Lower large intestine wall	2.73E—17*	4.10E—17†	2.02E—11†	2.65E—08†	3.34E—07	8.76E—07
Kidneys	8.39E—07*	1.26E—06	4.50E—06	1.57E—05	1.95E—05	1.58E—05
Liver	5.36E—04	4.96E—04	4.34E—04	2.97E—04	1.52E—04	9.14E—05
Lungs	7.76E—08	3.19E—06	9.46E—06	1.64E—05	1.45E—05	9.92E—06
"Other tissues" (suggested for muscle)	6.16E—07⊗	1.92E—06⊗	3.55E—06⊗	5.52E—06⊗	5.25E—06⊗	4.09E—06⊗
Ovaries	9.82E—15*	1.47E—14†	5.55E—10†	1.63E—07†	1.51E—06†	1.63E—06
Pancreas	1.47E—06*	2.21E—06*	2.94E—06	1.67E—05	2.18E—05	1.77E—05
Skeleton (suggested for total endosteal cells)	5.83E—08	5.37E—07	2.08E—06	5.80E—06	7.80E—06	4.93E—06
Red marrow	6.03E—08	5.27E—07	1.96E—06	5.68E—06	9.33E—06	7.14E—06
Skin	5.76E—08*	8.64E—08	4.77E—07	1.64E—06	1.89E—06	1.81E—06
Spleen	7.27E—13*	1.09E—12†	4.57E—09†	8.07E—07	2.93E—06	3.56E—06
Testes	2.17E—26*	3.26E—26†	9.37E—16†	2.77E—10†	3.42E—08†	1.90E—07†
Thymus	6.31E—16*	9.47E—16†	1.59E—10†	1.20E—07†	1.67E—06	2.50E—06
Thyroid	1.29E—24*	1.94E—24†	1.12E—14†	1.21E—09†	8.81E—08†	3.80E—07†
Uterus	1.75E—16*	2.62E—16†	8.22E—11†	8.39E—08†	9.07E—07	1.51E—06
Total body	1.43E—05	1.43E—05	1.42E—05	1.29E—05	9.48E—06	6.54E—06

Target	Energy (MeV) 0.200	0.500	1.000	1.500	2.000	4.000
Bladder wall	5.60E—07	1.21E—06	5.80E—07	8.48E—07	9.02E—07	9.93E—07
Stomach wall	6.96E—06	6.50E—06	6.44E—06	6.00E—06	6.11E—06	5.17E—06
Small intestine plus contents	6.01E—06	5.44E—06	5.16E—06	5.10E—06	4.64E—06	4.16E—06
Upper large intestine wall	8.71E—06	8.84E—06	7.71E—06	6.52E—06	6.96E—06	5.21E—06
Lower large intestine wall	8.55E—07	1.12E—06	8.88E—07	1.24E—06	9.53E—07	1.40E—06
Kidneys	1.36E—05	1.29E—05	1.18E—05	1.14E—05	1.10E—05	8.23E—06
Liver	8.82E—05	8.85E—05	8.07E—05	7.48E—05	6.86E—05	5.58E—05
Lungs	8.84E—06	8.23E—06	7.90E—06	7.72E—06	6.96E—06	5.60E—06
"Other tissues" (suggested for muscle)	3.82E—06⊗	3.85E—06⊗	3.69E—06⊗	3.46E—06⊗	3.30E—06⊗	2.80E—06⊗
Ovaries	1.80E—06	6.53E—07	2.49E—06†	3.44E—06	2.22E—06†	1.92E—06†
Pancreas	1.35E—05	1.66E—05	1.36E—05	1.21E—05	9.99E—06	8,75E—06
Skeleton (suggested for total endosteal cells)	3.17E—06	2.53E—06	2.30E—06	2.26E—06	2.20E—06	1.85E—06
Red marrow	4.64E—06	3.72E—06	3.21E—06	3.26E—06	3.17E—06	2.62E—06
Skin	1.89E—06	2.08E—06	2.08E—06	2.02E—06	2.10E—06	1.75E—06
Spleen	3.34E—06	3.44E—06	3.81E—06	2.95E—06	3.14E—06	2.14E—06
Testes	3.05E—07†	3.92E—07†	8.76E—07	4.70E—07†	4.79E—07†	4.62E—07†
Thymus	2.27E—06	4.64E—06	2.54E—06	2.17E—06	2.65E—06	3.92E—06
Thyroid	8.23E—07	6.32E—07†	6.81E—07†	6.87E—07†	6.90E—07†	6.40E—07†
Uterus	1.40E—06	1.52E—06	1.28E—06	2.07E—06	1.81E—06	1.38E—06
Total body	5.94E—06	5.86E—06	5.49E—06	5.16E—06	4.86E—06	4.06E—06

Specific Absorbed Fraction of Photon Energy (*cont.*)

† Build-up factor method. ⊕ Adjustment on density or composition. * Extrapolation from higher energy. ⊗ Calculated by difference.

Source in Lungs

Target	Energy (MeV) 0.010	0.015	0.020	0.030	0.050	0.100
Bladder wall	3.53E—28*	5.30E—28⊕	2.65E—16⊕	2.75E—10⊕	5.21E—08⊕	1.05E—07†
Stomach wall	5.37E—07*	8.05E—07	3.45E—06	8.04E—06	9.44E—06	7.19E—06
Small intestine plus contents	1.65E—17*	2.47E—17⊕	2.92E—11⊕	4.67E—08	3.60E—07	7.11E—07
Upper large intestine wall	2.53E—17*	3.80E—17⊕	4.52E—11⊕	7.70E—08	5.12E—07	8.35E—07
Lower large intestine wall	6.26E—20*	9.39E—20⊕	1.13E—12⊕	7.29E—09⊕	8.32E—08	1.78E—07
Kidneys	6.42E—11*	9.64E—11⊕	5.81E—08⊕	1.24E—06	2.97E—06	3.25E—06
Liver	7.16E—08	2.28E—06	7.25E—06	1.40E—05	1.29E—05	9.52E—06
Lungs	8.17E—04	6.58E—04	4.71E—04	2.30E—04	8.99E—05	5.05E—05
"Other tissues" (suggested for muscle)	3.89E—06⊗	6.82E—06⊗	9.44E—06⊗	1.00E—05⊗	7.05E—06⊗	4.87E—06⊗
Ovaries	5.25E—24*	7.88E—24⊕	3.48E—14⊕	3.13E—09⊕	2.30E—07⊕	3.02E—07†
Pancreas	8.53E—09*	1.28E—08⊕	6.38E—07	9.00E—06	1.27E—05	1.04E—05
Skeleton (suggested for total endosteal cells)	2.71E—08	7.11E—07	3.56E—06	1.04E—05	1.20E—05	6.56E—06
Red marrow	2.81E—08	7.53E—07	3.97E—06	1.20E—05	1.46E—05	8.36E—06
Skin	4.70E—08*	7.05E—08	6.44E—07	1.94E—06	2.19E—06	1.92E—06
Spleen	6.63E—07*	9.95E—07	4.70E—06	1.27E—05	1.10E—05	9.05E—06
Testes	8.51E—36*	1.28E—35⊕	5.76E—20⊕	5.56E—12⊕	5.03E—09⊕	1.92E—08†
Thymus	2.22E—07*	3.33E—07⊕	3.35E—06	1.78E—05	2.09E—05	1.79E—05
Thyroid	4.05E—12*	6.07E—12⊕	2.29E—08⊕	8.09E—07	2.30E—06	3.85E—06
Uterus	3.81E—25*	5.72E—25⊕	1.11E—14⊕	2.07E—09⊕	1.86E—07⊕	2.80E—07
Total body	1.43E—05	1.43E—05	1.41E—05	1.26E—05	8.78E—06	5.70E—06

Target	Energy (MeV) 0.200	0.500	1.000	1.500	2.000	4.000
Bladder wall	1.83E—07†	2.55E—07†	3.09E—07†	3.32E—07†	3.43E—07†	3.43E—07†
Stomach wall	6.61E—06	6.32E—06	6.34E—06	6.14E—06	6.45E—06	3.05E—06
Small intestine plus contents	7.14E—07	9.57E—07	1.08E—06	1.04E—06	1.16E—06	1.11E—06
Upper large intestine wall	8.57E—07	1.32E—06	1.38E—06	1.45E—06	1.37E—06	1.75E—06
Lower large intestine wall	4.05E—07	2.69E—07	3.21E—07	2.13E—07	2.34E—07	2.69E—07
Kidneys	3.14E—06	3.61E—06	3.22E—06	3.61E—06	2.31E—06	2.82E—06
Liver	8.81E—06	8.18E—06	7.72E—06	7.20E—06	6.52E—06	5.78E—06
Lungs	5.00E—05	5.01E—05	4.55E—05	4.32E—05	3.92E—05	3.08E—05
"Other tissues" (suggested for muscle)	4.61E—06⊗	4.62E—06⊗	4.31E—06⊗	4.04E—06⊗	3.81E—06⊗	3.17E—06⊗
Ovaries	4.41E—07†	5.28E—07†	5.74E—07†	5.87E—07†	5.94E—07†	5.58E—07†
Pancreas	8.94E—06	9.63E—06	7.11E—06	7.54E—06	5.86E—06	6.10E—06
Skeleton (suggested for total endosteal cells)	4.08E—06	3.45E—06	3.17E—06	2.87E—06	2.93E—06	2.40E—06
Red marrow	5.21E—06	4.35E—06	4.02E—06	3.69E—06	3.70E—06	3.02E—06
Skin	2.10E—06	2.34E—06	2.30E—06	2.25E—06	2.16E—06	1.73E—06
Spleen	8.18E—06	7.31E—06	7.18E—06	5.31E—06	6.23E—06	5.61E—06
Testes	4.62E—08†	8.43E—08†	1.22E—07†	1.43E—07†	1.58E—07†	1.74E—07†
Thymus	1.42E—05	1.26E—05	1.37E—05	1.35E—05	1.17E—05	8.45E—06
Thyroid	2.97E—06	4.15E—06	4.66E—06	2.75E—06	3.00E—06	2.64E—06†
Uterus	3.53E—07	3.29E—07	4.05E—07	9.28E—07	5.55E—07†	9.39E—07
Total body	5.11E—06	4.99E—06	4.66E—06	4.36E—06	4.12E—06	3.41E—06

SPECIFIC ABSORBED FRACTION OF PHOTON ENERGY (*cont.*)

†Build-up factor method. ⊕ Adjustment on density or composition. * Extrapolation from higher energy. ⊗ Calculated by difference.

Source in Muscle ⊗

Target	Energy (MeV) 0.010	0.015	0.020	0.030	0.050	0.100
Bladder wall	2.00E—07	2.02E—06	5.42E—06	9.69E—06	9.26E—06	6.86E—06
Stomach wall	3.98E—08	1.15E—06	3.92E—06	7.64E—06	7.31E—06	4.60E—06
Small intestine plus contents	6.96E—07	2.14E—06	4.24E—06	7.31E—06	7.49E—06	5.95E—06
Upper large intestine wall	4.38E—06	4.81E—06	6.20E—06	8.22E—06	7.74E—06	6.12E—06
Lower large intestine wall	6.79E—08	1.35E—06	4.12E—06	7.92E—06	8.26E—06	6.51E—06
Kidneys	1.42E—06	4.42E—06	7.39E—06	8.82E—06	6.78E—06	5.12E—06
Liver	6.16E—07	1.92E—06	3.55E—06	5.52E—06	5.25E—06	4.09E—06
Lungs	3.89E—06	6.82E—06	9.44E—06	1.00E—05	7.05E—06	4.87E—06
"Other tissues" (suggested for muscle)	1.81E—05	1.70E—05	1.52E—05	1.15E—05	7.13E—06	5.08E—06
Ovaries	4.11E—06	9.31E—06	1.19E—05	1.23E—05	1.00EE—05	7.57E—06
Pancreas	2.41E—06	6.73E—06	1.04E—05	1.21E—05	9.45E—06	6.87E—06
Skeleton (suggested for total endosteal cells)	3.85E—07	9.43E—07	1.63E—06	3.06E—06	3.82E—06	3.66E—06
Red marrow	2.01E—06	4.60E—06	8.73E—06	1.59E—05	1.56E—05	8.81E—06
Skin	4.00E—06	6.06E—06	6.46E—06	5.41E—06	3.48E—06	2.60E—06
Spleen	1.18E—06	3.78E—06	6.79E—06	9.42E—06	7.63E—06	5.48E—06
Testes	4.99E—08	7.21E—07	2.66E—06	5.71E—06	5.52E—06	4.28E—06
Thymus	1.74E—05	2.52E—05	2.59E—05	1.31E—05	1.23E—05	5.93E—06
Thyroid	3.35E—06	8.66E—06	1.19E—05	1.16E—05	7.22E—06	4.81E—06
Uterus	2.02E—05	1.76E—05	1.80E—05	1.42E—05	1.16E—05	8.77E—06
Total body	1.43E—05	1.37E—05	1.25E—05	9.93E—06	6.60E—06	4.86E—06

Target	Energy (MeV) 0.200	0.500	1.000	1.500	2.000	4.000
Bladder wall	6.28E—06	6.13E—06	5.71E—06	5.42E—06	5.11E—06	4.27E—06
Stomach wall	4.96E—06	4.84E—06	4.60E—06	4.38E—06	4.10E—06	3.45E—06
Small intestine plus contents	5.51E—06	5.33E—06	5.00E—06	4.73E—06	4.43E—06	3.77E—06
Upper large intestine wall	5.73E—06	5.61E—06	5.31E—06	5.00E—06	4.68E—06	3.97E—06
Lower large intestine wall	6.01E—06	5.84E—06	5.48E—06	5.18E—06	4.89E—06	4.09E—06
Kidneys	4.95E—06	4.93E—06	4.71E—06	4.47E—06	4.22E—06	3.55E—06
Liver	3.82E—06	3.85E—06	3.69E—06	3.46E—06	3.30E—06	2.80E—06
Lungs	4.61E—06	4.62E—06	4.31E—06	4.04E—06	3.81E—06	3.17E—06
"Other tissues" (suggested for muscle)	5.03E—06	5.22E—06	4.99E—06	4.65E—06	4.36E—06	3.66E—06
Ovaries	7.02E—06	6.84E—06	6.40E—06	6.00E—06	5.65E—06	4.75E—06
Pancreas	6.32E—06	6.20E—06	5.71E—06	5.43E—06	5.17E—06	4.20E—06
Skeleton (suggested for total endosteal cells)	3.74E—06	3.83E—06	3.67E—06	3.54E—06	3.33E—06	2.83E—06
Red marrow	5.58E—06	4.75E—06	4.26E—06	3.85E—06	3.95E—06	2.98E—06
Skin	2.74E—06	3.14E—06	3.15E—06	2.96E—06	2.81E—06	2.37E—06
Spleen	5.14E—06	5.07E—06	4.79E—06	4.55E—06	4.23E—06	3.60E—06
Testes	4.14E—06	4.27E—06	4.08E—06	3.97E—06	4.59E—06	3.20E—06
Thymus	1.03E—05	5.77E—06	4.35E—06	7.16E—06	5.70E—06	6.00E—06
Thyroid	4.63E—06	4.76E—06	4.45E—06	4.19E—06	3.94E—06	3.27E—06
Uterus	8.30E—06	6.46E—06	8.93E—06	5.29E—06	7.01E—06	6.15E—06
Total body	4.82E—06	4.97E—06	4.75E—06	4.45E—06	4.18E—06	3.51E—06

SPECIFIC ABSORBED FRACTION OF PHOTON ENERGY (*cont.*)

† Build-up factor method. ⊕ Adjustment on density or composition. * Extrapolation from higher energy. ⊗ Calculated by difference.

Source in Ovaries

Target	Energy (MeV) 0.010	0.015	0.020	0.030	0.050	0.100
Bladder wall	1.79E—08*	2.69E—08†	3.42E—06	2.65E—05	3.99E—05	2.91E—05
Stomach wall	8.09E—14*	1.21E—13†	2.07E—09†	5.22E—07	2.40E—06	3.24E—06
Small intestine plus contents	2.87E—06	2.23E—05	5.91E—05	9.45E—05	7.22E—05	4.80E—05
Upper large intestine wall	8.94E—06	6.76E—05	1.08E—04	1.03E—04	6.38E—05	4.16E—05
Lower large intestine wall	9.94E—06	8.15E—05	1.46E—04	1.45E—04	8.86E—05	5.62E—05
Kidneys	3.71E—14*	5.56E—14†	1.90E—09†	5.45E—07	2.64E—06	3.53E—06
Liver	1.05E—14*	1.58E—14†	5.74E—10†	2.42E—07	1.34E—06	2.00E—06
Lungs	5.12E—26*	7.68E—26†	1.02E—14†	9.26E—10†	8.58E—08	2.06E—07
"Other tissues" (suggested for muscle)	4.11E—06⊗	9.31E—06⊗	1.19E—05⊗	1.23E—05⊗	1.00E—05⊗	7.57E—06⊗
Ovaries	9.68E—02	5.91E—02	3.26E—02	1.15E—02	3.58E—03	2.22E—03
Pancreas	4.74E—18*	7.10E—18†	3.03E—11†	7.93E—08†	8.55E—07	1.99E—06
Skeleton (suggested for total endosteal cells)	2.70E—08*	4.06E—08	1.39E—06	7.90E—06	1.14E—05	6.94E—06
Red marrow	1.14E—07*	1.71E—07	5.85E—06	3.28E—05	4.51E—05	2.50E—05
Skin	8.61E—12*	1.29E—11†	1.85E—08	5.06E—07	1.35E—06	1.51E—06
Spleen	1.02E—16*	1.53E—16†	7.80E—11†	9.57E—08	1.03E—06	1.77E—06
Thymus	3.01E—33*	4.51E—33†	1.05E—18†	1.83E—11†	8.00E—09†	7.15E—08†
Thyroid	2.30E—23*	3.45E—23*	4.61E—23†	2.50E—13†	5.98E—10†	1.09E—08†
Uterus	1.82E—06*	2.74E—06	5.12E—05	1.44E—04	1.26E—04	8.03E—05
Total body	1.43E—05	1.43E—05	1.43E—05	1.39E—05	1.13E—05	8.10E—06

Target	Energy (MeV) 0.200	0.500	1.000	1.500	2.000	4.000
Bladder wall	2.45E—05	2.29E—05	2.28E—05	2.04E—05	1.72E—05	1.47E—05
Stomach wall	2.82E—06	2.90E—06	2.48E—06	3.25E—06	3.12E—06	3.13E—06
Small intestine plus contents	4.19E—05	4.07E—05	3.57E—05	3.34E—05	3.15E—05	2.51E—05
Upper large intestine wall	3.99E—05	3.82E—05	3.36E—05	3.19E—05	2.97E—05	2.21E—05
Lower large intestine wall	5.07E—05	4.88E—05	4.25E—05	4.02E—05	3.84E—05	3.37E—05
Kidneys	3.47E—06	3.92E—06	3.63E—06	3.32E—06	3.62E—06	2.98E—06
Liver	2.11E—06	2.11E—06	2.17E—06	2.14E—06	2.30E—06	1.76E—06
Lungs	2.56E—07	3.88E—07	5.40E—07	5.62E—07	5.86E—07	7.01E—07
"Other tissues" (suggested for muscle)	7.02E—06⊗	6.84E—06⊗	6.40E—06⊗	6.00E—06⊗	5.65E—06⊗	4.75E—06⊗
Ovaries	2.48E—03	2.62E—03	2.42E—03	2.28E—03	2.12E—03	1.72E—03
Pancreas	1.77E—06	1.90E—06	3.18E—06	1.76E—06	9.78E—07	2.66E—06
Skeleton (suggested for total endosteal cells)	4.18E—06	3.18E—06	2.88E—06	2.69E—06	2.60E—06	2.16E—06
Red marrow	1.43E—05	1.06E—05	9.34E—06	8.52E—06	8.06E—06	6.32E—06
Skin	1.63E—06	1.81E—06	1.90E—06	1.75E—06	1.56E—06	1.52E—06
Spleen	1.99E—06	2.40E—06	2.05E—06	1.62E—06	2.52E—06	1.98E—06
Thymus	1.40E—07†	2.11E—07†	2.66E—07†	2.92E—07†	3.08E—07†	3.13E—07†
Thyroid	3.01E—08†	6.22E—08†	9.61E—08†	1.17E—07†	1.31E—07†	1.51E—07†
Uterus	7.51E—05	6.18E—05	5.32E—05	4.86E—05	4.77E—05	3.91E—05
Total body	7.16E—06	6.88E—06	6.38E—06	5.97E—06	5.65E—06	4.70E—06

SPECIFIC ABSORBED FRACTION OF PHOTON ENERGY (*cont.*)

† Build-up factor method. ⊕ Adjustment on density or composition. * Extrapolation from higher energy. ⊗ Calculated by difference.

Source in Pancreas

Target	Energy (MeV) 0.010	0.015	0.020	0.030	0.050	0.100
Bladder wall	9.31E—22*	1.40E—21†	3.01E—13†	7.08E—09†	1.68E—07	5.22E—07
Stomach wall	2.26E—06	3.91E—05	1.10E—04	1.57E—04	1.13E—04	6.73E—05
Small intestine plus contents	2.59E—11*	3.89E—11†	4.01E—08†	1.87E—06	6.39E—06	7.11E—06
Upper large intestine wall	3.88E—11*	5.82E—11†	6.06E—08†	2.79E—06	8.27E—06	8.32E—06
Lower large intestine wall	1.41E—14*	2.12E—14†	8.81E—10†	3.08E—07	1.73E—06	2.22E—06
Kidneys	2.26E—08*	3.39E—08†	3.23E—06	2.50E—05	3.37E—05	2.65E—05
Liver	1.54E—07*	2.30E—07	3.61E—06	1.64E—05	2.24E—05	1.76E—05
Lungs	2.75E—09*	4.12E—09†	1.16E—06	8.87E—06	1.29E—05	1.02E—05
"Other tissues" (suggested for muscle)	2.41E—06⊗	6.73E—06⊗	1.04E—05⊗	1.21E—05⊗	9.45E—06⊗	6.87E—06⊗
Ovaries	5.28E—18*	7.92E—18†	3.07E—11†	7.99E—08†	1.29E—06†	1.82E—06
Pancreas	1.47E—02	1.09E—02	7.14E—03	3.16E—03	1.10E—03	6.51E—04
Skeleton (suggested for total endosteal cells)	3.77E—08*	5.66E—08	7.68E—07	5.02E—06	9.58E—06	6.58E—06
Red marrow	7.85E—08*	1.18E—07	1.46E—06	9.15E—06	1.78E—05	1.27E—05
Skin	7.33E—09*	1.10E—08*	1.47E—08	5.52E—07	1.39E—06	1.50E—06
Spleen	1.45E—06	4.48E—05	1.31E—04	1.79E—04	1.25E—04	7.62E—05
Testes	9.66E—30*	1.45E—29†	4.87E—17†	1.17E—10†	2.50E—08†	1.64E—07†
Thymus	1.99E—18*	2.98E—18†	1.78E—11†	5.77E—08†	1.70E—06	1.73E—06
Thyroid	1.21E—26*	1.81E—26†	2.08E—15†	7.82E—10†	7.99E—08†	3.78E—07†
Uterus	1.65E—18*	2.48E—18†	1.75E—11†	1.10E—07	1.01E—06	1.79E—06
Total body	1.43E—05	1.43E—05	1.43E—05	1.39E—05	1.11E—05	7.92E—06

Target	Energy (MeV) 0.200	0.500	1.000	1.500	2.000	4.000
Bladder wall	5.61E—07	6.82E—07	4.03E—07	1.02E—06	1.51E—06	1.19E—06
Stomach wall	6.54E—05	5.84E—05	5.40E—05	4.96E—05	4.54E—05	4.02E—05
Small intestine plus contents	6.67E—06	6.06E—06	5.90E—06	5.62E—06	5.27E—06	4.52E—06
Upper large intestine wall	7.94E—06	7.43E—06	7.18E—06	6.76E—06	5.91E—06	5.04E—06
Lower large intestine wall	2.06E—06	1.84E—06	1.85E—06	2.33E—06	1.54E—06	1.91E—06
Kidneys	2.27E—05	2.13E—05	1.94E—05	1.82E—05	1.63E—05	1.47E—05
Liver	1.56E—05	1.41E—05	1.29E—05	1.21E—05	1.08E—05	9.50E—06
Lungs	8.92E—06	8.21E—06	7.43E—06	6.77E—06	6.54E—06	5.20E—06
"Other tissues" (suggested for muscle)	6.32E—06⊗	6.20E—06⊗	5.71E—06⊗	5.43E—06⊗	5.17E—06⊗	4.20E—06⊗
Ovaries	1.14E—06	1.28E—06	3.50E—06	2.37E—06†	2.25E—06†	1.94E—06†
Pancreas	6.88E—04	7.29E—04	6.74E—04	6.33E—04	5.72E—04	4.58E—04
Skeleton (suggested for total endosteal cells)	4.00E—06	3.17E—06	2.86E—06	2.79E—06	2.60E—06	2.14E—06
Red marrow	7.70E—06	5.97E—06	5.34E—06	5.14E—06	4.76E—06	3.87E—06
Skin	1.57E—06	1.74E—06	1.90E—06	1.77E—06	1.69E—06	1.58E—06
Spleen	6.70E—05	6.59E—05	5.78E—05	5.34E—05	5.07E—05	3.75E—05
Testes	2.76E—07†	2.15E—07	4.28E—07†	4.50E—07†	4.61E—07†	4.48E—07†
Thymus	3.07E—06	1.83E—06	1.75E—06	2.15E—06	2.05E—06†	1.77E—06†
Thyroid	5.52E—07†	6.49E—07†	9.16E—08	7.11E—07†	7.11E—07†	6.61E—07†
Uterus	2.32E—06	2.17E—06	2.01E—06	1.87E—06	2.05E—06	1.42E—06
Total body	6.99E—06	6.74E—06	6.22E—06	5.90E—06	5.55E—06	4.53E—06

SPECIFIC ABSORBED FRACTION OF PHOTON ENERGY (*cont.*)

† Build-up factor method. ⊕ Adjustment on density or composition. * Extrapolation from higher energy. ⊗ Calculated by difference.
Source in Skeleton (suggested for Cancellous Bone, Cortical Bone, Red Marrow, and Yellow Marrow)

Target	Energy (MeV) 0.010	0.015	0.020	0.030	0.050	0.100
Bladder wall	2.27E—11*	3.40E—11†	2.00E—08†	2.96E—07	1.51E—06	2.12E—06
Stomach wall	1.45E—07*	2.18E—07*	2.90E—07	1.06E—06	1.73E—06	2.11E—06
Small intestine plus contents	1.52E—07*	2.28E—07	5.01E—07	1.69E—06	3.09E—06	2.81E—06
Upper large intestine wall	2.12E—07*	3.17E—07	6.44E—07	1.38E—06	2.57E—06	2.56E—06
Lower large intestine wall	6.48E—07*	9.71E—07	1.84E—06	3.46E—06	4.10E—06	3.97E—06
Kidneys	3.02E—07*	4.54E—07*	6.05E—07	1.81E—06	3.26E—06	3.01E—06
Liver	1.24E—07*	1.87E—07	4.79E—07	1.67E—06	2.37E—06	2.55E—06
Lungs	1.69E—07*	2.54E—07	1.04E—06	3.20E—06	4.42E—06	3.63E—06
"Other tissues" (suggested for muscle)	3.85E—07⊗	9.43E—07⊗	1.63E—06⊗	3.06E—06⊗	3.82E—06⊗	3.66E—06⊗
Ovaries	2.12E—08*	3.18E—08†	9.03E—07†	3.85E—06†	3.20E—06	2.81E—06
Pancreas	3.91E—07*	5.87E—07*	7.83E—07†	1.39E—06	3.22E—06	2.78E—06
Skeleton (suggested for total endosteal cells)	9.34E—05	8.97E—05	8.42E—05	6.84E—05	4.10E—05	1.81E—05
Red marrow	9.42E—05	8.85E—05	8.31E—05	6.45E—05	3.77E—05	1.65E—05
Skin	4.95E—07	1.18E—06	1.85E—06	2.27E—06	2.40E—06	2.35E—06
Spleen	6.52E—07*	9.78E—07*	1.30E—06*	1.96E—06	2.76E—06	2.09E—06
Testes	1.81E—11*	2.71E—11†	2.22E—08†	8.29E—07†	1.73E—06	2.59E—06
Thymus	2.85E—07*	4.27E—07*	5.69E—07†	7.64E—07	1.87E—06	2.19E—06
Thyroid	1.32E—07*	1.98E—07*	2.64E—07†	1.94E—06†	2.91E—06	2.76E—06
Uterus	9.28E—11*	1.39E—10†	4.96E—08†	1.23E—06	1.63E—06	2.11E—06
Total body	1.43E—05	1.42E—05	1.40E—05	1.28E—05	9.31E—06	5.73E—06

Target	Energy (MeV) 0.200	0.500	1.000	1.500	2.000	4.000
Bladder wall	1.68E—06	2.28E—06	1.37E—06	2.56E—06	1.87E—06	1.94E—06
Stomach wall	2.07E—06	2.02E—06	1.76E—06	1.90E—06	1.72E—06	1.48E—06
Small intestine plus contents	2.73E—06	2.77E—06	2.79E—06	2.36E—06	2.35E—06	1.98E—06
Upper large intestine wall	2.67E—06	2.40E—06	2.61E—06	2.33E—06	2.32E—06	1.77E—06
Lower large intestine wall	3.82E—06	3.24E—06	3.58E—06	3.03E—06	2.72E—06	2.54E—06
Kidneys	3.29E—06	3.12E—06	4.02E—06	3.16E—06	2.43E—06	2.32E—06
Liver	2.39E—06	2.44E—06	2.41E—06	2.30E—06	2.50E—06	2.05E—06
Lungs	3.50E—06	3.38E—06	3.29E—06	3.03E—06	3.09E—06	2.30E—06
"Other tissues" (suggested for muscle)	3.74E—06⊗	3.82E—06⊗	3.67E—06⊗	3.54E—06⊗	3.33E—06⊗	2.83E—06⊗
Ovaries	2.46E—06	3.75E—06	3.16E—06†	2.79E—06†	2.75E—06†	2.22E—06†
Pancreas	3.90E—06	3.11E—06	3.29E—06	3.71E—06	3.50E—06	1.59E—06
Skeleton (suggested for total endosteal cells)	1.29E—05	1.19E—05	1.11E—05	1.01E—05	9.39E—06	7.79E—06
Red marrow	1.15E—05	1.06E—05	1.01E—05	9.19E—06	8.37E—06	7.06E—06
Skin	2.63E—06	3.00E—06	2.87E—06	2.85E—06	2.62E—06	2.46E—06
Spleen	2.32E—06	2.90E—06	2.96E—06	2.02E—06	2.17E—06	1.23E—06
Testes	2.23E—06	1.98E—06	2.24E—06	2.37E—06	3.88E—06	2.13E—06
Thymus	1.65E—06	2.64E—06	2.84E—06	1.41E—06	2.17E—06†	2.78E—06
Thyroid	3.31E—06	3.86E—06	1.60E—06	3.82E—06	1.82E—07	1.91E—06†
Uterus	2.24E—06	1.97E—06	3.15E—06	1.98E—06	1.57E—06	1.65E—06
Total body	5.01E—06	4.93E—06	4.70E—06	4.43E—06	4.16E—06	3.50E—06

SPECIFIC ABSORBED FRACTION OF PHOTON ENERGY (*cont.*)

† Build-up factor method. ⊕ Adjustment on density or composition. * Extrapolation from higher energy. ⊗ Calculated by difference.

Source in Skin

Target	Energy (MeV) 0.010	0.015	0.020	0.030	0.050	0.100
Bladder wall	2.14E—07*	3.20E—07*	4.27E—07	1.91E—06	1.94E—06	1.79E—06
Stomach wall	2.74E—07*	4.11E—07*	5.48E—07	2.02E—06	2.58E—06	2.06E—06
Small intestine plus contents	6.45E—08*	9.68E—08*	1.29E—07	8.64E—07	1.72E—06	1.69E—06
Upper large intestine wall	1.26E—07*	1.89E—07*	2.52E—07	1.19E—06	1.95E—06	1.83E—06
Lower large intestine wall	5.72E—08*	8.59E—08*	1.14E—07	9.04E—07	1.65E—06	1.84E—06
Kidneys	1.82E—07*	2.72E—07	1.18E—06	2.81E—06	2.91E—06	2.08E—06
Liver	4.77E—08*	7.15E—08	4.86E—07	1.78E—06	2.32E—06	2.01E—06
Lungs	8.04E—08*	1.21E—07	7.45E—07	2.54E—06	2.94E—06	2.17E—06
"Other tissues" (suggested for muscle)	4.00E—06⊗	6.06E—06⊗	6.46E—06⊗	5.41E—06⊗	3.48E—06⊗	2.60E—06⊗
Ovaries	8.61E—12*	1.29E—11†	1.93E—08†	7.68E—07†	1.86E—06	1.22E—06
Pancreas	1.41E—08*	2.12E—08*	2.83E—08†	9.03E—07	1.45E—06	1.39E—06
Skeleton (suggested for total endosteal cells)	1.85E—06	4.30E—06	6.58E—06	8.80E—06	8.06E—06	4.19E—06
Red marrow	1.61E—06	3.90E—06	5.97E—06	7.98E—06	7.39E—06	4.02E—06
Skin	2.05E—04	1.03E—04	5.57E—05	2.24E—05	8.07E—06	5.33E—06
Spleen	1.67E—07*	2.50E—07*	3.34E—07	1.57E—06	2.49E—06	2.03E—06
Testes	8.27E—07	5.82E—06	8.59E—06	9.45E—06	4.85E—06	3.04E—06
Thymus	1.88E—08*	2.82E—08†	5.47E—07	1.56E—06	2.27E—06	2.22E—06
Thyroid	1.25E—06*	1.88E—06*	2.50E—06	5.24E—06	3.33E—06	2.50E—06
Uterus	5.17E—08*	7.75E—08*	1.03E—07	6.49E—07	1.84E—06	1.61E—06
Total body	1.13E—05	9.06E—06	7.77E—06	6.17E—06	4.21E—06	2.87E—06

Target	Energy (MeV) 0.200	0.500	1.000	1.500	2.000	4.000
Bladder wall	1.91E—06	2.16E—06	2.65E—06	2.69E—06	2.28E—06	1.50E—06
Stomach wall	1.97E—06	2.24E—06	2.23E—06	2.04E—06	1.94E—06	1.61E—06
Small intestine plus contents	1.73E—06	1.90E—06	2.01E—06	1.93E—06	1.80E—06	1.56E—06
Upper large intestine wall	1.66E—06	2.00E—06	2.25E—06	1.78E—06	1.86E—06	1.44E—06
Lower large intestine wall	1.78E—06	2.08E—06	2.25E—06	1.91E—06	2.08E—06	1.45E—06
Kidneys	2.29E—06	2.59E—06	2.60E—06	2.38E—06	2.30E—06	1.98E—06
Liver	2.02E—06	2.34E—06	2.25E—06	2.20E—06	2.21E—06	1.73E—06
Lungs	2.27E—06	2.34E—06	2.37E—06	2.33E—06	2.23E—06	1.84E—06
"Other tissues" (suggested for muscle)	2.74E—06⊗	3.14E—06⊗	3.15E—06⊗	2.96E—06⊗	2.81E—06⊗	2.37E—06⊗
Ovaries	1.77E—06	1.02E—06	2.11E—06	1.75E—06	2.10E—06	1.22E—06
Pancreas	2.07E—06	2.02E—06	1.63E—06	1.78E—06	1.87E—06	1.35E—06
Skeleton (suggested for total endosteal cells)	2.99E—06	2.93E—06	2.85E—06	2.74E—06	2.51E—06	2.18E—06
Red marrow	2.85E—06	2.72E—06	2.63E—06	2.59E—06	2.32E—06	2.06E—06
Skin	6.05E—06	6.93E—06	6.93E—06	6.38E—06	6.01E—06	4.78E—06
Spleen	2.06E—06	2.27E—06	2.67E—06	1.87E—06	1.82E—06	1.76E—06
Testes	3.82E—06	2.86E—06	3.10E—06	3.06E—06	2.97E—06	3.20E—06
Thymus	2.49E—06	2.99E—06	2.43E—06	1.94E—06	2.72E—06	2.13E—06
Thyroid	2.73E—06	2.89E—06	2.40E—06	2.31E—06	2.82E—06	2.18E—06
Uterus	1.39E—06	1.96E—06	2.58E—06	2.11E—06	1.74E—06	2.19E—06
Total body	2.81E—06	3.15E—06	3.15E—06	2.97E—06	2.81E—06	2.35E—06

Specific Absorbed Fraction of Photon Energy (*cont.*)

† Build-up factor method. ⊕ Adjustment on density or composition. * Extrapolation from higher energy. ⊗ Calculated by difference.

Source in Spleen

Target	Energy (MeV) 0.010	0.015	0.020	0.030	0.050	0.100
Bladder wall	6.70E—22*	1.00E—21†	2.23E—13†	5.34E—09†	2.28E—07†	3.15E—07
Stomach wall	8.63E—07*	1.29E—06	1.34E—05	5.33E—05	5.86E—05	4.09E—05
Small intestine plus contents	1.77E—10*	2.65E—10†	9.39E—08	1.49E—06	5.08E—06	5.60E—06
Upper large intestine wall	1.18E—10*	1.77E—10†	6.70E—08†	2.18E—06	5.92E—06	6.20E—06
Lower large intestine wall	1.82E—12*	2.73E—12†	5.78E—09†	4.55E—07	2.01E—06	2.35E—06
Kidneys	7.02E—06*	1.05E—05	3.63E—05	6.65E—05	5.30E—05	3.61E—05
Liver	7.20E—13*	1.08E—12†	4.25E—09†	5.99E—07	2.93E—06	3.78E—06
Lungs	6.50E—07*	9.76E—07	4.97E—06	1.25E—05	1.23E—05	8.97E—06
"Other tissues" (suggested for muscle)	1.18E—06⊗	3.78E—06⊗	6.79E—06⊗	9.42E—06⊗	7.63E—06⊗	5.48E—06⊗
Ovaries	9.82E—17*	1.47E—16†	7.94E—11†	8.80E—08†	1.19E—06†	1.14E—06
Pancreas	1.56E—06	4.29E—05	1.26E—04	1.87E—04	1.23E—04	7.35E—05
Skeleton (suggested for total endosteal cells)	6.83E—09	2.92E—07	1.70E—06	5.79E—06	8.30E—06	5.06E—06
Red marrow	7.07E—09	3.02E—07	1.74E—06	6.07E—06	1.05E—05	7.47E—06
Skin	7.80E—09*	1.17E—08	3.10E—08	1.54E—06	1.98E—06	1.74E—06
Spleen	5.44E—03	4.66E—03	3.59E—03	1.86E—03	7.24E—04	4.21E—04
Testes	1.22E—29*	1.83E—29†	4.62E—17†	9.74E—11†	2.09E—08†	1.42E—07†
Thymus	5.50E—20*	8.25E—20†	2.18E—12†	1.83E—08†	2.51E—07	9.39E—07
Thyroid	6.30E—27*	9.45E—27†	9.31E—16†	4.27E—10†	5.15E—08†	2.70E—07†
Uterus	7.54E—19*	1.13E—18†	9.33E—12†	4.07E—08†	6.87E—07	1.61E—06
Total body	1.43E—05	1.43E—05	1.42E—05	1.30E—05	9.47E—06	6.52E—06

Target	Energy (MeV)					
	0.200	0.500	1.000	1.500	2.000	4.000
Bladder wall	6.27E—07	4.52E—07	1.25E—06	8.91E—07	9.65E—07	9.26E—07†
Stomach wall	3.36E—05	3.39E—05	2.93E—05	3.14E—05	2.65E—05	2.06E—05
Small intestine plus contents	5.08E—06	4.77E—06	4.57E—06	4.67E—06	4.10E—06	3.55E—06
Upper large intestine wall	4.36E—06	4.81E—06	4.58E—06	4.63E—06	4.65E—06	3.03E—06
Lower large intestine wall	2.27E—06	2.41E—06	2.06E—06	2.11E—06	2.22E—06	1.81E—06
Kidneys	3.14E—05	2.93E—05	2.59E—05	2.55E—05	2.33E—05	1.79E—05
Liver	3.67E—06	3.69E—06	3.53E—06	3.45E—06	3.27E—06	2.82E—06
Lungs	7.91E—06	7.56E—06	6.87E—06	6.21E—06	6.38E—06	5.23E—06
"Other tissues" (suggested for muscle)	5.14E—06⊗	5.07E—06⊗	4.79E—06⊗	4.55E—06⊗	4.23E—06⊗	3.60E—06⊗
Ovaries	2.05E—06	3.70E—06	1.70E—06	2.17E—06†	3.46E—07	9.91E—08
Pancreas	6.58E—05	6.70E—05	5.80E—05	5.13E—05	4.94E—05	3.75E—05
Skeleton (suggested for total endosteal cells)	3.25E—06	2.65E—06	2.48E—06	2.36E—06	2.18E—06	1.94E—06
Red marrow	5.04E—06	3.90E—06	3.64E—06	3.35E—06	3.19E—06	2.82E—06
Skin	1.81E—06	2.18E—06	2.16E—06	2.05E—06	1.80E—06	1.66E—06
Spleen	4.32E—04	4.49E—04	4.10E—04	3.77E—04	3.55E—04	2.73E—04
Testes	2.45E—07†	3.30E—07†	3.88E—07†	4.14E—07†	4.26E—07†	4.17E—07†
Thymus	1.33E—06	1.65E—06	2.42E—06	4.14E—07	2.24E—06	1.27E—06†
Thyroid	4.16E—07†	5.10E—07†	5.68E—07†	5.83E—07†	5.88E—07†	5.58E—07†
Uterus	1.42E—06	1.61E—06	2.11E—06	6.25E—07	1.76E—06	2.23E—06
Total body	5.93E—06	5.82E—06	5.46E—06	5.16E—06	4.81E—06	4.02E—06

ITHP-GG

SPECIFIC ABSORBED FRACTION OF PHOTON ENERGY (*cont.*)

† Build-up factor method. ⊕ Adjustment on density or composition. * Extrapolation from higher energy. ⊗ Calculated by difference.

Source in Testes

Target	Energy (MeV) 0.010	0.015	0.020	0.030	0.050	0.100
Bladder wall	5.00E—09*	7.50E—09†	2.26E—06	1.74E—05	2.51E—05	1.89E—05
Stomach wall	1.39E—25*	2.08E—25†	3.73E—15†	6.66E—10†	6.18E—08	8.57E—08
Small intestine plus contents	2.67E—18*	4.01E—18†	1.62E—11†	4.66E—08	8.04E—07	1.25E—06
Upper large intestine wall	1.35E—17*	2.03E—17†	2.75E—11†	4.78E—08†	6.70E—07	1.06E—06
Lower large intestine wall	1.30E—09*	1.95E—09†	7.90E—07	6.85E—06	1.31E—05	1.06E—05
Kidneys	4.67E—27*	7.01E—27†	9.01E—16†	4.33E—10†	3.81E—08	1.16E—07
Liver	2.58E—26*	3.87E—26†	9.61E—16†	2.74E—10†	2.73E—08	9.67E—08
Lungs	2.24E—36*	3.36E—36†	1.83E—20†	1.65E—12†	1.49E—09†	1.23E—08
"Other tissues" (suggested for muscle)	4.99E—08⊗	7.21E—07⊗	2.66E—06⊗	5.71E—06⊗	5.52E—06⊗	4.28E—06⊗
Pancreas	9.64E—30*	1.45E—29†	4.97E—17†	1.17E—10†	2.49E—08†	1.63E—07†
Skeleton (suggested for total endosteal cells)	7.33E—11*	1.10E—10⊕	2.40E—08	1.15E—06	4.36E—06	4.01E—06
Red marrow	3.24E—09*	4.86E—09*	6.49E—09	6.46E—07	3.22E—06	3.12E—06
Skin	1.47E—06	9.12E—06	1.53E—05	1.39E—05	7.51E—06	4.93E—06
Spleen	1.31E—29*	1.97E—29†	4.57E—17†	9.84E—11†	2.10E—08†	6.08E—08
Testes	2.43E—02	1.86E—02	1.23E—02	5.34E—03	1.77E—03	1.07E—03
Thymus	2.20E—24*	3.30E—24*	4.40E—24†	5.01E—14†	2.17E—10†	5.16E—09†
Thyroid	2.50E—28*	3.75E—28*	5.00E—28†	7.32E—16†	1.68E—11†	7.89E—10†
Total body	1.43E—05	1.38E—05	1.26E—05	1.00E—05	6.84E—06	5.03E—06

Target	Energy (MeV) 0.200	0.500	1.000	1.500	2.000	4.000
Bladder wall	1.70E—05	1.75E—05	1.74E—05	1.38E—05	1.35E—05	9.58E—06
Stomach wall	1.71E—07	4.07E—07	2.71E—07	4.06E—07	4.80E—07	4.52E—07
Small intestine plus contents	1.50E—06	1.85E—06	1.78E—06	1.78E—06	1.86E—06	1.58E—06
Upper large intestine wall	1.33E—06	1.10E—06	1.40E—06	1.16E—06	2.30E—06	1.42E—06
Lower large intestine wall	9.91E—06	9.83E—06	7.24E—06	7.98E—06	7.32E—06	7.40E—06
Kidneys	2.04E—07	3.51E—07	5.07E—07	5.63E—07	4.05E—07	7.27E—07
Liver	1.48E—07	1.77E—07	3.61E—07	4.16E—07	4.07E—07	3.97E—07
Lungs	4.38E—08	4.94E—08	1.01E—07	1.60E—07	1.20E—07	1.68E—07
"Other tissues" (suggested for muscle)	4.14E—06⊗	4.27E—06⊗	4.08E—06⊗	3.97E—06⊗	4.59E—06⊗	3.20E—06⊗
Pancreas	2.76E—07†	1.28E—07	3.61E—07	4.50E—07†	6.34E—07	4.48E—07†
Skeleton (suggested for total endosteal cells)	2.73E—06	2.28E—06	2.15E—06	2.00E—06	1.95E—06	1.71E—06
Red marrow	2.25E—06	1.81E—06	1.69E—06	1.51E—06	1.61E—06	1.41E—06
Skin	5.13E—06	5.48E—06	5.15E—06	5.11E—06	4.59E—06	3.93E—06
Spleen	7.38E—08	4.55E—07	2.12E—07	2.38E—07	5.73E—07	2.33E—07
Testes	1.17E—03	1.23E—03	1.16E—03	1.04E—03	9.82E—04	7.66E—04
Thymus	1.62E—08†	3.82E—08†	6.43E—08†	8.18E—08†	9.46E—08†	1.14E—07†
Thyroid	3.53E—09†	1.17E—08†	2.46E—08†	3.51E—08†	4.35E—08†	6.00E—08†
Total body	4.82E—06	4.91E—06	4.69E—06	4.47E—06	4.23E—06	3.54E—06

Specific Absorbed Fraction of Photon Energy (*cont.*)

† Build-up factor method. ⊕ Adjustment on density or composition. * Extrapolation from higher energy. ⊗ Calculated by difference.

Source in Thyroid

Target	Energy (MeV)					
	0.010	0.015	0.020	0.030	0.050	0.100
Bladder wall	8.78E—25*	1.32E—24*	1.76E—24†	3.15E—14†	1.61E—10†	4.12E—09†
Stomach wall	2.07E—25*	3.11E—25†	4.50E—15†	7.26E—10†	2.01E—07	1.90E—07
Small intestine plus contents	4.58E—35*	6.87E—35†	1.08E—19†	5.26E—12†	3.49E—09†	1.97E—08
Upper large intestine wall	9.33E—35*	1.40E—34†	1.88E—19†	7.08E—12†	4.20E—09†	4.35E—08†
Lower large intestine wall	8.33E—38*	1.25E—37†	2.75E—21†	5.66E—13†	6.64E—10†	1.00E—08†
Kidneys	8.87E—30*	1.33E—29†	3.98E—17†	9.31E—11†	2.06E—08†	1.20E—07
Liver	1.26E—24*	1.90E—24†	1.16E—14†	1.19E—09†	1.93E—07	3.05E—07
Lungs	1.52E—13*	2.28E—12†	3.30E—08	1.23E—06	3.89E—06	3.67E—06
"Other tissues" (suggested for muscle)	3.35E—06⊗	8.66E—06⊗	1.19E—05⊗	1.16E—05⊗	7.22E—06⊗	4.81E—06⊗
Ovaries	2.33E—23*	3.50E—23*	4.66E—23†	2.50E—13†	5.97E—10†	1.09E—08†
Pancreas	1.24E—26*	1.86E—26†	2.08E—15†	7.80E—10†	7.98E—08†	1.91E—07
Skeleton (suggested for total endosteal cells)	1.79E—09*	2.69E—09	2.84E—07	2.99E—06	6.35E—06	4.45E—06
Red marrow	2.68E—09*	4.02E—09	3.14E—07	2.74E—06	6.46E—06	4.87E—06
Skin	5.06E—07*	7.59E—07	3.26E—06	5.35E—06	3.66E—06	2.64E—06
Spleen	6.81E—27*	1.02E—26†	9.64E—16†	4.29E—10†	1.16E—07	3.75E—07
Testes	2.48E—28*	3.72E—28*	4.96E—28†	7.32E—16†	1.67E—11†	7.87E—10†
Thymus	1.23E—09*	1.84E—09†	2.53E—07	1.13E—05	2.28E—05	1.65E—05
Thyroid	4.29E—02	2.93E—02	1.81E—02	7.41E—03	2.42E—03	1.44E—03
Uterus	5.98E—24*	8.97E—24*	1.20E—23†	2.03E—13†	5.28E—10†	9.99E—09†
Total body	1.43E—05	1.42E—05	1.35E—05	1.08E—05	6.95E—06	4.71E—06

Target	Energy (MeV) 0.200	0.500	1.000	1.500	2.000	4.000
Bladder wall	1.35E—08†	3.30E—08†	5.71E—08†	7.36E—08†	8.56E—08†	1.05E—07†
Stomach wall	1.45E—07	4.44E—07	4.62E—07	6.42E—07	3.70E—07	5.20E—07
Small intestine plus contents	5.80E—08	2.47E—08	1.38E—07	1.91E—07	1.92E—07	2.04E—07
Upper large intestine wall	4.03E—08	3.63E—08	1.51E—07	2.29E—07	1.85E—07	3.92E—07
Lower large intestine wall	2.59E—08†	5.22E—08†	8.09E—08†	9.91E—08†	1.11E—07†	1.30E—07†
Kidneys	1.40E—07	1.69E—07	4.12E—07	5.96E—07	4.35E—07	3.09E—07
Liver	4.28E—07	5.49E—07	6.54E—07	7.28E—07	8.78E—07	7.72E—07
Lungs	3.38E—06	3.67E—06	3.83E—06	3.35E—06	3.17E—06	3.22E—06
"Other tissues" (suggested for muscle)	4.63E—06⊗	4.76E—06⊗	4.45E—06⊗	4.19E—06⊗	3.94E—06⊗	3.27E—06⊗
Ovaries	3.00E—08†	6.21E—08†	9.62E—08†	1.17E—07†	1.31E—07†	1.51E—07†
Pancreas	3.98E—07	2.34E—07	1.88E—07	1.49E—06	7.11E—07	6.61E—07
Skeleton (suggested for total endosteal cells)	2.98E—06	2.61E—06	2.39E—06	2.33E—06	2.23E—06	1.97E—06
Red marrow	3.27E—06	2.89E—06	2.57E—06	2.56E—06	2.47E—06	2.11E—06
Skin	2.87E—06	2.97E—06	3.07E—06	2.96E—06	2.76E—06	2.41E—06
Spleen	4.36E—07	4.40E—07	7.21E—07	6.30E—07	4.40E—07	6.85E—07
Testes	3.52E—09†	1.17E—08†	2.46E—08†	3.51E—08†	4.36E—08†	6.00E—08†
Thymus	1.52E—05	1.52E—05	1.23E—05	1.32E—05	1.46E—05	9.37E—06
Thyroid	1.55E—03	1.66E—03	1.54E—03	1.45E—03	1.31E—03	1.05E—03
Uterus	2.80E—08†	5.89E—08†	9.19E—08†	1.12E—07†	1.27E—07†	1.46E—07†
Total body	4.42E—06	4.50E—07	4.26E—06	4.03E—06	3.78E—06	3.18E—06

Specific Absorbed Fraction of Proton Energy (*cont.*)

† Build-up factor method. ⊕ Adjustment on density or composition. * Extrapolation from higher energy. ⊗ Calculated by difference.

Source in Total Body

Target	Energy (MeV) 0.010	0.015	0.020	0.030	0.050	0.100
Bladder wall	1.16E—05	1.43E—05	1.84E—05	1.09E—05	1.04E—05	6.41E—06
Stomach wall	1.38E—05	1.41E—05	1.46E—05	1.14E—05	8.61E—06	5.76E—06
Small intestine plus contents	1.44E—05	1.45E—05	1.39E—05	1.34E—05	9.92E—06	7.56E—06
Upper large intestine wall	1.47E—05	1.51E—05	1.34E—05	1.36E—05	9.72E—06	7.39E—06
Lower large intestine wall	1.46E—05	1.21E—05	1.37E—05	1.26E—05	1.04E—05	6.69E—06
Kidneys	1.40E—05	1.45E—05	1.28E—05	1.06E—05	8.19E—06	6.00E—06
Liver	1.44E—05	1.42E—05	1.42E—05	1.23E—05	8.99E—06	6.23E—06
Lungs	1.32E—05	1.34E—05	1.43E—05	1.26E—05	8.40E—06	5.41E—06
"Other tissues" (suggested for muscle)	1.43E—05⊗	1.37E—05⊗	1.25E—05⊗	9.93E—06⊗	6.60E—06⊗	4.86E—06⊗
Ovaries	1.66E—05	2.47E—05	1.42E—05	1.23E—05	8.16E—06	7.01E—06
Pancreas	1.49E—05	1.43E—05	1.68E—05	1.24E—05	1.09E—05	6.65E—06
Skeleton (suggested for total endosteal cells)	1.51E—06	1.61E—05	1.82E—05	2.01E—05	1.64E—05	8.44E—06
Red marrow	1.56E—05	1.69E—05	1.95E—05	2.29E—05	1.90E—05	1.01E—05
Skin	1.14E—05	9.15E—06	7.62E—06	5.38E—06	3.49E—06	2.66E—06
Spleen	1.70E—05	1.20E—05	1.34E—05	1.26E—05	8.43E—06	6.16E—06
Testes	1.55E—05	1.27E—05	9.36E—06	8.90E—06	6.16E—06	4.36E—06
Thymus	1.37E—05	1.97E—05	2.03E—05	1.06E—05	1.02E—05	5.29E—06
Thyroid	1.97E—05	1.30E—05	1.54E—05	8.65E—06	7.45E—06	3.06E—06
Uterus	1.58E—05	1.37E—05	1.44E—05	1.23E—05	1.04E—05	7.93E—06
Total body	1.42E—05	1.38E—05	1.32E—05	1.14E—05	8.13E—06	5.41E—06

Target	Energy (MeV) 0.200	0.500	1.000	1.500	2.000	4.000
Bladder wall	6.16E—06	6.24E—06	6.78E—06	5.62E—06	4.76E—06	5.41E—06
Stomach wall	6.74E—06	6.14E—06	5.48E—06	4.09E—06	4.60E—06	3.98E—06
Small intestine plus contents	7.01E—06	6.81E—06	6.31E—06	5.85E—06	5.40E—06	4.82E—06
Upper large intestine wall	6.31E—06	6.75E—06	6.02E—06	5.34E—06	5.15E—06	4.23E—06
Lower large intestine wall	5.92E—06	6.83E—06	6.38E—06	5.89E—06	7.11E—06	4.03E—06
Kidneys	6.10E—06	6.06E—06	5.92E—06	4.88E—06	4.70E—06	3.94E—06
Liver	5.78E—06	5.99E—06	5.92E—06	5.02E—06	4.74E—06	4.28E—06
Lungs	4.97E—06	5.36E—06	4.58E—06	4.05E—06	3.85E—06	3.45E—06
"Other tissues" (suggested for muscle)	4.82E—06⊗	4.97E—06⊗	4.75E—06⊗	4.45E—06⊗	4.18E—06⊗	3.51E—06⊗
Ovaries	6.51E—06	5.84E—06	4.20E—06	6.57E—06	6.14E—06	6.47E—06
Pancreas	7.65E—06	5.95E—06	5.30E—06	4.90E—06	8.38E—06	3.84E—06
Skeleton (suggested for total endosteal cells)	5.75E—06	5.16E—06	4.69E—06	4.33E—06	4.22E—06	3.37E—06
Red marrow	6.57E—06	5.68E—06	5.17E—06	4.71E—06	4.65E—06	3.64E—06
Skin	2.81E—06	3.09E—06	3.15E—06	3.14E—06	2.81E—06	2.42E—06
Spleen	6.25E—06	6.16E—06	5.74E—06	4.66E—06	4.24E—06	3.59E—06
Testes	3.82E—06	6.49E—06	5.71E—06	4.90E—06	4.58E—06	4.31E—06
Thymus	8.53E—06	5.22E—06	4.11E—06	6.07E—06	5.03E—06	5.34E—06
Thyroid	4.35E—06	4.94E—06	5.24E—06	4.03E—06	4.12E—06	4.91E—06
Uterus	7.52E—06	5.97E—06	8.02E—06	4.98E—06	6.21E—06	5.48E—06
Total body	4.95E—06	4.97E—06	4.72E—06	4.40E—06	4.16E—06	3.48E—06

From ICRP Report No. 23, *Report on the Tests Group on Reference Man.*, Pergamon Press, Oxford, 1975.

ANSWERS TO PROBLEMS

2.1 0.53 m/s
2.2 a. 6.25×10^5 m/s^2
b. 3.12×10^4 N
c. 25 (kg-m)/s
d. 25 N-s
2.3 6.05×10^{24} kg
2.4 17.4 km/hr at 9.5° to direction of auto's path
2.5 0.0017 radian
2.6 2×10^{-4}C
2.7 98×10^{-9}C
2.8 14.5 V
2.9 a. 1.03×10^7 m/s,
0.73 m/s
b. 1.17×10^{-6} N
2.10 electron:
$v/c = 0.941$
$m/m_0 = 2.95$
proton:
$v/c = 0.046$
$m/m_0 = 1.001$
2.11 2.74×10^{-13} m
2.12 1.42×10^{-8} m
2.13 3.72 eV
2.14 2.94×10^{-10} m
2.15 a. 838 V
b. 0.83 MeV
c. 1.24 MeV
2.16 1.92×10^{-5} m/s
2.17 a. 1.88×10^7 m/s
b. 1.07×10^{-9} s
2.18 a. 1.88×10^7 m/s
b. 1.62×10^{-9} s
2.19 1.04×10^{-18} J
2.20 1,500 W
2.21 a. 1×10^{-4} C, 2×10^{-4} C
b. 0.33×10^{-4} C,
0.67×10^{-4} C
c. 33 V
2.22 a. 45.5 V
b. 4×10^6 m/s
2.23 1.42×10^{-12}
2.24 a. 0.8 kg
b. 2.4×10^6 tonnes
2.25 a. 3.83×10^{26} W
b. 4.26×10^6 tonnes/s
2.26 10.52 MeV
2.27 a. 0.39 Å
b. 6.63×10^{-24} m
c. 2.86×10^{-14} m
2.28 a. 1055 J/Btu
b. 4187 J/(kg-°C)
2.29 a. 0.73 A/m
b. 194 V/m
c. 10 mW/cm^2
2.30 a. 1.33×10^{-6} mW/cm^2
b. 2.65×10^{-4} A/m
2.31 a. 1.6×10^{-7} mW/cm^2
b. 3.5×10^{-2} V/m
c. 9.2×10^{-5} A/m
2.32 714 V/m, 1.89 A/m
3.1 4.3×10^{-12} cm
3.2 3.29×10^{22} atoms/cm^3
3.3 a. 2.83×10^{-11} cm
b. 2, 540 eV
3.4 54.4 eV
3.5 1.14 mA
3.6 0.0073
3.7 109,676 per cm
3.8 1.47×10^{-15} (gm—cm)/s
3.9 193 nm
3.10 a. 5.7×10^{11} J
b. 0.76 MeV
3.11 Be: 2.3×10^{14} gm/cm^3,
Pb: 2.3×10^{14} gm/cm^3
3.13 13.6 eV
3.14 0.421 eV
3.15 a. 2.93×10^6 m/s
b. 50.6 nm
3.16 3.2×10^{16} phot/s
4.1 0.156 MeV
4.2 1.3 MBq, 35.1 μCi
4.3

	Au-198	I-131
t(0)	65.2%	34.8%
t(3)	52.6%	47.4%
t(8)	33.5%	66.5%
t(16)	10.9%	89.1%

4.4 3.8×10^4
4.5 1.2 μg
4.6 299 yrs.
4.7 36.8%
4.8 3.14×10^{15}
4.9 9.84×10^5
4.10 a. 598 days
b. 86.6 days
4.11 38.8 metric tons
4.12 4.2 g
4.14 1.3°C
4.15 23 μCi/metric ton
4.16 6.28
4.17 103.5 hrs.
4.18 ~7 hrs
4.19 2.6 μg
4.20 0.27 g ^{238}U, 0.27 g ^{235}U,
0.58 g ^{232}U
4.21 70 yrs., 1.1×10^{14} Bq
4.22 26.1 min.
4.23 5.2×10^{16} Bq/m^3,
1.4×10^6 μCi/cm^3
4.24

	^{35}S	^{14}C
a.	7.8×10^{-9} W/MBq	7.9×10^{-9} W/MBq
b.	1.3×10^4 W/kg	1.3 W/kg

4.25 a. 1.8×10^{-7} W/MBq
b. 967 W/kg
4.26 a. 4.3×10^8 W/kg
b. 4.3×10^8 W/kg
4.27 5×10^9
4.28 6.4×10^{-5} cm^3
4.29 25 atmospheres
4.30 5.7×10^{-3} Mev/t
4.31 a. 136.5 hrs.
b. 3.5×10^{-8} g
4.32 a. 3×10^4 Ci
b. Yes
4.33 1.15×10^{19} Bq
4.34 7.1×10^{16} Bq/yr
5.1 4.08×10^{22}
5.2 Si: 2.66×10^{22},
O: 5.32×10^{22}
5.3 Al: 3.92×10^{23} electrons/cm^2
Fe: 3.8×10^{23} electrons/cm^2
5.4 ^{32}P
5.5 1.17 MeV, ^{210}Bi
5.6 a. 1.1 MeV
b. 1.31 MeV
5.7 a. 0.077 MeV
b. 0.023 MeV
5.8 a. 0.22 MeV
b. 0.22 MeV
5.9 0.48 MeV
5.10 0.61 MeV
5.11 a. 0.047 MeV
b. 0.617 MeV
c. 1.12 MeV
5.12 a. 0.63 per cm,
0.056 cm^2/g, 19.2 barns
b. 1.25 MeV
5.13 a. beta and gamma
b. ~0.6 MeV
c. ~0.4 MeV
d. ^{125}Sb
e. $Y = 900 \exp(-23.6t) + 100 \exp(-0.25t)$
5.14 41.3
5.15 1:0.8:1.2
5.16 a. HVL
0.1 MeV {Al 1.59 cm; Cu 0.18}
b. TVL
0.1 MeV {Al 5.29 cm; Cu 0.61}

a. HVL

0.8 MeV {Al 3.75; Cu 1.19}

b. TVL

0.8 MeV {Al 12.43; Cu 3.96}

c. TVL = 3.3 × HVL

5.17 a. 1.24 cm
b. 0.063
5.18 2.83 MeV
5.19 0.37 mW/cm^2
5.20 100,000 V
5.21 3,900
5.22 1.2×10^5 n/(sec-Bq), 4.3×10^3 n/(sec-μCi)
5.23 4.4×10^5 cm/s, 886°C
5.24 6.4×10^8 n/(cm^2-sec)
5.25 0.0055 cm
5.26 1.9×10^6 n/s
5.27 50.6 days
5.28 4.4×10^{-7} μCi
5.29 1.25×10^{-11}
5.30 1.008908 amu
5.31 a. 0.00185:1
b. 96.7:1
5.32 0.0174 cm
5.33 a. 4.82×10^{-4} cm
b. 2.075 m
5.34 a. 44
b. 117
5.35 6.57 MBq, 177.6 μCi
5.36 a. 0.249 per cm
b. 1.96×10^{10} n/s
6.1 146 V
6.2 0.33 μC/kg per hr, 1.29 mR/hr
6.3 0.082
6.4 1.2×10^{-3} degree C
6.5 a. 0.036 mGy/hr
b. 4.3 mGy/hr
6.6 0.96 mGy/hr
6.7 5.2 rads to body, 630 rads to kidneys
6.8 3.6 hr
6.9 111 MBq
6.10 0.382 mGy/yr, 38.2 mrads/yr
6.11 0.14 mGy/yr
6.12 284 mGy/sec
6.13 8.9 mGy
6.14 a. 0.77 Gy
b. 3.9×10^{-7} C
6.15 1.24×10^4 Gy, 2.56×10^{18} ion pairs/gm
6.16 3 mGy, 4.54 mGy
6.17 a. $A = 1.4 \exp - 0.58t + 2.3 \exp - 0.11t$
b. 0.19 mGy, 0.26 mGy
c. 0.32 mGy
6.18 5.75×10^5 Gy/sec
6.19 a. 4.2×10^{-6} W/kg
b. 1.1×10^{-6} W/kg
6.20 10.4 mGy/hr, 0.46 mGy/hr
6.21 0.15 Gy gamma, 0.26 Gy neutron
6.22 a. 0.047 mGy/hr
b. 0.31 mGy
6.23 188 mrads, 304 mrads
9.1 a. 0.02
b. 0.38
9.2 a. 2.37, 1.32 cpm
b. 2.71 cpm
c. 30.0 ± 4.5, 30.0 ± 7.0 cpm
9.3 a. Yes
b. No
9.4 Yes
9.5 Impossible
9.6 a. 17.0 ± 3.7
b. 11%
9.7 85 min.
9.8 a. 50 ± 20
b. 50 ± 8
9.9 a. 0.271
b. 0.090
c. 0.135
9.10 No
9.11 a. 0.056
b. 0.68
c. 0.16
9.12 7.8×10^{-13} amp
9.13 122 mR/hr
9.14 a. 1×10^{10} ohms
b. 2.5 sec
c. 11.5 sec
9.15 40 V
9.16 a. 2×10^6 ohms
b. 0.64 V
9.17 12%
9.18 2×10^{-5} V
9.19 3.95 cps per n/cm^2-sec
9.20 4.08×10^{-5} amp
9.21 892 years
9.22 6.4×10^{-16}, 1.4×10^{-15} amp/neut per cm^2 per sec
9.23 25.3 m
9.24 2000 neut/cm^2-sec
9.25 143 cps
10.1 a. 44 cm
b. 0.08 n/cm^2 per sec
c. 0.45 mrad per hr
10.2 2.38 mm
10.3 North wall: 49.5 in., West wall: 15.5 in concrete
10.4 71 cm
10.5 0.32
10.6 a. 3.19 cm^2/g, 33.16 per cm
b. 0.097 cm
10.7 6 mm
10.8 31 in.
10.9 0.79 g/cm^2
10.10 1,090 MBq
10.11 4 hr 17 min
10.12 3.53 cm
10.13 9.1 mR/hr (2.4 μC/kg·hr)
10.14 1.2 mSv/hr
10.15 0.53 μSv/hr
10.16 14 cm
10.17 436 cm
10.18 1.6%
10.19 0.16 μCi/cm^3
10.20 a. I-125: 2.44, Au-198: 6.4, Na-24: 25.6 (μX-m^2)/hr
b. 12.4 cm
10.21 1.1 cm polyethylene plus 4.2 cm Pb
11.1 38, 52 mGy
11.2 a. 7.4×10^3 rads/hr
b. 1.74 rad
11.3 282 rads, 0.066 rad
11.4 a. No
b. 3.7×10^5 Bq
c. 43 days
d. 1.5×10^{-2} Sv
e. 2.6×10^{-2} Sv
f. 2.6×10^{-2} Sv
11.5 Yes
11.6 a. No
b. 0.12 ppm, 50 ppm
11.7 a. 1.36, 2.71
b. 1.00002
c. 3.9 μCi/millimol
11.8 a. 4.9×10^7 Bq/sec
b. 3.1×10^3 m
11.9 a. 1.3×10^{-10} Gy, 2.5×10^{-13} Gy
b. 1.6×10^{-11} Gy, 5.5×10^{-14} Gy
11.10 3.1×10^5 μCi/gm
11.11 3×10^9 Bq/m^3 at 800 m
11.12 a. 8.2×10^{16} Bq
b. 3.8×10^4 years
11.13 3.2×10^{-7} Sv
11.14 14.4 m^3/min
12.1 96 Bq/l Na-24, 2.3×10^{-5} Bq/l Cl-36, 6×10^3 Bq/l Cl-38, 0.17 Bq/l S-35
12.2 1.25×10^3 Bq/l Na-24, 2.3×10^{-3} Bq/l Cl-36 3.3×10^3 Bq/l Cl-38, 16 Bq/l S-35
12.3 2 μGy/hr
12.4 696 Ci
12.5 3.5, 3×10^{-2}, 6×10^{-4}, 6×10^{-5} Gy/hr on surface
2×10^{-3}, 1×10^{-5}, 3×10^{-7}, 3×10^{-8} Gy/hr at 25 feet
12.6 6,560
12.7 a. 186.2 gm UO_2SO_4/l
b. 19.8 gm UO_2SO_4/l
12.8 1.88
12.9 a. 4×10^{-6} Gy
b. 3×10^{-5} Gy
12.10 a. 1023
b. 279
12.11 710 minutes
12.12 25 mGy gamma, 22.5 mGy neutrons
12.13 0.011 per cm

13.1 5.25×10^3 Gy/hr
13.2 No
13.3 5.6 cpm
13.4 a. 58.5 Bq-days
b. 27.4 μGy/hr
13.5 555 tpm
13.6 1 Bq/m^3: 3.6 Bq/l
13.7 3×10^3, 5×10^2 Bq
13.8 a. 1.3 MBq
b. 2.8 μGy
c. 81 μGy
13.9 0.520 ± 0.033 Bq/l
13.10 1.05 ± 0.23 Bq/m^3, $(28.4 \pm 6.1) \times 10^{-12}$ μCi/ml
13.11 1.58 ± 0.34 mSv
13.12 a. 3.6×10^{-12} Sv to lung
4×10^{-10} Sv to skeleton
b. 6×10^{-14} Sv to lung
7.5×10^{-10} Sv to skeleton
13.13 b. log-normal
c. CMD = $2.5\,\mu \times\!\!\div 2$
SMD = $8.2\,\mu \times\!\!\div 2$
MMD = $10.6\,\mu \times\!\!\div 2$
13.14 a. 6.5×10^4 Bq
b. 0.4 Gy
13.15 1.9×10^3 Bq
13.16 a. $A(u) = 25.4\,e^{-0.017t}$
b. 2.1×10^4 Bq
c. 38 mGy
13.17 0.33 Sv
13.18 a. 787.5 Bq
b. 42 mSv
13.19 a. 19.7 Ci ^{60}Co, 154 Ci ^{137}Cs, 690 Ci ^{144}Ce
b. 5 R/hr
c. 4,700 R/hr
13.20 3×10^{-8} μCi/cm^3
13.21 a. 1.77×10^4 Bq/m^3
b. 3.6×10^{-6} Sv
c. 1.25×10^3 Bq/l
14.1 3×10^6
14.2 a. 2.8 mm
b. 5.3 cm
c. 396 mW/cm^2
14.3 a. 0.088, 0.044 J/cm^2
b. 7 m
c. Yes
14.4 a. 5×10^{-4}
b. 2×10^4 W
c. 0.001 J
14.5 1.8 mW/cm^2
14.6 10.7 mW/cm^2
14.7 a. 1.72
b. 2.73
14.8 a. 3.3
b. 1.3
c. 1.91
14.9 a. 6.4×10^{-5} sec
b. 6.4×10^{-7} J/cm^2
c. 6.4×10^{-6} W/cm^2
d. class II
14.10 a. 10^{-4}
b. 100 W
14.11 a. 10^{-3}
b. 20,000 W
14.12 a. 88.8
b. 19.5
14.13 a. E = 122.8 V/m
b. H = 0.33 A/m
c. 0.133 pJ/cm^3
14.14 30 dB
14.15 0.27 A/m
14.16 a. 51 mW/cm^2
b. 204 mW/cm^2
c. 79 cm
d. 126 mW/cm^2
e. 355 cm
14.17 a. 500 mW/cm^2
b. 2000 mW/cm^2
c. 8 cm
d. 0.5 mW/cm^2
e. 358 cm
14.18 a. 10^{-3}
b. 2000 W
c. 42 dB
d. 320 m
14.19 a. 1.2×10^{-3}
b. 1.2×10^3 mW/cm^2
c. 21 mW/cm^2
d. Yes
14.20 derivation
14.21 27 cm, 98 cm
14.22 6.7 cm, 47.4 cm
14.23 29 dB
14.24 a. 4.2 m
b. 22.4 mW/cm^2
c. 14.2 m
14.25 0.11 mW/cm^2
14.26 175 mW/cm^2
14.27 12 seconds
14.28 a. 10^{-3}
b. 2000 W
c. 26 mW/cm^2
d. Yes

AUTHOR INDEX

SUBJECT INDEX

ABOUT THE AUTHOR

HERMAN CEMBER received a B.S. degree from City College of New York and his M.S. and PH.D. from the University of Pittsburgh. He was assistant professor then associate professor of Industrial Hygiene at the University of Pittsburgh Graduate School of Public Health for ten years, an associate professor of Industrial Health at the University of Cincinnati Medical School for five years, and professor of Environmental Health Engineering at the Northwestern University Technological Institute since 1964. He spent a year as a technical expert in Occupational Health at the International Labour Office in Geneva and was also a visiting professor (Fulbright Fellow) at the Hebrew University Hadassah Medical School. His main research interests are in radiogenic lung cancer and heavy metal toxicology, and he consults in the fields of health physics and industrial hygiene.